WASTEWATER

MICROBIOLOGY

Third Edition

WASTEWATER
MICROBIOLOGY

Third Edition

GABRIEL BITTON

Department of Environmental Engineering Sciences
University of Florida, Gainesville, Florida

WILEY-LISS

A JOHN WILEY & SONS, INC., PUBLICATION

Library of Congress Cataloging-in-Publication Data:

Bitton, Gabriel.
   Wastewater microbiology / Gabriel Bitton. -- 3rd ed.
     p. ; cm.
   Includes bibliographical references and index.
   ISBN 0-471-65071-4 (cloth)
  1. Sanitary microbiology. 2. Water--Microbiology. 3. Sewage--Microbiology.
   [DNLM: 1. Sewage--microbiology. 2. Water Microbiology. 3. Safety
  Management. 4. Water Purification--methods. QW 80 B624w 2005] I. Title.
  QR48.B53 2005
  628.3'01'579--dc22
                                 2004019935

Printed in the United States of America

10   9   8   7   6   5   4   3   2   1

to Julie and Natalie

CONTENTS

PREFACE

I would like to mention some of the changes and additions that have been included in the third edition of *Wastewater Microbiology*. In general, every chapter of the book has been revised (up to July 2004) to include the latest developments in the field, and I will highlight only the major ones.

A review of the most important molecular techniques has been added to Chapter 1, while the most recent methodology for measuring microbial biomass in environmental samples is described in Chapter 2. New developments in enhanced biological phosphorus removal (EBPR) are covered in Chapter 3. Chapter 4 covers new findings on old and emerging (e.g., *Helicobacter pylori*, *Cyclospora*, Microsporidia) microbial pathogens and parasites. Much progress has been made concerning the detection of *Cryptosporidium* and *Giardia* in environmental samples, including wastewater. The improved methodology is also covered in Chapter 4. As regards disinfection of water and wastewater, research efforts are now focusing on UV disinfection in industrialized countries and on the use of solar radiation in developing countries (Chapter 6).

Armed with new molecular tools and microsensor/microelectrode technology, investigators are making progress in understanding the microbial ecology and the surface properties of activated sludge flocs. The methodology used is similar to that used in biofilms. These advances will help us to better understand the flocculation process in activated sludge (Chapter 8). Concerning bulking and foaming in activated sludge plants, most of the recent studies have focused on the characterization and phylogeny of filamentous microorganisms (Chapter 9).

In the last few years we have witnessed an increased interest in biofilm microbiology. Biofilms develop on biological and nonbiological surfaces and are ubiquitous in natural aquatic environments and engineered systems (e.g., fixed-film bioreactors). Their beneficial role in fixed-film bioreactors has been known for years (chapter 10). However, the impact of biofilms on drinking water distribution systems has been the subject of increased research activity around the world (chapter 16). This interest is further heightened by the findings that biofilms are the source of medical problems such as dental plaques or colonization of artificial implants, leading to increased rate of infection in patients. The discovery of communication among members of the biofilm community (i.e., quorum sensing using signaling chemicals such as homoserine lactones) may lead to potential means of controlling biofouling of surfaces.

Chapter 13 shows that new procedures, particularly molecular techniques, have helped shed light on the phylogeny of methanogens and other Archaea.

Part D (Microbiology of Drinking Water Treatment) of the third edition now comprises three chapters instead of two as in the second edition. The third chapter (Chapter 17) introduces the reader to bioterrorism microbial agents and their potential impact on drinking water safety.

In Chapter 18 (Biotechnology of Waste Treatment: Pollution Control Biotechnology), I have added some information about membrane bioreactors (MBR technology), while in Chapter 21, new developments in the area of bioremediation have been included. Finally, in Chapter 23 (Wastewater Reuse), I have made an attempt to introduce the reader to the microbiological aspects of the treatment of wastewater effluents by natural and constructed wetlands and by the use of attached algae for polishing wastewater effluents.

Since the World Wide Web is increasingly becoming an integral part of the learning process at education institutions, I have added some Web resources to each chapter of the book to help students increase their knowledge or satisfy their curiosity about topics discussed in a given chapter. I have also included questions at the end of each chapter. These questions can help students in studying the material or can be used as homework.

I thank Jorge Gomez Moreno for drawing several of the new figures for the third edition of this book. His attention to detail is much appreciated.

I am grateful to Nancy, Julie, Natalie, Jonathan, my entire family, and friends for their love and moral support.

GABRIEL BITTON
Gainesville, Florida

PREFACE TO THE FIRST EDITION

Numerous colleagues and friends have encouraged me to prepare a second edition of *Introduction to Environmental Virology*, published by Wiley in 1980. Instead, I decided to broaden the topic by writing a text about the role of *all* microorganisms in water and wastewater treatment and the fate of pathogens and parasites in engineered systems.

In the 1960s, the major preoccupation of sanitary engineers was the development of wastewater treatment processes. Since then, new research topics have emerged and emphasis is increasingly placed on the biological treatment of hazardous wastes and the detection and control of new pathogens. The field of wastewater microbiology has blossomed during the last two decades as new modern tools have been developed to study the role of microorganisms in the treatment of water and wastewater. We have also witnessed dramatic advances in the methodology for detection of pathogenic microorganisms and parasites in environmental samples, including wastewater. New genetic probes and monoclonal antibodies are being developed for the detection of pathogens and parasites in water and wastewater. Environmental engineers and microbiologists are increasingly interested in toxicity and the biodegradation of xenobiotics by aerobic and anaerobic biological processes in wastewater treatment plants. Their efforts will fortunately result in effective means of controlling these chemicals. The essence of this book is an exploration of the interface between engineering and microbiology, which will hopefully lead to fruitful interactions between biologists and environmental engineers.

The book is divided into five main sections, which include fundamentals of microbiology, elements of public health microbiology, process microbiology, biotransformations and toxic impact of chemicals in wastewater treatment plants, and the public health aspects of the disposal of wastewater effluents and sludges on land and in the marine environment. In the process microbiology section, each biological treatment process is covered from both the process microbiology and public health viewpoints.

This book provides a useful introduction to students in environmental sciences and environmental engineering programs and a source of information and references to research workers and engineers in the areas of water and wastewater treatment. It should serve as a reference book for practicing environmental engineers and scientists and for public health microbiologists. It is hoped that this information will be a catalyst for scientists and engineers concerned with the improvement of water and wastewater treatment and with the quality of our environment.

I am very grateful to all my colleagues and friends who kindly provided me with illustrations for this book and who encouraged me to write *Wastewater Microbiology*. I will always be indebted to them for their help, moral support, and good wishes. I am indebted to my graduate students who have contributed to my interest and knowledge in the microbiology of engineered systems. Special thanks are due to Dr. Ben Koopman for lending a listening ear to my book project and to Dr. Joseph Delfino for his moral support. I thank Hoa Dang-Vu Dinh for typing the tables for this book. Her attention to detail is much appreciated.

Special thanks to my family, Nancy, Julie, and Natalie, for their love, moral support, and patience, and for putting up with me during the preparation of this book.

GABRIEL BITTON
Gainesville, Florida

PREFACE TO THE SECOND EDITION

The second edition of *Wastewater Microbiology* incorporates the latest findings in a field covering a wide range of topics.

During the past few years, we have witnessed significant advances in molecular biology, leading to the development of genetic probes, particularly the ribosomal RNA (rRNA) oligonucleotide probes, for the identification of wastewater microorganisms. The road is now open for a better identification of the microbial assemblages in domestic wastewater and their role in wastewater treatment.

The use of genetic tools has also been expanded as regards the detection of pathogens and parasites (Chapter 4), and biotechnological applications for wastewater treatment (Chapter 17). Chapter 4 has been expanded due to the emergence of new pathogens and parasites in water and wastewater. The topic of drinking water microbiology has been expanded, and two chapters are now devoted to this subject. Chapter 15 deals with water treatment and Chapter 16 covers the microbiology of water distribution systems. New methodology that shows the heterogeneous structure of biofilms and their complex biodiversity includes nondestructive confocal laser-scanning microscopy in conjunction with 16S rRNA-targeted oligonucleotide probes (Chapter 16). The topic of wastewater and biosolids disposal on land and in receiving waters has also been expanded and is now covered in two chapters (Chapters 20 and 21).

New figures and tables have been added to further enhance the illustration of the book. Many old figures and graphs were redrawn to improve the visual aspect of the book.

I am very grateful to the colleagues who reviewed the book proposal for their valuable suggestions concerning the second edition of *Wastewater Microbiology*. I am particularly grateful to my mentor and friend, Professor Ralph Mitchell, of Harvard University. As editor of the Wiley series in Ecological and Applied Microbiology, he offered me his full support in the undertaking of this project. I thank Dr. Charles Gerba of the University of Arizona for his continuous moral support and enthusiasm. I thank Dr. Robert Harrington, senior editor at Wiley, for enthusiastically endorsing this second edition of *Wastewater Microbiology*.

A picture is worth a thousand words. I thank Dr. Christopher Robinson of the Oak Ridge Institute of Science and Education, and Dr. H.D. Alan Lindquist of the U.S. EPA for promptly and kindly sending me photomicrographs of *Cryptosporidium parvum*. I am grateful to Dr. Rudolf Amann of the Max-Planck Institute for Marine Microbiology, Bremen, Germany, for allowing me to use his excellent color pictures on the use of rRNA probes in wastewater microbiology, and to Dr. Trello Beffa of the Universite de Neufcha-tel, Switzerland, for his scanning electron micrograph of compost microorganisms. Many thanks to Dr. Samuel Farrah and his students, Fuha Lu and 'Jerzy Lukasik, for supplying a scanning electron micrograph of *Zooglea*.

I am grateful to Nancy, Julie, Natalie, my entire family, and friends for their love and moral support.

GABRIEL BITTON
Gainesville, Florida

PART A

FUNDAMENTALS OF MICROBIOLOGY

1

THE MICROBIAL WORLD

Wastewater Microbiology, *Third Edition*, by Gabriel Bitton
Copyright © 2005 John Wiley & Sons, Inc.

1.1 INTRODUCTION

The three domains of life are *bacteria*, *archaea*, and *eukarya* (Fig. 1.1) (Rising and Reysenbach, 2002; Woese, 1987). Bacteria, along with actinomycetes and cyanobacteria (blue-green algae) belong to the *prokaryotes* while *eukaryotes* or eukarya include fungi, protozoa, algae, plant, and animal cells.

Viruses are obligate intracellular parasites that belong to neither of these two groups.

The main characteristics that distinguish prokaryotes from eukaryotes are the following (Fig. 1.2):

1. Eukaryotic cells are generally more complex than prokaryotic cells.
2. DNA is enclosed in a nuclear membrane and is associated with histones and other proteins only in eukaryotes.
3. Organelles are membrane-bound in eukaryotes.
4. Prokaryotes divide by binary fission whereas eukaryotes divide by mitosis.
5. Some structures are absent in prokaryotes: for example, Golgi complex, endoplasmic reticulum, mitochondria, and chloroplasts.

Other differences between prokaryotes and eukaryotes are shown in Table 1.1.

We will now review the main characteristics of prokaryotes, archaea, and eukaryotes. Later, we will focus on their importance in process microbiology and public health. We will also introduce the reader to environmental virology and parasitology, the study of the fate of viruses, and protozoan and helminth parasites of public health significance in wastewater and other fecally contaminated environments.

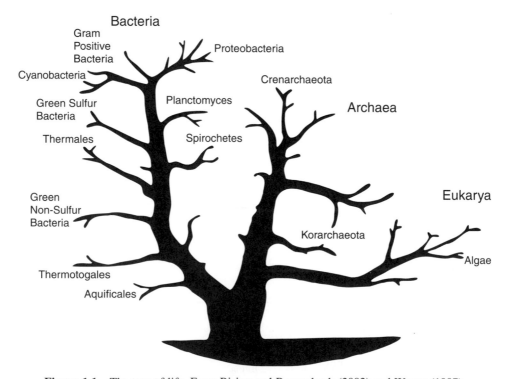

Figure 1.1   The tree of life. From Rising and Reysenbach (2002) and Woese (1987).

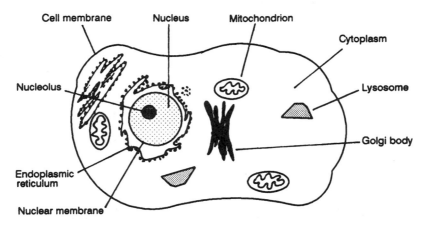

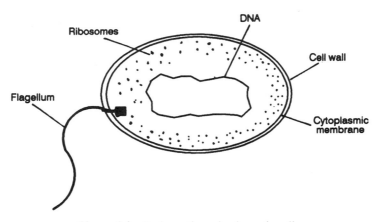

Figure 1.2 Prokaryotic and eukaryotic cells.

1.2 CELL STRUCTURE

1.2.1 Cell Size

Except for filamentous bacteria, prokaryotic cells are generally smaller than eukaryotic cells. Small cells have a higher growth rate than that of larger cells. This may be explained by the fact that small cells have a higher surface-to-volume ratio than larger cells. Thus, the higher metabolic activity of small cells is due to additional membrane surface available for transport of nutrients into and waste products out of the cell.

1.2.2 Cytoplasmic Membrane (Plasma Membrane)

The cytoplasmic membrane is a 40–80 Å-thick semipermeable membrane that contains a phospholipid bilayer with proteins embedded within the bilayer (fluid mosaic model) (Fig. 1.3). The phospholipid bilayer is made of hydrophobic fatty acids oriented towards the inside of the bilayer and hydrophilic glycerol moieties oriented towards the

TABLE 1.1. Comparison of Prokaryotes and Eukaryotes

Feature	Prokaryotes (Bacteria)	Eukaryotes (Fungi, Protozoa, Algae, Plants, Animals)
Cell wall	Present in most prokaryotes (absent in mycoplasma); made of peptidoglycan	Absent in animal; present in plants, algae, and fungi
Cell membrane	Phospholipid bilayer	Phospholipid bilayer + sterols
Ribosomes	70S in size	80S in size
Chloroplasts	Absent	Present
Mitochondria	Absent; respiration associated with plasma membrane	Present
Golgi complex	Absent	Present
Endoplasmic reticulum	Absent	Present
Gas vacuoles	Present in some species	Absent
Endospores	Present in some species	Absent
Locomotion	Flagella composed of one fiber	Flagella or cilia composed of microtubules; amoeboid movement
Nuclear membrane	Absent	Present
DNA	One single molecule	Several chromosomes where DNA is associated with histones
Cell division	Binary fission	Mitosis

outside of the bilayer. Cations such as Ca^{2+} and Mg^{2+} help stabilize the membrane structure. Sterols are other lipids that enter into the composition of plasma membranes of eukaryotic cells as well as some prokaryotes, such as mycoplasma (these bacteria lack a cell wall).

Chemicals cross biological membranes by diffusion, active transport, and endocytosis.

Diffusion.   Because of the hydrophobic nature of the plasma membrane, lipophilic compounds diffuse better through the membrane than ionized compounds. The rate of diffusion across cell membranes depends on their lipid solubility and concentration gradient across the membrane.

Active transport.   Hydrophilic compounds (that is, lipid insoluble) may be transferred through the membrane by active transport. This transport involves highly specific

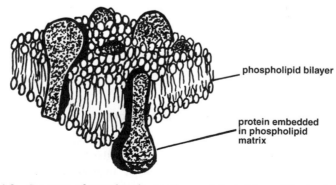

phospholipid bilayer

protein embedded in phospholipid matrix

Figure 1.3   Structure of cytoplasmic membrane. Adapted from Alberts et al. (1989).

carrier proteins, requires energy in the form of adenosine triphosphate (ATP) or phosphoenol-pyruvate (PEP) and allows cells to accumulate chemicals against a concentration gradient. There are specific active transport systems for sugars, amino acids, and ions. Toxic chemicals gain entry into cells mainly by diffusion and some may use active transport systems similar to those used for nutrients.

Endocytosis. In eukaryotic cells, substances can cross the cytoplasmic membranes by endocytosis, in addition to diffusion and active transport. Endocytosis includes *phagocytosis* (uptake of particles) and *pynocytosis* (uptake of dissolved substances).

1.2.3   Cell Wall

All bacteria, except mycoplasma, have a cell wall. This structure confers rigidity to cells and maintains their characteristic shape, and it protects them from high osmotic pressures. It is composed of a mucopolysaccharide called peptidoglycan or murein (glycan strands cross-linked by peptide chains). Peptidoglycan is composed of N-acetylglucosamine and N-acetylmuramic acid and amino acids. A cell wall stain, called the Gram stain differentiates between *gram-negative* and *gram-positive* bacteria on the basis of cell wall chemical composition. Peptidoglycan layers are thicker in gram-positive bacteria than in gram-negative bacteria. In addition to peptidoglycan, gram-positive bacteria contain teichoic acids made of alcohol and phosphate groups.

Animal cells do not have cell walls; however, in other eukaryotic cells, the cell walls may be composed of cellulose (e.g., plant cells, algae), chitin (e.g., fungi), silica (e.g., diatoms), or polysaccharides such as glucan and mannan (e.g., yeasts).

1.2.4   Outer Membrane

The outermost layer of gram-negative bacteria contains phospholipids, lipopolysaccharides (LPS), and proteins (Fig 1.4). Lipopolysaccharides constitute about 20 percent of

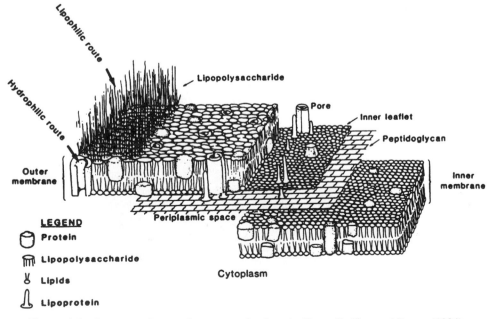

Figure 1.4   Outer membrane of gram-negative bateria. From Godfrey and Bryan (1984).

the outer membrane by weight and consist of a hydrophobic region bound to an oligosaccharide core. The LPS molecules are held together with divalent cations. Proteins constitute about 60 percent of the outer membrane weight and are partially exposed to the outside. Some of the proteins form water-filled pores, *porins*, for the passage of hydrophilic compounds. Other proteins have a structural role, as they help anchor the outer membrane to the cell wall. The outer membrane of gram-negative bacteria is an efficient barrier against hydrophobic chemicals, namely some antibiotics and xenobiotics, but is permeable to hydrophilic compounds, some of which are essential nutrients.

Chemical (e.g., Ethylendiamine tetraacetic acid, polycations) and physical (e.g., heating, freeze-thawing, drying, and freeze-drying) treatments, as well as genetic alterations, can increase the permeability of outer membranes to hydrophobic compounds.

1.2.5 Glycocalyx

The *glycocalyx* is made of extracellular polymeric substances (EPS), which surround some microbial cells and are composed mainly of polysaccharides. In some cells, the glycocalyx is organized as a capsule (Fig. 1.5). Other cells produce loose polymeric materials that are dispersed in the growth medium.

Extracellular polymeric substances are important from medical and environmental viewpoints: (1) capsules contribute to pathogen virulence; (2) encapsulated cells are protected from phagocytosis in the body and in the environment; (3) EPS helps bacteria adsorb to surfaces such as teeth, mucosal surfaces, and environmentally important surfaces such as water distribution pipes (see Chapter 16); (4) capsules protect cells against desiccation; (5) they play a role in metal complexation, particularly in wastewater treatment plants (see Chapters 20 and 21); and (6) they play a role in microbial flocculation in the activated sludge process (see Chapter 8).

1.2.6 Cell Motility

Microbial cells can move by means of flagella, cilia, or *pseudopods*. Bacteria display various flagellar arrangements ranging from *monotrichous* (polar flagellum; e.g., *Vibrio coma*), *lophotrichous* (bundle of flagella at one end of the cell; e.g., *Spirillum volutans*), to *peritrichous* (several flagella distributed around the cell; e.g., *Escherichia coli*)

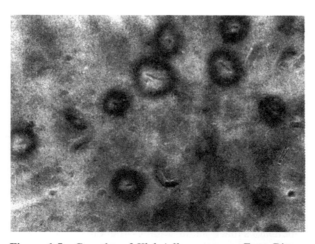

Figure 1.5 Capsules of *Klebsiella aerogenes*. From Bitton.

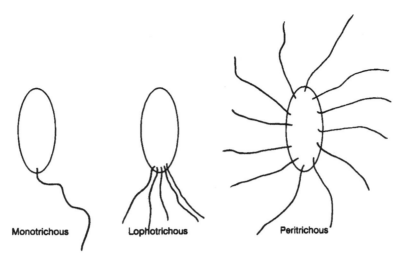

Figure 1.6 Flagellar arrangements in bacteria.

(Fig. 1.6). The flagellum is composed of a protein called *flagellin* and is anchored by a hook to a basal body located in the cell envelope. Flagella enable bacteria to attain speeds of 50–100 μm/s. They enable cells to move toward food (chemotaxis), light (phototaxis), or oxygen (aerotaxis). *Chemotaxis* is the movement of a microorganism toward a chemical, generally a nutrient. It also enables the movement away from a harmful chemical (negative chemotaxis). Chemotaxis can be demonstrated by placing a capillary containing a chemical attractant into a bacterial suspension. Bacteria, attracted to the chemical, swarm around the tip and move inside the capillary. Two sets of proteins, chemoreceptors and transducers, are involved in triggering flagellar rotation and subsequent cell movement. From an ecological viewpoint, chemotaxis provides a selective advantage to bacteria, allowing them to detect carbon and energy sources. Toxicants (e.g., hydrocarbons, heavy metals) inhibit chemotaxis by blocking chemoreceptors, thus affecting food detection by motile bacteria as well as predator–prey relationships in aquatic environments.

Eukaryotic cells move by means of flagella, cilia, or cytoplasmic streaming (i.e., amoeboid movement). The flagella have a more complex structure than that of prokaryotic flagella. Cilia are shorter and thinner than flagella. Ciliated protozoa use cilia for locomotion and for pushing food inside the *cytostome*, a mouthlike structure. Some eukaryotes (e.g., amoeba, slime molds) move by amoeboid movement by means of pseudopods (i.e., false feet), which are temporary projections of the cytoplasm.

1.2.7 Pili

Pili are structures that appear as short and thin flagella, attached to cells in a manner similar to that of flagella. They play a role in cell attachment to surfaces, conjugation (involvement of a sex pilus), and act as specific receptors for certain types of phages.

1.2.8 Storage Products

Cells may contain inclusions that contain storage products serving as a source of energy or building blocks. These inclusions may be observed under a microscope, using special stains, and include the following:

1. Carbon storage in the form of *glycogen*, *starch*, and poly-β-hydroxybutyric acid (PHB), which stains with Sudan black, a fat-soluble stain. The PHB occurs exclusively in prokaryotic microorganisms.

2. *Volutin granules*, which contain polyphosphate reserves. These granules, also called *metachromatic granules*, appear red when specifically stained with basic dyes such as toluidine blue or methylene blue.

3. *Sulfur granules* are found in sulfur filamentous bacteria (e.g., *Beggiatoa*, *Thiothrix*) and purple photosynthetic bacteria, which use H_2S as an energy source and electron donor. H_2S is oxidized to S^0, which accumulates inside sulfur granules, readily visible under a light microscope. Upon depletion of the H_2S source, the elemental sulfur is further oxidized to sulfate.

1.2.9 Gas Vacuoles

Gas vacuoles are found in cyanobacteria, halobacteria (i.e., salt-loving bacteria), and photosynthetic bacteria. Electron microscopic studies have shown that gas vacuoles are made of gas vesicles, which are filled with gases and surrounded by a protein membrane. Their role is to regulate cell buoyancy in the water column. Owing to this flotation device, cyanobacteria and photosynthetic bacteria sometimes form massive blooms at the surface of lakes or ponds.

1.2.10 Endospores

Endospores are formed inside bacterial cells and are released when cells are exposed to adverse environmental conditions. The location of the spore may vary. There are central, subterminal, and terminal spores. Physical and chemical agents trigger spore germination to form vegetative cells. Bacterial endospores are very resistant to heat and this is probably due to the presence of a dipicolinic acid–Ca complex in endospores. Endospores are also quite resistant to desiccation, radiation, and harmful chemicals. This is significant from a public health viewpoint because they are much more resistant to chemical disinfectants than vegetative bacteria in water and wastewater treatment plants (see Chapters 5 and 6).

1.2.11 Eukaryotic Organelles

Specialized structures, called organelles, are located in the cytoplasm of eukaryotic cells and carry out several important cell functions. We will now briefly review some of these organelles.

1.2.11.1 Mitochondria. *Mitochondria* (singular: mitochondrion) are oval or spherical structures surrounded by a double membrane. The outer membrane is very permeable to the passage of chemicals, and the inner membrane is folded and forms shelves called *cristae* (singular: crista) (Fig. 1.7). They are the site of cell respiration and ATP production in eukaryotic cells (see Chapter 2). The number of mitochondria per cell varies with the type and metabolic level of the cells.

1.2.11.2 Chloroplasts. Chloroplasts are relatively large chlorophyll-containing structures found in plant and algal cells and are also surrounded by a double membrane. They are made of units called *grana*, interconnected by lamellae. Each granum consists

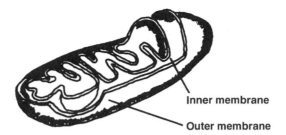

Figure 1.7 Mitochondrion structure.

of a stack of disks called thylakoids bathing in a matrix called stroma (Fig. 1.8). Chloroplasts are the sites for photosynthesis in plant and algal cells. The light and dark reactions of photosynthesis occur in the thylakoids and stroma, respectively (see Chapter 2).

1.2.11.3 Other Organelles. Other important organelles that are found in eukaryotic cells, but not in prokaryotic cells, are the following:

- The *Golgi complex* consists of a stack of flattened membranous sacs, called *saccules*, which form vesicles that collect proteins, carbohydrates, and enzymes.
- The *endoplasmic reticulum* is a system of folded membranes attached to both the cell membrane and the nuclear membrane. The rough endoplasmic reticulum is associated with ribosomes and is involved in protein synthesis. The smooth endoplasmic reticulum is found in cells that make and store hormones, carbohydrates, and lipids.
- *Lysosomes* are sacs that contain hydrolytic (digestive) enzymes and help in the digestion of phagocytized cells by eukaryotes.

1.3 CELL GENETIC MATERIAL

1.3.1 DNA in Prokaryotes and Eukaryotes

In prokaryotes, DNA occurs as a single circular molecule, which is tightly packed to fit inside the cell and is not enclosed in a nuclear membrane. Prokaryotic cells may also contain small circular DNA molecules called *plasmids.*

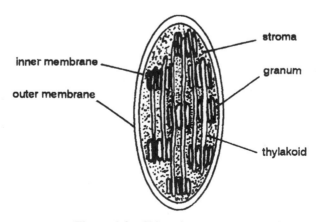

Figure 1.8 Chloroplast structure.

Eukaryotes have a distinct nucleus surrounded by a nuclear membrane with very small pores that allow exchanges between the nucleus and the cytoplasm. The DNA is present as chromosomes consisting of DNA associated with histone proteins. Cells divide by mitosis, which leads to a doubling of the chromosome numbers. Each daughter cell has a full set of chromosomes.

1.3.2 Nucleic Acids

Deoxyribonucleic acid (DNA) is a double-stranded molecule that is made of several millions of units (e.g., approximately 4 M base pairs (bp) in the *Escherichia coli* chromosome) called *nucleotides*. The double-stranded DNA is organized into a *double helix* (Fig. 1.9). Each nucleotide is made of a five-carbon sugar (deoxyribose), a phosphate group, and a nitrogen-containing base linked to the C-5 and C-1 of the deoxyribose molecule, respectively. The nucleotides on a strand are linked together via a phosphodiester bridge. The hydroxyl group of a C-3 of a pentose ($3'$ carbon) is linked to the phosphate group on the C-5 ($5'$ carbon) of the next pentose. There are four different bases in DNA, two purines (adenine and guanine), and two pyrimidines (cytosine and thymine). A base on one strand pairs through hydrogen bonding with another base on the complementary strand. Guanine always pairs with cytosine, while adenine always pairs wth thymine (Fig. 1.10). One strand runs in the $5' \rightarrow 3'$ direction, while the complementary strand runs in the $3' \rightarrow 5'$ direction. Physical and chemical agents cause DNA to unwind, leading to the separation of the two strands.

Ribonucleic acid (RNA) is generally single-stranded (some viruses have double-stranded RNA), contains ribose in lieu of deoxyribose, and uracil in lieu of thymine.

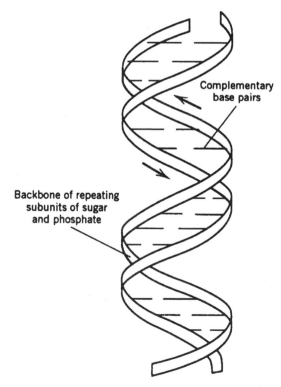

Complementary
base pairs

Backbone of repeating
subunits of sugar
and phosphate

Figure 1.9 DNA structure.

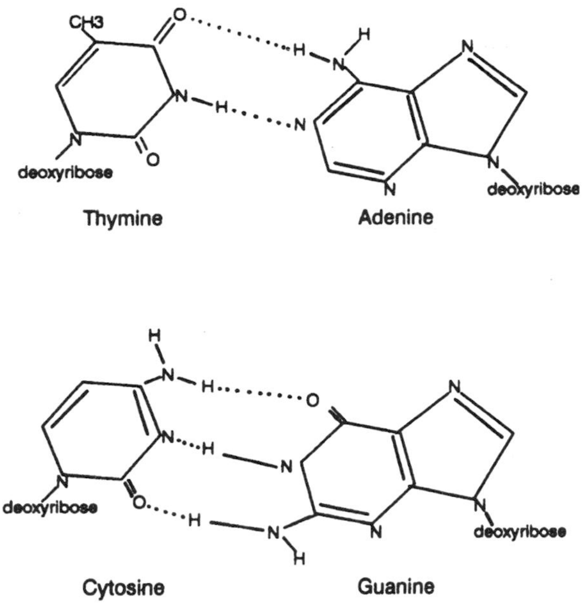

Figure 1.10 Base-pairing in DNA.

In RNA, guanine binds to cytosine while adenine binds to uracil. There are three forms of RNA: messenger RNA (mRNA), ribosomal RNA (rRNA), and transfer RNA (tRNA).

1.3.2.1 DNA Replication and Protein Synthesis.

Replication: The DNA molecule can make an exact copy of itself. The two strands separate and new complementary strands are formed. The double helix unwinds and each of the DNA strands acts as a template for a new complementary strand. Nucleotides move into the *replication fork* and align themselves against the complementary bases on the template. The addition of nucleotides is catalyzed by an enzyme called *DNA polymerase.*

Transcription: Transcription is the process of transfer of information from DNA to RNA. The complementary single-stranded RNA molecule is called *messenger RNA* (mRNA). mRNA carries the information from the DNA to the ribosomes where it controls protein synthesis. Transcription is catalyzed by an enzyme called *RNA polymerase.* Enzyme regulation (repression or induction) occurs at the level of transcription. Sometimes, the product formed through the action of an enzyme represses the synthesis of that enzyme. The enzyme product acts as a co-repressor, which, along with a *repressor,* combines with the operator gene to block transcription and, therefore, enzyme synthesis. The synthesis of other enzymes, called *inducible enzymes,* occurs only when the substrate is present in the medium. Enzyme synthesis is induced because the substrate, the inducer, combines with the repressor to form a complex that has no affinity for the operator gene.

Translation: mRNA controls protein synthesis in the cytoplasm. This process is called *translation.* Another type of RNA is the *transfer RNA* (tRNA), which has attachment sites for both mRNA and amino acids and brings specific amino acids to the ribosome.

Each combination of three nucleotides on the mRNA is called a *codon* or *triplet.* Each of these triplets codes for a specific amino acid. The sequence of codons determines that of amino acids in a protein. Some triplets code for the initiation and termination of amino acid sequences. There are 64 possible codons.

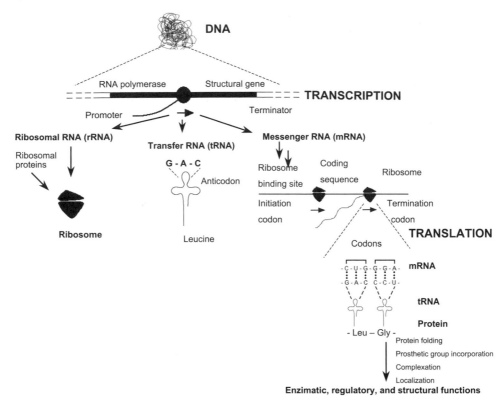

Figure 1.11 Protein synthesis: transcription and translation. From MacGregor (2002).

The sequence of events involved in protein synthesis is illustrated in Fig. 1.11 (MacGregor, 2002).

1.3.3 Plasmids

A *plasmid* is a circular extrachromosomal circular DNA containing from 1000 to 200,000 bp, and reproducing independent from the chromosomal DNA. Plasmids are inherited by daughter cells after cell division. Plasmid replication can be inhibited by *curing* the cells with compounds such as ethidium bromide. Some of the plasmids may exist in a limited number (1–3) of copies (stringent plasmids) or relatively large number (10–220) of copies (relaxed plasmids). Relaxed plasmids are most useful as cloning vectors. Some plasmids cannot coexist, making them incompatible with other plasmids in the same cell.

There are several categories of plasmid:

- *Conjugative plasmids*, which carry genes that code for their own transfer to other cells. *F factors* or *sex factors* are conjugative plasmids that can become integrated into the chromosomes. *E. coli* strains that possess the chromosome-integrated F factors are called *Hfr* (high frequency of recombination).
- *Resistance transfer factors* (*R factors*) are plasmids that confer upon the host cell resistance to antibiotics (e.g., tetracycline, chloramphenicol, streptomycin) and

heavy metals (e.g., mercury, nickel, cadmium). There is a great concern over these plasmids by the medical profession. The widespread use of antibiotics in medicine and agriculture results in the selection of multiple drug-resistant bacteria with R factors (see Chapter 4).

- *Col factors* are plasmids that code for production of colicins, which are proteinaceous bacterial inhibiting substances.

- *Catabolic plasmids* code for enzymes that drive the degradation of unusual molecules such as camphor, naphthalene, and other xenobiotics found in environmental samples. They are important in the field of pollution control. Plasmids can be engineered to contain desired genes and can be replicated by introduction into an appropriate host (see Section 1.3.6).

1.3.4 Mutations

Mutations, caused by physical and chemical agents, change the DNA code and impart new characteristics to the cell, allowing it, for example, to degrade a given xenobiotic or survive under high temperatures. Spontaneous mutations occur in one of 10^6 cells. However, the DNA molecule is capable of self-repair.

Conventional methods are used to obtain desired mutations in a cell. The general approach consists of exposing cells to a mutagen (e.g., ultraviolet [UV] light, chemical) and then expose them to desired environmental conditions. These conditions select for cells having the desired traits.

1.3.5 Genetic Recombinations

Recombination is the transfer of genetic material (plasmid or chromosomal DNA) fron a donor cell to a recipient cell. There are four means by which DNA is transferred to recipient cells (Brock and Madigan, 1991) (Fig. 1.12).

1.3.5.1 *Transformation.* Exogenous DNA enters a recipient cell and becomes an integral part of a chromosome or plasmid. A cell capable of transformation by exogenous DNA is called a *competent* cell. Cell competence is affected by the growth phase of the cells (i.e., physiological state of bacteria), as well as the composition of the growth medium. During transformation, the transforming DNA fragment attaches to the competent cell, is incorporated into the cell, becomes single-stranded, and one strand is integrated into the recipient cell DNA, while the other strand is broken down. Transformation efficiency is increased by treating cells with high concentrations of calcium under cold conditions. The widespread occurrence of DNAses in the environment, particularly in wastewater, affects the transformation frequency.

If the transforming DNA is extracted from a virus, the process is called *transfection*. DNA can be introduced into eukaryotic cells by electroporation (the use of an electric field to produce pores in the cell membrane) or through the use of a particle gun to shoot DNA inside the recipient cell.

1.3.5.2 *Conjugation.* This type of genetic transfer necessitates cell-to-cell contact. The genetic material (plasmid or a fragment of a chromosome mobilized by a plasmid) is transferred upon direct contact between a donor cell (F^+ or male cells) and a recipient cell (F^- or female cells). A special surface structure, called *sex pilus*, of the donor cell

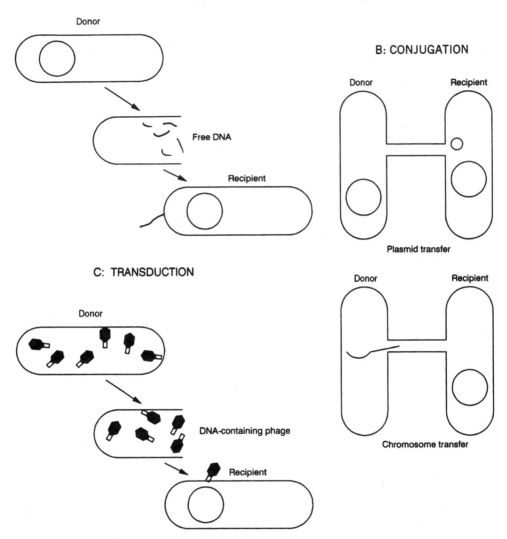

Figure 1.12 Genetic recombinations among bacteria: (a) conjugation; (b) transformation; (c) transduction. Adapted from Brock and Madigan (1988).

triggers the formation of a conjugation bridge that allows the transfer of the genetic material from the donor to the recipient cell. The conjugative pili are encoded by the *tra* genes. Certain bacteria (e.g., enterococci) have a specialized conjugation system induced by signal peptides secreted by the recipient cells. The signal peptides induce the synthesis by the donor cells of proteins involved in cell clumping (Wuertz, 2002).

Gene transfer, through conjugation, has been demonstrated in natural environments and engineered systems, including wastewater, freshwater, seawater, sediments, leaf surfaces, soils, and in the intestinal tract. The transfer rates vary between 10^{-2} and 10^{-8} transconjugants per recipient cell. Plasmids coding for antibiotic resistance can be transferred from environmental isolates to laboratory strains. Biotic and abiotic factors (e.g., cell type and density, temperature, oxygen, pH, surfaces) affect gene transfer by conjugation, but their impact under environmental conditions is not well known.

1.3.5.3 Transduction. This is the transfer of genetic material from a donor to a recipient cell using a bacterial phage as a carrier. A small fragment of the DNA from a donor cell is incorporated into the phage particle. Upon infection of a donor cell, the DNA of the transducing phage particle may be integrated into the recipient cell DNA. Contrary to conjugation, transduction is specific, due to the limited host range of the phage. There are reports on the occurrence of transduction in freshwater and wastewater treatment plants.

1.3.5.4 Transposition. Another recombination process is *transposition*, which consists of the movement (i.e., jumping) of small pieces of plasmid or chromosomal DNA, called *transposons* (jumping genes), from one location to another on the genome. Transposons, which can move from one chromosome to another or from one plasmid to another, carry genes that code for the enzyme transposase, which catalyzes their transposition.

1.3.6 Recombinant DNA Technology: Construction of a Genetically Engineered Microorganism (GEM)

Recombinant DNA technology, commonly known as genetic engineering or gene cloning, is the deliberate manipulation of genes to produce useful gene products (e.g., proteins, toxins, hormones). There are two categories of recombination experiments: (1) *in vitro recombination*, which consists of using purified enzymes to break and rejoin isolated DNA fragments in test tubes; and (2) *in vivo recombination*, which consists of encouraging DNA rearrangements that occur in living cells.

A typical gene cloning experiment consists of the following steps (Fig. 1.13):

1. *Isolation of the source DNA.* Several methods are used for the isolation of DNA from a wide range of cells.

2. *DNA fragmentation or splicing.* Restriction endonucleases are used to cleave the double-stranded DNA at specific sites. These enzymes normally help cells cope with foreign DNA and protect bacterial cells against phage infection. They are named after the microorganism from which they were initially isolated. For example, the restriction enzyme EcoRI was isolated from *E. coli*, whereas HindII enzyme was derived from *Haemophilus influenza*. EcoRI recognizes the following sequence on the double-stranded DNA:

   -G-A-A-T-T-C-
   -C-T-T-A-A-G-

   and produces the following fragments:

   -G-　　　　　A-A-T-T-C-
   -C-T-T-A-A　　　　-G-

   The DNA fragments can be separated according to their size by electrophoresis.

3. *DNA ligation.* DNA fragments are joined to a cloning vector using another enzyme called *DNA ligase*. Ligation is possible because both the source DNA and the cloning vector DNA were cut with the same restriction enzyme. Commonly used cloning vectors are plasmids (e.g., pBR322) or phages (e.g., phage λ).

4. *Incorporation of the recombinant DNA into a host.* The recombinant DNA is introduced into a cell for replication and expression. The recombinant DNA may be introduced into the host cell by transformation, for example. The most popular hosts are prokaryotes such as *E. coli* or eukaryotes such as *Saccharomyces*

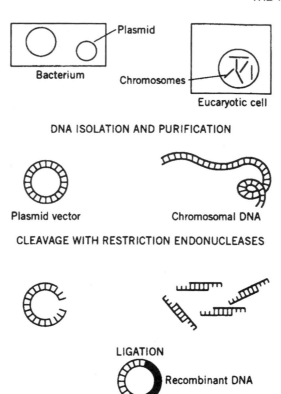

DNA ISOLATION AND PURIFICATION

CLEAVAGE WITH RESTRICTION ENDONUCLEASES

Figure 1.13   Steps involved in gene cloning.

cerevisiae. The host microorganism now containing the recombinant DNA will divide and make clones.

5. *Selection of the desirable clones.* Clones that have the desired recombinant DNA can be screened using markers like antibiotic resistance, that indicate the presence of the cloning vector in the cells. However, the selection of clones having the desired gene can be accomplished by using nucleic acid probes (see Section 1.3.7) or by screening for the gene product. If the gene product is an enzyme (e.g., β-galactosidase), clones are selected by looking for colonies that have the enzyme of interest (hosts cells are grown in the presence of the enzyme substrate).

Biotechnological applications of genetically engineered microorganisms (GEMs) have been realized in various fields, including the pharmaceuticals industry, agriculture, medicine, food industry, energy, and pollution control (see Chapter 18). Notorious applications are the production of human insulin and viral vaccines. In agriculture, research is focusing on the production of *transgenic plants* (i.e., genetically altered whole plants) that are resistant to insects, herbicides, or diseases.

The potential use of GEMs in pollution control is becoming increasingly attractive. It has been proposed to use GEMs to clean up hazardous waste sites and wastewaters by constructing microbial strains capable of degrading recalcitrant molecules. However, there are potential problems associated with the deliberate release of GEMs into the environment. This is because, unlike chemicals, GEMs have the potential ability to grow and reproduce under in situ environmental conditions.

1.3.7 Review of Selected Molecular Techniques

1.3.7.1 DNA Probes. DNA isolated from an environmental sample may be hybridized with a labeled probe, cloned into a plasmid, or may be amplified by polymerase chain reaction (PCR) (MacGregor, 2002). Nucleic acid probes help detect specific microorganisms in an environmental mixture of cells. They are based on nucleic acid hybridization. The two strands of DNA are said to be complementary. Single complementary strands are produced via DNA denaturation, using heat or alkali. Under appropriate conditions, the complementary strands *hybridize* (i.e., bind to each other). A DNA probe is a small piece of DNA (oligonucleotide) that contains specific sequences that, when combined with single-stranded target DNA, will hybridize (i.e., become associated) with the complementary sequence in the target DNA and form a double-stranded structure (Sayler and Blackburn, 1989). For easy detection, the probe is labeled with a radioisotope (e.g., ^{32}P), enzymes (e.g., β-galactosidase, peroxidase, or alkaline phosphatase), fluorescent compounds (e.g., fluorescein isothiocyanate). Gene probes can be used to detect gene sequences in bacterial colonies growing on a solid medium. This technique is called *colony hybridization. Dot hybridization* consists of spotting nucleic acid on a filter and then probing to show the presence/absence of a given sequence. Gene databases are available for checking the gene sequences of microorganisms.

The use of the *polymerase chain reaction* (PCR) greatly enhances the sensitivity of DNA probes. Some of these probes that can be combined with PCR technology, are available for detecting bacterial, viral, and protozoan pathogens and parasites (Sayler and Layton, 1990). However, DNA probes cannot be relied upon to evaluate the safety of disinfected water since they cannot distinguish between infectious and noninfectious pathogens (Moore and Margolin, 1994).

The following are a few applications of nucleic acid probes:

1. *Detection of pathogens in clinical samples.* Probes have been developed for clinically important microorganisms such as *Legionella* spp., *Salmonella* species, enteropathogenic *E. coli*, *Neisseria gonorrhoea*, human immunodeficiency virus (HIV), herpes viruses, or protozoan cysts of *Cryptosporidium* and *Giardia*. Probe sensitivity can be increased by using PCR technology (see Section 1.3.7.5).

2. *Detection of metal resistance genes* in environmental isolates. For example, a probe was constructed for detecting the *mer* operon, which controls mercury detoxification.

3. *Tracking of specific bacteria in the environment.* Probes are useful in following the fate of specific environmental isolates (e.g., nitrogen fixing bacteria or bacteria capable of degrading a specific substrate) and genetically engineered microbes in water, wastewater, biosolids, and soils.

1.3.7.2 RNA-Based Methods. Ribosomal RNA (rRNA) is a good target to probe because of the large number of ribosomes in living cells. RNA-based methods provide information about the activity of microbial communities. These methods aim at detecting rRNA or messenger RNA (mRNA).

rRNA probes are hybridized with extracted target RNA that has been blotted onto positively charged membranes (membrane hybridization) or with fixed target cells (fluorescent in situ hybridization or FISH). rRNA probes give information on community activity and are used for the identification and classification of indigenous microorganisms in environmental samples. These probes can be designed to target specific groups of microorganisms ranging from the subspecies to the domain level (DeLong, 1993; Head et al., 1998; Hofle, 1990; MacGregor, 2002). Their sensitivity is much greater than that of DNA probes. One major advantage of rRNA probes is the availability of more than 15,000 RNA sequences in public databases (Wilderer et al., 2002). However, because indigenous bacteria have a lower number copies of ribosomes than cultured bacteria, signal (e.g., fluorescence) amplification is sometimes needed. For example, up to 20-fold amplification of the signal was observed when using thyramide (Lebaron et al., 1997; Schönhuber et al., 1997). rRNA probes are quite helpful to microbial ecologists in their quest to identify nonculturable bacteria and to gain knowledge on the composition of complex microbial communities (Amann and Ludwig, 2000). The application of rRNA probes to the identification of filamentous bacteria in activated sludge will be discussed in Chapter 9.

mRNA carries information from the DNA to the ribosome. The detection of mRNA, present at much lower number of copies than rRNA, gives information on gene expression and function.

1.3.7.3 Fluorescent In Situ Hybridization (FISH).

We have seen that ribosomal RNA (rRNA) is a convenient target molecule occurring as thousands of copies inside a target cell. Gene probes, labeled with a fluorescent compound (called reporter), hybridize with whole cells in situ. Intact fluorescent cells with the desirable sequence are counted under a fluorescence microscope (Wilderer et al., 2002). Fluorescent in situ hybridization (FISH) can provide information about cell activity in environmental samples. This approach was useful in the identification of filamentous microorganisms in bulking sludge (Kanagawa et al., 2000; Wagner et al., 1994), nitrifying activated sludge (Juretschko et al., 1998), or sulfate-reducing biofilms (Ramsing et al., 1993). It also provides information on the interactions between different microorganisms in an environmental sample.

1.3.7.4 Nucleic Acid Fingerprinting.

This approach involves the use of PCR (see Section 1.3.7.5) to amplify specific fragments of DNA to be analyzed by denaturing gradient gel electrophoresis (DGGE), which allows the separation of fragments, based on length but with different sequences. Thus, this approach gives information concerning the microbial diversity in a wastewater microbial community.

1.3.7.5 Polymerase Chain Reaction (PCR).

The PCR technique was developed in 1986 at Cetus Corporation by Mullis and collaborators (Mullis and Fallona, 1987). This technique essentially simulates in vitro the DNA replication process occurring in vivo; it consists of amplifying discrete fragments of DNA by generating millions of copies of the target DNA (Atlas, 1991; Oste, 1988).

During cell division, two new copies of DNA are made and one set of genes is passed on to each daughter cell. Copies of genes increase exponentially as the number of generations increases. Polymerase chain reaction simulates in vitro the DNA duplication process and can create millions of copies of the target DNA sequence. It consists of three steps that constitute one cycle in DNA replication (Fig. 1.14):

1. *DNA denaturation (strand separation)*: When incubated at high temperature, the target double-stranded DNA fragment is denatured and dissociates into two strands.

Denature DNA and anneal primers

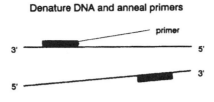

First cycle: Primer extension (complementary strand synthesis)

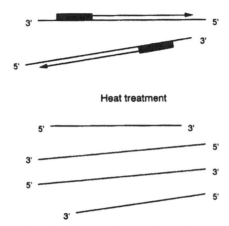

Heat treatment

Second cycle: Primer extension (complementary strand synthesis)

Figure 1.14   Polymerase chain reaction (PCR). Adapted from Brown (1990).

2. *Annealing of primers*: When the temperature is lowered, the target DNA fragment anneals to synthetic nucleotides primers made of 18–28 nucleotides that flank the target DNA fragment. These primers are complementary to the section of the DNA to be replicated. The primers target phylogenetic groups ranging from universal to subspecies level.

3. *Primer extension or amplification step*: The primers are extended with a thermostable DNA polymerase, the enzyme responsible for DNA replication in cells. This thermostable enzyme (*Taq* DNA polymerase) is extracted from *Thermus aquaticus*, a bacterium that is found in hot springs.

After approximately 30 cycles lasting approximately 3 hours, the target DNA fragment is amplified and accumulates exponentially. The PCR technique can be automated by using a DNA thermal cycler that automatically controls the temperatures necessary for the denaturation and annealing steps.

This procedure is very useful in cloning, DNA sequencing, tracking genetic disorders, and forensic analysis. It is a powerful tool in diagnostic microbiology, virology, and parasitology. The tests presently used to detect individuals who have been exposed to the human immunodeficiency virus (HIV) detect only antibodies to the virus and not the virus itself. However, HIV was identified directly, using PCR, in the blood of acquired immunodeficiency syndrome (AIDS) patients from whom the virus was also isolated by more traditional tissue culture techniques. Virus identification via PCR is relatively rapid as compared with traditional culture techniques (Ou et al., 1988). The following are some environmental applications of PCR technology (Atlas, 1991):

1. *Detection of specific bacteria.* Specific bacteria in environmental samples, including wastewater, effluents and sludges can be detected by PCR.

2. *Environmental monitoring of GEMs.* Genetically engineered bacteria that perform certain useful functions (e.g., pesticide or hydrocarbon degradation) can be tracked using PCR technology. The target DNA sequence would be amplified in vitro and then hybridized to a constructed DNA probe.

3. *Detection of indicator and pathogenic microorganisms.* Methodology for the detection of pathogens has shifted from cell cultures to molecular-based techniques because some enteric viruses grow poorly or fail to grow on tissue cultures (Metcalf et al., 1995). Hence, PCR technology has been considered for the detection of foodborne and waterborne pathogens and parasites in water, wastewater, and food. Examples of pathogens and parasites detected by PCR are invasive *Shigella flexneri*, enterotoxigenic *E. coli*, *Legionella pneumophila*, *Salmonella*, *Pseudomonas aeruginosa* (Khan and Cerniglia, 1994), *Yersinia enterocolitica* (Kapperud et al., 1993; Koide et al., 1993; Lampel et al., 1990; Tsai et al., 1993), hepatitis A virus (Divizia et al., 1993; Le Guyader et al., 1994; Prevot et al., 1993; Altmar et al., 1995; Morace et al., 2002), Norwalk virus (DeLeon et al., 1992; Altmar et al., 1995), rotaviruses (Gajardo et al., 1995; Le Guyader et al., 1994), adenoviruses (Girones et al., 1993), astroviruses (Abad et al., 1997), enteroviruses (Abbaszadegan et al., 1993), human immunodeficiency virus (Ansari et al., 1992), *Giardia* (Mahbubani et al., 1991), *Cryptosporidium* (Johnson et al., 1993), and indigenous and nonindigenous microorganisms in pristine environments (Baker et al., 2003). Multiplex PCR using several sets of primers allows the simultaneous detection of gene sequences of several pathogens or different genes within the same organism. For example, a triplex reverse transcriptase (RT)-PCR method was proposed for the simultaneous detection of poliovirus, hepatitis A virus and rotavirus in wastewater (Tsai et al., 1994; Way et al., 1993). Polymerase chain reaction was also used to amplify *lacZ* and *uidA* genes for the detection of total coliforms (β-galactosidase producers) and *E. coli* (β-glucuronidase producer), respectively. An advantage of PCR is the detection of phenotypically negative *E. coli* in environmental samples (see more details in Chapter 4). Improvements of PCR technology for environmental monitoring are needed. Environmental samples as well as chemicals used for virus concentration contain substances (e.g., humic and fulvic acids, heavy metals, beef extract used in virus concentration, and other unknown substances) that interfere with pathogen or parasite detection via PCR. DNA or rRNA purification by gel filtration (using Sephadex®) followed by treatment by an ion exchange resin (use of Chelex®) helps remove the interference due to humic substances and heavy metals, respectively (Abbaszadegan et al., 1993; Straub et al., 1995; Tsai and Olson, 1992; Tsai et al., 1993). Inhibitors found in shellfish extracts can be removed by treatment with

cetyltrimethylammonium bromide prior to PCR (Atmar et al., 1993; Moran et al., 1993) or granular cellulose (Le Guyader et al., 1994). The interference can also be removed by using a magnetic-antibody capture method, as shown for the detection by PCR of *Giardia* and *Cryptosporidium* in environmental samples (Bifulco and Schaefer, 1993; Johnson et al., 1995a, b) or by using an electropositive membrane (ZP60S; AMF Cuno Division, Meriden, CT) (Queiroz et al., 2001).

Polymerase chain reaction does not give, however, an indication of the viability of the pathogens and parasites detected in environmental samples, although some investigators reported a relationship between detection of viral RNA by PCR and the presence of infectious viruses or bacterial phages (Graff et al., 1993; Limsawat and Ohgaki, 1997).

For samples containing RNA viruses, an enzyme called *reverse transcriptase* is used to translate RNA into DNA. Then, the sample is subjected to PCR. The technique combining PCR with reverse transcriptase is called RT-PCR.

Competitive PCR (cPCR) involves the co-amplification of the target DNA and an internal standard, or competitor DNA, which is similar to but distinguishable from the target DNA. Estimation of the number of target sequences is achieved by comparison of ratios between target and competitor sequences with those of a standard curve generated by the amplification of competitor DNA with a range of target DNA concentrations. This standard curve is used to calculate the concentration of the target DNA in the sample (Phillipsa et al., 2000; Zimmermann and Mannhalter, 1996).

The proper decontamination of equipment and surfaces is necessary to avoid false positives by PCR. Among several disinfectants tested, chlorine appears to be the most efficient for degrading nucleic acid sequences contaminants (Ma et al., 1994).

The molecular-based techniques must be validated in order to be considered by regulatory agencies. A major concern is whether the microbial pathogens detected in environmental samples are infectious, thus posing a threat to public health (Metcalf et al., 1995).

1.3.7.6 *Microarrays.*

1.3.7.6 *Microarrays.* Microarrays, also called gene chips, consist of large sets of DNA sequences (probes) or oligonucleotides attached to a nonporous solid support and are hybridized with target fluorescently labeled sequences that have been isolated from environmental samples. Cy-3 and Cy-5 fluorescent dyes are generally used as labels, but alternative fluorescent dyes have been utilized to detect targets in microarrays. The probes, made of PCR products or oligonucleotides, are attached to the solid support by three main printing technologies such as photolithography, ink-jet ejection, or mechanical microspotting. Following hybridization of the probes with the targets, the microarrays are scanned with a high-resolution scanner and the digital images are analyzed, using commercially available software (Call et al., 2003; Zhou and Thompson, 2002). Figure 1.15 (Zhou and Thompson, 2002) shows the various steps involved in microarray preparation and utilization.

Microarrays offer a powerful tool for monitoring gene expression and function as well as the detection and characterization of pathogens in environmental samples. They offer several advantages, including the ability to attach thousands of probes over a very small surface area, high sensitivity, ability to detect several target sequences labeled with different fluorescent tags, low background fluorescence, amenability to automation, and potential for use in field-based studies (Bavykin et al., 2001; Wu et al., 2001; Zhou and Thompson, 2002).

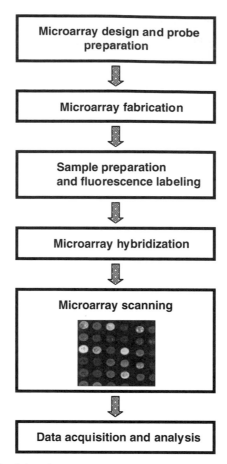

Figure 1.15 Steps involved in microarray preparation and utilization. Adapted from Zhou and Thompson (2002). Figure also appears in color figure section.

1.4 BRIEF SURVEY OF MICROBIAL GROUPS

We will now discuss the various microbial groups encountered in environmental samples. They include the bacteria (prokaryotes), the archaea eukaryotes (algae, protozoa, fungi), and viruses.

1.4.1 Bacteria

1.4.1.1 Bacterial Size and Shape. Except for filamentous bacteria (size may be greater than 100 μm) or cyanobacteria (size range approximately 5–50 μm), bacterial cell size generally ranges between 0.3 μm (e.g., *Bdellovibrio bacteriovorus*; *Mycoplasma*) and 1–2 μm (e.g., *E. coli*; *Pseudomonas*). Australian researchers have discovered an unusually large gram-positive bacterium, *Epulopiscium fishelsoni*, that can reach several hundred μm in length. It lives in a symbiotic relationship in the gut of surgeonfish (Clements and Bullivant, 1991).

Bacteria occur in three basic shapes: cocci (spherical shape; e.g., *Streptococcus*), bacilli (rods; e.g., *Bacillus subtilis*), and spiral forms (e.g., *Vibrio cholera*; *Spirillum volutans*)

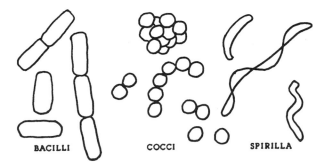

Figure 1.16 Bacterial cell shapes. Adapted from Edmonds (1978).

(Fig. 1.16). Because of their relatively small size, bacteria have a high surface-to-volume ratio, a critical factor in substrate uptake.

1.4.1.2 *Unusual Types of Bacteria (Fig. 1.17).*
• *Sheathed Bacteria*

These bacteria are filamentous microorganisms surrounded by a tubelike structure called a sheath. The bacterial cells inside the sheath are gram-negative rods that become flagellated (swarmer cells) when they leave the sheath. The swarmer cells produce a new sheath at a relatively rapid rate. They are often found in polluted streams and in wastewater treatment plants. This group includes three genera: *Sphaerotilus*, *Leptothrix*, and *Crenothrix*. These bacteria have the ability to oxidize reduced iron to ferric hydroxide (e.g., *Sphaerotilus natans*, *Crenothrix*) or manganese to manganese oxide (e.g., *Leptothrix*). In Chapter 9, we will discuss the role of *Sphaerotilus natans* in activated sludge bulking.

• *Stalked Bacteria*

Stalked bacteria are aerobic, flagellated (polar flagellum) gram-negative rods that possess a stalk, a structure that contains cytoplasm and is surrounded by a membrane and a wall. At the end of the stalk is a holdfast that allows the cells to adsorb to surfaces. Cells may adhere to one another and form rosettes. *Caulobacter* is a typical stalked bacterium that is found in aquatic environments with low organic content. *Gallionella* (e.g., *G. ferruginea*) is another stalked bacterium that makes a twisted stalk, sometimes called "ribbon," consisting of an organic matrix surrounded by ferric hydroxide. These bacteria are present in iron-rich waters and oxidize Fe^{2+} to Fe^{3+}. They are found in metal pipes in water distribution systems (see Chapter 15).

• *Budding Bacteria*

After attachment to a surface, budding bacteria multiply by budding. They make filaments or hyphae at the end of which a bud is formed. The bud acquires a flagellum (the cell is now called a swarmer), settles on a surface, and forms a new hypha with a bud at the tip. *Hyphomicrobium* is widely distributed in soils and aquatic environments and requires one-carbon (e.g., methanol) compounds for growth. A phototrophic bacterium, *Rhodomicrobium*, is another example of budding bacteria.

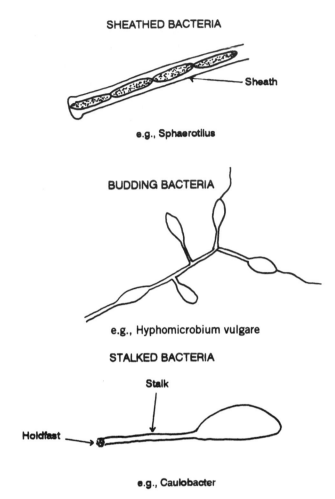

Figure 1.17   Some unusual types of bacteria. Adapted from Lechevalier and Pramer (1971).

• Gliding Bacteria

These filamentous gram-negative bacteria move by gliding, a slow motion on a solid surface. They resemble certain cyanobacteria except that they are colorless. *Beggiatoa* and *Thiothrix* are gliding bacteria that oxidize H_2S to S^0, which accumulates as sulfur granules inside the cells. *Thiothrix* filaments are characterized by their ability to form rosettes (more details are given in Chapters 3 and 9). Myxobacteria are another group of gliding microorganisms. They feed by lysing bacterial, fungal, or algal cells. Vegetative cells aggregate to make "fruiting bodies," which lead to the formation of resting structures called myxospores. Under favorable conditions, myxospores germinate into vegetative cells.

• Bdellovibrio (B. bacteriovorus)

These small (0.2–0.3 μm), flagellated (polar flagellum) bacteria are predatory on gram-negative bacteria. After attaching to the bacterial prey, *Bdellovibrio* penetrates the cells and multiplies in the periplasmic space (space between the cell wall and the plasma membrane). Because they lyse their prey, they are able to form plaques on a lawn of the host bacterium. Some *Bdellovibrio* can grow independently on complex organic media.

• *Actinomycetes*

Actinomycetes are gram-positive filamentous bacteria characterized by mycelial growth (i.e., branching filaments), which is analogous to fungal growth. However, the diameter of the filaments is similar in size to bacteria (approximately 1 μm). Most actinomycetes are strict aerobes, but a few of them require anaerobic conditions. Most of these microorganisms produce spores, and their taxonomy is based on these reproductive structures (e.g., single spores in *Micromonospora* or chains of spores in *Streptomyces*). They are commonly found in water, wastewater treatment plants, and soils (with preference for neutral and alkaline soils). Some of them (e.g., *Streptomyces*) produce a characteristic "earthy" odor that is due to the production of volatile compounds called *geosmins* (see Chapter 16). They degrade polysaccharides (e.g., starch, cellulose), hydrocarbons, and lignin. Some of them produce antibiotics (e.g., streptomycin, tetracycline, chloramphenicol). Two well-known genera of actinomycetes are *Streptomyces* and *Nocardia* (now called *Gordonia*) (Fig. 1.18). *Streptomyces* forms a mycelium with conidial spores at the tip of the hyphae. These actinomycetes are important industrial microorganisms that produce hundreds of antibiotic substances. *Gordonia* is commonly found in water and wastewater and degrades hydrocarbons and other recalcitrant (i.e., hard to degrade) compounds. *Gordonia* is a significant constituent of foams in activated sludge units (see Chapter 9).

• *Cyanobacteria*

Often referred to as blue-green algae, cyanobacteria are prokaryotic organisms that differ from photosynthetic bacteria in the fact that they carry out oxygenic photosynthesis (see Chapter 2) (Fig. 1.19). They contain chlorophyll *a* and accessory pigments such as *phycocyanin* (blue pigment) and *phycoerythrin* (red pigment). The characteristic blue-green color exhibited by these organisms is due to the combination of chlorophyll *a* and phycocyanin. Cyanobacteria occur as unicellular, colonial, or filamentous organisms.

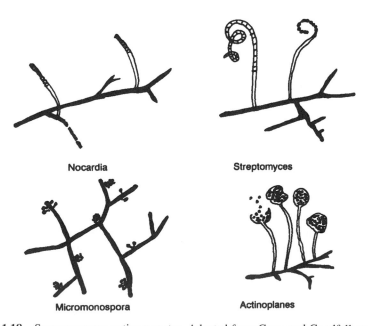

Nocardia   Streptomyces

Micromonospora   Actinoplanes

Figure 1.18 Some common actinomycetes. Adapted from Cross and Goodfellow (1973).

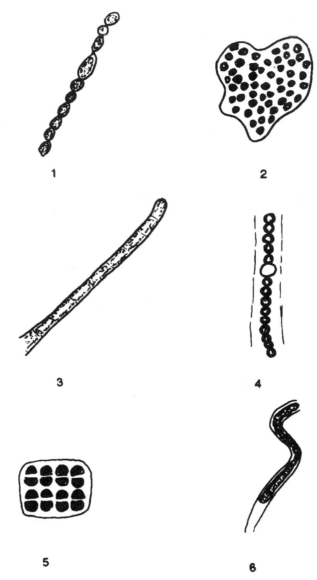

Figure 1.19 Some common cyanobacteria (blue-green algae): (1) *Anabaena*, (2) *Anacystis*, (3) *Oscillatoria*, (4) *Nostoc*, (5) *Agmenellum*, (6) *Lyngbya*. Adapted from Benson (1973).

They propagate by binary fission or fragmentation, and some may form resting structures, called akinetes, which, under favorable conditions, germinate into a vegetative form. Many contain gas vacuoles that increase cell buoyancy and help the cells float to the top of the water column where light is most available for photosynthesis. Some cyanobacteria (e.g., *Anabaena*) are able to fix nitrogen; the site of nitrogen fixation is a structure called a heterocyst.

Cyanobacteria are ubiquitous and, owing to their resistance to extreme environmental conditions (e.g., high temperatures, desiccation), they are found in desert soils and hot springs. They are responsible for algal blooms in lakes and other aquatic environments, and some are quite toxic (see Chapter 16).

1.4.2 Archaea

In the late 1970s, Carl Woese and his collaborators at the University of Illinois proposed the *Archaea* as the third domain of living organisms. This domain is subdivided into three kingdoms: Crenarchaeotes, Euryarchaeotes, and Korarchaeotes. The Euryarchaeota kingdom include the methanogens, which will be discussed in Chapter 13.

Although the archaea are considered to be prokaryotic cells, they possess certain characteristics that are different from those of bacteria or eukaryotes.

- Their membranes are made of branched hydrocarbon chains attached to glycerol by ether linkages.
- Their cell walls do not contain peptidoglycan.
- Their rRNA is different from eukaryotic and prokaryotic rRNA.

It appears that archaea are more closely related to eukaryotes than to bacteria. As regards their metabolism, archaea may range from organotrophs (use of organic compounds as a source of carbon and energy) to chemoautotrophs (use of CO_2 as a carbon source).

Most of the archaea live in extreme environments and are called *extremophiles*. They include the thermophiles, hyperthermophiles, psychrophiles, acidophiles, alkaliphiles, and halophiles. Thus, their unique products are of great interest to biotechnologists. Archaeal enzymes display attractive properties such as tolerance to high and low temperatures, high salt concentrations, high hydrostatic pressures, and organic solvents. The enzymes of interest to biotechnologists include the glycosyl hydrolases (e.g., cellulases, amylases, xylanases), proteases, DNA polymerases, and restriction endonucleases (Cowan and Burton, 2002).

1.4.3 Eukaryotes

1.4.3.1 Fungi. Fungi are eukaryotic organisms that produce long filaments called hyphae, which form a mass called mycellium. Chitin is a characteristic component of the cell wall of hyphae. In most fungi, the hyphae are septate and contain crosswalls that divide the filament into separate cells containing one nucleus each. In some others, the hyphae are nonseptate and contain several nuclei. They are called coenocytic hyphae.

Fungi are heterotrophic organisms that include both macroscopic and microscopic forms. They use organic compounds as a carbon source and energy and thus play an important role in nutrient recycling in aquatic and soil environments. Some fungi form traps that capture protozoa and nematodes. They grow well under acidic conditions (pH 5) in foods, water, or wastewater. Most fungi are aerobic, although some (e.g., yeast) can grow under facultatively anaerobic conditions. Fungi are significant components of the soil microflora, and a great number of fungal species are pathogenic to plants, causing significant damage to agricultural crops. A limited number of species are pathogenic to humans and cause fungal diseases called mycoses. Airborne fungal spores are responsible for allergies in humans. Fungi are implicated in several industrial applications, such as fermentation processes and antibiotic production (e.g., penicillin). In Chapter 12 we will discuss their role in composting.

Identification of fungi is mainly based on the type of reproductive structure. Most fungi produce spores (sexual or asexual spores) for reproduction, dispersal, and resistance to extreme environmental conditions. Asexual spores are formed from the mycelium and germinate, producing organisms identical to the parent. The nuclei of two mating strains fuse to give a diploid zygote, which gives haploid sexual spores following meiosis. There are four major groups of fungi (Fig. 1.20).

• *Phycomycetes*

These fungi are known as the *water molds* and occur on the surface of plants and animals in aquatic environments. They have nonseptate hyphae and reproduce by forming a sac called *sporangium*, which eventually ruptures to liberate *zoospores*, which settle and form a new organism. Some phycomycetes produce sexual spores. There are also terrestrial phycomycetes, such as the common bread mold (*Rhizopus*), which reproduces asexually as well as sexually.

• *Ascomycetes*

Ascomycetes have septate hyphae. Their reproduction is carried out by sexual spores (*ascospores*) contained in a sac called an *ascus* (eight or more ascopores in an ascus), or asexual spores called *conidia*, which are often pigmented. *Neurospora crassa* is a typical ascomycete. Most of the yeasts (e.g., baker's yeast *Saccharomyces cerevisiae*) are classified as ascomycetes. They form relatively large cells that reproduce asexually via budding or fission, and sexually by conjugation and sporulation. Some of these organisms (e.g., *Candida albicans*) are pathogenic to humans. Yeasts, especially the

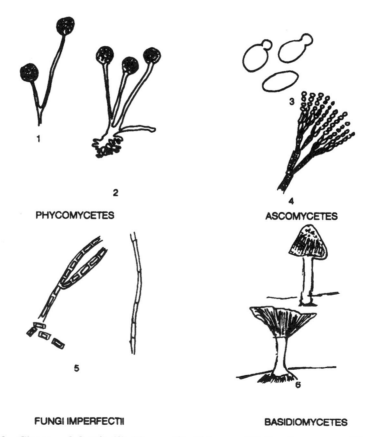

Figure 1.20 Classes of fungi: (1) *Mucor*, (2) *Rhizopus*, (3) *Saccharomyces*, (4) *Penicillium*, (5) *Geotrichum*, (6) fruiting bodies of mushrooms (basidiomycetes).

genus *Saccharomyces*, are important industrial microorganisms involved in bread, wine, and beer making.

• *Basidiomycetes*

Basidiomycetes also have a septate mycelium. They produce sexual spores called *basidiospores* on the surface of a structure called a *basidium*. Four basidiospores are formed on the surface of each basidium. Certain basidiomycetes, the *wood-rotting fungi* play a significant role in the decomposition of cellulose and lignin. Common edible mushrooms (e.g., *Agaricus*) belong to the basidiomycete group. Unfortunately, some of them (e.g., *Amanita*) are quite poisonous.

• *Fungi imperfectii*

Fungi imperfectii have septate hyphae but no known sexual stage. Some of them (e.g., *Penicillium*) are used for the commercial production of important antibiotics. These fungi cause plant diseases and are responsible for mycoses in animals and humans (e.g., athlete's foot).

1.4.3.2 Algae.

Most of algae are floating unicellular microorganisms and are called phytoplankton. Many of them are unicellular, some are filamentous (e.g., *Ulothrix*), and others are colonial (e.g., *Volvox*). Most are free-living organisms, but some form symbiotic associations with fungi (lichens), animals (corals), protozoa, and plants.

Algae play the role of primary producers in aquatic environments, including oxidation ponds for wastewater treatment. Most algae are phototrophic microorganisms (see Chapter 2). They all contain chlorophyll *a*; some contain chlorophyll *b* and *c* as well as other pigments such as xanthophyll and carotenoids. They carry out oxygenic photosynthesis (i.e., they use light as a source of energy and H_2O as electron donor) and grow in mineral media with vitamin supplements and with CO_2 as the carbon source. Under environmental conditions, vitamins are generally provided by bacteria. Some algae (e.g., euglenophyta) are heterotrophic and use organic compounds (simple sugars and organic acids) as a source of carbon and energy. Algae have either asexual or sexual reproduction.

The classification of algae is based mainly on the type of chlorophyll, cell wall structure, and nature of carbon reserve material produced by algae cells (Fig. 1.21).

- *Phylum Chlorophyta* (green algae): These algae contain chlorophylls *a* and *b*, have a cellulosic cell wall, and produce starch as a reserve material.
- *Phylum Chrysophyta* (golden-brown algae): This phylum contains an important group, the diatoms. They are ubiquitous, found in marine and freshwater environments, sediments, and soils. They contain chlorophylls *a* and *c*, and their cell wall typically contains silica (they are responsible for geological formations of diatomaceous earth); they produce lipids as reserve materials.
- *Phylum Euglenophyta*: The euglenophytes contain chlorophylls *a* and *b*, have no cell wall, and store reserves of paramylon, a glucose polymer. *Euglena* is a typical euglenophyte.
- *Phylum Pyrrophyta* (dinoflagellates): The pyrrophyta contain chlorophylls *a* and *c*, have a cellulosic cell wall, and store starch.
- *Phylum Rhodophyta* (red algae): They are found exclusively in the marine environment. They contain chlorophylls *a* and *d* and other pigments such as phycoerythrin; they store starch, and their cell wall is made of cellulose.

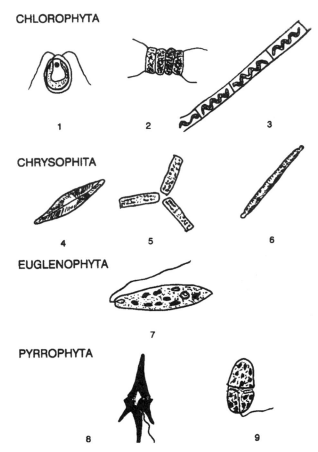

Figure 1.21   Some algae found in water and wastewater: (1) *Chlamydomonas*, (2) *Scenedesmus*, (3) *Spyrogira*, (4) *Navicula*, (5) *Tabellaria*, (6) *Synedra*, (7) *Euglena*, (8) *Ceratium*, (9) *Gymnodynium*. Adapted from Benson (1973).

- *Phylum Phaeophyta* (brown algae): The cells of these exclusively marine algae contain chlorophylls *a* and *c* and xanthophylls; they store laminarin (β 1,3-glucan) as reserve materials, and their cell walls are made of cellulose.

1.4.3.3   *Protozoa.* Protozoa are unicellular organisms that are important from public health and process microbiology standpoints in water and wastewater treatment plants. Cells are surrounded by a cytoplasmic membrane covered by a protective structure called a pellicle. They form *cysts* under adverse environmental conditions. These cysts are quite resistant to desiccation, starvation, high temperatures, lack of oxygen, and chemical insult, namely disinfection in water and wastewater treatment plants (see Chapter 6). Protozoa are found in soils and aquatic environments, including wastewater. Some are parasitic to animals, including humans.

Protozoa are heterotrophic organisms that can absorb soluble food that is transported across the cytoplasmic membrane. Others, the holozoic protozoa, are capable of engulfing particles such as bacteria. Ciliated protozoa use their cilia to move particles toward a mouthlike structure called a *cytostome*. They reproduce by binary fission, although sexual reproduction occurs in some species of protozoa (e.g., *Paramecium*).

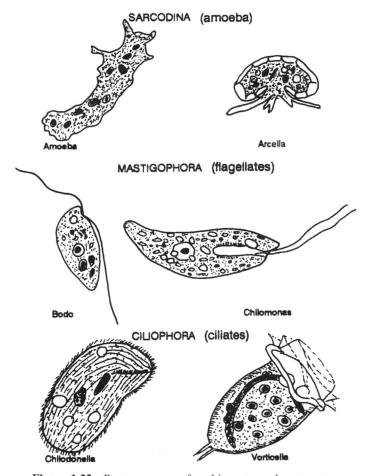

Figure 1.22   Protozoa groups found in water and wastewater.

The type of locomotion is a basis for classification of protozoa (Fig. 1.22). The medically important protozoa that may be transmitted through water and wastewater will be discussed in more detail in Chapter 4.

• *Sarcodina (amoeba)*

These protozoa move by means of *pseudopods* (i.e., false feet). Movement by pseudopods is achieved by changes in the viscosity of the cytoplasm. Many of the amoeba are free-living, but some are parasitic (e.g., *Entamoeba histolytica*). Amoeba feed by absorbing soluble food or by phagocytosis of the prey. The foraminifera are sarcodina, found in the marine environment, that have shells called tests. They can be found as fossils in geologic formations.

• *Mastigophora (flagellates)*

The mastigophora move by means of flagella. Protozoologists include in this group the phytomastigophora, which are photosynthetic (e.g., *Euglena*; this alga is also heterotrophic). Another example of flagellate is the human parasite, *Giardia lamblia* (see Chapter 4). Still another flagellate, *Trypanosoma gambiense*, transmitted to humans by the Tsetse fly, causes the African sleeping sickness, which is characterized by neurological disorders.

• *Ciliophora (ciliates)*

These organisms use cilia for locomotion, but they also help in feeding. A well-known large ciliate is *Paramecium*. Some are parasitic to animals and humans. For example, *Balantidium coli* causes dysentery when cysts are ingested.

• *Sporozoa*

The sporozoa have no means of locomotion and are exclusively parasitic. They feed by absorbing food and produce infective spores. A well-known sporozoan is *Plasmodium vivax*, which causes malaria.

1.4.4 Viruses

Viruses belong neither to prokaryotes nor to eukaryotes; they carry out no catabolic or anabolic function. Their replication occurs inside a host cell. The infected cells may be animal or plant cells, bacteria, fungi, or algae. Viruses are very small colloidal particles (25–350 nm) and most of them can be observed only with an electron microscope. Figure 1.23 shows the various sizes and shapes of some viruses.

1.4.4.1 Virus Structure. A virus is made of a core of nucleic acid (double-stranded or single-stranded DNA; double-stranded or single-stranded RNA) surrounded by a protein coat called a capsid. Capsids are composed of arrangements of various numbers of protein subunits known as capsomeres. The combination of capsid and nucleic acid core is called nucleocapsid. There are two main classes of capsid symmetry. In helical symmetry, the capsid is a cylinder with a helical structure (e.g., tobacco mosaic virus). In polyhedral symmetry, the capsid is an icosahedron, consisting of 20 triangular faces, 12 corners, and 30 edges (e.g., poliovirus). Some viruses have more complex structures (e.g., bacterial phages), and some (e.g., influenza or herpes viruses) have an envelope composed of lipoproteins or lipids.

1.4.4.2 Virus Replication. Bacterial phages have been used as models to elucidate the phases involved in virus replication. The various phases are as follows (Fig. 1.24):

1. *Adsorption.* This is the first step in the replication cycle of viruses. In order to infect the host cells, the virus particle must adsorb to receptors located on the cell surface. Animal viruses adsorb to surface components of the host cell. The receptors may be polysaccharides, proteins, or lipoproteins.
2. *Entry.* This step involves the entry of a virus particle or its nucleic acid into the host cell. Bacteriophages "inject" their nucleic acid into the host cell. For animal viruses the whole virion penetrates the host cell by endocytosis.
3. *Eclipse.* During this step, the virus particle is "uncoated" (i.e., stripping of the capsid), and the nucleic acid is liberated.
4. *Replication.* This step involves the replication of the viral nucleic acid.
5. *Maturation.* The protein coat is synthesized and is assembled with the nucleic acid to form a nucleocapsid.
6. *Release of mature virions.* Virus release is generally attributable to the rupture of the host cell membrane.

1.4.4.3 Virus Detection and Enumeration. There are several approaches to virus detection and enumeration.

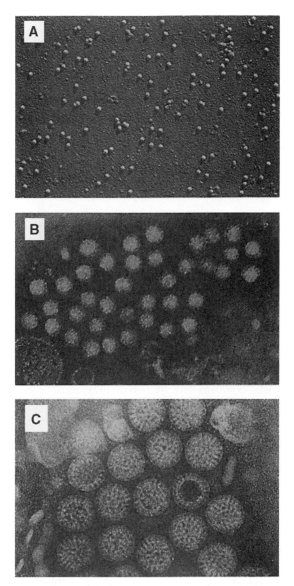

Figure 1.23 Some enteric viruses of public health importance: (a) poliovirus 1; (b) hepatitis A virus; (c) rotavirus. Courtesy of R. Floyd and J. E. Banatvala.

• *Animal Inoculation*

This was the traditional method for detecting viruses before the advent of tissue cultures. Newborn mice are infected with the virus and are observed for symptoms of disease. Animal inoculation is essential for the detection of some enteroviruses such as Coxsackie A viruses.

• *Tissue Cultures*

Viruses are quantified by measuring their effect on established host cell lines, which, under appropriate nutritional conditions, grow and form a monolayer on the inner

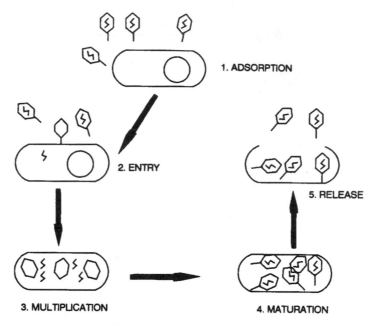

Figure 1.24   Viral lytic cycle.

surface of glass or plastic bottles. There are two main types of host cell lines: (1) *Primary cell lines*, which are cells that are removed directly from the host tissues and can be sub-cultured only a limited number of times; and (2) *Continuous cell lines*, which are animal cells that, after serial subculturing, acquire characteristics that are different from the original cell line, allowing them to be subcultured indefinitely; they are derived from normal or cancerous tissues. Cell lines traditionally used in water virology laboratories include HEp-2, HeLa, VERO, and BGM cells. The Buffalo Green Monkey cell line (BGM) is the most popular and perhaps the most sensitive of all the ones used for the detection of enteroviruses.

Many viruses (enteroviruses, reoviruses, adenoviruses) infect host cells and display a *cytopathic effect*. Others (e.g., rotaviruses, hepatitis A virus) multiply in the cells but do not produce a cytopathic effect. The presence of the latter needs to be confirmed by other tests, including immunological procedures, monoclonal antibodies (MAbs), or nucleic acid probes. Other viruses (e.g., Norwalk type virus) cannot yet be detected by tissue cultures.

• *Plaque Assay*

A viral suspension is placed on the surface of a cell monolayer and, after adsorption of viruses to the host cells, an overlay of soft agar or carboxymethylcellulose is poured on the surface of the monolayer. Virus replication leads to localized areas of cell destruction called *plaques*. The results are expressed in numbers of *plaque-forming units* (PFU). Bacteriophages are also assayed by a plaque assay method based on similar principles. They form plaques that are zones of lysis of the host bacterial lawn (Fig. 1.25).

• *Serial Dilution Endpoint*

Aliquots of serial dilutions of a viral suspension are inoculated into cultured host cells and, after incubation, viral cytopathic effect (CPE) is recorded. The titer or endpoint is

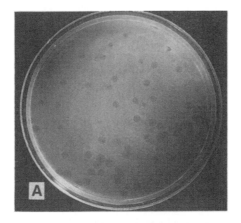

Figure 1.25 Viral enumeration by plaque assay: (a) bacterial phage; (b) animal virus (poliovirus 1).

the highest viral dilution (i.e., smallest amount of viruses) capable of producing CPE in 50 percent of cultures and is referred to as $TCID_{50}$ (tissue culture infectious dose).

• *Most Probable Number (MPN)*

Virus titration is carried out in tubes or 96-well microplates, using three dilutions of the viral suspension. Virus-positive tubes or wells are recorded and the most probable number (MPN) is computed from MPN tables.

1.4.4.4 Rapid Detection Methods

• *Immunoelectron Microscopy*

Viruses are incubated with specific antibodies and examined by electron microscopy for the presence of virus particles aggregated by the antibody. This is a useful technique for examining viruses such as the Norwalk type agent.

• *Immunofluorescence*

A fluorescent dye-labeled antibody is combined with the viral antigen, and the complex formed is observed with a fluorescence microscope. This approach enables the detection of rotaviruses as fluorescent foci in MA-104 or CaCo-2 cultured cells. This method can be accelerated by using flow cytometry. Immunomagnetic separation has been used to detect rotavirus and hepatitis A virus (HAV) in environmental samples.

• *Enzyme Linked Immunosorbent Assay (ELISA)*

A specific antibody is fixed on a solid support and the antigen (virus) is added to form an antigen–antibody complex. An enzyme-labeled specific antibody is then added to the fixed antigen. The presence of the virus is detected by the formation of a colored product upon addition of the enzyme substrate. This enzymatic reaction can be conveniently quantified with a spectrophotometer.

• *Radioimmunoassay (RIA)*

This assay is also based on the binding of an antigen by a specific antibody. The antigen is quantified by labeling the antibody with a radioisotope (e.g., ^{125}I) and measuring the radioactivity bound to the antigen–antibody complex. When viruses growing in host cells are treated with a ^{125}I-labeled antibody, the radioactive foci can be enumerated following contact with a special film. This test, the *radioimmunofocus assay* (RIFA), is used for the detection of hepatitis A virus.

• *Nucleic Acid Probes*

Molecular-based methodology has helped in the detection of viruses that show no growth or grow marginally in tissue cultures (e.g., Norwalk-like viruses, rotaviruses).

Gene probes are pieces of nucleic acid that help identify unknown microorganisms by hybridizing (i.e., binding) to the homologous organism's nucleic acid. For easy detection, the probes can be labeled with radioactive isotopes such as ^{32}P or with enzymes such as alkaline phosphatase, peroxidase, or β-galactosidase. Nucleic acid probes have been used for the detection of viruses (e.g., polioviruses, hepatitis A virus) in environmental samples (water, sediments, shellfish). Single-stranded RNA (ssRNA) probes were used for the detection of hepatitis A virus in shellfish concentrates. Unfortunately, these probes are not sensitive and detect a minimum of 10^6 HAV particles.

Amplification of the target viral sequences by PCR has also been considered. Presently, the most popular method for detecting viruses (enteroviruses, adenoviruses, rotaviruses, astroviruses, Norwalk-like viruses) in environmental samples is the reverse transcriptase-PCR (RT-PCR) method.

1.4.4.5 Virus Classification.

Viruses may be classified on the basis of the host cell they infect. We will briefly discuss the classification of animal, algal, and bacterial phages.

• *Animal Viruses*

Animal viruses are classified mainly on the basis of their genetic material (DNA or RNA), presence of an envelope, capsid symmetry, and site of capsid assembly. All

DNA viruses have double-stranded DNA except members of the parvovirus group, and all RNA viruses have single-stranded RNA, except members of the reovirus group. Tables 1.2 and 1.3 show the major groups of animal viruses. Of great interest to us in this book is the *enteric virus* group, the members of which may be encountered in water and wastewater. This group will be discussed in more detail in Chapter 4.

Retroviruses are a special group of RNA viruses. These viruses make an enzyme, called *reverse transcriptase*, which converts RNA into double-stranded DNA, which integrates into the host genome and controls viral replication. A notorious retrovirus is the human immunodeficiency virus (HIV), which causes acquired immunodeficiency syndrome (AIDS).

• *Algal Viruses*

A wide range of viruses or "viruslike particles" of eukaryotic algae have been isolated from environmental samples. The nucleic acid is generally double-stranded DNA, but is unknown in several of the isolates. Viruses infecting *Chlorella* cells are large particles (125–200 nm diameter), with an icosahedral shape and a linear double-stranded DNA.

Cyanophages were discovered during the 1960s. They infect a number of cyanobacteria (blue-green algae). They are generally named after their hosts. For example, LPP1 cyanophage has a series of three cyanobacterial hosts: *Lyngbia, Phormidium*, and *Plectonema*. They range in size from 20 to 250 nm, and all contain DNA. Cyanophages have been isolated around the world from oxidation ponds, lakes, rivers, and fish ponds. Cyanophages have been proposed as biological control agents for the overgrowth (i.e., blooms) of cyanobacteria. Although some of the experiments were successful on a relatively small scale, cyanobacteria control by cyanophages under field conditions remains to be demonstrated.

• *Bacterial Phages*

Bacteriophages infect a wide range of bacterial types. A typical T-even phage is made of a *head* (capsid), which contains the nucleic acid core, a *sheath* or "tail," which is attached to the head through a "neck," and *tail fibers*, which help in the adsorption of the phage to its host cell. The genetic material is mostly double-stranded DNA, but it may be single-stranded DNA (e.g., ΦX174) or single-stranded RNA (e.g., f2, MS2). Phages adsorb to the host bacterial cell and initiate the lytic cycle, which results in the production of phage progeny and the destruction of the bacterial host cell. Sometimes, the phage becomes incorporated in the host chromosome as a prophage. This process,

TABLE 1.2. Major Groups of Animal DNA Viruses[a]

Group	Parvoviruses	Papovaviruses	Adenoviruses	Herpesviruses	Poxviruses
Capsid symmetry	Cubic	Cubic	Cubic	Cubic	Complex
Virion: naked or enveloped	Naked	Naked	Naked	Enveloped	Complex coat
Site of capsid assembly	Nucleus	Nucleus	Nucleus	Nucleus	Cytoplasm
Reaction to ether (or other liquid solvent)	Resistant	Resistant	Resistant	Sensitive	Resistant
Diameter of virion (nm)	18–26	45–55	70–90	100	230–300

Adapted from Melnick (1976).

[a]All DNA viruses of vertebrates have double-stranded DNA, except members of the parvoviruses, which have single-stranded DNA.

TABLE 1.3. Major Groups of Animal RNA Viruses[a]

Group	Picornavirus	Reovirus	Rotavirus	Rubella	Arbovirus	Myxovirus	Paramyxovirus	Rhabdovirus
Capsid symmetry	Cubic	Cubic	Cubic	Cubic	Cubic	Helical (or unknown)	Helical (or unknown)	Helical (or unknown)
Virion: naked or enveloped	Naked	Naked	Naked	Enveloped	Enveloped	Enveloped	Enveloped	Enveloped
Site of capsid assembly / Reaction to ether (or to other liquid solvent)	Cytoplasm Resistant	Cytoplasm Resistant	Cytoplasm Resistant	Cytoplasm Sensitive	Cytoplasm Sensitive	Cytoplasm Sensitive	Cytoplasm Sensitive	Cytoplasm Sensitive
Diameter of virion (nm)	20–30	75	64–66	60	40	80–120	150–300	60–180

Adapted from Melnick (1976).

[a]All RNA viruses of vertebrates have single-stranded DNA, except members of the reovirus group, which are double-stranded.

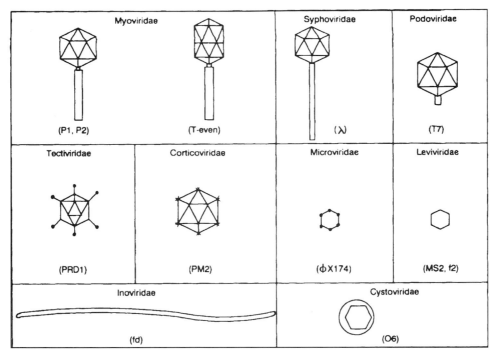

Figure 1.26   Major groups of bacteriophages. Adapted from Jofre (1991) and Coetzee (1987).

called *lysogeny*, does not lead to the destruction of the host cell. Phages infecting *E. coli* are called *coliphages*. These phages are considered as potential indicators of fecal contamination (see Chapter 5). The various families of phages are illustrated in Figure 1.26.

1.5 WEB RESOURCES

General

http://www.cat.cc.md.us/~gkaiser/goshp.html (microbiology lecture material)
http://www.microbes.info/resources/Environmental_Microbiology/   (collection   of websites on microbes)
http://home.att.net/~gallgosp/bacteria.htm (collection of websites on microbes)

Bacteria

http://commtechlab.msu.edu/sites/dlc-me/zoo/ (various good pictures of bacteria, from the Digital Center for Microbial Ecology)
http://commtechlab.msu.edu/sites/dlc-me/zoo/microbes/microbemonth.htm   (various good pictures of bacteria, from the Digital Center for Microbial Ecology)
http://www.micrographia.com/specbiol/bacteri/bacter/bact0100.htm (some good pictures of bacteria and cyanobacteria)
http://www.ulb.ac.be/sciences/biodic/EPageImages.html (good pictures of bacteria from Universite Libre de Bruxelles)
http://www.biology.pl/bakterie_sw/index_en.html (luminescent bacteria)
http://www.lifesci.ucsb.edu/~biolum/ (bioluminescence)

Cyanobacteria

http://www-cyanosite.bio.purdue.edu/ (excellent cyanobacteria and general pictures from Purdue University)

Algae

http://www.dipbot.unict.it/sistematica/Index.html (good pictures, in Italian)
http://www.nmnh.si.edu/botany/projects/algae/Alg-Menu.htm (Smithsonian, decent pictures and basic explanations)
http://www.whoi.edu/redtide/ (Woods Hole Institute Harmful algae page)
http://www.tmcwebatlas.com/WebSites/Disciplinary_and_Academic_Fields/
Life_Science/Algae_and_Fungi/default.asp (web atlas on algae and fungi)
http://www.nwfsc.noaa.gov// (NOAA harmful algae page)
http://www.nhm.ac.uk/hosted_sites/ina/ (nanoplankton web page)
http://www.geo.ucalgary.ca/~macrae/palynology/dinoflagellates/dinoflagellates.html
(University of Calgary page on dinoflagellates)
http://rbg-web2.rbge.org.uk/algae/index.htm?publications/
refs_mann_subset5.htm&main (Royal Botanic Garden of Edinburgh, about diatoms)
http://www.psaalgae.org/(Phycological Society of America)
http://seawifs.gsfc.nasa.gov/SEAWIFS/sanctuary_4.html (NASA SeaWiFS web page)
http://microscope.mbl.edu/baypaul/microscope/general/page_01.htm (good collection of pictures of microorganisms from the Astrobiology Institute, Marine Biological Laboratory, Woods Hole, MA)
http://www.botany.uwc.ac.za/algae/

Protozoa

http://www.uga.edu/~protozoa/ (pictures and information from the Society of Protozoologists)
http://www.life.sci.qut.edu.au/LIFESCI/darben/protozoa.htm (picture collection of protozoa)
http://www3.baylor.edu/~Darrell_Vodopich/welcome.html (color pictures on protozoa and helminthes from Darrell Vodopich, Baylor U. pictures)
http://zoology.okstate.edu/zoo_lrc/zool1604/lab/protozoa.htm (excellent collection of webpages on protozoa; good collection of pictures)
http://www.microscopy-uk.org.uk/micropolitan/index.html (excellent pictures, from Microscopy-UK)
http://www.ulb.ac.be/sciences/biodic/EPageImages.html (good pictures of protozoa from Universite Libre de Bruxelles)

Archaea

http://www.ucmp.berkeley.edu/alllife/threedomains.html (tree of life: the three domains)
http://www.ucmp.berkeley.edu/archaea/archaea.html (introduction to archaea; Berkeley)
http://www.earthlife.net/prokaryotes/archaea.html
http://www3.ncbi.nlm.nih.gov/Taxonomy/Browser/wwwtax.cgi?name=Archaea
(archaea classification)
http://www.daviddarling.info/encyclopedia/A/archaea.html (archaea: general)

http://faculty.washington.edu/leighj/mmarchaea.html
http://www.microbe.org/microbes/archaea.asp

Microbial Genetics

http://www.cat.cc.md.us/courses/bio141/lecguide/unit4/index.html#conjugation
(several animation diagrams)
http://evolution.genetics.washington.edu/phylip/software.html (phylogeny software packages)
http://www.ncbi.nlm.nih.gov/About/primer/microarrays.html (microarray primer)
http://www.ncbi.nlm.nih.gov (National Center for Biotechnology Information)
http://www.tigr.org (Institute for Genomic Research)

1.6 QUESTIONS AND PROBLEMS

1. State whether the following characteristics are seen in prokaryotes or eukaryotes:
   (a) circular DNA
   (b) nuclear membrane
   (c) presence of histones
   (d) mitosis
   (e) binary fission
   (f) produce gametes through meiosis

2. An electron micrograph of a cell shows a cell wall, cytoplasmic membrane, nuclear body without a nuclear membrane, and no endoplasmic reticulum or mitochondria. The cell is:
   (a) a plant cell
   (b) an animal cell
   (c) a bacterium
   (d) a fungus
   (e) a virus

3. Match the following descriptions with the best answer.
   - The movement of water across a membrane from an area of higher water concentration (lower solute concentration) to lower water concentration (higher solute concentration).
   - The net movement of small molecules or ions from an area of higher concentration to an area of lower concentration. No energy is required.
   - Also known as facilitated diffusion, the transport of substances across the membrane. The transport is from an area of higher concentration to lower concentration and no energy is required.
   - If the net flow of water is out of a cell, the cell is in _____ environment.
   - If the net flow of water is into a cell, the cell is in _____ environment.
     A passive transport
     B active transport
     C simple diffusion

D osmosis

E a hypotonic

F a hypertonic

G an isotonic

4. In the complement base pairing of nucleotides, adenine can form hydrogen bonds with _____ and guanine can form hydrogen bonds with _____ .

5. Copies the genetic information in the DNA by complementary base pairing and carries this "message" to the ribosomes where the proteins are assembled. This best describes:

   (a) tRNA

   (b) mRNA

   (c) rRNA

6. Transfer RNA picks up specific amino acids, transfers the amino acids to the ribosomes, and inserts the correct amino acids in the proper place according to the mRNA message. This best describes:

   (a) tRNA

   (b) mRNA

   (c) rRNA

7. Give the ecological role of the cell wall, outer membrane, glycocalyx, and gas vacuoles.

8. What type of plasmid is of interest to bioremediation experts?

9. As compared to DNA, why is RNA a good target to probe?

10. Give and explain the stages involved in protein synthesis.

11. Explain the main features of fluorescent in situ hybridization (FISH).

12. What problems are encountered in PCR when handling environmental samples?

13. What is, essentially, RT-PCR?

14. What are some differences between Archaea and Bacteria?

15. Give the different groups of algae and point out the groups that are exclusively of marine origin?

16. Give the steps involved in the viral lytic cycle.

17. Compare lytic to lysogenic cycles in virus replication.

2

MICROBIAL METABOLISM AND GROWTH

2.1 INTRODUCTION

In Chapter 1 we examined the microbial world, with particular emphasis on the structure of microbial cells as well as the range of microorganisms found in the environment. In the present chapter we will examine enzyme kinetics, and the metabolism and growth kinetics of microbial populations.

Wastewater Microbiology, *Third Edition*, by Gabriel Bitton
Copyright © 2005 John Wiley & Sons, Inc.

2.2 ENZYMES AND ENZYME KINETICS

2.2.1 Introduction

Enzymes are protein molecules that serve as catalysts of biochemical reactions in animal, plant, and microbial cells. Commercial applications have been found for enzymes in the food (e.g., wine, cheese, beer), detergent, medical, pharmaceutical, and textile industries. However, no significant application has been proposed yet for wastewater treatment (see Chapter 18).

Enzymes do not undergo structural changes after participating in chemical reactions and can be used repeatedly. They lower the activation energy and increase the rate of biochemical reactions. Enzymes may be intracellular or extracellular and are generally quite specific for their substrates. The substrate combines with the active site of the enzyme molecule to form an enzyme–substrate complex (ES). A new product (P) is formed and the unchanged enzyme E is ready to react again with its substrate:

$$E + S \underset{k_{-1}}{\overset{k_1}{\rightleftharpoons}} ES \overset{k_2}{\rightarrow} E + P$$

Some nonprotein groups or cofactors may become associated with enzyme molecules and participate in the catalytic activity of the enzyme. These include coenzymes (e.g., nicotinamide adenine dinucleotide, coenzyme A, flavin–adenine dinucleotide, or flavin mononucleotide) and metallic activators (e.g., K, Mg, Fe, Co, Cu, Zn, Mn, Mo). Dehydrogenase enzymes require coenzymes (e.g., FAD, FMN, NAD, coenzyme A, biotin), which accept the hydrogen removed from the substrate.

Enzymes are currently subdivided into six classes:

1. *Oxidoreductases*: responsible for oxidation and reduction processes in the cell.
2. *Transferases*: responsible for the transfer of chemical groups from one substrate to another.
3. *Hydrolases*: hydrolyze carbohydrates, proteins, and lipids into smaller molecules (e.g., β-galactosidase hydrolyses lactose into glucose and galactose).
4. *Lyases*: catalyze the addition or removal of substituent groups.
5. *Isomerases*: catalyze isomer formation.
6. *Ligases*: catalyze the joining of two molecules, using an energy source such as ATP.

2.2.2 Enzyme Kinetics

The most important factors controlling enzymatic reactions are substrate concentration, pH, temperature, ionic strength, and the presence of toxicants.

At low substrate concentration, the enzymatic reaction rate V is proportional to the substrate concentration (first-order kinetics). At higher substrate concentrations, V reaches a plateau (zero-order kinetics). The enzymatic reaction rate V as a function of substrate concentration is given by the *Michaelis–Menten equation* (2.1):

$$V = \frac{V_{max}[S]}{K_m + [S]} \tag{2.1}$$

where V = reaction rate (units/time), V_{max} = maximum reaction rate (units/time), $[S]$ = substrate concentration (mol/L), and K_m = half saturation constant (Michaelis constant), which is the substrate concentration at which V is equal to $V_{max}/2$.

The hyperbolic curve shown in Figure 2.1a can be linearized, using the Lineweaver–Burke plot (Fig. 2.1b), which uses the reciprocals of both the substrate concentration and the reaction rate V:

$$\frac{1}{V} = \frac{K_m}{V_{max}} \frac{1}{[S]} + \frac{1}{V_{max}} \qquad (2.2)$$

Plotting $1/V$ versus $1/[S]$ gives a straight line with a slope of K_m/V_{max}, y intercept of $1/V_{max}$ and an x intercept of $-1/K_m$ (Fig. 2.1b). V_{max} and K_m can be obtained directly from the Lineweaver–Burke plot.

2.2.3 Effect of Inhibitors on Enzyme Activity

A wide range of toxicants are commonly found in industrial and municipal wastewater treatment plants. These inhibitors decrease the activity of enzyme-catalyzed reactions. There are three types of enzyme inhibition: competitive, noncompetitive, and uncompetitive (Fig. 2.2).

2.2.3.1 Competitive Inhibition. In competitive inhibition, the inhibitor (I) and the substrate (S) compete for the same reactive site on the enzyme (E). In the presence of a

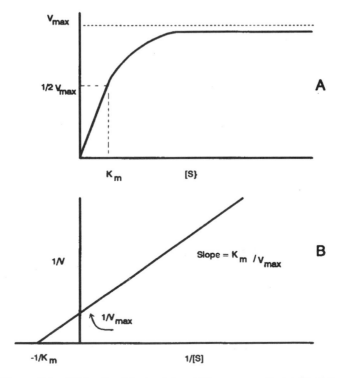

Figure 2.1 Michaelis–Menten (a) and Lineweaver–Burke (b) plots.

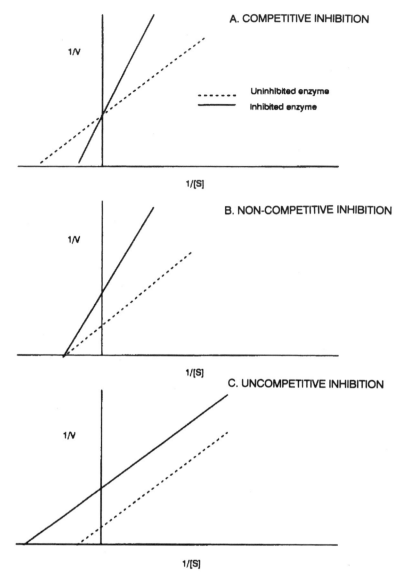

Figure 2.2 Types of enzyme inhibition: (a) competitive inhibition; (b) noncompetitive inhibition; (c) uncompetitive inhibition. Adapted from Bitton and Koopman (1986) and Marison (1988b).

competitive inhibitor I, the reaction rate V is given by Eq. (2.3):

$$V = \frac{V_{\max}[S]}{[S] + K_m(1 + [I]/K_i)} \tag{2.3}$$

where $[S]$ = substrate concentration (mol/L), $[I]$ = inhibitor concentration (mol/L), K_i = inhibition coefficient, $V_{\max}$ = maximum enzyme reaction rate (t^{-1}), and K_m = Michaelis constant (mol/L).

The Lineweaver–Burke representation of Eq. (2.3) is as follows:

$$\frac{1}{V} = \frac{K_m}{V_{max}}(1 + [I]/K_i)\frac{1}{[S]} + \frac{1}{V_{max}} \tag{2.4}$$

In competitive inhibition, V_{max} is unaffected whereas K_m is increased by a factor of $(1 + [I]/K_i)$ (Fig. 2.2a).

2.2.3.2 *Noncompetitive Inhibition.* In noncompetitive inhibition, the inhibitor I can bind to both the enzyme E and the ES complex (Fig. 2.2b). According to Michaelis–Menten kinetics, the reaction rate is given by Eq. (2.5):

$$V = \frac{V_{max}[S]}{(K_m + [S])(1 + [I]/K_i)} \tag{2.5}$$

The Lineweaver–Burke transformation of Eq. (2.5) is as follows:

$$\frac{1}{V} = \frac{K_m}{V_{max}}(1 + [I]/K_i)\frac{1}{[S]} + \frac{1}{V_{max}}(1 + [I]/K_i) \tag{2.6}$$

In the presence of the inhibitor, the slope of the double reciprocal plot is increased whereas V_{max} is decreased. K_m remains unchanged.

2.2.3.3 *Uncompetitive Inhibition.* In uncompetitive inhibition the inhibitor binds to the enzyme–substrate complex (ES) but not to the free enzyme (Fig. 2.2c).

$$V = \frac{V_{max}[S]}{K_m + [S](1 + [I]/K_i)} \tag{2.7}$$

The Lineweaver–Burke plot transformation gives Eq. (2.8):

$$\frac{1}{V} = \frac{K_m}{V_{max}}\frac{1}{[S]} + \frac{1}{V_{max}}(1 + [I]/K_i) \tag{2.8}$$

In the presence of an inhibitor, the slope of the reciprocal plot remains the same, but both V_{max} and K_m are affected by the inhibitor concentration (Fig. 2.2c).

2.3 MICROBIAL METABOLISM

2.3.1 Introduction

Metabolism is the sum of biochemical transformations that includes interrelated catabolic and anabolic reactions. Catabolic reactions are *exergonic* and release energy derived from organic and inorganic compounds. Anabolic reactions (i.e., biosynthetic) are *endergonic*: they use the energy and chemical intermediates provided by catabolic reactions for biosynthesis of new molecules, cell maintenance, and growth. The relationship between catabolism and anabolism is shown in Figure 2.3.

The energy generated by catabolic reactions is transferred to energy-rich compounds such as adenosine triphosphate (ATP) (Fig. 2.4). This phosphorylated compound is

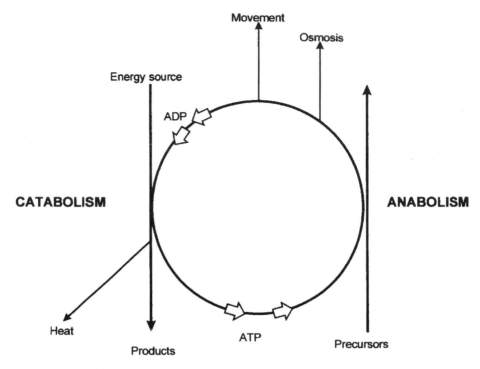

Figure 2.3   Relationship between anabolic and catabolic reactions. From Scragg (1988).

composed of adenine, ribose (a five-carbon sugar), and three phosphates; it has two high-energy bonds that release chemical energy when hydrolyzed to adenosine diphosphate (ADP). Upon hydrolysis under standard conditions, each molecule of ATP releases approximately 7500 calories.

$$A - P \sim P \sim P + H_2O \longleftrightarrow A - P \sim P + Pi + Energy \qquad (2.9)$$

where A = Adenine + Ribose.

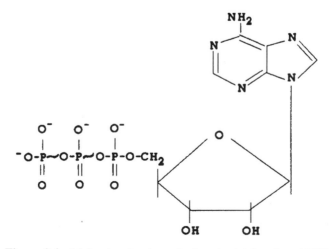

Figure 2.4   Molecular structure of adenosine triphosphate (ATP).

The energy released is used for biosynthetic reactions, active transport, or movement, and some of it is dissipated as heat. Other energy-rich phosphorylated compounds are phosphoenolpyruvate (PEP) ($\Delta G^0 = -14.8$ kcal/mol) and 1,3-diphosphoglycerate ($\Delta G^0 = -11.8$ kcal/mol). Adenosine triphosphate is generated by three mechanisms of phosphorylation.

2.3.1.1 Substrate-Level Phosphorylation.
Substrate-level phosphorylation is the direct transfer of high-energy phosphate to ADP from an intermediate in the catabolic pathway. It produces all the energy in fermentative microorganisms but only a small portion of the energy in aerobic and anaerobic microorganisms. In fermentation, glucose is transformed to pyruvic acid through the Embden–Meyerhof pathway. Pyruvic acid is further transformed to several byproducts by various microorganisms (e.g., alcohol by yeasts or lactic acid by *Streptococcus lactis*). Substrate-level phosphorylation results in ATP formation via the transfer of phosphate groups from a high-energy phosphorylated compound such as 1,3-diphosphoglyceric acid, to ADP. In a typical fermentation, only two molecules of ATP (equivalent to approximately 15,000 calories) are released per molecule of glucose. The free energy ΔG^0 released by the combustion of a molecule of glucose being $-686,000$ calories, the efficiency of the process is only 2 percent ($14,000/686,000 \times 100 = 2$ percent).

2.3.1.2 Oxidative Phosphorylation (Electron Transport System).
Adenosine triphosphate can also be generated by oxidative phosphorylation where electrons are transported through the *electron transport system* (ETS) from an electron donor to a final electron acceptor, which may be oxygen, nitrate, sulfate, or CO_2. Further details are given in Section 2.3.2.

The electron transport system is located in the cytoplasmic membrane of prokaryotes and in mitochondria of eukaryotes. Although the electron carriers used by bacteria are similar to those utilized in mitochondria, the former can use alternate final electron acceptors (e.g., nitrate or sulfate). It is assumed that the ATP yield for oxidative phosphorylation using alternate final electron acceptors is lower than in the presence of oxygen.

2.3.1.3 Photophosphorylation.
Photophosphorylation is a process by which light energy is converted to chemical energy (ATP) and occurs in both eukaryotes (e.g., algae) and prokaryotes (e.g., cyanobacteria, photosynthetic bacteria). The ATP generated via photophosphorylation drives the reduction of CO_2 during the dark phase of photosynthesis. In photosynthetic organisms CO_2 serves as the source of carbon. The electron transport system of these organisms supplies both ATP and NADPH for the synthesis of cell food. Green plants and algae use H_2O as the electron donor and release O_2 as a byproduct. Photosynthetic bacteria do not produce oxygen from photosynthesis and use H_2S as the electron donor (more details are given in Section 2.3.2).

2.3.2 Catabolism

2.3.2.1 Aerobic Respiration.
Respiration is an ATP-generating process that involves the transport of electrons through an electron transport system. The substrate is oxidized, O_2 being used as the terminal electron acceptor. As indicated earlier, a small portion of ATP is also generated via substrate-level phosphorylation. The electron donor may be an organic compound (e.g., oxidation of glucose by heterotrophic microorganisms) or an

inorganic compound (e.g., oxidation of H_2, Fe(II), NH_4, or S^0 by chemoautotrophic micro-organisms). The breakdown of glucose involves the following steps:

• *Glycolysis (Embden−Meyerhof−Parnas Pathway) (Fig. 2.5)*

In glycolysis, glucose is first phosphorylated and cleaved into two molecules of a key intermediate compound, glyceraldehyde-3-phosphate, which is then converted into pyruvic acid. Thus, glycolysis is the oxidation of one molecule of glucose to two molecules of pyruvic acid, a three-carbon compound. This pathway results in the production of two molecules of NADH and a net gain of two molecules of ATP (4 ATP molecules produced − 2 molecules ATP used) per molecule of glucose.

Some microorganisms use the pentose phosphate pathway to oxidize pentose. Glucose oxidation via this pathway produces 12 molecules of NADPH and one molecule of ATP.

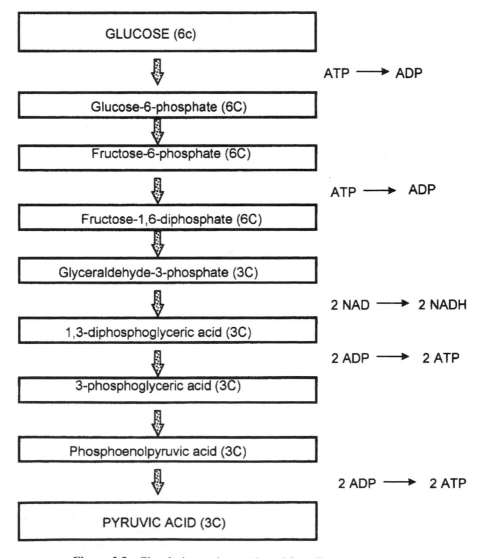

Figure 2.5 Glycolysis reactions. Adapted from Tortora et al. (1989).

Some prokaryotic microorganisms use the Entner–Doudoroff pathway, which produces two molecules of NADPH and one molecule of ATP per molecule of glucose oxidized.

• Transformation of Pyruvic Acid to Acetyl Coenzyme A

Pyruvic acid is decarboxylated (i.e., loss of one CO_2) to an acetyl group that combines with coenzyme A to give acetyl-CoA, a two-carbon compound. During this process NAD^+ is reduced to NADH.

$$\text{Pyruvic acid} + NAD^+ + \text{Coenzyme A} \longrightarrow \text{acetyl-CoA} + NADH + CO_2 \qquad (2.10)$$

• Krebs Cycle

The Krebs cycle (citric acid cycle or tricarboxylic acid cycle) (Fig. 2.6), in which pyruvate is completely oxidized to CO_2, releases more energy than glycolysis; it occurs in mitochondria in eukaryotes but is associated with the cell membrane in prokaryotes. Acetyl-CoA enters the Kreb's cycle and combines with oxaloacetic acid (4-C compound) to form the six-carbon citric acid. The cycle consists of a series of biochemical reactions, each one being catalyzed by a specific enzyme. It results in the production of two molecules of CO_2 per molecule of acetyl-CoA entering the cycle. The oxidation of compounds in the cycle releases electrons that reduce NAD^+ to NADH or flavin–adenine dinucleotide (FAD) to FADH. For each acetyl CoA, 3 NADH and 1 FADH are formed. Moreover, one GTP (guanosine triphosphate) is formed by substrate-level phosphorylation during the oxidation of α-glutaric acid to succinic acid.

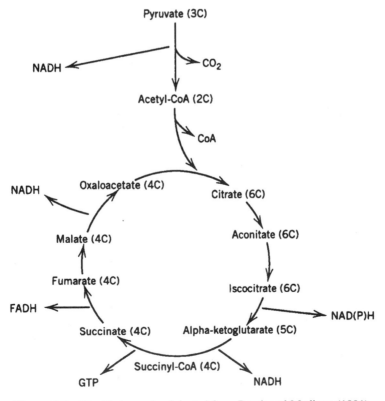

Figure 2.6 The Krebs cycle. Adapted from Brock and Madigan (1991).

• *Electron Transport System*

The electron transport system (ETS) is a chain of electron carriers associated with the cytoplasmic membrane in prokaryotes and mitochondria in eukaryotes. This system transfers the energy stored in NADH and FADH into ATP. The chain consists of electron carriers such as flavin mononucleotide (FMN), coenzyme Q, iron–sulfur proteins, and cytochromes. For efficient trapping of energy, there is a stepwise release of ATP as the electrons are transported from one carrier to another to reach the final electron acceptor. Each molecule of NADH and FADH generates three and two molecules of ATP, respectively. A series of steps are involved in the electron transport system (ETS) (Fig. 2.7):

1. Electrons originating from the substrate are transferred to NAD (precursor: nicotinic acid), which is reduced to NADH.

2. Flavoproteins such as FMN and FAD accept hydrogen atoms and donate electrons.

3. Quinones (coenzyme Q) are lipid-soluble carriers that also accept hydrogen atoms and donate electrons.

4. Cytochromes are proteins with porphyrin rings containing ferric iron, which is reduced to Fe^{2+}. There are several classes of cytochromes designated by letters. The sequence of electron transport is the following: cytochrome b → cytochrome c → cytochrome a.

5. Oxygen is the final electron acceptor in aerobic respiration. Anaerobic respiration involves final electron acceptors other than oxygen. These electron acceptors may be NO_3, SO_4, CO_2, or some organic compounds. Anaerobic respiration releases less energy than aerobic respiration.

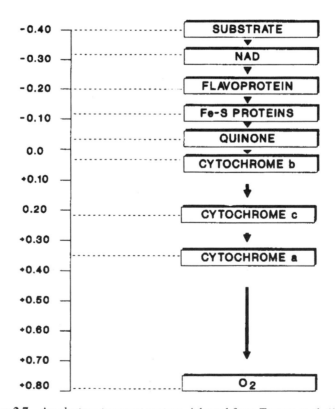

Figure 2.7   An electron transport system. Adapted from Tortora et al. (1989).

Inhibitors (e.g., cyanide, carbon monoxide), acting at various points of the ETS, can inhibit the electron carriers and thus interfere with both the electron flow and ATP synthesis. Other inhibitors (e.g., dinitrophenol), called *uncouplers*, inhibit only ATP synthesis.

· *Summary of ATP Release by Aerobic Respiration*
Aerobic respiration releases 38 ATP molecules per molecule of glucose oxidized to CO_2. A breakdown of ATP molecules released is as follows:

Glycolysis: Oxidation of glucose to pyruvic acid
- Yield of substrate-level phosphorylation .... 2 ATP
- Yield of two molecules of NADH going through ETS if oxygen is available .... 6 ATP

Transformation of pyruvic acid to acetyl-Coenzyme A
- Yields 2 molecules of NADH (there are two molecules of pyruvic acid/molecule glucose)
- Yield of 2 NADH .... 6 ATP

Krebs Cycle
- For each acetyl-coenzyme A going through the cycle, 12 ATP are generated (9 from NADH, 2 from FADH, and 1 from GTP)
- 2 acetyl-coenzyme A molecules .... 24 ATP

Total ATP/molecule glucose .... 38 ATP

The efficiency of complete oxidation of glucose via aerobic respiration to CO_2 is approximately 38 percent.

2.3.2.2 Fermentation.
Fermentation is the transformation of pyruvic acid to various products in the absence of a terminal electron acceptor. Both the electron donor and acceptor are organic compounds and ATP is generated solely via substrate-level phosphorylation. Fermentation releases little energy (2 ATP/molecule of glucose) and most of it remains in fermentation products. The latter depends on the type of microorganism involved in fermentation. For example, ethanol, lactic acid, and propionic acid are formed by *Saccharomyces*, *Lactobacillus*, and *Propionobacterium*, respectively. Alcoholic fermentation is a well-known fermentation carried out by yeasts. Alcoholic fermentation proceeds as follows:

$$\text{Pyruvic acid} \xrightarrow[\text{NADH}_2 \rightarrow \text{NAD}]{\nearrow^{CO_2}} \text{Ethyl alcohol} \tag{2.11}$$
$$\underset{\text{CH}_3\text{COCOOH}}{} \qquad\qquad \underset{\text{CH}_3\text{CH}_2\text{OH}}{}$$

It involves the two following steps:

1. Decarboxylation step (i.e., removal of carboxyl group)

$$\text{Pyruvic acid} \xrightarrow{\nearrow^{CO_2}} \text{acetaldehyde} \tag{2.12}$$

2. Reduction of acetaldehyde

$$\text{Acetaldehyde} \longrightarrow \text{ethyl alcohol} \tag{2.13}$$

The overall reaction for glucose fermentation to ethanol by yeasts is as follows:

$$\underset{\text{glucose}}{C_6H_{12}O_6} \longrightarrow \underset{\text{ethanol}}{2\,C_2H_5OH} + \underset{\text{carbon dioxide}}{2\,CO_2} \qquad (2.14)$$

Yeasts can switch from aerobic respiration to fermentation, depending upon growth conditions.

Fermenting bacteria include strict anaerobes (e.g., *Clostridium* spp.), facultative anaerobes, which can carry out oxidative phosphorylation (*Pseudomonas* spp.) and aerotolerant anaerobes (e.g., *Lactobacillus* spp.) (Leis and Flemming, 2002).

2.3.3 Anabolism

Anabolism (biosynthesis) includes all the energy-consuming processes that result in the formation of new cells. It is estimated that 3000 μmoles of ATP are required to make 100 mg of dry mass of cells. Moreover, most of this energy is used for protein synthesis (Brock and Madigan, 1991). Cells use energy (ATP) to make building blocks, synthesize macromolecules, repair damage to cells (maintenance energy), and maintain movement and active transport across the cell membrane. Most of the ATP generated by catabolic reactions is used for biosynthesis of biological macromolecules such as proteins, lipids, polysaccharides, purines, and pyrimidines. Most of the precursors of these macromolecules (amino acids, fatty acids, monosaccharides, nucleotides) are derived from intermediates formed during glycolysis, the Krebs cycle, and other metabolic pathways (Entner–Doudoroff and pentose phosphate pathways). These precursors are linked together by specific bonds (e.g., peptide bond for proteins, glycoside bond for polysaccharides, phosphodiester bond for nucleic acids) to form cell biopolymers.

2.3.4 Photosynthesis

Photosynthesis is a process that converts light energy into chemical energy, using CO_2 as a carbon source, and light as an energy source. Light is absorbed by chlorophyll molecules that are found in algae, plants, and photosynthetic bacteria.

The general equation for photosynthesis is:

$$6\,CO_2 + 12\,H_2O + \text{Light} \longrightarrow C_6H_{12}O_6 + 6\,H_2O + 6\,O_2 \qquad (2.15)$$

Oxygenic photosynthesis consists of two types of reactions: (1) *light reactions*, which result essentially in the conversion of light energy into chemical energy (ATP) and the production of NADPH; and (2) *dark reactions*, in which NADPH is used to reduce CO_2.

2.3.4.1 Light Reactions.   Most of the energy captured by photosynthetic pigments is contained in wavelengths between 400 nm (visible light) and 1100 nm (near-infrared light). Algae, green photosynthetic bacteria, and purple photosynthetic bacteria absorb light at 670–685 nm, 735–755 nm, and 850–1000 nm, respectively.

Light is absorbed by chlorophyll *a* (Fig. 2.8), a compound made up of pyrrole rings surrounding a Mg atom, and displaying two absorption peaks, the first peak being at 430 nm and the second at 675 nm (Fig. 2.9). The pigments are organized in clusters called *photosystem I* (p700) and *photosystem II* (p680), which play a role in the transfer of electrons from H_2O to NADP (photosynthetic bacteria have only photosystem I).

The flow of electrons in oxygenic photosynthesis has a path that resembles the letter "Z" and is called the *noncyclic electron flow* or *Z scheme* (Fig. 2.10; Boyd, 1988). Upon

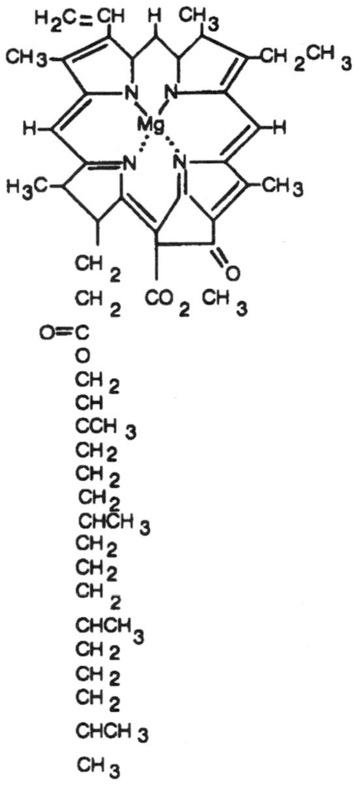

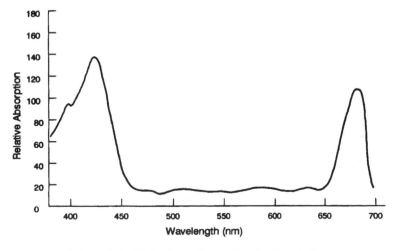

Figure 2.8 Chemical structure of chlorophyll *a*.

exposure of photosystem II to light, the energy released boosts an electron to a higher energy level as the oxidation–reduction potential becomes more negative. The electron hole is filled by an electron produced through photolysis of H_2O. Adenosine triphosphate is produced as the boosted electron travels downhill, via electron carriers (e.g., quinones, cytochromes, plastocyanin), and is captured by photosystem I (p700). Upon exposure of

Figure 2.9 Light absorption peaks of chlorophyll *a*.

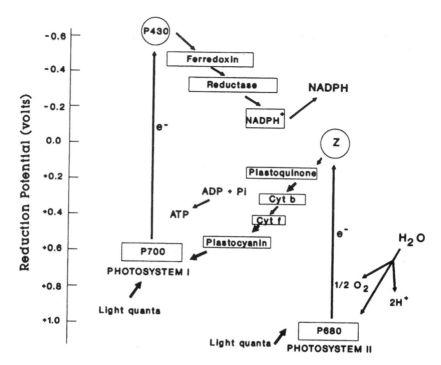

Figure 2.10   Electron transfer during light reactions in oxygenic photosynthesis: The Z scheme. Adapted from Boyd (1988).

p700 to light, the electron released is again boosted to a higher energy level. The NADP is reduced to NADPH after downhill electron flow, via other electron carriers (e.g., ferredoxin). Oxygen is a byproduct of light reactions carried out in oxygenic photosynthesis by algae, cyanobacteria, and plants. Sometimes, the Z scheme displays a cyclic flow of electrons, which results in the production of ATP but not NADPH.

2.3.4.2 Dark Reactions (Calvin–Benson Cycle).

The endergonic dark reactions, collectively called the Calvin cycle or Calvin–Benson cycle, take place in the stroma of chloroplasts and use products of the light reactions to fix carbon. The reducing compound (NADPH) and the energy (ATP) produced during the light reactions are used to reduce CO_2 to organic compounds during the Calvin–Benson cycle, the first step of which is the combination of CO_2 with ribulose-1,5-diphosphate (RuDP), a reaction catalyzed by the enzyme ribulose bisphosphate carboxylase. A subsequent series of reactions leads to the formation of one hexose molecule, fructose-6-phosphate, which can be converted to glucose.

The overall equation for the dark reactions is given by

$$6\,CO_2 + 18\,ATP + 12\,H_2O + 12\,NADPH \longrightarrow C_6H_{12}O_6 + 12\,Pi$$
$$+ 18\,ADP + 12\,NADP^+ \tag{2.16}$$

Thus, the fixation of one molecule of CO_2 requires three molecules of ATP and two molecules of NADPH.

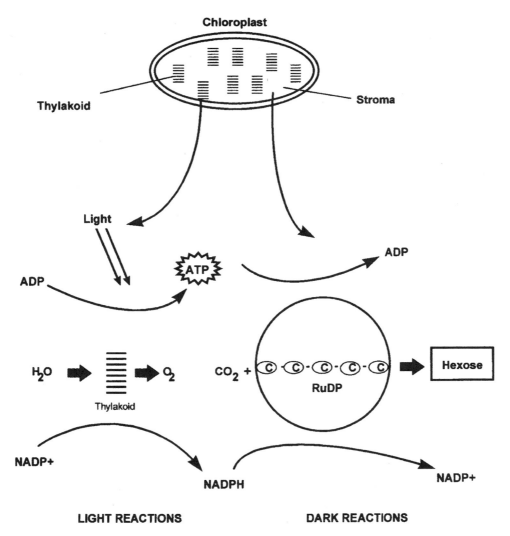

Figure 2.11 Summary of light and dark reactions in photosynthesis. Adapted from Norton (1986).

The coupling of light and dark reactions in photosynthesis is summarized in Figure 2.11 (Norton, 1986).

2.3.5 Metabolic Classification of Microorganisms

The major elements that enter in the composition of a microbial cell are carbon, oxygen, nitrogen, hydrogen, phosphorus, and sulfur. Other nutrients necessary for biosynthesis of cell components include cations (e.g., Mg^{2+}, Ca^{2+}, Na^+, K^+, Fe^{2+}), anions (e.g., Cl^-, SO_4^{2-}), trace elements (e.g., Co, Cu, Mn, Mo, Zn, Ni, Se), which serve as components or cofactors of several enzymes, and growth factors such as vitamins (e.g., riboflavin, thiamin, niacin, vitamin B_{12}, folic acid, biotin, vitamins B_6). A typical *E. coli* cell contains approximately 70 percent water, 3 percent carbohydrates, 3 percent amino acids, nucleotides, and lipids, 22 percent macromolecules (mostly proteins as well as RNA, DNA), and 1 percent inorganic ions.

Microorganisms need a carbon source (CO_2 or organic carbon) and an energy source (light or energy derived from the oxidation of inorganic or organic chemicals). The metabolic classification of microorganisms is based on two main criteria: energy source and carbon source.

2.3.5.1 Phototrophs.

These microorganisms use light as the source of energy. They are subdivided into photoautotrophs and photoheterotrophs.

• *Photoautotrophs*

This group includes algae, cyanobacteria (blue-green algae), and photosynthetic bacteria also called phototrophic bacteria. Photoautotrophs uses CO_2 as a carbon source, and H_2O, H_2, or H_2S as electron donors. Photosynthetic bacteria carry out *anoxygenic photosynthesis*, and most of them require anaerobic conditions. Oxygen is detrimental to the synthesis of their photosynthetic pigments, bacteriochlorophylls, and carotenoids. For cyanobacteria and algae, the electron donor is H_2O, whereas for photosynthetic bacteria the donor is H_2S (some cyanobacteria are also able to use H_2S as electron donor, resulting in the deposition of S^0 outside their cells). The general taxonomy of phototrophic bacteria is shown in Table 2.1. There are approximately 60 species of phototrophic bacteria broadly grouped into *purple* and *green* bacteria. Purple bacteria contain bacteriochlorophyll *a* (maximum absorption at 825–890 nm) and *b* (maximum absorption at about 1000 nm), whereas green bacteria contain bacteriochlorophyll *c*, *d*, and *e*, which absorb light at wavelengths between 705 and 755 nm. These phototrophic bacteria (e.g., chromatiaceae, chlorobiaceae) use CO_2 as a carbon source, light as an energy source, and reduced sulfur compounds (e.g., H_2S, S^0) as electron donors.

Anoxygenic photosynthesis, using H_2S as the reductant source, is summarized as follows:

$$12\,H_2S + 6\,CO_2 \longrightarrow C_6H_{12}O_6 + 6\,H_2O + 12\,S^0 \tag{2.17}$$

S^0 is deposited inside (in the case of purple bacteria) or outside (in the case of green bacteria) the cells of photosynthetic bacteria.

• *Photoheterotrophs (or photoorganotrophs)*

This group comprises all the facultative heterotrophs that derive energy from light or from organic compounds that serve as carbon sources and electron donors. The purple nonsulfur bacteria, the Rhodospirillaceae, use organic compounds as electron donors.

2.3.5.2 Chemotrophs.

These microorganisms obtain their energy via oxidation of inorganic or organic compounds. They are subdivided into lithotrophs (chemoautotrophs) and heterotrophs (organotrophs).

• *Lithotrophs (Chemoautotrophs)*

Lithotrophs use carbon dioxide as a carbon source (carbon fixation) and derive their energy (ATP) needs by oxidizing inorganic compounds such as NH_4, NO_2, H_2S, Fe^{2+}, or H_2. Most of them are aerobic.

Nitrifying bacteria.   These are widely distributed in soils, water, and wastewater, and oxidize ammonium to nitrate (see Chapter 3 for more details).

$$NH_4^+ \xrightarrow{\textit{Nitrosomonas}} NO_2^- \xrightarrow{\textit{Nitrobacter}} NO_3^- \tag{2.18}$$

TABLE 2.1. Recognized Genera of Anoxygenic Phototrophic Bacteria[a]

Taxonomic Group	Morphology
Purple bacteria	
Purple sulfur bacteria (Chromatiaceae and Ectothiorhodospiraceae)	
Amoebobacter	Cocci embedded in slime; contain gas vesicles
Chromatium	Large or small rods
Lamprocystis	Large cocci or ovoids with gas vesicles
Lamprobacter	Large ovals with gas vesicles
Thiocapsa	Small cocci
Thiocystis	Large cocci or ovoids
Thiodictyon	Large rods with gas vesicles
Thiospirillum	Large spirilla
Thiopedia	Small cocci with gas vesicles; cells arranged in flat sheets
Ectothiorhodospira	Small spirilla; do not store sulfur inside the cell
Purple nosulfur bacteria (Rhodospirillaceae)	
Rhodocyclus	Half-circle or cirle
Rhodomicrobium	Ovoid with stalked budding morphology
Rhodopseudomonas	Rods, dividing by budding
Rhodobacter	Rods and cocci
Rhodopila	Cocci
Rhodospirillum	Large or small spirilla
Green bacteria	
Green sulfur bacteria (Chlorobiaceae)	
Anacalochloris	Prosthecate spheres with gas vesicles
Chlorobium	Small rods or vibrios
Pelodictyon	Rods or vibrios, some form three-dimensional net; contain gas vesicles
Prosthecochloris	Spheres with prosthecae
Green gliding bacteria (Chloroflexaceae)	
Chloroflexus	Narrow filaments (multicellular), up to 100 μm long
Chloroherpeton	Short filaments (unicellular)
Chloronema	Large filaments (multicellular), up to 250 μm long; contain gas vesicles
Oscillochloris	Very large filaments, up to 2500 μm long, contain gas vesicles

[a]From Madigan (1988).

Sulfur oxidizing bacteria. These bacteria use hydrogen sulfide (H_2S), elemental sulfur (S^0), or thiosulfate ($S_2O_3^{2-}$) as energy sources. They are capable of growth in very acidic environments (pH $\leq$ 2). The oxidation of elemental sulfur to sulfate is given by:

$$S + 3\,O_2 + 2\,H_2O \xrightarrow[\substack{\textit{Thiobacillus thiooxidans} \\ \textit{(Acidithiobacillus thiooxidans)}}]{} 2\,H_2SO_4 + \text{energy} \qquad (2.19)$$

Iron bacteria. These include bacteria that are acidophilic, derive energy via oxidation of Fe^{2+} to Fe^{3+}, and are also capable of oxidizing sulfur (e.g., *Thiobacillus ferrooxidans*).

Others oxidize ferrous iron at neutral pH (e.g., *Sphaerotilus natans*, *Leptothrix ochracea*, *Crenothrix*, *Clonothrix*, *Gallionella ferruginea*).

Hydrogen bacteria. These (e.g., *Hydrogenomonas*) use H_2 as the energy source and CO_2 as the carbon source. Hydrogen oxidation is catalyzed by a hydrogenase enzyme. These bacteria are facultative lithotrophs since they can grow also in the presence of organic compounds.

· *Heterotrophs (Organotrophs)*

This is the most common nutritional group among microorganisms and includes mostly bacteria, fungi, and protozoa. Heterotrophic microorganisms obtain their energy via oxidation of organic matter. Organic compounds serve both as energy source and carbon source. This group includes the majority of bacteria, fungi, and protozoa in the environment.

2.4 MICROBIAL GROWTH KINETICS

Prokaryotic organisms such as bacteria reproduce mainly by binary fission (i.e., each cell gives two daughter cells). Growth of a microbial population is defined as an increase in numbers or an increase in microbial mass. *Growth rate* is the increase in microbial cell numbers or mass per unit time. The time required for a microbial population to double in numbers is the *generation time* or *doubling time*, which may vary from minutes to days.

Microbial populations can grow as batch cultures (closed systems) or as continuous cultures (open systems) (Marison, 1988a).

2.4.1 Batch Cultures

When a suitable medium is inoculated with cells, the growth of the microbial population follows the growth curve displayed in Figure 2.12, which shows four distinct phases.

2.4.1.1 Lag Phase. The lag phase is a period of cell adjustment to the new environment. Cells are involved in the synthesis of biochemicals and undergo enlargement. The duration of the lag phase depends on the cells' prior history (age, prior exposure to

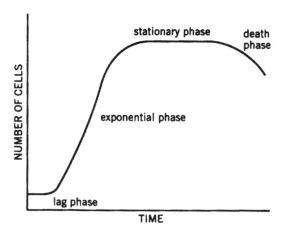

Figure 2.12 Microbial growth curve.

damaging physical or chemical agents, culture medium). For example, no lag phase is observed when an exponentially growing culture is transferred to a similar medium with similar growth conditions. Conversely, a lag period is observed when damaged cells are introduced into the culture medium.

2.4.1.2 *Exponential Growth Phase (Log Phase).* The number of cells increase exponentially during the log phase. The exponential growth varies with the type of microorganism and growth conditions (e.g., temperature, medium composition). Under favorable conditions, the number of bacterial cells (e.g., *Escherichia coli*) double every 15–20 min. The growth follows a geometric progression ($2^0 \rightarrow 2^1 \rightarrow 2^2 \rightarrow 2^n$).

$$X_t = X_0 e^{\mu t} \tag{2.20}$$

where μ = specific growth rate (h^{-1}), X_t = cell biomass or numbers after time t, and X_0 = initial number or biomass of cells.

Using the natural logarithms on both sides of Eq. (2.20), we obtain

$$\ln X_t = \ln X_0 + \mu t \tag{2.21}$$

where μ is given by

$$\mu = \frac{\ln X_t - \ln X_0}{t} \tag{2.22}$$

If n is the number of population doublings (i.e., number of generations) after time t, the doubling time t_d is given by

$$t_d = \frac{t}{n} \tag{2.23}$$

μ is related to the doubling time t_d by

$$\mu = \frac{\ln 2}{t_d} = \frac{0.693}{t_d} \tag{2.24}$$

Cells in the exponential growth phase are more sensitive to physical and chemical agents than those in the stationary phase.

2.4.1.3 *Stationary Phase.* The cell population reaches the stationary phase because microorganisms cannot grow indefinitely, mainly because of lack of nutrients and electron acceptors, and the production and the accumulation of toxic metabolites. Secondary metabolites (e.g., certain enzymes, antibiotics) are produced during the stationary phase. There is no net growth (cell growth is balanced by cell death or lysis) of the population during the stationary phase.

2.4.1.4 *Death Phase.* During this phase, the death (decay) rate of the microbial population is higher than the growth rate. Cell death may be accompanied by cell lysis. The viable count of microorganisms decreases, although the turbidity of the microbial suspension may remain constant.

2.4.2 Continuous Culture of Microorganisms

So far, we have described the growth kinetics of batch cultures. Maintenance of microbial cultures at the exponential growth phase over a long period of time can be achieved by growing continuously the cells in a completely mixed reactor in which a constant volume is maintained. The most commonly used device is the *chemostat* (Fig. 2.13), which is essentially a complete-mix bioreactor without recycle. In addition to the flow rate of growth-limiting substrate, environmental parameters such as oxygen level, temperature, and pH are also controlled. The substrate is added continuously at a flow rate Q to a reactor with a volume V containing concentration X of microorganisms. The dilution rate D, the reciprocal of the hydraulic retention time t, is given by

$$D = \frac{Q}{V} = \frac{1}{t} \tag{2.25}$$

where D = dilution rate (time^{-1}), V = reactor volume (L), Q = flow rate of substrate S (L/time), and t = time.

In continuous-flow reactors, microbial growth is described by

$$\frac{dX}{dt} = \mu X - DX = X(\mu - D) \tag{2.26}$$

$$X_t = X_0 e^{(\mu - D)t} \tag{2.27}$$

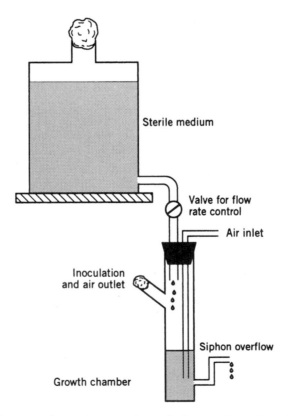

Figure 2.13 Chemostat for continuous culture of microorganisms. From Marison (1988a).

Equations (2.26) and (2.27) show that the supply rate of the limiting substrate controls the specific growth rate μ. At $D > \mu_{max}$, we observe a decrease in cell concentration and a washout of the population. Cell washout starts at the critical dilution rate D_c, which is approximately equal to μ_{max}.

The mass balance for X is given by

$$V\frac{dX}{dt} = \mu XV - QX \tag{2.28}$$

$$= \frac{\mu_{max}S}{K_s + S}XV - QX \tag{2.29}$$

At steady state

$$\frac{dX}{dt} = 0 \;\rightarrow\; \mu = D = \frac{Q}{V} = \frac{\mu_{max}S}{K_s + S} \tag{2.30}$$

At steady state, the substrate concentration S and the cell concentration X in the reactor are given by Eqs (2.30) and (2.31), respectively:

$$S = K_s\frac{D}{\mu_{max} - D} \tag{2.31}$$

$$X = Y(S_i - S_e) \tag{2.32}$$

where Y = growth yield, S_i = influent substrate concentration, and S_e = effluent substrate concentration.

The steady state breaks down at very low or very high dilution rates.

2.4.3 Other Kinetic Parameters

There are three important parameters in microbial growth kinetics: growth yield Y, specific growth rate μ, and specific substrate uptake rate q.

The rate of increase of microorganisms in a culture (dX/dt) is proportional to the rate of substrate uptake/removal (dS/dt) by microbial cells:

$$\frac{dX}{dt} = Y\frac{dS}{dt} \tag{2.33}$$

where Y = growth yield coefficient expressed as mg cells formed per mg of substrate used, dX/dt = rate of increase in microorganism concentration (mg/L/day), and dS/dt = rate of substrate removal (mg/L/day).

A more simplified equation showing the relationship between the three parameters is the following:

$$\mu = Yq \tag{2.34}$$

where μ = specific growth rate (time^{-1}), Y = growth yield (mg cells formed per mg of substrate removed), and q = substrate uptake rate (mg/L/day).

2.4.3.1 Growth Yield. As shown above, growth yield is the amount of biomass formed per unit of amount of substrate removed. It reflects the efficiency of conversion of substrate to cell material. The yield coefficient Y is obtained as

$$Y = \frac{X - X_0}{S_0 - S} \tag{2.35}$$

where S_0 and S = initial and final substrate concentrations, respectively (mg/L or mol/L), X_0 and X = initial and final microbial concentrations, respectively.

Several factors influence the growth yield: type of microorganisms, growth medium, substrate concentration, terminal electron acceptor, pH, and incubation temperature. Yield coefficients for several bacterial species are within the range 0.4–0.6 (Heijnen and Roels, 1981).

For a pure microbial culture growing on a single substrate, the growth yield Y is assumed to be constant. However, in the environment, particularly in wastewater, there is a wide range of microorganisms, few of which are in the logarithmic phase. Many are in the stationary or in the declining phase of growth. Some of the energy will be used for cell maintenance. Thus, the growth yield Y must be corrected for the amount of cell decay occurring during the declining phase of growth. This correction will give the true growth yield coefficient, which is lower than the measured yield. Equation (2.34) becomes:

$$\mu = Yq - k_d \tag{2.36}$$

where k_d is the endogenous decay coefficient (day^{-1})

2.4.3.2 Specific Substrate Uptake Rate q. The specific substrate uptake (removal) is given by

$$q = dS/dt/X \tag{2.37}$$

where q (time^{-1}) is given by the Monod's equation:

$$q = q_{max} \frac{[S]}{K_s + [S]} \tag{2.38}$$

2.4.3.3 Specific Growth Rate μ. This is given by:

$$\mu = \frac{dX/dt}{X} \tag{2.39}$$

where μ (day^{-1}) is given by Monod's equation:

$$\mu = \mu_{max} \frac{[S]}{K_s + [S]} \tag{2.40}$$

The in situ specific growth rate of bacteria in wastewater was measured using the labeled thymidine growth assay (thymidine is a precursor of DNA in cells). In an aerobic tank, the specific growth rate μ was 0.5 d^{-1} (doubling time t_d = 1.4 d) whereas in an anaerobic tank μ was equal to 0.2 d^{-1} (t_d = 3.9 d) (Pollard and Greenfield, 1997).

In waste treatment, the reciprocal of μ is the *biological solid retention time* θ_c, that is

$$\mu = 1/\theta_c \tag{2.41}$$

Thus

$$1/\theta_c = Yq - k_d \tag{2.42}$$

2.4.4 Physical and Chemical Factors Affecting Microbial Growth

2.4.4.1 *Substrate Concentration.* The relationship between the specific growth rate μ and substrate concentration S is given by the Monod's equation (Fig. 2.14a):

$$\mu = \mu_{max} \frac{[S]}{K_s + [S]} \tag{2.43}$$

where μ_{max} = maximum specific growth rate (h^{-1}), S = substrate concentration (mg/L), K_s = half-saturation constant (mg/L). This is the substrate concentration at which the specific growth rate is equal to $\mu_{max}/2$. K_s represents the affinity of the microorganism for the substrate. μ_{max} and K_s are influenced by temperature, type of carbon source, and other factors.

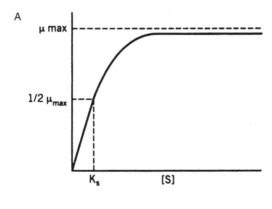

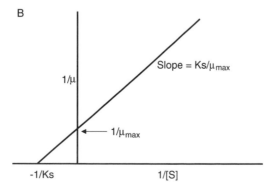

Figure 2.14 Relationship between the specific growth rate μ and substrate concentration S: (a) Monod's saturation curve; (b) Lineweaver–Burke plot.

Monod's equation can be linearized using the Lineweaver–Burke equation:

$$\frac{1}{\mu} = \frac{K_s}{\mu_{max}[S]} + \frac{1}{\mu_{max}} \tag{2.44}$$

Figure 2.14b shows a plot of $1/\mu$ vs. $1/S$. The slope, y-intercept, and x-intercept are (K_s/μ_{max}), $(1/\mu_{max})$, and $(-1/K_s)$, respectively. This plot allows the computation of K_s and μ_{max}. K_s values for individual chemicals found in wastewater are between 0.1 and 1.0 mg/L (Hanel, 1988).

2.4.4.2 Temperature.

This is one of the most important factors affecting microbial growth and survival. Microbial growth can occur at temperatures varying from below freezing to more than 100°C. Based on the optimum temperature for growth, microorganisms are classified as mesophiles, psychrophiles, thermophiles, or extreme thermophiles. Microbial growth rate is related to temperature by the Arrhenius equation:

$$\mu = Ae^{-E/RT} \tag{2.45}$$

where A = constant, E = activation energy (kcal/mole), R = gas constant, and T = absolute temperature (K).

Psychrophiles can grow at low temperatures because their cell membrane has a high content of unsaturated fatty acids, which helps maintain membrane fluidity, whereas a high content of saturated fatty acids help thermophiles function at high temperatures. The decreased μ at high temperatures is due to the thermal denaturation of proteins, particularly enzymes, as well as changes in membrane structure, leading to alterations in cell permeability.

2.4.4.3 pH.

Biological treatment of wastewater occurs generally at neutral pH. In general, the optimum pH for bacterial growth is around 7, although some may be obligately acidophilic (e.g., *Thiobacillus*, *Sulfolobus*) and thrive at pH < 2. Fungi prefer acidic environments with a pH of 5 or lower. Cyanobacteria grow optimally at pH higher than 7. Bacterial growth generally results in a decrease of the pH of the medium through the release of acidic metabolites (e.g., organic acids, H_2SO_4). Conversely, some microorganisms can increase the pH value of their surrounding milieu (e.g., denitrifying bacteria, algae).

pH affects the activity of microbial enzymes. It affects the ionization of chemicals and thus plays a role in the transport of nutrients and toxic chemicals into the cell.

2.4.4.4 Oxygen Level.

Microorganisms can grow in the presence or in the absence of oxygen. There are divided into strict aerobes, facultative anaerobes (can grow in the presence or in the absence of oxygen), and strict anaerobes. Aerobic microorganisms use oxygen as the terminal electron acceptor in respiration. Anaerobic counterparts use other electron acceptors such as sulfate, nitrate, or CO_2. Some microorganisms are micro-aerophilic and require low levels of oxygen for growth. Through their metabolism, aerobes may render the environment suitable for anaerobes by using oxygen.

Upon reduction, oxygen forms toxic products such as superoxide (O_2^-), hydrogen peroxide (H_2O_2), or hydroxyl radicals. However, microorganisms have acquired enzymes to deactivate them. For example, H_2O_2 is destroyed by catalase and peroxidase enzymes, whereas O_2^- is deactivated by superoxide dismutase. Catalase and superoxide dismutase-catalyzed reactions are represented by

$$2\,O_2^- + 2\,H^+ \xrightarrow[\substack{\text{superoxide} \\ \text{dismutase}}]{} O_2 + H_2O_2 \tag{2.46}$$

$$2\,H_2O_2 \xrightarrow[\text{catalase}]{} 2\,H_2O + O_2 \tag{2.47}$$

2.4.5 Measurement of Microbial Biomass

Various approaches are available for measuring microbial biomass in laboratory cultures or in environmental samples (Sutton, 2002).

2.4.5.1 *Total Number of Microbial Cells.*

Total number of cells (live and dead cells) can be measured by using special counting chambers such as the Petroff–Hauser chamber for bacterial counts or the Sedgewick–Rafter chamber for algal counts. The use of a phase-contrast microscope is required when nonphotosynthetic microorganisms are under consideration. Presently, the most popular method consists of retaining the cells on a membrane filter treated to suppress autofluorescence (use of polycarbonate filters treated with Irgalan Black) and staining the cells with fluorochromes such as acridine orange (AO) or 4′,6-diamidino-2-phenylindol (DAPI). The microorganisms are subsequently counted using an epifluorescence microscope (Kepner and Pratt, 1994). An advantage of DAPI is its stable fluorescence. A wide range of other fluorochromes are available for many applications in environmental microbiology studies. These include, among others, PicoGreen, SYBR-Green 1 and 2, Hoechst 33342, YOYO-1, and SYTO dyes (green, red, and blue) (Neu and Lawrence, 2002).

Scanning electron microscopy (SEM) has also been considered for measuring total microbial numbers.

Electronic particle counters are also used for determining the total number of microorganisms in a sample. These instruments do not differentiate, however, between live and dead microorganisms, and very small cells may be missed.

Flow cytometers are fluorescence-activated cell sorters and include a light source (argon laser or a mercury lamp) and a photodetector, which measures fluorescence (use correct excitation wavelength) and scattering of the cells. They sort and collect cells with predefined optical parameters. They are often used in the biomedical and aquatic microbiology fields (Paul, 1993). They have been used to sort algal cells and to distinguish between cyanobacteria from other algae, based on phycoerythrin (orange) and chlorophyll (red) fluorescence. They can help identify microorganisms when combined with fluorescent antibodies.

2.4.5.2 *Measurement of the Number of Viable Microbes on Solid Growth Media.*

This approach consists of measuring the number of viable cells capable of forming colonies on a suitable growth medium.

Plate count is determined by using the pour plate method (0.1–1 mL of microbial suspension is mixed with molten agar medium in a petri dish), or the spread plate method (0.1 mL of bacterial suspension is spread on the surface of an agar plate). The results of plate counts are expressed as *colony forming units* (CFU). The number of CFU per plate should be between 30 and 300. Membrane filters can also be used to determine microbial numbers in dilute samples. The sample is filtered and the filter is placed directly on a suitable growth medium.

Culture-based methods have been routinely used in soil, aquatic, and wastewater microbiology, but they reveal only about 0.1–10 percent of the total bacterial counts in most environments (Pickup, 1991). Indeed, some microorganisms (e.g., *E. coli, Salmonella typhimurium, Vibrio* spp.) can enter into the *viable but nonculturable* (VBNC) state and are not detected by plate counts, especially when using selective growth media (Koch, 2002; Roszak and Colwell, 1987). The VBNC state can be triggered by factors such as nutrient deprivation or exposure to toxic chemicals. This phenomenon is particularly important for pathogens that may remain viable in the VBNC state for longer periods of time than previously thought. The VBNC pathogens may remain virulent and cause disease in humans and animals.

2.4.5.3 *Measurement of Active Cells in Environmental Samples.*

Several approaches have been considered for assessing microbial viability/activity in environmental samples (Fig. 2.15). Epifluorescence microscopy, in combination with the use of oxido-reduction dyes, is used to determine the percent of active cells in aquatic environments. The most popular oxido-reduction dyes are INT (2-(*p*-iodophenyl)-3-(*p*-nitrophenyl)-5-phenyl tetrazolium chloride) and CTC (cyanoditolyl tetrazolium chloride) (Posch et al., 1997; Pyle et al., 1995a). A good correlation was found between the number of CTC-positive *E. coli* cells and the CFU (colony forming units) count, regardless of the growth phase (Créach et al., 2003).

The direct viable count (DVC) method was pioneered by Kogure and his collaborators in Japan (Kogure et al., 1984). The sample is incubated with trace amounts of yeast extract

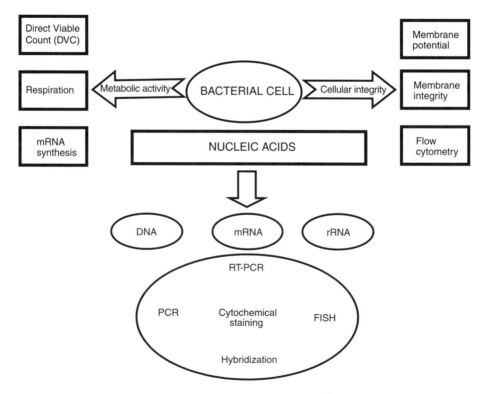

Figure 2.15 Approaches used for assessing microbial viability/activity. From Keer and Birch (2000).

and nalidixic acid. The latter blocks DNA replication but not RNA synthesis. This leads to cell elongation of active cells, which are counted using epifluorescence microscopy. Some methods allow the detection in aquatic samples of specific bacterial pathogens, including those in the VBNC state. One such method combines fluorescent in situ hybridization (FISH; see Chapter 1 for more details) with DVC, followed by cell enumeration using a laser scanning cytometer. This approach gives information on the identity of the bacteria (Baudart et al., 2002). To detect specific bacteria in aquatic environments, the cells can be simultaneously labeled with a fluorescent antibody (FA technique) in combination with viability/activity markers such as cyanoditolyl tetrazolium chloride or propidium iodide, which is an indicator of membrane integrity (Caruso et al., 2003).

Fluorescein diacetate (FDA) is transformed by esterase enzymes to a fluorescent compound, fluorescein, which accumulates inside the cells (FDA is a nonpolar compound and fluorescein a polar compound). The active fluorescent cells are counted under a fluorescent microscope. This method is best suited for active fungal filaments but can be applied to bacterial cells.

2.4.5.4 *Determination of Cell Biochemicals.*
Microbial biomass can also be measured by determination of specific cell biochemical constituents such as ATP, DNA, RNA, proteins, phospholipids, bacterial cell wall components, or photosynthetic pigments (Sutton, 2002).

· *ATP*
Adenosine triphosphate has often been used to determine live microbial biomass in environmental samples, using a ratio of C/ATP = 250 for aquatic samples. However, the ATP content of cells varies with the growth rate and metabolic state of microorganisms and nutrient limitation. A better measure is the *total adenylate pool* A_T (A_T = ATP + ADP + AMP) because it does not change greatly with changes in metabolic activities of the microorganisms. The *adenylate energy charge* (EC) ratio provides information on growth potential of naturally occurring microbial populations.

$$EC = \frac{ATP + \frac{1}{2}ADP}{ATP + ADP + AMP} \qquad (2.48)$$

An EC of 0.5–0.6 indicates senescence of the microbial population, whereas an EC of 0.8–0.9 indicates active microbial growth. Adenosine triphosphate determination has been applied to wastewater treatment, disinfection control, and pollution assessment.

· *Thymidine and Leucine Incorporation into Cells*
Bacterial biomass and production can also be estimated by measuring the incorporation of tritiated thymidine into DNA or radioactive leucine into bacterial proteins (bacterial biomass comprises 60 percent proteins) (Kirchman and Ducklow, 1993).

· *Lipid Biomarkers*
Cell lipids can be classified into neutral lipids (NL), glycolipids (GL), and polar lipids (PL). They serve as storage material (e.g., poly-β-hydroxybutyrate), electron acceptors in the electron transport chain in respiration (e.g., quinones), and components of membranes (e.g., phospholipids) or outer membranes of gram-negative bacteria (e.g., lipopolysaccharides) (Tunlid, 2002).

Lipid biomarkers give the following information:

Microbial biomass. Lipid biomarkers can help in the determination of microbial biomass. For example, phospholipid analysis can determine either the phospholipid-bound phosphate (PLP) or the ester-linked phospholipid fatty acids (PLFA), which serve as biomarkers for estimating microbial biomass and community composition. Suggested conversion factors are 190 μmol P/g C, 100 μmol P/g C, and 50 μmol P/g C for aerobic bacteria, anaerobic bacteria, and eukaryotes, respectively. Phospholipid analysis is unfortunately complex and requires sophisticated equipment. Ergosterol can serve as a biomarker for living fungal biomass.

Community composition. For example, detection of signature PLFAs indicates the presence of specific groups of microorganisms in an environmental sample. Phytanyl-ether lipids indicate the presence of archaea.

Metabolic activity of the microbial community. This is accomplished by measuring the uptake of ^{14}C-labeled substrates into lipid biomarkers such as PLFAs or poly-β-hydroxybutyrate.

• *Bacteria Cell Wall Components*

Cell wall components such as muramic acid and lipopolysaccharide can serve as biomarkers for estimating bacterial biomass.

• *Molecular Techniques for the Determination of Cell Viability/Activity*

Both mRNA and rRNA are well correlated with cell viability. However, due to its shorter half-life in the environment, mRNA is generally preferred over rRNA for indicating cell viability. Reverse transcriptase-PCR (RT-PCR) can also give an indication of cell viability. It consists of transcribing a target RNA sequence into a complimentary DNA (cDNA) sequence, which is then amplified using PCR. Some of these methods have been used to monitor the viability of bacterial pathogens and protozoan parasites (Keer and Birch, 2003). Chapter 1 gives more details on molecular techniques in microbiology.

2.5 WEB RESOURCES

Bacterial Metabolism and Growth

http://www.cat.cc.md.us/courses/bio141/lecguide/unit4/index.html#conjugation
(good animated diagrams)

Photosynthesis

http://www.emc.maricopa.edu/faculty/farabee/BIOBK/BioBookPS.html#Table of Contents
http://www.biology-online.org/1/4_photosynthesis.htm
http://mike66546.tripod.com/biology/Chapter_5/photosynthesis.htm

Respiration

http://www.biology-online.org/1/3_respiration.htm (cell respiration)
http://scidiv.bcc.ctc.edu/rkr/Biology201/lectures/Respiration/Respiration.html

http://pcist2.pc.cc.va.us/energetics/CellRespiration.htm (good site on cell respiration with reference to several other sites)

http://cja.cj.cnd.pvt.k12.oh.us/janderson/Pages/hb_cellrespiration.html/ (slide show on cell respiration and photosynthesis)

http://photoscience.la.asu.edu/photosyn/photoweb/ (photosynthesis and the web: good diagrams)

2.6 QUESTIONS AND PROBLEMS

1. If we start with a bacterial culture of 10^3 cells per mL and if the doubling time t_d is 1.5 h, calculate the final cell concentration (cells/mL) after 16 h.

2. If 58 g of biomass was formed from one mole of glucose substrate, calculate the cell yield Y of the microbial culture.

3. In a bioreactor the specific growth rate μ of a nitrifying bacteria population is $0.02 \ h^{-1}$. What is the minimum mean cell residence time required?

4. Assuming an exponential growth of a bacterial population with an initial number of cells $X_0 = 500$ cells/L and a specific growth rate $\mu = 0.5 \ h^{-1}$, determine:

   (a) the cell concentration after 6 h

   (b) the doubling time t_d

5. Write the Monod's equation and show its graph. How would you go about computing the μ_{max} and K_s?

6. Compare microbial growth in batch and continuous culture. Give the equations and show the differences between the two.

2.7 FURTHER READING

Bitton, G. Editor-in-chief. 2002. *Encyclopedia of Environmental Microbiology*, Wiley-Interscience, N.Y., 3527 pp.

Gaudy, A.F., Jr. and E.T. Gaudy. 1988. *Elements of Bioenvironmental Engineering*, Engineering Press Inc., San Jose, CA, 592 pp.

Grady, C.P.L., Jr. and H.C. Lim. 1980. *Biological Wastewater Treatmnent: Theory and Applications*, Marcel Dekker, New York, 963 pp.

Hammer, M.J. 1986. *Water and Wastewater Technology*, John Wiley and Sons, New York, 536 pp.

Heijnen, J.J. and J.A. Roels. 1981. A macroscopic model describing yield and maintenance relationships in aerobic fermentation. Biotech. Bioeng. 23: 739–741.

Hurst, C.J., Editor-in-chief. 2002. *Manual of Environmental Microbiology*, ASM Press, Washington, D.C. 1138 pp.

Kepner, R.L., and J.R. Pratt. 1994. Use of fluorochromes for direct enumeration of total bacteria in environmental samples: past and present. Microbiol. Reviews 58: 603–615.

Madigan, M. 2003. *Brock Biology of Microorganisms*, 10th ed., Prentice-Hall, Englewood Cliffs, N.J.

Marison, L.W. 1988a. Growth kinetics. In: *Biotechnology for Engineers: Biological Systems in Technological Processes*, A. Scragg, Ed., Ellis Horwood, Chichester, U.K. (pp. 184–217).

Marison, L.W. 1988b. Enzyme kinetics. In: *Biotechnology for Engineers: Biological Systems in Technological Processes*, A. Scragg, Ed., Ellis Horwood, Chichester, U.K. (pp. 96–119).

Metcalf and Eddy, Inc. 1991. *Wastewater Engineering: Treatment, Disposal and Reuse*, McGraw-Hill, New York, 1334 pp.

Michal, G. 1978. Determination of Michaelis constant and inhibitor constants. pp. 29–42. In: *Principles of Enzymatic Analysis*, H.U. Bergmeyer, Ed., Verlag Chemie, Wienheim,

Norton, C.F. 1986. *Microbiology*, 2nd Ed., Addison-Wesley, Reading, MA, 860 pp.

Rittmann, B.E., D.E. Jackson and S.L. Storck. 1988. Potential for treatment of hazardous organic chemicals with biological processes, pp. 15–64, In: *Biotreatment Systems*, Vol. 3, D.L Wise, Ed., CRC Press, Boca Raton, FL.

Tortora, G.J., B.R. Funke and C.L. Case. 1998. *Microbiology: An Introduction*, Benjamin/ Cummings, Menlo Park, CA, 832 pp.

3

ROLE OF MICROORGANISMS IN BIOGEOCHEMICAL CYCLES

In this chapter we will examine the biogeochemical cycles of nitrogen, phosphorus, and sulfur. We will discuss the microbiology of each cycle and emphasize the biotransformations and subsequent biological removal of these nutrients in wastewater treatment plants. Public health and technological aspects of each cycle will also be addressed.

3.1 NITROGEN CYCLE

3.1.1 Introduction

Nitrogen is essential to life as it is a component of proteins and nucleic acids in microbial, animal, and plant cells. It is ironic that nitrogen gas is the most abundant gas (79 percent of the Earth atmosphere) in the air we breathe and yet it is a limiting nutrient in aquatic environments and in agricultural lands, leading to protein deficiency being experienced by millions of people in developing countries. Unfortunately, nitrogen gas cannot be used by most organisms unless it is first converted to ammonia. This is because N_2 is a very stable molecule that will undergo changes only under extreme conditions (e.g., electrical discharge, high temperatures and pressures) (Barnes and Bliss, 1983).

3.1.2 Microbiology of the Nitrogen Cycle

Microorganisms play a major role in nitrogen cycling in the environment. The nitrogen cycle is displayed in Figure 3.1. We will now discuss the microbiology of the five steps involved in nitrogen cycling: nitrogen fixation, assimilation, mineralization, nitrification, and denitrification (Alexander, 1977; Atlas and Bartha, 1987; Barnes and Bliss, 1983; Grady and Lim, 1980).

3.1.2.1 Nitrogen Fixation.   Chemical reduction of nitrogen is very much energy-intensive and expensive. As regards biological reduction of nitrogen, only a few species

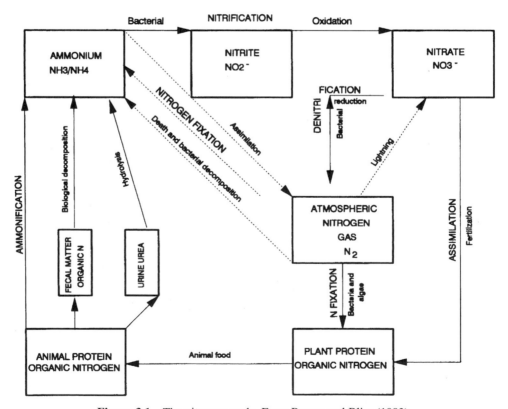

Figure 3.1   The nitrogen cycle. From Barnes and Bliss (1983).

of bacteria and cyanobacteria (blue-green algae) are capable of carrying out nitrogen fixation, which ultimately results in the production of ammonia. The global biological nitrogen fixation is approximately 2×10^8 metric tonnes of N_2/year. Agronomists have put considerable efforts toward exploiting biological nitrogen fixation for maximizing crop yields.

• Nitrogen-Fixing Microorganisms

There are two categories of nitrogen-fixing microorganisms (Table 3.1): nonsymbiotic nitrogen-fixing microorganisms and symbiotic nitrogen-fixing microorganisms.

Nonsymbiotic nitrogen-fixing microorganisms include *Azotobacter* (e.g., *A. agilis*, *A. chroococcum*, *A. vinelandii*), a gram-negative bacterium that forms cysts and fixes nitrogen in soils and other environments. Other nitrogen-fixing microorganisms are *Klebsiella*, *Clostridium* (anaerobic, spore-forming bacteria active in sediments), and cyanobacteria (e.g., *Anabaena*, *Nostoc*). The latter fix nitrogen in natural waters and soils, and their fixation rate is 10 times higher than that of other free-nitrogen-fixing microorganisms in soils. The site of nitrogen fixation in cyanobacteria is a special cell called a *heterocyst* (Fig. 3.2). Cyanobacteria sometimes form associations with aquatic plants (e.g., the *Anabaena–Azolla* association).

Symbiotic nitrogen-fixing microorganisms include some prokaryotes that may enter in a symbiotic relationship with higher plants to fix nitrogen. An example of significant agronomic importance is the legume–*Rhizobium* association. Upon infection of the root, *Rhizobium* form a *nodule*, which is the site of nitrogen fixation. Other examples are the association between *Frankia* and roots of woody perennial plants, and the association (with no nodule formation) between *Azospirillum* and the roots of maize and tropical grasses.

• Nitrogenase: The Enzyme Involved in Nitrogen Fixation

Nitrogen fixation is driven by an enzyme called nitrogenase, which is made of iron sulfide and molybdo-iron proteins, both of which are sensitive to oxygen. This enzyme has the ability to reduce the triple bonded molecule N_2 to NH_4^+, and requires Mg^{2+} and energy in the form of ATP (15–20 ATP/N_2). The biosynthesis of this enzyme is controlled by the *nif* genes. Microorganisms have developed means to avoid inactivation of

TABLE 3.1. Nitrogen-Fixing Microorganisms[a]

Category	Microorganisms
(A) *Free-living nitrogen-fixing microorganisms*	
• Aerobes	*Azotobacter*
	Beijerinckia
• Microaerophilic	*Azospirillum*
	Corynebacterium
• Facultative anaerobes	*Klebsiella*
	Erwinia
• Anaerobes	*Clostridium*
	Desulfovibrio
(B) *Symbiotic associations*	
• Microbe–higher plants	Legume + *Rhizobium*
• Cyanobacteria–aquatic weeds	*Anabaena–Azolla*
• Others	Termites + enterobacteria

[a]From Grant and Long (1981).

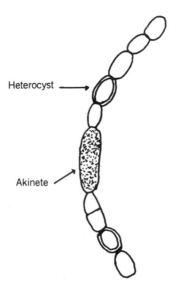

Figure 3.2   Heterocysts, sites of nitrogen fixation in cyanobacteria (e.g., *Anabaena flos-aquae*).

the oxygen-sensitive nitrogenase enzyme. For example, *Azotobacter* produces a copious amount of polysaccharides, which help reduce O_2 diffusion, protecting the enzyme from inactivation. *Azotobacter vinelandii* is protected by an alginate capsule, which forms an effective barrier to O_2 transfer into the cell (Sabra et al., 2000).

· *Nitrogen Fixation Methodology*

Nitrogen fixation is determined using the acetylene reduction technique. This technique consists of measuring the reduction of C_2H_2 (acetylene) to C_2H_4 (ethylene). Ethylene production can be monitored using gas chromatography or laser photoacoustic detection (conversion of light into acoustical signals) (Zuckermann et al., 1997).

The amount of nitrogen fixed is given by Eqs. (3.1) and (3.2).

$$\text{Moles of N}_2 \text{ fixed} = \frac{\text{Moles C}_2\text{H}_2 \rightarrow \text{C}_2\text{H}_4}{3} \tag{3.1}$$

$$\text{g N}_2 \text{ fixed} = \frac{\text{Moles C}_2\text{H}_2 \rightarrow \text{C}_2\text{H}_4}{3} \times 28 \tag{3.2}$$

3.1.2.2 Nitrogen Assimilation. Heterotrophic and autotrophic microorganisms uptake and assimilate NH_4^+ and NO_3^- after reduction to NH_4 (for more details see Section 3.1.2.4). Assimilation is responsible for some nitrogen removal in wastewater treatment plants. Plant and algal cells take up nitrogen preferably in the form of NH_4^+. In soils, NH_4^+-based fertilizers are preferred to NO_3^--based ones (N-SERVE is used to inhibit nitrification in soils).

Cells convert NO_3^- or NH_4^+ to proteins and grow until nitrogen becomes limiting. For each 100 units of carbon assimilated, cells need approximately 10 units of nitrogen (C/N ratio = 10).

3.1.2.3 Nitrogen Mineralization (Ammonification). *Ammonification* is the transformation of organic nitrogenous compounds to inorganic forms. This process is driven by a wide variety of microorganisms (bacteria, actinomycetes, fungi). In soils, some organic nitrogenous compounds become resistant to biodegradation because they form complexes with either phenols and polyphenols, or both.

Proteins are mineralized to NH_4^+ according to the following sequence:

$$proteins \longrightarrow amino\ acids \longrightarrow deamination\ to\ NH_4$$

An example is the transformation of urea to ammonium:

$$O{=}C{\overset{\textstyle NH_2}{\underset{\textstyle NH_2}{\diagup}}} + H_2O \xrightarrow{\text{Urease enzyme}} 2\,NH_3 + CO_2 \tag{3.3}$$

Proteins are converted to peptides and amino acids by extracellular proteolytic enzymes. Ammonium is produced after *deamination* (e.g., oxidative or reductive deaminations) of the amino acids according to reactions 3.4 and 3.5:

Oxidative deamination:

$$\underset{\substack{NH_2 \\ \textit{amino acid}}}{R{-}CH{-}COOH} + \tfrac{1}{2}O_2 \longrightarrow \underset{\substack{O \\ \textit{keto acid}}}{R{-}C{-}COOH} + NH_4 \tag{3.4}$$

Reductive deamination

$$\underset{NH_2}{R{-}CH{-}COOH} + 2\,H \longrightarrow \underset{\textit{acid}}{R{-}CH_2{-}COOH} + NH_4 \tag{3.5}$$

NH_4^+ predominates in acidic and neutral aquatic environments. As pH increases, NH_3 predominates and is released to the atmosphere:

$$NH_4^+ \longrightarrow NH_3 + H^+ \tag{3.6}$$

3.1.2.4 *Nitrification*

• *Microbiology of Nitrification*

Nitrification is the conversion of ammonium to nitrate by microbial action. This process is carried out by two categories of microorganisms, in two stages: conversion of ammonia to nitrite, and conversion of nitrite to nitrate.

Conversion of ammonia to nitrite.   The conversion of ammonia to nitrite is carried out by the *ammonia oxidizing bacteria* (AOB), which belong to the β- and γ-subdivisions of the proteobacteria. *Nitrosomonas* (e.g., *N. europaea*, *N. oligocarbogenes*) is an autotrophic bacterium that oxidizes ammonia to nitrite via hydroxylamine (NH_2OH). Other AOBs are *Nitrosospira*, *Nitrosococcus*, *Nitrosolobus*, and *Nitrosovibrio* (Focht and Verstraete, 1977). In most wastewater environments, the dominant AOB belong to the genus *Nitrosomonas*.

$$NH_3 + O_2 + 2\,H^+ \xrightarrow[\substack{\text{Ammonia} \\ \text{monoxygenase}}]{} \underset{\text{Hydroxylamine}}{NH_2OH} + H_2O \tag{3.7}$$

Ammonia monooxygenase is encoded by the *AmoA* gene.

$$NH_2OH + H_2O \longrightarrow NO_2 + 5\ H^+ \tag{3.8}$$

The overall reaction is (Ward, 2002)

$$NH_3 + \tfrac{3}{2}O_2 \xrightarrow[\substack{\text{Hydroxylamine} \\ \text{oxidoreductase}}]{} NO_2^- + H^+ + H_2O \tag{3.9}$$

Detection of nitrifying bacteria in environmental samples includes most probable number (MPN) methods using selective growth media, as well as immunofluorescent and molecular-based techniques. The latter showed a higher diversity of ammonium oxidizing bacteria than cultivation-based methods, and showed that these bacteria belong to the α-, β-, and γ-subdivisions of proteobacteria (Head et al., 1993; 1998; Wolfe and Lieu, 2002). A 16S rRNA-targeted oligonucleotide probe was used to detect AOB in activated sludge. Microscopic examination showed that these bacteria occur as densely packed aggregates in activated sludge (Wagner et al., 1995). Thus, probing the *amoA* gene gives good information about the nitrifying capacity of a microbial community. *AmoB* can also serve as a molecular marker for AOB (Calvo and Garcia-Gil, 2004).

Conversion of nitrite to nitrate. The conversion of nitrite to nitrate is carried out by the *nitrite oxidizing bacteria* (NOB), which belong to the α-subdivision of the proteobacteria and are obligate autotrophs except for *Nitrobacter*, which can grow heterotrophically in the presence of acetate, formate, or pyruvate. For example, *Nitrobacter* (e.g., *N. agilis*, *N. winogradski*) converts nitrite to nitrate:

$$NO_2^- + \tfrac{1}{2}O_2 \xrightarrow{\text{Nitrite oxidoreductase}} NO_3^- \tag{3.10}$$

Other chemolithotrophic nitrite oxidizers are *Nitrospina*, *Nitrospira*, and *Nitrococcus* (Wolfe and Lieu, 2002). Although *Nitrobacter* has been the most studied in wastewater treatment plants and other environments, *Nitrospira* is often detected in nitrifying biofilms and activated sludge samples, using fluorescence in situ hybridization (FISH) and confocal laser scanning microscopy (Coskuner and Curtis, 2002; Daims et al., 2001) and is sometimes the dominant NOB genus found in wastewater environments.

The oxidation of NH_3 to NO_2^- and then to NO_3^- is an energy-yielding process. Microorganisms use the generated energy to assimilate CO_2. Carbon requirements for nitrifiers are satisfied by carbon dioxide, bicarbonate, or carbonate. Nitrification is favored by the presence of oxygen and sufficient alkalinity to neutralize the hydrogen ions produced during the oxidation process. Theoretically, the oxygen requirement is 4.6 mg O_2/mg ammonia oxidized to nitrate (U.S. EPA, 1975). Although they are obligate aerobes, nitrifiers have less affinity for oxygen than aerobic heterotrophic bacteria. The optimum pH for growth of *Nitrobacter* is 7.2–7.8. Acid production resulting from nitrification can cause problems in poorly buffered wastewaters.

Heterotrophic nitrification. Although autotrophic nitrifiers are predominant in nature, nitrification may be carried out by heterotrophic bacteria (e.g., *Arthrobacter*) and fungi (e.g., *Aspergillus*) (Falih and Wainwright, 1995; Verstraete and Alexander, 1972). These microorganisms utilize organic carbon sources and oxidize ammonium to nitrate. However, heterotrophic nitrification requires energy and is much slower than autotrophic nitrification and probably does not have any significant contribution. In a forest soil, heterotrophic nitrification was found to be less than 10 percent of total nitrification (Barraclough and Puri, 1995).

• *Nitrification Kinetics*

The growth rate of *Nitrobacter* is higher than that of *Nitrosomonas*. Thus, the rate-limiting step in nitrification is the conversion of ammonia to nitrite by *Nitrosomonas*. The following equation describes the growth according to Monod's model (Barnes and Bliss, 1983):

$$\mu = \mu_{max} \frac{[NH_4^+]}{K_s + [NH_4^+]} \quad (3.11)$$

where μ = specific growth rate (days^{-1}), μ_{max} = maximum specific growth rate (days^{-1}), $[NH_4^+]$ = ammonium concentration (mg/L), and K_s = half saturation constant (ammonium substrate) (mg/L).

The ammonium oxidation rate q is related to the specific growth rate μ by the following equation (U.S. EPA, 1975):

$$q = \frac{\mu}{Y} \quad (3.12)$$

where Y = yield coefficient (see Chapter 2).

In wastewater treatment plants, oxygen is a limiting factor controlling the growth of nitrifiers. Therefore, Eq. (3.11) is modified to take into account the effect of oxygen concentration:

$$\mu = \mu_{max} \frac{[NH_4^+]}{K_s + [NH_4^+]} \frac{[DO]}{K_0 + [DO]} \quad (3.13)$$

where [DO] = dissolved oxygen concentration (mg/L), and K_0 = half saturation constant (oxygen) (mg/L). K_0 has been estimated to be 0.15–2 mg/L, depending on temperature (U.S. EPA, 1975). Others have reported a range of $K_0 = 0.25$–1.3 mg/L (Hawkes, 1983; Stenstrom and Song, 1991; Verstraete and van Vaerenbergh, 1986).

More complex equations expressing the growth kinetics of nitrifiers take into account the substrate $[NH_4^+]$ concentration as well as environmental factors such as temperature, pH, and dissolved oxygen (Eq. (3.13); Barnes and Bliss, 1983):

$$\mu_n = \mu_{max} \frac{[NH_4]}{0.4e^{0.118(T-15)} + (NH_4)} \times \frac{[DO] \cdot e^{0.095(T-15)}}{1 + [DO]} \times (1.83)(pH_{opt} - pH) \quad (3.14)$$

where $\mu_n = \mu$ of nitrifiers, T = temperature (°C), pH_{opt} = optimum pH = 7.2, and $\mu_{max} = 0.3$ day^{-1}.

The minimum residence time is a function of μ_n:

$$\text{Minimum residence time} = 1/\mu_n \quad (3.15)$$

The μ_{max} of nitrifiers (0.006–0.035 h^{-1}) is much lower than the μ_{max} of mixed cultures of heterotrophs using glucose as a substrate (0.18–0.38 h^{-1}) (Grady and Lim, 1980; Christensen and Harremoes, 1978; U.S. EPA, 1977). The cell yield of nitrifiers is also lower than that of heterotrophic microorganisms (Rittmann, 1987). The maximum cell yield for *Nitrosomonas* is 0.29; the cell yield of *Nitrobacter* is much lower and is around 0.08. However, experimental yield values are much lower and vary from 0.04 to

0.13 for *Nitrosomonas* and from 0.02 to 0.07 for *Nitrobacter* (Edeline, 1988; Painter, 1970). Thus, in nitrifying environments, *Nitrosomonas* is present in higher numbers than *Nitrobacter*. These relatively low yields have many implications for nitrification in wastewater treatment plants. The half-saturation constant (K_s) for the energy substrate is 0.05–5.6 mg/L for *Nitrosomonas* and 0.06–8.4 mg/L for *Nitrobacter* (Verstraete and van Vaerenbergh, 1986).

Nitrification is favored in biologically treated effluents with low BOD and high ammonia content. A suspended-growth aeration process is the most favorable to nitrification of effluents. Nitrification occurs in the aeration tank (4–6 h detention time), and sludge containing high numbers of nitrifiers is recycled to maintain high nitrifier activity.

• *Factors Controlling Nitrification*

Several factors control nitrification in wastewater treatment plants. These factors are ammonia/nitrite concentration, oxygen concentration, pH, temperature, BOD_5/TKN ratio, and the presence of toxic chemicals (Grady and Lim, 1980; Hawkes, 1983; Metcalf and Eddy, 1991).

Ammonia/nitrite concentration. Growth of *Nitrosomonas* and *Nitrobacter* follows Monod's kinetics and depends on ammonia and nitrite concentration, respectively (see previous subsection).

Oxygen level. Dissolved oxygen (DO) concentration remains one of the most important factors controlling nitrification. The half-saturation constant for oxygen (K_0) is 1.3 mg/L (Metcalf and Eddy, 1991). For nitrification to proceed, the oxygen should be well distributed in the aeration tank of an activated sludge system and its level should not be less than 2 mg/L.

$$NH_3 + 2O_2 \longrightarrow NO_3^- + H^+ + H_2O \qquad (3.16)$$

To oxidize 1 mg of ammonia, 4.6 mg of O_2 are needed (Christensen and Harremoes, 1978). Pure culture studies have demonstrated the possible growth of *Nitrobacter* in the absence of dissolved oxygen, with NO_3 used as electron acceptor and organic substances as the source of carbon (Bock et al., 1988; Ida and Alexander, 1965; Smith and Hoare, 1968). Furthermore, nitrifiers may behave as microaerophiles in aquatic environments such as the sediment–water interface. The nitrate produced helps support denitrification in subsurface sediments (Ward, 2002).

Temperature. The growth rate of nitrifiers is affected by temperature in the range 8–30°C. The optimum temperature has been reported to be in the range 25–30°C

pH. The optimum pH for *Nitrosomonas* and *Nitrobacter* lies between 7.5 and 8.5 (U.S. EPA, 1975). Nitrification ceases at or below pH 6.0 (Painter, 1970; Painter and Loveless, 1983). Alkalinity is destroyed as a result of ammonia oxidation by nitrifiers. Theoretically, nitrification destroys alkalinity, as $CaCO_3$, in amounts of 7.14 mg/1 mg of NH_4^+–N oxidized (U.S. EPA, 1975). Therefore, there should be sufficient alkalinity in wastewater to balance the acidity produced by nitrification. The pH drop that results from nitrification can be minimized by aerating the wastewater to remove CO_2. Lime is sometimes added to increase wastewater alkalinity.

BOD_5/TKN ratio. The fraction of nitrifying organisms decreases as the BOD_5/TKN ratio increases. In combined carbon oxidation–nitrification processes this ratio is greater than 5, whereas in separate stage nitrification processes, the ratio is lower than 3 (Metcalf and Eddy, 1991).

Toxic inhibition. Nitrifiers are subject to product and substrate inhibition and are also quite sensitive to several toxic compounds found in wastewater (Bitton, 1983; Bitton et al.,

1989). It appears that many of those compounds are more toxic to *Nitrosomonas* than to *Nitrobacter*. Organic matter in wastewater is not directly toxic to nitrifiers. Apparent inhibition by organic matter may be indirect and may be due to O_2 depletion by heterotrophs (Barnes and Bliss, 1983). The most toxic compounds to nitrifiers are cyanide, thiourea, phenol, anilines, and heavy metals (silver, mercury, nickel, chromium, copper, and zinc). The toxic effect of copper to *Nitrosomonas europea* increases as the substrate (ammonia) concentration is raised from 3 to 23 mg/L as N (Sato et al., 1988).

Table 3.2 summarizes the conditions necessary for optimal growth of nitrifiers (U.S. EPA, 1977).

3.1.2.5 Denitrification.
Nitrification exerts an oxygen demand in a receiving body of water. Therefore, nitrate must be removed before discharge to receiving waters, particularly if the receiving stream serves as a source of drinking water.

• *Microbiology of Denitrification*

The two most important mechanisms of biological reduction of nitrate are assimilatory and dissimilatory nitrate reduction (Tiedje, 1988).

Assimilatory nitrate reduction. By this mechanism, nitrate is taken up and converted to nitrite and then to ammonium by plants and microorganisms. It involves several enzymes that convert NO_3^- to NH_3, which is then incorporated into proteins and nucleic acids. Nitrate reduction is driven by a wide range of assimilatory nitrate reductases, the activity of which is not affected by oxygen. Certain microorganisms (e.g., *P. aeruginosa*) possess both an assimilatory nitrate reductase as well as a dissimilatory nitrate reductase that is oxygen-sensitive. The two enzymes are encoded by different genes (Sias et al., 1980).

Dissimilatory nitrate reduction (denitrification). This is an anaerobic respiration where NO_3 serves as the terminal electron acceptor. NO_3^- is reduced to nitrous oxide N_2O and nitrogen gas N_2. N_2 liberation is the predominant output of denitrification. However, N_2 has a low water solubility and thus tends to escape as rising bubbles. The bubbles may interfere with sludge settling in a sedimentation tank (Dean and Lund, 1981). The microorganisms involved in denitrification are aerobic autotrophic or hetero-

TABLE 3.2. Optimal Conditions for Nitrification[a]

Characteristic	Design Value
Permissible pH range (95% nitrification)	7.2–8.4
Permissible temperatures (95% nitrification) (°C)	15–35
Optimum temperature, °C (approximately)	30°
DO level at peak flow, mg/L	>1.0
MLVSS, mg/L	1200–2500
Heavy metals inhibiting nitrification (Cu, Zn, Cd, Ni, Pb, Cr)	<5 mg/L
Toxic organics inhibiting nitrification	
Halogen-substituted phenolic compounds	0 mg/L
Halogenated solvents	0 mg/L
Phenol and cresol	<20 mg/L
Cyanides and all compounds from which hydrocyanic acid is liberated on acidification	<20 mg/L
Oxygen requirement (stoichiometric, Ib O_2/lb NH_3–N, plus carbonaceous oxidation demand)	4.6

[a]Adapted from the U.S. EPA (1977).

trophic microorganisms that can switch to anaerobic growth when nitrate is used as the electron acceptor.

Denitrification is carried out according to the following sequence:

$$NO_3 \xrightarrow[\text{reductase}]{\text{Nitrate}} NO_2 \xrightarrow[\text{reductase}]{\text{Nitrite}} NO \xrightarrow[\text{reductase}]{\text{Nitric oxide}} N_2O \xrightarrow[\text{reductase}]{\text{Nitrous oxide}} N_2$$

Denitrifiers belong to several physiological (organotrophs, lithotrophs, and phototrophs) and taxonomic groups (Tiedje, 1988) and can use various energy sources (organic or inorganic chemicals or light). Microorganisms capable of denitrification belong to the following genera: *Pseudomonas, Bacillus, Spirillum, Hyphomicrobium, Agrobacterium, Acinetobacter, Propionobacterium, Rhizobium, Corynebacterium, Cytophaga, Thiobacillus, and Alcaligenes.* Using 16S rRNA-targeted probes, *Paracoccus* sp., another denitrifier, has been isolated in a denitrifying sand filter (Neef et al., 1996). The most widespread genera are probably *Pseudomonas (P. fluorescens, P. aeruginosa, P. denitrificans)* and *Alcaligenes,* which are frequently found in soils, water, and wastewater (Painter, 1970; Tiedje, 1988). Nitrous oxide (N_2O) may be produced during denitrification in wastewater, leading to incomplete removal of nitrate. This gas is a major air pollutant, the production of which must be prevented or at least reduced. Under certain conditions, up to 8 percent of nitrate is converted to N_2O; favorable conditions for its production are low COD/NO_3, short solid retention time, and low pH (Hanaki et al., 1992).

Nitrate respiration has also been observed under aerobic conditions and is driven by a nitrate reductase located in the bacterial periplasmic space. Several bacterial species capable of aerobic nitrate respiration were isolated from soils and sediments (Carter et al., 1995) and from an upflow anaerobic filter (Patureau et al., 1994).

In the absence of oxygen and available organic matter, autotrophic ammonia oxidizers can carry out denitrification by using NH_4 as the electron donor and NO_2 as the electron acceptor:

$$NH_4^+ + NO_2^- \longrightarrow N_2 + 2H_2O \tag{3.17}$$

The Anammox process, described in Chapter 8, is based on the above reaction.

• *Conditions for Denitrification*

The main factors controlling denitrification in wastewater treatment plants and other environments are the following (Barnes and Bliss, 1983; Hawkes, 1983).

Nitrate concentration.   Since nitrate serves as an electron acceptor for denitrifying bacteria, the growth rate of denitrifiers depends on nitrate concentration and follows Monod-type kinetics (see next subsection).

Anoxic conditions.   Oxygen competes effectively with nitrate as a final electron acceptor in respiration. Glucose oxidation in the presence of oxygen releases more free energy (686 kcal/molecule glucose) than in the presence of nitrate (570 kcal/molecule glucose) (Delwiche, 1970). That is why denitrification must be conducted in the absence of oxygen. Denitrification may occur inside activated sludge flocs and biofilms despite relatively high levels of oxygen in the bulk liquid. Thus, the presence of oxygen in wastewater may not preclude denitrification at the microenvironment level (Christensen and Harremoes, 1978).

Presence of organic matter.   Denitrifying bacteria must have an electron donor to carry out the denitrification process. Several sources of electrons have been suggested and studied. The sources include pure compounds (e.g., acetic acid, citric acid, methanol,

ethanol), raw domestic wastewater, wastes from food industries (brewery wastes, molasses), biosolids, or ammonium as carried out in the Anammox process (see Section 8.4.1.2 in Chapter 8). The preferred source of electrons, although more expensive, is methanol, which is used as a carbon source to drive denitrification (Christensen and Harremoes, 1978). Biogas, containing approximately 60 percent methane, can also serve as a sole carbon source in denitrification (Werner and Kaiser, 1991). It has long been known that methane can be used as a carbon source in denitrification (Harremoes and Christensen, 1971), as methanotrophic bacteria oxidize methane to methanol, which is used as a carbon source in denitrification (Mechsner and Hamer, 1985; Werner and Kaiser, 1991).

$$6\,NO_3 + 5\,CH_3OH \longrightarrow 3\,N_2 + 5\,CO_2 + 7\,H_2O + 6\,(OH)^- \qquad (3.18)$$

Thus, $5/6$ mole of methanol is necessary for denitrifying one mole of NO_3. However, some of the methanol is used for cell respiration and cell synthesis. The maximum removal of nitrate is achieved when the ratio CH_3OH/NO_3 is approximately 2.5 (Fig. 3.3; Christensen and Harremoes, 1977). In an anaerobic upflow filter, near complete nitrate removal (99.8 percent) was achieved at a ratio ≥ 2.65 (Hanaki and Polprasert, 1989). It has been suggested that a value of 3.0 should ensure complete denitrification (U.S. EPA, 1975). Using ethanol as the carbon source, the maximum specific denitrification rate in drinking water was obtained at a C/N ratio around 2.2 (Nuhoglu et al., 2002).

pH.   In wastewater, denitrification is most effective at pH values between 7.0 and 8.5 and the optimum is around 7.0 (Christensen and Harremoes, 1977; Metcalf and Eddy, 1991). Alkalinity and pH increase following denitrification. Theoretically, denitrification produces 3.6 mg alkalinity as $CaCO_3$ per mg nitrate reduced to N_2. However, in practice, this value is lower and a value of 3.0 has been suggested for design purpose (U.S. EPA, 1975). Thus, denitrification replaces about half of the alkalinity consumed during nitrification.

Temperature.   Denitrification may occur at 35–50°C. It also occurs at low temperatures (5–10°C) but at a slower rate.

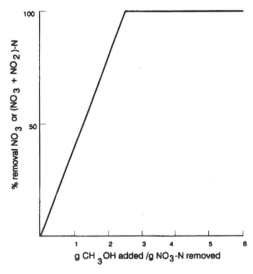

Figure 3.3   Percent removal of nitrate as a function of methanol/nitrate ratio. Adapted from Christensen and Harremoes (1977).

Effect of trace metals.   Denitrification is readily stimulated in the presence of molybdenum and selenium, which are active in the formation of formate dehydrogenase, one of the enzymes implicated in the metabolism of methanol. Molybdenum is essential to the synthesis of nitrate reductase (Chakrabarti and Jones, 1983).

Toxic chemicals.   Denitrifying organisms are less sensitive to toxic chemicals than are nitrifiers.

· Denitrification Kinetics

The environmental factors discussed above have an effect on the growth kinetics of denitrifiers, as shown in Eq. (3.19) (U.S. EPA, 1975):

$$\mu_D = \mu_{max} \frac{D}{K_d + D} \frac{M}{K_m + M} \tag{3.19}$$

where μ = growth rate of denitrifiers, μ_{max} = maximum growth rate of denitrifiers (as affected by nitrate and methanol concentrations, temperatuture and pH), D = nitrate concentration (mg/L), K_d = half-saturation constant for nitrate (mg/L), M = methanol concentration (mg/L), and K_m = half-saturation constant for methanol (mg/L).

Denitrification rate is related to growth rate through Eq. (3.20):

$$q_d = \mu_D/Y_d \tag{3.20}$$

where q_d = nitate removal rate (mg NO_3-N/mg VSS/day), Y_d-growth yield (mg VSS/mg NO_3 removed).

In a completely mixed reactor, the solids retention time is given by Eq. (3.21):

$$1/q = Y_d q_d - K_d \tag{3.21}$$

where q = solids retention time (days), and K_d = decay coefficient (day^{-1}).

· Denitrification Methodology

Denitrification can be measured by determining NO_3 disappearance, N_2 or N_2O formation, or by using ^{15}N. A popular method is the acetylene inhibition method, by which all nitrogen released is thus in the form of N_2O because of the specific inhibition of N_2O reductase by acetylene (Smith et al., 1978). Chemical analysis is greatly simplified since N_2O is a minor atmospheric constituent and can be assayed by gas chromatography.

3.1.3   Nitrogen Removal in Wastewater Treatment Plants

In domestic wastewater, nitrogen is found mostly in the form of organic nitrogen and ammonia. The average total nitrogen concentration in domestic sewage is approximately 35 mg/L. In this section we will mention only briefly the extent of, and means for, nitrogen removal in wastewater treatment plants. Specific processes for nitrogen removal will be covered in greater detail in Chapters 8, 10, and 11.

3.1.3.1   *Extent of Nitrogen Removal.*   Primary treatment of domestic wastewater removes approximately 15 percent of total nitrogen, mainly solids-associated organic nitrogen. Conventional biological treatment removes approximately another 10 percent of nitrogen that is associated with cell biomass, which settles in the sedimentation tank. In biologically treated effluents, approximately 90 percent of the nitrogen is in the form

of ammonia. Nitrogen removal may be increased by the recycling of supernatants from sludge digesters to the wastewater treatment plant (Hammer, 1986). We will now discuss briefly additional means for nitrogen removal.

3.1.3.2 *Means for Removal of Nitrogen.* In sewage treatment plants, there are several means of removing nitrogen from incoming wastewater:

- *Microbiological means: Nitrification–denitrification.* Denitrification provides a means of removing nitrogen from a well-nitrified wastewater effluent. The overall efficiency of nitrification–denitrification can be as high as 95 percent (U.S. EPA, 1975).
- *Chemical and physical means.* Liming results in a high pH (pH $= 10$ or 11), which converts NH_4^+ into NH_3, which can be removed from solution by air stripping in packed cooling towers:

$$NH_4^+ + OH^- \longrightarrow NH_4OH \xrightarrow{\text{Air stripping}} NH_3 + H_2O \qquad (3.22)$$

A first full-scale stripping tower was installed in the late 1970s in south Lake Tahoe in the United States. Some of the problems encountered were freezing in cold weather (ice formation on the packing material), scale formation (calcium carbonate scale), and air pollution with ammonia gas (Dean and Lund, 1981). Ammonia can, however, be recovered as an ammonium salt fertilizer.

Breakpoint chlorination or superchlorination oxidizes ammonium to nitrogen gas according to the reaction shown below (see Chapter 6 for more details):

$$3\,Cl_2 + 2\,NH_4^+ \longrightarrow N_2 + 6\,HCl + 2H^+ \qquad (3.23)$$

Breakpoint chlorination can remove from 90 to 100 percent ammonium (Metcalf and Eddy, 1991; U.S. EPA, 1975). Nitrogen can also be removed via selective ion exchange, filtration, dialysis, or reverse osmosis.

3.1.4 Adverse Effects of Nitrogen Discharges from Wastewater Treatment Plants

Wastewater treatment plants may discharge effluents with high ammonium or nitrate concentrations into receiving aquatic environments. This may lead to several environmental problems, which are summarized as follows (Bitton, 1980a; Hammer, 1986; U.S EPA, 1975).

3.1.4.1 *Toxicity.* Un-ionized ammonia is toxic to fish. At neutral pH, 99 percent of the ammonia occurs as NH_4^+, whereas NH_3 concentration increases at pH > 9.

$$NH_4^+ + OH^- \rightleftarrows NH_3 + H_2O \qquad (3.24)$$

Therefore, ammonia toxicity is particularly important after discharge of alkaline wastewaters or rapid algal photosynthesis, which leads to high pH.

3.1.4.2 Oxygen Depletion in Receiving Waters.

Ammonia may result in oxygen demand in receiving waters (recall that 1 mg ammonia exerts an oxygen demand of 4.6 mg O_2). The oxygen demand exerted by nitrifiers is called *nitrogenous oxygen demand* (NOD). Oxygen depletion adversely affects aquatic life.

3.1.4.3 Eutrophication of Surface Waters.

Discharge of nitrogen into receiving waters may stimulate algal and aquatic plant growth. These, in turn, exert a high oxygen demand at night time, which adversely affects fish and other aquatic life, and has a negative impact on the beneficial use of water resources for drinking or recreation. Nitrogen and phosphorus often are limiting nutrients in aquatic environments. Algal assay procedures help determine which of these two nutrients is the limiting one.

3.1.4.4 Effect of Ammonia on Chlorination Efficiency.

Chlorine combines with ammonium to form chloramines, which have a lower germicidal effect than free chlorine (see Chapter 6 for more details).

3.1.4.5 Corrosion.

Ammonia, at concentrations exceeding 1 mg/L may cause corrosion of copper pipes (Dean and Lund, 1981).

3.1.4.6 Public Health Aspects of Nitrogen Discharges.

Nitrate may be the cause of *methemoglobinemia* in infants and certain susceptible segments of the adult population (e.g., Navajo Indians, Inuits, and people with genetic deficiency of glucose-6-phosphate dehydrogenase or methemoglobin reductase have a higher incidence of methemoglobinemia), and it can lead to the formation of carcinogenic compounds (Bouchard et al., 1992; Craun, 1984a).

Methemoglobinemia ("blue babies" syndrome) is due to the conversion of nitrate to nitrite by nitrate-reducing bacteria in the gastrointestinal tract. Hemoglobin is converted to a brown pigment, methemoglobin, after oxidation, by nitrite, of Fe^{2+} in hemoglobin to Fe^{3+}.

$$\text{Hemoglobin (Fe}^{2+}) + O_2 \xrightarrow[\text{(red pigment)}]{} \text{Oxyhemoglobin} \tag{3.25}$$

$$\text{Hemoglobin (Fe}^{2+}) + NO_2^- \xrightarrow[\text{(brown pigment)}]{} \text{Methemoglobin (Fe}^{3+}) \tag{3.26}$$

Since methemoglobin is incapable of binding molecular oxygen, the ultimate result is suffocation. Babies are more susceptible to methemoglobinemia because the higher pH in their stomach allows a higher reduction of nitrate to nitrite by nitrate-reducing bacteria. Vitamin C offers a protective effect and helps maintain lower levels of methemoglobin. An enzyme, *methemoglobin reductase*, keeps methemoglobin at 1–2 percent of the total hemoglobin in healthy adults.

Nitrite can also combine with secondary amines in the diet to form nitrosamines, which are known to be mutagenic and carcinogenic:

$$\underset{\text{Nitrite}}{NO_2^-} + \underset{\substack{\text{Secondary}\\\text{amine}}}{\overset{R}{\underset{R'}{>}}NH} \rightarrow \underset{\text{Nitrosamine}}{\overset{R}{\underset{R'}{>}}N-N{=}O} \tag{3.27}$$

where R and R' are alkyl and aryl groups.

There is also a Centers for Disease Control (CDC) report of an association between high nitrate concentrations in water and miscarriages in women drinking water that exceeded (19–26 mg/L) the government standard of 10 mg NO_3-N/L (CDC, 1996).

The U.S. EPA interim drinking water standard for nitrate is 10 mg/L as NO_3-N. The World Health Organization (WHO) has also adopted a similar guideline. Public water supplies serving 1 percent of the U.S. population exceed the EPA limit (e.g., some Texas wells contain 110–690 mg/L NO_3-N) (Craun, 1984b). The European Community's standards for nitrogen compounds in potable water are even lower (EC 80/179, 1980): 25 mg/L for nitrate (NO_3^-), 0.1 mg/L of nitrite (NO_2^-), and 0.05 mg/L of ammonia (NH_4^+).

3.2 PHOSPHORUS CYCLE

3.2.1 Introduction

Phosphorus is a macronutrient necessary to all living cells. It is an important component of adenosine triphosphate (ATP), nucleic acids (DNA and RNA), and phospholipids in cell membranes. It may be stored in intracellular volutin granules as polyphosphates in both prokaryotes and eukaryotes. It is a limiting nutrient for algal growth in lakes. The average concentration of total phosphorus (inorganic and organic forms) in wastewater is in the range 10–20 mg/L.

Approximately 15 percent of the U.S. population contributes wastewater effluents to lakes, resulting in eutrophication of these water bodies. Eutrophication leads to significant changes in water quality and lowers the value of surface waters for fishing as well as for industrial and recreational uses. This can be controlled by reducing P inputs to receiving waters (Hammer, 1986).

3.2.2 Microbiology of the Phosphorus Cycle

The major transformations of phosphorus in aquatic environments are described below (Ehrlich, 1981).

3.2.2.1 Mineralization. Organic phosphorus compounds (e.g., phytin, inositol phosphates, nucleic acids, phospholipids) are mineralized to orthophosphate by a wide range of microorganisms that include bacteria (e.g., *B. subtilis, Arthrobacter*), actinomycetes (e.g., *Streptomyces*), and fungi (e.g., *Aspergillus, Penicillium*). Phosphatases are the enzymes responsible for degradation of phosphorus compounds.

3.2.2.2 Assimilation. Microorganisms assimilate phosphorus, which enters in the composition of several macromolecules in the cell. Some microorganisms have the ability to store phosphorus as polyphosphates in special granules. This topic will be covered in more detail in Section 3.2.4.

3.2.2.3 Precipitation of Phosphorus Compounds. The solubility of orthophosphate is controlled by the pH of the aquatic environment and by the presence of Ca^{2+}, Mg^{2+}, Fe^{3+}, and Al^{3+}. When precipitation occurs, there is formation of insoluble compounds such as hydroxyapatite ($Ca_{10}(PO_4)_6(OH)_2$, vivianite $Fe_3(PO_4)_2 \cdot 8H_2O$ or variscite $AlPO_4 \cdot 2H_2O$ (Ehrlich, 1981).

3.2.2.4 Microbial Solubilization of Insoluble Forms of Phosphorus. Through their metabolic activity, microorganisms help in the solubilization of P compounds. The mechanisms of solubilization are metabolic processes involving enzymes, production of organic and inorganic acids by microorganisms (e.g., succinic acid, oxalic acid, nitric acid, and sulfuric acid), production of CO_2, which lowers pH, production of H_2S, which may react with iron phosphate and liberate orthophosphate, and the production of chelators, which can complex Ca, Fe, or Al.

3.2.3 Phosphorus Removal in Wastewater Treatment Plants

The average concentration of total phosphorus (inorganic and organic forms) in wastewater is within the range 10–20 mg/L, much of which comes from phosphate builders in detergents. Common forms of phosphorus in wastewater are orthophosphate (PO_4) (50–70 percent of phosphorus), polyphosphates, and phosphorus tied to organic compounds. Orthophosphate comprises approximately 90 percent of phosphorus in biologically treated effluents (Meganck and Faup, 1988). Since phosphorus is a limiting nutrient and is mainly responsible for eutrophication of surface waters, it must be removed by wastewater treatment processes before discharge of the effluents into surface waters. Most often, phosphorus is removed from wastewater by chemical means but can also be removed by biological processes (see Section 3.2.4). Several biological and chemical mechanisms are responsible for phosphorus removal in wastewater treatment plants (Arvin, 1985; Arvin and Kristensen, 1983):

1. Chemical precipitation, which is controlled by pH and cations such as Ca, Fe and Al;
2. Phosphorus assimilation by wastewater microorganisms;
3. Enhanced biological phosphorus removal (EBPR);
4. Microorganism-mediated enhanced chemical precipitation.

Primary treatment of wastewaters removes only 5–15 percent of phosphorus associated with particulate organic matter, and conventional biological treatment does not remove a substantial amount of phosphorus (approximately 10–25 percent; Metcalf and Eddy, 1991). Most of the retained phosphorus is transferred to biosolids (Hammer, 1986). Additional phosphorus removal can be achieved by adding iron and aluminum salts or lime to wastewater. Commercially available aluminum and iron salts are alum, ferric chloride, ferric sulfate, ferrous sulfate, and waste pickle liquor from the steel industry. These are generally added in excess to compete with natural alkalinity. Lime is less frequently used for phosphorus removal because of increased production of sludge as well as operation and maintenance problems associated with its use (U.S. EPA, 1987b).

Aluminum reacts with phosphorus to form aluminum phosphate:

$$Al^{3+} + PO_4^{3-} \longrightarrow AlPO_4 \tag{3.28}$$

Ferric chloride reacts with phosphorus to form ferric phosphate:

$$FeCl_3 + PO_4^{3-} \longrightarrow FePO_4 + 3\ Cl^- \tag{3.29}$$

Other treatments for phosphorus removal include adsorption to activated alumina, ion exchange, electrochemical methods, and deep-bed filtration (Meganck and Faup, 1988).

3.2.4 Biological Phosphorus Removal

In addition to chemical precipitation, phosphorus may also be removed by biological means. During the 1960s, Shapiro and collaborators demonstrated the role of microorganisms in the uptake and release of phosphorus in the activated sludge system (Shapiro, 1967; Shapiro et al., 1967). Since then, two approaches have been taken to explain the enhanced biological phosphorus removal in wastewater treatment plants: microorganism-mediated chemical precipitation and microorganism-mediated enhanced uptake of phosphorus.

3.2.4.1 *Microorganism-Mediated Chemical Precipitation.* According to this approach, phosphate precipitation and subsequent removal from wastewater is mediated by microbial activity in the aeration tank of the activated sludge process. At the head of a plug-flow aeration tank, microbial activity leads to low pH, which solubilizes phosphate compounds. At the end of the tank, a biologically mediated pH increase leads to phosphate precipitation and incorporation into the sludge (Menard and Jenkins, 1970).

Biologically mediated phosphate precipitation also occurs inside denitrifying biofilms. Since denitrification is an alkalinity-producing process, denitrifier activity leads to pH increase and subsequent calcium phosphate precipitation inside biofilms (Arvin, 1985; Arvin and Kristensen, 1983). Precipitation of phosphorus may also be induced by the increase in phosphate concentration resulting from P release from the polyphosphate pool under anaerobic conditions.

3.2.4.2 *Enhanced Biological Phosphorus Removal (EBPR).* Enhanced biological phosphorus removal or *luxury uptake* of phosphorus is the result of microbial action in the activated sludge process (Toerien et al., 1990). Several mechanisms have been proposed to explain the enhanced uptake of phosphorus by microorganisms in wastewater.

Several microorganisms, called poly P bacteria or *polyphosphate accumulating organisms* (PAOs), have the ability to accumulate phosphorus in excess of the normal cell requirement, which is around $1-3$ percent of the cell dry weight. Phosphorus is accumulated intracellularly inside polyphosphate granules (i.e., volutin granules or metachromatic granules), which can be easily observed under brightfield, phase-contrast, or fluorescence microscopy. Stains like methylene blue and toluidine blue are often used to show these granules. Staining with DAPI imparts a yellow fluorescence to these granules. Methods for quantification and visualization of PAO storage polymers include light and epifluorescence microscopy, chemical extraction and quantification of the polymers, and in vivo nuclear magnetic resonance (NMR) for online monitoring of intracellular reserves. (Serafim et al., 2002). Nuclear magnetic resonance has also been used to detect and locate polyphosphate granules in wastewater microorganisms (Florentz and Granger, 1983; Randall et al., 1997a).

An enzyme, polyphosphate kinase (PPK), catalyzes polyphosphate biosynthesis in the presence of Mg^{2+} ions by transferring the terminal phosphoryl group from ATP to the polyphosphate chain. Polyphosphate kinase homologs are found in several human pathogens (e.g., *Mycobacterium tuberculosis*, *Helicobacter pylori*, *Vibrio cholerae*, *Salmonella typhimurium*, *Shigella flexneri*) and play a role in bacterial physiological functions such as motility, virulence, or biofilm formation (Kornberg et al., 1999; McMahon et al., 2002).

Polyphosphate degradation is driven by several enzymes (van Groetnestijn et al., 1988b; Kornberg, 1957; Meganck and Faup, 1988; Ohtake et al., 1985; Szymona and

Ostrowski, 1964) according to the following reactions:

$$(\text{Polyphosphate})_n + \text{AMP} \xrightarrow{\substack{\text{Polyphosphate AMP} \\ \text{phosphotransferase}}} (\text{Polyphosphate})_{n-1} + \text{ADP} \qquad (3.30)$$

$$2\ \text{ADP} \xrightleftharpoons{\text{adenylate kinase}} \text{ATP} + \text{AMP} \qquad (3.31)$$

Polyphosphatases are other hydrolytic enzymes that are involved in polyphosphate degradation. In some bacteria, the hydrolysis of polyphosphates is driven by polyphosphate glucokinase and polyphosphate fructokinase, resulting in the phosphorylation of glucose and fructose, respectively.

Other microorganisms, known as the *glycogen accumulating organisms* (GAOs), are also part of the EBPR microbial assemblages. These bacteria (e.g., *Candidatus Competibacter phosphatis*) were always found in an examination of six full-scale plants in Australia (Saunders et al., 2003). Although GAOs carry out carbon transformations similar to those of PAOs, they do not take up or release phosphorus (Blackall et al., 2002).

For the successful operation of the EBPR process one must create conditions for the predominance of PAOs over GAOs. One such condition is the operation of the EBPR process at pH higher than 7 (Filipe et al., 2001a; 2001b) since GAOs predominate at lower pHs. Low temperature is another factor controlling the predominance of PAOs over GAOs. The psychrophilic PAOs are generally favored at low temperatures, leading to an increase in EBPR efficiency (Erdal et al., 2003).

In an anaerobic–aerobic activated sludge unit, inorganic phosphate is released under anaerobic conditions and taken up by microorganisms under aerobic conditions (Barnard, 1975). Figure 3.4 shows the release and uptake of phosphorus in a laboratory anaerobic-aerobic activated sludge unit (Hiraishi et al., 1989). Polyphosphate accumulating organisms

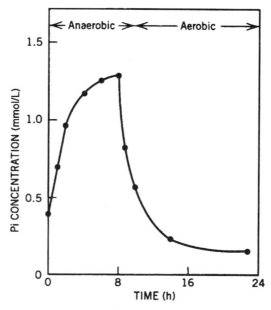

Figure 3.4 P_i release and uptake by a laboratory anaerobic–aerobic activated sludge. From Hiraishi et al. (1989).

such as *Acinetobacter* take up phosphorus under aerobic conditions, accumulate it as polyphosphate in granules, and release it under anaerobic conditions (Fuhs and Chen, 1975). For example, *Acinetobacter calcoaceticus* takes up phosphorus under aerobic conditions at a rate of 0.4–0.5 mmole P/g dry cells per hour and releases it under anaerobic conditions at a rate of 0.015 mmole/g dry cells per hour (Ohtake et al., 1985). Another P-accumulating bacterium found in wastewater is *Acinetobacter Johnsonii* (Wiedmann-Al-Ahmad et al., 1994). Magnesium plays a role in EBPR stability and acts as an important counterion of polyphosphates and is taken up and released simultaneously with phosphate (van Groenestijn et al., 1988a). Others have shown that, in addition to Mg^{2+}, K^+ and Ca^{2+} are also co-transported with phosphate (Comeau et al., 1987).

There is a controversy over whether *Acinetobacter* is the predominant microorganism involved in enhanced phosphorus uptake. In an early study, a correlation was found between the number of *Acinetobacter* and the extent of phosphorus removal (Cloete and Steyn, 1988). However, the use of respiratory quinone profiles to characterize the bacterial population structure of anaerobic–aerobic activated sludge system showed that *Acinetobacter* species were not particularly important in these systems (Hiraishi et al., 1989; 1998). Cultivation on nutrient-rich media tends to favor the growth of *Acinetobacter* spp. (more than 30 percent of the isolates), which belongs to the gamma subclass of proteobacteria. The use of rRNA oligonucleotide probes in activated sludge has indeed confirmed that *Acinetobacter* constitutes only 3–9 percent of the isolates, and does not play a major role in enhanced biological phosphate removal processes and that gram-positive bacteria with a high G+C DNA content (e.g., *Nocardia*, *Arthrobacter*, *Rhodococcus*) are the predominant phosphate-removing bacteria (Eschenhagen et al., 2003; Snaidr et al., 1997; Wagner et al., 1994). However, it was shown that *Acinetobacter*, as detected by the biomarker diaminopropane, was the dominant organism only in wastewater treatment plants with low organic loading (Auling et al., 1991). Other poly P bacteria (*Pseudomonas*, *Aeromonas*, *Moraxella*, *Klebsiella*, *Enterobacter*, *Tetrasphaera* spp.) also contribute to the removal of phosphorus in activated sludge.

Molecular methods (e.g., FISH probes consisting of fluorescently-labeled 16S rRNA or 23S rRNA sequences) have essentially shown that β-proteobacteria and actinobacteria are the predominant microorganisms in the EBPR community (Seviour et al., 2003). More specifically, *Candidatus Accumulibacter phosphatis* and *Rhodocyclus*-like organisms were the PAOs found in EBPR systems (Dabert et al., 2001; Eschenhagen et al., 2003; Hesselmann et al., 1999; McMahon et al., 2002; Saunders et al., 2003). In full-scale EBPR plants in South Africa, *Rhodocyclus*-like organisms within the beta subclass of Proteobacteria represented up to 73 percent of the PAOs in one plant (Zilles et al., 2002). Similar results were found in a sequencing batch reactor (SBR) sludge (Jeon et al., 2003).

3.2.4.3 *Biochemical Model of Enhanced Phosphorus Uptake.* Biochemical models have been proposed to explain the enhanced removal of phosphorus in activated sludge (Comeau et al., 1986; Wentzel et al., 1986). We will discuss the model of Comeau and collaborators, which is illustrated in Figure 3.5 (Comeau et al., 1986):

• *Phosphorus Release under Anaerobic Conditions.*
Under anaerobic conditions (Fig. 3.5*a*), bacteria use energy derived from polyphosphate hydrolysis to takeup carbon substrates that are stored as poly-β-hydroxybutyrate (PHB) reserves, and to regulate the pH gradient across the cytoplasmic membrane. This phenomenon leads to the release of inorganic phosphorus. Except for propionic acid, C_2 to C_5 volatile fatty acids (e.g., acetic acid) are taken up by microorganisms and

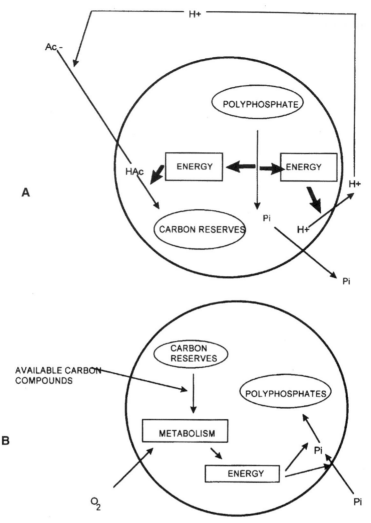

Figure 3.5 Biochemical model explaining the enhanced P removal in activated sludge: (a) anaerobic conditions; (b) aerobic conditions. From Comeau et al. (1986).

stored intracellularly as poly-β-hydroxyalkanoates (PHAs), of which PHB is a well-known representative. PHB is formed following acetate assimilation while propionate results in the formation of mainly poly-3-hydroxy-butyrate as the predominant form of PHA. These compounds are subsequently used as carbon sources during the aerobic phase. Acetate is converted to acetyl-CoA and the reaction is driven by energy supplied by the hydrolysis of accumulated intracellular polyphosphates. NADH, which provides the reducing power for PHB synthesis, is derived from the consumption of intracellular carbohydrate (Arun et al., 1988). Some investigators have suggested that the anaerobic zone serves as a fermentation milieu in which microorganisms (Brodisch and Joyner, 1983; Meganck et al., 1984) produce volatile fatty acids (VFA) such as acetate, which are taken up by PAOs and stored as PHAs (Fig. 3.6) (Meganck and Faup, 1988). Glycogen is synthesized and stored by PAOs under aerobic conditions and is depleted in the anaerobic zone (Blackall et al., 2002). However, PAOs must, under certain conditions (e.g., presence of glucose as a substrate), compete for VFA with the "G" bacteria, which do

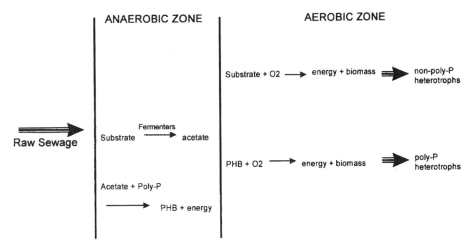

Figure 3.6 Acetate formation by fermentation and subsequent storage as PHB by Poly P bacteria. Adapted from Meganck and Faup (1988).

not accumulate polyphosphates (Cech et al., 1993). These "G" bacteria are sometimes referred to as GAOs and dominate in systems with poor phosphorus removal. They assimilate glucose under anaerobic conditions and synthesize glycogen under aerobic conditions. They share the same ecological niche with the PAOs, but their role in EBPR systems needs further investigation (Seviour, 2002).

During the anaerobic phase, the uptake of acetic and isovaleric acids results in the formation of poly-3-hydroxy-butyrate as the predominant form of PHA. It was shown that poly-3-hydroxy-butyrate results in higher uptake of phosphorus during the aerobic phase (Randall and Liu, 2002).

To maximize carbon storage under anaerobic conditions, Comeau et al. (1987) suggested increasing the addition of simple carbon sources (e.g., septic wastewater, fermented primary sludge supernatant, acetate) and minimizing the addition of electron acceptors such as O_2 and NO_3. Readily biodegradable short chain carbon substrates (butyric and isobutyric acids, valeric and isovaleric acids, ethanol, acetic acid, methanol) are responsible for an enhanced removal of phosphorus (Abu-Ghararah and Randall, 1990; Jones et al., 1987; Randall et al., 1997b).

· Enhanced P Uptake under Aerobic Conditions

Under aerobic conditions (Fig. 3.5b), the energy derived from the metabolism of stored PHAs is used for the accumulation of polyphosphates inside the cells (sometimes surpassing 15 percent of cell dry weight) while PHA cell concentration decreases. Under these conditions, inorganic phosphorus is taken up by the cells and stored as polyphosphates.

It was found that, under laboratory conditions, reducing the pH of the mixed liquor to values between 5.5 and 6.5 can greatly increase phosphorus uptake by microorganisms, as indicated by an increase in poly P granules (McGrath et al., 2001).

Toxicants such as 2,4-dinitrophenol and H_2S have an adverse effect on phosphorus uptake under aerobic conditions (Comeau et al., 1987). Methods have been developed for distinguishing between intracellular polyphosphate from extracellular precipitated orthophosphate. The use of these methods has indicated that polyphosphate is the predominant form of bioaccumulated phosphorus in activated sludge (de Haas, 1989).

Several proprietary processes based on release of phosphorus under anaerobic conditions and its uptake under aerobic conditions have been developed and marketed for phosphate removal in wastewater treatment plants. These processes will be discussed in Chapter 8.

3.3 THE SULFUR CYCLE

3.3.1 Introduction

Sulfur is relatively abundant in the environment, and seawater is the largest reservoir of sulfate. Other sources include sulfur-containing minerals (e.g., pyrite, FeS_2, and chalcopyrite $CuFeS_2$), fossil fuels, and organic matter. It is an essential element for microorganisms and enters in the composition of amino acids (cystine, cysteine, and methionine), cofactors (thiamine, biotin, and coenzyme A), ferredoxins and enzymes (—SH groups).

The sources of sulfur in wastewaters are organic sulfur found in excreta, and sulfate, which is the most prevalent anion in natural waters.

3.3.2 Microbiology of the Sulfur Cycle

The steps involved in the sulfur cycle are summarized in Fig. 3.7 (Sawyer and McCarty, 1967) and described in the following.

3.3.2.1 *Mineralization of Organic Sulfur.* Several types of microorganisms mineralize organic sulfur compounds, through aerobic and anaerobic pathways (Paul and Clark, 1989). Under aerobic conditions, sulfatase enzymes are involved in the degradation of sulfate esters to SO_4^{2-}:

$$R—O—SO_3^- + H_2O \xrightarrow{\text{sulfatase}} ROH + H^+ + SO_4^{2-} \qquad (3.32)$$

Under anaerobic conditions, sulfur-containing amino acids are degraded to inorganic sulfur compounds or to mercaptans, which are odorous sulfur compounds.

3.3.2.2 *Assimilation.* Microorganisms assimilate oxidized as well as reduced forms of sulfur. Anaerobic microorganisms assimilate reduced forms such as H_2S, whereas aerobes use the more oxidized forms. The carbon to sulfur ratio is $100:1$.

3.3.2.3 *Oxidation Reactions.* Several groups of microorganisms are involved in sulfur oxidation.

· *H_2S Oxidation*

H_2S is oxidized to elemental sulfur under aerobic and anaerobic conditions. Under aerobic conditions, *Thiobacillus thioparus* oxidizes S^{2-} to S^0:

$$S^{2-} + \tfrac{1}{2} O_2 + 2 H^+ \longrightarrow S^0 + H_2O \qquad (3.33)$$

Under anaerobic conditions, oxidation is carried out by photoautotrophs (e.g., photosynthetic bacteria) and a chemoautotroph, *Thiobacillus denitrificans*. Photosynthetic bacteria use H_2S as an electron donor and oxidize H_2S to S^0, which is stored within the cells of chromatiaceae (purple sulfur bacteria) or outside the cells of chlorobiaceae (green sulfur

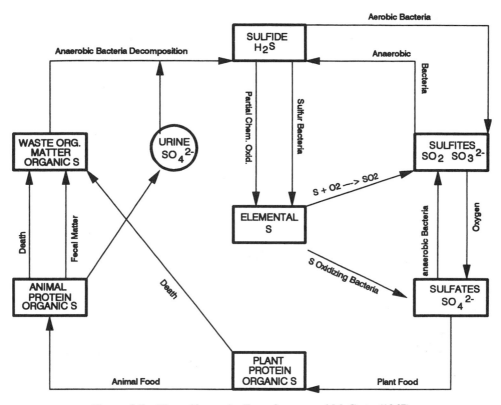

Figure 3.7   The sulfur cycle. From Sawyer and McCarty (1967).

bacteria). Filamentous sulfur bacteria (e.g., *Beggiatoa*, *Thiothrix*) also carry out H_2S oxidation to S, which is deposited in S granules.

• *Oxidation of Elemental Sulfur*

This is carried out mainly by the aerobic, gram-negative, nonspore forming acidithio-bacilli (e.g., *Acidithiobacillus thiooxidans*, formerly *Thiobacillus thiooxidans*), which grow at very low pH.

$$2\,S + 3\,O_2 + 2\,H_2O \;\longrightarrow\; 2\,H_2SO_4 \qquad\qquad (3.34)$$

$$\underset{\text{Thiosulfate}}{Na_2S_2O_3} + 2\,O_2 + H_2O \;\longrightarrow\; Na_2SO_4 + H_2SO_4 \qquad\qquad (3.35)$$

Another sulfur oxidizer is *Sulfolobus* (e.g., *S. metallicus*), which is a thermophilic acidophilic, and autotrophic archae found in hot acidic spring waters (pH = 2–3; temperature = 55–85°C) and grows by oxidizing ferrous iron, reduced inorganic sulfur compounds, or sulfide ores (Rawlings, 2002).

• *Sulfur Oxidation by Heterotrophs*

Heterotrophs (e.g., *Arthrobacter*, *Micrococcus*, *Bacillus*, *Pseudomonas*) may also be responsible for sulfur oxidation in neutral and alkaline soils (Paul and Clark, 1989).

3.3.2.4   *Sulfate Reduction.* Sulfides are produced by assimilatory and dissimilatory sulfate reduction.

• *Assimilatory Sulfate Reduction*

H_2S may result from the anaerobic decomposition by proteolytic bacteria (e.g., *Clostridia*, *Vellionella*) of organic matter containing S amino acids such as methionine, cysteine, and cystine (Bowker et al., 1989).

• *Dissimilatory Sulfate Reduction*

Sulfate reduction is the most important source of H_2S in wastewater. It is the reduction of sulfate by strict anaerobes, the sulfate reducing bacteria.

$$SO_4^= + \text{organic compounds} \longrightarrow S^= + H_2O + CO_2 \qquad (3.36)$$

$$S^= + 2H^+ \longrightarrow H_2S \qquad (3.37)$$

H_2S is toxic to animals and plants. It causes problems in paddy rice fields and is toxic to wastewater treatment plant operators (see Chapter 14).

In 1895, Beijerinck first isolated a bacterium capable of reducing sulfate to hydrogen sulfide. Sulfate-reducing bacteria (SRB) belonging to the following genera have since been isolated from environmental samples (anaerobic sludge digesters, aquatic sediments, gastrointestinal tract): *Desulfovibrio*, *Desulfotomaculum*, *Desulfobulbus*, *Desulfomonas*, *Desulfobacter*, *Desulfococcus*, *Desulfonema*, *Desulfosarcina*, *Desulfobacterium*, and *Thermodesulfobacterium*. *Desulfotomaculum* is the only spore-forming genus among sulfate-reducing bacteria (Hamilton, 1985; Widdel, 1988).

In the absence of oxygen and nitrate, these strict anaerobic bacteria use sulfate as the terminal electron acceptor. They use low-molecular weight carbon sources (e.g., electron donors) produced via the fermentation of carbohydrates, proteins, and other compounds. These carbon sources include lactate, pyruvate, acetate, propionate, formate, fatty acids, alcohols (ethanol, propanol), dicarboxylic acids (succinic, fumaric, and malic acids), and aromatic compounds (benzoate, phenol). H_2 is also used as electron donor. These bacteria have very low cell yields (Hamilton, 1985; Rinzema and Lettinga, 1988; Widdel, 1988). The detection of sulfate-reducing bacteria in aerobic wastewater treatment processes (activated sludge flocs, trickling filter biofilms) suggest that these bacteria may be regarded as microaerophiles and may tolerate some oxygen in their environment (Cypionka et al., 1985; Lens et al., 1995; Manz et al., 1998).

Detection of SRB in the environment is accomplished by using lactate-sulfate growth media, enzymatic test kits for hydrogenase and sulfite reductase, radiometric tests that determine the production of labeled H_2S, immunoassays, and molecular techniques based on the use of 16S rRNA probes (Barton and Plunkett, 2002).

3.3.3 Environmental and Technological Problems Associated with Sulfur Oxidation or Reduction

3.3.3.1 *Problems Associated with Sulfate-Reducing Bacteria.*

• *Corrosion Problems*

Biocorrosion or microbially influenced corrosion (MIC) is the deterioration of metals by microorganisms. This problem is important to the shipping, power, oil, and gas industries, as well as water and wastewater treatment industries (Beech, 2002).

Pitting corrosion develops under strictly anaerobic conditions and involves sulfate-reducing bacteria such as *Desulfovibrio desulfuricans*. The pits are filled with iron sulfide. There are still questions regarding the exact role of sulfate-reducing bacteria in metal corrosion. Corrosion may be inherently linked to the metabolism of sulfate reducers or may be due to hydrogen sulfide or iron sulfides (Odom, 1990).

The development of biofilms on pipe surfaces leads to anaerobic conditions that are ideal for the growth of obligate anaerobic sulfate-reducing bacteria. These conditions are achieved when the biofilm reaches a certain thickness, varying from 10 to 100 µm. A model for anaerobic microbial corrosion is displayed in Figure 3.8 (Hamilton, 1985).

The most accepted theory of anaerobic corrosion by sulfate-reducing bacteria is the *cathodic depolarization theory* of von Wolzogen-Kuhr and van der Vlugt (1934) (Fig. 3.9; Widdel, 1988). A net oxidation of the metal is catalyzed by the activity of sulfate-reducing bacteria. At the anode, the metal is polarized by losing Fe^{2+}. The electrons released at the anode reduce protons (H^+) from water to molecular hydrogen (H_2) at the cathode. The molecular hydrogen is removed from the metal surface by sulfate-reducing bacteria (cathodic depolarization). The latter produce hydrogen sulfide, which combines with Fe^{2+} to form FeS, which accumulates on the metal surface.

Anaerobic corrosion of iron is summarized as follows (Ford and Mitchell, 1990; Hamilton, 1985):

$$4\,Fe(0) \longrightarrow 4\,Fe(II) + 8\,e^- \qquad \text{(reaction at the anode)} \qquad (3.38)$$

$$8\,H_2O \longrightarrow 8\,H^+ + 8\,OH^- \qquad \text{(H}_2\text{O dissociation)} \qquad (3.39)$$

$$8\,H + 8\,e^- \longrightarrow 4\,H_2 \qquad (3.40)$$

$$4\,H_2 + SO_4 \longrightarrow S^{2-} + 4\,H_2O \qquad \text{(cathodic depolarization)} \qquad (3.41)$$

The presence of organic substrates (i.e., electron donors) helps in the production of additional sulfide. Hydrogenases (three known types) of sulfate-reducing bacteria drive the reversible oxidation of hydrogen. They remove cathodic hydrogen, and are involved

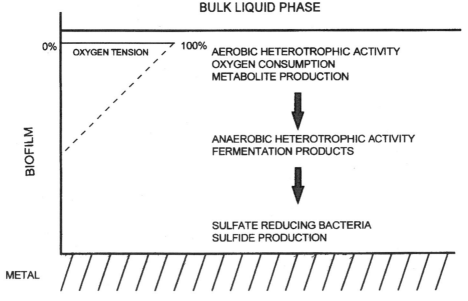

Figure 3.8 Model for anaerobic microbial corrosion. From Hamilton (1985). Courtesy of American Society of Microbiology.

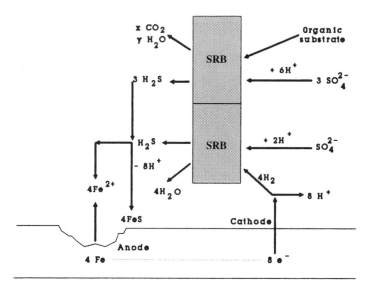

Figure 3.9 Reactions during anaerobic corrosion as suggested by cathodic depolarization theory of von Wolzogen-Kuhr and van der Vlugt. From Widdel (1988).

in the initiation of the biocorrosion process. It was suggested that their activity is a good indicator of biocorrosion (Bryant and Laishley, 1990; Bryant et al., 1991). A commercial kit is available for measuring hydrogenase activity (Costerton et al., 1989). The genes encoding *Desulfovibrio* periplasmic hydrogenase enzymes have been cloned and sequenced. This has led to the construction of hydrogenase gene probes for the detection of *Desulfovibrio* species in environmental samples (Voordouw, 1992; Voordouw et al., 1990). Both hydrogenase and iron sulfide can serve as depolarizing agents. It has also been suggested that a phosphorus compound produced by sulfur-reducing bacteria could also be responsible for corrosion (Beech, 2002).

Sulfate-reducing bacteria can be controlled by preventing or reducing the input of organic matter or sulfate, changing environmental conditions by aeration or pH change and by adding appropriate biocides (Widdel, 1988).

• Impact of Sulfate Reduction on Anaerobic Digesters

In anaerobic digesters treating wastewaters with high sulfate concentrations (e.g., molasses-based fermentation and paper and board industry), H_2S produced via sulfate reduction is transferred to the biogas and causes corrosion problems, unpleasant odors, and inhibition of methanogenic bacteria (see Chapter 13) (Colleran et al., 1995; Rinzema and Lettinga, 1988). Sulfate reducers and methanogens share similar characteristics. They both live under strict anaerobic conditions with similar pH and temperature ranges. Like methanogens, some sulfate reducers are able to oxidize H_2 and acetate and thus may compete with methanogens for these substrates (Fig. 3.10) (Rinzema and Lettinga, 1988). Competition for these substrates has been studied mostly in estuarine sediments where sulfate supply is more abundant than in freshwater sediments (Abram and Nedwell, 1978; Oremland and Polcin, 1982; Sorensen et al., 1981). Kinetic studies have shown that sulfate reducers generally have higher maximum growth rates and higher affinity for substrates (i.e., lower half-saturation constants K_s) than methanogens. The half-saturation constant K_s for hydrogen is 6.6 μM for methanogens as compared

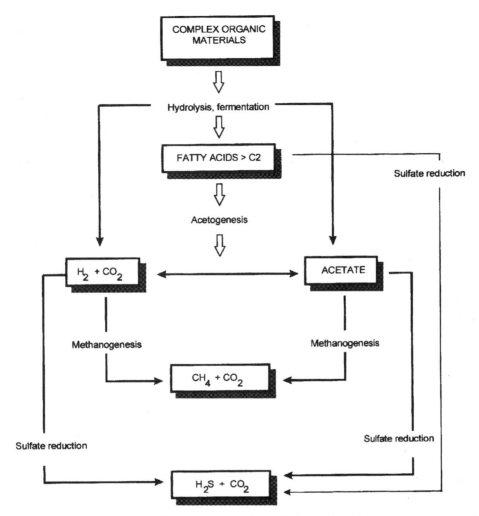

Figure 3.10 Substrate competition between sulfate-reducing and methanogenic or acetogenic bacteria. From Rinzema and Lettinga (1988).

to 1.3 μM for sulfate reducers. Similarly, the K_s for acetate is 3 μM and 0.2 μM for methanogens and sulfate reducers, respectively (Speece, 1983). Thus, sulfate reducers may predominate over methanogens provided the sulfate supply is not limiting. However, despite their kinetic advantages, sulfate-reducing bacteria rarely predominate in anaerobic wastewater treatment. Future research should provide a satisfactory explanation for this phenomenon (Rinzema and Lettinga, 1988).

• *Aquatic Environments and Flooded Soils*

The excessive production of hydrogen sulfide in anoxic waters and sediments may result in the poisoning of plants and animals, including fish. This detrimental effect is also observed in waterlogged soils (paddy fields) (Widdel, 1988).

• *Sulfate Reduction and Oil Technology and Paper-Making Industries*

Both environments are conducive to the formation of electron donors for sulfate-reducing bacteria. In the paper-making industry, cellulose fermentation produces electron donors for sulfate reduction. Furthermore, aerobic degradation of hydrocarbons may release organic acids, which can be used by sulfate reducers. The resulting hydrogen sulfide combines with iron and other heavy metals to form sulfides, which tend to plug injection wells. H_2S gas may also contaminate fuel gas and oil (Widdel, 1988).

• *Activity of Sulfate-Reducing Bacteria in Municipal Solid Waste Landfills*
Gypsum drywall is a major component of construction and demolition wastes disposed into landfills. Gypsum drywall contains about 90 percent of calcium sulfate ($CaSO_4 \cdot 2H_2O$), which, under anaerobic conditions, serves as an electron acceptor for sulfate-reducing bacteria, which produce high concentrations of H_2S. H_2S levels as high as 12,000 ppm were reported following monitoring of ten landfills in Florida, although some dilution with ambient air is expected (Lee, 2000). In addition to the rotten egg odor of H_2S, there are some public health concerns over its release around solid wastes landfills.

3.3.3.2 Environmental Aspects of Sulfur-Oxidizing Bacteria.
• *Acidic Mine Drainage*
Thousands of miles of U.S. waterways, particularly in Appalachia, are affected by acidity. Acid mine drainage results from water flowing through mines, or runoff from mounds of mine tailings. Iron pyrite (FeS_2) is commonly found associated with coal deposits. Under acidic conditions, iron- and sulfur-oxidizing bacteria drive the oxidation of pyrite to ferric sulfate and sulfuric acid, which causes the degradation of water quality and fish kills (Dugan, 1987a; b).

$$FeS_2 + 3O_2 + 2H_2O \longrightarrow 2H_2SO_4 + Fe^{3+} \tag{3.42}$$

Fe^{2+} is oxidized to Fe^{3+}, which forms ferric hydroxide and more acidity according to Eq. (3.41):

$$Fe^{3+} + 3H_2O \longrightarrow Fe(OH)_3 + 3H^+ \tag{3.43}$$

One of the means of control of acidic mine drainage is inhibition of iron- and sulfur-oxidizing bacteria, using anionic surfactants, benzoic acid, organic acids, alkyl benzene sulfonates, and sodium dodecyl sulfate (SDS) (Dugan, 1987a). Some of these chemicals or their combinations (e.g., combination of sodium dodecyl sulfate and benzoic acid) were able to reduce acidic drainage from coal refuse under simulated field conditions (Dugan, 1987b). However, the application of these techniques under field conditions has not been attempted. Acid mine drainage can also be treated, using sulfate-reducing bacteria. Their activity produces alkalinity and H_2S, which help raise the pH and remove metal contaminants (Dvorak et al., 1992; Webb et al., 1998).

• *Removal of Pyritic Sulfur from Coal*
Iron- and sulfur-oxidizing bacteria (*Acidithiobacillus ferrooxidans*, *Acidithiobacillus thiooxidans*) have been considered for removing sulfur from coal. The rate of biological sulfur removal increases as the particle size of the coal decreases (Tillet and Myerson, 1987).

• *Microbial Leaching (or Biomining)*
Sulfur-oxidizing microorganisms participate in metal recovery (Cu, Ni, Zn, Pb) from low-grade ores. Mine tailings are piled up and the effluents from the piles are recirculated

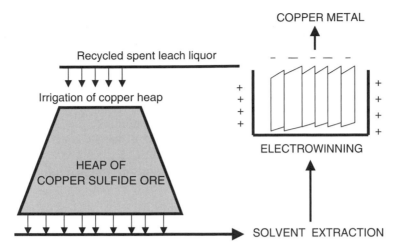

Figure 3.11 Microbial leaching or biomining. From Rawlings (2002).

and become enriched with metals (they are called "pregnant" solutions). Figure 3.11 shows the heap-leaching of copper-containing ore (Rawlings, 2002). The microorganisms involved in biomining include *Acidithiobacillus thiooxidans*, *Acidithiobacillus ferrooxidans*, *Acidithiobacillus caldus*, and *Leptospirillum ferrooxidans*. Extreme thermophilic archaea such as *Sulfolobus* can also be used in metal recovery from mining wastes. These microorganisms are difficult to grow on solid growth media, and molecular-based techniques have been used for their identification.

The advantages of biomining over physico-chemical processes are (Rawlings, 2002):

- Biomining is environmentally friendly and does not result in harmful gas emissions.
- Biomining can extract metals from low-grade ores.
- Biomining is less costly than physico-chemical processes.

- *Role in Corrosion*

Chemical corrosion is due mainly to oxidation and to the corrosive action of chlorine, acids, alkalis, and metal salts. Bacteria are also responsible for corrosion by producing inorganic (e.g., H_2SO_4) or organic acids (e.g., acetic acid, butyric acid) (Ford and Mitchell, 1990). Sulfuric acid, formed by oxidation of H_2S by sulfide-oxidizing bacteria (e.g., *Thiobacillus concretivorus*), is a major corrosive agent in distribution pipes, particularly concrete pipes. Another bacterium, *Acidithiobacillus ferrooxidans* is responsible for the corrosion of iron pipes and pumps. Other iron-oxidizing bacteria, *Gallionella* and *Sphaerotilus natans*, are associated with tubercle formation and corrosion of distribution pipes (Hamilton, 1985). H_2S can be removed by activated carbon, addition of iron salts, or by using oxidants such as chlorine and hydrogen peroxide (Hamilton, 1985; Odom, 1990).

3.4 WEB RESOURCES

Nitrogen Cycle

http://www.sp.uconn.edu/~terry/229sp02/lectures/Ncycleanim.html (animation of the nitrogen cycle; University of Connecticut)

http://www.sp.uconn.edu/~terry/229sp02/lectures/Ncycleanim.html (University of Missouri-Columbia)
http://www.neuse.ncsu.edu/nitrogen/index.htm (nitrogen cycle in soils)
http://ridge.icu.ac.jp/genedN2.htm (Powerpoint presentation)

Phosphorus Cycle

http://soils1.cses.vt.edu/ch/biol_4684/cycles/pcycle.html
http://www.sdsmt.edu/online-courses/geology/geol299/lect29.htm
http://www.enviroliteracy.org/article.php/480.html
http://www.epa.gov/owow/monitoring/volunteer/stream/vms56.html

Sulfur Cycle

http://www.asmusa.org/edusrc/library/images/Tterry/anim/Scycleanim.html (animated sulfur cycle; University of Connecticut)
http://www.biosci.ohio-state.edu/~mgonzalez/Micro521/23.html (Source: Ohio State University)

Iron and Mercury Cycle

http://www.biosci.ohio-state.edu/~mgonzalez/Micro521/24.html (Ohio State University)

Corrosion

http://www.corrosion-doctors.org/Microbial/MIC-References.htm (several good links to corrosion problems)
http://microscope.mbl.edu/reflections/scripts/microscope.php?func=imgDetail&imageID=108 (pictures of Begiatoa)

Acid Mine Drainage

http://www.mines.edu/fs_home/jhoran/ch126/index.htm (site on acid mine drainage from Colorado School of Mines)
http://www.splammo.net/bact102/102ironbact.html (iron bacteria)
http://www.im.nbs.gov/blitz/bionorslide.html (iron bacteria; USGS)
http://soils1.cses.vt.edu/ch/biol_4684/Microbes/Leptothrix.html (iron bacteria: *Leptothrix*)
http://pubs.usgs.gov/publications/text/Norrielist.html (iron bacteria photos from USGS)

3.5 QUESTIONS AND PROBLEMS

1. What is the importance of biological nitrogen fixation? If did not exist, how would we be able to get nitrogen?
2. CO_2 and N_2O are two greenhouse gases. What are the human activities that generate these gases?
3. What microorganisms in the oceans help mitigate the increase of CO_2 in the atmosphere?
4. What are the factors favoring nitrification in water and wastewater?

5. Why does methemoglobinemia affect mostly infants?

6. What is the difference between assimilatory and dissimilatory nitrate reduction?

7. What are the conditions favoring denitrification in wastewater treatment plants?

8. How is nitrogen removed in wastewater treatment plants?

9. What is the principle behind enhanced biological phosphorus removal? How is it accomplished in wastewater treatment plants?

10. Why do sulfate-reducing bacteria compete with methanogens?

11. What causes acidic mine drainage?

12. Discuss the anaerobic corrosion of iron pipes.

3.6 FURTHER READING

Barnes, D., and P.J. Bliss. 1983. *Biological Control of Nitrogen in Wastewater Treatment*, E. & F.N. Spon, London.

Barton, L.L., and R.M. Plunkett. 2002. Sulfate reducing bacteria: Environmental and technological aspects, pp. 3087–3096, In: *Encyclopedia of Environmental Microbiology*, Gabriel Bitton, editor-in-chief, Wiley-Interscience, N.Y.

Beech, I.B. 2002. Biocorrosion: Role of sulfate reducing bacteria, pp. 465–475, In: *Encyclopedia of Environmental Microbiology*, Gabriel Bitton, editor-in-chief, Wiley-Interscience, N.Y.

Christensen, M.H., and P. Harremoes. 1978. Nitrification and denitrification in wastewater treatment, pp. 391–414, In: *Water Pollution Microbiology*, Vol. 2, R. Mitchell, Ed., John Wiley & Sons, New York, N.Y.

Ehrlich, H.L. l98l. *Geomicrobiology*, Marcel Dekker, New York.

Ford, T., and R. Mitchell. 1990. The ecology of microbial corrosion. Adv. Microb. Ecol. 11: 231–262.

Maier, R.M. Biogeochemical cycling, pp. 319-346, In: *Environmental Microbiology*, R.M. Maier, I.L. Pepper, and C.P. Gerba, Eds., Academic Press, San Diego, CA, 585 pp.

Meganck, M.T.J., and G.M. Faup. 1988. Enhanced biological phosphorus removal from waste waters, pp. 111–203, In: *Biotreatment Systems*, Vol. 3, D.L. Wise, Ed., CRC Press, Boca Raton, FL.

Rawlings, D.E. 2002. Heavy metal mining using microbes. Ann. Rev. Microbiol. 56: 65-91.

Seviour, R.J., T. Mino, and M. Onuki. 2003. The microbiology of biological phosphorus removal in activated sludge systems. FEMS Microbiol. Rev. 27: 99–127.

Widdel, F. 1988. Microbiology and ecology of sulfate- and sulfur-reducing bacteria. pp. 469–585, In: *Biology of Anaerobic Microorganisms*, A.J.B. Zehnder, Ed., John Wiley & Sons, New York.

PART B

PUBLIC HEALTH MICROBIOLOGY

4

PATHOGENS AND PARASITES IN DOMESTIC WASTEWATER

4.1 ELEMENTS OF EPIDEMIOLOGY

4.1.1 Some Definitions

Epidemiology is the study of the spread of infectious diseases in populations. *Infectious* diseases are those that can be spread from one host to another. Epidemiologists play an important role in the control of these diseases.

Incidence of a disease is the number of individuals with the disease in a population, whereas *prevalence* is the percentage of individuals with the disease at a given time. A disease is *epidemic* when the incidence is high and *endemic* when the incidence is low. *Pandemic* refers to the spread of the disease across continents.

Infection is the invasion of a host by an infectious microorganism. It involves the entry (e.g., through the gastrointestinal and respiratory tracts, skin) of the pathogen into the host and its multiplication and establishment inside the host. *Inapparent infection* (or covert infection) is a subclinical infection with no apparent symptoms (i.e., the host reaction is not clinically detectable). It does not cause disease symptoms but confers the same degree of immunity as an overt infection. For example, most enteric viruses cause inapparent infections. A person with inapparent infection is called a healthy carrier. Carriers constitute, however, a potential source of infection for others in the community (Finlay and Falkow, 1989; Jawetz et al., 1984). *Nosocomial infections* are hospital-acquired infections, which affect approximately 2.5 million patients annually in the United States. This represents approximately 5 percent of patients with a documented infection acquired in a hospital. Intensive care unit patients represent about 20 percent of hospital-acquired infections. In developing countries, the overall rates vary between 3 and 13.5 percent. The patients most at risk for nosocomial infections are elderly patients, patients with intravenous devices or urinary catheters, those on parenteral nutrition or antimicrobial chemotherapy, and HIV and cancer patients (see review of Mota and Edberg, 2002).

Pathogenicity is the ability of an infectious agent to cause disease and injure the host. Pathogenic microorganisms may infect susceptible hosts, leading sometimes to *overt* disease, which results in the development of clinical symptoms that are easily detectable. The development of the disease depends on various factors, including infectious dose, pathogenicity, and host and environmental factors. Some organisms, however, are *opportunistic pathogens* that cause disease only in compromised individuals.

4.1.2   Chain of Infection

The potential for a biological agent to cause infection in a susceptible host depends on the various factors described in the following.

4.1.2.1   *Type of Infectious Agent.*   Several infectious organisms may cause diseases in humans. These agents include bacteria, fungi, protozoa, metazoa (helminths), rickettsiae, and viruses (Fig. 4.1).

Evaluation of infectious agents is based on their virulence or their potential for causing diseases in humans. *Virulence* is related to the dose of infectious agent necessary for infecting the host and causing disease. The potential for causing illness also depends on the stability of the infectious agent in the environment. The *minimal infective dose* (MID) varies widely with the type of pathogen or parasite. For example, for *Salmonella typhi* or enteropathogenic *E. coli*, thousands to millions of organisms are necessary to establish infection, whereas the MID for *Shigella* can be as low as 10 cells. A few protozoan cysts or helminth eggs may be sufficient to establish infection. For some viruses, only one or a few particles are sufficient for infecting individuals. For example, 17 infectious particles of echovirus 12 are sufficient for establishing infection (Table 4.1; Bitton, 1980a; Bryan, 1977; Gunnerson et al., 1984; Schiff et al., 1984a; 1984b).

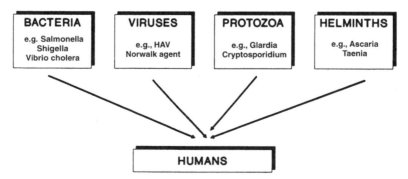

Figure 4.1   Categories of organisms of public health significance.

TABLE 4.1.　Minimal Infective Doses for Some Pathogens and Parasites

Organism	Minimal Infective Dose
Salmonella spp.	$10^4 - 10^7$
Shigella spp.	$10^1 - 10^2$
Escherichia coli	$10^6 - 10^8$
Escherichia coli O157:H7	<100
Vibrio cholerae	10^3
Campylobacter jejuni	about 500
Giardia lamblia	$10^1 - 10^2$ cysts
Cryptosporidium	10^1 cysts
Entamoeba coli	10^1 cysts
Ascaris	1–10 eggs
Hepatitis A virus	1–10 PFU

4.1.2.2　*Reservoir of the Infectious Agent.* A *reservoir* is a living or nonliving source of the infectious agent and allows the pathogen to survive and multiply. The human body is the reservoir for numerous pathogens; person-to-person contact is necessary for maintaining the disease cycle. Domestic and wild animals also may serve as reservoirs for several diseases (e.g., rabies, brucellosis, turbeculosis, anthrax, leptospirosis, toxoplasmosis) called *zoonoses*, that can be transmitted from animals to humans. Table 4.2 shows a list of animal reservoirs for waterborne human pathogens (Berger and Oshiro, 2002). Nonliving reservoirs such as water, wastewater, food, or soils can also harbor infectious agents.

TABLE 4.2.　Examples of Animal Reservoirs for Waterborne Human Pathogens and Parasites[a]

Organism	Major Disease	Animal
Bacteria		
E. coli O157:H7	Hemolytic uremic syndrome	Cattle and other ruminants
Campylobacter jejuni	Gastroenteritis	Poultry, pigs, sheep, dogs, cats
Helicobacter pylori	Peptic ulcers; stomach cancer	Not known
Shigella spp.	Bacillary dysentery	Nonhuman primates
Salmonella typhi	Typhoid fever	Not known
Enteric Viruses (human strains)		
Caliciviruses	Gastroenteritis	Humans only
Rotaviruses	Gastroenteritis	Humans only
Polioviruses	Polio	Humans only
Hepatitis A virus	Infectious hepatitis	Nonhuman primates rarely
Protozoa		
Cryptosporidium parvum	Gastroenteritis	Many mammals, especially calves
Giardia lamblia	Gastroenteritis	Muskrats, beavers, small rodents, many domestic and wild animals
Naegleria fowleri	Primary amoebic meningoencephalitis	None (free-living)
Toxoplasma gondii	Flu-like symptoms	Cats

[a]Adapted from Berger and Oshiro (2002).

4.1.2.3 Mode of Transmission.   Transmission involves the transport of an infectious agent from the reservoir to the host. It is the most important link in the chain of infection. Pathogens can be transmitted from the reservoir to a susceptible host by various routes (Sobsey and Olson, 1983).

• *Person-to-Person Transmission*

The most common route of transmission of infectious agents is from person to person. The best examples of direct contact transmission are the sexually transmitted diseases such as syphilis, gonorrhea, herpes, or acquired immunodeficiency syndrome (AIDS). Coughing and sneezing discharge very small droplets containing pathogens within a few feet of the host (droplet infection). Transmission by these infectious droplets is sometimes considered as an example of direct contact transmission.

• *Waterborne Transmission*

The waterborne transmission of cholera was established in 1854 by John Snow, an English physician who noted a relationship between a cholera epidemics and consumption of water from the Broad Street well in London. The waterborne route is not, however, as important as the person-to-person contact route for the transmission of fecally transmitted diseases.

The World Health Organization (WHO) reported that diarrhoeal diseases contracted worldwide mainly by contaminated water or food, killed 3.1 million people, most of them children (WHO, 1996). In the United States, waterborne disease outbreaks are reported to the U.S. Environmental Protection Agency (U.S. EPA) and the Centers for Disease Control and Prevention (CDC) by local epidemiologists and health authorities; the system was started in the 1920s (Craun, 1986a, b; 1988). During the period 1971–1985, 502 waterborne outbreaks and 111,228 cases were reported. Figure 4.2 (Craun, 1988) shows that about three-quarters of the outbreaks were due to untreated or inadequately treated groundwater and surface waters. Gastrointestinal illnesses of unidentified etiology and giardiasis are the most common waterborne diseases for groundwater and surface water systems (Table 4.3; Craun, 1988). The outbreak rate (expressed as the number of outbreaks/1000 water systems) and the illness rate (expressed as

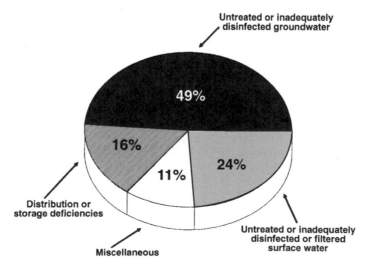

Figure 4.2   Causes of 502 waterborne disease outbreaks: 1971–1985. From Craun (1988).

TABLE 4.3. Etiology of Waterborne Outbreaks for Groundwater and Surface Water Systems: 1971–1985[a]

Illness	No. of Outbreaks	Cases of Illness
Gastroenteritis, undefined	251	61,478
Giardiasis	92	24,365
Chemical poisoning	50	3,774
Shigellosis	33	5,783
Hepatitis A	23	737
Gastroenteritis, viral	20	6,254
Campylobacterosis	11	4,983
Salmonellosis	10	2,300
Typhoid	5	282
Yersiniosis	2	103
Gastroenteritis, toxigenic *E. coli*	1	1,000
Cryptosporidiosis	1	117
Cholera	1	17
Dermatitis	1	31
Amoebiasis	1	4
Total	502	111,228

[a]From Craun (1988). (Courtesy of the American Water Works Association.)

numbers of cases/million-person year) decrease as the raw water is filtered and disinfected (Table 4.4) (Craun, 1988).

• *Foodborne Transmission*

Food may serve as a vehicle for the transmission of numerous infectious diseases caused by bacteria, viruses, protozoa, and helminth parasites. The World Health Organization estimates that accidental food poisoning kills up to 1.5 million people per year. In the United States, it is estimated that foodborne illnesses affect some 6 to 80 million persons/year, leading to approximately 9000 deaths. Emerging foodborne pathogens reported in the United States are shown in Table 4.5 (Altekruse et al., 1997). Food contamination results from unsanitary practices during production or preparation. Several pathogens and parasites have been detected in risky foodstuffs such as shellfish, vegetables, raw milk, runny eggs or pink chicken, turkey, ground beef and ground pork, alfalfa sprouts, and unpasteurized apple juice/cider. Their presence is of public health significance, particularly for foods that are eaten raw (e.g., shellfish, fresh produce). There is also an increased risk among the elderly and immunocompromised people (HIV and leukemia patients, and those taking immunosuppressive drugs such as steroids, cyclosporine, and radiation therapy).

TABLE 4.4. Effect of Water Treatment on Outbreak and Illness Rates: 1971–1985[a]

Type of Community Water System	Waterborne Outbreaks per 1000 Water Systems	Waterborne Illness per Million Person-Years
Untreated surface water	32.5	370.9
Disinfected-only surface water	40.5	66.3
Filtered and disinfected surface water	5.0	4.7

[a]From Craun (1988). (Courtesy of the American Water Works Association.)

TABLE 4.5.   Estimated Number of Cases per Year Caused by Infection with Selected Foodborne Bacterial Pathogens in the United States[a]

Foodborne Pathogen	Estimated Cases (10^3)	Commonly Implicated Foods
Campylobacter jejuni	4000	Poultry, raw milk, untreated water
Salmonella (nontyphoid)	2000	Eggs, poultry, meat, fresh produce, other raw foods
Escherichia coli O157:H7	25	Ground beef, raw milk, lettuce, untreated water, unpasteurized cider/apple juice
Listeria monocytogenes	1.5	Ready-to-eat foods (e.g., soft cheese, deli foods, pâté)
Vibrio sp. (e.g., *Vibrio vulnificus*)	10	Seafood (e.g., molluscan, crustacean shellfish) raw, undercooked, cross-contaminated

[a]Adapted from Altekruse et al. (1997).

Vegetables contaminated with wastewater effluents are also responsible for disease outbreaks (e.g., typhoid fever, salmonellosis, amebiasis, ascariasis, viral hepatitis, gastroenteritis). Raw vegetables and fruit become contaminated as a result of being handled by an infected person during processing, storage, distribution or final preparation, or following irrigation with fecally contaminated water (Seymour and Appleton, 2001). Vomitus (estimation of 20 to 30 million virus particles released during vomiting) from infected food handlers can also contaminate exposed food and surfaces via production of bioaerosols. In England and Wales, viruses accounted for 4.3 percent of all foodborne outbreaks for the period 1992–1999, with Norwalk-like viruses (NLVs) being the most commonly found agents (O'Brien et al., 2000). Outbreaks of hepatitis were associated with fresh produce (e.g., salads, iceberg lettuce, diced tomatoes, frozen raspberries).

Shellfish (e.g., oysters, clams, mussels) are significant vectors of human diseases of bacterial, viral, and protozoan origin. Several surveys have been carried out worldwide to show the presence of pathogens in shellfish samples. The use of molecular techniques (RT-PCR) has helped in the detection of enteric viruses (hepatitis A virus, Norwalk-like virus, enterovirus, rotavirus, and astrovirus) in oyster and mussel samples in France (Le Guyader et al., 2000). In Switzerland, 8 of 87 imported oyster samples were positive for Norwalk-like viruses (Beuret et al., 2003). Moreover, infectious oocysts of *Cryptosporidium* were detected in mussels and cockles in Spain (Gomez-Bautista et al., 2000). Enteric viruses (enteroviruses, Norwalk-like viruses, and adenoviruses) were detected in 50–60 percent of mussel samples in two sites in the west coast of Sweden (Fig. 4.3; Hernroth et al., 2002). Of 36 mussel samples from the Adriatic Sea, 13 were contaminated with hepatitis A virus and 5 samples with enteroviruses (Croci et al., 2000). In several surveys, *E. coli* was not found to be a good indicator of virus presence in mussels.

Shellfish are important in disease transmission for the following reasons (Bitton, 1980a; Doré and Lees, 1995; Poggi, 1990):

- They live in estuarine environments, which are often contaminated by domestic wastewater effluents.
- As filter-feeders, they concentrate pathogens and parasites by pumping large quantities of estuarine water (4–20 L/h). The accumulation occurs mainly in the digestive tissues (mostly in the digestive gland as demonstrated for male-specific phage accumulation in mussels). The bioaccumulation of enteric bacteria and viruses by bivalve mollusks vary with the species of shellfish, type of microorganism, environmental conditions, and season. In oysters (*Crassostrea virginica*), the average

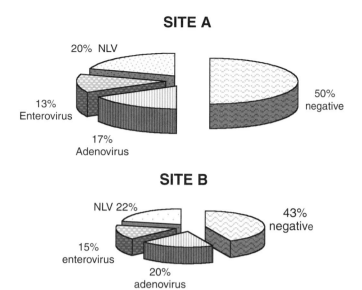

Figure 4.3 Distribution of human viruses in mussels in Scandinavia (February 2000–July 2001). Adapted from Hernroth et al. (2002).

accumulation factors for *E. coli* and F+ coliphage were 4.4 and 19.0 (range from < 1 to 99), respectively (Burkhardt and Calci, 2000).

- They are often eaten raw or insufficiently cooked. It has been estimated that only one-third of shellfish consumed every year in France are sufficiently cooked. In the United States, this habit has led to 2100 cases in the period 1991–1998. Sixty percent of the cases were caused by enteric viruses, particularly Norwalk-like viruses (NLV) (Glatzer, 1998; Shieh et al., 2003). Shellfish steaming or cooking do not seem to prevent viral-associated gastroenteritis (Kirkland et al., 1996; McDonnell et al., 1997). A temperature of about 90°C is necessary to inactivate hepatitis A virus in shellfish.
- Depuration of shellfish, while effective for bacterial contaminants, is not successful for enteric viruses such as hepatitis A virus (Franco et al., 1990).
- Other health hazards associated with shellfish consumption result from the ability of these mollusks to concentrate dinoflagellate toxins, heavy metals, hydrocarbons, pesticides, and radionuclides.

• Airborne Transmission

Some diseases (e.g., Q fever, some fungal diseases) can be spread by airborne transmission. This route is important in the transmission of biological aerosols generated by wastewater treatment plants or spray irrigation with wastewater effluents. This topic is discussed in depth in Chapter 14.

• Vector-Borne Transmission

The most common vectors for disease transmission are arthropods (e.g., fleas, insects) or vertebrates (e.g., rodents, dogs, and cats). The pathogen may or may not multiply inside the arthropod vector. Some vector-borne diseases are malaria (caused by *Plasmodium*), yellow fever, or encephalitis (both due to arboviruses) and rabies (from virus transmitted by the bite of rabid dogs or cats).

• *Fomites*

Some pathogens may be transmitted by nonliving objects or *fomites* (e.g., clothes, utensils, toys, environmental surfaces).

4.1.2.3 Portal of Entry. Pathogenic microorganisms can gain access to the host mainly through the gastrointestinal tract (e.g., enteric viruses and bacteria), the respiratory tract (e.g., *Klebsiella pneumonae*, *Legionella*, myxoviruses) or the skin (e.g., *Aeromonas*, *Clostridium tetani*, *Clostridium perfringens*). Although the skin is a formidable barrier against pathogens, wounds or abrasions may facilitate their penetration into the host.

4.1.2.4 Host Susceptibility. Both the immune system and nonspecific factors play a role in the resistance of the host to infectious agents. Immunity to an infectious agent may be natural or acquired. *Natural immunity* is genetically specified and varies with species, race, age (the young and the elderly are more susceptible to infection), hormonal status, and the physical and mental health of the host. People in poor health and the elderly are more susceptible to infectious agents than healthy adults. *Acquired immunity* develops as a result of exposure of the host to an infectious agent. Acquired immunity can be *passive* (e.g., fetus acquiring the mother's antibodies) or *active* (e.g., active production of antibodies through contact with the infectious agent) (Jawetz et al., 1984).

The nonspecific factors include physiological barriers at the portal of entry (e.g., unfavorable pH, bile salts, production of digestive enzymes and other chemicals with antimicrobial properties, competition with the natural microflora in the colon) and destruction of the invaders by phagocytosis.

4.2 PATHOGENS AND PARASITES FOUND IN DOMESTIC WASTEWATER

Several pathogenic microorganisms and parasites are commonly found in domestic wastewater as well as in effluents from wastewater treatment plants. The three categories of pathogens encountered in the environment are (Leclerc et al., 2002):

• *Bacterial pathogens*: Some of these pathogens (e.g., *Salmonella*, *Shigella*) are enteric bacteria. Others (e.g., *Legionella*, *Mycobacterium avium*, *Aeromonas*) are indigenous aquatic bacteria.

• *Viral pathogens*: These are also released into aquatic environments but are unable to multiply outside their host cells. Their infective dose is generally lower than for bacterial pathogens.

• *Protozoan parasites*: These are released into aquatic environments as cysts or oocysts, which are quite resistant to environmental stress and to disinfection, and do not multiply outside their hosts.

4.2.1 Bacterial Pathogens

Fecal matter contains up to 10^{12} bacteria per gram. The bacterial content of feces represents approximately 9 percent by wet weight (Dean and Lund, 1981). 16S rRNA-gene-targeted group-specific primers were used to detect and identify the predominant bacteria in human feces. Of 300 bacterial isolates from feces of six healthy individuals, 74 percent belonged to the *Bacteroides fragilis* group, *Bifidobacterium*, *Clostridium coccoides* group, and *Prevotella* (Matsuki et al., 2002).

Wastewater bacteria have been characterized and belong to the following groups (Dott and Kampfer, 1988):

- Gram-negative facultatively anaerobic bacteria: e.g., *Aeromonas*, *Plesiomonas*, *Vibrio*, *Enterobacter*, *Escherichia*, *Klebsiella*, and *Shigella*.
- Gram-negative aerobic bacteria: e.g., *Pseudomonas*, *Alcaligenes*, *Flavobacterium*, *Acinetobacter*.
- Gram-positive spore forming bacteria: e.g., *Bacillus* spp.
- Nonspore-forming gram-positive bacteria: e.g., *Arthrobacter*, *Corynebacterium*, *Rhodococcus*.

A compilation of the most important bacteria that may be pathogenic to humans and that can be transmitted directly or indirectly by the waterborne route is shown in Table 4.6 (Dart and Stretton, 1980; Dean and Lund, 1981; Feachem et al., 1983; Sobsey and Olson, 1983; Theron and Cloete, 2002). These pathogens cause enteric infections such as typhoid fever, cholera, or shigellosis.

We will now review some of the most important bacterial pathogens found in wastewater.

4.2.1.1 *Salmonella*.

Salmonellae are enterobacteriaceae and are widely distributed in the environment and include more than 2000 serotypes. The *Salmonella* numbers in wastewater range from a few to 8000 organisms/100 mL, are the most predominant pathogenic bacteria in wastewater, and cause typhoid and paratyphoid fever, and gastroenteritis. *Salmonella* numbers in wastewater in the United States range from 10^2 to 10^4 organisms/100 mL, but much higher concentrations (up to 10^9/100 mL) have been reported in

TABLE 4.6. Major Waterborne Bacterial Diseases[a]

Bacterial Agent	Major Disease	Major Reservoir	Principal Site Affected
Salmonella typhi	Typhoid fever	Human feces	Gastrointestinal tract
Salmonella paratyphi	Paratyphoid fever	Human feces	Gastrointestinal tract
Shigella	Bacillary dysentery	Human feces	Lower intestine
Vibrio cholerae	Cholera	Human feces	Gastrointestinal tract
Pathogenic *E. coli*	Gastroenteritis Hemolytic uremic syndrome	Human feces	Gastrointestinal tract
Yersinia enterocolitica	Gastroenteritis	Human/animal feces	Gastrointestinal tract
Campylobacter jejuni	Gastroenteritis	Human/animal feces	Gastrointestinal tract
Legionella pneumophila	Acute respiratory illness (legionnaires' disease)	Thermally enriched waters	Lungs
Mycobacterium tuberculosis	Tuberculosis	Human respiratory exudates	Lungs
Leptospira	Leptospirosis (Weil's disease)	Animal feces and urine	Generalized
Opportunistic bacteria	Variable	Natural waters	Mainly gastrointestinal tract

[a]Adapted from Sobsey and Olson (1983).

developing countries (Jimenez and Chavez, 2000). It is estimated that two to four million human *Salmonella* infections occur each year in the United States (Feachem et al., 1983). An estimated 0.1 percent of the population excretes *Salmonella* at any given time. In the United States, salmonellosis is primarily due to food contamination, but its transmission through drinking water is still of great concern. *Salmonella typhi* is the etiological agent for typhoid fever, a deadly disease that was brought under control as a result of the development of adequate water treatment processes (e.g., chlorination, filtration). This pathogen produces an endotoxin that causes fever, nausea, and diarrhea and may be fatal if not properly treated by antibiotics (Sterritt and Lester, 1988). Species implicated in food contamination are *S. enteriditis* and *S. typhimurium*. These species can grow readily in contaminated foods and cause food poisoning, leading to diarrhea and abdominal cramps. *Salmonella enteriditis*, through consumption of contaminated poultry and eggs, emerged as a public health problem in the 1980s in Europe and in the United States. Of 371 outbreaks of *S. enteriditis* that occurred in the United States, 80 percent were associated with consumption of insufficiently cooked eggs (Patrick et al., 2004). *Salmonella enteriditis* is transmitted directly from breeding flocks to egg-laying hens. In Israel, about 30 percent of the 35 flocks of laying hens examined were contaminated with *S. enteriditis*. This has led to programs to control this pathogen (Shimshony, 1997).

4.2.1.2 *Shigella*.
Shigella is the causative agent of bacillary dysentery or shigellosis, an infection of the large bowel that leads to cramps, diarrhea, and fever. This disease produces bloody stools as a result of inflammation (induction and release of proinflammatory cytokines) and ulceration of the intestinal mucosa. Globally, this pathogen is responsible for approximately one million deaths and 163 million cases of bacillary dysentery annually, with 99 percent of the cases occurring in developing countries (Torres, 2002). There are four pathogenic species of *Shigella*: *S. dysenteriae* (13 serotypes), *S. flexneri* 15 serotypes), *S. boydii* (18 serotypes) and *S. sonnei.* (1 serotype). *S. dysenteriae*, serotype 1, produces a potent toxin called the Shiga toxin. Infection with bacteria producing this toxin may lead to hemolytic uremic syndrome which results in kidney failure (Torres, 2002). No current vaccine is available for protection against *Shigella*.

This pathogen is transmitted by direct contact with an infected individual, who may excrete up to 10^9 shigellae per gram of feces. The infectious dose for *Shigella* is relatively small and can be as low as 10 organisms. Although person-to-person contact is the main mode of transmission of this pathogen, food-borne (via salads and raw vegetables) and waterborne transmissions have also been documented. At least three large epidemics caused by *Shigella dysenteriae* type 1 have occurred in Bangladesh between 1972 and 1994, and the pathogen was isolated from surface waters, using genetic probes and culturing followed by biochemical tests. The environmental isolates shared some virulence genes with clinical isolates and were resistant to one or more antibiotics (Faruque et al., 2002). Groundwater was found to be responsible for a shigellosis outbreak in Florida that involved 1200 people. However, *Shigella* persists less in the environment than fecal coliforms. Few quantitative data are available on its occurrence and removal in water and wastewater treatment plants.

4.2.1.3 *Vibrio cholerae*.
Vibrio cholera is a gram-negative curved rod bacterium that is an autochtonous member of the aquatic microbial community. This bacterium is the causative agent of cholera. In 1854, John Snow first demonstrated that contaminated drinking water was the cause of cholera. This pathogen releases an enterotoxin that causes mild to profuse diarrhea, vomiting, and a very rapid loss of fluids, which may result in death in a relatively short period of time. The infectious dose for *V. cholera* is

between 10^4 and 10^6 cells. Among about 200 known serogroups of *Vibrio cholerae*, only two (O1 and O139) are known to cause disease and can be detected using serum agglutination assays or monoclonal antibodies (Huk et al., 2002). Although rare in the United States and Europe, this disease appears to be endemic in various areas throughout Asia, particularly in Bangladesh. This pathogen is found in wastewater, and levels of $10-10^4$ organisms/100 mL during a cholera epidemic have been reported (Kott and Betzer, 1972). Explosive epidemics of cholera and typhoid fever have been documented in Peru and Chile and have been associated with the consumption of sewage-contaminated vegetables (Shuval, 1992). *Vibrio cholerae* can be detected in environmental samples, using immunological or molecular methods. Although the nucleic acid-based methods are rapid and relatively easy, they do not allow a differentiation between viable and nonviable cells.

Vibrios are naturally present in many aquatic environments and survive by attaching to solids, including zooplankton (e.g., copepods), cyanobacteria (e.g., *Anabaena*), and phytoplankton (e.g., *Volvox*) cells. These plankton-associated bacteria may occur in the viable but nonculturable state (VBNC) and can be observed under the microscope by means of the fluorescent-monoclonal antibody technique (Brayton et al., 1987; Huk et al., 1990).

4.2.1.4 *Escherichia Coli*.

Several strains of *E. coli*, many of which are harmless, are found in the gastrointestinal tract of humans and warm-blooded animals. There are several categories of *E. coli* strains, however, that bear virulence factors and cause diarrhea. There are enterotoxigenic (ETEC), enteropathogenic (EPEC), enterohemorrhagic (EHEC), enteroinvasive (EIEC), and enteroaggregative (EAggEC) types of *E. coli* (Guerrant and Thielman, 1995; Levine, 1987). Enterotoxigenic *E. coli* causes gastroenteritis with profuse watery diarrhea accompanied with nausea, abdominal cramps, and vomiting. Approximately 2–8 percent of the *E. coli* present in water were found to be enteropathogenic *E. coli*, which causes Traveler's diarrhea. Food and water are important in the transmission of this pathogen. However, the infective dose for this pathogen is relatively high, within the range of 10^6-10^9 organisms. Distinct features of enteroaggregative *E. coli* are its adherence to Hep2 cells in tissue culture in an aggregative pattern and its ability to cause persistent diarrhea.

Some of these diarrheagenic strains of *E. coli* have been detected in treated water with genetic probes, and they can represent a health risk to consumers (Martins et al., 1992). During a 1989–1990 survey of waterborne disease outbreaks in the United States, the etiologic agent in one out of 26 outbreaks was enterohemorrhagic *E. coli* O157:H7, an agent that produces shiga-like toxins (SLT) I and/or II, has a relatively low infectious dose (<100 organisms) and causes bloody diarrhea, particularly among the very young and very old members of the community (Herwaldt et al., 1992). Infections, if left untreated, may lead to hemolytic uremic syndrome, a leading cause of kidney damage and possible failure in children, with a 3–5 percent death rate in patients (Boyce et al., 1995). In the United States, it is estimated that *E. coli* O157:H7 causes more than 20,000 infections and as many as 250 deaths each year (Boyce et al., 1995). Twenty outbreaks of *E. coli* O157:H7 infections reported in the United States between 1982 and 1993 led to 1557 cases, 358 hospitalizations, and 19 deaths (Griffin, 1995). Food (e.g., fresh or undercooked ground meats) appears to be the primary source of infections (Doyle and Schoeni, 1987). Other sources include cider, raw milk, lettuce, and contaminated waters.

Enterohemorrhagic *E. coli* was also isolated from a water reservoir in Philadelphia, PA (McGowan et al., 1989), untreated and treated drinking water (Martins et al., 1992), but surprisingly was not detected in primary and secondary wastewater effluents (Grant et al., 1996). Several outbreaks of *E. coli* O157:H7 were shown to be associated with waterborne transmission from sources such as groundwater (Nataro and Kaper, 1998), recreational waters (Keene et al., 1994), and municipal water systems. One outbreak occurred in Cabool, Missouri, in the winter of 1990 after disturbances in the water distribution network; it resulted in 243 documented cases of diarrhea and four deaths among elderly citizens (Geldreich et al., 1992; Swerdlow et al., 1992). A more recent outbreak occurred in May 2000 in Canada and resulted in 2000 infections and six deaths (ProMed, 2000).

A rapid method has been developed for detecting respiring *E. coli* O157:H7 in water. This membrane filtration method is based on the use of a specific fluorescent antibody combined with cyanoditolyl tetrazolium chloride (CTC) to assess the respiratory activity of the cells (Pyle et al., 1995a). In food, such as hamburger meat, this pathogen can be detected by combining immunomagnetic separation with a sandwich enzyme-linked immunosorbent assay (ELISA) (Tsai et al., 2000). A new proposed methodology detects *E. coli* O157:H7 by using green fluorescent-labeled PP01 specific bacteriophage, which, following adsorption to the host cells, leads to the visualization of the cells under a fluorescence microscope. This method detects *E. coli* O157:H7 in both the culturable and the culturable but not viable (VBNC) states, (Oda et al., 2004).

4.2.1.5 *Yersinia.*

Yersinia enterocolitica is responsible for acute gastroenteritis with invasion of the terminal ileum. Swine are the major animal reservoir, but many domestic and wild animals can also serve as reservoirs for this pathogen. Food-borne (e.g., milk, tofu) outbreaks of yersiniosis have also been documented in the United States. The role of water is uncertain, but there are instances in which this pathogen was suspected to be the cause of waterborne transmission of gastroenteritis (Schiemann, 1990). This psychrotrophic organism thrives at temperatures as low as 4°C, is mostly isolated during the cold months, but is poorly correlated with traditional bacterial indicators (Sobsey and Olson, 1983; Wetzler et al., 1979). This organism has been isolated from wastewater effluents, river water, and from drinking water (Bartley et al., 1982; Meadows and Snudden, 1982; Stathopoulos and Vayonas-Arvanitidou, 1990; Wetzler et al., 1979).

4.2.1.6. *Campylobacter.*

This pathogen (e.g., *C. fetus* and *C. jejuni*) is known to infect humans as well as wild and domestic animals. It is ubiquitous in domestic wastewater and in effluents from abattoirs and poultry processing plants. In Lancaster, United Kingdom, the occurrence of Campylobacters in wastewater showed a seasonal trend similar to that of the incidence of infections in humans (Jones, 2001). Its relatively low infectious dose of approximately 500 organisms (for *C. jejuni*) makes it the leading cause of food-borne infections, with approximately 4 million *C. jejuni* infections/year in the United States. Campylobacter is the most frequent cause of diarrhea in the United States and United Kingdom (Schwartzbrod et al., 2002). These infections may lead in one of 1000 patients to Guillain–Barré syndrome, which is an acute paralytic illness (Altekruse et al., 1997). It is also a common cause of acute gastroenteritis (fever, nausea, abdominal pains, bloody diarrhea, vomiting) and is transmitted to humans through contaminated food, mainly undercooked poultry, unpasteurized milk, contaminated drinking water, and water from mountain streams (Mentzing, 1981; Palmer et al., 1983; Taylor et al., 1983). This pathogen has been the cause of several outbreaks of gastroenteritis in the United States and worldwide (Blaser and Peller, 1981; Blaser et al., 1983; Hänninen et al., 2003; Jones, 2001; Miettinen et al., 2001). The first outbreak in the United States was reported in 1978 in Vermont, where 2000 out of a population of 10,000 were affected. The seasonal occurrence of *Campylobacter* in surface waters has been documented (Carter et al., 1987). *Campylobacter* has been detected in surface waters, potable water, and wastewater, but no organisms have been recovered from digested sludge (Andrin and Schwartzbrod, 1992; Stathopoulos and Vayonas-Arvanitidou, 1990; Stelzer, 1990). Recovery from surface waters was highest in the fall (55 percent of samples positive) and winter (39 percent of samples positive). Numbers of *Campylobacter* did not display any correlation with heterotrophic plate counts, total and fecal coliforms, or fecal streptococci. Owing to their sensitivity to oxygen and inability to grow at temperatures below 30°C (optimum temperature is 42°C), *C. jejuni* survives but does not grow in the environment (Park, 2002).

Campylobacters can be detected in contaminated natural waters by using selective growth agar media or molecular methods following an enrichment step in a selective broth medium (Moreno et al., 2003; Sails et al., 2002).

4.2.1.7. *Leptospira.*

Leptospira is a small spirochete that can gain access to the host through abrasions of the skin or through mucous membranes. It causes leptospirosis, which is characterized by the dissemination of the pathogen in the patient's blood and the subsequent infection of the kidneys and the central nervous system. The disease can be transmitted from animals (rodents, domestic pets, and wildlife) to humans coming into contact (e.g., bathing) with waters polluted with animal wastes. This *zoonotic* disease may strike sewage workers. This pathogen is not of major concern because it does not appear to survive well in wastewater (Rose, 1986).

4.2.1.8 *Legionella pneumophila.*

This bacterial pathogen is the etiological agent of *Legionnaires' disease*, first described in a 1976 outbreak in Philadelphia, PA. It is estimated that there are 10,000–25,000 cases of Legionnaires' disease/year in the United States (Harb and Kwaik, 2002).

This disease is a type of acute pneumonia with a relatively high fatality rate; it may also involve the gastrointestinal and urinary tracts as well as the nervous system. *Legionella pneumophila* causes pneumonia as a result of its ability to multiply within alveolar macrophages. *Pontiac fever* is a milder nonfatal form of Legionnaires' disease associated with *Legionella* infection. People manifesting this syndrome have fever, headaches, and muscle aches, but may recover without any treatment. Aquatic environments and soils can act as natural reservoirs for pathogenic species of Legionellae. This organism is transmitted mainly by aerosolization of contaminated water or soil (Harb and Kwaik, 2002; Muraca et al., 1988).

- *Aerosolization.* Outbreaks of legionnaire's disease are associated with exposure to *L. pneumophila* aerosols from cooling towers, evaporative condensers, humidifiers, shower heads, air conditioning systems, whirlpools, mist machines in produce departments of grocery stores, mechanical aerosolization of soil particles during gardening, or dental equipment. A study of public showers in Bologna, Italy, showed that 22 of 48 samples were positive for *L. pneumophila* (Table 4.7; Leoni et al., 2001). Natural draft cooling towers are used to cool the hot water generated by power-generating plants. These towers generate microbial aerosols, including *Legionella* at the top of the structure. It is postulated that the source of *Legionella* is the water drawn from nearby surface waters or the potable water supply to replace the moisture lost during the cooling cycle (Muraca et al., 1988). Perhaps the world largest epidemics of Legionnaire's disease occurred in Murcia, Spain, in 2001, with 449 confirmed cases out of 800 suspected cases. The outbreak was due to contamination from the hospital cooling towers (Garcia-Fulgueiras et al., 2003).
- *Ingestion. Legionella pneumophila* serogroup 1 has been detected in drinking water systems (Hsu et al., 1984; Tobin et al., 1986) but, so far, no outbreak has been attributed to consumption of contaminated drinking water.

TABLE 4.7. Frequency of Isolation of *Legionella* spp. from Shower Waters[a]

Legionella	Number of Positive Samples ($n = 48$)	Range (CFU/L)
Legionella pneumophila (SG1)	6	20–8,700
Legionella pneumophila (SG3)	5	10–650
Legionella pneumophila (SG4)	1	1530
Legionella pneumophila (SG5)	2	15,150–15,440
Legionella pneumophila (SG6)	8	20–19,250
Legionella bozemanii	7	100–6,000
Legionella dumofii	4	20–510
Legionella gormanii	3	600–4,000
Legionella micdadei	4	30–1,800
All the species	27	10–25,250

[a]Adapted from Leoni et al. (2001).

Surveys have shown that hospitals are the setting for many outbreaks of *Legionella* infections. The sources of nosocomial Legionnaire's disease can be traced back to hospital cooling towers and to potable water distribution system in hospitals (Best et al., 1984; Harb and Kwaik, 2002). The number of cases increase after a pressure drop in the distribution system, which probably causes the release of *Legionella* cells associated with biofilms growing in the distribution pipes. Biofilms appear to provide a protective environment to *Legionella* in distribution systems. However, the number of cases dropped after hyperchlorination (>2 mg/L free residual chlorine) (Shands et al., 1985) (Fig. 4.4). This bacterium appears to be ubiquitous in the environment and has been isolated from wastewater, soil, and natural aquatic environments including tropical waters (Fliermans et al., 1979; 1981; Ortiz-Roque and Hazen, 1987). Its presence in wastewater has been linked, on one occasion at least, with increased levels of antibodies among wastewater irrigation workers (Bercovier et al., 1984). However, the epidemiological significance of this finding remains unclear. In the natural environment, this pathogen can thrive in association with other bacteria which may provide the L-cysteine required by *Legionella*, green and blue-green algae (Berendt, 1980; Pope et al., 1982; Stout et al., 1985; Tison et al., 1980), amoeba (e.g., *Acanthamoeba*, *Naegleria*), and other protozoa (e.g., *Hartmanella*, *Tetrahymena*) (Barbaree et al., 1986; Barker et al., 1992; Neumeister et al., 1997; Thyndall and Domingue, 1982), or ciliates (Fields et al., 1984; Smith-Somerville et al., 1991). Conversely, 16–32 percent of heterotrophic plate count bacteria isolated from chlorinated drinking water were found to inhibit *Legionella* species (Toze et al., 1990).

Legionella association with protozoa provides increased resistance to biocides such as chlorine, low pH, and high temperatures (Barker et al., 1992; States et al., 1990). Protozoa are able to sustain the intracellular growth of *L. pneumophila* and have an effect on its virulence to mammalian cells. It was found that the same virulence genes are required to infect both protozoa and mammalian cells (Harb et al., 2000).

Control of *Legionella* in water distribution systems includes thermal treatment (e.g., increase of water temperature to 60–70°C followed by flushing), treatment with bactericidal agents (e.g., copper, silver), and hyperchlorination up to 50 mg/L (Harb and Kwaik, 2002; Kim et al., 2002). However, because biofilms protect bacterial pathogens from inactivation by free chlorine, treatment with monochloramine provides a better control of *Legionella* in water distribution lines. An epidemiological study showed that 10 times fewer outbreaks of Legionnaires' disease occurred in hospitals having monochloramine as disinfectant residual than those using free chlorine as a residual (Kool et al., 2000).

4.2.1.9 *Bacteroides fragilis.*

Bacteroides species are a major part of the microorganisms in the human colon and account for approximately 25 percent of all colonic isolates (Wilson et al., 1997). This pathogen has been found in wastewater at levels ranging from 6.2×10^4 to 1.1×10^5 colony forming units/mL, 9.3 percent of which were enterotoxigenic (Shoop et al., 1990). Enterotoxin-producing strains of this anaerobic bacterium may be involved in causing diarrhea in humans.

4.2.1.10 *Opportunistic Bacterial Pathogens.*

This group includes heterotrophic gram-negative bacteria belonging to the following genera (Sobsey and Olson, 1983): *Pseudomonas, Aeromonas, Klebsiella, Flavobacterium, Enterobacter, Citrobacter, Serratia, Acinetobacter, Proteus* and *Providencia*, and nontubercular mycobacteria. Segments of the population particularly at risk of infection with opportunistic pathogens are newborn babies, and elderly and sick people. These organisms may occur in high numbers in institutional (e.g., hospital) drinking water and attach to water distribution pipes, and some of them may grow in finished drinking water (see Chapter 16). However, their public health significance with regard to the population at large is not well known.

Pseudomonas aeruginosa is ubiquitous in the environment and is frequently found in water, wastewater, soils and plants. Although it poses no risks in drinking water, it is responsible for 10–20 percent of nosocomial (i.e., hospital-acquired) infections.

Other opportunistic pathogens are the nontubercular mycobacteria, which cause pulmonary infections and other diseases. The most frequently isolated nontubercular mycobacteria belong to the species of *Mycobacterium avium* complex (MAC) (i.e., *M. avium* and *M. intracellulare*), which infect humans (mostly AIDS and other immunocompromised patients) and animals (e.g., pigs) (Falkinham, 2002; Falkinham et al., 2001; Wolinsky, 1979). Another member of the MAC complex is *M. avium* subspecies *paratuberculosis* (MAP), which causes inflammation. There is now evidence that infection with MAP is one of the causes of Crohn's disease, a chronic inflammation of the intestine affecting animals and humans. A new generation of antibiotics (rifabutin, clarithromycin, azithromycin) has been proposed to control or reduce MAP activity in the gastrointestinal tract (Gui et al., 1997; Hermon-Taylor et al., 2000).

Mycobacteria are ubiquitous and are found in environmental waters, including drinking water, ice, and hospital hot water systems, soils, plants, air–water interface of aquatic environments, biofilms in water distribution systems, medical instruments, and aerosols (Covert et al., 1999; duMoulin et al., 1988; Le Dantec et al., 2002; Schulze-Robbecke et al., 1992). Owing to the hydrophobic nature of the cell surface of mycobacteria, a major route for their transmission is via aerosolization. This also explains their resistance to commonly used disinfectants and to antibiotics (Falkinham, 2002; Taylor et al., 2000). Mycobacteria persist well and grow under environmental conditions. Potable water, particularly hospital water supplies, can support the growth of these bacteria, which may be linked to nosocomial infections (duMoulin et al., 1981). Their growth in water distribution systems was correlated with assimilable organic carbon and biodegradable organic carbon levels (Falkinham et al., 2001).

Mycobacterium avium and *M. intracellulare* can be detected using cultural, biochemical, and molecular-based methods (commercially available DNA probes and PCR amplification).

4.2.1.11 *Helicobacter pylori.*

Helicobacter pylori is a bacterial agent that is responsible for peptic ulcers (chronic gastritis), stomach cancer, lymphoma, and adenocarcinoma.

In the United States alone, over 5 million people are diagnosed annually with peptic ulcers, and some 40,000 of them undergo surgery, sometimes leading to death in complicated cases (Levin et al., 1998; Sonnenberg and Everhart, 1997).

There are indications of person-to-person as well as waterborne and food-borne transmissions of this pathogen (Brown, 2000; Hopkins et al., 1993; Klein et al., 1991). It has been suggested that *H. pylori* may be transmitted via four routes: the fecal–oral route, oral–oral route (person-to-person transmission via saliva), gastric–oral route (e.g., contaminated vomit in children), and by endoscopic procedures in the hospital setting (Schwartzbrod et al., 2002). This pathogen has been associated with an increased risk of gastric cancer among sewage workers (Friis et al., 1993; Lafleur and Vena, 1991). *Helicobacter pylori* infection is generally treated via administration of antibiotics such metronidazole, tetracycline, amoxycillin, clarithromycin, and azithromycin. The treatment is complicated by the appearance of antibiotic resistance in this pathogen (Kusters and Kuipers, 2001).

Helicobacter pylori has been detected in wastewater (Sutton et al., 1995), seawater (Cellini et al., 2004), and drinking water (Handwerker et al., 1995; Hegarty et al., 1999). When exposed to environmental conditions, *H. pylori* enters a viable but nonculturable (VBNC) state that allows its persistence in aquatic environments (Adams et al., 2003). Although Johnson and colleagues (1997) reported that *H. pylori* was readily inactivated by free chlorine, it is, however, more resistant than *E. coli* to chorine (Fig. 4.5; Baker et al., 2002). This higher resistance to disinfectants may allow this pathogen to persist in water distribution systems.

Helicobacter pylori was detected in environmental samples, using culture-based methods as well as immunological, autoradiography, and molecular-based methods. Although no standard method is available for *H. pylori* detection, this pathogen was recently isolated from wastewater using fluorescent in situ hybridization (FISH), a combination of immunomagnetic separation (IMS) and culturing techniques, and identified

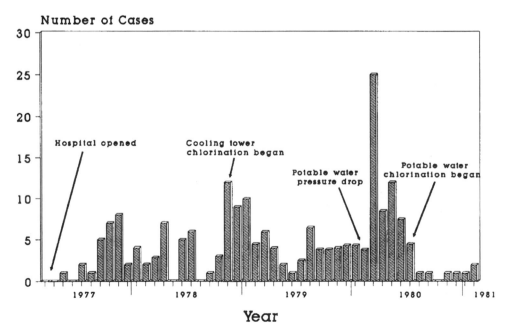

Figure 4.4 Reduction of Legionnaire's disease cases by water superchlorination. From Shands et al. (1985). (Courtesy of the American Medical Association.)

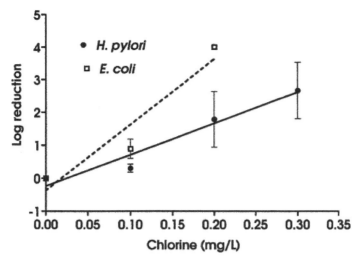

Figure 4.5 Effect of chlorine on *Helicobacter pylori* and *Escherichia coli*. From Baker et al. (2002).

using PCR-based 16S rRNA sequences (Engstrand, 2001; Lu et al., 2002; Moreno et al., 2003). *Helicobacter* DNA was also isolated from biofilms in municipal water distribution systems (Park et al., 2001).

4.2.1.12 Antibiotic-Resistant Bacteria.
Antibiotics act on microorganisms by inhibiting peptidoglycan synthesis, protein synthesis, and nucleic acid synthesis (interruption of nucleotide metabolism, inhibitition of RNA polymerase or DNA gyrase) or by affecting the cell membrane integrity (Russel, 2002). Increased use of antibiotics is often associated with an increased resistance of bacteria to these chemicals, especially in the hospital setting (Swartz, 1997). A global rise in antibiotic resistance has been reported. A comparison of pre-antibiotic era strains of *E. coli* and *Salmonella enterica* to contemporary strains showed that the former were susceptible to antibiotics, whereas 20 percent of the latter displayed resistance to at least one of the antibiotics (Houndt and Ochman, 2000). In the United States, 46 percent of *Streptococcus pneumoniae* isolates are now resistant to penicillin, and methicillin-resistant *Staphylococcus aureus* accounts for 30 percent of nosocomial infections with this pathogen.

An investigation of the antibiotic resistance pattern of *E. coli* strains in a wastewater treatment plant in Austria showed that these strains were resistant to 16 of 24 tested antibiotics. Antibiotic-resistant strains numbers increase when the influent to the wastewater treatment plant is from a hospital source (Reinthaler et al., 2003). Resistance to vancomycin has also been documented (Fox, 1997). Vancomycin-resistant enterococci (VRE) were isolated from 60 percent of raw wastewater and 36 percent of wastewater effluents (Iversen et al., 2002). As regards resistance to vancomycin, the minimum inhibitory concentrations (MIC) of VRE from hospital wastewaters were found to be much higher than those from residential wastewaters (Harwood et al., 2001; Table 4.8). Using a real-time PCR assay, the antibiotic resistance genes *vanA* of VRE, and *ampC* (resistance gene for the synthesis of β-lactamase) of *Enterobacteriaceae* were detected in 20 and 78 percent of wastewater samples, respectively (Volkmann et al., 2004).

Patients receiving antibiotic therapy harbor a large number of antibiotic-resistant bacteria in their intestinal tract. These bacteria are excreted in large numbers in feces and eventually reach the community wastewater treatment plant. The genes coding for

TABLE 4.8.   **Minimum Inhibitory Concentrations (MICs)
of Vancomycin for Vancomycin-Resistant Enterococci from
Residential and Hospital Wastewaters**[a]

Source	Species	MIC[b] (μg/mL)
Hospital	*E. avium*	32
Hospital	*E. avium*	20
Hospital	*E. avium*	20
Hospital	*E. faecalis*	>100
Hospital	*E. faecalis*	>100
Residential	*E. gallinarum*	3
Residential	*E. gallinarum*	3
Residential	*E. gallinarum*	3

[a]Adapted from Harwood et al. (2001).
[b]MIC = minimum inhibitory concentration.

antibiotic resistance are often located on plasmids (R factors) and, under appropriate conditions, can be transferred to other bacteria through conjugation that requires cell-to-cell contact, or through other modes of recombination. If the recipient bacteria are potential pathogens, they may be of public health concern as a result of their acquisition of antibiotic resistance. Drug-resistant microorganisms produce nosocomial and community-acquired human infections, which can lead to increased morbidity, mortality, and disease incidence. Resistance to antibiotics, including quinolones (e.g., nalidixic acid, ciprofloxacin), can in turn complicate and increase the cost of therapy based on administration of antibiotics to patients exposed to pathogens of environmental origin. Patients infected with antibiotic-resistant bacterial strains are likely to require hospitalization, sometimes for long periods (Lee et al., 1994). Bacterial resistance to antibiotics has been demonstrated in terrestrial and aquatic environments, particularly those contaminated with wastes from hospitals (Grabow and Prozesky, 1973).

Gene transfer between microorganisms is known to occur in natural environments as well as in engineered systems such as wastewater treatment plants (Colwell and Grimes, 1986; McClure et al., 1990). Investigators have used survival chambers to demonstrate the transfer of R plasmids among bacteria in domestic wastewater. The mean transfer frequency in wastewater varied between 4.9×10^{-5} and 7.5×10^{-5}. The highest transfer frequency (2.7×10^{-4}) was observed between *Salmonella enteritidis* and *E. coli* (Mach and Grimes, 1982). Nonconjugative plasmids (e.g., pBR plasmids) can also be transferred and this necessitates the presence of a mobilizing bacterial strain to mediate the transfer (Gealt et al., 1985). Several indigenous mobilizing strains have been isolated from raw wastewater. Each of these strains is capable of aiding in the transfer of the plasmid pBR325 to a recipient *E. coli* strain (McPherson and Gealt, 1986). Under laboratory conditions, plasmid mobilization from genetically engineered bacteria to environmental strains was also demonstrated under low temperature and low nutrient conditions in drinking water (Sandt and Herson, 1989).

The occurrence of multiple-antibiotic resistant (MAR) indicator and pathogenic (e.g., *Salmonella*) bacteria in water and wastewater treatment plants has been documented (Alcaid and Garay, 1984; Armstrong et al., 1981; 1982; Walter and Vennes, 1985). In untreated wastewater, the percentage of multiple-antibiotic resistant coliforms varies between less than 1 to about 5 percent of the total coliforms (Walter and Vennes, 1985) (Fig. 4.6). Chlorination appears to select for resistance to antibiotics in wastewater treatment plants (Staley et al., 1988). However, others (Murray et al., 1984) observed that

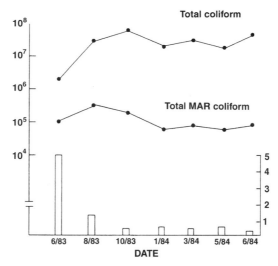

Figure 4.6 Multiple-antibiotic resistant (MAR) bacteria in domestic wastewater. From Walter and Vennes (1985). (Courtesy of the American Society for Microbiology.)

chlorination increased the bacterial resistance to some antibiotics (e.g., ampicillin, tetracycline) but not to others (e.g., chloramphenicol, gentamicin). The proportion of bacteria carrying R factors seems to increase after water and wastewater treatment (Armstrong et al., 1981; Bell, 1978; Bell et al., 1981; Calomiris et al., 1984). For example, in one study, MAR was expressed by 18.6 percent of heterotrophic plate count bacteria in untreated water as compared to 67.8 percent for bacteria in the distribution system (Armstrong et al., 1981). Similarly, in a water treatment plant in Oregon, the percentage of MAR bacteria rose from 15.8 percent in untreated (river) water to 57.1 percent in treated water (Armstrong et al., 1982). Multiple-antibiotic resistance is furthermore associated with resistance to heavy metals (e.g., Cu^{2+}, Pb^{2+}, Zn^{2+}). This phenomenon was observed both in drinking water (Calomiris et al., 1984) and wastewater (Varma et al., 1976). The public health significance of this phenomenon deserves further study.

Strategies for tackling the serious problem of drug resistance include the reduced use of antibiotics in humans and animals, preventive measures for the transmission of infectious diseases, and increased efforts by the scientific community to better understand the mechanisms of drug resistance in microorganisms (Cohen, 1992; McKeegan et al., 2002; Neu, 1992).

4.2.2 Viral Pathogens

Water and wastewater may become contaminated by approximately 140 types of enteric viruses. These viruses enter into the human body orally, multiply in the gastrointestinal tract, and are excreted in large numbers in the feces of infected individuals. Table 4.9 (Bitton, 1980a; Jehl-Pietri, 1992; Schwartzbrod, 1991; Schwartzbrod et al., 1990) lists the enteric viruses found in aquatic environments that are pathogenic to humans. Many of the enteric viruses cause non apparent infections that are difficult to detect. They are responsible for a broad spectrum of diseases ranging from skin rash, fever, respiratory infections, and conjunctivitis to gastroenteritis and paralysis. It was estimated that nonpolio enteroviruses cause 10–15 million symptomatic infections/year in the United States (Zaoutis and Klein, 1998). Virus presence in the community wastewater reflects virus infections among the population. Enteric viruses are present in relatively small numbers

TABLE 4.9. **Some Human Enteric Viruses**

Virus Group	Serotypes	Some Diseases Caused
A. *Enteroviruses*		Paralysis
Poliovirus	3	Aseptic meningitis
Coxsackievirus		
A	23	Herpangia
		Aseptic meningitis
		Respiratory illness
		Paralysis
		Fever
B	6	Pleurodynia
		Aseptic meningitis
		Pericarditis
		Myocarditis
		Congenital heart
		Anomalies
		Nephritis
		Fever
Echovirus	34	Respiratory infection
		Aseptic meningitis
		Diarrhea
		Pericarditis
		Myocarditis
		Fever, rash
Enteroviruses (68–71)	4	Meningitis
		Respiratory illness
Hepatitis A virus (HAV)		Infectious hepatitis
Hepatitis E virus (HEV)		Hepatitis
B. *Reoviruses*	3	Respiratory disease
C. *Rotaviruses*	4	Gastroenteritis
D. *Adenoviruses*	41	Respiratory disease
		Acute conjunctivitis
		Gastroenteritis
E. Norwalk agent (calicivirus)	1	Gastroenteritis
F. Astroviruses	5	Gastroenteritis

in water and wastewater. Therefore, environmental samples of 10–1000 L must be concentrated in order to detect these pathogens. An ideal method should fulfil the following criteria: applicability to a wide range of viruses, processing of large sample volumes with small-volume concentrates, high recovery rates, reproducibility, rapidity, and low cost.

A number of approaches have been considered for accomplishing this task (Farrah and Bitton, 1982; Gerba, 1987b; Goyal and Gerba, 1982a; Wyn-Jones and Sellwood, 2001). The most widely used approach is based on the adsorption of viruses to electronegative and electropositive microporous filters of various compositions (e.g., nitrocellulose, fiberglass, charge-modified cellulose, epoxy-fiberglass, cellulose + glass fibers, positively charged nylon membranes). This step is followed by elution of the adsorbed viruses from the filter surface. Further concentration of the sample can be obtained by membrane filtration, organic flocculation (using beef extract or casein) or aluminum hydroxide hydroextraction. The concentrate is then assayed using animal tissue cultures, immunological or genetic probes (see Chapter 1). Other adsorbents considered for virus concentration include glass powder, glass wool, bituminous coal, bentonite, iron oxide, modified diatomaceous earth or pig erythrocyte membranes. Table 4.10 (Gerba, 1987b) shows a compilation of most of the methods available for concentrating viruses from water and wastewater.

From an epidemiological standpoint, enteric viruses are mainly transmitted via person-to-person contacts. However, they may also be communicated by water transmission either directly (drinking water, swimming, aerosols) or indirectly via contaminated food (e.g., shellfish, vegetables). Waterborne transmission of enteric viruses is illustrated in Figure 4.7 (Gerba et al., 1975a). Some enteric viruses (e.g., hepatitis A virus) persist on environmental surfaces, which may serve as vehicles for the spread of viral infections in day-care centers or hospital wards (Abad et al., 1994). The infection process depends on the *minimal infectious dose* (MID) and on host susceptibility, which involves host factors (e.g., specific immunity, sex, age) and environmental factors (e.g., socioeconomic level, diet, hygienic conditions, temperature, humidity) factors. Although the MID for viruses is controversial, it is generally relatively low as compared with bacterial pathogens. Experiments with human volunteers have shown an MID of 17 PFU (plaque-forming units) for echovirus 12 (Schiff et al., 1984a, b). Several epidemiological surveys have shown that enteric viruses are responsible for 4.7–11.8 percent of waterborne epidemics (Cliver, 1984; Craun, 1988; Lippy and Waltrip, 1984).

TABLE 4.10. Methods Used for Concentrating Viruses from Water[a]

Method	Initial Volume of Water	Applications	Remarks
Filter adsorption–elution Negatively charged filters	Large	All but the most turbid waters	Only system shown useful for concentrating viruses from large volumes of tapwater, sewage, sea water, and other natural waters; cationic salt concentration and pH must be adjusted before processing.
Positively charged filters	Large	Tap water, sewage, seawater	No preconditioning of water necessary at neutral or acidic pH level.
Adsorption to metal salt precipitate, aluminum hydroxide, ferric hydroxide	Small	Tapwater, sewage	Have been useful as reconcentration methods.
Charged filter aid	Small	Tapwater, sewage	40 L volumes tested, low cost; used as a sandwich between prefilters.
Polyelectrolyte PE60	Large	Tapwater, lake water, sewage	Because of its unstable nature and lot-to-lot variation in efficiency for concentrating viruses, method has not been used in recent years.
Bentonite	Small	Tapwater, sewage	
Iron oxide	Small	Tapwater, sewage	
Talcum powder	Large	Tapwater, sewage	Can be used to process up to 100 L volumes as a sandwich between filter paper support.

(continued)

TABLE 4.10. *Continued*

Method	Initial Volume of Water	Applications	Remarks
Gauze pad	Large		First method developed for detection of viruses in water, but not quantitative or very reproducible.
Glass powder	Large	Tapwater, seawater	Columns containing glass powder have been made that are capable of processing 400 L volumes.
Organic flocculation	Small	Reconcentration	Widely used method for reconcentrating viruses from primary filter eluates.
Protamine sulfate	Small	Sewage	Very efficient method for concentrating reoviruses and adenoviruses from small volumes of sewage.
Polymer two-phase	Small	Sewage	Processing is slow; method has been used to reconcentrate viruses from primary eluates.
Hydroextraction	Small	Sewage	Often used as a method for reconcentrating viruses from primary eluates.
Ultrafiltration			
Soluble filters	Small	Clean waters	Clogs rapidly even with low turbidity.
Flat membranes	Small	Clean waters	Clogs rapidly even with low turbidity.
Hollow fiber or capillary	Large	Tapwater, lake water	Up to 100 L may be processed, but water must often be prefiltered.
Reverse osmosis	Small	Clean waters	Also concentrates cytotoxic compounds that adversely affect assay methods.

[a]Adapted from Gerba (1987b).

Epidemiological investigations have definitely proved the waterborne and food-borne transmission of viral diseases such as hepatitis and gastroenteritis.

4.2.2.1 *Hepatitis.* Hepatitis is caused mainly by the following viruses (Jehl-Pietri, 1992; Pilly, 1990):

- *Infectious hepatitis* is caused by hepatitis A virus (HAV), a 27 nm RNA enterovirus (enterovirus type 72 belonging to the family picornaviridae) with a relatively short incubation period (2–6 weeks) and displaying a fecal–oral transmission route. Although it can be replicated on primary and continuous human or animal tissue cultures, it is hard to detect because it does not always display a cytopathic effect. Other means of detection of HAV include genetic probes, use of PCR (Altmar et al., 1995), and immunological methods (immunoelectron microscopy, radioimmunoassay, enzyme immunoassay, radioimmuno-focus assay; see Chapter 1 for more details).

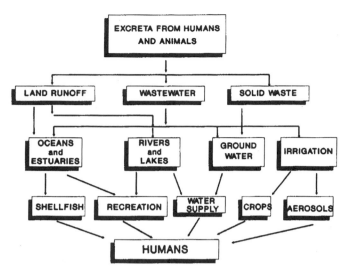

Figure 4.7 Waterborne transmission of enteric viruses. From Gerba et al. (1975). (Courtesy of the American Chemical Society.)

- *Serum hepatitis* is caused by hepatitis B virus (HBV), a 42 nm DNA virus displaying a relatively long incubation time (4–12 weeks). This virus is transmitted by contact with infected blood or by sexual contact. The mortality rate (1–4 percent) is higher than for infectious hepatitis (<0.5 percent). Hepatitis B virus is responsible for approximately 60 percent of the 434,000 cases of liver cancer worldwide (WHO, 1996).
- *Non-A, non-B infectious hepatitis* is caused by hepatitis E virus (HEV).

Hepatitis A virus causes liver damage with necrosis and inflammation. After the onset of infection, the incubation period may last up to 6 weeks. One of the most characteristic symptoms is jaundice. In the United States, approximately 140,000 persons are infected annually with HAV (Berge et al., 2000).

Hepatitis A is transmitted via the fecal–oral route either by person-to-person direct contact, waterborne, or food-borne transmission (Jehl-Pietri, 1992; Myint et al., 2002). Concentration of HAV in feces can reach 10^7 to $10^9/g$. This disease is distributed worldwide and the prevalence of HAV antibodies is higher among lower socioeconomic groups and increases with the age of the infected individuals. Direct contact transmission has been documented mainly in nurseries (especially among infants wearing diapers), mental institutions, prisons, or military camps.

Waterborne transmission of infectious hepatitis has been conclusively demonstrated and documented worldwide on several occasions. It has been estimated that 4 percent of hepatitis cases observed during the period 1975–1979 in the United States were transmitted through the waterborne route (Cliver, 1985). The hepatitis cases are due to consumption of improperly treated water or contaminated well water (De Serres et al., 1999). Hepatitis A outbreaks were also associated with swimming in lakes or public pools (Mahoney et al., 1992). Food-borne transmission of HAV appears to be more important than waterborne transmission. Consumption of shellfish grown in wastewater-contaminated waters accounts for numerous hepatitis and gastroenteritis outbreaks documented worldwide (Jehl-Pietri, 1992; Schwartzbrod, 1991) (Table 4.11).

TABLE 4.11.   Some Viral Hepatitis Outbreaks Due to Shellfish Consumption[a]

Year	Shellfish	Country	No. of Cases
1953	Oysters	USA	30
1955	Oysters	Sweden	600
1961	Oysters	USA	84
1962	Clams	USA	464
1963/1966	Clams/oysters	USA	180
1964	Clams	USA	306
1964	Oysters	USA	3
1966	Clams	USA	4
1968/1971	Clams/oysters	Germany	34
1971	Clams	USA	17
1972	Mussels	France	13
1973	Oysters	USA	265
1976	Mussels	Australia	7
1978	Mussels	England	41
1979	Oysters	USA	8
1980	Oysters	Philippines	7
1980/1981	Cockles	England	424
1982	Various shellfish	England	172
1982	Oysters	USA	204
1982	Clams	USA	150
1984	Cockles	Malaysia	322
1984	Mussels	Yugoslavia	51
1985	Clams	USA	5
1988	Clams	China	292,000

[a]Adapted from Jehl-Pietri (1992) and Schwartzbrod (1991).

Passive immunization by means of pooled immunoglobulin is used for the prevention of infectious hepatitis. Vaccines against hepatitis A are available around the world.

Hepatitis E virus (HEV) is a single-stranded RNA virus, the classification of which is not known. It is believed that this virus may be a calicivirus. This virus is not well characterized due to the lack of a tissue culture cell line for its assay. Unlike HAV, it is mainly transmitted via fecally contaminated water, while the person-to-person transmission is very low (Myint et al., 2002).

Hepatitis E virus epidemics generally involve thousands of cases. A notorious non-A, non-B hepatitis (now recognized as hepatitis E virus) epidemic broke out in 1956 in New Delhi, India, and resulted in approximately 30,000 cases. A more recent outbreak involved approximately 79,000 people in Kampur (Ray et al., 1991). It attacks mostly young adults and pregnant women and has clinical symptoms similar to those of HAV (Moe, 1997). The fatality rates are 2–3 percent in the general population and as high as 32 percent for pregnant women in their third trimester (Bader, 1995; Haas et al., 1999). A new recombinant HEV vaccine is under development by the U.S. Army (Myint et al., 2002).

4.2.2.2 *Viral Gastroenteritis.*

Gastroenteritis is probably the most frequent waterborne illness; it is caused by protozoan parasites, and by bacterial and viral pathogens (e.g., rotaviruses, Norwalk-like agents, adenoviruses, astroviruses) (Williams and Akin, 1986). In this section we will examine rotaviruses, caliciviruses genetically related to Norwalk-type virus, and enteric adenoviruses as causal agents of gastroenteritis.

• *Rotaviruses*

Rotaviruses, belonging to the family Reoviridae, are 70 nm particles containing double-stranded RNA surrounded with a double-shelled capsid (Fig. 4.8). Rotaviruses are the major cause of infantile acute gastroenteritis in children less than two years of age (Gerba et al., 1985). This disease largely contributes to childhood mortality in developing countries, and is responsible for millions of childhood deaths per year in Africa, Asia, and Latin America. It is also responsible for outbreaks among adult populations (e.g., the elderly) and is a major cause of travelers' diarrhea. Up to 10^{11} rotavirus particles can be detected in patient stools. The virus is spread mainly by the fecal–oral route, but a respiratory route has also been suggested (Flewett, 1982; Foster et al., 1980). There have been several outbreaks of gastroenteritis where rotaviruses originating from wastewater have been implicated. Some waterborne outbreaks associated with rotaviruses are summarized in Table 4.12 (Gerba et al., 1985; Williams and Akin, 1986).

Detection of rotaviruses in wastewater, drinking water, and other environmental samples is accomplished by using electron microscopy, enzyme-linked immunosorbent assays (ELISA kits are commercially available), reverse transcription-PCR method (RT-PCR), which helped identify types 1, 2, and 3 in wastewater (Gajardo et al., 1995), or tissue cultures (a popular cell line is MA-104, which is derived from fetal rhesus monkey kidney). Detection in cell cultures include methods such as plaque assay, cytopathic effect (CPE), or immunofluorescence. Information on the fate of rotaviruses in the environment is available mostly for simian rotaviruses (e.g., strain SA-11) and little is known concerning the four known human rotavirus serotypes. A recent molecular epidemiological survey for rotaviruses in wastewater showed that the environmental rotavirus isolates from wastewater effluents displayed profiles similar to human rotaviruses isolated from fecal samples (Dubois et al., 1997). Using RT-PCR, rotavirus RNA was detected in drinking water in homes with children suffering from rotaviral acute gastroenteritis, but the sequences found in drinking water were different from those found in the patients' feces (Gratacap-Cavallier et al., 2000).

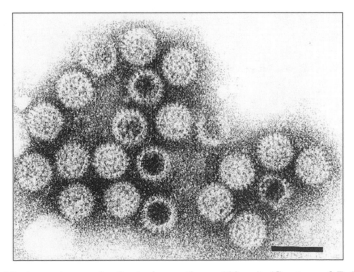

Figure 4.8   Electron micrograph of rotaviruses (bar = 100 nm). (Courtesy of F. M. Williams, U.S. EPA.)

TABLE 4.12. Some Rotavirus Waterborne Outbreaks[a]

Year	Location	No. Ill	Remarks
1977	Sweden	3,172	Small town water supply contaminated with sewage effluent
1980	Brazil	~900	Contamination of private school's water
1980	Norfolk Island	–	Contamination of community water supply
1981	Russia	173	Contamination of groundwater system
1981	Colorado	1,500	Source contamination (chlorinator and filtration failure)
1982	Israel	~2,000	Reservoir contaminated by children
1981–1982	East Germany	11,600	Floodwater contamination of wells
1982–1983	China	13,311	Contaminated water supply
1991	Arizona	900	Well water contaminated by sewage at a resort

[a]Adapted from Gerba et al. (1985) and Williams and Akin (1986).

· *Human Caliciviruses*

Gastroenteritis is often caused by small round-structured viruses that have been characterized as caliciviruses genetically related to the Norwalk-like virus (NLV) (Estes and Hardy, 1995; Le Guyader et al., 1996). The latter is a small 27 nm virus (Fig. 4.9), first discovered in 1968 in Norwalk, Ohio, and is a major cause of waterborne disease and is also implicated in food-borne outbreaks, especially those associated with shellfish consumption. The virus causes diarrhea and vomiting and appears to attack the proximal

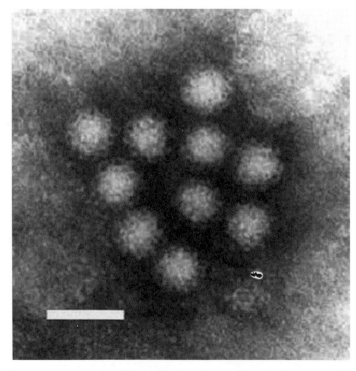

Figure 4.9   Electron micrograph of Norwalk virus (bar = 50 nm). (Courtesy of F. M. Williams, U.S. EPA.)

small intestine, but the mechanism of pathogenesis is poorly understood as the exact site of virus replication has not been identified. The Norwalk virus plays a major role in water-borne gastroenteritis (Table 4.13; Gerba et al., 1985; Williams and Akin, 1986) but also appears to play a role in travelers' diarrhea (Keswick et al., 1982). There is a lack of long-term immunity to NLVs. These viruses are stable under environmental conditions and appear to have a low infectious dose (estimated at 10–100 virus particles). Although person-to-person transmission is the prevalent mode of transmission of NLVs, food-borne (e.g., via shellfish consumption, contaminated fruits and vegetables, food handlers) and waterborne transmission (e.g., tapwater, ice, well water, bottled water) have been well documented around the world (Fig. 4.10; Lopman et al., 2003; Parshionikar et al., 2003). Approximately half of the reported outbreaks of gastroenteritis in the US, Japan and Europe are caused by Norwalk-like viruses (NLV) (Lopman et al., 2003).

Because NLV cannot be propagated in tissue cultures, the tools mostly used for their detection in clinical samples, immune electron microscopy and radioimmunoassay techniques, are not sensitive enough for environmental monitoring. New methodology includes immunomagnetic separation followed by RT-PCR or an RT-PCR-DNA enzyme immunoassay for detection of NLV in stools and shellfish (Altmar et al., 1993; De Leon et al., 1992; Jiang et al., 1992; Meschke and Sobsey, 2002; Schwab et al., 2001) but more work is needed for the development of rapid diagnostic assays for environmental monitoring.

· *Enteric Adenoviruses*

Enteric adenoviruses belong to subgroup F adenoviruses, which comprises 2 serotypes types 40 and 41, and are commonly found in stools ($>10^{11}$ viruses/g feces) of children with gastroenteritis (Cruz et al., 1990). These nonenveloped double-stranded DNA viruses have an 80-nm diameter. The most frequent symptoms of infection by these viruses are diarrhea and vomiting (Herrmann and Blacklow, 1995). Enteric adenoviruses can be detected using a commercial monoclonal ELISA method (Wood et al., 1989a) or PCR technique (Allard et al., 1992). They are found in raw wastewater and effluents,

TABLE 4.13. Some Examples of Waterborne Outbreaks of Norwalk-Like Virus[a]

Year	Location	Number Ill	Remarks
1978	Pennsylvania	350	Drinking water with insufficient chlorination (attack rate 17–73%).
1978	Tacoma, WA	~600	Well of 51.4 m depth (attack rate = 72%).
1979	Arcata, CA	30	Sprinkler irrigation system not for human consumption.
1980	Maryland	126	Well 95 ft. (attack rate 64%) disinfected water.
1980	Rome, GA	~1500	Spring (attack rate 72%).
1982	Tate, GA	~500	Springs and well. Springs possibly contaminated from rainfall events from the surface.
1998	Finland	1500–3000	Contaminated drinking water. Well contaminated by contaminated river water by back flow. Finding of NLV genotype 2 in tap water.
2000	Italy	344	Outbreak of gastroenteritis at a tourist resort. NLV found in 22 out of 28 stool specimens. Contaminated water and ice.

[a]Adapted from Boccia et al., 2002; Gerba et al., 1985; Kukkula et al., 1999; Williams and Akin, 1986. NLV, Norwalk-like virus.

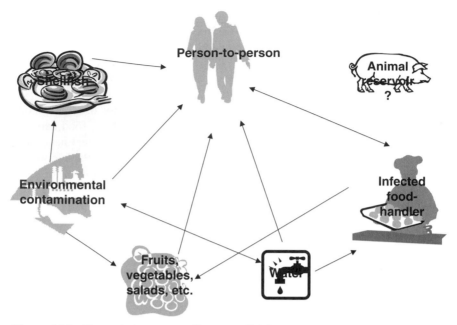

Figure 4.10 Transmission routes of human caliciviruses. From Lopman et al. (2002).

and are suspected to also be transmitted by the water route (Enriquez et al., 1995). They are quite resistant to UV irradiation (see Chapter 6).

• *Astroviruses*

Astroviruses are 27–34 nm spherical nonenveloped single-stranded RNA viruses with a characteristic starlike appearance (Fig. 4.11). Seven human serotypes have been identified to date. Astroviruses mostly affect children and immunocompromised adults and are transmitted by the fecal–oral route, spreading via person-to-person contacts and via contaminated food or water. Astroviruses are the second leading cause of viral gastroenteritis in children and adults (Willcocks et al., 1995). The mild, watery diarrhea lasts for 3–4 days, but can be long-lasting in immunocompromised patients (Matsui, 1995). Astroviruses are traditionally detected with immune electron microscopy but molecular probes (e.g., RT-PCR, RNA probes) and immunoassays (e.g., monoclonal antibodies) are now available for their diagnosis (Jonassen et al., 1993; Metcalf et al., 1995; Nadan et al., 2003). After treatment with trypsin, astroviruses can grow in human embryonic kidney cultures or in Caco cell cultures. Infectious astroviruses can be detected in environmental samples through growth on tissue cultures followed by hybridization with a specific cDNA probe (Pinto et al., 1996).

4.2.3 Protozoan Parasites

Most of the protozoan parasites produce *cysts*, which are able to survive outside their host under adverse environmental conditions. Encystment is triggered by factors such as lack of nutrients, accumulation of toxic metabolites, or host immune response. Under appropriate conditions, a new *trophozoite* is released from the cyst. This process is called excystment (Rubin et al., 1983).

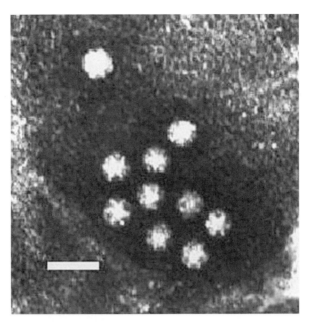

Figure 4.11   Electron micrograph of astroviruses (note the star-like appearance) (bar = 50 nm). (Courtesy of F. M. Williams, U.S. EPA.)

The major waterborne pathogenic protozoa affecting humans are presented in the following subsections (Table 4.14).

4.2.3.1 Giardia.   This flagellated protozoan parasite has a pear-shaped trophozoite (9–21 μm long) and an ovoid cyst stage (8–12 μm long and 7–10 μm wide) (Figs 4.12 and 4.13). An infected individual may shed up to $1-5 \times 10^6$ cysts/g feces (Jakubowski and Hoff, 1979; Lin, 1985). Domestic wastewater is a significant source of *Giardia*, and wild and domestic animals act as important reservoirs of *Giardia* cysts. This parasite is endemic in mountainous areas in the United States and infects both humans and domestic and wild animals (e.g., beavers, muskrats, dogs, cats). Infection is caused by ingestion of

TABLE 4.14.   Major Waterborne Diseases Caused by Protozoa

Organism	Disease (Site Affected)	Major Reservoir
Giardia lamblia	Giardiasis (G.I. tract)	Human and animal feces
Entamoeba histolytica	Amoebic disentery (G.I. tract)	Human feces
Acanthamoeba castellani	Amoebic meningoencephalitis (central nervous system)	Soil and water
Naeleria gruberi	Amoebic meningoencephalitis (central nervous system)	Soil and water
Balantidium coli	Dysentery/intestinal ulcers (G.I. tract)	Human feces
Cryptosporidium	Profuse and watery diarrhea; weight loss; nausea; low-grade fever (G.I. tract)	Human and animal feces
Cyclospora	Watery diarrhea alternating with constipation	Feces, contaminated fruits and vegetables
Microsporidia	Chronic diarrhea, dehydration, weight loss	Feces

G.I., gastrointestinal tract.

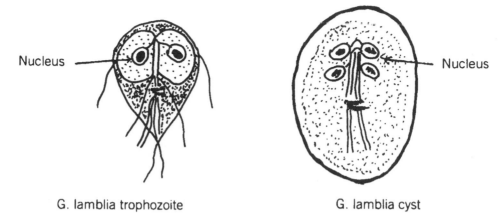

Nucleus

Nucleus

G. lamblia trophozoite

G. lamblia cyst

Figure 4.12 *Giardia lamblia* trophozoite and cyst. From Lin (1985). (Courtesy of the American Water Works Association.)

the cysts found in water. Passage through the stomach appears to promote the release of *trophozoites*, which attach to the epithelial cells of the upper small intestine and reproduce by binary fission. They may coat the intestinal epithelium and interfere with absorption of fats and other nutrients. They encyst as they travel through the intestines and reach the large intestine (AWWA, 1985a). The life cycle of *Giardia* is shown in Figure 4.14 (Rochelle, 2002). In humans, infections may last months to years. The infectious dose for Mongolian gerbils is more than 100 cysts (Schaefer et al., 1991), and is generally between 25 and 100 in human volunteers, but may be as low as 10 cysts (Rendtorff, 1979).

Giardia has an incubation period of 1–8 weeks and causes diarrhea, abdominal pains, nausea, fatigue, and weight loss. However, giardiasis is rarely fatal. Although its usual mode of transmission is the person-to-person or food routes, *Giardia* is recognized as one of the most important etiological agents in waterborne disease outbreaks (Craun, 1979; 1984b). The first major documented outbreak of giardiasis in the United States occurred in 1974 in Rome, New York, and was associated with the presence of *Giardia* in the water supply. It affected approximately 5000 people (10 percent of the town's

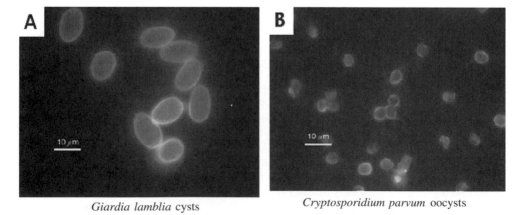

Giardia lamblia cysts

Cryptosporidium parvum oocysts

Figure 4.13 Immunofluorescence image of *Giardia lamblia* cysts and *Cryptosporidium parvum* oocysts. (Courtesy of H. D. A. Lindquist, U.S. EPA.) Figure also appears in color figure section.

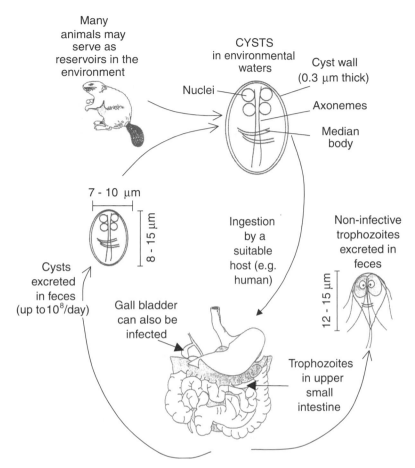

Figure 4.14   Life cycle of *Giardia*. Adapted from Rochelle (2002).

population). The outbreak occurred as a result of consumption of water that has been chlorinated but not filtered. Other outbreaks have been reported in Colorado, New Hampshire, Pennsylvania, South Dakota, Tennessee, Utah, Vermont, and Washington. In the United States, during the 1971–1985 period, more than 50 percent of the outbreaks resulting from the use of surface waters were caused by *Giardia* (Craun, 1988) (Fig. 4.15). Approximately 80 giardiasis outbreaks were recorded in the United States from 1965 to 1983 (AWWA, 1985a). In the United States, *Giardia* was the etiologic agent responsible for 27 percent of the outbreaks reported in 1989–1990, and for 16.6 percent of the outbreaks reported in 1993–1994 (Herwaldt et al., 1992; Kramer et al., 1996). A survey of giardiasis cases in the United States between 1992 and 1997 indicated 0.9–42.3 cases of giardiasis per 100,000 population, with a national average of 9.5 cases per 100,000 population. It is estimated that as many as 2.5 million cases of giardiasis occur annually in the United States (Furness et al., 2000). It is estimated that *Giardia* is responsible for 100 million mild cases and one million severe cases per year, worldwide (Smith, 1996).

   Most of the giardiasis outbreaks are associated with the consumption of untreated or unsuitably treated water (e.g., water chlorinated but not filtered, interruption of disinfection) and with recreational activities during the summer season. Amendments of the Safe Drinking Water Act (Surface Water Treatment Rule) in the United States mandates the U.S. EPA to require filtration and disinfection for all surface waters and groundwater

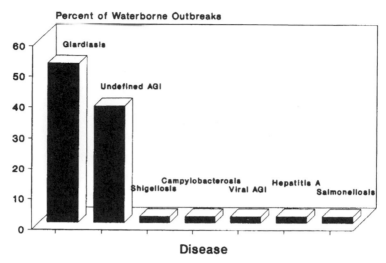

Figure 4.15 Etiology of waterborne disease outbreaks: surface water (1971–1985). From Craun (1988). (Courtesy of American Water Works Association.)

under the direct influence of surface water to control the transmission of *Giardia* spp. and enteric viruses. However, exceptions (e.g., effective disinfection) to this requirement have been considered (Clark et al., 1989; U.S. EPA, 1989d). Faulty design or construction of filters may lead to breakthrough of *Giardia lamblia* and subsequent contamination of drinking water. Traditional bacterial indicators are not suitable as surrogates for the presence of *Giardia* cysts in water and other environmental samples (Rose et al., 1991). A good correlation has been reported between the removal of *Giardia* cysts, *Cryptosporidium* oocysts and bacterial pathogens, and some traditional parameters of water quality such as turbidity (LeChevallier et al., 1991a, b; Xagoraraki et al., 2004). Microbiological quality can be significantly improved at a turbidity below 0.2 NTU (Xagoraraki et al., 2004). Oocyst-sized polystyrene microspheres also appear to be reliable surrogates for *C. parvum* oocyst removal by filtration (Emelko and Huck, 2004).

Giardia cysts generally occur in low numbers in aquatic environments and must be concentrated from water and wastewater, by means of ultrafiltration cassettes, vortex flow filtration, or adsorption to polypropylene or yarn wound cartridge filters (APHA, 1999; Hibler and Hancock, 1990; Isaac-Renton et al., 1986; Mayer and Palmer, 1996). Because *Giardia lamblia* cannot be cultured in the laboratory, the detection of the cysts necessitates other approaches such as immunofluorescence, using polyclonal or monoclonal antibodies, or by phase-contrast microscopy (Sauch, 1989). Moreover, cysts exposed to chlorine concentration from 1 to 11 mg/L, although fluorescing, could not be confirmed by phase contrast microscopy because they lost their internal structures (Sauch and Berman, 1991). Cysts can also be selectively concentrated from waters samples by an antibody-magnetite procedure. Cysts, following exposure to a mouse anti-*Giardia* antibody, are allowed to react with an anti-mouse antibody-coated magnetite particles and then concentrated by high-gradient magnetic separation (Bifulco and Schaefer, 1993).

U.S. EPA methods 1622 and 1623 (McCuin and Clancy, 2003; U.S. EPA, 2001a) consist of filtration of the sample through a pleated membrane capsule (1 μm) or a compressed foam filter, followed by elution, purification by immunomagnetic separation (IMS) of the cysts, and observation under a florescence microscope following staining with fluorescein isothiocyanate (FITC) conjugated monoclonal antibody (FAb) and counterstaining with DAPI. Cyst recovery is significantly improved by heating at 80°C for 10 min prior to DAPI staining (Ware et al., 2003).

In vivo infectivity assays, using gerbils, give information on both the viability and infectivity of the cysts. Alternative methods include the use of cell cultures (e.g., caco-2 cells), in vitro excystation, and fluorogenic dyes. The latter include a propidium iodide (PI) in combination with fluorescein diacetate (FDA) or DAPI. Fluorescein diacetate uptake and degradation releases fluorescein, which renders the cysts fluorescent. Although cysts' response to FDA sometimes correlates well with infectivity to animals, this stain may overestimate their viability (Campbell and Wallis, 2002; Labatiuk et al., 1991). Conversely, there is a negative correlation between cyst staining with propidium iodide and infectivity, indicating that this stain can be used to determine the number of nonviable cysts (Schupp and Erlandsen, 1987; Sauch et al., 1991). Alternative vital dyes include SYTO-9 and SYTO-59 fluorogenic dyes (Bukhari et al., 2000). It was proposed to combine the use of these fluorogenic dyes with Nomarski differential interference contrast microscopy for examination of morphological features of the cysts (Smith, 1996).

Molecular-based methods have also been applied for the detection of cysts and oocyts in water and wastewater. A cDNA probe was constructed for the detection of *Giardia* cysts in water and wastewater concentrates (Abbaszadegan et al., 1991; Nakhforrosh and Rose, 1989). However, this method does not provide information on cyst viability (Rose et al., 1991). Amplification of the *giardin* gene by PCR has been used to detect *Giardia*, and several primer pairs have been developed to detect this parasite at the genus or species level (Cacciò et al., 2003; Rochelle, 2002). Distinction of live from dead cysts was made possible by measuring the amount of RNA before and after excystation (Mahbubani et al., 1991). Viable *Giardia* cysts can also be detected by PCR amplification of heat-shock-induced mRNA that codes for heat shock proteins (Abbaszadegan et al., 1997). Several substances (e.g., humic acids) that are present in environmental samples interfere with pathogen and parasite detection by PCR technology (Rodgers et al., 1992). However, several methods have been proposed to remove this interference.

A survey of raw wastewater from several states in the United States showed that *Giardia* cysts numbers varied from hundreds to thousands of cysts per liter (Casson et al., 1990; Sykora et al., 1990), but cyst concentration may be as high as $10^5/L$ (Jakubowski and Eriksen, 1979). In Italy, *Giardia* cysts were found in raw wastewater throughout the year at concentrations ranging from 2.1×10^3 to 4.2×10^4 cysts/L (Cacciò et al., 2003). In Arizona, *Giardia* was detected at concentrations of 48 cysts/40 L of activated sludge effluent (Rose et al., 1989a). This concentration decreased to 0.3 cysts/40 L after sand filtration. It was suggested that wastewater examination for *Giardia* cysts may serve as a means for determining the prevalence of giardiasis in a given community (Jakubowski et al., 1990). A survey of a wastewater treatment plant in Puerto Rico showed that 94–98 percent of these parasites are removed following passage through the plant (Correa et al., 1989). This parasite is more resistant to chlorine than bacteria (Jarrol et al., 1984). Furthermore, *Giardia* cysts have been detected in 16 percent of potable water supplies (lakes, reservoirs, rivers, springs, groundwater) in the United States at an average concentration of 3 cysts/100 L (Rose et al., 1991). Another survey of surface water supplies in the United States and Canada showed that cysts occurred in 81 percent of the samples (LeChevallier et al., 1991b).

A survey of drinking water plants in Canada indicated that 18 percent of the treated water samples were positive for *Giardia* cysts, but viable cysts were found in only 3 percent of the samples. A total of 80 percent of the plants treated their water solely by chlorination without any filtration (Wallis et al., 1996). In Japan, *Giardia* was detected in 12 percent of drinking water samples with a mean concentration of 0.8 cyts/1000 L (Hashimoto et al., 2002).

4.2.3.2 Cryptosporidium.   The coccidian protozoan parasite *Cryptosporidium* was first described at the beginning of the twentieth century. It was known to infect mostly animal species (calves, lambs, chicken, turkeys, mice, pigs, dogs, cats), but infection of humans was reported only during the 1970s in an immunocompetent child. *Cryptosporidium parvum* is the major species responsible for infections in humans and animals (Adal et al., 1995; Current, 1987; Rose et al., 1985; Rose, 1990).

The infective stage of this protozoan is a thick-walled *oocyst* (5–6 µm in size) (Fig. 4.13), which readily persists under environmental conditions. An infected individual may release up to 10^9 oocysts per day. Following ingestion by a suitable host, the oocysts undergo excystation and release infective *sporozoites*, which parasitize epithelial cells mainly in the host's gastrointestinal tract. The life cycle of *Cryptosporidium* is illustrated in Fig. 4.16 (Fayer and Ungar, 1986). Animal models showed that as few as one to ten oocysts may initiate infection (Kwa et al., 1993; Miller et al., 1986; Rose, 1988), while a study with 29 healthy human volunteers showed a minimum infective dose of 30 oocysts and a median infective dose of 132 *C. parvum* oocysts (DuPont et al., 1995). The parasite causes a profuse and watery diarrhea that typically lasts for 10 to 14 days in immunocompetent hosts and is often associated with weight loss and sometimes nausea, vomiting, and low-grade fever (Current, 1988). The duration of the symptoms and the outcome depend on the immunological status of the patient. The diarrhea generally lasts 1–10 days in immunocompetent patients, but may persist for longer periods (more than 1 month) in immunodeficient patients (e.g, AIDS patients, cancer patients undergoing chemotherapy). Drug therapy to control this parasite is not yet available. Examination of thousands of human fecal

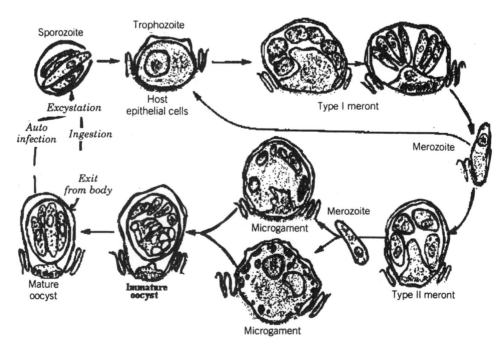

Figure 4.16 Diagrammatic representation of the life cycle of *Cryptosporidium*. From Fayer and Ungar (1986). (Courtesy of American Society of Microbiology.)

samples in the United States, Canada, and Europe has shown that the prevalence of human cryptosporidiosis is within a range of 1–5 percent (Ongerth and Stibbs, 1987).

It is estimated that *C. parvum* is responsible for 250–500 million infections per year in developing countries (Current and Garcia, 1991). Person-to-person, waterborne, food-borne, and zoonotic routes are all involved in the transmission of *Cryptosporidium*. Person-to-person transmission is the major route, especially in day-care centers. The zoonotic route, the transmission of the pathogen from infected animals to humans, is suspected to be greater for *Cryptosporidium* than for *Giardia* (AWWA, 1988). Molecular analysis of isolates from human and animal sources showed that there are two *Cryptosporidium* genotypes, with genotype 1 found in humans, and genotype 2 found in both animals and humans. This supports the existence of two separate transmission cycles of this parasite (Peng et al., 1997).

Studies have been conducted on the prevalence of this protozoan parasite in the environment since some of the outbreaks are waterborne. Ten of the 30 drinking-water-associated outbreaks reported in 1993–1994 in the United States were attributed to *Cryptosporidium* or *Giardia* (Kramer et al., 1996). Cryptosporidiosis outbreaks occurred in Georgia, Minnesota, Nevada, Texas, Washington, and Wisconsin. The outbreak in Carrollton, Georgia, affected approximately 13,000 people and was epidemiologically associated with consumption of drinking water from a water treatment plant where rapid sand filtration was part of the treatment processes. Problems discovered in the plant included ineffective flocculation and restarting of the sand filter without backwashing. *Cryptosporidium* was identified in 39 percent of the stools of patients during the outbreak and in samples of treated water, while no other traditional indicator was identified in the samples (Hayes et al., 1989). The largest documented waterborne disease outbreak occurred in Milwaukee, Wisconsin, where 403,000 people became ill, of whom 4400 were hospitalized and 54 died (Kaminski, 1994; MacKenzie et al., 1994). A retrospective cost-of-illness analysis showed that the total cost of that outbreak was $96.2 million in medical costs and productivity losses (Corso et al., 2003). There are also reports of swimming-associated cases of cryptosporidiosis (Sorvillo et al., 1992). Other outbreaks of cryptosporidiosis have been reported in Europe (Rush et al., 1990).

Cryptosporidium is not efficiently removed or inactivated by traditional water treatment processes such as sand filtration or chlorination, although lime treatment for water softening can partially inactivate *Cryptosporidium* oocysts (Robertson et al., 1992; Rose, 1990), and short-term (15 s) pasteurization at 71.7°C is able to destroy the infectivity of *C. parvum* oocysts (Harp et al., 1996). Compliance with U.S. EPA standards does not guarantee protection from infection with *Cryptosporidium*.

Concentration techniques have been developed for the recovery of this parasite but they are still at the developmental stage (Schaefer, 1997). The methods used involve, for example, the retention of oocysts on polycarbonate filters (Ongerth and Stibbs, 1987), polypropylene cartridge filters (Musial et al., 1987), vortex flow filtration (Mayer and Palmer, 1996), hollow fiber ultrafiltin (Kuhn and Oshima, 2001), or concentration via passage through a membrane filter that is dissolved in acetone and then centrifuged to pelletize the oocysts (Graczyk et al., 1997c). Although the recovery efficiency of these concentration techniques is relatively low, methodology improvements have been reported (Kuhn and Oshima, 2001; Mayer and Palmer, 1996).

The oocysts are detected in the concentrates using polyclonal or monoclonal antibodies combined with epifluorescence microscopy (Rose et al., 1989b), flow cytometry (Vesey et al., 1993), genetic probes in combination with PCR, or electronic imaging of the fluorescent oocysts, using cooled charge couple devices (CCD) (Campbell et al., 1992b; Johnson et al., 1993; 1995a; Leng et al., 1996; Mayer and Palmer, 1996; Rochelle et al., 1997a, b; Webster et al., 1993). The method currently used by the U.S. EPA is the immunomagnetic separation (IMS) fluorescent antibody (FA) detection technique (method 1623; U.S. EPA, 1999). Method 1623 and cell culture-PCR gave similar recovery efficiencies as regards the detection of *Cryptosporidium* in water. The comparison between the two methods revealed that about 37 percent of the *Cryptosporidium* oocysts detected by Method 1623 are infectious (LeChevallier et al., 2003). The recovery efficiency of the method can be also be improved by using a polysulfone hollow-fiber single-use filter instead of the pleated polyethersulfone filter used in the EPA method (Simmons et al., 2001). Immunomagnetic separation was also combined with PCR for oocyst detection (Hallier-Soulier and Guillot, 2000; Sturbaum et al., 2002). Viable oocysts can be detected with IMS followed by RT-PCR, which targets the *hsp70* heat-shock-induced mRNA. The assay did not give any signal following heat treatment (20 min at 95°C) of the oocysts, confirming that the method detects only viable oocysts (Hallier-Soulier and Guillot, 2003). Oocyst viability and infectivity is generally determined by in vitro excystation, mouse infectivity assay, in vitro cell culture infectivity assays, or staining with fluorogenic vital dyes such as DAPI (4,6-diamino-2-phenylindole), propidium iodide, SYTO-9 or SYTO-59 (Bukhari et al., 2000; Campbell et al., 1992a; Shin et al., 2001). However, the vital-dye-based assays and in vitro excystation were found to overestimate oocyst viability. Although the mouse infectivity assay is considered the method of choice for assessing oocyst infectivity (Bukhari et al., 2000), the in vitro cell culture assay should be considered as a practical and accurate alternative (Rochelle et al., 2002).

A PCR method, based on the amplification of an 873 bp gene fragment, detects the excysted sporozoite, and therefore allows a distinction between live and dead oocysts (Wagner-Wiening and Kimmig, 1995). An immunomagnetic capture PCR method was proposed to detect viable *Cryptosporidium parvum* in environmental samples. The procedure consists of capturing *Cryptosporidium* oocysts on IgG-coated magnetite particles followed by excystation, PCR amplification, and identification of the PCR products (Deng et al., 1997). Oocyst infectivity can also be determined by combining infectivity assays on cell cultures with RT-PCR, which targets a heat shock protein 70 (*hsp70*) gene (Rochelle et al., 1997a, b).

Figure 4.17 exhibits micrographs of *Cryptosporidium parvum* oocysts, using a combination of a fluorescent antibody and a fluorescent in situ hybridization (FISH) probe.

It was also suggested that the Asian benthic freshwater clam (*Corbicula fluminea*) or the marine mussel (*Mytilus edulis*), be used as biomonitors for the presence of *Cryptosporidium* oocysts in water and wastewater (Chalmers et al., 1997; Graczyk et al., 1997a, b). Immunofluoresence microscopy showed that the clam concentrates the oocysts in the hemolymph, where they are phagocytosed by the hemocytes. Oocysts uptake and phagocytosis by hemocytes was also demonstrated for the Eastern oyster, *Crassostrea virginica* (Fayer et al., 1997; Graczyk et al., 1997b).

These methodologies allow the detection of this hardy parasite in wastewater, surface, and drinking water and show that oocysts occur in raw wastewater at levels varying between 850 and 13,700 oocysts/L. The range of oocyst concentrations in wastewater effluents varies between 4 and 3960 cysts/L (Madore et al., 1987; Musial et al., 1987). A survey of potable water supplies in the United States showed that oocysts were present in 55 percent of the samples at an average concentration of 43 oocysts/100 L (Rose et al., 1991). Other surveys indicated the presence of oocysts in up to 87 percent of surface water samples (Chauret et al., 1995; LeChevallier et al., 1991a, b), 8.7 percent of coastal waters, and 16.7 percent of cistern waters (Johnson et al., 1995a, b). Analysis of river water in the western United States showed *Cryptosporidium* oocysts in each of the 11 rivers examined, at concentrations ranging from 2 to 112 oocysts/L

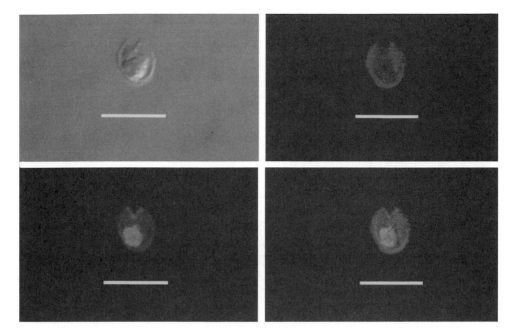

Figure 4.17 Oocysts of *Cryptosporidium parvum* viewed with a combination of immunological and genetic tools. (Courtesy of Christopher Robinson, Oak Ridge Institute of Science and Education, and H. D. Alan Lindquist, U.S. EPA.) The gray frame on the top left shows a *Cryptosporidium parvum* oocyst with some apparent damage to the oocyst wall, and some material remaining inside. The red image on the right is of the same oocyst as detected by an anti-*C. parvum* red fluorescent antibody (*Fab*) staining procedure. Only the surface features of the oocyst are evident with this type of staining. The image on the bottom left is of the same oocyst, hybridized to a *C. parvum* specific anti-ribosomal RNA fluorescent in situ hybridization (FISH) probe labeled with a green fluorescent compound. There is some nonspecific staining with this probe, but the majority of the probe appears to be hybridized within one region, indicating the presence of ribosomal RNA in this structure. The image on the bottom right is an overlay of Fab and FISH staining of the oocyst. The scale in all images represents 5 μm. Figure also appears in color figure section.

(Ongerth and Stibbs, 1987). In Japan, a survey of 18 rivers that serve as sources of water supply, showed that 47 percent of samples were positive for *C. parvum* oocysts (Ono et al., 2001). Another study showed that 13 of 13 samples of source water in Japan were positive for *Cryptosporidium* oocysts (Hashimoto et al., 2002). This parasite has also been detected in finished drinking water (Hashimoto et al., 2002; Rose et al., 1986).

4.2.3.3 Cyclospora. *Cyclospora cayetanensis* is another emerging diarrhea-causing coccidian parasite, which was first reported in 1986, and is often mentioned in the literature as a "cyanobacterium-like body" (Adal et al., 1995; Soave, 1996). *Cyclospora* oocysts are spheroidal with an 8–10 μm diameter and contain two sporocysts, with two sporozoites per sporocyst. Cyclosporiasis has an incubation period of approximately one week, and clinical symptoms include long-lasting watery diarrhea, sometimes alternating with constipation, abdominal cramps, nausea, weight loss, sometimes vomiting, anorexia, and fatigue. Infections may last up to four months in AIDS patients. This parasite infects epithelial cells of the duodenum and jejunum. Diagnosis of infection is based on microscopic examination of stool specimens. (Ortega et al., 2002; Soave, 1996). The treatment of cyclosporiasis necessitates the use of trimethropim sulfamethoxazole (TMX-SMX).

Cyclospora outbreaks have been reported in developing countries, and the parasite is endemic in certain countries such as Nepal, Haiti, and Peru. Some water-associated cases have been documented in the United States (CDC, 1991). Even after chlorination, drinking water appears to be implicated in outbreaks of diarrhea associated with *Cyclospora* (Huang et al., 1996; Rabold et al., 1994). In the United States, most infections are associated with the consumption of contaminated fruits and vegetables. An outbreak of cyclosporiasis in 1996 in the United States and Canada resulted in 1465 cases, 67 percent of which were confirmed by various laboratories. The outbreak was associated with consumption of raspberries imported from Guatemala (Herwaldt et al., 1997). Other outbreaks have been linked to the consumption of contaminated basil (Lopez et al., 2001).

Cyclospora oocysts are concentrated in environmental samples by filtration methods similar to those employed for *Cryptosporidium*. Microscopic examination is used to detect oocysts in sample concentrates. A distinct feature of *Cyclospora* oocysts is their auto-fluorescence, as they appear as blue circles when examined under a fluorescence microscope (365 nm excitation filter). Polymerase chain reaction can detect less than 40 oocysts per 100 g of raspberries or basil (Steele et al., 2003). Fluorogenic probes, in conjunction with real-time PCR, have been used to detect *Cyclospora* oocysts (Varma et al., 2003). A method based on PCR-restriction fragment length polymorphism (PCR-RFLP) allows the differentiation of *Cyclospora cayetanensis* from other *Cyclospora* species, as well as other coccidian parasites such as *Eimeria* (Shields and Olson, 2003). Unfortunately, there is a lack of in vivo or in vitro culture assays to assess the viability of *Cyclospora* oocysts.

No data are available on the removal/inactivation of this parasite following water and wastewater treatment.

4.2.3.4 Microsporidia.
Microsporidia are obligate intracellular protozoan parasites that cause infections in humans, especially AIDS patients. Prevalence in patients with chronic diarrhea varies between 10 and 50 percent (Cotte et al., 1999). Individuals ingesting the small (1–5 μm) resistant spores experience chronic diarrhea, dehydration, and significant weight loss. Some microsporidia species may be transmitted by the water route (Schaefer, 1997).

4.2.3.5 Entamoeba histolytica.
Entamoeba histolytica forms infective cysts (10–15 μm in diameter) that are shed for relatively long periods by asymptomatic carriers; it persists well in water and wastewater and may be subsequently ingested by new hosts. Level of cysts in raw wastewater may be as high as 5000 cysts/L.

This protozoan parasite is transmitted to humans mainly via contaminated water and food. It causes amoebiasis or amoebic dysentery, which is a disease of the large intestine. Symptoms vary from diarrhea alternating with constipation to acute dysentery. It may cause ulceration of the intestinal mucosa, resulting in diarrhea and cramps. It is a cause of morbidity and mortality mostly in developing countries and is acquired mostly via consumption of contaminated drinking water in tropical and subtropical areas. Waterborne transmission of this protozoan parasite is, however, rare in the United States.

4.2.3.6 Naegleria.
Naegleria are free-living protozoa that have been isolated from wastewater, surface waters, swimming pools, soils, domestic water supplies, thermal spring waters, and thermally polluted effluents (Marciano-Cabral, 1988). *Naegleria fowleri* is the causative agent for *primary amoebic meningoencephalitis* (PAME), first reported in Australia in 1965. It is most often fatal, 4–5 days after entry of the amoeba into the body. The protozoan enters the body via the mucous membranes of the nasal

cavity and migrates to the central nervous system. The disease has been associated with swimming and diving mostly in warm lakes in southern states of the United States (Florida, South Carolina, Georgia, Texas, and Virginia). Using an animal model, it was estimated that PAME risk to humans, as a function of *N. fowleri* concentration in water, is 8.5×10^{-8} at a concentration of 10 amoebae per liter (Cabanes et al., 2001). Another concern is the fact that *Naegleria* may harbor *Legionella pneumophila* and other pathogenic microorganisms (Newsome et al., 1985). The implications of this association with regard to human health are not well known.

There are now rapid identification techniques (e.g., cytometry, API ZYM system, which is based on detection of enzyme activity), monoclonal antibodies, DNA probes, and PCR methods that can distinguish *Naegleria fowleri* from other free-living amoebas in the environment (Behets et al., 2003; Kilvingston and Beeching, 1995a, b; Kilvington and White, 1985; Visvesvara et al., 1987). A recent method for detecting *N. fowleri* is solid-phase cytometry, which uses fluorescent labeling of microorganisms on a membrane filter followed by an automated counting system (Pougnard et al., 2002).

4.2.3.7 *Toxoplasma gondii.* *Toxoplasma gondii* is a coccidian parasite that uses cats as a host (Dubey, 2002). It also causes parasitic infections in humans worldwide. Many infections are congenitally acquired and cause ocular disease in children. Others are postnatally acquired and lead to enlargement of the lymph nodes. This parasite causes most damage among AIDS patients and other immunosuppressed individuals. Humans become infected following ingestion of uncooked or undercooked meat or water contaminated with *T. gondii* oocysts. An outbreak of toxoplasmosis was linked to the consumption of water from a reservoir in Canada (Isaac-Renton et al., 1998). Preventive measures are directed mostly to pregnant women, who should avoid contact with cats as well as contaminated meat (Dubey, 2002).

Following concentration and purification (immunomagnetic separation can be used) of environmental samples, *T. gondii* oocysts can be detected using light or fluorescence microscopy or PCR-based methods. The molecular methods are hampered by the multi layered nature of the oocyst walls, which cause difficulties in DNA extraction. Oocyst viability is assessed via mouse bioassay (Dumètre and Dardé, 2003).

4.2.4 Helminth Parasites

Although helminth parasites are not generally studied by microbiologists, their presence in wastewater, along with bacterial and viral pathogens and protozoan parasites, is nonetheless of great concern as regards human health. It is estimated that about 63 percent of the Chinese population is infected with one or more helminth parasites, particularly with *Ascaris lumbricoides*, *Trichuris trichiura*, and hookworms *Ancylostoma duodenale* and *Necator americanus*. Most of these infections are acquired by the food-borne route (Hotez et al., 1997; Xu et al., 1995).

The ova (eggs) constitute the infective stage of parasitic helminths; they are excreted in feces and spread via wastewater, soil, or food. The ova are very resistant to environmental stresses and to chlorination in wastewater treatment plants (Little, 1986). There are seasonal fluctuations in the number of helminth eggs in wastewater. Egg concentration in raw wastewater from Marrakech, Morocco, varied between 0 and 120 eggs/L with an annual mean of 32 eggs/L (Mandi et al., 1992). The most important helminth parasites are presented in the following subsections (Table 4.15).

TABLE 4.15. Major Parasitic Helminths

Organism	Disease (Main Site Affected)
Nematodes (roundworms)	
Ascaris lumbricoides	Ascariasis – intestinal obstruction in children (small intestine)
Trichuris trichiura	Whipworm – (trichuriasis) (intestine)
Hookworms	
Necator americanus	Hookworm disease (GI tract)
Ancylostoma duodenale	Hookworm disease (GI tract)
Cestodes (tapeworms)	
Taenia saginata	Beef tapeworm – abdominal discomfort, hunger pains, chronic indigestion (GI tract)
Taenia solium	Pork tapeworm (GI tract)
Trematodes (flukes)	
Schistosoma mansoni	Schistosomiasis (complications in liver [cirrhosis], bladder, and large intestine)

***4.2.4.1 Taenia* spp.** *Taenia saginata* (beef tapeworm) and *Taenia solium* (pig tapeworm) are now relatively rare in the United States. These parasites develop in an intermediate host to reach a larval stage called *cysticercus* and may finally reach humans, which serve as final hosts. Cattle ingest the infective ova while grazing, and serve as intermediate hosts for *Taenia saginata*, pigs being the intermediate hosts for *Taenia solium.* The cysticerci invade muscles, eyes, and brain. These parasites cause enteric disturbances, abdominal pains, and weight loss.

4.2.4.2 Ascaris lumbricoides (Roundworms). The life cycle of this helminth (Fig. 4.18) includes a phase in which the larvae migrate through the lungs and cause pneumonitis (Loeffler's syndrome). This disease can be acquired through ingestion of only a few infective eggs. Infected individuals excrete a large quantity of eggs and each female *Ascaris* can produce approximately 200,000 eggs/day (Little, 1986). The eggs are dense and are well removed via sedimentation in wastewater treatment plants. Although they are effectively removed by the activated sludge treatment, they are quite resistant to chlorine action (Rose, 1986).

4.2.4.3 Toxocara canis. This parasite infects mainly children with habits of eating dirt. In addition to causing intestinal disturbances, the larvae of this parasite can migrate into the eyes, causing severe ocular damage, sometimes resulting in loss of the eye.

4.2.4.4 Trichuris trichiura. *Trichuris trichiura* causes whipworm infections in humans. The eggs are dense and settle quite well in sedimentation tanks.

4.2.5 Other Problem-Causing Microorganisms

Surface waters feeding water treatment plants may harbor large concentrations of blue-green algae (cyanobacteria) such as *Anabaena flos aquae*, *Microcystis aeruginosa*, and *Schizothrix calcicola.* These algae produce exotoxins (peptides and alkaloids) as well as endotoxins (lypopolysaccarides) that may be responsible for syndromes such as gastroenteritis (Carmichael, 1981a, b; 1989). Presently, studies are being undertaken to gain

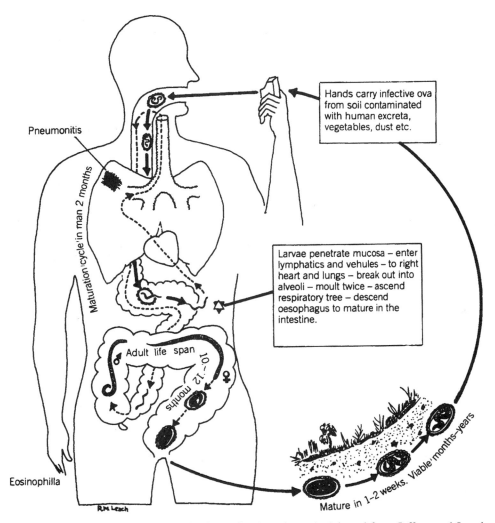

Figure 4.18 Life cycle of *Ascaris lumbricoides* (roundworm). Adapted from Jeffrey and Leach (1972).

knowledge about the occurrence and removal potential of these toxins in water and waste-water treatment plants. Their health risks have not yet been fully evaluated (see Chapter 16 for more details).

4.3 WEB RESOURCES

http://www.swbic.org/products/bioinfo/envirusat/main.html (very good web page focused on enteric viruses, excellent pictures)

http://www.bact.wisc.edu/Bact330/330Lecturetopics (microbiology course, University of Michigan, Madison)

http://web.uct.ac.za/depts/mmi/stannard/linda.html (lectures notes on viral diseases and EM pictures of viruses from the University of Capetown)

http://www.ncbi.nlm.nih.gov/ICTVdb/Images/index2.htm (excellent EM pictures and diagrams of viruses and comprehensive description of phylogeny from the Universal Virus Database of the International Committee on Taxonomy of Viruses)

http://www.biosci.ohio-state.edu/~parasite/a-z.html (good selection of good pictures of protozoan and other parasites)

http://www.epa.gov/microbes/ (EPA methods on pathogens, parasites, and indicator organisms)

http://www.cdc.gov/ (Center for Disease Control and Prevention)

4.4 QUESTIONS AND PROBLEMS

1. What is the difference between a primary pathogen and an opportunistic one?
2. Why are shellfish important in disease transmission?
3. Discuss the problems associated with antibiotic-resistant bacteria.
4. Give the causative agents for the following:
   - peptic ulcers;
   - pontiac fever;
   - serum hepatitis;
   - viral gastroenteritis;
   - Primary amoebic meningoencephalitis (PAME).
5. Give two examples of bacterial pathogens that may be transmitted via aerosolization.
6. Which hepatitis viruses are transmitted via the waterborne route?
7. Give the categories of pathogenic *E. coli* strains.
8. Give an example of an aquatic free-living protozoan parasite.
9. What is the most important transmission route for viruses?
10. How would you remove *Cryptosporidium* oocysts in water treatment plants?
11. Give the major factors affecting the chain of infection.

4.5 FURTHER READING

Altekruse, S.F., M.L. Cohen, and D.L. Swerdlow. 1997. Emerging foodborne diseases. Emerging Infect. Dis. 3: 285–293.

Bitton, G. 1980. *Introduction to Environmental Virology*, John Wiley & Sons, New York. 326 pp.

Cliver, D.O. 1984. Significance of water and environment in the transmission of virus disease. Monog. Virol. 15: 30–42.

Craun, G.F., Ed. 1986. *Waterborne Diseases in the United States*, CRC Press, Boca Raton, FL., 295 pp.

Feachem, R.G., D.J. Bradley, H. Garelick, and D.D. Mara. 1983. *Sanitation and Disease: Health Aspect of Excreta and Wastewater Management*, John Wiley & Sons, Chichester, U.K.

Gerba, C.P., S.N. Singh, and J.B. Rose. 1985. Waterborne gastroenteritis and viral hepatitis. CRC Crit. Rev. Environ. Contam. 15: 213–236.

Jones, K. 2001. Campylobacters in water, sewage and the environment. J. Appl. Microbiol. 90: 68S–79S.

Leclerc, H., L. Schwartzbrod, and E. Dei-Cas. 2002. Microbial agents associated with waterborne disease. Crit. Rev. Microbiol. 28: 371–409.

Rusin, P., C.E. Enriquez, D. Johnson, and C.P. Gerba. 2000. Environmentally transmitted pathogens, pp. 447–489, In: *Environmental Microbiology*, R.M. Maier, I.L. Pepper, and C.P. Gerba, Eds., Academic Press, San Diego, CA, 585 pp.

Rose, J.B. 1990. Occurrence and Control of *Cryptosporidium* in drinking water, pp. 294–321, In: *Drinking Water Microbiology*, G.A. McFeters, Ed., Springer Verlag, New York.

Schwartzbrod, L. Ed. 1991. *Virologie des milieux hydriques*, TEC & DOC Lavoisier, Paris, France, 304 pp.

Seymour, I.J., and H. Appleton. 2001. Foodborne viruses and fresh produce. J. Appl. Microbiol. 91: 759–773.

Theron, J., and T.E. Cloete. 2002. Emerging waterborne infections: Contributing factors, agents, and detection tools. Crit. Rev. Microbiol. 28: 1–26.

Wyn-Jones, A.P., and J. Sellwood. 2001. Enteric viruses in the aquatic environment. J. Appl. Microbiol. 91: 945–962.

5

MICROBIAL INDICATORS OF FECAL CONTAMINATION

5.1 INTRODUCTION

The direct detection of pathogenic bacteria and viruses, and cysts of protozoan parasites (see Chapter 4) requires costly and time-consuming procedures, and well-trained labor. These requirements led to the concept of indicator organisms of fecal contamination.

Wastewater Microbiology, *Third Edition*, by Gabriel Bitton
Copyright © 2005 John Wiley & Sons, Inc.

As early as 1914, the U.S. Public Health Service (U.S.P.H.S.) adopted the coliform group as an indicator of fecal contamination of drinking water. Later on, various microorganisms have been proposed and used for indicating the occurrence of fecal contamination, treatment efficiency in water and wastewater treatment plants, and the deterioration and post-contamination of drinking water in distribution systems.

The criteria for an ideal indicator organism are:

1. It should be one of the intestinal microflora of warm-blooded animals.
2. It should be present when pathogens are present, and absent in uncontaminated samples.
3. It should be present in greater numbers than the pathogen.
4. It should be at least equally resistant as the pathogen to environmental insults and to disinfection in water and wastewater treatment plants.
5. It should not multiply in the environment.
6. It should be detectable by means of easy, rapid, and inexpensive methods.
7. The indicator organism should be nonpathogenic.

5.2   REVIEW OF INDICATOR MICROORGANISMS

Proposed or commonly used microbial indicators are discussed below (APHA, 1989; Berg, 1978; Ericksen and Dufour, 1986; Leclerc et al., 2000; Olivieri, 1983) (Fig. 5.1).

5.2.1   Total Coliforms

The total coliform group belongs to the family *enterobacteriaceae* and includes the aerobic and facultative anaerobic, gram-negative, nonspore-forming, rod-shaped bacteria that ferment lactose with gas production within 48 hours at 35°C (APHA, 1989).

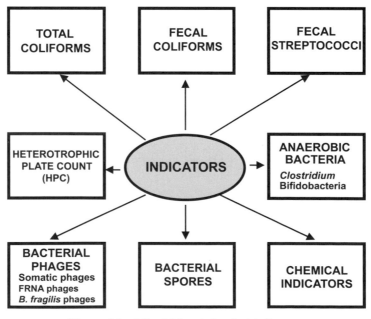

Figure 5.1   Microbial and chemical indicators.

This group includes *Escherichia coli*, *Enterobacter*, *Klebsiella*, and *Citrobacter*. These coliforms are discharged in high numbers (2×10^9 coliforms/day/capita) in human and animal feces, but not all of them are of fecal origin. These indicators are useful for determining the quality of potable water, shellfish-harvesting waters, and recreational waters. They are less sensitive, however, than viruses or protozoan cysts to environmental factors and to disinfection. Some members (e.g., *Klebsiella*) of this group may sometimes grow under environmental conditions in industrial and agricultural wastes. In water treatment plants, total coliforms are one of the best indicators of treatment efficiency of the plant. For example, this group has been found useful for assessing the safety of reclaimed wastewater in the Windhoek reclamation plant in Namibia (Grabow, 1990).

5.2.2 Fecal Coliforms

Fecal coliforms or thermotolerant coliforms include all coliforms that can ferment lactose at 44.5°C. The fecal coliform group comprises bacteria such as *Escherichia coli* or *Klebsiella pneumonae*. The presence of fecal coliforms indicates the presence of fecal material from warm-blooded animals. However, human and animal sources of contamination cannot be differentiated. Some investigators have suggested the sole use of *E. coli* as an indicator of fecal pollution as it can be easily distinguished from the other members of the fecal coliform group (e.g., absence of urease and presence of β-glucuronidase). Fecal coliforms display a survival pattern similar to that of bacterial pathogens, but their usefulness as indicators of protozoan or viral contamination is limited. They are much less resistant to disinfection than viruses or protozoan cysts. Coliform standards are thus unreliable with regard to contamination of aquatic environments with viruses and protozoan cysts. They may also regrow in water and wastewater under appropriate conditions. Several methodological modifications have been proposed to improve the recovery of these indicators, particularly injured fecal coliforms (Ericksen and Dufour, 1986). Moreover, the growth and detection of *E. coli* in pristine sites in a tropical rain forest suggest that it may not be a reliable indicator of fecal pollution in tropical environments (Bermudez and Hazen, 1988; Hazen, 1988). The distinction between human and animal sources of *E. coli* may be accomplished by using methods such as antibiotic resistance patterns or DNA fingerprinting (see Section 5.2.3).

5.2.3 Fecal Streptococci

This group comprises *Streptococcus faecalis*, *S. bovis*, *S. equinus*, and *S. avium*. Because they commonly inhabit the intestinal tract of humans and warm-blooded animals, they are used to detect fecal contamination in water. Members of this group survive longer than other bacterial indicators, but do not reproduce in the environment. A subgroup of the fecal streptococci group, the enterococci (*S. faecalis* and *S. faecium*), has been suggested as useful for indicating the presence of viruses, particularly in biosolids and seawater.

The fecal coliform to fecal streptococci ratio (FC/FS ratio) has served for many years as an indicator of the origin of pollution of surface waters. A ratio of ≥ 4 indicates a contamination of human origin, whereas a ratio of <0.7 is indicative of animal pollution (Geldreich and Kenner, 1969). This ratio is only valid, however, for recent (24 hours) fecal pollution and is unreliable for chlorinated effluents, and some investigators have questioned its usefulness (Pourcher et al., 1991). The source of fecal contamination can also be identified, based on the antibiotic resistance pattern of coliforms (Kaspar et al., 1990) or fecal streptococci (Hagedorn et al., 1999; Knudson and Hartman, 1993;

Wiggins, 1996), multiple antibiotic resistance (MAR) profiles (Burnes, 2003; Parveen et al., 1997), ribotyping, pulsed-field gel electrophoresis, biochemical finger-printing (Manero et al., 2002), phenotypic fingerprinting with carbon source utiliza-tion profiles, using the Biolog system (Hagedorn et al., 2003) or amplified fragment length polymorphism (AFLP), which can distinguish between nonpathogenic and pathogenic strains of *E. coli* (Leung et al., 2004).

Phages of *Bacteroides fragilis* can also serve as indicators of human fecal contami-nation (see Section 5.2.5).

5.2.4 Anaerobic Bacteria

The main anaerobic bacteria that have been considered as indicators are given in the following subsections.

5.2.4.1 *Clostridium Perfringens.*   Clostridia are mostly opportunistic pathogens, but are also implicated in human diseases such as gas gangrene (*C. perfringens*), tetanus (*C. tetani*), botulism (*C. botulinum*), or acute colitis (*C. difficile*) (Payment et al., 2002). *Clostridium perfringens* is an anaerobic gram-positive, endospore-forming, rod-shaped, sulfite-reducing bacterium found in the colon and represents approximately 0.5 percent of the fecal microflora. It produces spores that are quite resistant to environ-mental stresses and to disinfection. Since it is a member of the sulfite reducing clostridia (SRC) group, it is detected in growth media containing sulfite. It is commonly found in human and animal feces and in wastewater-contaminated aquatic environments. In Europe, SRC have been traditionally used as indicators of water quality, but new European Union (EU) regulations consider more specifically *C. perfringens* as the indicator of choice. The EU standard was set at 0/100 mL of drinking water supply (European Union, 1998). The hardy spores make this bacterium too resistant to be useful as an indi-cator organism. It has been suggested nonetheless to use this microorganism as an indi-cator of past pollution and as a tracer to follow the fate of pathogens. *Clostridium perfringens* was also proposed as a suitable indicator for viruses and protozoan cysts in water treatment plants (Payment and Franco, 1993), *Cryptosporidium parvum* oocysts after mixed-oxidant disinfection (Venczel et al., 1997), and the quality of recreational waters (Fujioka, 1997). This bacterium is generally much more resistant to oxidizing agents and to UV than bacterial and phage indicators. It also appears to be a reliable indi-cator for tracing fecal pollution in the marine environment (e.g., marine sediments impacted by sludge dumping) and survives in sediments for long periods (>1 year) after cessation of sludge dumping (Burkhardt and Watkins, 1992; Hill et al., 1993; 1996).

5.2.4.2 *Bifidobacteria.*   These anaerobic, nonspore-forming, gram-positive bacteria live in human and animal guts and have been suggested as fecal indicators. *Bifidobacter-ium* is the third most common genus found in the human intestinal microflora. Because some of them (e.g., *B. bifidum*, *B. adolescentis*, *B. infantis*, *B. dentium*) are primarily associated with humans, they may help distinguish between human and animal contami-nation sources. Their detection is now made possible through the use of rRNA probes (Bonjoch et al., 2004; Langendijk et al., 1995). Enrichment for bifidobacteria followed by detection of *Bifidobacterium adolescentis* via colony hybridization have shown that this bacterium can serve as a specific indicator of human fecal contamination (Lynch et al., 2002). A multiplex PCR approach showed that *B. adolescentis* and *B. dentium* were found exclusively in human sewage (Bonjoch et al., 2004).

5.2.4.3 *Bacteroides* spp.

These anaerobic bacteria occur in the intestinal tract at concentrations in the order of 10^{10} cells of feces. The survival of *B. fragilis* in water is lower than that of *E. coli* or *S. faecalis*. A fluorescent antiserum test for this bacterium was suggested as a useful method for indicating the fecal contamination of water (Fiksdal et al., 1985; Holdeman et al., 1976).

5.2.5 Bacteriophages

Three groups of bacteriophages have been proposed as indicator organisms: somatic coliphages, male-specific RNA coliphages (FRNA phages), and phages infecting *Bacteroides fragilis* (Berger and Oshiro, 2002; Leclerc et al., 2000).

5.2.5.1 *Somatic Coliphages.*

They infect mostly *E. coli*, but some can infect other enterobacteriaceae. Coliphages are similar to the enteric viruses, but are more easily and rapidly detected in environmental samples and are found in higher numbers than enteric viruses in wastewater and other environments (Bitton, 1980a; Goyal et al., 1987; Grabow, 1986). Several investigators have suggested the potential use of coliphage as water quality indicators in estuaries (O'Keefe and Green, 1989), seawater (good correlation between coliphage and *Salmonella*; Borrego et al., 1987), freshwater (Dutka et al., 1987; Skraber et al., 2002), potable water (Ratto et al., 1989) biosolids and wastewater (Mocé-Llivina et al., 2003). Phages can also serve as biotracers to identify pollution sources in surface waters and aquifers (Harvey, 1997b; McKay et al., 1993; Paul et al., 1995). Genetically modified phages have been proposed to avoid interference with indigenous phages present in environmental samples. A unique DNA sequence was inserted into the phage genome, which then can be detected, using PCR or plaque hybridization (Daniell et al., 2000).

Of all the indicators examined, coliphages exhibited the best correlation with enteric viruses in polluted streams in South Africa. The incidences of both enteric viruses and coliphages were inversely correlated with temperature (Geldenhuys and Pretorius, 1989). They may also serve as indicators for assessing the removal efficiency of water and wastewater treatment plants (Bitton, 1987) as well as water treatment plants where they help provide information concerning the performance of water treatment processes such as coagulation, flocculation, sand filtration, adsorption to activated carbon, or disinfection (Payment, 1991) (Table 5.1). In an activated sludge system, coliphages forming plaques >3 mm were significantly correlated with numbers of enteroviruses (Funderburg and Sorber, 1985).

5.2.5.2 *Male-specific RNA phages (FRNA phages).*

These are single-stranded RNA phages (FRNA phage) with a cubic capsid measuring 24–27 nm. All FRNA phages belong to the family Leviviridae (see Fig. 1.26 in Chapter 1). They enter a host bacterial cell by adsorbing to the F or sex pilus of the cell. Because F-specific phages are infrequently detected in human fecal matter and show no direct relationship with fecal pollution level, they cannot be considered as indicators of fecal pollution (Havelaar et al., 1990a; Leclerc et al., 2000; Morinigo et al., 1992b). Their presence in high numbers in wastewaters and their relatively high resistance to chlorination contribute to their consideration as indicators of wastewater contamination (Debartolomeis and Cabelli, 1991; Havelaar et al., 1990a; Nasser et al., 1993; Yahya and Yanko, 1992). There are four genotypes of F-specific RNA phages. With a few exceptions, genotypes II and III are generally associated with human feces while genotypes I and IV are associated with

TABLE 5.1.   Monitoring of Coliphages, Clostridia, and Viruses at Various Stages of the Pont-Viau, Canada, Water Filtration Plant

Organisms	Type of Water	Geometric Mean	Percentage Reduction	Percentage Positive
Human enteric viruses (mpniu/100 L)	Raw[a]	79	NA	91.0
	Settled[a]	0	>99	0.0
	Filtered[b]	0	>99.9	6.5
	Finished[b]	0	>99.9	0.0
Coliphages (PFU/100 L)	Raw[a]	565	NA	100.0
	Settled[a]	3.1	99.953	50.0
	Filtered[b]	0.5	99.992	30.3
	Finished[b]	0.0	99.99997	0.6
Clostridia (CFU/100 L)	Raw[a]	11,349	NA	100.0
	Settled[a]	83.8	99.262	93.8
	Filtered[b]	1.2	99.989	51.5
	Finished[b]	0.0	99.9982	1.9

[a]100 L sample.
[b]100 L sample.
CFU, colony-forming units; PFU, plaque-forming units; NA, not available.
Adapted from Payment (1991).

animal feces. It was suggested that they could be used as indicators of the source (human vs. animal sources) of fecal contamination (Schaper et al., 2002).

Monitoring of postchlorinated effluents after a rainfall showed that fecal coliforms and enterococci were much more sensitive to chlorine than are male-specific bacteriophages (Rippey and Watkins, 1992). As regards shellfish contamination and depuration, male-specific phages provide a suitable model for studying the fate of animal viruses in shellfish (Doré and Lees, 1995). They survive at least seven days in hard-shelled clams at ambient seawater temperatures and do not undergo replication with or without added host cells (Burkhardt et al., 1992). They appear to be suitable indicators for viral contamination in the marine environment.

5.2.5.3   *Phages Infecting Bacteroides Fragilis.*   The potential of bacteriophages of *Bacteroides* spp. to serve as indicators of viral pollution was also explored (Tartera and Jofre, 1987). Phages active against *Bacteroides fragilis* HSP 40 were detected in feces (found in 10 percent of human fecal samples but not in animal feces), sewage, and other polluted aquatic environments (riverwater, seawater, groundwater, sediments) and were absent in nonpolluted sites (Cornax et al., 1990; Tartera and Jofre, 1987) (Table 5.2). These indicators do not appear to multiply in environmental samples (Tartera et al., 1989), are more resistant to chlorine than bacterial indicators (*S. faecalis*, *E. coli*) or viruses (poliovirus type 1, rotavirus SA11 and coliphage f2) (Fig. 5.2; Abad et al., 1994; Bosch et al., 1989) but are less resistant than coliphage f2 to UV irradiation (Bosch et al., 1989). The higher resistance to chlorine of bacterial phages as compared to bacterial indicators was confirmed for sewage effluents (Table 5.3; Durán et al., 2003).

TABLE 5.2. Levels of Bacteriophages Active Against *B. fragilis* HSP 40 in Water and Sediments

Samples	No. of Samples	% Samples Positive for Phages	Maximum Value/ 100 mL	Minimum Value/ 100 mL	Mean Value/ 100 mL
Sewage	33	100	1.1×10^5	7	6.2×10^3
River water[a]	22	100	1.1×10^5	93	1.6×10^4
River sediment[a]	5	100	4.6×10^5	90	1.08×10^5
Seawater[a]	22	77.2	1.1×10^3	<3	1.2×10^2
Marine sediment[a]	12	91.0	43	<3	13.4
Groundwater[a]	19	21.0	Unknown	0	
Nonpolluted[b]	50	0			

[a]Samples from areas with sewage pollution.
[b]Water and sediments from areas without known sewage pollution.
Adapted from Tartera and Jofre (1987).

Phage chlorine resistance was ranked as follows:

$$\text{Phages infecting } \textit{Bacteroides fragilis} > \text{F-RNA bacteriophages} > \text{Somatic coliphages}$$

Moreover, *B. fragilis* phage displayed a similar reduction to enteroviruses. Thus, these organisms may be suitable indicators of human fecal pollution and their use enables the distinction between human and animal fecal pollution. They display a positive correlation with enteroviruses and rotaviruses (Jofre et al., 1989) and their persistence is similar to that of enteric viruses (e.g., hepatitis A virus) in seawater and shellfish (Chung and Sobsey, 1993a; Lucena et al., 1994). Phages infecting *Bacteroides fragilis* are more resistant to treatment processes than bacterial (fecal coliforms and streptococci, clostridia), phage (somatic and male-specific phages) indicators, and enteroviruses, as shown in three water treatment plants in Spain (Jofre et al., 1995). They are also more resistant to

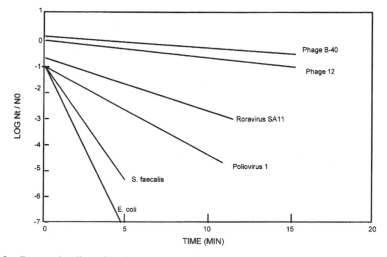

Figure 5.2 Regression line of the inactivation of bacteriophages active against *Bacteroides fragilis* (B-40) and other microorganisms by chlorine. Adapted from Bosch et al. (1989).

TABLE 5.3.   Log$_{10}$ Reduction (SD) of Naturally Occurring Bacterial Indicators and Bacteriophages by Chlorination of a Secondary Sewage Effluent[a]

Microorganism	Chlorine[b] 20 mg/L	Chlorine[b] 40 mg/L
Fecal coliforms	>4.1	>4.7
Enterococci	>3.2	>3.5
Somatic phages	1.3 (0.4)	2.6 (0.2)
F-specific RNA phages	0.5 (0.5)	1.1 (0.4)
B. fragilis phages	0.3 (0.2)	0.5 (0.3)

[a]Adapted from Durán et al. (2003).
[b]Contact time = 30 min.
SD, standard deviation.

natural inactivation in freshwater than fecal coliforms and other phages (FRNA and somatic phages) (Durán et al., 2002).

It is doubtful that bacterial phages can be used as indicators for enteric viruses in all situations (Gerba, 1987a).

5.2.6   Yeasts and Acid-Fast Organisms

Some investigators have proposed yeasts and acid-fast mycobacteria (*Mycobacterium fortuitum* and *M. phlei*) as indicators of disinfection efficiency (Grabow et al., 1980; Haas et al., 1985). The acid-fast bacterium, *Mycobacterium fortuitum*, is more resistant to free chlorine and ozone than is *E. coli* or poliovirus type 1 (Engelbrecht and Greening, 1978; Engelbrecht et al., 1974; Farooq and Akhlaque, 1983). Little is known about the use of these microorganisms as indicators in field studies.

5.2.7   Bacterial Spores

Aerobic spores are nonpathogenic, ubiquitous in aquatic environments, occur at much higher concentrations than the parasitic protozoan cysts, do not grow in environmental waters, and their assay is simple, inexpensive, and relatively quick. Since turbidity is not an adequate surrogate, it was suggested to use *Bacillus* spores as surrogates to assess water treatment plant performance as regards the removal of *Cryptosporidium* or *Giardia* cysts (Nieminski, 2002; Nieminski et al., 2000; Rice et al., 1996), and to assess disinfection efficiency (Chauret et al., 2001; Radziminsk et al., 2002). Aerobic spore-forming bacilli are present in surface and treated waters in much higher concentrations than *Cl. perfringens*. Since they occur in high concentrations in source waters and are also found in finished waters, they allow the calculation of treatment efficiencies up to 5 log. However, no relationship was found between *Cryptosporidium* and *Giardia* cysts removal and spores removal due to methodological problems associated with the detection of parasites (Nieminski et al., 2000).

5.2.8   Heterotrophic Plate Count

The heterotrophic plate count (HPC) represents the aerobic and facultative anaerobic bacteria that derive their carbon and energy from organic compounds. The number of recovered bacteria depends on medium composition, period of incubation (1–7 days), and temperature of incubation (20–35°C) (Reasoner, 1990). A low-nutrient medium, R2A, is used to determine bacterial numbers in water distribution systems. Plate counts in R2A medium are higher than those obtained on plate count agar or sheep blood agar (Fig. 5.3; Carter et al., 2000). This group includes gram-negative bacteria belonging to the following genera: *Pseudomonas, Aeromonas, Klebsiella, Flavobacterium, Enterobacter, Citrobacter, Serratia, Acinetobacter, Proteus, Alcaligenes, Enterobacter,* and *Moraxella.* The HPC microorganisms found in chlorinated distribution water are shown in Table 5.4 (LeChevallier et al., 1980). Some members of this group are opportunistic pathogens (e.g., *Aeromonas, Flavobacterium* (see Chapter 4), but little is known about the effects of high numbers of HPC bacteria on human health. In drinking water, the number of HPC bacteria may vary from <1 to $>10^4$ CFU/mL and they are influenced mainly by temperature, presence of a chlorine residual, and level of assimilable organic matter. Heterotrophic plate count level should not exceed 500 organisms/mL (LeChevallier et al., 1980). The HPC group was found to be the most sensitive indicator for the removal and inactivation of microbial pathogens in reclaimed wastewater.

Heterotrophic plate count is useful to water treatment plant operators with regard to the following (AWWA, 1987; Grabow, 1990; Reasoner, 1990):

1. Assessing the efficiency of various treatment processes, including disinfection, in a water treatment plant;
2. Monitoring the bacteriological quality of the finished water during storage and distribution;
3. Determining bacterial growth on surfaces of materials used in treatment and distribution systems;
4. Determining the potential for regrowth or aftergrowth in treated water in distribution systems.

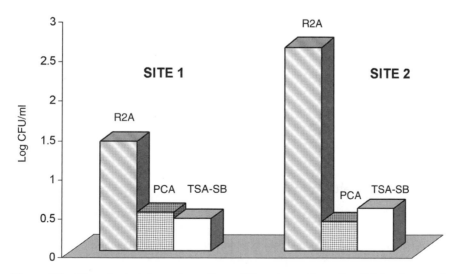

Figure 5.3 Mean bacterial counts on three different growth media: (R2A = low-nutrient agar; PCA = plate count agar; TSA-SB = Tryptic soy agar + 5% sheep's blood). Adapted from Carter et al. (2000).

TABLE 5.4. Identification of HPC Bacteria in Distribution Water and Raw Water[a]

Organism	Distribution Water		Raw Water	
	Total	% of Total	Total	% of Total
Actinomycete	37	10.7	0	0
Arthrobacter spp.	8	2.3	2	1.3
Bacillus spp.	17	4.9	1	0.6
Corynebacterium spp.	31	8.9	3	1.9
Micrococcus luteus	12	3.5	5	3.2
Staphylococcus aureus	2	0.6	0	0
S. epidermidis	18	5.2	8	5.1
Acinetobacter spp.	19	5.5	17	10.8
Alcaligenes spp.	13	3.7	1	0.6
F. meningosepticum	7	2.0	0	0
Group IVe	4	1.2	0	0
Group M5	9	2.6	2	1.3
Group M4	8	2.3	2	1.3
Moraxella spp.	1	0.3	1	0.6
Pseudomonas alcaligenes	24	6.9	4	2.5
P. cepacia	4	1.2	0	0
P. fluorescens	2	0.6	0	0
P. mallei	5	1.4	0	0
P. maltophilia	4	1.2	9	5.7
Pseudomonas spp.	10	2.9	0	0
Aeromonas spp.	33	9.5	25	15.9
Citrobacter freundii	6	1.7	8	5.1
Enterobacter agglomerans	4	1.2	18	11.5
Escherichia coli	1	0.3	0	0
Yersinia enterocolitica	3	0.9	10	6.4
Group IIK biotype I	0	0	1	0.6
Hafnia alvei	0	0	9	5.7
Enterobacter aerogenes	0	0	1	0.6
Enterobacter cloacae	0	0	1	0.6
Klebsiella pneumoniae	0	0	0	0
Serratia liquefaciens	0	0	1	0.6
Unidentified	65	18.7	28	17.8
Total	347	100.0	157	99.7

[a]Adapted from LeChevallier et al. (1980).
HPC, heterotrophic plate count.

5.2.9 Chemical Indicators of Water Quality

5.2.9.1 Fecal Sterols. These include coprostanol, coprosterol, cholesterol, and coprostanone. Some investigators have reported a correlation between fecal sterols and fecal contamination. A good relationship was found between levels of *E. coli* and coprostanol concentrations in tropical regions (R^2 varying from 0.81 and 0.92) but was affected by temperature (Isobe et al., 2004). However, fecal sterols may be degraded after water and wastewater treatment operations and may not be affected by chlorination. Bile acids (e.g., deoxycholic and lithocholic acids) are also potential useful indicators of wastewater pollution.

They are more resistant to degradation than coprostanol and can help in distinguishing between human and animal contamination sources (Elhmmali et al., 1997; 2000).

5.2.9.2 *Free Chlorine Residual.* This is a good indicator for drinking water quality.

5.2.9.3 *Levels of Endotoxins.* Endotoxins are lipopolysaccharides present in the outer membrane of gram-negative bacteria and some cyanobacteria. Some endotoxin-related symptoms are fever, diarrhea, and vomiting. Exposure to endotoxins is of particular concern to dialysis patients. Endotoxin concentration in environmental samples is conveniently measured via the *Limulus* amoebocyte lysate (LAL) assay. This test is based on the reaction of white blood cells of the horseshoe crab with endotoxins. This reaction leads to increased sample turbidity, which is measured with a spectrophotometer. Endotoxin concentration in untreated water generally ranges between 1 and 400 endotoxin units/mL (Anderson et al., 2003). A number of studies have been undertaken to establish a relationship between endotoxin levels in wastewater and drinking water with the levels of total and fecal coliform bacteria (Evans et al., 1978; Haas et al., 1983; Jorgensen et al., 1979). Although a statistically significant association was noted, endotoxin level was not recommended as a surrogate indicator (Haas et al., 1983).

5.2.9.4 *Fluorescent Whitening Agents.* Wastewater from human sources often contains fluorescent whitening agents that are included in detergents and washing powders. They have been used to indicate contamination from septic tanks (Close et al., 1989).

In conclusion, there is no ideal indicator as no-one fulfills all the criteria presented at the beginning of this chapter. In the end, we may have to resort to direct tests for detecting certain important pathogens (e.g., HAV or Norwalk viruses) or parasites (e.g., *Cryptosporidium*). We might also consider the use of an enterovirus indicator (e.g., poliovirus) for enteric viruses, a protozoan indicator (e.g., *Giardia* or *Cryptosporidium*) for cysts of parasitic protozoa, and coliforms for bacterial pathogens.

5.3 DETECTION METHODOLOGY FOR SOME INDICATOR MICROORGANISMS

Several methods are available for the detection of indicator microorganisms in environmental samples, including wastewater (Ericksen and Dufour, 1986; Seidler and Evans, 1983). Some of the procedures have been standardized and are routinely used by government and private laboratories (APHA, 1989). In this section we will focus on the detection of total and fecal coliforms, heterotrophic plate count, bacterial spores, and bacteriophages, stressing only some of the methodological advances made in the past few years.

5.3.1 Standard Methods for the Detection of Total and Fecal Coliforms

As noted above, the total coliform group includes all the aerobic and facultative anaerobic, gram-negative, nonspore-forming, rod-shaped bacteria that ferment lactose with gas production within 48 hours at 35°C. Total coliforms are detected via the most probable numbers (MPN) technique or via the membrane filtration method. These procedures are described in detail in *Standard Methods for the Examination of Water and Wastewater* (APHA, 1999). Most probable numbers generally overestimates coliform numbers in tested samples; the overestimation depends on the number of total coliforms present in the water sample, and on the number of tubes per dilution.

Fecal coliforms are defined as those bacteria that produce gas when grown in EC broth at 44.5°C or blue colonies when grown in m-FC agar at 44.5°C. A 7 h test is also available for detecting this indicator group (Reasoner et al., 1979). An evaluation of this test showed more than 90 percent agreement between this method and the traditional MPN test (Barnes et al., 1989).

Several factors influence the recovery of coliforms, among them the type of growth medium, the diluent, and membrane filter used, the presence of noncoliforms, and the sample turbidity (one should use the MPN approach when turbidity exceeds 5 Nephelometric Turbidity Units; NTU). Heterotrophic plate count bacteria are also able to reduce the number of coliform bacteria, presumably by competing successfully for limiting organic carbon (LeChevallier and McFeters, 1985a; McFeters et al., 1982). Another important factor affecting the detection of coliforms in water and wastewater is the occurrence of injured bacteria in environmental samples. Injury is due to physical (e.g., temperature, light), chemical (e.g., toxic metals and organic toxicants, chlorination), and biological factors. These debilitated bacteria do not grow well in the selective detection media used (presence of selective ingredients such as bile salts and deoxycholate) under temperatures much higher than those encountered in the environment (Bissonnette et al., 1975; 1977; Domek et al., 1984; McFeters et al., 1982; Zaske et al., 1980). In gram-negative bacteria, injury causes damage to the outer membrane, which becomes more permeable to selective ingredients such as deoxycholate. The injured cells can undergo repair, however, when grown on a nonselective nutrient medium. The low recovery of injured coliforms in environmental samples may underestimate the presence of fecal pathogens in the samples. Copper- and chlorine-induced injuries have been studied under in vitro and in vivo conditions to learn more about the pathogenicity of injured pathogens (LeChevallier et al., 1985; Singh et al., 1985; 1986a). Sublethally injured pathogens display a temporary reduction in virulence, but under suitable in vivo conditions, they may regain their pathogenicity. Copper- and chlorine-stressed cells retain their full pathogenic potential even after exposure to the low pH of the stomach of orally infected mice (Singh and McFeters, 1987).

A growth medium, m-T7 agar, was proposed for the recovery of injured microorganisms (LeChevallier et al., 1983). The recovery of fecal coliforms on m-T7 agar is greatly improved when the samples are preincubated at 37°C for 8 h. Recovery of fecal coliforms from wastewater samples was three times higher than with the standard m-FC method (LeChevallier et al., 1984b). Chlorine-stressed *E. coli* displays reduced catalase, leading to its inhibition from the accumulated hydrogen peroxide. Thus, improved detection of chlorine-stressed coliform bacteria can also be accomplished by incorporating catalase or pyruvate, or both in the growth medium to block and degrade hydrogen peroxide (Calabrese and Bissonnette, 1989; 1990; McDonald et al., 1983).

5.3.2 Rapid Methods for Coliform Detection

5.3.2.1 Enzymatic Assays. Enzymatic assays constitute an alternative approach for detecting indicator bacteria, namely total coliforms and *E. coli*, in water and wastewater. These assays, are specific, sensitive, and rapid. In most tests, the detection of total coliforms consists of observing β-galactosidase activity, which is based on the hydrolysis of chromogenic substrates such as ONPG (*o*-nitrophenyl-β-D-galactopyranoside), CPRG (chlorophenol red-β-D-galactopyranoside), X-GAL (5-bromo-4-chloro-3-indolyl-β-D-galactopyranoside) or cyclohexenoesculetin-β-D-galactoside to colored reaction products.

Other substrates used for β-galactosidase assay are fluorogenic compounds, such as 4-methylumbelliferone-β-D-galactoside (MUGA) or fluorescein-di-β-galactopyranoside (FDG) (Berg and Fiksdal, 1988; Bitton et al., 1995; James et al., 1996; Plovins et al., 1994), or chemiluminescent compounds such as phenyl galactose-substituted 1,2-dioxethane derivative, which considerably increases the sensitivity of the assay (Bronstein et al., 1989; van Poucke and Nelis, 1995). The detection of total coliforms by β-galactosidase assay can be improved by incorporating isopropyl-β-D-thiogalactopyranoside (IPTG), a gratuitous inducer of β-galactosidase production, in the growth medium (Diehl, 1991).

Rapid assays for detection of *E. coli* are based on the hydrolysis of a fluorogenic substrates, 4-methylumbelliferone glucuronide (MUG) by β-glucuronidase, an enzyme found in *E. coli*. The endproduct is fluorescent and can be easily detected with a long-wave UV lamp. These tests have been used for the detection of *E. coli* in clinical and environmental samples (Berg and Fiksdal, 1988; Trepeta and Edberg, 1984). β-glucuronidase is an intracellular enzyme found in *E. coli* as well as some *Shigella* species (Feng and Hartman, 1982). A most probable number–fluorogenic assay based on β-glucuronidase activity has been used for the detection of *E. coli* in water and food samples (Feng and Hartman, 1982; Robison, 1984). The assay consists of incubating the sample in lauryltryptose broth amended with 100 mg/L MUG, and observing the development of fluorescence within 24 h incubation at 35°C. This assay can be adapted to membrane filters since β-glucuronidase-positive colonies are fluorescent or have a fluorescent halo when examined under a long-wave UV light. This test can detect the presence of one viable *E. coli* cell within 24 h. A similar miniaturized fluorogenic assay, using MUG as the substrate, was considered for the determination of *E. coli* numbers in marine samples. This assay displayed a 87.3 percent confirmation rate (Hernandez et al., 1990; 1991).

A commercial test, the Autoanalysis Colilert (AC) test, also called the minimal media ONPG-MUG (MMO-MUG) was developed to enumerate simultaneously in 24 h both total coliforms and *E. coli* in environmental samples (Covert et al., 1989; Edberg et al., 1988; 1989; 1990). The test is performed by adding the sample to tubes that contain powdered ingredients consisting mainly of salts and specific enzyme substrates, which also serve as the only carbon source for the target microorganisms. The enzyme substrates are *o*-nitrophenyl-β-D-galactopyranoside (ONPG) for detecting total coliforms, and 4-methylumbelliferyl-β-D-glucuronide (MUG) for specifically detecting *E. coli*. Thus, according to the manufacturer, ONPG and MUG serve as enzyme substrates as well as a food source for the microorganisms. After 24 h incubation, samples positive for total coliforms turn yellow, whereas *E. coli*-positive samples fluoresce under long-wave UV illumination in the dark. It appears that *Escherichia* species other than *E. coli* are not detected by the Colilert test (Rice et al., 1991). Examination of human and animal (cow, horse) fecal samples revealed that 95 percent of *E. coli* isolates were β-glucuronidase positive after 24 h incubation (Rice et al., 1990). Several surveys concerning coliform detection in drinking water have shown that the AC test had a similar sensitivity as the standard multiple tube fermentation method, or the membrane filtration method for drinking water (Edberg et al., 1988; Katamay, 1990). This test yielded numbers of chlorine-stressed *E. coli* in wastewater that were similar to or higher than U.S. EPA-approved EC-MUG tests (Covert et al., 1992; McCarty et al., 1992). A version of Colilert, Colilert-MW, was developed to detect total coliforms and *E. coli* in marine waters (Palmer et al., 1993).

Some problems have been signaled regarding the AC test:

- Some *E. coli* strains are nonfluorogenic; one-third of *E. coli* isolates from fecal samples from human volunteers were found to be nonfluorogenic (Chang et al., 1989).

- A certain percentage of *E. coli* isolates producing virulence factors (e.g., enterotoxigenic or enterohemorrhagic *E. coli*) are not recovered on AC medium (Martins et al., 1992).

- Only 26 percent of *uidA* gene-bearing *E.coli* isolated from treated drinking water expressed β-glucuronidase when incubated in AC medium (Martins et al., 1993).

- Some microalgae and macrophytes can produce β-galactosidase and β-glucuronidase. Laboratory studies showed that their presence at high concentrations in aquatic environments may interfere with the detection of total coliforms and *E. coli*. The significance of this interference under field conditions remains to be investigated (Davies-Colley et al., 1994).

Thus, some investigators do not recommend the implementation of the AC test as a routine procedure for *E. coli* in environmental samples (Lewis and Mak, 1989). Furthermore, the AC test disagreed with the standard membrane filtration fecal coliform test for treated, but not for untreated water samples. This was due to the presence of false-negative results obtained via the AC test (Clark et al., 1991).

ColiPAD™ is another detection test for total coliforms and *E.coli* in environmental samples. It is based on the hydrolysis of chlorophenol red-β-D-galacto-pyranoside (CPRG) and 4-methylumbelliferone glucuronide (MUG) for the rapid detection on an assay pad of total coliforms (purple spots) and *E. coli* (fluorescent spots), respectively. Monitoring of wastewater effluents and lake water showed a good correlation between results obtained by ColiPAD and the standard multiple tube fermentation method (Bitton et al., 1995) (Fig. 5.4).

Early indication of fecal contamination can be obtained by using rapid tests (25 min) based on the activity rate of β-galactosidase and β-glucuronidase. Although the correlation between enzyme activity and culturable fecal coliforms is fairly good, the test sensitivity is relatively low, however (Fiksdal et al., 1994). A rapid (<4 h) enzymatic test was developed to measure *E. coli* numbers directly on membrane filters. The test consists of passing a given sample through a membrane filter, inducing *E. coli* to produce β-glucuronidase directly on the filter surface, fluorescence labeling with the substrate fluorescein-di-β-D-glucuronide, and laser scanning of the membrane surface. This procedure showed a good agreement with reference methods. Although the method is relatively complex, it can be used in emergency situations (van Poucke and Nelis, 2000).

An MUG-based solid medium was proposed for the detection of *E. coli* after only a 7.5 h incubation. Testing this method in water gave a specifity of 96.3 percent (Sarhan and Foster, 1991). Chromogenic substrates such as indoxyl-β-D-glucuronide (IBDG) and 5-bromo-4-chloro-3-indolyl-β-D-glucuronide (X-Gluc), are also useful for the rapid and specific identification of *E. coli* on solid media (Gaudet et al., 1996; Watkins et al., 1988). *Escherichia coli* colonies turn blue following incubation of samples for 22–24 h at 44.5°C. When IBDG was used as a substrate, 99 percent of the blue colonies were also MUG positive (Watkins et al., 1988), and 93 percent of the colonies were confirmed as *E. coli* in a survey of surface waters in Ohio (Haines et al., 1993). The modified mTEC method proposed by EPA uses a medium that contains the chromogen 5-bromo-6-chloro-3-indolyl-β-D-glucuronide. Following sample filtration, the filters are placed on modified mTEC medium and incubated for 2 h at 35°C and then for 20–22 h at 44.5°C. Magenta colonies are counted as *E. coli* (Francy and Darner, 2000). Brenner and colleagues (1996) developed an agar method based on the use of two enzyme substrates, 4 methylumbelliferyl-β-D-galactopyranoside and indoxyl-β-D-glucuronide for the detection of total coliforms and *E. coli*, respectively. Assays based on the presence of the enzyme

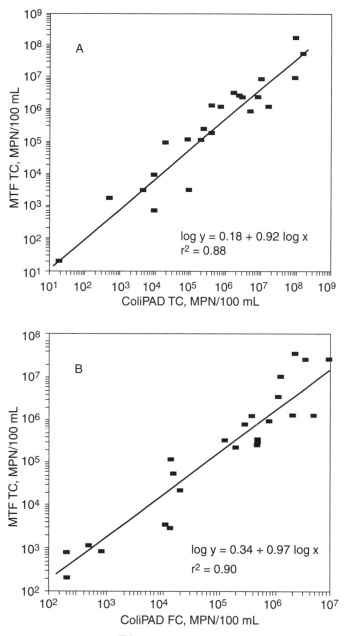

Figure 5.4 Comparison of ColiPAD^{TM} to the standard multiple tube fermentation (MTF) method for the detection of total coliforms (TC) and *Escherichia coli* (EC): (a) total coliforms; (b) fecal coliforms. From Bitton et al. (1995). (With permission from the publisher.)

glutamate decarboxylase are also very selective for *E. coli* (the reagent turns from yellow to blue) (Rice et al., 1993).

5.3.2.2 Other Methods for Coliform Detection
• *Monoclonal Antibodies*

Escherichia coli can be detected, using monoclonal antibodies directed against outer membrane proteins (e.g., *OmpF* protein) or alkaline phosphatase, an enzyme localized in the cell periplasmic space (Joret et al., 1989). Although some monoclonal antibodies are specific for *E. coli* and *Shigella*, some investigators have questioned their specificity and affinity, and further research is needed to demonstrate the application of this tool to routine *E. coli* detection in field samples (Kfir et al., 1993).

• *Polymerase Chain Reaction (PCR)-Gene Probe Detection Method*

In this method, specific genes (e.g., *LacZ*, *lamB* genes) in *E. coli* are amplified via PCR and subsequently detected with a gene probe. With this method one can detect 1–5 cells of *E. coli* per 100 mL of water (Atlas et al., 1989; Bej et al., 1990). Another genetic probe involves the *uidA* gene, which codes for β-glucuronidase in *E. coli* and *Shigella* species and which has been detected in 97.7 percent of *E. coli* isolates from drinking water and raw water sources (Martins et al., 1993). The *uidA* probe, when combined with PCR, can detect as few as 1–2 cells, but is unable to distinguish between *E. coli* and *Shigella* (Bej et al., 1991a; Cleuziat and Robert-Baudouy, 1990). The PCR-gene probe method appears to be more sensitive than the AC method as regards the detection of *E. coli* in environmental samples (Bej et al., 1991b). This is probably due to the presence of approximately 15 percent of β-glucuronidase-negative strains in environmental samples.

Escherichia coli can also be concentrated on a membrane filter and be allowed to grow on a solid growth medium for 5 h. This is followed by detection of microcolonies by in situ hybridization with a peroxidase-labeled nucleic acid (PNA) probe that targets the 16S rRNA of *E. coli*.

5.3.3 Fecal Streptococci/Enterococci

Fecal streptococci/enterococci can be detected using selective growth media in most probable numbers or membrane filtration formats. Enzymatic methods have been developed for the detection of fecal streptococci, including the enterococci group. These indicators can be detected by incorporating fluorogenic (MUD = 4-methylumbelliferone β-D-glucoside) or chromogenic (indoxyl-β-D-glucoside) substrates into selective media (Messer and Dufour, 1998). Miniaturized tests, using microtitration plates and MUD, were successful in the selective detection of this group in fecal, freshwater, wastewater, and marine samples (Hernandez et al., 1990; 1993; Pourcher et al., 1991). The enterococci group can be rapidly detected via fluorogenic or chromogenic enzymatic assays. These tests are based on the detection of the activity of two specific enzymes, pyroglutamyl aminopeptidase and β-D-glucosidase (Manafi and Sommer, 1993). Enterolert was marketed as a 24 h MPN test for the detection of enterococci, and is based, as shown for Colilert, on the use of a methylumbelliferyl substrate (Budnick et al., 1996). It has been argued, however, that Enterolert may give false negative and false positive results. Thus, an alternative use of Enterolert substrate is to incorporate it in a bacteriological agar for the rapid (4 h) confirmation of presumptive enterococci isolated on m-Enterococcus agar (MEA) (Adcock and Saint, 2001).

5.3.4 Heterotrophic Plate Count

The heterotrophic plate count in water and wastewater is defined as the total number of bacteria that can grow after incubation of the sample on plate count agar at 35°C for 48 h. These bacteria may interfere with the detection of coliforms in water samples. Heterotrophic plate count is greatly influenced by the temperature and length of incubation, the growth medium, and the plating method (pour plate vs. spread plate). A growth medium, designated R2A, was developed for use in heterotrophic plate counts.

This medium is recommended with an incubation period of 5–7 days at 28°C (Reasoner and Geldreich, 1985). A significant relationship was found between ATP concentration of surface waters and HPC count using R2A agar (Delahaye et al., 2003).

Heterotrophic plate count should not exceed 500 colonies/mL. Numbers above this limit generally signal a deterioration of water quality in distribution systems.

An original approach is the use of recombinant *lux*+ phages for the detection of indicator bacteria within 1–5 h. Enteric bacteria become bioluminescent following infection with recombinant phage and can be measured with a bioluminometer. This procedure awaits further development (Kodikara et al., 1991).

5.3.5 Bacteriophages

Domestic wastewater harbors a wide range of phage strains that can be detected using a variety of host bacteria. Their levels in raw wastewater are within the range of $10^5 – 10^7$ phage particles/L, but decrease significantly following waste treatment operations (Bitton, 1987). The detection of phage in water or wastewater effluents comprises the following steps (Goyal, 1987).

5.3.5.1 *Phage Concentration.* Phage can be concentrated from large volumes of water by adsorption to negatively or positively charged membrane filters. This step is followed by elution of adsorbed phage from the membrane surface by using glycine at high pH (pH 11.5), beef extract, or casein at pH 9. If necessary, a reconcentration step is included to obtain a low-volume concentrate (Goyal and Gerba, 1983; Goyal et al., 1980; Logan et al., 1980). Phage can also be concentrated from 2 to 4 L samples by Magnetite-organic flocculation (Bitton et al., 1981a). The sample, after addition of casein and magnetite, is flocculated at pH 4.5–4.6. The flocs, with the trapped viruses, are pulled down with a magnet, resolubilized, and assayed for phage.

5.3.5.2 *Decontamination of Concentrates.* Indigenous bacteria interfering with the bacterial phage assay must be inactivated or removed from the concentrate by chloroform extraction, membrane filtration, addition of antibiotics, or use of selective media (e.g., nutrient broth amended with sodium dodecyl sulfate) (Goyal, 1987; Kennedy et al., 1985). Treatment of the concentrate with hydrogen peroxide followed by plating on a media supplemented with crystal violet is also useful for the inactivation of interfering bacteria (Ashgari et al., 1992).

5.3.5.3 *Phage Assay.* Concentrates are assayed for phage, by the double-layer-agar method (Adams, 1959) or single-layer-agar procedures (APHA, 1999; Grabow and Coubrough, 1986; Havelaar and Hogeboom, 1983). The APHA method employs a tetrazolium salt, 2,3,5-triphenyltetrazolium chloride, which aids in visualizing the plaques, which appear as clear zones in a pink bacterial lawn. The phage numbers may also be obtained by a most probable number (MPN) procedure (Kott, 1966). Somatic coliphage can be assayed on an *E. coli* C host, while the assay of male-specific phage requires the use of specific host cells such as *Salmonella typhimurum* strain WG49 or *Escherichia coli* strain *HS[pFamp]R*, but may be complicated by the growth of somatic phages on the host cells. Somatic phages can be suppressed by treating the sample with lipopolysaccharides (LPS) from the host cells (Handzel et al., 1993). A novel approach to phage assay is based on the ability of phages to cause lysis of their bacterial hosts with subsequent release of induced β-galactosidase, which is detected by using an appropriate substrate. In an agar-based format, plaques appear as blue zones when X-gal is used as the substrate

(Ijzerman and Hagedorn, 1992). A liquid colorimetric presence-absence assay is also available (Ijzerman and Hagedorn, 1992). A field evaluation of these phage assays showed they were as sensitive as the APHA method (Ijzerman et al., 1994). Finally, bacterial phages can also be detected by the RT-PCR technique as shown for F^+-specific coliphage in fecally contaminated marine waters (Rose et al., 1997).

The U.S. EPA has published two methods (methods # 1601 and 1602) to detect somatic coliphages (host is *E. coli* CN-13) and F-specific coliphages (host is *E. coli* F-amp) in aquatic environments. Method 1601 includes an overnight enrichment step (water is supplemented with the host, $MgCl_2$, and tryptic soy broth) followed by "spotting" onto a host bacterial lawn. In method 1602, a 100 mL water sample is supplemented with $MgCl_2$, host bacteria, and double-strength molten agar. The mixture is poured onto Petri dishes and the plaques are counted after overnight incubation (U.S. EPA, 2001b, c).

5.3.6 Bacterial Spores

The detection of bacterial spores is relatively simple and consists of pasteurizing the sample ($60°C$ for 20 min), and passing it through a membrane filter that is incubated on nutrient agar supplemented with 0.005 percent bromothymol blue (Francis et al., 2001).

We have reviewed the characteristics and detection methodology for the traditional and less traditional microbial indicators used for assessing contamination of aquatic and other environments by pathogenic microorganisms. Most, if not all, of these indicators are not ideal because some are more sensitive (e.g., vegetative bacterial indicators) or more resistant (e.g., bacterial spores) than viruses or protozoan parasites to environmental stresses and disinfection. There is probably no universal ideal indicator microorganism that is suitable for various environments under variable conditions. A search for the elusive ideal indicator is still under way in many laboratories the world over. The advent of gene probes and PCR technology have given hope for the development of rapid and simple methods for detecting small numbers of bacterial or viral pathogens and protozoan parasites in wastewater, wastewater effluents, biosolids, food, drinking water, and other environmental samples. Furthermore, multiplex PCR can be used to detect a wide range of pathogenic microorganisms and parasites in the same sample. The road is open to direct, rapid, and possibly inexpensive methods for detecting pathogens and parasites in wastewater and other environmental samples.

5.4 WEB RESOURCES

http://www.epa.gov/microbes/ (EPA methods on pathogens, parasites and indicator organisms)
http://bcn.boulder.co.us/basin/data/FECAL/info/FColi.html (fecal coliforms)
http://oh.water.usgs.gov/micro/qcmanual/manual.pdf (methodology for indicators and pathogens from USGS)

5.5 QUESTIONS AND PROBLEMS

1. What are some reasons for using coliphages as indicator microorganisms?
2. Give two enzymes that serve as a basis for the detection of total and fecal coliforms.
3. What are the criteria for an ideal microbial indicator?

4. What are some of the reasons for using bacterial spores as indicators?

5. Why do we use R2A growth medium for determining the heterotrophic plate count in water distribution systems?

6. Give examples of fecal indicators of contamination.

7. What is the basis and the use of the following tests for indicator organisms?
   - Colilert
   - ColiPAD
   - Enterolert

8. What are the different categories of phage presently used as indicator organisms?

9. Discuss in detail U.S. EPA method # 1602 for detecting phage in environmental samples (please go to the original reference).

5.6 FURTHER READING

APHA, 1999. *Standard Methods for the Examination of Water and Wastewater.* 20th Ed. American Public Health Association, Washington, D.C.

Berg, G., Ed. 1978. *Indicators of Viruses in Water and Food*, Ann Arbor Scientific Publications, Ann Arbor, MI.

Ericksen, T.H., and A.P. Dufour. 1986. Methods to identify water pathogens and indicator organisms, pp. 195–214, In: *Waterborne Diseases in the United States*, G.F. Craun, Ed., CRC Press, Boca Raton, Fla.

Fujioka, R.S. 1997. Indicators of marine recreational water quality, pp. 176–183, In: *Manual of Environmental Microbiology*, C.J. Hurst, G.R. Knudsen, M.J. McInerney, L.D. Stetzenbach, and M.V. Walter, Eds., ASM Press, Washington, D.C.

Gerba, C.P. 1987. Phage as indicators of fecal pollution. pp. 197–209, In: *Phage Ecology*, S.M. Goyal, C.P. Gerba and G. Bitton, Eds., Wiley Interscience., New York.

Gerba, C.P. 2000a. Indicator organisms, pp. 491–503, In: *Environmental Microbiology*, R.M. Maier, I.L. Pepper, and C.P. Gerba, Eds., Academic Press, San Diego, CA, 585 pp.

Goyal, S.M., C.P. Gerba, and G. Bitton, Eds. 1987. *Phage Ecology*, Wiley Interscience, New York, 321 pp.

Leclerc, H. S. Edberg, V. Pierzo, and J. M. Delattre. 2000. Bacteriophages as indicators of enteric viruses and public health risk in groundwaters. J. Appl. Microbiol. 88: 5–21.

Toranzos, G.A. and G.A. McFeters. 1997. Detection of indicator microorganisms in environmental freshwaters and drinking waters, pp. 184–194, In: *Manual of Environmental Microbiology*, C.J. Hurst, G.R. Knudsen, M.J. McInerney, L.D. Stetzenbach, and M.V. Walter, Eds. ASM Press, Washington, D.C.

6

WATER AND WASTEWATER DISINFECTION

Wastewater Microbiology, Third Edition, by Gabriel Bitton
Copyright © 2005 John Wiley & Sons, Inc.

6.1   INTRODUCTION

Disinfection is the destruction of microorganisms capable of causing diseases. Disinfection is an essential and final barrier against human exposure to disease-causing pathogenic microorganisms, including viruses, bacteria, and protozoan parasites. Chlorination was initiated at the beginning of the twentieth century to provide an additional safeguard against pathogenic microorganisms. The destruction of pathogens and parasites by disinfection helped considerably in the reduction of waterborne and food-borne diseases. However, in recent years, the finding that chlorination can lead to the formation of byproducts that can be toxic or genotoxic to humans and animals, has led to a quest for safer disinfectants. It was also realized that some pathogens or parasites are indeed quite resistant to disinfectants and that the traditional indicator microorganisms are sometimes not suitable for ensuring safe water.

In addition to their use for pathogen and parasite destruction, some of the disinfectants (e.g., ozone, chlorine dioxide) are also employed for oxidation of organic matter, iron, and manganese and for controlling taste and odor problems and algal growth. This chapter deals with the disinfectants most frequently used by the water and wastewater treatment industries. It also deals with the physical removal of pathogens by membranes in treatment plants.

6.2   FACTORS INFLUENCING DISINFECTION

Several factors control the disinfection of water and wastewater (Lippy, 1986; Sobsey, 1989).

6.2.1   Type of Disinfectant/Biocide

Biocides, including disinfectants, exert cidal or inhibiting effects by interacting with one or more targets in microbial cells (Russel et al., 1997) (Fig. 6.1). The target sites include the peptidoglycan layer, cytoplasmic membrane, outer membrane, structural proteins, thiol groups of enzymes, nucleic acids, viral envelopes, capsids or nucleic acids, and bacterial spore coats (Russel et al., 1997). The efficacy of water and wastewater disinfection depends on the type of chemical used. Some disinfectants (e.g., ozone, chlorine dioxide) are stronger oxidants than others (e.g., chlorine).

TARGET SITES OF BIOCIDES IN MICROBIAL CELLS

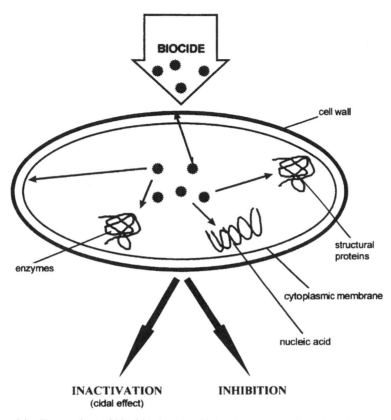

Figure 6.1   Target sites of biocides in microbial cells. Adapted from Russel et al. (1997).

6.2.2   Type of Microorganism

There is a wide variation between the various microbial pathogens as regards their resistance to disinfectants. Spore-forming bacteria are generally more resistant to disinfectants than are vegetative bacteria. Resistance to disinfectants varies also among nonspore-forming bacteria and among strains belonging to the same species (Ward et al., 1984). For example, *Legionella pneumophila* is much more resistant to chlorine than *Escherichia coli* (Kuchta et al., 1983; Muraca et al., 1987). In general, resistance to disinfection goes along the following order:

Nonspore-forming bacteria < Enteric viruses

< Spore-forming bacteria < Protozoan cysts

6.2.3   Disinfectant Concentration and Contact Time

Inactivation of pathogens with disinfectants increases with time and, ideally, should follows first-order kinetics. Inactivation vs. time follows a straight line when data are plotted on a log–log scale.

$$N_t/N_0 = e^{-k_t} \tag{6.1}$$

where N_0 = number of microorganisms at time 0; N_t = number of microorganisms at time t; k = decay constant (time^{-1}); and t = time.

However, field inactivation data actually show a deviation from first-order kinetics (Hoff and Akin, 1986) (Fig. 6.2). Curve C in Figure 6.2 shows deviation from first-order kinetics. The tailing off of the curve results from the survival of a resistant sub-population within a heterogeneous population or from protection of the pathogens by interfering factors (see below). Microbial clumping may explain the "shoulder" of survival curves obtained when exposing microorganisms to chlorine action (Rubin et al., 1983).

Disinfectant effectiveness may be expressed as Ct, C being the disinfectant concentration, and t the time required to inactivate a certain percentage of the population under specific conditions (pH and temperature). The relationship between disinfectant concentration and contact time is given by the Watson's law (Clark et al., 1989):

$$K = C^n t \tag{6.2}$$

where K = constant for a given microorganism exposed to a disinfectant under specific conditions; C = disinfectant concentration (mg/L); t = time required to kill a certain percentage of the population (min); and n = constant, also called the "coefficient of dilution."

When t is plotted against C on a double logarithmic paper, n is the slope of the straight line (see Fig. 6.3; Clark et al., 1989). The value of n determines the importance of the disinfectant concentration or contact time in microorganism inactivation. If $n < 1$, disinfection is more affected by contact time than by disinfectant concentration. If $n > 1$, the disinfectant level is the predominant factor controlling disinfection (Rubin et al., 1983). However, the value of n is often close to unity.

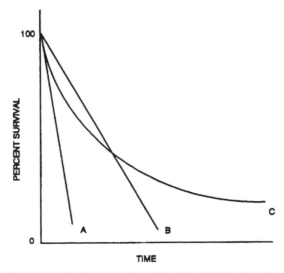

Figure 6.2 Inactivation curves of microorganisms following disinfection: A, sensitive homogeneous population; B, more resistant homogeneous population; C, heterogeneous population or partially protected by aggregation. From Holf and Akin (1986). (With permission of the publisher.)

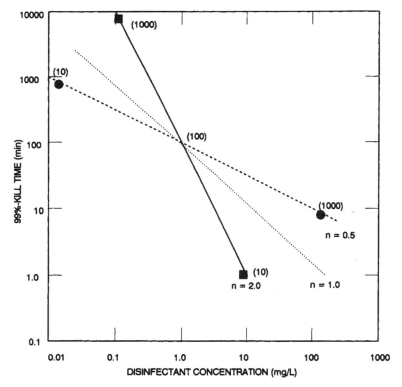

Figure 6.3 Effect of n values on Ct at various disinfectant concentrations (Ct values given in parentheses). From Clark et al. (1989). (With permission of the publisher.)

Determination of Ct can also take into account the temperature and pH value of the suspending medium. For example, an equation was developed to predict *Giardia lamblia* cysts inactivation after chlorine treatment (Clark et al., 1989; Hibler et al., 1987).

$$Ct = 0.9847 \, C^{0.1758} \, \text{pH}^{2.7519} \, T^{-0.1467} \tag{6.3}$$

where C = chlorine concentration ($C \leq 4.23$ mg/L); t = time to inactivate 99.99 percent of the cysts; pH range is 6–8; T = temperature range, 0.5–5.0°C.

Ct values for a range of pathogenic microorganisms are shown in Table 6.1 (Hoff, 1986; Hoff and Akin, 1986). The order of resistance to chlorine is in the following order:

Protozoan cysts > Viruses > Nonspore-forming bacteria

Another way to express the efficiency of a given disinfectant is the lethality coefficient given by the equation (Morris, 1975):

$$\lambda = 4.6/Ct_{99} \tag{6.4}$$

where 4.6 = natural log of 100; C = residual concentration of disinfectant (mg/L); and t_{99} = contact time (minutes) for 99 percent inactivation of microorganisms.

The values of λ for destroying 99 percent of a range of microorganisms by ozone in 10 min at 10–15°C vary from 5 for *Entamoeba histolytica* to 500 for *E. coli* (Chang, 1982).

TABLE 6.1.   Microbial Inactivation by Chlorine: *Ct* Values
(Temperature = 5°C; pH = 6.0)[a]

Microorganism	Chlorine Concentration (mg/L)	Inactivation Time (min)	*Ct*
Escherichia coli	0.1	0.4	0.04
Poliovirus I	1.0	1.7	1.7
Entamoeba histolytica cysts	5.0	18	90
Giardia lamblia cysts	1.0	50	50
	2.0	40	80
	2.5	100	250
Giardia muris cysts	2.5	100	250

[a]Adapted from Hoff and Akin, 1986.

6.2.4   Effect of pH

As regards disinfection with chlorine, pH controls the amount of HOCl (hypochlorous acid) and OCl$^-$ (hypochorite ion) in solution (see Section 6.3.1.). HOCl is 80 times more effective than OCl$^-$ for *E. coli*. For disinfection with chlorine, *Ct* increases with pH (Lippy, 1986). Conversely, bacterial, viral, and protozoan cyst inactivation by chlorine dioxide is generally more efficient at higher pH values (Berman and Hoff, 1984; Chen et al., 1985; Sobsey, 1989). The effect of pH on microbial inactivation by chloramine is not well established because of conflicting results.

6.2.5   Temperature

Pathogen and parasite inactivation increases (i.e., *Ct* decreases) as temperature increases.

6.2.6   Chemical and Physical Interference with Disinfection

Chemical compounds interfering with disinfection are inorganic and organic nitrogenous compounds, iron, manganese, and hydrogen sulfide. Dissolved organic compounds also exert a chlorine demand; their presence results in reduced disinfection efficiency (see Section 6.3.7).

Turbidity in water is composed of inorganic (e.g., silt, clay, iron oxides) and organic matter as well as microbial cells. It is measured by determining light scattering by particulates present in water. It interferes with the detection of coliforms in water (Geldreich et al., 1978; LeChevallier et al., 1981), but it can also reduce the disinfection efficiency of chlorine and other disinfectants. The need to remove turbidity is based on the fact that particle-associated microorganisms are more resistant to disinfection than freely suspended microorganisms. The total organic carbon (TOC) associated with turbidity exerts a chlorine demand and thus interferes with the maintenance of a chlorine residual in water. Microorganisms associated with fecal material, cell debris, or wastewater solids are also protected from disinfection (Berman et al., 1988; Foster et al., 1980; Harakeh, 1985; Hejkal et al., 1979; Hoff, 1978; Narkis et al., 1995). These findings are particularly important for communities that treat their water solely by chlorination. Figure 6.4 illustrates the protective effect of turbidity toward coliform bacteria (LeChevallier et al., 1981). It was also shown that the protective effect of particulates in water and wastewater depends on the nature and the size of the particles. Hence, cell-associated poliovirus is protected

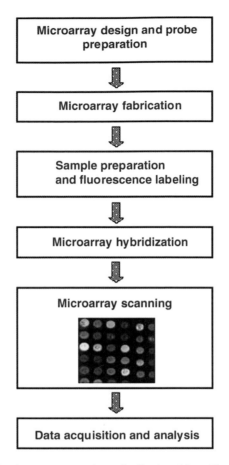

Figure 1.15 Steps involved in microarray preparation and utilization. Adapted from Zhou and Thompson (2002).

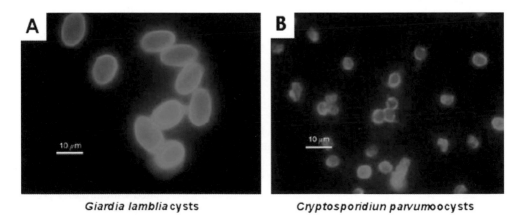

Giardia lamblia cysts

Cryptosporidiun parvum oocysts

Figure 4.13 Immunofluorescence image of *Giardia lamblia* cysts and *Cryptosporidium parvum* oocysts. (Courtesy of H. D. A. Lindquist, U.S. EPA.)

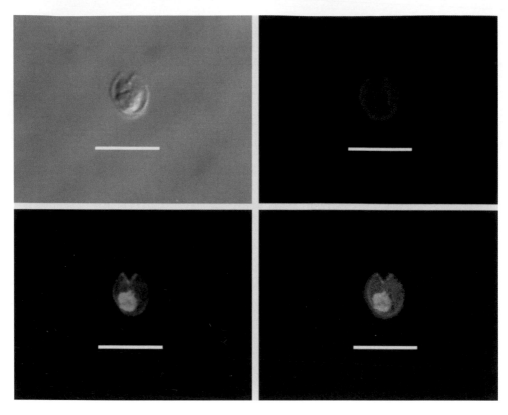

Figure 4.17 Oocysts of *Cryptosporidium parvum* viewed with a combination of immunological and genetic tools. (Courtesy of Christopher Robinson, Oak Ridge Institute of Science and Education, and H. D. Alan Lindquist, U.S. EPA.) The gray frame on the top left shows a *Cryptosporidium parvum* oocyst with some apparent damage to the oocyst wall, and some material remaining inside. The red image on the right is of the same oocyst as detected by an anti-*C. parvum* red fluorescent antibody (*Fab*) staining procedure. Only the surface features of the oocyst are evident with this type of staining. The image on the bottom left is of the same oocyst, hybridized to a *C. parvum* specific anti-ribosomal RNA fluorescent in situ hybridization (FISH) probe labeled with a green fluorescent compound. There is some nonspecific staining with this probe, but the majority of the probe appear to be hybridized within one region, indicating the presence of ribosomal RNA in this structure. The image on the bottom right is an overlay of Fab and FISH staining of the oocyst. The scale in all images represents 5μm.

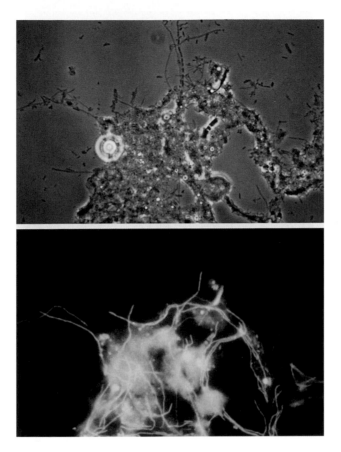

Figure 9.7 *Haliscomenobacter hydrossis* in activated sludge, as shown with a fluorescein-labeled oligonucleotide probe: (a) phase contrast micrograph; (b) epifluorescence micrograph. From Wagner et al. (1994).

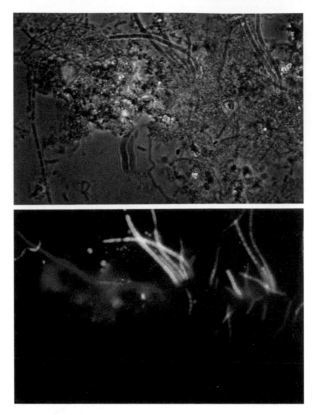

Figure 9.8 *Thiothrix* sp. and type 021N in activated sludge, as shown with labeled oligonucleotide probes: (a) phase contrast micrograph; (b) epifluorescence micrograph (*Thiothrix* is green, and type 021N is red). From Wagner et al. (1994).

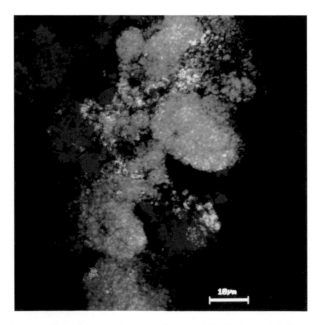

Figure 10.7 Confocal laser-scanning microscopy in conjunction with 16S rRNA-targeted oligo-nucleotide probes showing clusters of nitrifiers (red cells) in close proximity to more dense clusters of denitrifiers (green cells) in a 20 m thick section of biofilm. From Schramm et al. (1996). (Courtesy of R. Amann and with permission of the publisher.)

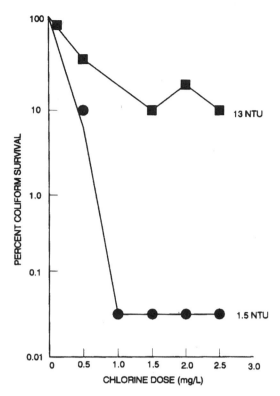

Figure 6.4 Effect of turbidity on coliform persistence in chlorinated water. Adapted from LeChevallier et al. (1981).

from chlorine inactivation, whereas bentonite or aluminum phosphate do not offer such protection to the virus (Hoff, 1978). Viruses and bacterial indicators are not protected from ozone inactivation by bentonite (Boyce et al., 1981). A study of solid-associated viruses under field conditions showed that they are more resistant to chlorination than are "free" viruses. Reducing turbidity to less than 0.1 NTU could be a preventive measure for counteracting the protective effect of particulate matter during disinfection.

6.2.7 Protective Effect of Macroinvertebrates

Macroinvertebrates may enter and colonize water distribution systems (Levy et al., 1984; Small and Greaves, 1968). The public health significance of the presence of the organisms in water distribution systems has been addressed. Nematodes may ingest viral and bacterial pathogens and thus protect these microorganisms from chlorine action (Chang et al., 1960). *Hyalella azteca*, an amphipod, protects *E. coli* and *Enterobacter cloacae* from chlorination. In the presence of 1 mg/L chlorine the decay rate of macroinvertebrate-associated *E. cloacae* was $k = 0.022\,h^{-1}$, whereas the decay rate of unassociated *E. cloacae* was $k = 0.93\,h^{-1}$ (Levy et al., 1984) (Fig. 6.5). Enteropathogenic bacteria are also protected from chlorine action when ingested by protozoa (King et al., 1988).

6.2.8 Other Factors

Several studies have shown that laboratory-grown pathogenic and indicator bacteria are more sensitive to disinfectants than those that occur in the natural aquatic environment.

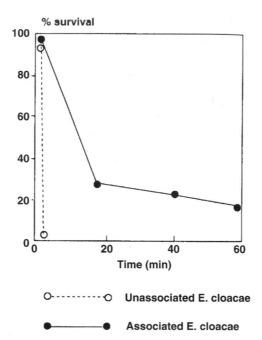

Figure 6.5 Effect of chlorination on *Enterobacter cloacae* associated with a macroinvertebrate. Adapted from Levy et al. (1984).

Hence, naturally occurring *Flavobacterium* spp. was 200 times more resistant to chlorine than when subcultured under laboratory conditions (Wolfe and Olson, 1985). *Klebsiella pneumoniae* was more resistant to chloramine when grown under low nutrient conditions (Stewart and Olson, 1992a). The increased resistance to chloramine is due to several physiological factors: increased cell aggregation and extracellular polymer production, alteration of membrane lipids, and decreased oxidation of sulfhydryl groups (Stewart and Olson, 1992b). The greater resistance of "environmental" bacterial strains to nutrient limitation and deleterious agents such as disinfectants may also be due to the synthesis of stress proteins, the role of which is not well understood (Matin and Harakeh, 1990). This phenomenon raises questions regarding the usefulness of laboratory disinfection data to predict pathogen inactivation under field conditions (Hoff and Akin, 1986).

Prior exposure may also increase microbial resistance to disinfectants. Repeated exposure of microorganisms to chlorine results in the selection of bacteria and viruses that are resistant to disinfection (Bates et al., 1977; Leyval et al., 1984; Ridgeway and Olson, 1982). Clumping or aggregation of pathogenic microorganisms generally reduces the disinfectant efficiency. Bacterial cells, viral particles, or protozoan cysts inside the aggregates are well protected from disinfectant action (Chen et al., 1985; Sharp et al., 1976).

6.3 CHLORINE

6.3.1 Chlorine Chemistry

Chlorine gas (Cl_2) introduced in water, hydrolyzes according to the following equation:

$$\underset{\substack{\text{Chlorine}\\\text{gas}}}{Cl_2} + H_2O \rightleftharpoons \underset{\substack{\text{Hypochlorous}\\\text{acid}}}{HOCl} + H^+ + Cl^- \qquad (6.5)$$

Hypochlorous acid dissociates in water according to the following:

$$HOCl \rightleftharpoons H^+ + OCl^- \qquad (6.6)$$

$\underset{\text{Hypochlorous acid}}{} \qquad \underset{\text{Hypochlorite ion}}{}$

Figure 6.6 shows that the proportion of HOCl and OCl$^-$ depends on the pH of the water. Chlorine, as HOCl or OCl$^-$, is defined as free available chlorine. HOCl combines with ammonia and organic nitrogen compounds to form chloramines that are combined available chlorine (see Section 6.4).

6.3.2 Inactivation of Microorganisms by Chlorine

Of the three chlorine compounds (HOCl, OCl$^-$, and NH$_2$Cl), hypochlorous acid is the most effective for the inactivation of microorganisms in water and wastewater. The presence of interfering substances in wastewater reduces the disinfection efficacy of chlorine, and relatively high concentrations of chlorine (20–40 ppm) are required for adequate reduction of viruses. In wastewater effluents, no free chlorine species are available after a few seconds of contact.

Chlorine, specifically HOCl, is generally quite efficient in inactivating pathogenic and indicator bacteria. Water treatment with ≤1 mg/L for about 30 min is generally efficient in significantly reducing bacterial numbers. For example, *Campylobacter jejuni* displays more than

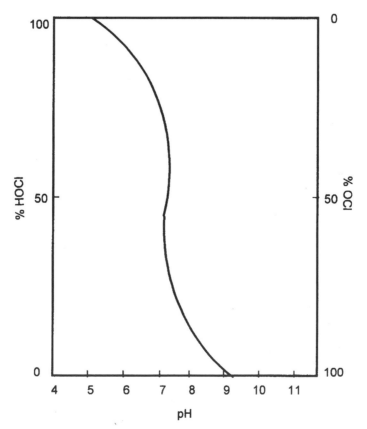

Figure 6.6 Distribution of HOCl and OCl$^-$ in water as a function of pH. From Bitton (1980). (Courtesy of John Wiley & Sons, Inc.)

99 percent inactivation in the presence of 0.1 mg/L free chlorine (contact time = 5 min) (Blaser et al., 1986). Although there is wide variation in the resistance of enteric viruses to chlorine, these pathogens are generally more resistant to this disinfectant than are vegetative bacteria. That explains why viruses are frequently detected in chlorinated secondarily treated effluents. Chloramines are much less efficient than free residual chlorine (about 50 times less efficient) as regards viral inactivation. Protozoan cysts (e.g., *Giardia lamblia*, *Entamoeba histolytica*, *Naegleria gruberi*) are more resistant to chlorine than are both bacteria and viruses. In the presence of HOCl at pH 6, the *Ct* for *E. coli* is 0.04, compared with a *Ct* value of 1.05 for poliovirus type 1 and 80 for *G. lamblia* (Logsdon and Hoff, 1986).

Cryptosporidium oocysts are extremely resistant to disinfection but can be inactivated by ammonia at concentrations found in natural environments (Jenkins et al., 1998). A chlorine or monochloramine concentration of 80 mg/L is necessary to cause 90 percent inactivation after 90 min contact time (Korich et al., 1990). This parasite is not completely inactivated in a 3 percent solution of sodium hypochlorite (Campbell et al., 1982), and the oocysts can remain viable for 3–4 months in 2.5 percent potassium dichromate solution (Current, 1988). The *Ct* value for *Cryptosporidium* is in the 1000s, as shown in Figure 6.7 (Driedger et al., 2000). This parasite would thus be extremely resistant to disinfection as carried out in water and wastewater treatment plants (Korich et al., 1989).

6.3.3 Cell Injury by Chlorine

Physical (heat, freezing, sunlight) and chemical agents (chlorine, sublethal levels of heavy metals) can cause injury to bacterial cells (LeChevallier and McFeters, 1985b). Injury

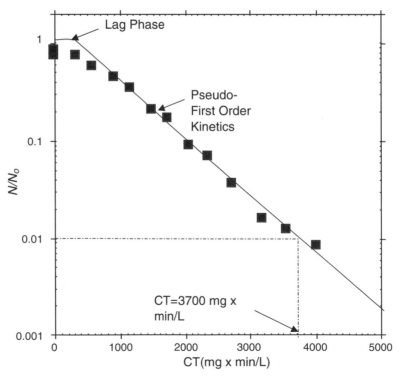

Figure 6.7   Primary inactivation of *Cryptosporidium parvum* oocysts with free chlorine at pH 6 and 20°C. From Driedger et al. (2000).

caused by environmental agents can lead to cell size reduction, damage to cell barriers, as well as altered cell physiology and virulence (Singh and McFeters, 1990).

Chlorine and copper appear to cause significant injury to coliform bacteria in drinking water (Camper and McFeters, 1979; Domek et al., 1984). The injured bacteria fail to grow in the presence of selective agents (e.g., sodium lauryl sulfate, sodium deoxycholate) traditionally incorporated in growth media designed for the isolation of indicator and pathogenic bacteria (Bissonnette et al., 1975; Busta, 1976). However, chlorine- and copper-injured pathogens (e.g., enterotoxigenic *E. coli*) retain their potential for enterotoxin production (Singh and McFeters, 1986) and are able to recover in the small intestine of animals, retaining their pathogenicity. This finding suggests that cells injured by chlorine treatment still have a potential health significance (Singh et al., 1986a). Injury by chlorine can affect a wide variety of pathogens, including enterotoxigenic *E. coli*, *Salmonella typhimurium*, *Yersinia enterocolitica*, and *Shigella* spp. The extent of injury by chlorine depends on the type of microorganism involved.

6.3.4 Potentiation of the Cidal Effect of Free Chlorine

The killing action of free chlorine can be potentiated by adding salts such as KCl, NaCl, or CsCl (Berg et al., 1990; Haas et al., 1986; Sharp et al., 1980). After chlorination, viruses are more effectively inactivated in drinking water (e.g., Cincinnati drinking water) than in purified water (Berg et al., 1989) (Fig. 6.8). The mechanism of the potentiating effect of salts is not fully understood.

The disinfecting ability of chlorine can also be enhanced in the presence of heavy metals. The inactivation rate of pathogenic (e.g., *Legionella pneumophila*) bacteria and viruses (e.g., poliovirus) is increased when free chlorine is amended with electrolytically generated copper and silver (400 and 40 mg/L, respectively) (Landeen et al., 1989; Yahya and Gerba, 1990) (Fig. 6.9). This phenomenon was also demonstrated for indicator

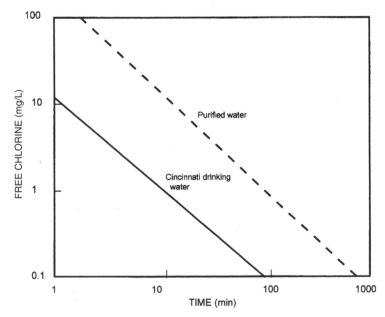

Figure 6.8 Relative rates of inactivation of 99.99 percent of poliovirus 1 at 5°C by free chlorine at pH 9.0 in purified water and in Cincinnati drinking water. From Berg et al. (1989). (Courtesy of the American Society of Microbiology.)

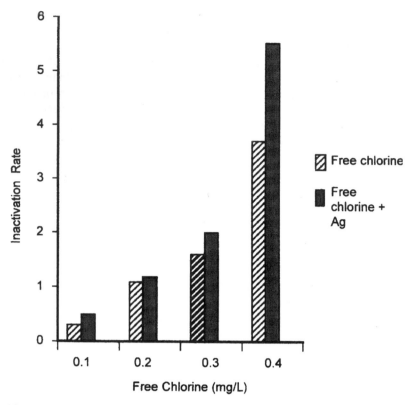

Figure 6.9 Inactivation of *Legionella pneumophila* by exposure to electrolytically generated copper and silver and/or various concentrations of free chlorine. From Landeen et al. (1989). (Courtesy of the American Society of Microbiology.)

bacteria and *Naegleria fowleri* ameba in water (Cassells et al., 1995; Yahya et al., 1989). This process does not, however, completely eliminate enteric viruses of medical importance such as hepatitis A virus (HAV) or human rotavirus (Abad et al., 1994; Bosch et al., 1993). Adaptation of *Salmonella typhimurium* to acid also enhances its sensitivity to hypochlorous acid (Leyer and Johnson, 1997), but the practical application of this finding to water and wastewater treatment remains unknown.

Inactivation of *Cryptosporidium* oocysts with free chlorine can also be enhanced by ozone pretreatment. A 4- to 6-fold increase in free chlorine efficacy was obtained with ozone pretreatment. The sequential inactivation using ozone followed by free chlorine was shown to increase the inactivation of *Salmonella*, bacterial phage, poliovirus type 1, and parasites such *Cryptosporidium* and *Giardia* (Finch et al., 1998; Kott et al., 1980).

6.3.5 Mechanism of Action of Chlorine

Chlorine causes considerable damage to bacterial cells:

1. Disruption of cell permeability. Free chlorine disrupts the integrity of the bacterial cell membrane, thus leading to loss of cell permeability and to the disruption of other cell functions (Berg et al., 1986; Haas and Engelbrecht, 1980; Leyer and Johnson, 1997; Venkobachar et al., 1977). Exposure to chlorine leads to a leakage of proteins, RNA and DNA (Venkobachar et al., 1977). Cell death is the

result of release of TOC and UV-absorbing materials, decrease in potassium uptake, and reduction in protein and DNA synthesis (Haas and Engelbrecht, 1980). Permeability disruption was also implicated as the cause of damage of chlorine to bacterial spores (Kulikovsky et al., 1975).

2. Damage to nucleic acids and enzymes. Chlorine also damages bacterial nucleic acids (Hoyana et al., 1973; Shih and Lederberg, 1974) as well as enzymes (e.g. catalase, dehydrogenases). One of the consequences of reduced catalase activity is inhibition by the accumulated hydrogen peroxide (Calabrese and Bissonnette, 1990).

3. Other effects. Hypochlorous acid oxidizes sulfhydryl groups, damage iron-sulfur centers, disrupts nutrient transport, inhibits cell respiration, and impairs the ability of cells to maintain an adequate adenylate energy charge to remain viable (Barrette et al., 1989; Leyer and Johnson, 1997).

As for viruses, the mode of action of chlorine may depend on the type of virus. Nucleic acid damage is the primary mode of inactivation for bacterial phage f2 (Dennis et al., 1979; Olivieri et al., 1980) or poliovirus type 1 (Nuanualsuwan and Cliver, 2003; O'Brien and Newman, 1979). The protein coat appears to be the target site for other types of viruses (e.g., rotaviruses) (Vaughn and Novotny, 1991).

6.3.6 Toxicology of Chlorine and Chlorine Byproducts

In general, the risks from chemicals in water are not as well defined as those from pathogenic microorganisms and parasites. This is because of the lack of data on disinfection byproducts (DBP). The toxicology of chlorine and byproducts is of obvious importance since it is estimated that 79 percent of the U.S. population is exposed to chlorine (U.S. EPA, 1989f). There is evidence of an association between chlorination of drinking water and increased risk of bladder, kidney, and colorectal cancers. This association is stronger for consumers who have been exposed to chlorinated water for more than 15 years (Craun, 1988; Jolley et al., 1985; Larson, 1989). Epidemiological studies have also linked disinfection byproducts with reproductive and developmental effects but were considered inadequate for showing evidence of this relationship (Reif et al., 1996).

Disinfection byproducts (DBP) are formed following the reaction of chlorine with precursors such as natural organic matter (chiefly humic and fulvic acids) and microorganisms such as algal cells (especially blue-green algae and diatoms) and their extracellular products (Plummer and Edzwald, 2001). There is a good relationship between total THM formation potential (TTHMFP) and total organic carbon (TOC) in water (Fig. 6.10; LeChevallier et al., 1992). Disinfection byproducts include trihalomethanes (THM) such as chloroform ($CHCl_3$), bromodichloromethane ($CHBrCl_2$), dibromochloromethane ($CHBr_2Cl$), and bromoform ($CHBr_3$), haloacetic acids (monochloroacetic acid, monobromoacetic acid, dichloroacetic acid, dibromoacetic acid, trichloroacetic acid) and halocetonitriles. The DBPs are suspected mutagens/carcinogens and teratogens. Chlorohydroxyfuranones, especially 3-chloro-4-(dichloromethyl)-5-hydroxy-2(5H)-furanone (designated as MX) have also been identified as chlorination byproducts. The MX is a potent mutagen and a suspected carcinogen, but its effect on human health remains to be elucidated (Huixian et al., 2002; Meier et al., 1987). There is also the possibility of an association of water chlorination with increased risk of cardiovascular diseases (Craun, 1988). Swimmers can also become exposed to trihalomethanes via dermal absorption or inhalation. The estimated risk due to swimming in a pool contaminated with DBP such as chloroform is 2.5 excess cancer cases per 100,000 exposed bathers as compared to 0.6 cases per 100,000 from

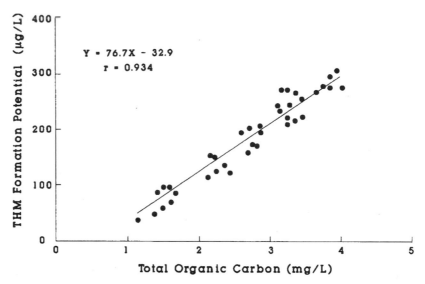

Figure 6.10 Relationship between total THM formation potential (TTHMFP) and total organic carbon (TOC). From LeChevallier et al. (1992). (Courtesy of the American Water Works Association.)

ingestion of contaminated drinking water (D. Ambroise, T. Francais, M. Joyeux and M. Morlot, unpublished data). These findings led the U.S. EPA to establish a maximum contaminant level (MCL) of 80 µg/L for THM and 60 µg/L for haloacetic acids in finished drinking water. Because water treatment with chloramines does not produce any trihalomethanes, consumers drinking chloraminated water appear to experience less bladder cancers than do those consuming chlorinated water (Zierler et al., 1987).

There are three main approaches for reducing or controlling DBP in drinking water (Wolfe et al., 1984):

1. Removal or reduction of DBP precursors (e.g., natural organic matter, including humic substances, algae and their extracellular products) before disinfection: Organic carbon concentrations can be reduced by enhanced coagulation, granular activated carbon or membrane filtration, but this practice can also lead to an increase in brominated THMs (Black et al., 1996). A combination of coagulation, ozonation, and biofiltration can effectively reduce trihalomethane and haloacetic acid formation potential (Chaiket et al., 2002). Natural organic matter (NOM), the main precursor of DBPs, can also be removed by iron oxide coated filtration media (Chang et al., 1997).

2. Removal of DBPs in the water treatment plant. In homes, THMs can be volatilized upon boiling tap water. Laboratory tests have shown that the level of haloacetic acids can be reduced via biodegradation involving aquatic bacteria, and adsorption to granular activated carbon. An increase in the number of halogen atoms decreases biodegradation but enhances adsorption (Zhou and Xie, 2002).

3. Use of alternative disinfectants that do not generate THMs (e.g., chloramination, ozone, or UV irradiation).

6.3.7 Chloramination

Chloramination is the disinfection of water with chloramine in lieu of free chlorine.

6.3.7.1 *Chloramine Chemistry.*

In aqueous solutions, HOCl reacts with ammonia and forms inorganic chloramines according to the following equations (Snoeyink and Jenkins, 1980):

$$NH_3 + HOCl \longrightarrow \underset{\text{monochloramine}}{NH_2Cl} + H_2O \tag{6.7}$$

$$NH_2Cl + HOCl \longrightarrow \underset{\text{dichloramine}}{NHCl_2} + H_2O \tag{6.8}$$

$$NHCl_2 + HOCl \longrightarrow \underset{\text{trichloramine}}{NCl_3} + H_2O \tag{6.9}$$

The proportion of the three forms of chloramines greatly depends on the pH of the water. Monochloramine is predominant at pH > 8.5. Monochloramine and dichloramine coexist at a pH value of 4.5–8.5 and trichloramine occurs at pH < 4.5. Figure 6.11 (Wolfe et al., 1984) shows the effect of pH on the distribution of chloramines. Monochloramine is the predominant chloramine formed at the pH range usually encountered in water and wastewater treatment plants (pH = 6–9). In water treatment plants the formation of monochloramine is desirable because dichloramine and trichloramines impart an unpleasant taste to the water.

The mixing of chlorine and ammonia produces a chlorine dose-residual curve displayed in Figure 6.12 (Kreft et al., 1985). In the absence of chlorine demand, a chlorine dose of 1 mg/L produces a chlorine residual of 1 mg/L. However, in the presence of ammonia, the chlorine residual reaches a peak (formation of mostly monochloramine; chlorine to ammonia-N ratio = 4:1 to 6:1) and then decreases to a minimum called the breakpoint.

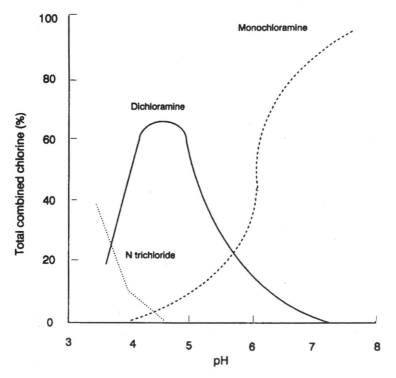

Figure 6.11 Distribution of chloramine species with pH. From Wolfe et al. (1984). (Courtesy of the American Water Works Association.)

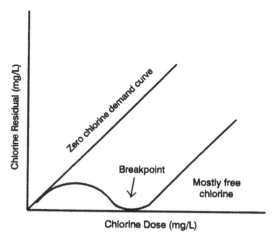

Figure 6.12 Dose–demand curve for chlorine–ammonia reaction after 1 hour at pH 7–8. From Kreft et al. (1985). (Courtesy of the American Water Works Association.)

The breakpoint, where the chloramine is oxidized to nitrogen gas, occurs when the chlorine to ammonia-N ratio is between 7.5 : 1 and 11 : 1.

$$2\,NH_3 + 3\,HOCl \longrightarrow N_2 + 3\,H_2O + 3\,HCl \tag{6.10}$$

Addition of chlorine beyond *breakpoint chlorination* ensures the existence of a free available residual.

6.3.7.2 Biocidal Effect of Inorganic Chloramines.

During the 1940s, Butterfield and collaborators showed that free chlorine inactivated enteric bacteria much faster than did inorganic chloramines. Furthermore, the bactericidal activity of chloramines increased with temperature and hydrogen ion concentration (Wattie and Butterfield, 1944). Similar observations were made with regard to viruses (Kelly and Sanderson, 1958) and protozoan cysts (Stringer and Kruse, 1970). Mycobacteria, some enteric viruses (e.g., hepatitis A virus, rotaviruses), and protozoan cysts are, however, quite resistant to chloramines (Pelletier and DuMoulin, 1988; Rubin, 1988; Sobsey et al., 1988). Thus, it was recommended that drinking water should not be disinfected with only chloramines unless the source water is of good quality (Sobsey, 1989). The inactivation of pathogens and parasites with chloramines is summarized in Table 6.2 (Sobsey, 1989).

6.3.7.3 Advantages and Disadvantages of Chloramination.

The Denver Water Department has been successfully using chloramination for more than 70 years for water treatment (Dice, 1985), and many plants with high coliform counts in the distribution system have switched from free chlorine to chloramines. Approximately 30 percent of U.S. drinking water plants use chloramines for disinfection (AWWA, 2000; Betts, 2002). Chloramines offer several benefits, including lower THMs such as trihalomethanes and haloacetic acids, lower coliform regrowth in distribution pipes, and improved maintenance of disinfectant residual (Norton and LeChevallier, 1997). Although less effective disinfectants than free chlorine, they appear to be more effective in controlling biofilm microorganisms because they interact poorly with capsular polysaccharides. Hospitals using free chlorine as residual disinfectant are 10 times more likely to experience outbreaks of Legionnaires' disease than those using monochloramine (Kool et al., 2000). Thus, it has been suggested to use free chlorine as a primary disinfectant in water distribution systems and to convert the residual to monochloramine if biofilm control is the goal (LeChevallier et al., 1990).

TABLE 6.2.   Inactivation of Health-Related Microorganisms in Water by Chloramines: Ct Values[a]

Microbe	Water	Temp. (°C)	pH	Est Ct
Bacteria				
Escherichia coli	BDF	5	9.0	113
Coliforms	Tap + 1%	20	6.0	8.5
Salmonella typhimurium, Shigella sonnei	Sewage			
Mycobacterium fortuitum	BDF	20	7.0	2,667
M. avium	BDF	17	7.0	ND
M. intracellulare	BDF	17	7.0	ND
Viruses				
Polio I	BDF	5	9.0	1,420
Polio I	Primary effluent	25	7.5	345
Hepatitis A	BDF	5	8.0	592
Coliphage MS2	BDF	5	8.0	2,100
Rotavirus SA11				
Dispersed	BDF	5	8.0	4,034
Cell-associated	BDF	5	8.0	6,124
Protozoan cysts				
Giardia muris	BDF	3	6.5–7.5	430–580
Giardia muris	BDF	5	7.0	1,400

BDF, buffered-demand free water; ND, no data available.
[a]Adapted from Sobsey (1989).

However, chloramination may promote the growth of nitrifying bacteria, which convert ammonia to nitrite and nitrate (Wilczak et al., 1996). Ammonia oxidizing bacteria (AOB) produce nitrite, which may exert a chloramine demand, thus contributing to the deterioration of water quality in drinking water distribution systems. Using molecular techniques, Regan et al. (2002) studied the diversity of ammonia oxidizing bacteria (AOB) and nitrite oxidizing bacteria (NOB) in a pilot-scale chloraminated drinking water treatment system. The AOB found in this system were *Nitrosomonas* (*N. oligotropha* and *N. urea*) along some *Nitrosospira*-like organisms. The NOB community was comprised of *Nitrospira* and to a lesser extent *Nitrobacter*. Thus, the presence of *Nitrospira*-like NOBs may help lower the nitrite concentration and, hence, reduce the chloramine demand in the distribution system (see Section 6.3.7.4, Toxicology of Chloramines, for further details).

6.3.7.4 *Toxicology of Chloramines.*   As regards the toxicology of chloramines, dichloramine and trichloramine have offensive odors and have a threshold odor concentration of 0.8 and 0.02 mg/L, respectively (Kreft et al.,1985). Chloramines cause hemolytic anemia in kidney hemodialysis patients (Eaton et al., 1973), but no effect was observed in animals or humans ingesting chloramines. Chloramines are mutagenic to bacteria and initiate skin papillomas in mice. Monochloramine can also react with dimethylamine and other precursors to form *N*-nitrosodimethylamine (NDMA), which has been classified as a potential human carcinogen by the U.S. EPA (Choi and Valentine, 2002). Two NDMA appears to be a byproduct of the chloramination of water and wastewater (Najm and Trussell, 2001). It has been detected in drinking water wells at levels ranging from 70 to 3000 ng/L. In bench-scale tests it was found that up to 98 percent of NDMA can be removed in drinking water treated with pulsed UV at an applied dose of 11.2 kW · h/1000 gal (Liang et al., 2003).

In the aquatic environment, chloramines are toxic to fish and invertebrates. At $20°C$, the 96 h LC_{50} of monochloramine ranges from 0.5 to 1.8 mg/L. One of the mechanisms of toxicity to fish is the irreversible oxidation of hemoglobin to methemoglobin, which has a lower oxygen-carrying capacity (Groethe and Eaton, 1975; Wolfe et al., 1984).

6.4 CHLORINE DIOXIDE

6.4.1 Chemistry of Chlorine Dioxide Byproducts

Chlorine dioxide (ClO_2) use as a disinfectant in water treatment is becoming widespread because it does not appear to form trihalomethanes, nor does it react with ammonia to form chloramines. Because it cannot be stored in compressed form in tanks, chlorine dioxide must be produced at the site; it is generated from chlorine gas reacting with sodium chlorite:

$$2\,NaClO_2 + Cl_2 \longrightarrow 2\,ClO_2 + 2\,NaCl \tag{6.11}$$

The ClO_2 does not hydrolyze in water but exists as a dissolved gas. In alkaline solutions, it forms chlorite and chlorate:

$$2\,ClO_2 + 2\,OH^- \longrightarrow ClO_2^- + ClO_3^- + H_2O \tag{6.12}$$

Chlorite is the predominant species formed in water treatment plants. The U.S. EPA maximum contaminant level for chlorite is 1 mg/L. To reduce THM formation, ClO_2 can be used as a preoxidant and a primary disinfectant and is followed by the addition of chlorine to maintain a residual (Aieta and Berg, 1986).

6.4.2 Effect of Chlorine Dioxide on Microorganisms

Chlorine dioxide is a fast and effective microbial disinfectant and is equal or superior to chlorine in inactivating bacteria and viruses in water and wastewater (Aieta and Berg, 1986; Bitton, 1980a; Longley et al., 1980; Narkis and Kott, 1992). It is also effective in the destruction of cysts of pathogenic protozoa such as *Naegleria gruberi* (Chen et al., 1985). As shown for human and simian rotaviruses, the virucidal efficiency of chlorine dioxide increases as the pH is increased from 4.5 to 9.0 (Chen and Vaughn, 1990). Bacteriophage f2 inactivation is also much higher at pH 9.0 than at pH 5.0 (Noss and Olivieri, 1985) (Fig. 6.13). Inactivation of health-related microorganisms (Ct values) by chlorine dioxide is summarized in Table 6.3. The $Ct_{99.9\%}$ for *Mycobacterium avium* ranged from 2 to 11, depending on the strain under consideration (Taylor et al., 2000). As regards *C. parvum*, Ct values ranging from 75 to 1000 mg.min/L to obtain a $2-\log_{10}$ inactivation were recently reported for oocysts obtained from different suppliers (the suspending medium was deionized water at $21°C$ and pH 8) (Fig. 6.14; Chauret et al., 2001). *Giardia lamblia* was more sensitive to ClO_2 and displayed a $3-\log_{10}$ inactivation at a Ct value of 15. As regards *Bacillus subtilis* spores, a $2.0\log_{10}$ inactivation required Ct values of 100 and 25 at pH 6 and 8, respectively. It was suggested that bacterial spores could be used as indicators for *C. parvum* inactivation (Radziminski et al., 2002).

6.4.3 Mode of Action of Chlorine Dioxide

The primary mode of action of chlorine dioxide involves the disruption of protein synthesis in bacterial cells (Bernarde et al., 1967; Russel et al., 1997). The disruption

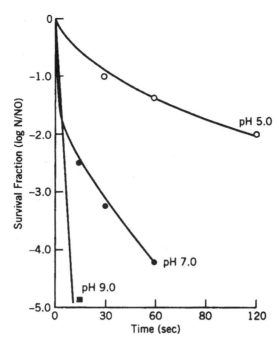

Figure 6.13 Inactivation of bacteriophage f2 by chlorine dioxide as a function of pH. Adapted from Noss and Olivieri (1985).

of the outer membrane of gram-negative bacteria has also been observed (Berg et al., 1986).

An examination of the mechanisms of viral inactivation by chlorine dioxide has revealed contradicting results (Vaughn and Novotny, 1991). Work with f2 bacterial phage has shown that the protein coat is the primary target of the lethal action of chlorine dioxide (Noss et al., 1986; Olivieri et al., 1980). The loss of attachment of this phage to its host cell paralleled virus inactivation (Noss et al., 1986). Specifically, degradation of tyrosine residues within the protein coat appears to be the primary site of action of chlorine dioxide in f2 phage (Noss et al., 1986). The disruption of the viral protein coat was suggested in regard to other viruses such as poliovirus (Brigano et al., 1979). Some other investigators have concluded that the primary site of action of chlorine dioxide is the viral genome (Alvarez and O'Brien, 1982; Taylor and Butler, 1982).

6.4.4 Toxicology of Chlorine Dioxide

Chlorine dioxide interferes with the thyroid function and produces high serum cholesterol in animals fed a diet low in calcium and rich in lipids (Condie, 1986). Chlorine dioxide has two inorganic byproducts, chlorite (ClO_2^-) and chlorate (ClO_3^-). Chlorite is of greater health concern than chlorate and both may combine with hemoglobin to cause methemoglobinemia. The U.S. EPA limits the chlorite concentration to 1 mg/L. Reduced sulfur compounds and ferrous ions can be used to remove chlorite from water and wastewater. The reduction of chlorite to chloride ion by ferrous ions is shown in Eq. 6.13 (Katz and Narkis, 2001).

$$4\,Fe^{2+} + ClO_2^- + 10\,H_2O \longrightarrow 4\,Fe(OH)_{3(S)} + Cl^- + 8\,H^+ \qquad (6.13)$$

TABLE 6.3. Inactivation of Health-Related Microorganisms in Water by Chlorine Dioxide[a]: Ct values

Microbe	Medium	ClO_2 Residual (mg/L)	Temp. (°C)	pH	Temp. (min)	% Reduction	Ct
Bacteria							
E. coli	BDF[b]	0.3–0.8	5	7.0	0.6–1.8	99	0.48
Fecal coliforms	Effl.	1.9	?	?	10	99.94	ND
Fecal streptococci	Effl.	1.9	?	?	10	99.5	ND
C. perfringens	Effl.	1.9	?	?	10	0	ND
L. pneumophila	BDF	0.35–0.5	23	?	15	99.9–99.99	ND
K. pneumonia	BDF	0.12	23	?	15	99.3–99.7	ND
B. subtilis spores			6			99	100
B. subtilis spores			8			99	25
Viruses							
Coliphage f2	BDF	1.5	5	7.2	2	99.994	ND
Polio 1	BDF	0.4–14.3	5	7.0	0.2–11.2	99	0.2–6.7
Polio 1	Effl.	1.9	?	?	10	99.4	ND
Rota SA11:							
Dispersed	BDF	0.5–1.0	5	6.0	0.2–0.6	99	0.2–0.3
Cell-assoc.	BDF	0.45–1.0	5	6.0	1.2–4.8	99	1.0–2.1
Hepatitis A	BDF	0.14–0.23	5	6.0	8.4	99	1.7
Hepatitis A	BDF	0.2	5	9.0	<0.33	>99.9	<0.04
Coliphage MS2	BDF	0.15	5	6.0	34	99	5.1
Coliphage MS2	BDF	0.15	5	—	<0.33	>99.95	<0.03
Protozoan cysts							
N. gruberi	BDF	0.8–1.95	5	7.0	7.8–19.9	99	15.5
N. gruberi	BDF	0.46–1.0	25	5.0	5.4–13.2	99	6.35
N. gruberi	BDF	0.42–1.1	25	9.0	2.5–6.7	99	2.91
G. muris	BDF	0.1–5.55	5	7.0	1.3–168	99	10.7
G. muris	BDF	0.26–1.2	25	5.0	4.0–24	99	5.8
G. muris	BDF	0.21–1.12	25	7.0	3.3–28.8	99	5.1
G. muris	BDF	0.15–0.81	25	9.0	2.1–19.2	99	2.7
G. lamblia	Deionized water		21	8.0		99.9	15
C. parvum	Deionized water		21	8.0		99	75–1000

[a]Adapted from Sobsey (1989), Chauret et al. (2001) and Radziminski et al. (2002).
[b]BDF = buffered-demand free water.

Chlorination of urban wastewater with chlorine dioxide was shown to lead to the formation of mutagenic compounds, as demonstrated by the Ames mutagenicity test (Monarca et al., 2000).

6.5 OZONE

6.5.1 Introduction

Ozone is produced by passing dried air between electrodes separated by an air gap and a dielectric and by applying an alternating current with the voltage ranging from 8000 to

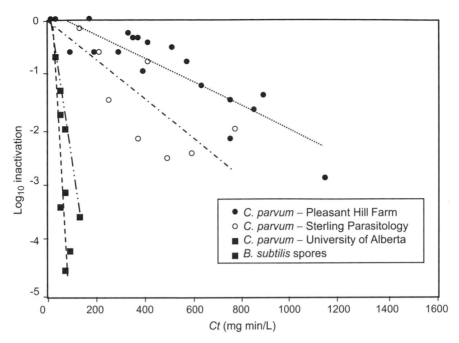

Figure 6.14 Chlorine dioxide inactivation of *Cryptosporidium parvum* oocysts from three different sources (comparison with *Bacillus subtilis* spores). Adapted from Chauret et al. (2001).

20,000 V (Fig. 6.15). Ozone was first introduced as a strong oxidizing agent for the removal of taste, color, and odors. The first water treatment plant using ozone started operations in 1906 in Nice, France. This oxidant is now used as a primary disinfectant to inactivate pathogenic microorganisms and for the oxidation of iron and manganese, taste- and odor-causing compounds, color, refractory organic compounds, and THM precursors. Preozonation also lowers THM formation potential, and promotes particle coagulation during water treatment (Chang and Singer, 1991); it is also used in combination with activated carbon treatment. Ozone breaks down complex compounds into simpler ones, some of

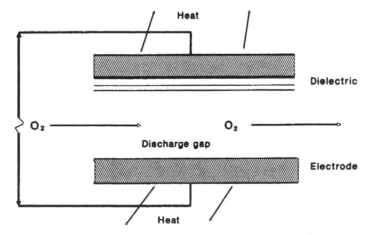

Figure 6.15 Basic ozonator configuration. From McCarthy and Smith (1974). (Courtesy of the American Water Works Association.)

which may serve as substrates for microbial growth in water distribution systems (Bancroft et al., 1984). In the United States, several wastewater treatment plants are using ozone mostly as an oxidant and a few of them use it as a disinfectant. Ozone can be applied at various points of a conventional water treatment plant, depending on the type of use (AWWA, 1985b; Rice, 1989). Its effectiveness as a disinfectant is not controlled by pH and it does not interact with ammonia (Driedger et al., 2001).

Ozonation is more expensive than chlorination or even UV disinfection. Power use is the most expensive operation cost item. Because ozone does not leave any residual in water, ozone treatment is sometimes combined with postchlorination. A synergistic effect was observed in sequential disinfection schemes that use ozone as a primary disinfectant, and free chlorine or monochloramine as a secondary disinfectant. Ozone pretreatment increases the inactivation rate and removes the characteristic lag phase associated with monochloramine or free chlorine disinfection (Rennecker et al., 2000). With the sequential addition of ozone and free chlorine, the enhancing effect decreases with an increase in pH (Driedger et al., 2000).

6.5.2   Effect of Ozone on Indicator and Pathogenic Microorganisms

Ozone is a much more powerful oxidant than chlorine. The standard oxidation potential of ozone is 2.07 eV as compared with a potential of 1.36 eV for chlorine or 0.95 eV for chlorine dioxide (Acher et al., 1997). The threshold ozone concentration above which bacterial inactivation is very rapid is only 0.1 mg/L. The Ct values for 99 percent inactivation are very low and range between 0.001 and 0.2 for *E. coli* and from 0.04 to 0.42 for enteric viruses (Engelbrecht, 1983; Hall and Sobsey, 1993).

Ozone appears to be more effective against human and simian rotaviruses than chlorine, monochloramine, or chlorine dioxide (Chen and Vaughn, 1990; Korich et al., 1990). The ozone concentration required to inactivate 99.9 percent of enteroviruses in water (25°C, pH 7.0) in 10 min varied between 0.05 and 0.6 mg/L (Engelbrecht, 1983). The $Ct_{99.9\%}$ for *Mycobacterium avium* exposed to ozone ranged from 0.10 to 0.17, depending on the strain (Taylor et al., 2000). The resistance of a number of microorganisms to ozone was as follows: *Mycobacterium fortuitum* > poliovirus type 1 > *Candida parapsilosis* > *E. coli* > *Salmonella typhimurium* (Fig. 6.16; Farooq and Akhlaque, 1983). Suspended solids (e.g., clays, sludge solids) significantly reduce viral inactivation by ozone. The protective effect of solids is illustrated in Figure 6.17 (Kaneko, 1989).

We have seen that *Cryptosporidium* oocysts are very resistant to chlorination. Ozone, at a concentration of 1.1 mg/L, totally inactivates *Cryptosporidium parvum* oocysts in 6 min at levels of 10^4 oocysts/mL (Peeters et al., 1989). The extent of inactivation of *C. parvum* with ozone is displayed in Figure 6.18 (Driedger et al., 2000). Cysts of *Giardia lamblia* and *G. muris* are also effectively inactivated by ozone (Fig. 6.19; Wickramanayake et al., 1985). At pH 7, at 5°C, more than 2–log reduction of the viability of *G. lamblia* cysts is achieved in a few minutes with <0.5 mg ozone/L.

Temperature greatly affects the effectiveness of ozonation. The resistance of *G. lamblia* cysts to ozone increased when the temperature was lowered from 25 to 5°C (Wickramanayake et al., 1985). A similar phenomenon was observed for *Cryptosporidium* oocysts (Joret et al., 1992; Rennecker et al., 2000; Fig. 6.20) and for *B. subtilis* spores (Driedger et al., 2001).

The PEROXONE process, using a mixture of ozone and hydrogen peroxide, has been investigated for controlling taste and odors, disinfection byproducts, and microbial pathogens. It appears that the inactivation efficiency of PEROXONE ($H_2O_2 : O_3 \leq 0.3$) is

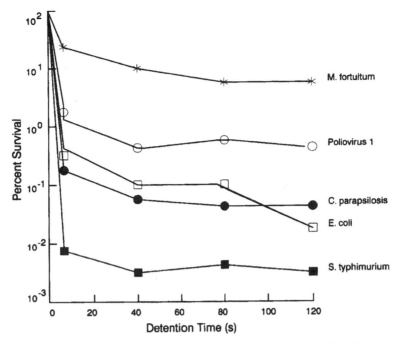

Figure 6.16 Inactivation of various microorganisms by ozone in an activated sludge effluent. Adapted from Farooq and Akhlaque (1983).

similar to that of ozone alone. However, PEROXONE is superior to ozone for the oxidation of taste- and odor-causing compounds (Ferguson et al., 1990).

6.5.3 Mechanisms of Ozone Action

In aqueous media, ozone produces free radicals that inactivate microorganisms. Ozone affects the permeability, enzymatic activity, and DNA of bacterial cells (Hamelin et al., 1978; Ishizaki et al., 1987), and guanine or thymine, or both, appear to be the most susceptible targets of ozone (Ishizaki et al., 1984). As shown in poliovirus, ozone inactivates viruses by damaging the nucleic acid core (Roy et al., 1981). The protein coat is also affected (De Mik and De Groot, 1977; Riesser et al., 1977; Sproul et al., 1982), but the damage may be small and may not significantly affect the adsorption of poliovirus to its the host cell (VP4, a capsid polypeptide responsible for attachment to the host cell, was not affected by ozone). For rotaviruses, ozone alters both the capsid and RNA core (Chen et al., 1987).

Ozone does not appear to cause damage to bacterial spore DNA, but rather to the spore's inner membrane (Young and Setlow, 2004).

6.5.4 Toxicology of Ozonation Byproducts

We have discussed the formation of mutagenic/carcinogenic compounds after chlorination of water and wastewater. Less is known regarding ozonation byproducts. Byproducts of concern are bromate (BrO_3^-), an animal carcinogen produced by the reaction of bromide ion with molecular ozone and hydroxyl radicals as well as aldehydes and keto

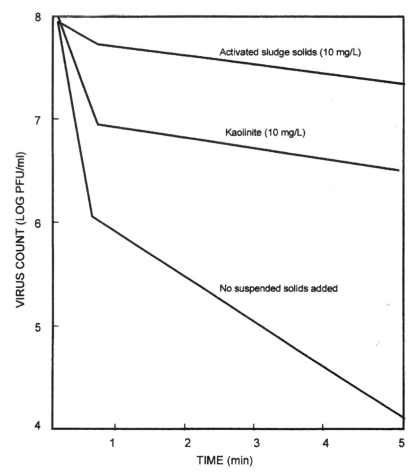

Figure 6.17 Effect of suspended solids on poliovirus inactivation by ozone. Adapted from Kaneko (1989).

acids resulting from the reaction of ozone with natural organic matter (von Gunten, 2003). The current limit of bromate in the United States and European Union (EU) is 10 μg/L. The health significance of ozonation byproducts remains largely unknown (U.S. EPA, 1989e). Preozonation of algal suspensions with 1–3 mg/L ozone increased the formation of disinfection byproducts (trihalomethanes and haloacetic acid) (Plummer and Edzwald, 2001). Recent studies show, using the Ames test for mutagenicity, that the ozonation of urban wastewater produces mutagenic compounds (Monarca et al., 2000). These compounds can be removed by granular activated carbon (GAC) treatment (Bourbigot et al., 1986; van Hoof et al., 1985; Matsuda et al., 1992; Rice, 1989). Ozonation may also increase effluent toxicity, as shown with the *Ceriodaphnia dubia* toxicity assay, but the alteration appears to be site-specific (Blatchley et al., 1997).

Another approach to oxidant disinfection is the use of a mixture of oxidants (free chlorine, chlorine dioxide, ozone, hydrogen peroxide, and other unknown oxidants) generated on site by the electrolysis of a sodium chloride solution. In laboratory studies, the mixed oxidant solution was shown to be more effective than free chlorine and inactivated more than 99.9 percent of *Cryptosporidium parvum* oocysts in 4 h (Venczel et al., 1997).

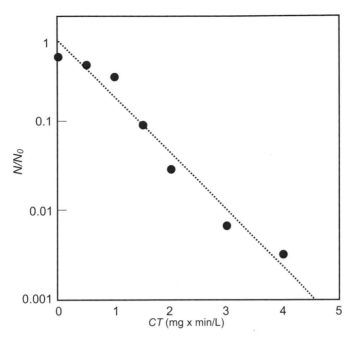

Figure 6.18 Primary inactivation of *Cryptosporidium parvum* with ozone at pH 7 and 20°C. From Driedger et al. (2000).

6.6 ULTRAVIOLET LIGHT

6.6.1 Introduction

Ultraviolet (UV) light is the portion of the electromagnetic spectrum that is located between visible light and X-rays. It is broken down into UV-A (315–400 nm), UV-B (280–315 nm), UV-C (200–280 nm), and vacuum UV (100–200 nm). Germicidal activity is due to radiation at wavelengths ranging from 245 to 285 nm (U.S. EPA, 1999).

Ultraviolet biocidal properties were discovered at the beginning of the twentieth century. Ultraviolet disinfection was then considered for treating water in Henderson, Kentucky, but was abandoned in favor of chlorination. Owing to technological improvements, this disinfection alternative is now regaining popularity, particularly in Europe (Wolfe, 1990).

There are two categories of UV lamps (Mofidi et al., 2001):

- Continuous-wave emission lamps, which include low-pressure (10^{-3} to 10^{-2} torr) mercury lamps with a peak emission at 253.7 nm, and medium-pressure (10^2 to 10^4 torr) mercury lamps, which emit light at 185 nm to more than 300 nm.
- Pulsed emission lamps, which produce pulsed light (30 pulses/s) over a wide spectrum (polychromatic) ranging from 185 nm to more than 800 nm.

Ultraviolet disinfection systems use mercury lamps enclosed in quartz tubes. The tubes are immersed in flowing water in a tank and allow passage of UV radiation at the germicidal wavelength of 253.7 nm. However, transmission of UV by quartz decreases upon continuous use. Therefore, the quartz lamps must be regularly cleaned, using mechanical,

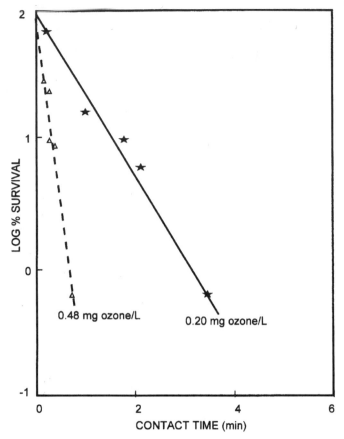

Figure 6.19 Inactivation of *Giardia lamblia* cysts by ozone. Adapted from Wickramanayake et al. (1985).

chemical, and ultrasonic cleaning methods. Teflon® has been proposed as an alternative to quartz, but its transmission of UV radiation is lower that in quartz systems.

6.6.2 Mechanism of UV Damage

Ultraviolet radiation damages microbial DNA at a wavelength of approximately 260 nm. It causes thymine and cytosine dimerization, which blocks DNA replication and effectively inactivates microorganisms. Thymine dimers are more prevalent due to the higher absorbance of thymine in the germicidal range. Studies with viruses have demonstrated that the initial site of UV damage is the viral genome, followed by structural damage to the virus coat (Nuanualsuwan and Cliver, 2003; Rodgers et al., 1985). The damage caused by UV radiation is repaired by bacteria, using two basic repair systems: excision repair which occurs in the dark, and photoreactivation which requires light (see Section 6.6.4 for more details on photoreactivation).

6.6.3 Inactivation of Pathogens and Parasites by UV Radiation

Microbial inactivation is proportional to the UV dose, which is expressed in $\mu W \cdot s/cm^2$. The inactivation of microorganisms by UV radiation can be represented by the following

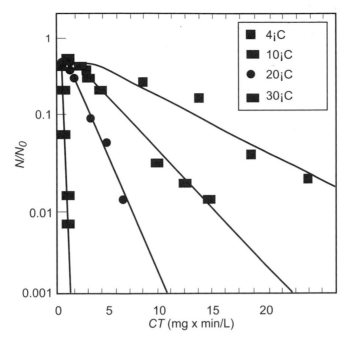

Figure 6.20 Effect of temperature on inactivation of *Cryptosporidium parvum* with ozone at pH 7. From Renneker et al. (2000).

equation (Luckiesh and Holladay, 1944; Severin, 1980).

$$N_t/N_0 = e^{-KP_d t} \tag{6.14}$$

where N_0 = initial number of microorganisms (no. per mL); N_t = number of surviving microorganisms (no. per mL); K = inactivation rate constant (time^{-1}); P_d = UV light intensity reaching the organisms (μW/cm^2); and t = exposure time in seconds.

The above equation is subject to several assumptions, one of which is that the logarithm of the survivor fraction should be linear with regard to time (Severin, 1980). In environmental samples, however, the inactivation kinetics is not linear with time, which may be due to resistant organisms among the natural population and to differences in flow patterns.

The efficacy of UV disinfection depends on the applied dose and the type of microorganism under consideration. The UV dose is given by the following equation (U.S. EPA, 1999):

$$D = I.t \tag{6.15}$$

where D = UV dose in mW $\cdot$ s/cm^2 (1 mW $\cdot$ s = 1 mJ); I = intensity in mW/cm^2; and t = exposure time in seconds.

In general, the resistance of microorganisms to UV follows the same pattern as with chemical disinfectants and is as follows (Chang et al., 1985) (Fig. 6.21):

Protozoan cysts > Bacterial spores > Viruses > Vegetative bacteria

This trend is supported by Table 6.4 (Wolfe, 1990), which shows the approximate UV dose in μW $\cdot$ s/cm^2 (equivalent μJ/cm^2) for 90 percent inactivation of microorganisms.

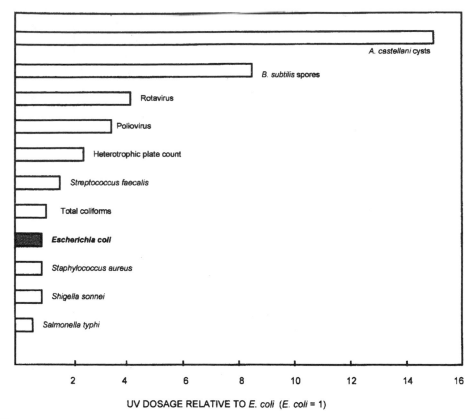

Figure 6.21 Relative UV dose required to inactivate 99.9 percent of various microorganisms compared to that for *Escherichia coli*. From Chang et al. (1985). (Courtesy of the American Society of Microbiology.)

The UV doses required for 4–log reduction of bacterial phages were $750 \, J/m^2$, $100 \, J/m^2$, and $290 \, J/m^2$ for MS2, PhiX174, and B40-8 (phage infecting *Bacteroides fragilis*), respectively (Sommer et al., 2001). It appears that phage MS2 can be considered as a conservative indicator of virus or parasite inactivation by UV irradiation. A 3–log inactivation of enteroviruses (echovirus 1, echovirus 2, poliovirus 1, coxsackievirus B3 and B5) occurs at UV doses of $20–27 \, mW \cdot s/cm^2$. However, adenovirus type 2 displayed the highest resistance to UV, requiring a dose of $119 \, mW \cdot s/cm^2$ for a 3–log inactivation (Gerba et al., 2002). This was confirmed in a pilot-scale study reporting adenovirus persistence in UV-treated wastewater tertiary effluents. A dose of approximately $170 \, mW \cdot s/cm^2$ was necessary for a 4–log inactivation of adenoviruses 2 and 15 (Thompson et al., 2003). Hepatitis A virus requires a UV dose of $3.7 \, mW \cdot s/cm^2$ for 1–log inactivation (Wolfe, 1990), but necessitates $20 \, mW \cdot s/cm^2$ to achieve a 3–log reduction (Battigelli et al., 1993). A 4–log reduction of poliovirus requires a dose ranging between 20 and $35 \, mJ/cm^2$ (Battigelli et al., 1993; Harris et al., 1987; Meng and Gerba, 1996; Sommer et al., 1989).

As regards protozoan parasites, a $3.9–\log_{10}$ inactivation was obtained for *Cryptosporidium* at a dose of $19 \, mJ/cm^2$, using a medium pressure UV lamp and animal infectivity assays (Bukhari et al., 1999). A UV dose of $11 \, mJ/cm^2$ led to a 2–log inactivation of *C. parvum* (genotype 2 Iowa isolate) using pulsed UV light or medium-pressure UV lamps and cell culture infectivity assays (Mofidi et al., 2001).

**TABLE 6.4. Approximate Dosage for 90%
Inactivation of Selected Microorganisms by UV**

Microorganism	Dosage ($\mu W \cdot s/cm^2$)
Bacteria	
Escherichia coli	3,000
Salmonella typhi	2,500
Pseudomonas aeruginosa	5,500
Salmonella enteritis	4,000
Shigella dysenteriae	2,200
Shigella paradysenteriae	1,700
Shigella flexneri	1,700
Shigella sonnei	3,000
Staphylococcus aureus	4,500
Legionella pneumonphila	380
Vibrio cholerae	3,400
Viruses	
Poliovirus I	5,000
Coliphage	3,600
Hepatitis A Virus	3,700
Rotavirus SA 11	8,000
Protozoan cysts	
Giardia muris	82,000
Cryptosporidium parvum	80,000
Giardia lamblia	63,000
Acanthamoeba castellanii	35,000

[a]Adapted from Wolfe (1990), and Rice and Hoff (1981).

Cryptosporidium parvum oocysts were found to be sensitive to both low-pressure and medium-pressure UV radiation, displaying a $3-\log_{10}$ inactivation at a UV dose of $3\,mJ/cm^2$ (Clancy et al., 2000; Shin et al., 2001). Similar results were obtained for *Giardia lamblia* cysts (Mofidi et al., 2002). In another study with *Giardia lamblia*, it was found that 10 and $20\,mJ/cm^2$ were required to obtain $2-\log_{10}$ and $3-\log_{10}$ inactivation, respectively (Campbell and Wallis, 2002). A comparison of the sensitivity of five strains of *C. parvum* oocysts (Iowa, Moredun, Texas A&M strain, Maine, and Glasgow) showed that a UV dose of $10\,mJ/cm^2$ was sufficient to cause at least $4-\log_{10}$ inactivation in all strains (Clancy et al., 2004).

Encephalitozoon intestinalis (microsporidia) spores, at a UV dose of $3\,mJ/cm^2$, showed an inactivation of $1.6-2.0\,\log_{10}$ whereas at $6\,mJ/cm^2$ more than $3.6-\log_{10}$ of microsporidia were inactivated (Huffman et al., 2002).

6.6.4 Variables Affecting UV Action

Many variables (e.g., suspended particles, organic matter, color) in wastewater effluents affect UV transmission in water and, thus, the dose necessary for disinfection (Harris et al., 1987; Severin, 1980). Several organic and inorganic chemicals (e.g., humic substances, phenolic compounds, lignin sulfonates from the pulp and paper mill industry, iron, manganese) interfere with UV transmission in water.

Microorganisms are partially protected from the harmful UV radiation when embedded within particulate matter (Oliver and Cosgrove, 1977; Harris et al., 1987a, b;

Qualls et al., 1983; 1985). Coliform inactivation tests with UV showed that the inactivation efficiency decreased when particles with size of 7 μm or larger were present in tertiary effluents (Jolis et al., 2001). Suspended solids only partially protect microorganisms from the lethal effect of UV radiation. This is because suspended particles in water and wastewater absorb only a portion of the UV light (Bitton et al., 1972). Wastewater solids absorb 75% of the light, and scattering accounts for the remaining 25% (Qualls et al., 1983; 1985). Most clay minerals do not exert much protection to microorganisms because they scatter most of the UV light. It is known that their protective effect depends on their specific absorption and scattering of UV radiation, and decrease with an increase of light scattering (Bitton et al., 1972). Thus, flocculation followed by filtration of effluents through sand or anthracite beds to remove interfering substances should improve UV disinfection efficiency (Dizer et al., 1993).

Microorganisms have DNA repair processes, including *photoreactivation* and *excision repair*. *Photoreactivation* (i.e., photo repair) may occur after exposure of UV-damaged microbial cells to visible light at wavelengths of 300–500 nm (Jagger, 1958). DNA damage can also be repaired in the dark by the cell excision repair system (i.e., dark repair). Several enzymes are involved in excision repair. The UV-damaged DNA segment is excised and replaced by a newly synthesized segment. The potential for UV-damaged bacteria to undergo repair after UV irradiation has been demonstrated (Carson and Petersen, 1975; Mechsner et al., 1990; Zimmer et al., 2003). Under laboratory conditions, it was shown that *E. coli* exposed to UV radiation at doses ranging from 3 to 10 mJ/cm^2 underwent photorepair when exposed to a low-pressure UV source, but no repair was detected following exposure to a medium-pressure UV source (Zimmer and Slawson, 2002). Figure 6.22 shows the photoreactivation of *E. coli* following exposure to UV at 10 mJ/cm^2 (Zimmer et al., 2003). No photo or dark repairs were observed in *Cryptosporidium parvum* exposed to low-pressure or medium-pressure UV radiation (Shin et al., 2001; Zimmer et al., 2003). It was demonstrated that, although photoreactivation and dark repair occurred in *Cryptosporidium* oocysts, their infectivity was not restored (Morita et al., 2002).

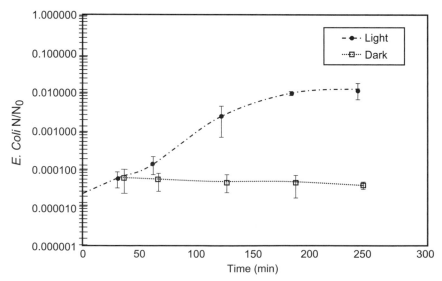

Figure 6.22 Photoreactivation of *Escherichia coli* following exposure to UV at 10 mJ/cm^2. From Zimmer et al. (2003).

Photoreactivation was demonstrated in a full-scale wastewater treatment plant using UV disinfection. However, although total and fecal coliforms were photoreactivated, fecal streptococci showed no or slight photoreactivation (Baron, 1997; Harris et al., 1987a,b; Whitby et al., 1984). *Legionella pneumophila* treated by UV irradiation was also photoreactivated following exposure to visible light (indirect sunlight). Photoreactivation has also been demonstrated for sunlight-damaged natural viruses in seawater, although no repair was observed for bacterial phages in wastewater (Baron, 1997; Weinbauer et al., 1997).

6.6.5 Ultraviolet Disinfection of Water and Wastewater

6.6.5.1 Potable Water. Ultraviolet disinfection is particularly useful for potable water (Wolfe, 1990). Because groundwater accounts for about 50 percent of waterborne diseases in the United States, UV is a potential useful disinfectant for this resource. Europe has more than 2000 drinking water facilities using UV disinfection as compared to 500 in North America (Parrotta and Bekdash, 1998). This disinfectant is particularly efficient against viruses that are major agents of waterborne diseases in groundwater (Craun, 1986a, b). A drinking water plant using UV disinfection in London treats up to 14.5 MGD. In hospitals, chlorine is added to maintain a disinfectant residual after UV irradiation. Continuous UV irradiation rapidly inactivates *Legionella* in plumbing systems and in circulating hot tubs and whirlpools (Gilpin, 1984; Muraca et al., 1987). At a dose of $30 \, \text{mW} \cdot \text{s/cm}^2$ ($30 \, \text{mJ/cm}^2$), UV reduces *Legionella pneumophila* by 4–5 logs within 20 min in a hospital water distribution system (Muraca et al., 1987) while a dose of $38,000 \, \text{W} \cdot \text{s/cm}^2$ is necessary for inactivating *Bacillus subtilis* spores (Parrotta and Bekdash, 1998). Although this dose is ineffective in inactivating *Giardia lamblia* cysts, which require a dose of $63 \, \text{mW} \cdot \text{s/cm}^2$ for 1–log reduction in cyst viability (Rice and Hoff, 1981), it is much higher than the minimum dose of $16 \, \text{mW} \cdot \text{s/cm}^2$ recommended by the U.S. Public Health Service for disinfection of water with ultraviolet irradiation (U.S. DHEW, 1967). In Europe some countries are recommending a minimum UV dose (radiant exposure) of $40 \, \text{mJ/cm}^2$ ($40 \, \text{mW} \cdot \text{s/cm}^2$) and the proposed guideline for North America may range between 50 and $100 \, \text{mJ/cm}^2$ for reclaimed drinking water (Anderson et al., 2003; National Water Research Institute, 2000).

Small portable UV disinfection devices, capable of treating drinking water at a flow rate of about 4 gal/min, are being considered in small villages in rural areas of developing countries (Reuther, 1996). These units provide a low-cost alternative to chlorination which requires more expert supervision.

6.6.5.2 Wastewater Effluents. Ultraviolet disinfection of wastewater effluents is now an economically competitive alternative to chlorination or ozonation and does not generate toxic or genotoxic byproducts as shown for chlorination. Several UV disinfection systems have been built or are in the planning stage. Problems encountered include the difficulty for measuring the UV dose necessary for disinfection of wastewater effluents. Models as well as bioassays using *Bacillus subtilis* spores or RNA phage MS2 have been proposed for the determination of the effective UV dose for disinfection (Havelaar et al., 1990b; Qualls and Johnson, 1983; Qualls et al., 1989; Wilson et al., 1992). MS2 displays higher resistance to UV irradiation than enteric viruses (e.g., hepatitis A virus, rotavirus) or *C. parvum* oocysts (Battigelli et al., 1993; Shin et al., 2001; Wiedenmann et al., 1993). However, as regards UV inactivation of *C. parvum* oocysts, MS2 is considered as too conservative (Mackey et al., 2002). Spores of the *B. subtilis* strain ATCC 6633 are currently the European dosimetry standards for UV disinfection of drinking

TABLE 6.5.   Inactivation of Microorganisms, Using the Multitubular Photoreactor

Microorganism (UV Dose Needed)	Initial Count (No. per 100 mL)	Final Count[a] (No. per 100 mL)
Escherichia coli	1.5×10^8	0
Streptococcus faecalis		
$(10,000 \ \mu W \cdot s/cm^2)$	7.6×10^6	0
Enterococci	9.3×10^5	0
Poliovirus		
$(21,000 \ \mu W \cdot s/cm^2)$	2×10^5	0

[a]The exposure time at a flow rate of 5 m^3 h^{-1} was 2 s.
Adapted from Acher et al. (1997).

water. The D value (UV dose to reduce spore viability by a factor of 10) for this strain is 120 J/m^2 (Hoyer, 2000; Nicholson and Galeano, 2003).

The U.S. Public Health Service minimum dose of 16 mW · s/cm^2 leads to more than 3–log reduction of coliforms in wastewater effluents. The wide variation observed in coliform inactivation by UV in wastewater is probably due to the varying proportions of solids-associated coliforms in the effluents (Qualls et al., 1985). Ultraviolet disinfection, investigated in a full-scale plant in Ontario, Canada, was shown to be as efficient as chlorination in respect to the inactivation of total coliforms, fecal coliforms, and fecal streptococci (Whitby et al., 1984). Moreover, it was superior to chlorination for the inactivation of *Clostridium perfringens* and coliphages. In tertiary treated wastewater effluents subjected to filtration, a UV dose of approximately 75 μW · s/cm^2 was sufficient to reduce the concentration of fecal coliforms, fecal streptococci, enterococci, MS2 phage, and poliovirus type 1 by 4 logs (Oppenheimer et al., 1997). New UV-based devices are being proposed to overcome the interference due to effluent turbidity, thus allowing the treatment of poor quality effluents. One of these devices is the patented multitubular photoreactor, which consists of a transparent quartz pipe (through which the wastewater is passed at a flow rate of ≤5 m^3/h) surrounded by eight UV lamps. Preliminary results show that this device efficiently inactivates microorganisms in wastewater effluents (Acher et al., 1997) (Table 6.5).

Simultaneous addition of ozone and UV is not additive with regard to inactivation of fecal coliforms in wastewater effluents. However, inactivation of the bacterial indicators is increased if UV irradiation precedes or follows ozonation (Venosa et al., 1984).

Ultraviolet disinfection can also be coupled with a novel photocatalytic oxidation (PCO) technology, which consists of using $TiO_2 - Fe_2O_3$ as photocatalysts and a ceramic membrane for the recovery of catalyst particles. A 2–log removal of *E. coli* was obtained at a DO level of 21.3 mg/L, a hydraulic retention time (HRT) of 60 s, and a bacterial concentration of 10^9 CFU/mL (Sun et al., 2003).

6.6.6   Some Advantages and Disadvantages of UV Disinfection

There are several advantages of disinfecting water and wastewater with UV irradiation (Sobsey, 1989; Wolfe, 1990):

1. Efficient inactivation of bacteria and viruses in potable water. Higher doses are required for protozoan cysts.

2. No production of known undesirable mutagenic/carcinogenic or toxic byproducts; no adverse effects were observed in rainbow trout exposed to UV-treated effluents

TABLE 6.6. Disinfection Byproducts in Chlorinated and UV-Irradiated Filtered Secondary Effluent[a]

Compound Detected	Filtered Effluent (μg/L)	Chlorinated Effluent (μg/L)	UV-Treated Effluent (μg/L)
Chloroform	2.7	21	2.9
Dibromochloromethane	0.9	22	0.8
Dichlorobromomethane	1.1	27	1.0
Bromoform	<0.5	2.7	<0.5
Acetaldehyde	<5	21	7.0
Formaldehyde	<0.5	24	8.0

[a]Adapted from Oppenheimer et al. (1997).

(Oliver and Carey, 1976). Similarly, fish and *Ceriodaphnia* acute and chronic toxicity assays did not show any toxicity for a UV-treated tertiary effluent in a full-scale wastewater treatment plant, but did produce chronic toxicity for a chlorinated effluent from the same plant (Oppenheimer et al., 1997). Depending on the concentration of natural organic matter (NOM), UV-treated water can become toxic due to the formation of carboxylic acids, ketoacids, and aldehydes, and to the release of NOM-associated heavy metals (e.g., copper) (Parkinson et al., 2001). Chemical analysis for disinfection byproducts after chlorination or UV irradiation showed that no trihalomethanes were produced in the UV-treated effluent, and aldehyde concentrations were higher in the chlorinated effluent than in the UV-irradiated effluent (Oppenheimer et al., 1997) (Table 6.6). The available data allow us to rank the effect of disinfectants on effluent toxicity in the following order: chlorination/dechlorination > ozonation > UV irradiation.

3. No taste and odor problems.

4. No need to handle and store toxic chemicals.

5. Small space requirement by UV units.

Disadvantages of UV disinfection include the following:

1. No disinfectant residual in treated water (therefore, a postdisinfectant such as chlorine should be added).

2. Difficulty in determining UV dose.

3. Biofilm formation on the lamp surface (however, modern UV units are designed to prevent fouling by microorganisms).

4. Lower disinfection in high-turbidity effluents.

5. Problems in maintenance and cleaning of UV lamps.

6. Potential problem due to photoreactivation of UV-treated microbial pathogens.

7. Cost of UV disinfection: UV disinfection is becoming competitive with chlorination at a dose of 40 mW · s/cm^2 (Parrotta and Bekdash, 1998).

6.6.7 Other Technologies Based on Photoinactivation

6.6.7.1 *Use of Solar Radiation for Disinfection of Drinking Water.* In remote areas with no access to treated drinking water, solar radiation has been considered for disinfecting water. A solar radiation intensity of $\geq$600 W/m^2 for 5 h is necessary for an adequate reduction of bacterial indicators and pathogens (e.g., *Vibrio cholerae,*

Salmonella typhi, and so on) (Odeymi, 1990). Transparent plastic bottles exposed to sunshine in equatorial climates for 7 h can raise the drinking water temperature to 55°C, causing the total eradication of *E. coli* (Joyce et al., 1996). No data are available so far on the inactivation of parasites.

The Solar Water Disinfection (SODIS) process, sponsored by EAWAG (Swiss Federal Institute for Environmental Science and Technology), is a point-of-use treatment system that uses solar radiation to inactivate waterborne pathogens. It has been adopted by several developing countries (e.g., Latin American countries, Sri Lanka, Indonesia, Thailand) and will probably be helpful in reducing waterborne diseases in those countries. The process is simple and consists of partially filling transparent plastic bottles (preferably made of polyethylene terephtalate), shaking the bottles to saturate the water with oxygen, and exposing them to solar radiation for 6 h if the sky is bright or partially cloudy, or for two consecutive days if the sky is 100 percent cloudy. The exposure time is also dependent on ambient temperature. Pathogen inactivation is due to the synergistic effects of UV-A (wavelength 150–400 nm) and increased water temperature. Furthermore, inactivation can also be due to the highly reactive species such as superoxides (O_2^-), hydrogen peroxides (H_2O_2), and hydroxyl radicals ($OH^\bullet$). However, SODIS efficacy can be reduced by low ambient air temperature, turbidity (should be less than 30 NTU) and topography (lower inactivation at higher elevations (Oates et al., 2003; www.sodis.ch; Wegelin et al., 1994; Wegelin and Sommer, 1998). A mathematical model, based on satellite-derived data, was developed to simulate monthly mean, minimum, and maximum 5 h averaged peak solar radiation intensities in Haiti. Based on the threshold of 3–5 h of solar radiation above 500 W/m^2, the model suggested that SODIS can be used all year round in Haiti (Oates et al., 2003).

A low-cost portable unit based on disinfection by solar radiation was recently tested for wastewater and river water. A 2–log coliform inactivation was obtained after about 40 min (Caslake et al., 2004). A new SODIS reflective solar disinfection pouch made of food-grade packaging material helped achieve a 5–log reduction of *E. coli* and a 3.5–log reduction of phage MS2 following 6 h-exposure to sunlight (Walker et al., 2004).

6.6.7.2 *Photodynamic Inactivation.* *Photodynamic inactivation* (photochemical disinfection) consists of the use of visible light or sunlight as the energy source, O_2, and a sensitizer dye such as methylene blue, rose bengal, or eosin. It is based on the transfer of sunlight energy to oxygen, producing oxidative species (e.g., singlet oxygen), which inactivate pathogens. This approach was studied under laboratory conditions (Acher and Juven, 1977; Gerba et al., 1977) and in pilot plants (Acher et al., 1990; 1994; Eisenberg et al., 1987). Photochemical disinfection of wastewater effluents under field conditions achieved microbial reductions of 1.8–log for poliovirus1, 3.0–log for coliforms, 3.1–log for fecal coliforms, and 3.7–log for enterococci in approximately 1 h under alkaline conditions. These reductions necessitate a light intensity of 700–2100 $\mu E \ m^{-2} \ s^{-1}$, a methylene blue level of 0.8–0.9 ppm, pH = 8.7–8.9, and a DO level of 4.5–5.5 mg/L (Acher et al., 1990). A pilot plant with a flow rate of 50 m^3/h and a disinfection time of 35 min, led to 4–5 log reduction of microorganisms. There is, however, no need to remove methylene blue after the oxidative treatment if the dye is immobilized on a support (Acher et al., 1994; 1997).

Titanium dioxide (TiO_2), in combination with fluorescent light or sunlight, has also been considered for the photocatalytic oxidation and hence inactivation of pathogenic microorganisms in water and wastewater. This disinfection process requires relatively long contact times since a 2–log reduction of poliovirus and coliform bacteria in wastewater is achieved after 30 min and 150 min, respectively (Watts et al., 1995).

TiO_2 photocatalysis not only results in the inactivation of *E. coli*, but also degrades the endotoxin released from the cells (Sunada et al., 1998).

6.7 WASTEWATER IRRADIATION AND OTHER EMERGING DISINFECTION TECHNOLOGIES

In Chapter 12, we will discuss the treatment of wastewater sludges by gamma radiation (cesium-137 or cobalt-60) or high energy electron beams. Research is now focusing on the effectiveness of this technology in the disinfection of wastewater as well as in the removal of organic matter (BOD and COD). Disinfection of raw wastewater with gamma radiation (^{60}Co source), at a dose of 463 krad, resulted in 3–log inactivation for coliphage and 4– to 5–log inactivation for coliforms and heterotrophic plate count (Farooq et al., 1992) (Fig. 6.23). Wastewater treatment with high-energy electrons was less effective than gamma radiation and resulted in 2– to 3–log inactivation for the three categories of microbial indicators (Farooq et al., 1992) (Fig. 6.24). Similarly, 4– to 5–log reduction of coliform bacteria was obtained following irradiation of raw sewage with a dose of 200 krad (Rawat et al., 1998).

Emerging disinfection/removal technologies include the use of pulsed UV irradiation, microfiltration, sonication, and electroporation. The latter involves the perforation of microbial cell walls in an electric field (Marshall, 1998).

Another oxidizing agent is peracetic acid, which has been used in the food and pharmaceutical industries as well as in wastewater treatment alone or in combination with other disinfectants. Its use in the disinfection of wastewater effluents in a full-scale plant has resulted in more than 97 percent reduction of total coliforms, *E. coli* and enterococci.

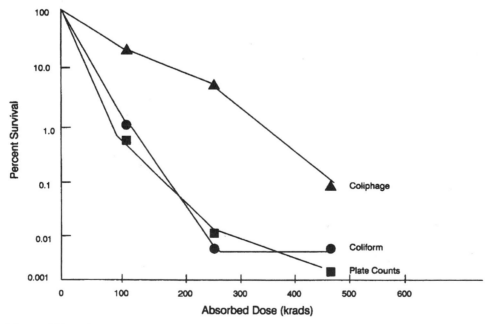

Figure 6.23 Disinfection of raw wastewater with gamma irradiation. From Farooq et al. (1992). (Courtesy of Pergamon Press.)

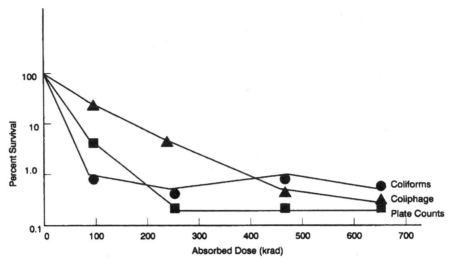

Figure 6.24 Disinfection of raw wastewater with electron beams. From Farooq et al. (1992). (Courtesy of Pergamon Press.)

Peracetic acid also led to a reduction in *Salmonella* levels. Inactivation increased with temperature but decreased with an increase in BOD level (Stampi et al., 2001).

Another approach is the physical removal of microorganisms by microfiltration using membrane bioreactors. Membranes remove efficiently the particulate matter in wastewater, and their small pore size also allows the physical removal of microorganisms. A set of three 0.4 μm pore size polyethylene membranes achieved up to a 7–log removal of fecal coliforms, 5–log removal of fecal streptococci, and 5.9–log removal of phage from settled wastewater. Phage removal efficiency increased as filtration resistance across the membrane increased. The increased resistance was due to biofilm accumulation on the membrane (Ueda and Horan, 2000).

Owing to toxicological problems linked to the use of chemical disinfectants (e.g., chlorine), disinfection using physical treatments is potentially attractive. Ultrasound is known to cause damage to microbial cells (e.g., cell lysis, damage to membrane integrity). A preliminary study on disinfection of wastewater has shown that a 20 kHz ultrasound unit, operated at 700 W/L, causes a 4–log units inactivation of fecal coliforms within 6 min. Approximately half of the inactivation was due to thermal effects resulting from ultrasound treatment (Madge and Jensen, 2002). A 20 kHz ultrasound inactivated more than 90 percent *Cryptosporidium* oocysts within 1.5 min under laboratory conditions (Ashokkumar et al., 2003).

6.8 WEB RESOURCES

http://www.epa.gov/enviro/html/icr/dbp.html (disinfection byproducts from U.S. EPA)
http://cfpub.epa.gov/ncea/cfm/recordisplay.cfm?deid=2829 (chloramines)
http://www.epa.gov/owmitnet/mtb/chlo.pdf (chlorine disinfection)
http://www.epa.gov/pesticides/factsheets/chemicals/chlorinedioxidefactsheet.htm
(disinfection with chlorine dioxide)
http://www.epa.gov/owm/mtb/uv.pdf (UV disinfection)
www.sodis.ch (Solar Disinfection (SODIS) process)

6.9 QUESTIONS AND PROBLEMS

1. Pathogens decrease exponentially when exposed to a disinfectant. The decrease is described by Chick's law:

$$\ln N_t/N_0 = -kt$$

where N_0 = initial number of pathogens (cells/L); N_t = number of pathogens after time t (cells/L); t = duration of disinfection; k = empirical constant (t^{-1}).

The following data were obtained in a disinfection experiment dealing with inactivation of poliovirus

Time (s)	N/N_0
4	1/13
8	1/158
12	1/2000

From Floyd et al. (1978).

(a) Plot $-\ln(N/N_0)$ vs. time

(b) Calculate the k value

(c) Calculate the time required for a 1/5000 reduction of poliovirus.

2. As regards the effect of pH on disinfection efficiency, what is the difference between chorine and chlorine dioxide?

3. Why is chloramination recommended for water distribution systems?

4. Why does chloramination promote the growth of nitrifying bacteria?

5. What are some advantages and disadvantages of disinfection with ozone?

6. What is the ozonation byproduct of most concern at the present time?

7. Give the categories of UV lamps used in disinfection studies.

8. What are the main factors affecting the efficiency of UV disinfection?

9. Define photodynamic inactivation.

10. Summarize the toxicological problems associated with disinfection (covers chlorine, chlorine dioxide, ozone, UV).

6.10 FURTHER READING

Aieta, E.M., and J.D. Berg. 1986. A review of chlorine dioxide in drinking water treatment. J. Amer. Water Works Assoc. 78: 62–72.

Gerba, C.P., D. Gramos, and N. Nwachuku. 2002. Comparative inactivation of enteroviruses and adenovirus 2 by UV light. Appl. Environ. Microbiol. 68: 5167–5169.

Hoff, J.C., and E.W. Akin. 1986. Microbial resistance to disinfectants: Mechanisms and significance. Environ. Health Perspect. 69: 7–13.

Oates, P.M., P. Shanahan, and M.F. Polz. 2003. Solar disinfection (SODIS): simulation of solar radiation for global assessment and application for point-of-use water treatment in Haiti. Water Res. 37: 47–54.

Radziminski, C., L. Ballantyne, J. Hodson, R. Creason, R.C. Andrews, and C. Chauret. 2002. Disinfection of *Bacillus subtilis* spores with chlorine dioxide: a bench-scale and pilot-scale study. Water Res. 36: 1629–1639.

Russel, A.D., J.R. Furr, and J.-Y. Maillard. 1997. Microbial susceptibility and resistance to biocides. Amer. Soc. Microbiol. News 63: 481–487.

Sobsey, M.D. 1989. Inactivation of health-related microorganisms in water by disinfection processes. Water Sci. Technol. 21: 179–195.

Thompson, S.S., J.L. Jackson, M. Suva-Castillo, W.A. Yanko, Z. El Jack, J. Kuo, C.-L. Chen, F.P. Williams, and D.P. Schnurr. 2003. Detection of infectious adenoviruses in tertiary-treated and ultraviolet-disinfected wastewater. Water Environ. Res. 75: 163–170.

U.S. EPA. 1999. *Alternative Disinfectants and Oxidants Guidance Manual*, Report # 815-R-99-014, U.S. EPA, Office of Water, Washington, DC.

Wolfe, R.L., N.R. Ward and B.H. Olson. 1984. Inorganic chloramines as drinking water disinfectants: A review. J. Amer. Water Works Assoc. 76: 74–88.

PART C

MICROBIOLOGY OF
WASTEWATER TREATMENT

7

INTRODUCTION TO WASTEWATER TREATMENT

7.1 INTRODUCTION

In mid-nineteenth century England, waterborne diseases such as cholera were rampant and several epidemics in London resulted in thousands of victims. Increasing awareness of the role of microorganisms in diseases led to an enhanced demand for wastewater treatment. This has resulted in the passage of pieces of legislation that encouraged the construction of wastewater treatment plants (Guest, 1987). The practice of wastewater treatment started at the beginning of the twentieth century. By the end of the nineteenth century, the British Royal Commission on Sewage Disposal proposed that the goal of wastewater treatment should be to produce a final effluent of 30 mg/L of suspended solids and 20 mg/L of biochemical oxygen demand (BOD) (Sterritt and Lester, 1988).

Today, there are more than 15,000 wastewater treatment facilities in the United States, 80 percent of which are small plants (<1 MGD). These plants treat approximately 37 billion gallons of wastewater per day. About 75 percent of the facilities have secondary treatment or greater (Ouellette, 1991; U.S. EPA, 1989f). These plants have been constructed to treat both domestic and industrial wastes. Nontoxic wastes are contributed mainly by the food industry and by domestic sewage, whereas toxic wastes are contributed by coal processing (phenolic compounds, ammonia, cyanide), petrochemical, (oil, petrochemicals, surfactants), pesticide, pharmaceutical, and electroplating (toxic metals such as cadmium, copper, nickel, zinc) industries (Kumaran and Shivaraman, 1988). Biodegradation and toxicity of these chemicals will be discussed in Chapters 19 and 20. As many of these plants may fail to meet the desired criteria, pretreatment steps are required at industrial sites before the entry of the waste into municipal wastewater treatment plants.

The major contaminants found in wastewater are biodegradable organic compounds, volatile organic compounds, recalcitrant xenobiotics, toxic metals, suspended solids, nutrients (nitrogen and phosphorus), and microbial pathogens and parasites (Fig. 7.1). In the beginning, the requirements for treatment plants were to remove organic matter and suspended solids. Research efforts are now being focused on the removal of nutrients (N, P), odors, volatile organic compounds, metals, and toxic organics after their passage through wastewater treatment plants.

There are several objectives of waste treatment processes:

1. Reduction of the organic content of wastewater (i.e., reduction of BOD).
2. Removal/reduction of trace organics that are recalcitrant to biodegradation and may be toxic or carcinogenic (see Chapters 18 and 19).
3. Removal/reduction of toxic metals.

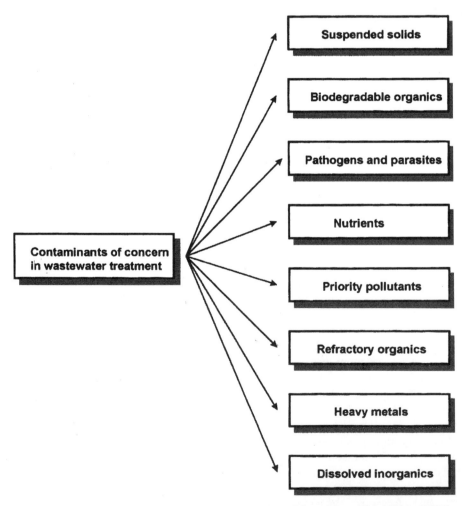

Figure 7.1   Major contaminants in wastewater. Adapted from Metcalf and Eddy (1991).

4. Removal/reduction of nutrients (N, P) to reduce pollution of receiving surface waters or groundwater if the effluents are applied onto land (see Chapters 8, 10, and 11).

5. Removal or inactivation of pathogenic microorganisms and parasites.

7.2 COMPOSITION OF DOMESTIC WASTEWATER

Domestic wastewater is a combination of human and animal excreta (feces and urine) and gray water resulting from washing, bathing, and cooking. People excrete 100–500 g wet weight of feces and between 1 and 1.3 L of urine per capita per day. Each person contributes 15–20 g BOD_5/day (Feachem et al., 1983; Gotaas, 1956; Sterritt and Lester, 1988). Other characteristics of human feces and urine are displayed in Table 7.1 (Feachem et al., 1983; Gotaas, 1956; Polprasert, 1989). The chemical characteristics of untreated domestic wastewater are displayed in Table 7.2 (Metcalf and Eddy, 1991).

Domestic wastewater is composed mainly of proteins (40–60 percent), carbohydrates (25–50 percent), fats and oils (10 percent), urea derived from urine, and a large number of trace organic compounds that include pesticides, surfactants, phenols, and priority pollutants. The latter category comprises nonmetals (As, Se), metals (e.g., Cd, Hg, Pb), benzene compounds (e.g., benzene, ethylbenzene), and chlorinated compounds (e.g., chlorobenzene, tetrachloroethene, trichloroethene) (Metcalf and Eddy, 1991). The bulk of organic matter in domestic wastewater is easily biodegradable and consists mainly of carbohydrates, amino acids, peptides and proteins, volatile acids, and fatty acids and their esters (Giger and Roberts, 1978; Painter and Viney, 1959).

In domestic wastewaters, organic matter occurs as dissolved organic carbon (DOC) and particulate organic carbon (POC). Particulate organic carbon represents approximately 60 percent of organic carbon and some of it is of sufficient size to be removed via sedimentation (Rickert and Hunter, 1971) (Fig. 7.2). In fixed-film processes, DOC is directly absorbed by the biofilm, whereas POC is adsorbed to the biofilm surface to be subsequently hydrolyzed by microbial action (Sarner, 1986).

Three main tests are used for the determination of organic matter in wastewater. These include biochemical oxygen demand (BOD), total organic carbon (TOC), and chemical oxygen demand (COD). Trace organics are detected and measured, using more sophisticated instruments such as gas chromatography and mass spectroscopy.

TABLE 7.1. Composition of Human Feces and Urine[a]

Component	Feces	Urine
Quantity (wet) per person per day	100–400 g	1.0–1.31 kg
Quantity (dry solids) per person per day	30–60 g	50–70 g
Moisture content	70–85%	93–96%
Approximate composition (percent dry weight) organic matter	88–97	65–85
Nitrogen (N)%	5.0–7.0	15–19
Phosphorus (as P_2O_5)%	3.0–5.4	2.5–5.0
Potassium (as K_2O)%	1.0–2.5	3.0–4.5
Carbon (C)%	44–55	11–17
Calcium (as CaO)%	4.5	4.5–6.0
C/N ratio	6–10	1
BOD_5 content per person per day	15–20	10 g

[a]From Polprasert (1989).

TABLE 7.2. Typical Characteristics of Domestic Wastewater[a]

| | Concentration | | |
Parameter	Strong (mg/L)	Medium (mg/L)	Weak (mg/L)
BOD_5	400	220	110
COD	1000	500	250
Organic N	35	15	8
NH_3-N	50	25	12
Total N	85	40	20
Total P	15	8	4
Total solids	1200	720	350
Suspended solids	350	220	100

[a]From Metcalf and Eddy Inc. (1991).

7.2.1 Biochemical Oxygen Demand

Biochemical oxygen demand (BOD) is the amount of dissolved oxygen (DO) consumed by microorganisms for the biochemical oxidation of organic (carbonaceous BOD) and inorganic matter (autotrophic or nitrogenous BOD). The sample must be diluted if the BOD exceeds 8 mg/L.

7.2.1.1 Carbonaceous BOD (CBOD). Carbonaceous BOD (CBOD) is the amount of oxygen used by a mixed population of heterotrophic microorganisms to oxidize organic compounds in the dark at 20°C over a period of 5 days. This is described by the following general equations:

$$\text{Organic compounds} \xrightarrow[\text{heterotrophs}]{O_2} CO_2 + H_2O + NH_4 + \text{bacterial mass}$$

$$\text{Bacterial biomass} \xrightarrow[\text{protozoa}]{O_2} \text{Protozoan biomass} + CO_2$$

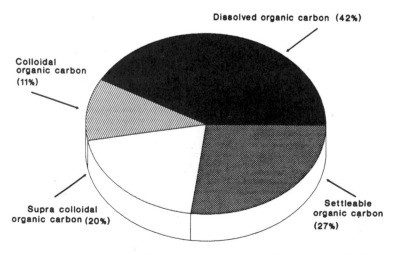

Figure 7.2 Total organic carbon (TOC) size fractions in domestic wastewater. Estimated particles sizes: settleable, >100 mm; supracooloidal, 1–100 mm; colloidal, 1 nm–1 mm; dissolved, <1 nm. Adapted from Rickert and Hunter (1971).

The conversion of bacterial biomass to protozoan biomass proceeds with a yield of 0.78 mg protozoa/mg bacteria (Edeline, 1988).

The BOD test was originally developed to predict the effect of wastewater on receiving streams and determine their capacity to assimilate organic matter (Gaudy, 1972; Gaudy and Gaudy, 1988). It is one of the most important tests considered by regulators to measure the effectiveness of wastewater treatment plants and assess the impact of the plant effluents on receiving waters. Much has been written on the kinetics of BOD exertion in wastewater. Figure 7.3 (Gaudy, 1972) shows the existence of a plateau in oxygen uptake and its length depends on the relationship between prey (bacteria) and predator (protozoa) microorganisms. Many reasons have been offered for the existence of a plateau during oxygen uptake: (1) predation by protozoa; (2) a lag period between exogenous respiration on the substrate and the onset of endogenous respiration; (3) substrate made of various carbon sources that are metabolized sequentially; and (4) biodegradation of a metabolic product only after an acclimation period.

Aliquots of wastewater are placed in a 300 mL BOD bottle and diluted in phosphate buffer (pH 7.2) containing other inorganic elements (N, Ca, Mg, Fe) and saturated with oxygen. Sometimes, acclimated microorganisms or dehydrated cultures of microorganisms sold in capsule form are added to municipal and industrial wastewaters that may not have a sufficient microflora to carry out the BOD test. The microbial seed is usually composed of microorganisms commonly found in the wastewater environment (e.g., *Pseudomonas, Bacillus, Nocardia, Streptomyces*). A nitrification inhibitor is sometimes added to the sample to determine the *carbonaceous BOD*, which is exerted solely by heterotrophic microorganisms (Fitzmaurice and Gray, 1989). Dissolved oxygen concentration is determined at time 0 and after five-day incubation, by means of an oxygen electrode, chemical procedures (e.g., the Winkler test), or a manometric BOD apparatus (Hammer, 1986). The BOD test is carried out on a series of dilutions of the sample, the dilution depending on the source of the sample. Other details concerning the BOD test are given in *Standard Methods for the Examination of Water and Wastewater* (APHA, 1999).

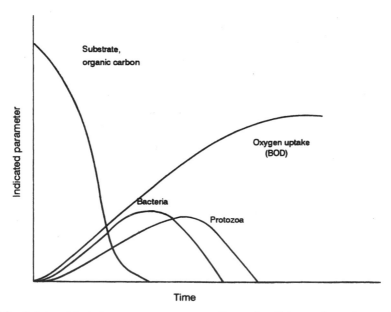

Figure 7.3 Relationship between oxygen consumption, microbial growth, and organic carbon removal. Adapted from Gaudy (1972).

When dilution water is not seeded, the BOD value is expressed in mg/L according to the following equation (APHA, 1999; Hammer, 1986):

$$BOD \ (mg/L) = \frac{D_1 - D_5}{P} \tag{7.1}$$

where D_1 = initial DO; D_5 = DO at day 5; and P = decimal volumetric fraction of wastewater used.

The microbial biomass (bacteria and protozoa) consumes oxygen during the five-day incubation period (carbonaceous BOD). Because of depletion of the carbon source, the carbonaceous BOD reaches a plateau (Fig. 7.4) called the *ultimate carbonaceous BOD*. The relationship between BOD at any time and ultimate carbonaceous BOD is given by the following equation (Hammer, 1986):

$$BOD_t = L(1 - 10^{-kt}) \tag{7.2}$$

where L = ultimate BOD (mg/L); k = BOD rate constant with a value of approximately 0.1/day for domestic wastewaters; and k depends on the following (Davis and Cornwell, 1985):

- *Type of waste. k* varies from 0.15–0.30 in raw wastewater to 0.05–0.10 in surface water.
- *Acclimation of microorganisms to the type of waste analyzed.* The BOD test should be conducted with acclimated microbial seeds.

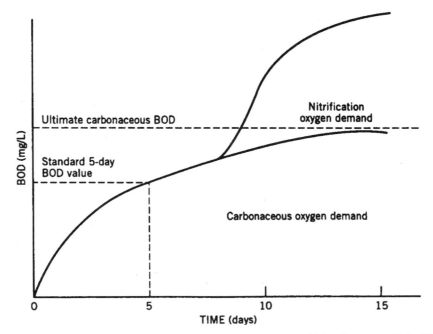

Figure 7.4 Carbonaceous and autotrophic BOD. From Hammer (1986). (Courtesy of John Wiley.)

- *Temperature*. The BOD test is conducted at the standard temperature of 20°C. At other temperatures, k is given by:

$$k_T = k_{20}\theta^{T-20} \tag{7.3}$$

where T = temperature in °C; k_T = BOD rate constant at a given temperature T (day^{-1}); k_{20} = rate constant at 20°C; and θ = temperature coefficient = 1.135 at $4 < T < 20$°C and = 1.056 at $20 < T < 30$°C.

Toxic substances present in wastewater interfere with the BOD test. Another disadvantage of the BOD test is that it overlooks recalcitrant compounds.

The above described method of BOD testing is time consuming and is not amenable to on-line monitoring. Biosensor-based methods (Liu and Mattiasson, 2002), which do not have these limitations, are described in Chapter 18.

7.2.1.2 Nitrogenous Oxygen Demand.

Autotrophic bacteria such as nitrifying bacteria also require oxygen to oxidize NH_4^+ to nitrate (see Chapter 3). The oxygen demand exerted by nitrifiers is called *autotrophic BOD* or *nitrogenous oxygen demand* (NOD) (Fig. 7.4). During the determination of BOD of wastewater samples, nitrifiers exert an oxygen demand and their activity leads to higher BOD values, sometimes resulting in noncompliance of the wastewater treatment plant with federal or state regulations. It has been estimated that about 60 percent of the compliance violations in the United States were due to nitrification occurring during the BOD test (Dague, 1981; Hall and Foxen, 1983). This phenomenon is significant in nitrified wastewater effluents where nitrifying bacteria may account for 24–86 percent of the total BOD (Hall and Foxen, 1983; Washington et al., 1983). The theoretical nitrogenous oxygen demand is 4.57 g oxygen used per g of ammonium oxidized to nitrate. However, the actual value is somewhat lower because some of the nitrogen is incorporated into the bacterial cells (Davis and Cornwell, 1985). The nitrogenous oxygen demand (NOD) is as follows (Verstraete and van Vaerenbergh, 1986):

$$NOD \text{ (mg/L)} = (\text{Available N} - \text{Assimilated N}) \times 4.33 \tag{7.4}$$

It is thus necessary to carry out an inhibited BOD test to distinguish between carbonaceous and nitrogenous BOD (APHA, 1989). It is recommended to add 2-chloro-6-(trichloromethyl) pyridine at a final concentration of 10 mg/L for nitrification inhibition. It has been shown that this chemical does not suppress the oxidation of organic matter (Young, 1983).

The BOD$_5$ of municipal wastewaters varies between 100 and 300 mg/L, and much higher values are found in some industrial wastewaters (Table 7.3). Removal of BOD may vary from 70 percent in high rate activated sludge to up to 95 percent in extended aeration activated sludge (see Chapter 8).

7.2.2 Chemical Oxygen Demand

Chemical oxygen demand (COD) is the amount of oxygen necessary to oxidize the organic carbon completely to CO_2, H_2O, and ammonia (Sawyer and McCarty, 1967). Chemical oxygen demand is measured via oxidation with potassium dichromate ($K_2Cr_2O_7$) in the presence of sulfuric acid and silver and is expressed in mg/L. Thus, COD is a measure of the oxygen equivalent of the organic matter as well as microorganisms in the wastewater (Pipes and Zmuda, 1997). If the COD value is much higher than the

TABLE 7.3.   COD, BOD$_5$, and BOD$_5$/COD Ratios of Selected Wastewaters[a]

	COD (mg/L)	BOD$_5$ (mg/L)	BOD$_5$/COD
Domestic sewage			
Raw	500	300	0.60
After biological treatment	50	10	0.20
Slaughterhouse wastewater	3,500	2,000	0.57
Distillery vinasse	60,000	30,000	0.50
Dairy wastewater	1,800	900	0.50
Rubber factory	5,000	3,300	0.66
Tannery wastewater	13,000	1,270	0.10
Textile dyeing			
Raw	1,360	660	0.48
After biological treatment	116	5	0.04
Draft mill effluent			
Raw	620	226	0.36
Biologically stabilized	250	30	0.12

[a]Adapted from Verstraete and van Vaerenbergh (1986).

BOD value, the sample contains large amounts of organic compounds that are not easily biodegraded.

In untreated domestic wastewater, COD ranges between 250 and 1000 mg/L. For typical untreated domestic wastewater, the BOD$_5$/COD ratio varies from 0.4 to 0.8 (Metcalf and Eddy, 1991). The BOD$_5$, COD, and BOD$_5$/COD ratios of some typical wastewaters are shown in Table 7.3 (Verstraete and van Vaerenbergh, 1986).

7.2.3   Total Organic Carbon

Total organic carbon (TOC) represents the total organic carbon in a given sample and is independent of the oxidation state of the organic matter. It is determined via oxidation of the organic matter with heat and oxygen (aeration step is eliminated if VOCs are present in the sample) or chemical oxidants, followed by the measurement of the CO_2 liberated with an infrared analyzer (Hammer, 1986; Metcalf and Eddy, 1991).

7.3   OVERVIEW OF WASTEWATER TREATMENT

Physical forces as well as chemical and biological processes drive the treatment of wastewater. Treatment methods that rely on physical forces are called *unit operations.* These include screening, sedimentation, filtration, or flotation. Treatment methods based on chemical and biological processes are called *unit processes.* Chemical unit processes include disinfection, adsorption, or precipitation. Biological unit processes involve microbial activity, which is responsible for organic matter degradation and removal of nutrients (Metcalf and Eddy, 1991).

Wastewater treatment comprises the following four steps (Fig. 7.5):

1. *Preliminary treatment.* The objective of this operation is to remove debris and coarse materials that may clog equipment in the plant.
2. *Primary treatment.* Treatment is brought about by physical processes (unit operations) such as screening and sedimentation.

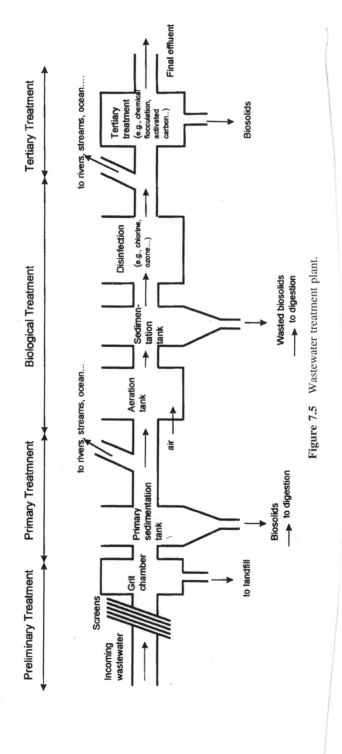

Figure 7.5 Wastewater treatment plant.

221

3. *Secondary treatment*. Biological (e.g., activated sludge, trickling filter, oxidation ponds) and chemical (e.g., disinfection) unit processes are used to treat wastewater. Nutrient removal also generally occurs during secondary treatment of wastewater.

4. *Tertiary or advanced treatment*. Unit operations and chemical unit processes are used to further remove BOD, nutrients, pathogens, and parasites, and sometimes toxic substances.

The third part of this book will address the following aspects for each unit process:

- *Process microbiology*. We will examine the types and activity of microorganisms involved in each process.
- *Public health microbiology*. Some of the unit processes were specifically designed to remove pathogenic microorganisms and parasites. We will examine the effectiveness of each process in removing or inactivation of microbial pathogens and parasites.

7.4 WEB RESOURCES

http://ga.water.usgs.gov/edu/wuww.html (beginner's wastewater treatment)

http://www.epa.gov/ebtpages/treawaterpwastewatertreatment.html (EPA wastewater treatment topics)

http://ei.cornell.edu/biodeg/wastewater/ (wastewater treatment from Cornell University)

http://www.dep.state.pa.us/dep/deputate/waterops/ (operator's information center, Pennsylvania Department of Environmental Protection)

http://www.scitrav.com/wwater/waterlnk.htm (collection of web pages on wastewater treatment)

http://www.scitrav.com/wwater/waterlnk.htm#microbiology (rich collection of sites on wastewater microbiology)

http://www.swopnet.com/engr//londonsewers/londontext1.html (a glimpse at London early sewers, Part 1)

http://www.swopnet.com/engr//londonsewers/londontext2.html (a glimpse at London early sewers, Part 2)

http://www.swopnet.com/engr//londonsewers/londontext3.html (a glimpse at London early sewers, Part 3)

http://www.college.ucla.edu/webproject/micro7/studentprojects7/Rader/asludge2.htm (pictures of the Hyperion Treatment Plant, Playa del Rey, California)

http://www.scitrav.com/wwater/waterlnk.htm (collection of web pages on activated sludge)

http://www.swbic.org/education/env-engr/index.html (online course on water and wastewater treatment)

http://www.rpi.edu/dept/chem-eng/Biotech-Environ/FUNDAMNT/streem/envmicro.htm (general text on waste treatment; diagrams)

http://www.rpi.edu/dept/chem-eng/Biotech-Environ/FUNDAMNT/streem/microbug.htm (song about wastewater bugs)

7.5 QUESTIONS AND PROBLEMS

1. BOD of a wastewater sample: 5 mL of wastewater is added to a 300 mL BOD bottle. Add dilution water to the bottle until obtaining a volume of 300 mL. The initial DO

concentration was 7.8 mg/L and the final DO concentration was 4.3 mg/L after five-day incubation.

    (a) Calculate the BOD_5 of the sample.

    (b) Calculate the ultimate BOD (assuming a k value of 0.1).

2. What are some problems of the BOD test?
3. Give the different steps involved in wastewater treatment.
4. Explain the difference between carbonaceous BOD and autotrophic BOD.
5. Give two examples of unit operations and two examples of unit processes.

7.6   FURTHER READING

Metcalf & Eddy, Inc. 2003. *Wastewater Engineering: Treatment and Reuse*, 4th ed./revised by George Tchobanoglous, Franklin L. Burton, H. David Stensel, McGraw-Hill, Boston.

Cheremisino, N.P. 2002. *Handbook of water and wastewater treatment technologies*, Butterworth-Heinemann, Boston, 636 pp.

Grady, C.P.L., Jr., and H.C. Lim. 1980. *Biological Waste Treatment: Theory and Applications*. Marcel Dekker, New York, 963 pp.

8

ACTIVATED SLUDGE PROCESS

8.1   INTRODUCTION

Activated sludge is a suspended-growth process that began in England at the turn of the century. This process has since been adopted worldwide as a secondary biological treatment for domestic wastewaters. This process consists essentially of an aerobic treatment that oxidizes organic matter to CO_2 and H_2O, NH_4, and new cell biomass. Air is provided by using diffused or mechanical aeration. The microbial cells form flocs that are allowed to settle in a clarification tank.

Wastewater Microbiology, Third Edition, by Gabriel Bitton
Copyright © 2005 John Wiley & Sons, Inc.

8.2 DESCRIPTION OF THE ACTIVATED SLUDGE PROCESS

8.2.1 Conventional Activated Sludge System

A conventional activated sludge process includes (Fig. 8.1) the following:

- *Aeration tank.* Aerobic oxidation of organic matter is carried out in this tank. Primary effluent is introduced and mixed with *return activated sludge* (RAS) to form the *mixed liquor*, which contains 1500–2500 mg/L of suspended solids. Aeration is provided by mechanical means. An important characteristic of the activated sludge process is the *recycling of a large portion of the biomass.* This makes the mean cell residence time (i.e., sludge age) much greater than the hydraulic retention time (Sterritt and Lester, 1988). This practice helps maintain a large number of microorganisms that effectively oxidize organic compounds in a relatively short time. The detention time in the aeration basin varies between 4 and 8 hours.
- *Sedimentation tank.* This tank is used for the sedimentation of microbial flocs (sludge) produced during the oxidation phase in the aeration tank. A portion of the sludge in the clarifier is recycled back to the aeration basin and the remainder is wasted to maintain a proper F/M (food to microorganisms ratio).

We will now define some operational parameters commonly used in activated sludge (Davis and Cornwell, 1985; Verstraete and van Vaerenbergh, 1986).

8.2.1.1 Mixed Liquor Suspended Solids (MLSS). The content of the aeration tank in an activated sludge system is called *mixed liquor.* The MLSS is the total amount of organic and mineral suspended solids, including microorganisms, in the mixed liquor. It is determined by filtering an aliquot of mixed liquor, drying the filter at 105°C, and determining the weight of solids in the sample.

8.2.1.2 Mixed Liquor Volatile Suspended Solids (MLVSS). The organic portion of MLSS is represented by MLVSS, which comprises nonmicrobial organic matter as well as dead and live microorganisms and cellular debris (Nelson and Lawrence, 1980). The MLVSS is determined after heating of dried filtered samples at 600–650°C, and represents approximately 65–75 percent of MLSS.

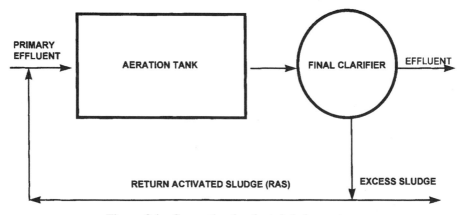

Figure 8.1 Conventional activated sludge system.

8.2.1.3 *Food-to-Microorganism Ratio (F/M).* The food-to-microorganisms (F/M) ratio indicates the organic load into the activated sludge system and is expressed in kilogram BOD per kilogram of MLSS per day (Curds and Hawkes, 1983; Nathanson, 1986). It is expressed as:

$$F/M = \frac{Q \times BOD}{MLSS \times V} \tag{8.1}$$

where Q = flow rate of sewage in million gallons per day (MGD); BOD = five-day biochemical oxygen demand (mg/L); $MLSS$ = mixed liquor suspended solids (mg/L); and V = volume of aeration tank (gallons).

The food-to-microorganism ratio is controlled by the rate of activated sludge wasting. The higher the wasting rate the higher the F/M ratio. For conventional aeration tanks the F/M ratio is 0.2–0.5 lb BOD_5/day/lb MLSS but it can be higher (≤ 1.5) for activated sludge using high purity oxygen (Hammer, 1986). A low F/M ratio means that the microorganisms in the aeration tank are starved, generally leading to a more efficient wastewater treatment.

8.2.1.4 *Hydraulic Retention Time (HRT).* Hydraulic retention time is the average time spent by the influent liquid in the aeration tank of the activated sludge process; it is the reciprocal of the dilution rate D (Sterritt and Lester, 1988).

$$HRT = \frac{1}{D} = \frac{V}{Q} \tag{8.2}$$

where V = volume of the aeration tank; Q = flow rate of the influent wastewater into the aeration tank; and D = dilution rate.

8.2.1.5 *Sludge Age.* Sludge age is the mean residence time of microorganisms in the system. While the hydraulic retention time may be in the order of hours, the mean cell residence time may be in the order of days. This parameter is the reciprocal of the microbial growth rate μ (see Chapter 2). Sludge age is given by the following formula (Hammer, 1986; Curds and Hawkes, 1983):

$$\text{Sludge age (days)} = \frac{MLSS \times V}{SS_e \times Q_e + SS_w \times Q_w} \tag{8.3}$$

where $MLSS$ = mixed liquor suspended solids (mg/L); V = volume of aeration tank (L); SS_e = suspended solids in wastewater effluent (mg/L); Q_e = quantity of wastewater effluent (m^3/day); SS_w = suspended solids in wasted sludge (mg/L); and Q_w = quantity of wasted sludge (m^3/day).

Sludge age may vary from 5 to 15 days in conventional activated sludge. It varies with the season of the year and is higher in the winter than in the summer season (U.S. EPA, 1987a).

The important parameters controlling the operation of an activated sludge are organic loading rates, oxygen supply, and control and operation of the final settling tank. This tank has two functions: clarification and thickening. For routine operation, one must measure sludge settleability by determining the *sludge volume index* (SVI) (Forster and Johnston, 1987) (see Chapter 9 for more details).

8.2.2   Some Modifications of the Conventional Activated Sludge Process

There are several modifications of the conventional activated sludge process (Nathanson, 1986; U.S. EPA, 1977) (Fig. 8.2). These are given in the following subsections.

8.2.2.1   *Extended Aeration System (Fig. 8.2a).*   This process, used in package treatment plants, has the following features:

1. The aeration time is much longer (about 30 h) than in conventional systems. The sludge age is also longer and can be extended to >15 days.
2. The wastewater influent entering the aeration tank has not been treated by primary settling.
3. The system operates at much lower F/M ratio (generally <0.1 lb BOD/day/lb MLSS) than conventional systems (0.2–0.5 lb BOD/day/lb MLSS).
4. This system requires less aeration than conventional treatment and is mainly suitable for small communities that use package treatment.

8.2.2.2   *Oxidation Ditch (Fig. 8.2b).*   The oxidation ditch consists of an aeration oval channel with one or more rotating rotors for wastewater aeration. This channel receives screened wastewater and has a hydraulic retention time of approximately 24 h.

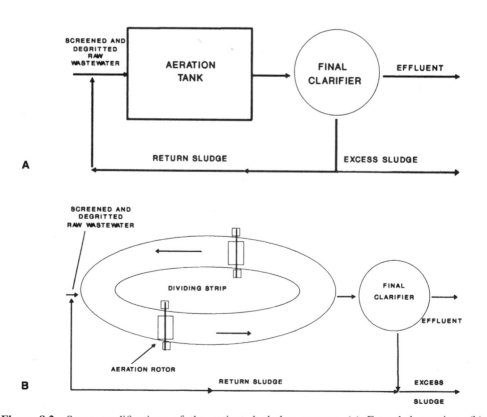

Figure 8.2   Some modifications of the activated sludge process: (a) Extended aeration; (b) oxidation ditch. From the U.S. EPA (1977).

8.2.2.3 Step Aeration. The primary effluent enters the aeration tank through several points, thus improving its distribution into the tank and making more efficient use of oxygen. This increases the treatment capacity of the system.

8.2.2.4 Contact Stabilization. After contact of the wastewater with the sludge in a small contact tank for a short period of time (20–40 min), the mixture flows to a clarifier and the sludge is returned to a stabilization tank with a retention time of 4–8 h. This system produces less sludge.

8.2.2.5 Completely Mixed Aerated System. A completely mixed aerated system allows a more uniform aeration of the wastewater in the aeration tank. This system can sustain shock and toxic loads.

8.2.2.6 High-Rate Activated Sludge. This system is used for the treatment of high-strength wastes and is operated at much higher BOD loadings than those encountered in the conventional activated sludge process. This results in shorter hydraulic retention periods (i.e., shorter aeration periods). The system is operated at higher MLSS concentrations.

8.2.2.7 Pure Oxygen Aeration. The pure oxygen aeration system is based on the principle that the rate of transfer of pure oxygen is higher than that of atmospheric oxygen. This results in higher availability of dissolved oxygen, leading to improved treatment and reduced production of sludge.

Table 8.1 summarizes the design and operational characteristics of some activated sludge processes.

TABLE 8.1. Design and Operational Characteristics of Some Activated Sludge Processes[a]

Process	Flow Regime	Aeration System	θ_c (days)	F/M	BOD Removal (%)
Conventional	Plug-flow	Diffused air, mechanical aerators	5–15	0.2–0.4	85–95
Contact stabilization	Plug-flow	Diffused air, mechanical aerators	5–15	0.2–0.6	80–90
Step aeration	Plug-flow	Diffused air	5–15	0.2–0.4	85–95
Extended aeration	Complete-mix	Diffused air, mechanical aerators	20–30	0.05–0.15	75–95
High-rate aeration	Complete-mix	Diffused air, mechanical aerators	5–10	0.4–1.5	75–90
Pure oxygen	Complete-mix	Mechanical aerators	8–20	0.25–1	85–95
Complete mix	Complete-mix	Diffused air, mechanical aerators	5–15	0.2–0.6	95–95

[a]Adapted from http://www.swbic.org/education/env-engr/secondary/principles/as
θ_c = mean cell residence time.
F/M = food-to-microorganisms ratio.

8.3 BIOLOGY OF ACTIVATED SLUDGE

There are two main goals of the activated sludge system:

1. Oxidation of the biodegradable organic matter in the aeration tank (soluble organic matter is thus converted to new cell mass).
2. Flocculation, that is, the separation of the newly formed biomass from the treated effluent.

8.3.1 Survey of Organisms Present in Activated Sludge Flocs

The activated sludge flocs contain mostly bacterial cells as well as other microorganisms, and inorganic and organic particles. Floc size varies between <1 μm (the size of some bacterial cells) to ≥ 1000 μm (Parker et al., 1971; U.S. EPA, 1987a) (Fig. 8.3). Figure 8.4 illustrates the main microorganisms in the activated sludge microbial community (Wagner and Amann, 1996).

Early studies showed that viable cells in the floc, as measured by ATP analysis and dehydrogenase activity, would account for 5–20 percent of the total cells (Weddle and Jenkins, 1971). Some investigators estimated that the active fraction of bacteria in activated sludge flocs represents only 1–3 percent of total bacteria (Hanel, 1988). However, fluorescently labeled oligonucleotide probes show that a higher percentage of the microbial biomass is metabolically active (Head et al., 1998; Wagner et al., 1993). Flow cytometry, in conjunction with fluorogenic viability/activity dyes (see Chapters 1 and 2), were also used to detect the viability and activity of microorganisms in activated sludge flocs. It was found that 62 percent of the total bacteria were active in the flocs

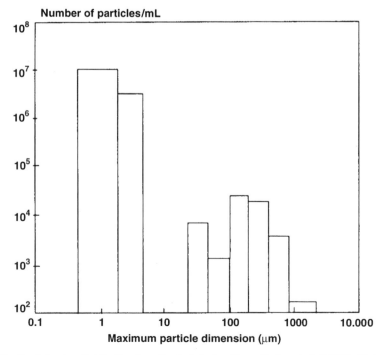

Figure 8.3 Particle size distribution in activated sludge. From Parker et al. (1971). (Courtesy of the Water Environment Federation.)

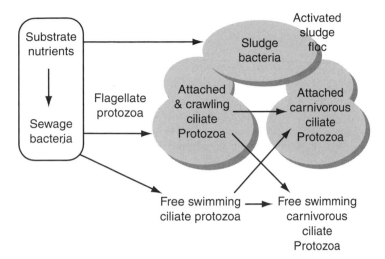

Figure 8.4 Microbial community in activated sludge flocs. From Wagner and Amann (1996).

(Ziglio et al., 2002). These techniques allow the distinction between viable cells, dead cells, and damaged cells. The development of microsensor technology has enabled the exploration of the floc microenvironment (see Chapter 18). Microelectrodes are now being used to determine oxygen, pH, redox potential, nitrate, ammonia or sulfide micro-profiles within activated sludge flocs. These microelectrodes give information on microbial activity within the floc (Li and Bishop, 2004).

Activated sludge flocs contain a wide range of prokaryotic and eukaryotic microorganisms, and many of them can be routinely observed with regular phase-contrast microscopy. A color atlas of wastewater organisms is available and should be consulted to become familiar with the most encountered organisms in activated sludge or trickling filters (Berk and Gunderson, 1993).

8.3.1.1 *Bacteria.* As the oxygen level in the flocs is diffusion-limited, the number of active aerobic bacteria decreases as the floc size increases (Hanel, 1988). Anoxic zones can occur within flocs, depending on the oxygen concentration in the tank. They disappear when the oxygen concentration exceeds 4 mg/L (Li and Bishop, 2004). The inner region of relatively large flocs favors the development of strictly anaerobic bacteria such as methanogens or sulfate-reducing bacteria (SRB) (Lens et al., 1995). The presence of methanogens and sulfate-reducing bacteria can be explained by the formation of several anaerobic pockets inside the flocs or by the tolerance of certain methanogens and SRB to oxygen (Wu et al., 1987) (Fig. 8.5). Thus, activated sludge could be a convenient and suitable seed material for starting anaerobic reactors.

Bacteria, particularly the gram-negative bacteria, constitute the major component of activated sludge flocs. Hundreds of bacterial strains thrive in activated sludge but only a relatively small fraction can be detected by culture-based techniques. They are responsible for the oxidation of organic matter and nutrient transformations, and produce polysaccharides and other polymeric materials that aid in the flocculation of microbial biomass.

The total aerobic bacterial counts in standard activated sludge are in the order of 10^8 CFU/mg of sludge. When using culture-based techniques, it was found that the major genera in the flocs are *Zooglea, Pseudomonas, Flavobacterium, Alcaligenes, Achromobacter, Corynebacterium, Comomonas, Brevibacterium, Acinetobacter, Bacillus*

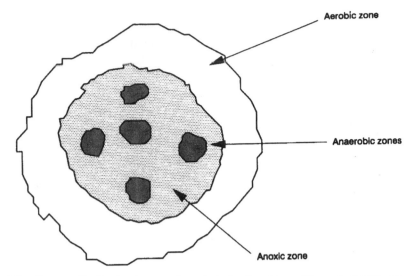

Figure 8.5   Oxygen distribution in an activated sludge floc. From Wu et al. (1987). (Courtesy of Pergamon Press.)

spp., as well as filamentous microorganisms. Some examples of filamentous microorganisms are the sheathed bacteria (e.g., *Sphaerotilus*) and gliding bacteria (e.g., *Beggiatoa*, *Vitreoscilla*), which are responsible for sludge bulking (see Chapter 9 for more details on these bacteria). Table 8.2 displays some bacterial genera found in standard activated sludge, using culture-based techniques. The majority of the bacterial isolates were identified as *Comamonas-Pseudomonas* species. Analysis of the quinone composition of activated sludge revealed that ubiquinone Q-8 was the predominant quinone (Hiraishi et al., 1989).

Culture-based techniques show, however, less than 10 percent of the total cell numbers obtained by direct microscopic counts. New approaches for characterizing the bacterial communities in activated sludge include 16S and 23S rRNA-targeted oligonucleotide fluorescent probes for in situ identification of bacteria (Manz et al., 1994;

TABLE 8.2.   Distribution of Aerobic Heterotrophic Bacteria in Standard Activated Sludge

Genus or Group	% Total Isolates
Comamonas-Pseudomonas	50.0
Alcaligenes	5.8
Pseudomonas (fluorescent group)	1.9
Paracoccus	11.5
Unidentified (gram-negative rods)	1.9
Aeromonas	1.9
Flavobacterium-Cytophaga	13.5
Bacillus	1.9
Micrococcus	1.9
Coryneform	5.8
Arthrobacter	1.9
Aureobacterium-Microbacterium	1.9

Wagner et al., 1993). These techniques give information about the bacterial community structure in activated sludge. This approach showed that the dominant group in activated sludge is the β subclass of proteobacteria (the dominant group within this subclass is the β1 group encompassing bacteria such as *Comamonas, Hydrogenophaga, Acidovorax, Sphaerotilus natans, Leptothrix discophora*, and several pseudomonads). Other groups found include the alpha (e.g., *Sphingomonas*) and gamma subclasses of proteobacteria, *Cytophaga-flavobacterium* cluster, gram-positive bacteria with high G+C DNA content, *Acinetobacter* and *Arcobacter*, a potential human pathogen (Snaidr et al., 1997).

Caulobacter, a stalked bacterium generally found in organically poor waters, has also been isolated from wastewater treatment plants in general and activated sludge in particular (MacRae and Smit, 1991). *Hyphomicrobium*, a gram-negative budding bacterium that produces swarmer cells was also found in activated sludge; some species carry out denitrification, using methanol as the carbon source (Holm et al., 1996). 16S RNA analysis showed the presence of *Hyphomicrobium* strains in an industrial activated sludge treating wastewater from chemical manufacturing. *Hyphomicrobium* 16S rRNA was approximately 5 percent of the 16S rRNA in the activated sludge (Layton et al., 2000).

Gram-negative cocci, known as the "G" bacteria, are found in bioreactors fed glucose and acetate (Cech et al., 1993). They are seen microscopically as tetrads or aggregates in activated sludge (Fig. 8.6; Seviour, 2002). They dominate in systems with poor phosphorus removal because they out-compete phosphorus-accumulating organisms (PAO) by accumulating polysaccharides instead of polyphosphates (see Chapter 3). Two strains of "G" bacteria were identified as *Tetracoccus cechii* and belong to the alpha group of proteobacteria (Blackall et al., 1997). A fluorescently-labelled rRNA-targeted oligonucleotide probe was used to detect the G-bacterium *Amaricoccus* in 46 activated sludge samples, thus showing that this bacterium is commonly found in activated sludge (Maszenan et al., 2000).

Zoogloea are exopolysaccharide-producing bacteria that produce typical finger-like projections and are found in wastewater and other organically enriched environments (Fig. 8.7) (Norberg and Enfors, 1982; Unz and Farrah, 1976; Williams and Unz, 1983).

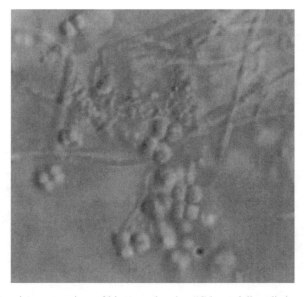

Figure 8.6 Light microscope view of biomass showing "G bacteria" as distinctive cocci in tetrads and clusters associated with the floc. From Seviour (2002).

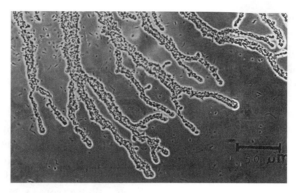

Figure 8.7 *Zooglea ramigera*. (Courtesy of Samuel R. Farrah.)

These finger-like projections consist of aggregates of *Zooglea* cells surrounded by a polysaccharide matrix (Fig. 8.8). They are isolated by using enrichment media containing *m*-butanol, starch, or *m*-toluate as the carbon source. They are found in various stages of wastewater treatment but their numbers comprise only 0.1–1 percent of the total bacterial numbers in the mixed liquor (Williams and Unz, 1983) although, through the use of 16S rRNA-targeted probes, their level was found to be as high as 10 percent (Rosellot-Mora et al., 1995). Other methods used to detect *Zooglea* in wastewater treatment plants and in surface waters include polyclonal antibodies, scanning electron microscopy, and RT-PCR (Lu et al., 2001). The relative importance of these bacteria in wastewater treatment needs further investigation.

Activated sludge flocs also harbor autotrophic bacteria such as nitrifiers (*Nitrosomonas, Nitrobacter*), which convert ammonium to nitrate (see Chapter 3). The use of 16S rRNA-targeted probes showed that *Nitrosomonas* and *Nitrobacter* species occur in clusters and are in close contact in activated sludge flocs and in biofilms (Mobarry et al., 1996; see Fig. 10.7 in Chapter 10). Phototrophic bacteria such as the purple nonsulfur bacteria

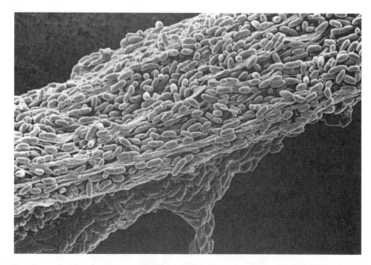

Figure 8.8 Portion of a finger-like projection of *Zooglea ramigera* 106 (ATTC 19544) from a 72 h culture grown on YP medium (yeast extract: 2.5 g; peptone: 2.5 g; water: 1 L) at 28°C. (Courtesy of Fuhu Lu, Jerzy Lukasik, and Samuel Farrah.)

(Rhodospirillaceae) are also detected at concentrations of approximately 10^5 cells/mL. The purple and green sulfur bacteria are found at much lower levels. However, phototrophic bacteria probably play a minor role in BOD removal in activated sludge (Madigan, 1988; Siefert et al., 1978). Figure 8.9 displays a scanning electron micrograph of an activated sludge floc.

8.3.1.2 Fungi.

Activated sludge does not usually favor the growth of fungi, although some fungal filaments are observed in activated sludge flocs. Fungi may grow abundantly under specific conditions of low pH, toxicity, and nitrogen-deficient wastes. The predominant genera found in activated sludge are *Geotrichum*, *Penicillium*, *Cephalosporium*, *Cladosporium*, and *Alternaria* (Pipes and Cooke, 1969; Tomlinson and Williams, 1975). Sludge bulking (see Chapter 9) may result from the abundant growth of *Geotrichum candidum*, which is favored by low pH from acid wastes.

Laboratory experiments have shown that fungi are also capable of carrying out nitrification and denitrification. This suggests that they could play a role in nitrogen removal in wastewater under appropriate conditions. Some advantages of a fungi-based treatment system are the ability of fungi to carry out nitrification in a single step, and their greater resistance to inhibitory compounds than bacteria (Guest and Smith, 2003).

8.3.1.3 Protozoa.

Protozoa are significant predators of bacteria in activated sludge as well as in natural aquatic environments (Curds, 1982; Drakides, 1980; Fenchel and Jorgensen, 1977; LaRiviere, 1977). Protozoan grazing on bacteria (i.e., bacterivory) can be experimentally determined by measuring the uptake of ^{14}C- or ^{35}S-labeled bacteria or fluorescently-labeled bacteria (Hoffmann and Atlas, 1987; Sherr et al., 1987). Protozoan predation on bacteria can also be studied by using bacteria tagged with the green fluorescent protein (GFP) derived from the jellyfish *Aequorea victoria*. This approach was used to follow bacterial ingestion and digestion by ciliates in activated sludge (Chalfie et al., 1994; Eberl et al., 1997). Such grazing can be significantly reduced in the presence

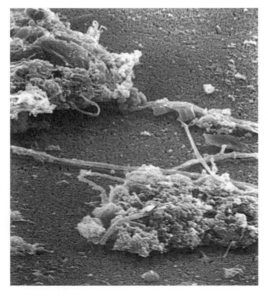

Figure 8.9 Activated sludge flocs observed via scanning electron microscopy. (Courtesy of R. J. Dutton and G. Bitton.)

of toxicants (e.g., heavy metals). For example, *Aspidisca costata* grazing on bacteria in activated sludge is reduced in the presence of cadmium (Hoffman and Atlas, 1987). Protozoa may also graze on *Cryptosporium* oocysts and thus help in the dispersion and transmission of protozoan parasites (Stott ct al., 2001; 2003b).

The protozoa most found in activated sludge have been thoroughly described (Fig. 8.10) (Abraham et al., 1997; Dart and Stretton, 1980; Edeline, 1988; Eikelboom and van Buijsen, 1981).

• *Ciliates*

The cilia that give the organisms their name are used for locomotion and for pushing food particles into the mouth. Ciliates appear to be the most abundant protozoa in activated sludge plants (Sudo and Aiba, 1984). They are subdivided into *free*, *creeping*, and *stalked* ciliates. Free ciliates feed on free-swimming bacteria. The most important genera found in activated sludge are *Chilodonella, Colpidium, Blepharisma, Euplotes, Paramecium, Lionotus, Trachelophyllum,* and *Spirostomum.* Creeping ciliates graze on bacteria on the surface of activated sludge flocs. Two important genera are *Aspidisca* and *Euplotes.* Stalked ciliates are attached by their stalk to the flocs. The stalk has a muscle (myoneme) that allows it to contract. The predominant stalked ciliates are *Vorticella* (e.g., *V. convallaria, V. microstoma*), *Carchesium, Opercularia,* and *Epistylis.*

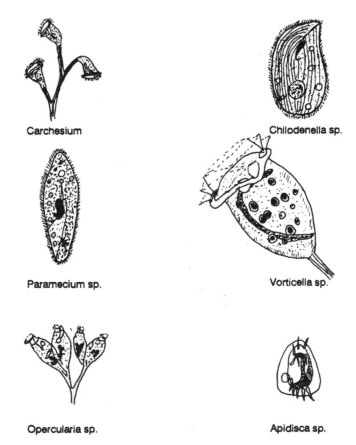

Carchesium

Chilodenella sp.

Paramecium sp.

Vorticella sp.

Opercularia sp.

Apidisca sp.

Figure 8.10   Some protozoa found in wastewater, particularly in activated sludge. Adapted from Dart and Stretton (1980).

• *Flagellates*

These protozoa move via one or several flagella. They take up food via the mouth or via absorption through their cell wall. Some important flagellates found in wastewater are *Bodo* ssp., *Pleuromonas* spp., *Monosiga* spp., *Hexamitus* spp., and a colonial protozoa *Poteriodendron* spp.

• *Rhizopoda (amebae)*

Amebae move slowly via pseudopods (false feet), which are temporary projections of the cells. This group is subdivided into ameba (e.g., *Amoeba proteus*) and thecameba, which are surrounded by a shell (e.g., *Arcella*).

Flagelated protozoa and free-swimming ciliates are usually associated with high bacterial concentrations ($>10^8$ cells/mL), whereas stalked ciliates occur at low bacterial concentrations ($<10^6$/mL). Protozoa contribute significantly to the reduction of BOD, suspended solids, and numbers of bacteria, including pathogens (Curds, 1975). There is an inverse relationship between the number of protozoa in mixed liquor and the COD and suspended solids concentration in activated sludge effluents (Sudo and Aiba, 1984). Changes in the protozoan community reflect those of the plant operating conditions, namely F/M ratio, nitrification, sludge age, or dissolved oxygen level in the aeration tank (Madoni et al., 1993). The protozoan species composition of activated sludge may indicate the BOD removal efficiency of the process. For example, the presence of large numbers of stalked ciliates and rotifers indicate a low BOD. The ecological succession of microorganisms during activated sludge treatment is illustrated in Figure 8.11 (Bitton, 1980a).

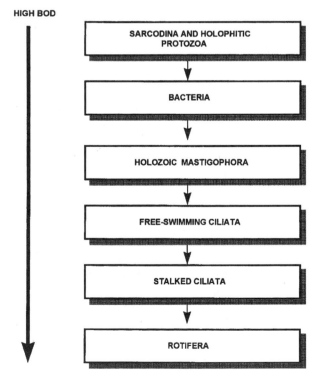

Figure 8.11 Ecological succession of microorganisms in activated sludge. From Bitton (1980).

• *Rotifers*

Rotifers are metazoa (i.e., multicellular organisms) with sizes varying from 100 μm to 500 μm. Their body, anchored to a floc particle, frequently "stretches out" from the floc surface (Doohan, 1975; Eikelboom and van Buijsen, 1981). The rotifers found in wastewater treatment plants belong to two main orders, Bdelloidea (e.g., *Philodina* spp., *Habrotrocha* spp.) and Monogononta (e.g., *Lecane* spp., *Notommata* spp.). The four most common rotifers found in activated sludge and trickling filters are illustrated in Figure 8.12 (Curds and Hawkes, 1975). The role of rotifers in activated sludge is twofold:

1. They help remove freely suspended bacteria (i.e., nonflocculated bacteria) and other small particles and contribute to the clarification of wastewater (Fig. 8.13; Lapinski and Tunnacliffe, 2003). They are also capable of ingesting *Cryptosporidium* oocysts in wastewater and can thus serve as vectors for the transmission of this parasite (Stott et al., 2003a).
2. They contribute to floc formation by producing fecal pellets surrounded by mucus.

The presence of rotifers at later stages of activated sludge treatment is due to the fact that these animals display a strong ciliary action that helps in feeding on reduced numbers of suspended bacteria (their ciliary action is stronger than that of protozoa).

8.3.2 Organic Matter Oxidation in the Aeration Tank

Domestic wastewater has a C:N:P ratio of 100:5:1, which satisfies the C, N, and P requirements of a wide variety of microorganisms. Organic matter in wastewater occurs as soluble, colloidal, and particulate fractions (see Chapter 7). The soluble organic matter serves as a food source for the heterotrophic microorganisms in the mixed liquor. It is quickly removed by adsorption, coflocculation, as well as absorption and oxidation by microorganisms. Aeration for only a few hours leads to the transformation of

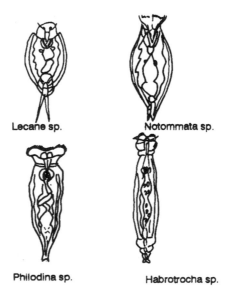

Figure 8.12 Rotifers most commonly found in activated sludge. From Curds and Hawkes (1975). (Courtesy of Academic Press, Orlando, FL.)

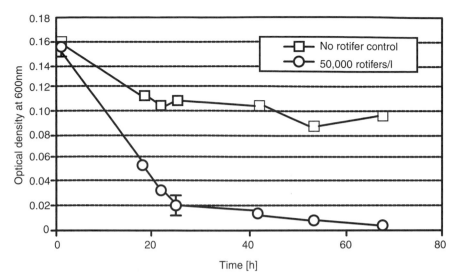

Figure 8.13 Wastewater clarification using bdelloid rotifers. From Lapinski and Tunnacliffe (2003).

soluble BOD into microbial biomass. Aeration serves two purposes: (1) supplying oxygen to the aerobic microorganisms, and (2) keeping the activated sludge flocs in constant agitation to provide an adequate contact between the flocs and the incoming wastewater. An adequate dissolved oxygen concentration is also necessary for the activity of heterotrophic and autotrophic microorganisms, especially nitrifying bacteria. The dissolved oxygen level must be in the 0.5–0.7 mg/L range. Nitrification ceases when DO is <0.2 mg/L (Dart and Stretton, 1980).

Figure 8.14 (Curds and Hawkes, 1983) summarizes the degradation and biosynthetic reactions that occur in the aeration tank of an activated sludge process.

8.3.3 Sludge Settling

The mixed liquor is transferred from the aeration tank to the settling tank where the sludge separates from the treated effluent. A portion of the sludge is recycled to the aeration tank and the remaining sludge is wasted and transferred to an aerobic or anaerobic digester for further treatment (see Chapter 12).

Microbial cells occur as aggregates or flocs, the density of which is sufficient for settling in the clarifying tank. Floc sedimentation is followed by "secondary clarification," which is due to the attachment of dispersed bacterial cells and small flocs to the settling flocs (Olofsson et al., 1998). Flocculation or aggregation of cells are generally a response of microorganisms to low nutrient conditions in their environment. They provide a more efficient utilization of food due to the close proximity of cells. Products released by one group of microorganisms can serve as a growth substrate for another group (McLoughlin, 1994). Thus, sludge settling depends on the F/M ratio and sludge age. Good settling occurs when the sludge microorganisms are in the endogenous phase, which occurs when carbon and energy sources are limited and when the microbial specific growth rate is low. Good sludge settling with subsequent efficient BOD removal occurs at low F/M ratio (i.e., high MLSS concentration). Conversely, a high F/M ratio is conducive to poor sludge settling. In municipal wastewaters, the optimum F/M ratio is 0.2–0.5

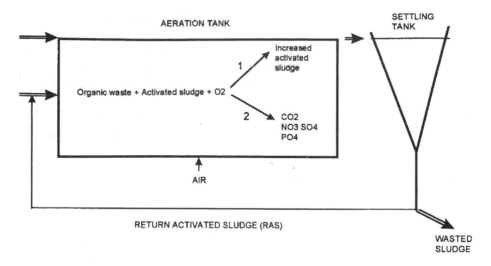

1 BIOSYNTHESIS

2 BIODEGRADATION

Figure 8.14 Organic matter removal by the activated sludge process. Adapted from Curds and Hawkes (1983).

(Gaudy and Gaudy, 1988; Hammer, 1986). A mean cell residence time of 3–4 days is necessary for effective settling (Metcalf and Eddy, 1991). Poor settling can also be caused by sudden changes in physical parameters (e.g., temperature, pH), absence of nutrients (e.g., N, P, micronutrients) and presence of toxicants (e.g., heavy metals), which may cause a partial deflocculation of activated sludge (Chudoba, 1989).

A model explaining the structure of an activated sludge floc was proposed. According to this model, filamentous microorganisms form backbones to which zoogleal (floc-forming) microorganisms attach to form strong flocs (Sezgin et al., 1978). This model does not explain however the absence of filamentous backbones in well flocculating activated sludges (Chudoba, 1989). The production of an intracellular storage product, poly-β-hydroxybutyric acid, was first thought to be responsible for bacterial aggregation. Extracellular polysaccharides in the form of capsules and loose extracellular slimes produced by *Zooglea ramigera* and other activated sludge microorganisms play a leading role in bacterial flocculation and floc formation (Friedman et al., 1969; Harris and Mitchell, 1973; Norberg and Enfors, 1982; Pavoni et al., 1972; Tenney and Stumm, 1965; Vallom and McLoughlin, 1984). It is now accepted that the extracellular polymeric substances (EPS) produced by some activated sludge microorganisms are mainly responsible for floc formation. The use of both transmission electron microscopy and scanning confocal laser microscopy showed that exopolymeric fibrils are important stabilizing components of the floc matrix (Liss et al., 1996). This fibrillar material appears to fill the void spaces between cells in the flocs. Exopolymers are produced during the endogenous phase of growth and help bridge the microbial cells to form a three-dimensional matrix (Parker et al., 1971; Pavoni et al., 1972; Tago and Aida, 1977). Figure 8.15 (Pavoni et al., 1972) shows the correlation between microbial flocculation and extracellular polymer production by activated sludge microorganisms. The polymer concentration increases as the solid retention time increases (Chao and Keinath, 1979). Extracellular polymeric substances production by activated sludge microorganisms has been studied in the laboratory under batch and continuous culture conditions.

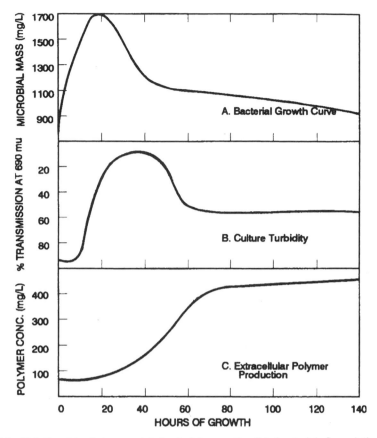

Figure 8.15 Relationship between (a) bacterial growth, (b) bacterial flocculation, and (c) polysaccharide production. Adapted from Pavoni et al. (1972).

These polymeric substances are composed of carbohydrates (e.g., glucose, galactose), amino sugars, uronic acids (glucuronic, galacturonic acids), proteins, lipids, and small amounts of nucleic acids, and are refractory to biodegradation (Flemming and Wingender, 2002; Hejzlar and Chudoba, 1986; Horan and Eccles, 1986; Liao et al., 2001). Recently, it was reported that proteins were the most important polymeric components found in extracted EPS (Wilén et al., 2003). Most of these components contribute to the negative surface charge of flocs. Various probes are available for EPS components (Neu and Lawrence, 2002). For example, specific stains (e.g., Calcofluor White, Congo Red) and lectins help demonstrate the presence of polysaccharides in EPS. Fluorescent markers are used for proteins and nucleic acids Figure 8.16 (Nielsen, 2002) shows the composition of organic matter in activated sludge, with emphasis on EPS.

Since EPS have both hydrophobic and hydrophilic properties, some investigators have demonstrated that hydrophobic interactions are also involved in microbial floc formation and in the adhesion of bacteria to the flocs (Urbain et al., 1993; Zita and Hermansson, 1997a, b). A positive correlation was found between relative hydrophobicity of the flocs and their flocculating ability (Liu and Fang, 2003; Wilén et al., 2003), suggesting that surface charge is less important than hydrophobic binding. Figure 8.17 shows that the adhesion of *E. coli* to activated sludge flocs increases with hydrophobicity of the cell surface (Zita and Hermansson, 1997a). These findings were confirmed by comparing the attachment of green fluorescent protein-marked bacteria to activated sludge flocs.

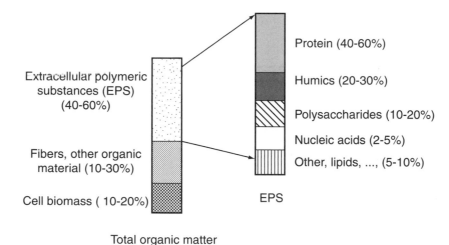

Figure 8.16 Composition of the organic part of sludge (EPS: extracellular polymeric substances). From Nielsen (2002).

Hydrophobic *Serratia marcescens* attached in higher numbers to activated sludge flocs than hydrophilic *E. coli* (Olofsson et al., 1998). Sludge retention time (SRT) influences the composition of EPS and the physicochemical properties (hydrophobicity and surface charge) of the flocs. As SRT increases, the floc surface is less negatively charged and more hydrophobic (Liao et al., 2001).

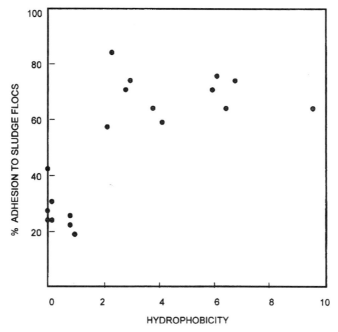

Figure 8.17 Effect of cell surface hydrophobicity on adhesion of bacteria to activated sludge flocs. From Zita and Hermansson (1997), with permission of the publisher.

Divalent cations, mostly Ca and Mg, bridge the negatively charged groups of EPS, play an important role in the flocculation of activated sludge and retain the biopolymers in the floc (Forster and Dallas-Newton, 1980; Kakii et al., 1985; Murthy and Novak, 2001; Wilén et al., 2003). Conversely, monovalent cations such as Na and NH_4 negatively affect the settling properties of activated sludge and help release biopolymers in the suspending medium (Novak, 2001; Murthy and Novak, 2001). It appears that floc stability is affected by the ionic strength of the influent wastewater (Zita and Hermansson, 1994). A suggested structure of the activated sludge floc is displayed in Figure 8.18 (Forster and Dallas-Newton, 1980).

However, excessive production of EPS can be responsible for *bulking* (see Chapter 9), a condition consisting of loose flocs that do not settle well. Poor sludge settleability, as indicated by a high sludge volume index (SVI) was associated with the amount of total EPS (Liao et al., 2001). This condition (nonfilamentous bulking), contrasts with filamentous bulking, which is caused by the excessive growth of filamentous bacteria. Microbial flocculation can be enhanced by adding commercial polyelectrolytes or by adding iron and aluminum salts as coagulants (see Chapter 9).

Extracellular polymers are also responsible for removing phosphorus in activated sludge. Scanning electron microscopy combined with energy dispersive spectrometry (EDS) showed that EPS alone contained, on average, between 27 and 30 percent phosphorus (Cloete and Oosthuizen, 2001).

The conventional way of monitoring for sludge settleability is by determining the *sludge volume index* (SVI). Mixed liquor drawn from the aeration tank is introduced into a 1 L graduated cylinder and allowed to settle for 30 min. Sludge volume is recorded. Sludge volume index, which is the volume occupied by 1 g of sludge, is given by:

$$SVI \ (\text{mL/g}) = \frac{SV \times 1000}{MLSS} \qquad (8.4)$$

where $SV = $ volume of the settled sludge in the graduated cylinder (mL), and $MLSS = $ mixed liquor suspended solids (mg/L).

In a conventional activated sludge plant (with MLSS <3500 mg/L) the normal range of SVI is $50-150$ mL/g. This topic will be discussed further in Chapter 9.

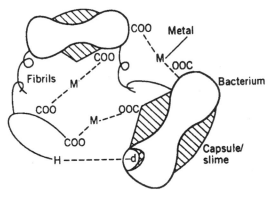

Figure 8.18 Possible structure of an activated sludge floc. From Forster and Dallas-Newton (1980). (Courtesy of the Water Environment Federation.)

8.4 NUTRIENT REMOVAL BY THE ACTIVATED SLUDGE PROCESS

8.4.1 Nitrogen Removal

We have seen in Chapter 3 that nitrogen can be removed from incoming wastewater by chemical–physical means (e.g., breakpoint chlorination or air stripping to remove ammonia) or by biological means, which consist of nitrification followed by denitrification. We will now discuss these two biological processes in suspended-growth bioreactors.

8.4.1.1 *Nitrification in Suspended-Growth Reactors.* The advent of molecular techniques has shed some light on the nitrifying bacterial populations in wastewater treatment plants. Competitive PCR (cPCR; see Chapter 1) enabled the enumeration of *Nitrosomonas* (an ammonia oxidizing bacterium or AOB) and *Nitrospira* (a nitrite oxidizing bacterium or NOB) in activated sludge. The target regions were the *AmoA* gene of *Nitrosomonas* and the 16S rDNA of *Nitrospira*. This showed that AOB represented 0.0033 percent of the total bacterial population as compared to 0.39 percent for NOB (Dionisi et al., 2002). This is much lower than the 2 to 3 percent reported for the nitrifying biomass in municipal wastewater (Koch et al., 2001). These data are important for design models of wastewater treatment plants.

Several factors control nitrification kinetics in activated sludge plants. These factors, discussed in detail in Chapter 3, are ammonia/nitrite concentration, oxygen concentration, pH, temperature, BOD_5/TKN ratio, and the presence of toxic chemicals (Grady and Lim, 1980; Hawkes, 1983a; Metcalf and Eddy, 1991).

The establishment of a nitrifying population in activated sludge depends on the wastage rate of the sludge and, therefore, on the BOD load, MLSS, and mean cell retention time. The growth rate of nitrifying bacteria (μ_n) must be higher than the growth rate (μ_h) of heterotrophs in the system (Barnes and Bliss, 1983; Christensen and Harremoes, 1978; Hawkes, 1983a).

$$\mu_h = 1/\theta \longrightarrow \mu_n \geq 1/\theta$$

where θ = detention time.

In reality, the growth rate of nitrifiers is lower than that of heterotrophs in sewage and, therefore, a long sludge age is necessary for the conversion of ammonia to nitrate. Nitrification is expected at a sludge age above 4 days (Hawkes, 1983a).

$$\mu_h = 1/\theta = Y_h q_h - K_d \tag{8.5}$$

where Y_h = heterotrophic yield coefficient (lb BOD obtained/lb BOD removed/day); q_h = rate of substrate removal (lb BOD removed/lb VSS/day); μ_h = specific growth rate of heterotrophs (day^{-1}); and K_d = decay coefficient (day^{-1}).

Y_h and K_d are assumed to be constant. Therefore, μ_h is reduced by decreasing q_h (U.S. EPA, 1977).

There are at least two approaches for enhancing nitrification in aerobic bioreactors:

- Immobilization of nitrifying bacteria (Tanaka et al., 1996).
- Using bioaugmentation of nitrifiers (Rittmann, 1996; Salem et al., 2003). This approach consists of growing the nitrifiers in a separate reactor fed with return sludge or with an external ammonium source. This bioaugmentation batch enhanced (BABE) process can be used to augment nitrification in activated sludge operating at

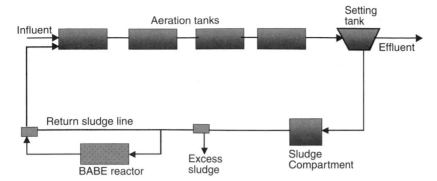

Figure 8.19 The bioaugmentation batch enhanced (BABE) process for bioaugmentation of nitrifiers in activated sludge. From Salem et al. (2003).

suboptimal solid retention times. Figure 8.19 (Salem et al., 2003) shows the configuration of a Dutch activated sludge process with a BABE reactor. Using FISH and microelectrodes, it was shown that bioaugmentation can also enhance nitrification in biofilm reactors (Satoh et al., 2003).

There are two nitrification systems in suspended growth reactors:

1. *Combined carbon oxidation-nitrification (single-stage nitrification system)*. This process is characterized by a high BOD_5/TKN ratio and has a low population of nitrifiers. Most of the oxygen requirement is exerted by heterotrophs (Fig. 8.20).
2. *Two-stage nitrification*. Nitrification proceeds well in two-stage activated sludge systems. In the first stage, BOD is removed, while nitrifiers are active in the second stage (Fig. 8.21*b*).

8.4.1.2 Denitrification in Suspended-Growth Reactors. Nitrification must be followed by denitrification to remove nitrogen from wastewater. Since the nitrate concentration is often much greater (>1 mg/L) than the half saturation constant for denitrification ($K_s = 0.08$ mg/L), the rate of denitrification is independent of nitrate concentration but depends on the concentration of biomass and electron donor (e.g., methanol) in wastewater

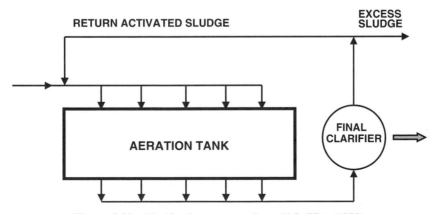

Figure 8.20 Nitrification systems. From U.S. EPA (1977).

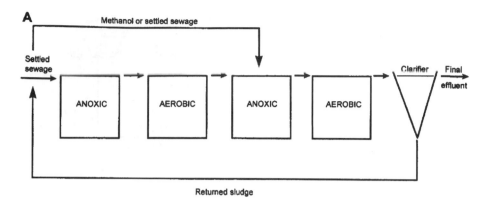

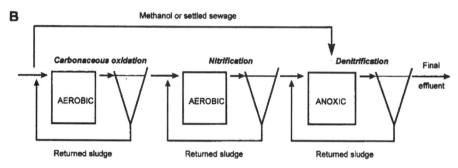

Figure 8.21 Denitrification systems: (A) single-sludge system; (B) multi-sludge system. From Curds and Hawkes (1983). (Courtesy of Academic Press, Orlando, FL.)

(Barnes and Bliss, 1983). The conventional activated sludge system can be modified to encourage denitrification. The systems used are the following (Curds and Hawkes, 1983) (Fig. 8.21).

• *Single Sludge System (Fig. 8.21a)*
This system comprises a series of aerobic and anoxic tanks in lieu of a single aeration tank.

• *Multisludge System (Fig. 8.21b)*
Carbonaceous oxidation, nitrification, and denitrification are carried out in three separate systems. Methanol or settled sewage can serve as sources of carbon for denitrifiers.

• *Bardenpho Process*
This process, illustrated in Figure 8.22, was developed by Barnard in South Africa (U.S. EPA, 1975). The process consists of two aerobic and two anoxic tanks followed by a sludge settling tank. Tank 1 is anoxic and is used for denitrification, using wastewater as a carbon source. Tank 2 is an aerobic tank utilized for both carbonaceous oxidation and nitrification. The mixed liquor from this tank, which contains nitrate, is returned to tank 1. The anoxic tank 3 removes by denitrification the nitrate remaining in the effluent. Finally, tank 4 is an aerobic tank used to strip the nitrogen gas that results from denitrification, thus improving mixed liquor settling.

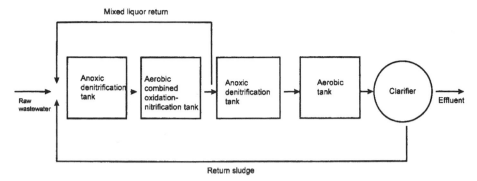

Figure 8.22 Denitrification systems: Bardenpho process. From U.S. EPA (1975).

• Sharon–Anammox Process

The *Sharon process* is a nitrification system that is suitable for wastewater with high ammonia concentrations (>0.5 g N/L). In this process, 50 percent of ammonium is converted to nitrite in the presence of oxygen. The alkalinity of the wastewater is enough to compensate for the production of acidity (Jetten et al., 1997). The Sharon process selects for the fast growing ammonium oxidizing bacteria over the nitrite oxidizers.

$$0.5\,NH_4^+ + HCO_3^- + 0.75\,O_2 \longrightarrow 0.5\,NO_2^- + CO_2 + 1.5\,H_2O \qquad (8.6)$$

The *Anammox process* consists of converting, under anaerobic conditions, nitrite into nitrogen gas, using NH_4^+ as an electron donor and NO_2^- as electron acceptor according to reaction (8.7):

$$NH_4^+ + NO_2^- \longrightarrow N_2 + 2\,H_2O \qquad (8.7)$$

The reaction is carried out by autotrophic bacteria, which do not need an external source of organic matter to produce N_2. These anaerobic ammonia-oxidizing bacteria (e.g., Candidatus *Brocadia anammoxidans*) belong to the Planctomycetales (Strous et al., 1999).

The combined Sharon–Anammox processes remove ammonium according to reaction (8.8):

$$2\,NH_4^+ + 2\,HCO_3^- + 1.5\,O_2 \longrightarrow N_2 + 2\,CO_2 + 5\,H_2O \qquad (8.8)$$

This combined process requires less oxygen and no organic matter (e.g., methanol) is needed to produce N_2 in denitrification.

• CANON Process

The CANON (completely autotrophic nitrogen removal over nitrite) process combines partial nitrification and anammox in a single, aerated reactor. The nitrifiers oxidize ammonia to nitrite and exert an oxygen demand, which leads to anoxic conditions necessary for the anammox bacteria (Schmidt et al., 2003).

The two bacterial groups cooperate according to the following reactions:

$$NH_4^+ + 1.5\,O_2 \longrightarrow NO_2^- + 2\,H^+ + H_2O \qquad [\Delta G^{0\prime} - 275\,kJ\,mol^{-1}] \qquad (8.9)$$

$$NH_4^+ + NO_2^- \longrightarrow N_2 + 2\,H_2O \qquad [\Delta G^{0\prime} - 357\,kJ\,mol^{-1}] \qquad (8.10)$$

8.4.2 Phosphorus Removal

In wastewater treatment plants, phosphorus is removed by chemical means (e.g., P precipitation using iron or aluminum) and by microbiological means. In Chapter 3, we discussed the mechanisms of microorganism-mediated chemical precipitation and enhanced uptake of phosphorus. We will now briefly describe some proprietary phosphate removal processes. All the processes incorporate aerobic and anaerobic stages and are based on phosphorus uptake during the aerobic stage and its subsequent release during the anaerobic stage (Manning and Irvine, 1985; Meganck and Faup, 1988; U.S. EPA, 1987b). The commercial processes can be divided into mainstream and sidestream processes. The most popular processes are described below.

8.4.2.1. Mainstream Processes. Mainstream P removal processes are illustrated in Figure 8.23.

• *A/O (Anaerobic/Oxic) Process*
The A/O process consists of a modified activated sludge system, which includes an anaerobic zone (detention time = 0.5–1 h) upstream of the conventional aeration tank (detention time = 1–3 h) (U.S. EPA, 1987b). Figure 8.24 (Deakyne et al., 1984) illustrates the microbiology of the A/O process. During the anaerobic phase, inorganic phosphorus is released from the cells as a result of polyphosphate hydrolysis. The energy liberated is used for the uptake of BOD from wastewater. Removal efficiency is high when the BOD/phosphorus ratio exceeds 10 (Metcalf and Eddy, 1991). During the aerobic phase, soluble phosphorus is taken up by bacteria that synthesize polyphosphates, using the energy released from BOD oxidation.

The A/O process results in phosphorus and BOD removal from effluents and produces a phosphorus-rich sludge. The key features of this process are the relatively low SRT (solid retention time) and high organic loading rates (U.S. EPA, 1987b).

• *Bardenpho Process*
This system, described in Section 8.4.1.2, removes nitrogen by nitrification–denitrification as well as phosphorus (Meganck and Faup, 1988; Barnard, 1973; 1975).

• *UCT (University of Capetown) Process*
The treatment train in the UCT process includes three tanks (anaerobic → anoxic → aerobic) followed by a final clarifier. In order to have strictly anaerobic conditions in the anaerobic tank, sludge is not recycled from the final clarifier to the first tank (Meganck and Faup, 1988; U.S. EPA, 1987b).

8.4.2.2 Sidestream Processes. *PhoStrip* is a sidestream process that is designed for phosphorus removal by biological as well by chemical means (Tetreault et al., 1986) (Fig. 8.25). A sidestream return activated sludge is diverted to an anaerobic tank called an anaerobic phosphorus stripper, where phosphorus is released from the sludge. The phosphorus-rich supernatant is treated with lime to remove phosphorus by chemical

A: A/O PROCESS

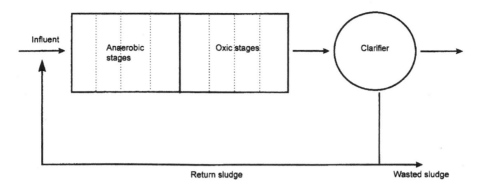

B: BARDENPHO PROCESS

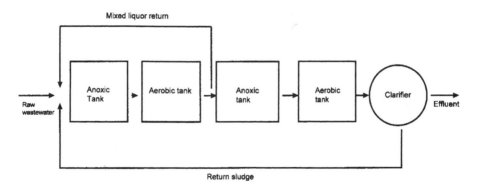

C: UCT PROCESS

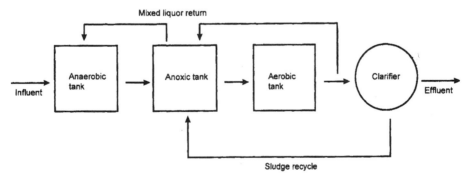

Figure 8.23 Propietary biological phosphorus removal processes: Mainstream processes. Adapted from Meganck and Faup (1988).

precipitation. The solid retention time in the anaerobic tank is 5–20 h. Phosphorus uptake in the aeration basin is ensured when the DO level is >2 mg/L. The PhoStrip process can help achieve an effluent total P concentration of <1 mg/L if the soluble BOD_5 : soluble P is low (12–15) (Tetreault et al., 1986).

The processes discussed above are used in nutrient removal plants to achieve an efficient removal of carbonaceous compounds, nitrogen, and phosphorus. An example of

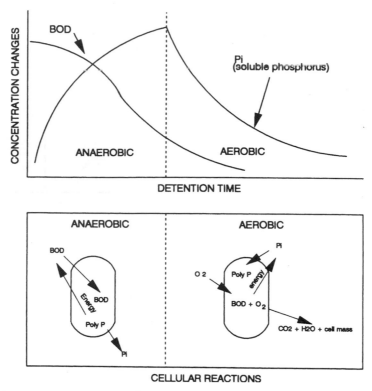

Figure 8.24 Microbiology of the A/O process. From Deakyne et al. (1984). (Courtesy of the Water Environment Federation.)

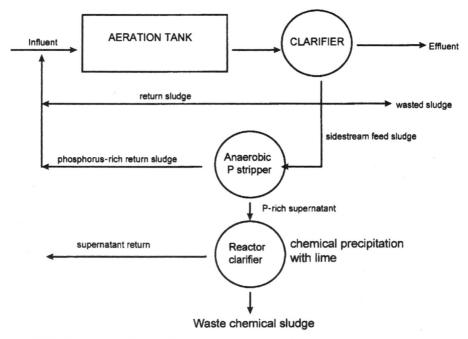

Figure 8.25 Propietary biological phosphorus removal processes: Sidestream processes (PhoStrip). From Tetreault et al. (1986). (Courtesy of the Water Environment Federation.)

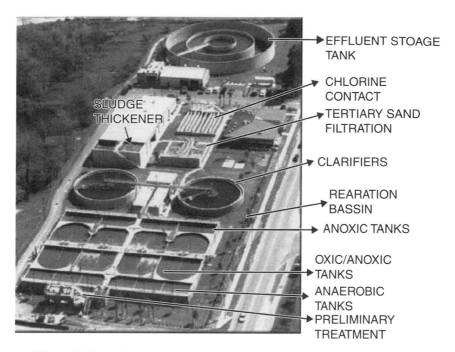

EFFLUENT STOAGE TANK

CHLORINE CONTACT

TERTIARY SAND FILTRATION

CLARIFIERS

REARATION BASSIN

ANOXIC TANKS

OXIC/ANOXIC TANKS

ANAEROBIC TANKS

PRELIMINARY TREATMENT

SLUDGE THICKENER

Figure 8.26 Aerial view of the University of Florida water reclamation plant.

such plants is illustrated in Figure 8.26, which shows an aerial view of the University of Florida water reclamation plant.

8.5 ACTIVATED SLUDGE MODELS

We have discussed the microbiology of the activated sludge process and we have learned that microorganisms, particularly bacteria, are implicated in carbon, nitrogen, and phosphorus transformations and removal (see also Chapter 3).

Activated sludge models are used to describe the main microbiological activities occurring in the system. The modeling has several objectives, including (Mino, 2002):

- prediction of wastewater effluent quality;
- estimation of oxygen demand;
- estimation of sludge produced;
- prediction of microbial behavior in the system.

The International Water Association (IWA) has proposed activated sludge models (ASM) that include several versions (e.g., ASM1, ASM2, ASM2d, ASM3) to predict the behavior of microorganisms in activated sludge (Henze et al., 2000). The main microbial groups that are considered in these models are the heterotrophs (C removal), nitrifiers (production of NO_3), and the phosphate accumulating organisms (PAO) (responsible for enhanced biological phosphorus removal). As an illustration, Figures 8.27 and 8.28 show heterotrophic and nitrifier activities according to the ASM2 model (Mino, 2002). These models will probably be fine tuned in the future as our knowledge of activated sludge microbiology increases.

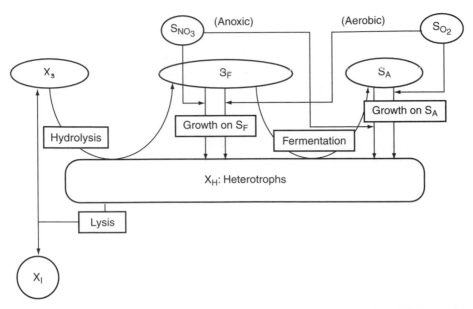

Figure 8.27 Transformation of organic substrates by heterotrophs in ASM 2 model. S_{NO_3}, nitrate and nitrite nitrogen; S_{O_2}, dissolved oxygen; S_F, readily fermentable biodegradable substrates; S_A, fermentation products; X_S, slowly biodegradable substrates; X_H, heterotrophic biomass; X_I, inert organics. From Mino (2002).

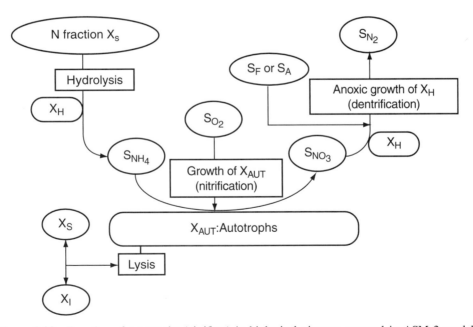

Figure 8.28 Function of autotrophs (nitrifiers) in biological nitrogen removal in ASM 2 model. S_{NH_4}, ammonia nitrogen; S_{N_2}, nitrogen gas; S_{NO_3}, nitrate and nitrite nitrogen; S_{O_2}, dissolved oxygen; S_F, readily fermentable biodegradable substrates; S_A, fermentation products; X_S, slowly biodegradable substrates; X_H, heterotrophic biomass; X_{AUT}, autotrophs (nitrifiers); X_I, inert organics. From Mino (2002).

8.6 PATHOGEN AND PARASITE REMOVAL BY ACTIVATED SLUDGE

Both components (aeration and sedimentation tanks) of the activated sludge process affect, to some extent, the removal/inactivation of pathogens and parasites. During the aeration phase, environmental (e.g., temperature, sunlight) and biological (e.g., inactivation by antagonistic microorganisms) factors, possibly aeration, have an impact on pathogen/parasite survival. Floc formation during the aeration phase is also instrumental in removing undesirable microorganisms. During the sedimentation phase, certain organisms (e.g., parasites) undergo sedimentation, while floc-entrapped microbial pathogens settle readily in the tank. As compared with other biological treatment processes, activated sludge is relatively efficient in removing pathogenic microorganisms and parasites from incoming primary effluents.

8.6.1 Bacteria

Activated sludge is generally more efficient than trickling filters for the removal of indicator (e.g., coliforms) and pathogenic (e.g., *Salmonella*) bacteria. The removal efficiency may vary from 80 percent to more than 99 percent. Bacteria are removed via inactivation, grazing by ciliated protozoa (grazing is particularly effective for free-swimming bacteria), and adsorption to sludge solids or encapsulation within sludge flocs, or both, followed by sedimentation (Feachem et al., 1983; Omura et al., 1989; Yaziz and Lloyd, 1979).

8.6.2 Viruses

The activated sludge process is the most efficient biological process for virus removal from sewage. It appears that most of the virus particles (>90 percent) are solids-associated and are ultimately transferred to sludge (Bitton, 1980a; Omura et al., 1989; Rao et al., 1986). The ability of activated sludge to remove viruses is related to the capacity to remove solids. Thus, many of the viruses found in the effluents are solids-associated (Gerba et al., 1978). Viruses are also inactivated by environmental and biological factors. Attempts have been made to estimate the contribution of both association to solids and inactivation to virus removal in activated sludge. After 10 h of aeration, 25 percent are removed by adsorption to sludge flocs, and 75 percent are removed by inactivation (Glass and O'Brien, 1980). Therefore, inactivation alone is not sufficient for removing most of the viruses with a retention time varying from 6 to 12 h.

Field studies in India have shown that 90–99 percent of enteric viruses were removed by a conventional activated sludge process (Rao et al., 1977) (Table 8.3). Rotaviruses are removed to the same extent as enteroviruses, and a 93–99 percent removal was observed in a 1.5 MGD plant in Houston, Texas (Rao et al., 1986). Lower removal of enteroviruses was observed in some instances. For example, enterovirus removal at Haut de Seine and Nancy wastewater treatment plants in France was only 48 and 69 percent, respectively (Schwartzbrod et al., 1985).

Coliphages have been used as model viruses for studying virus removal by the activated sludge process (Bitton, 1987). These studies generally show that the activated sludge process removes 90–99 percent of phage (Safferman and Morris, 1976). Studies on indigenous coliphage in wastewater treatment plants showed that while 12–30 percent of coliphage are associated with suspended solids in raw wastewater, more than 97 percent are solids-associated in the aeration tank. Most of the adsorbed coliphages were FRNA phages (Ketratanakul and Ohgaki, 1989).

TABLE 8.3. Removal of Enteric Viruses by Primary and Activated Sludge Treatments at the Dadat Sewage Treatment Plant, Bombay, India[a]

Season	Months	Raw Sewage (Virus Conc. PFU/L)	% Reduction after Primary Treatment	% Reduction after Activated Sludge Treatment
Rainy	June 1972	1000	33.5	97.9
	July 1972	1250	24.1	97.0
	June 1973	1200	29.7	95.5
	July 1973	837	29.8	98.9
Autumn	September 1972	300	64.7	98.0
	October 1972	312	66.0	91.7
	October 1973	572	73.0	96.4
	November 1973	1087	56.0	90.0
Winter	January 1973	587	41.4	96.8
	February 1973	468	83.4	95.5
	January 1974	605	47.0	99.6
Summer	March 1973	812	57.0	97.6
	April 1973	875	59.7	98.6
	May 1973	731	66.0	93.5
	March 1974	250	68.8	98.0
	June 1974	694	74.7	99.0

[a]Adapted from Rao et al. (1977).

In summary, virus removal/inactivation by activated sludge may be due to the following:

- virus adsorption to or encapsulation within sludge solids (this results in the transfer of viruses to sludge);
- virus inactivation by sewage bacteria (some activated sludge bacteria may have some antiviral activity);
- virus ingestion by protozoa (ciliates) and small metazoa (e.g., nematodes).

8.6.3 Parasitic Protozoa

Protozoan cysts such as those of *Entamoeba histolytica* are not inactivated under the conditions prevailing in the aeration tank of an activated sludge process. They are, however, entrapped in sludge flocs and are thus transferred to sludge after sedimentation. Similar removals were observed for both *Entamoeba histolytica* and *Giardia* cysts (Panicker and Krishnamoorthi, 1978). More than 98 percent of *Giardia* cysts are removed and become concentrated in sludge. Cysts numbers at various treatment stages of a California wastewater treatment plant are shown in Table 8.4. More than 99 percent of *Giardia* cysts were removed by the activated sludge treatment and most of them were transferred to sludge (Casson et al., 1990; Mayer and Palmer, 1996). A recent survey of four activated sludge plants in Italy showed that *Giardia* removal following activated sludge treatment and disinfection ranged from 87.0 to 98.4 percent (Cacciò et al., 2003). Under laboratory conditions, the activated sludge process removes 80–84 percent of *Cryptosporidium parvum* oocysts (Villacorta-Martinez de Maturana et al., 1992). In-plant reduction of *Cryptosporidium* by the activated sludge process varied from 84.6 percent in a California plant to 96.8 percent in a Canadian plant (Chauret et al., 1995; Mayer and Palmer, 1996). A recent survey of six wastewater treatment plants in the United States showed that the

TABLE 8.4. *Giardia* Cyst Removal in a California Wastewater Treatment Plant[a]

		Sampling Point						
		1	2	3	4	5	6	7
Date	Sampling Period	Raw Wastewater	Primary Effluent	Secondary Effluent	Postchlorination and Filtration	Chlorine Contact Chamber Effluent	Final Effluent	Return Activated Sludge
3/13/89	7:00–15:00	260	380	1	1	1	2	270
	15:00–23:00	360	380	4	1	1	4	530
	23:00–7:00	140	620	2	1	1	1	670
Calculated flow composite		276	427	2	1	1	3	450
3/14/89	7:00–15:00	310	100	2	1	1	1	270
	15:00–23:00	470	660	1	4	1	2	270
	23:00–7:00	110	220	1	1	1	1	530
Calculated flow composite		326	351	1	2	1	1	331

[a]From Casson et al. (1980). Courtesy of the Water Environment Federation.

removal of total or infectious *Cryptosporidium* oocysts varied between 94 and 99 percent following biological treatment. However, 40 percent of the samples were still positive for infectious oocysts following biological treatment, postfiltration, and final disinfection (Gennaccaro et al., 2003). A study in Scotland showed that biological treatment (activated sludge or trickling filter) was superior to primary treatment as regards the removal of *Giardia* cysts and *Cryptosporidium* oocysts. The removal efficiency was higher for *Giardia* than for *Cryptosporidium*. However, the removal of protozoan parasites by biological treatment is very variable. The removal efficiency is estimated at 0–90 percent for *Cryptosporidium* and 60–90 percent for *Giardia* cysts in a wastewater treatment plant that incorporates both primary and secondary stages. Oocysts, introduced into survival chambers that were immersed in wastewater, survived well in this environment. Thus, the removal observed in activated sludge was probably due to sedimentation along with entrapment inside the flocs (Robertson et al., 2000).

8.6.4 Helminth Eggs

Because of their size and density, eggs of helminth parasites (e.g., *Taenia*, *Ancylostoma*, *Necator*) are removed by sedimentation during primary treatment of wastewater and during the activated sludge treatment, thus they are largely concentrated in sludges (Dean and Lund, 1981). Following a survey of wastewater treatment plants in Chicago, parasite eggs were not detected in unchlorinated sewage effluents (Fitzgerald, 1982). Some investigators, however, have reported that activated sludge does not completely eliminate parasite eggs of *Ascaris* and *Toxocara* from wastewater (Schwartzbrod et al., 1989).

8.7 WEB RESOURCES

http://www.scitrav.com/wwater/waterlnk.htm (collection of web pages on activated sludge)
http://protist.i.hosei.ac.jp/PDB/images/Ciliophora/Vorticella/index.html   (page   on *Vorticella*)
http://www.linnbenton.edu/process1/resource/asbasic/asbasic1.html (slide show on activated sludge)
http://www.italocorotondo.it/tequila/ (European online course on water and wastewater treatment)
http://www.swbic.org/education/env-engr/secondary/principles/activated_sludge.html#as (activated sludge principles)
http://www.swbic.org/education/env-engr/secondary/asconfig/configuration.html#Top (activated sludge reactor configuration)
http://wwtp.org/index.php (the Activated Sludge Biomolecular Database, Syracuse University)

8.8 QUESTIONS AND PROBLEMS

1. Compare MLSS to MLVSS. Which one represents more accurately the microbial fraction of activated sludge?
2. Describe an activated sludge floc.
3. What modern tools are used to assess microbial activity in flocs?

4. Since the activated sludge is an aerobic process, is it possible to find anaerobic microorganisms inside the flocs?

5. Discuss the "G" bacteria and their role in activated sludge.

6. As regards the microbiology of activated sludge, what is the advantage(s) of molecular techniques over the older culture-based methods.

7. Summarize the roles of bacteria in an activated sludge system.

8. What is the role of protozoa in activated sludge?

9. What is the effect of F/M ratio on sludge settling?

10. Explain the role of extracellular polymeric substances (EPS) on floc formation in activated sludge.

11. What are some proposed approaches for enhancing nitrification in activated sludge?

12. Explain the Sharon–Anammox Process and the advantages it provides.

13. Explain the mechanisms involved in the removal of pathogens/parasites by activated sludge.

8.9 FURTHER READING

Barnes, D., and P.J. Bliss. 1983. *Biological Control of Nitrogen in Wastewater Treatment*, E. & F.N. Spon, London.

Berk, S.G., and J.H. Gunderson. 1993. *Wastewater Organisms: A Color Atlas*, Lewis Pub., Boca Raton, FL., 25 pp.

Curds, C.R., and H.A. Hawkes. 1983. *Ecological Aspects of Used-Water Treatment*, Vol.2, Academic Press, London.

Edeline, F. 1988. *L'epuration biologique des eaux residuaires: theorie et technologie*, Editions CEBEDOC, Liege, Belgium, 304 pp.

Flemming, H.-C., and J. Wingender. 2002. Extracellular polymeric substances (EPS): Structural, ecological and technical aspects pp. 1223–1231, In: *Encyclopedia of Environmental Microbiology*, Gabriel Bitton, editor-in-chief, Wiley-Interscience, N.Y.

Forster, C.F., and D.W.M. Johnston. 1987. Aerobic processes, pp. 15–56, In: *Environmental Biotechnology*, C.F. Forster and D.A.J. Wase, Eds., Ellis Horwood, Chichester, U.K.

Hanel, L. 1988. *Biological Treatment of Sewage by the Activated Sludge Process*, Ellis Horwood, Chichester, U.K.

Liu, Y., and H.P. Fang. 2003. Influences of extracellular polymeric substances (EPS) on flocculation, settling, and dewatering of activated sludge. Crit. Rev. Environ. Sci. & Technol., 33: 237–264.

Metcalf and Eddy, Inc. 1991. *Wastewater Engineering: Treatment, Disposal and Reuse*, McGraw-Hill, NewYork, 1334 pp.

BULKING AND FOAMING IN ACTIVATED SLUDGE PLANTS

Wastewater Microbiology, Third Edition, by Gabriel Bitton
Copyright © 2005 John Wiley & Sons, Inc.

9.1   INTRODUCTION

Since the introduction of continuous-flow reactors, sludge bulking has been one of the major problems affecting biological waste treatment (Sykes, 1989). There are several types of problems regarding solid separation in activated sludge. These problems are summarized in Table 9.1 (Eikelboom and van Buijsen, 1981; Jenkins, 1992; Jenkins et al., 1984).

9.1.1   Dispersed Growth

In a well-operated activated sludge, bacteria that are not associated with the flocs are generally consumed by protozoa. Their presence in high numbers as dispersed cells results in a turbid effluent. Dispersed growth is associated with the failure of floc-forming bacteria to flocculate. This phenomenon occurs as a result of high BOD loading and oxygen limitation. Toxicity (e.g., metal toxicity) may also cause a dispersed growth of activated sludge bacteria.

9.1.2   Nonfilamentous Bulking

This is sometimes called "zoogleal bulking" and is caused by excess production of exopolysaccharides by bacteria (e.g., *Zooglea*) found in activated sludge. This results in reduced settling and compaction. This type of bulking is rare and is corrected by chlorination (Chudoba, 1989). In industrial wastewater, nonfilamentous bulking may also be due to nitrogen deficiency in wastewater (Peng et al., 2003). Recently, molecular-based methods have shown that zoogleal clusters responsible for dewatering problems in an industrial activated sludge were composed of bacteria belonging to the genus *Thauera* (Lajoie et al., 2002).

9.1.3   Pinpoint Flocs

Pinpoint flocs are caused by the disruption of sludge flocs into very small fragments that may pass into the activated sludge effluent. It was argued that filamentous bacteria constitute the backbone of activated sludge flocs and therefore that their occurrence in low numbers may cause the flocs to lose their structure, resulting in poor settling and release of turbid effluents.

TABLE 9.1.   Causes and Effects of Activated Sludge Separation Problems[a]

Name of Problem	Cause of Problem	Effect of Problem
Dispersed growth	Microorganisms do not form flocs but are dispersed, forming only small clumps or single cells	Turbid effluent. No zone settling of sludge
Slime (jelly) Viscous bulking (also possibly has been referred to as nonfilamentous resulting bulking)	Microorganisms are present in large amounts of extracellular slime	Reduced settling and compaction rates. Virtually no solids separation, in severe cases in overflow of sludge blanket from secondary clarifier
Pin floc (or pinpoint floc)	Small, compact, weak, roughly spherical flocs are formed, the larger of which settle rapidly Smaller aggregates settle slowly	Low sludge volume index (SVI) and a cloudy, turbid effluent
Bulking	Filamentous organisms extend from flocs into the bulk solution and interfere with compaction and settling of activated sludge	High SVI; very clear supernatant
Rising sludge (blanket rising)	Denitrification in secondary clarifier releases poorly soluble N_2 gas, which attaches to activated sludge flocs and floats them to the secondary clarifier surface	A scum of activated sludge forms on the surface of the secondary clarifier
Foaming/scum formation	Caused by (1) nondegradable surfactants and (2) the presence of *Nocardia* sp. and sometimes (3) the presence of *Microthrix parvicella*	Foams float large amounts activated sludge solids to the surface of treatment units Foam accumulates and putrefies. Solids can overflow into secondary effluent or overflow onto walkways

[a]Adapted from Jenkins et al. (1984).

9.1.4   Rising Sludge

Rising sludge is the result of excess denitrification, which results from anoxic conditions in the settling tank. Sludge particles attach to rising nitrogen bubbles and form a sludge blanket at the surface of the clarifier. The final outcome is a turbid effluent with an increased BOD_5. One solution to rising sludge is to lower the sludge retention time (i.e., increase the recirculation rate of activated sludge) in the settling tank.

9.1.5   Filamentous Bulking

Bulking is a problem consisting of slow settling and poor compaction of solids in the clarifier of the activated sludge system (Jenkins and Richard, 1985). Filamentous bulking is usually caused by the excessive growth of filamentous microorganisms. This common problem will be discussed in more detail in Section 9.2.

9.1.6   Foaming/Scum Formation

This problem is due to the proliferation of *Gordonia* and *Microthrix* in aeration tanks of activated sludge units and will be discussed in Section 9.6.

9.2 FILAMENTOUS BULKING

Bulking is caused by the overgrowth of filamentous bacteria in activated sludge. These bacteria are normal components of activated sludge flocs but may outcompete the floc-forming bacteria under specific conditions.

9.2.1 Measurement of Sludge Settleability

Sludge settleability is determined by measuring the *sludge volume index* (SVI), which is given by:

$$SVI = \frac{V \times 1000}{MLSS} \tag{9.1}$$

where V = volume of settled sludge after 30 min (mL/L); and MLSS = mixed liquor suspended solids (mg/L).

The sludge volume index is expressed in mL per gram and is thus the volume occupied by one gram of sludge. A high SVI (>150 mL/g) indicates bulking conditions, whereas an SVI below 70 mL/g indicates the predominance of pin (small) flocs (U.S. EPA, 1987a). Image analysis of activated sludge flocs has helped establish a relationship between the total filaments length and SVI (da Motta et al., 2002).

Two other sludge settleability indices are the *stirred sludge volume index* (the cylinder has a rotating stirring device) and the *diluted sludge volume index* (the volume of settled sludge volume should not exceed 200 mL/L) (Wanner, 2002).

9.2.2 Relationship Between Filamentous and Floc-Forming Bacteria

Based on the relationship between floc-forming and filamentous bacteria, three types of flocs are observed in activated sludge: normal flocs, pin-point flocs, and filamentous bulking.

Normal flocs.   A balance between floc-forming and filamentous bacteria results in strong flocs that keep their integrity in the aeration basin and settle well in the sedimentation tank.

Pin-point flocs.   In these flocs, filamentous bacteria are absent or occur in low numbers. This results in small flocs that do not settle well. The secondary effluent is turbid despite the low SVI.

Filamentous bulking.   Filamentous bulking is caused by the predominance of filamentous organisms. The filaments interfere with sludge settling and compaction. Poor sludge settling, as expressed by SVI, is observed when the total extended filament length exceeds 10^7 μm/mL (Palm et al., 1980; Sezgin, 1982; Sezgin et al., 1978; 1980) (Fig. 9.1). Approximately 25 types of filamentous microorganisms have been identified in bulking activated sludge.

Major physiological differences between floc-forming and filamentous bacteria are summarized in Table 9.2 (Sykes, 1989). Filamentous bacteria have a higher surface-to-volume ratio than that of their floc-forming counterparts, which helps them survive under low oxygen concentration and low nutrient conditions. They also display a low half-saturation constant (K_s in Monod's equation; see Chapter 2) and have a high affinity for substrates, thus behaving as *oligotrophs* and surviving well under starvation conditions. Filamentous bacteria are able to predominate under low dissolved oxygen, low F/M, low nutrient conditions or high sulfide levels. However, it appears that low F/M is the predominant cause of bulking in wastewater treatment plants. These differences

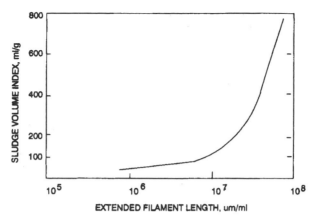

Figure 9.1 Relationship between numbers of filamentous bacteria and sludge volume index. From Sezgin (1982). (Courtesy of Pergamon Press.)

between filamentous and floc-forming bacteria can be exploited to control filamentous bulking in activated sludge (see Section 9.3.2).

9.2.3 Types of Filamentous Microorganisms

Some 20 to 30 types of filamentous microorganisms are known to be involved in activated sludge bulking. A survey of bulking activated sludge plants in the United States has revealed that approximately 15 major types of filamentous microorganisms are responsible for bulking, the most predominant ones being *Nocardia* (*Gordonia*), which is responsible for foaming (see Section 9.6) and type 1701 (Jenkins and Richard, 1985) (Table 9.3). Type 021N was found in 19 percent of more than 400 bulking sludge samples in the United States. Of all the bulking episodes attributable to type 021N, 70 percent were associated with the treatment of industrial wastes or mixtures of industrial and domestic wastes (Richard et al., 1985b). A survey of 167 plants in Italy revealed that two-thirds of the plants experienced bulking or foaming problems. The most encountered filamentous

TABLE 9.2. Comparison of Physiological Characteristics of Floc-Formers and Filamentous Organisms[a]

| Characteristic | Bacteria | |
	Floc-Formers	Filamentous
Maximum substrate uptake rate	High	Low
Maximum specific growth rate	High	Low
Endogenous decay rate	High	Low
Decrease in specific growth rate from low substrate concentration	Significant	Moderate
Resistance to starvation	Low	High
Decrease in specific growth rate from low DO	Significant	Moderate
Potential to sorb organics when excess is available	High	Low
Ability to use nitrate as an electron acceptor	Yes	No
Exhibits luxury uptake of phosphorus	Yes	No

[a]From Sykes (1989), with permission.
DO, dissolved oxygen.

TABLE 9.3. Filamentous Organisms Predominant in U.S. Bulking Activated Sludges

Rank	Filamentous Organism	Percentage of Treatment Plants with Bulking Sludge Where Filament was Observed to be Dominant[a]
1	*Nocardia* spp.	31
2	Type 1701	29
3	Type 021N	19
4	Type 0041	16
5	*Thiothrix* spp.	12
6	*Sphaerotilus natans*	12
7	*Microthrix parvicella*	10
8	Type 0092	9
9	Haliscomenobacter hydrossis	9
10	Type 0675	7
11	Type 0803	6
12	*Nostocoida limicola*	6
13	Type 1851	6
14	Type 0961	4
15	Type 0581	3
16	*Beggiatoa* spp.	<1
17	Fungi	<1
18	Type 0914	<1
	All others	<1

[a]Percentage of 525 samples from 270 treatment plant with bulking problems.
From Jenkins and Richard (1985).

microorganisms were *Microthrix parvicella* followed by Eikelboom types 0041, 021N, 0092, 0675, and *Thiothrix* (Madoni et al., 2000). Another survey of 17 wastewater treatment plants in Pennsylvania showed that the four most encountered filamentous microorganisms were type 0041, type 1701, *Haliscomenobacter hydrossis*, and type 021N (Williams and Unz, 1985b). Type 021N was also detected with relative frequencies of 13 and 21 percent in two wastewater treatment plants in Berlin, Germany (Ziegler et al., 1990). In activated sludge plants in Australia, the dominant organism was *Microthrix parvicella* (Seviour et al., 1994). Approximately 40 new morphotypes of filamentous bacteria have been observed in a recent survey of industrial activated sludge systems (Eikelboom and Geurkink, 2002).

A full description of filamentous organisms found in bulking sludges and in biological foams is given by Kämpfer and Wagner (2002).

9.2.4 Techniques for the Isolation and Identification of Filamentous Microorganisms

There are three approaches for the identification of filamentous bacteria: microscopy-based techniques, immunological techniques, and RNA chemotaxonomy.

9.2.4.1 *Microscopy-Based Techniques.* Filamentous bacteria have long been considered "unusual" microorganisms in classical microbiology textbooks. The pioneering work of Eikelboom and van Buijsen has led to the development of methods for the isolation and identification of these organisms in activated sludge (Eikelboom, 1975; Eikelboom and van Buijsen, 1981; Richard et al., 1985a; Williams and Unz, 1985b; 1989).

Some have isolated the filaments by micromanipulation followed by plating on specific growth media (Seviour et al., 1994). Although some gram-positive filamentous bacteria (e.g., *Gordonia*, *Rhodococcus rhodochrous*, *Sphaerotilus natans*) can grow on relatively rich media, most of the gram-negative filamentous organisms prefer nutrient-poor media. Ammonium and 2- to 4-carbon organic acids support the growth of all filamentous sulfur bacteria (Williams and Unz, 1989). *Microthrix parvicella* has also been cultivated successfully in a chemically defined medium containing mainly Tween 80, reduced nitrogen and sulfur compounds (Slijkhuis and Deinema, 1988).

The application of conventional techniques for the identification of filamentous microorganisms is both difficult and time-consuming. Other problems are their slow growth and difficulties in obtaining pure cultures from activated sludge samples (Wanner and Grau, 1989). Thus, filamentous microorganisms were first characterized mainly by microscopic examination, mostly with a phase contrast microscope. For such identification, information about the following characteristics should be obtained (Eikelboom and van Buijsen, 1981) (Fig. 9.2):

- *Filament shape.* The filaments can be straight, curved, mycelial, or twisted.
- *Size and shape of cells* within the filament (e.g., rods, cocci).
- *Branching.* Fungi and actinomycetes such as *Gordonia* have branched filaments. Some filamentous bacteria such as *Sphaerotilus natans* display false branching.
- *Filament motility.* For example, *Beggiatoa* move by *gliding* on a surface.
- *Presence of a sheath.* Some filamentous bacteria produce a *sheath* (see Chapter 1), which is a tubular structure that encloses the cells. Although difficult to see, this structure can be observed in preparations where cells are missing. Staining with 0.1 percent crystal violet is a common method for sheath detection. However, for some filamentous bacteria such as *Thiothrix*, sheaths may be present in some strains and absent in others (Williams et al., 1987).

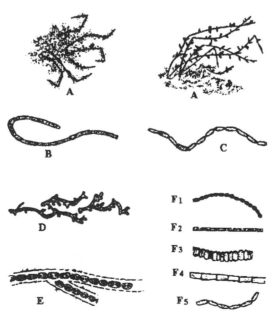

Figure 9.2 Some morphological characteristics of filamentous bacteria. Adapted from Eikelboom and van Buijsen (1981).

- *Presence of epiphytic bacteria on filament surfaces.*   Several bacterial cells become attached to the surface of some filamentous bacteria.

- *Filament size and diameter.*   These factors can be useful for distinguishing between fungi and branched filamentous actinomycetes such as *Nocardia (Gordonia)*.

- *Presence of granules.*   Granules are inclusions in filamentous bacteria and are used to store food. The presence of sulfur granules is indicated by the observation of bright yellow colored granules following addition of sodium sulfide or thiosulfate to the sample. *Beggiatoa, Thiothrix,* and type 021N have sulfur granules (Richard et al., 1985b). The dark-field micrographs displayed in Figure 9.3 show sulfur granules in filaments of *Thiothrix* (Brigmon and Bitton, unpublished). The sulfur granules are surrounded by a single-layered envelope within invaginations of the cytoplasmic membrane. *Thiothrix* was first observed by Winogradsky in 1888, but its isolation in pure culture was not undertaken until more than a hundred years later. This sheathed bacterium forms filaments that produce motile gonidia (Larkin, 1980).

Other tests help detect the presence of polyphosphate and polyhydroxybutyric (PHB) granules.

- *Staining of Activated Sludge Dry Smears*

The following tests are performed, using regular transmitted light.

- *Gram stain.*   This test distinguishes between gram-positive and gram-negative bacteria and is based on the chemical composition of bacterial cell walls.

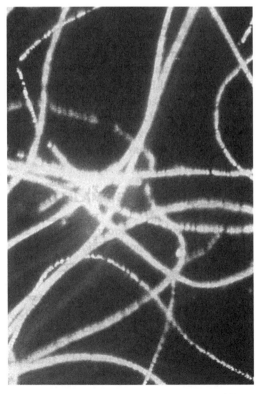

Figure 9.3   Sulfur granules of *Thiothrix* (dark-field microscopy; 1000×). From R. Brigmon and G. Bitton.

- *Neisser stain*. This staining technique helps in the observation of polyphosphate granules in filamentous bacteria. Cells are stained with a mixture of methylene blue and crystal violet, and are counterstained with chrysoidin Y (Eikelboom and van Buijsen, 1981). Neisser-negative bacteria appear as light brown to yellowish filaments, whereas Neisser-positive bacteria display dark polyphosphate granules.
- *Other characteristics*. *Rosettes* are sometimes observed in *Thiothrix* and type 021N (Fig. 9.4) (Brigmon and Bitton, unpublished). In both organisms, the basal cells at the end of the filaments are held together by a holdfast material, which is stained with ruthenium red, suggesting the presence of acidic polysaccharides (Costerton, 1980). Using these characteristics, one then attempts to identify the filamentous bacteria according to the dichotomous key shown in Figure 9.5 (Eikelboom, 1975; Eikelboom and van Buijsen, 1981; Jenkins et al., 1984). Figure 9.6 displays micrographs of some common filamentous bacteria.

There are, however, limitations to the morphology-based identification of filamentous bacteria. These limitations may include variable Gram stain reaction, nonfilamentous growth for some bacteria, loss of sheath-forming capacity, and difficulties in distinguishing between certain types of bacteria (Wagner et al., 1994). Since microscopic techniques are not always reliable for the identification of filamentous microorganisms, other approaches have been explored.

9.2.4.2 Fluorescent-Antibody Techniques.

A fluorescent-antibody technique has been used for the detection of *Sphaerotilus natans* in activated sludge. The antiserum did not display any reaction with other tested filamentous microorganisms in activated sludge (e.g., *Haliscomenobacter hydrossis*, *Microthrix parvicella*, type 1701, type 021N, type 0041, *Thiothrix*) (Howgrave-Graham and Steyn, 1988). Monoclonal antibodies may also be useful tools for the rapid identification of filamentous microorganisms in activated sludge. A monoclonal antibody has been made against *Thiothrix* and tested in wastewater treatment plants as well as in sulfur spring waters in Florida (Brigmon et al., 1995). Polyclonal and monoclonal antibodies were also useful for the detection of *Microthrix parvicella* in activated sludge (Connery et al., 2002). However, immunological methods are not suitable for the identification of many filamentous bacteria.

9.2.4.3 RNA Chemotaxonomy.

RNA chemotaxonomy is the use of ribosomal RNA (rRNA) for the taxonomic classification and identification of microorganisms

Figure 9.4 Rosette formation in *Thiothrix* (dark-field microscopy; 1000×). From R. Brigmon and G. Bitton.

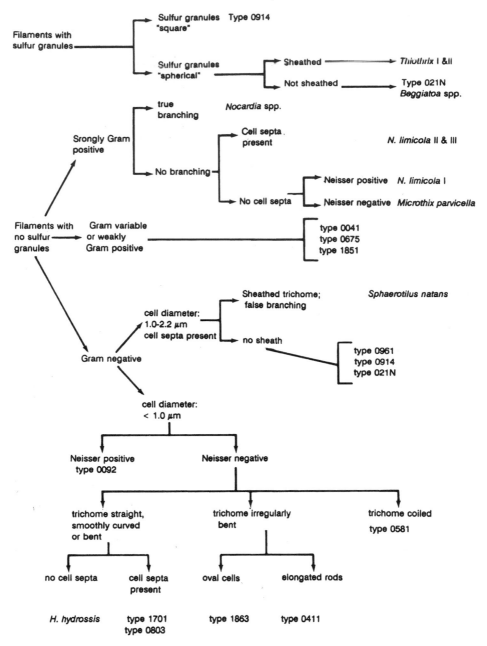

Figure 9.5 Simplified dichotomous key for identification of filamentous microorganisms. Adapted from Jenkins et al. (1984). (Courtesy of Pergamon Press.)

(Head et al., 1998; Hofle, 1990), avoiding a reliance on morphological characteristics or culture-based techniques. Sensitivity is much greater (actively growing cells may contain $\leq 10^4$ ribosomes/cell) than for the detection of DNA sequences. Among more than 80 morphotypes of filamentous bacteria, only about 20 species have been identified using FISH (fluorescent in situ hybridization) technology (Martins et al., 2004). Fluorescently labeled 16S rRNA-targeted oligonucleotide probes were considered for

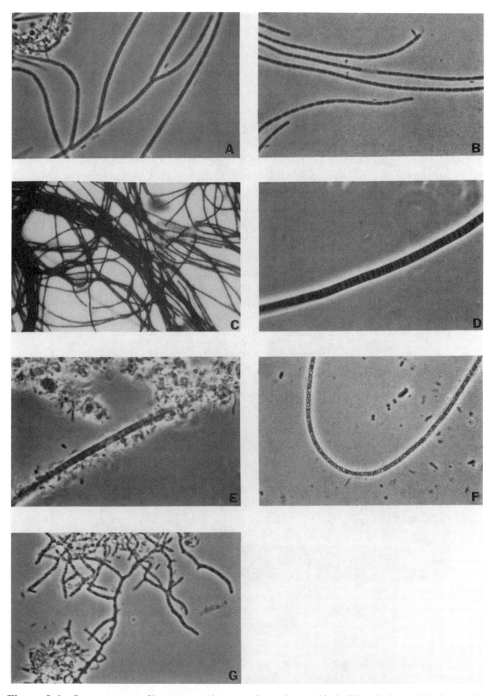

Figure 9.6   Some common filamentous microorganisms observed in bulking sludges: (a) *Sphaerotilus natans*; (b) type 1701; (c) *Microthrix parvicella*; (d) type 021N; (e) type 0041; (f) *Thiothrix* II; (g) *Nocardia* sp. (except for *M. parvicella*, which was gram-stained, all others were observed via phase contrast microscopy at 1000×). (Courtesy of M. Richards.)

in situ detection of filamentous bacteria (*Sphaerotilus* spp., *Thiothrix*, *Haliscomenobacter*, type 021N, *Gordonia amarae*, and other causative organisms of foaming) in activated sludge (De Los Reyes et al., 1997; Wagner et al., 1994). Figure 9.7 shows *Haliscomeno-bacter hydrossis*, which was detected in activated sludge with a fluorescein-labeled oligo-nucleotide probe (Wagner et al., 1994). The rRNA probes can help distinguish between two filamentous sulfur bacteria, *Thiothrix* sp. and Eikelboom type 021N. In Figure 9.8, *Thiothrix* sp. is stained in green, and type O21N is stained in red. The fluorescence signal can be improved by using a scanning confocal laser microscope. 16S ribosomal DNA (rDNA) sequence analysis showed three distinct groups among the Eikelboom type 021N isolates (Kanagawa et al., 2000). The use of the FISH technique has revealed the existence of five phylogenetic groups of *Thiothrix*. *Thiothrix eikelboomii* was the predominant *Thiothrix* found in activated sludge plants with poor sludge settleability (Kim et al., 2002b). This approach was taken for the in situ identification of filamentous bacteria in activated sludges from industrial wastewater treatment plants in Europe (Eikelboom and Geurkink, 2002; van der Waarde et al., 2002). RNA chemotaxonomy showed sometimes an affiliation between bacteria (e.g., *Sphaerotilus* spp. and Eikelboom type 1701) that are distinct according to the Eikelboom key (Wagner et al., 1994). Similarly, 16S ribosomal DNA (rDNA) sequence analysis showed a monophyletic cluster of 021N and *Thiothrix* species (Kanagawa et al., 2000).

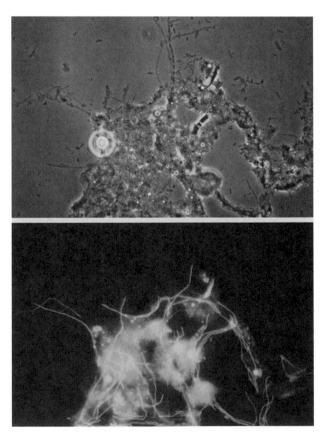

Figure 9.7   *Haliscomenobacter hydrossis* in activated sludge, as shown with a fluorescein-labeled oligonucleotide probe: (a) phase contrast micrograph; (b) epifluorescence micrograph. From Wagner et al. (1994). Figure also appears in color figure section.

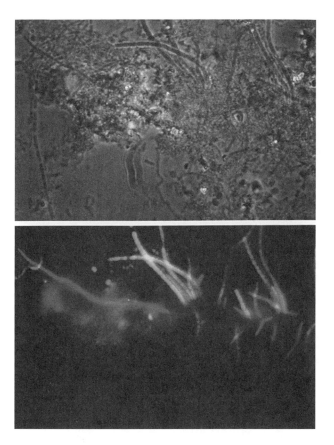

Figure 9.8 *Thiothrix* sp. and type 021N in activated sludge, as shown with labeled oligonucleotide probes: (a) phase contrast micrograph; (b) epifluorescence micrograph (*Thiothrix* is green, and type 021N is red). From Wagner et al. (1994). Figure also appears in color figure section.

Probe sequences are now available for many filamentous bacteria that cause bulking or foaming in municipal and industrial treatment plants. Table 9.4 displays the probe sequences and target sites for some important filamentous bacteria (Kämpfer and Wagner, 2002). The phylogenetic 16S rRNA-based trees for the gram-positive and gram-negative filamentous bacteria are displayed in Figures 9.9 and 9.10 (Kämpfer and Wagner, 2002).

9.3 SOME FACTORS CAUSING FILAMENTOUS BULKING

Filamentous microorganisms are normal components of the activated sludge flocs. Their overgrowth may be due to one or a combination of the following factors.

9.3.1 Wastewater Composition

High-carbohydrate wastes (e.g., brewery and corn wet-milling industries) appear to be conducive to sludge bulking. Carbohydrates composed of glucose, maltose, and lactose, but not galactose, support the growth of filamentous bacteria (Chudoba, 1985;

TABLE 9.4. Probe Sequences and Target Sites for Different Filamentous Bacteria[a]

Target Organism	Probe Name (ODP Name)[b]	Sequence	Target Site (16S RNA Position)
Microthrix parvicella	MPA60	5'-GGATGGCCGCGTTCGACT-3'	60–77
	MPA223	5'-GCCGCGAGACCCTCCTAG-3'	223–240
	MPA645	5'-CCGGACTCTAGTCAGAGC-3'	645–661
	MPA650	5'-CCCTACCGGACTCTAGTC-3'	650–666
Gordonia	(S-G-Gor-0596-a-A-22)	5'-TGCAGAATTTCACAGACGACGC-3'	596–617
Gordonia amarae group 1	(S-G-G.am1-0439-a-A-19)	5'-TCGCGCTTCGTCCCTGGTG-3'	439–457
Gordonia amarae group 2	(S-G-G.am2-0439-a-A-19)	3'-CGAAGCTTCGTCCCTGGCG-5'	439–456
Nostocoida limicola-like filaments	AHW183	5'-CCGACACTACCCACTCGT-3'	183–200
	Noli-644	5'-TCCGGTCTCCAGCCACA-3'	644–660
	PPx3-1428	5'-TGGCCCACCGGCTTCGGG-3'	1428–1447
	MC2-649	5'-CTCTCCCGGACTCGAGCC-3'	649–667
Sphaerotilus natans and some other bacteria	SNA	5'-CATCCCCCTCTACCGTAC-3'	665–673
Thiothrix spp.	TNI	5'-CTCCTCTCCCACATTCTA-3'	652–669
021N group 1	G1B (S-[c]021Ng1-1029-a-A-18)	5'-TGTGTTCGAGTTCCTTGC-3'	1029–1046
021N group 2	G2 M(S-[c] 021Ng2-842-a-A-18)	5'-GCACCACCGACCCCTTAG-3'	842–859
021N group 3	G3 M(S-[c]021Ng3-996-a-A-18)	5'-CTCAGGGATTCCTGCCAT-3'	996–1013
Haliscomenobacter hydrossis	HHY	5'-GCCTACCTCAACCTGATT-3'	655–672

[a]Adapted from a compilation by Kämpfer and Wagner (2002).
[b]OPD = oligonucleotide probe database.
[c]*E. coli* numbering.

Chudoba et al.,1985). Some filaments (e.g., *S. natans*, *Thiothrix* spp., type 021N) appear to be favored by readily biodegradable organic substrates (e.g., alcohols, volatile fatty acids, amino acids) (Wanner, 2002), while others (e.g., *M. parvicella*) are able to use slowly biodegradable substrates (Jenkins, 1992).

9.3.2 Substrate Concentration (F/M Ratio)

Low substrate concentration (i.e., low F/M ratio) appears to be the most prevalent cause of filamentous bulking. Filamentous microorganisms are slow-growing organisms and have lower half-saturation constant K_s and μ_{max} than floc-formers. A study of the interaction between type 021N (a filamentous bacterium) and *Zooglea ramigera* (a typical floc-forming bacterium) showed that, under low substrate concentration (low F/M ratio), type 021N outcompetes *Z. ramigera* due to its higher affinity for the substrate (i.e., low K_s). Conversely, under high substrate concentration, *Z. ramigera* outcompetes the

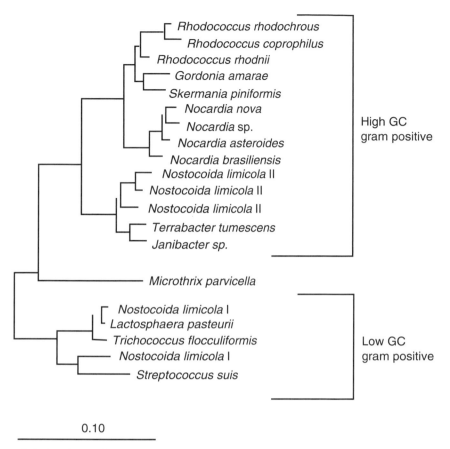

Rhodococcus rhodochrous
Rhodococcus coprophilus
Rhodococcus rhodnii
Gordonia amarae
Skermania piniformis
Nocardia nova
Nocardia sp.
Nocardia asteroides
Nocardia brasiliensis
Nostocoida limicola II
Nostocoida limicola II
Nostocoida limicola II
Terrabacter tumescens
Janibacter sp.

High GC
gram positive

Microthrix parvicella

Nostocoida limicola I
Lactosphaera pasteurii
Trichococcus flocculiformis
Nostocoida limicola I
Streptococcus suis

Low GC
gram positive

0.10

Figure 9.9 Phylogenetic 16S rRNA-based tree showing the affiliation of gram-positive filamentous bacteria. From Kämpfer and Wagner (2002).

filamentous bacterium because of its higher maximum growth rate (van Niekerk et al., 1987). Thus, at low substrate concentration, filamentous microorganisms have a higher substrate removal rate than that of floc-formers, which prevail at high substrate concentrations (Chudoba, 1985; 1989; Chudoba et al., 1973) (Fig. 9.11). *Microthrix parvicella*, a gram-positive filamentous bacterium found in both bulking and foaming sludges, is often associated with low F/M ratios. Although it is aerobic, it can grow for long periods of time under anaerobic conditions. Owing to its low K_s (i.e., high affinity for substrate) it can compete effectively with floc-forming microorganisms at low substrate concentrations (Rossetti et al., 2002). A high substrate concentration can be established by means of a *biological selector* (see Section 9.5.4).

9.3.3 Sludge Loading and Sludge Age

These two parameters are related by the following formula (Chudoba, 1985):

$$\frac{1}{\theta} = YB - K_d \tag{9.2}$$

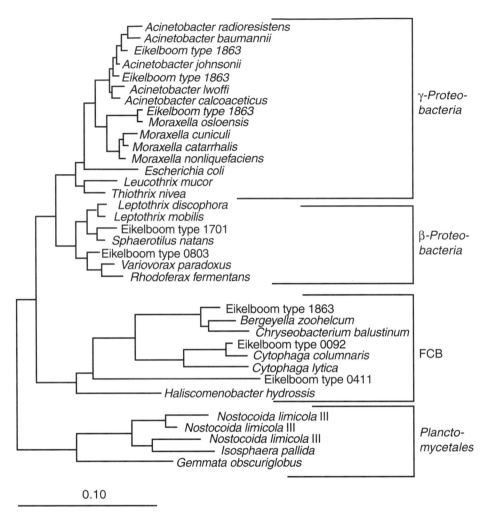

Figure 9.10 Phylogenetic 16S rRNA-based tree showing the affiliation of gram-negative filamentous bacteria; FCB, *Flexibacter-Cytophaga-Bacteroides* phylum. From Kämpfer and Wagner (2002).

where θ = sludge age; B = sludge loading; Y = yield coefficient; and K_d = decay rate of total biomass.

The relationship depends on whether the reactor is a completely mixed or plug flow system. In completely mixed systems, increasing sludge loading leads to a decrease of SVI and thus to a decrease of filamentous microorganisms. At high B values (low sludge age values), filamentous microorganisms are washed out and this leads to poor quality effluents. In the plug-flow pattern, floc-forming bacteria predominate at an optimum B value of approximately $0.3 \, \mathrm{g \, g^{-1} \, day^{-1}}$ (Chudoba, 1985).

Some filamentous microorganisms (e.g., *Thiothrix*, type 1701, *S. natans*) occur over a wide range of sludge age (i.e., MCRT = mean cell retention time) values while others occur only at low (e.g., type 1863) or high (e.g., *M. parvicella*, type 0092) values (Jenkins, 1992) (Fig. 9.12).

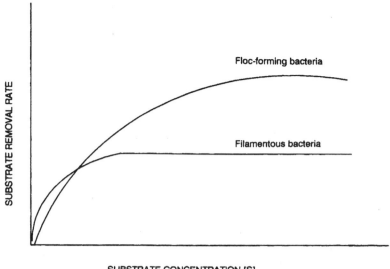

Figure 9.11   Substrate removal rate (r_x) of floc-forming and filamentous bacteria. From Chudoba (1985). (Courtesy of Pergamon Press.)

9.3.4 pH

The optimum pH in the aeration tank is 7–7.5. pH values below 6 may favor the growth of fungi (e.g., *Geotrichum*, *Candida*, *Trichoderma*) and cause filamentous bulking (Pipes, 1974). In laboratory activated sludge units, bulking caused by the excessive growth of fungi occurred after 30 days at pH 4.0 and 5.0 (Hu and Strom, 1991).

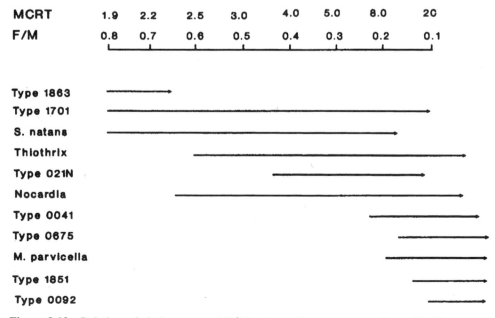

Figure 9.12   Relation of sludge age and F/M ratio to the occurrence of specific filamentous microorganisms in activated sludge. From Jenkins (1992). (Courtesy of Pergamon Press.)

A survey of wastewater treatment plants in Pennsylvania showed that fungi were encountered in 10 percent of samples (Williams and Unz, 1985b).

9.3.5 Sulfide Concentration

High sulfide concentration in the aeration tank causes the overgrowth of sulfur filamentous bacteria such as *Thiothrix*, *Beggiatoa*, or type 021N. These microorganisms use sulfide as a source of energy and oxidize it to elemental sulfur, which is stored as intracellular sulfur granules. *Beggiatoa* growth occurs mostly in fixed-film bioreactors (Jenkins, 1992).

9.3.6 Dissolved Oxygen Level

The growth of certain filamentous bacteria (e.g., *Sphaerotilus natans*, type 1701, *Haliscomenobacter hydrossis*) is favored by relatively low dissolved oxygen levels in the aeration tank (Lau et al., 1984a; Palm et al., 1980; Sezgin et al., 1978; Travers and Lovett, 1984). A substrate overload in the tank may induce oxygen deficiency. Aeration tanks should be operated with a minimum of 2 mg O_2/L to avoid a predominance of specific filamentous microorganisms, namely *Sphaerotilus natans* (Chudoba, 1985). The growth kinetics of *Sphaerotilus natans* and its interaction with a floc-forming bacterium (*Citrobacter* sp.) have been studied under laboratory conditions, using continuous culture techniques. It was essentially shown that low dissolved oxygen is a major factor contributing to the proliferation of this filamentous bacterium in activated sludge. *Sphaerotilus* has a lower K_{DO} ($K_{DO} = 0.01$ mg/L) than that of the floc-forming bacterium ($K_{DO} = 0.15$ mg/L) and, thus, would thrive at low DO in the mixed liquor (Lau et al., 1984a, b). However, no relationship between oxygen level and filament numbers was found when the dominant filamentous bacteria were *Microthrix parvicella* or type 0041 (Forster and Dallas-Newton, 1980).

9.3.7 Nutrient Deficiency

Deficiencies in nitrogen, phosphorus, iron, or trace elements may cause bulking. Some filamentous microorganisms display a high affinity for nutrients. The growth of *S. natans*, *Thiothrix*, and type 021N may be associated with nitrogen and phosphorus deficiencies. It was suggested that the C/N/P ratio should be 100/5/1 (U.S. EPA, 1987).

9.3.8 Temperature

Increased temperature supports the growth of filamentous bacteria associated with low dissolved oxygen concentrations. Moreover, there is a tendency of *Microthrix parvicella* to be the dominant filamentous microorganism during the winter season (Wanner, 2002). This tendency to thrive at low temperature was confirmed in a foaming activated sludge in Italy (Miana et al., 2002).

An integrated hypothesis for sludge bulking has been proposed. According to this hypothesis, activated sludge consists of three categories of "model" microorganisms (Chiesa and Irvine, 1985): (1) fast-growing zoogleal type microorganisms; (2) slow-growing filamentous organisms with high substrate affinity (i.e., low K_s); and (3) fast-growing filamentous organisms with a high affinity for dissolved oxygen (i.e., low K_{DO}). At high substrate concentrations, category 1 is favored as long as there is enough dissolved oxygen. Low substrate concentrations below a critical concentration S^* favor the proliferation of category 2. Category 3 prevails under low DO conditions. Intermittent feeding pattern creates favorable conditions for the development of nonfilamentous microorganisms that have high substrate uptake rates during periods of high substrate concentration and a capacity to store reserve materials during periods of starvation (endogenous metabolism).

Another hypothesis on filamentous bulking is based on the ability of filamentous bacteria to denitrify nitrate to only nitrite with no accumulation of toxic nitric oxide (NO) by the cells. This gives a competitive advantage over floc-forming bacteria (Casey et al., 1992). Further studies are needed to support this hypothesis.

9.4 USE OF FILAMENTOUS MICROORGANISM IDENTIFICATION AS A TOOL TO DIAGNOSE THE CAUSE(S) OF BULKING

Excessive growth of specific filamentous microorganisms is indicative of specific operational problems in the plant, such as low DO, low F/M ratio (i.e., low organic loading rate), high concentration of sulfides in wastewater, nitrogen and phosphorus deficiencies, and low pH (Richard et al., 1982; 1985a; Strom and Jenkins, 1984) (Table 9.5). Therefore, identification of the causative organism is recommended, although some investigators question their use as indicator organisms in the assessment of sludge bulking (Wanner and Grau, 1989; Williams and Unz, 1985b).

In a survey of 89 U.S. wastewater treatment plants with bulking problems, type 1701 was the predominant filamentous microorganism in 33 percent of plants (Richard et al., 1985a). The overgrowth of these filamentous bacteria in activated sludge systems is linked to low dissolved oxygen in the aeration tank. This bacterium has a $K_{DO} = 0.014$ mg/L as compared to $K_{DO} = 0.073$ mg/L for a floc-forming microorganism (Richard et al., 1985a). It thrives in activated sludge plants that treat complex carbohydrates such as brewing wastes and starches.

Excessive growth of fungi is indicative of a low pH level in the aeration basin. The predominance of *Sphaerotilus natans*, type 021N, type 1701, and *Thiothrix* (also an indicator of high sulfide levels) is indicative of low DO in the aeration basin.

Bulking due to sulfur-oxidizing *Thiothrix* is associated with the presence of sulfides in septic wastes (Farquhar and Boyle, 1972). Bulking caused by type 021N is associated with the treatment of septic domestic wastes containing readily degradable carbonaceous substrates such as simple sugars and organic acids and operated at a low F/M ratio (<0.3). This type of bulking is also associated with nutrient deficiencies (N or P deficiencies) or high sulfide levels (Richard et al., 1985a). The predominance of

TABLE 9.5. Dominant Filament Types Indicative of Activated Sludge Operational Problems[a]

Suggested Causative Conditions	Indicative Filament Types
Low F/M	*M. parvicella*, *Nocardia* sp., *H. hydrossis*, 0041, 0675, 0092, 0581, 0961, 0803
Low DO	1701, *S, natans*; possibly 021N and *Thiothrix* sp.
Presence of sulfides	*Thiothrix* sp., *Beggiatoa* sp., possibly 021N
Low pH	Fungi
Nutrient deficiency (N and/or P)	*Thiothrix* sp., possibly 021N

[a]From Richard et al. (1985).

Microthrix parvicella, types 0041, 0675, 0961, 0803, and 0092 is associated with low F/M ratios (Nowak et al., 1986; U.S. EPA, 1987; Daigger et al., 1985). These filamentous microorganisms prevail over floc-forming bacteria under alternating anoxic–aerobic conditions (Casey et al., 1992). The occurrence of *Nostocoida lumicola* is also associated with low organic loading (Nowak and Brown, 1990).

Low temperature encourages the excessive growth of *M. parvicella*. These micro-organisms also maintain a higher growth rate than do floc-forming microorganisms at low organic concentrations. One cure for this type of bulking is the incorporation of a selector that promotes the growth of floc-forming organisms at the expense of the filamentous types (see Section 9.5.4). *Microthrix parvicella* also displays excessive growth in systems (activated sludge or oxidation ditch) fed wastewater with excess long-chain fatty acids and low DO concentration. It displays a high affinity for oxygen (Andreasen and Nielsen, 2000; Slijkhuis and Deinema, 1988). The growth of type 0961 is affected by the feed pattern in the aeration basin. Higher organic loading leads to the loss of their selective advantage over floc-forming bacteria (Nowak et al., 1986).

A simple test was developed to predict the bulking potential of a given sludge. Briefly, the test sludge is amended with milk and its SV_{30} is measured following 24 h incubation (see more details of this methodology in Fig. 9.13; Seka et al., 2001). The ratio of the SV_{30} of the milk-amended sludge to the SV_{30} of the tapwater-amended sludge (control) gives an indication of the stability of the sludge sample. A ratio of 4 or more indicates an instable sludge, whereas a ratio of 2 or less indicates a stable sludge.

9.5 CONTROL OF SLUDGE BULKING

Various approaches are available for controlling sludge bulking in wastewater treatment plants. Some of them are described below.

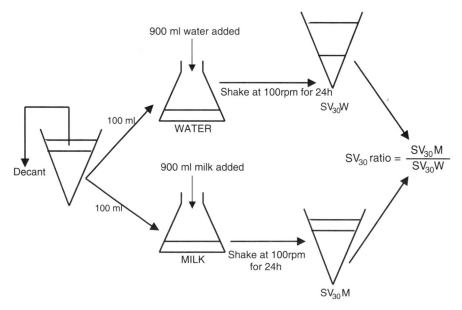

Figure 9.13 Test to predict the bulking potential of an activated sludge. Adapted from Seka et al. (2001).

9.5.1 Treatment with Oxidants

Filamentous bacteria can be controlled by treating the return sludge with chlorine or hydrogen peroxide to selectively kill filamentous microorganisms. This approach is based on the fact that filamentous microorganisms protruding from the flocs are more exposed to oxidants, whereas most of the floc-forming microorganisms embedded inside the flocs are protected from the lethal action of the oxidants (Wanner, 2002). Bulking control by chlorination was proposed over 50 years ago and this practice is probably the most widely used cost-effective and short-term method for controlling filamentous bacteria. Chlorine may be added to the aeration tank or to the return activated sludge (RAS) (Jenkins et al., 1984) (Fig. 9.14). The method of choice is the addition of chlorine to the RAS line as chlorine gas or sodium hypochlorite about three times per day. Chlorine concentration should be 10–20 mg/L (concentrations of >20 mg/L may cause deflocculation and formation of pin-point flocs). Chlorine dosage for bulking control can be rapidly estimated, using a short-term enzyme assay based on reduction of a tetrazolium salt by dehydrogenases. This test helps determine the inhibition of filamentous bacteria by chlorine, using a microscope (Logue et al., 1983). The LIVE/DEAD® BacLight™ stain mixture can also help distinguish live filamentous bacteria from dead ones, based on the effect of the oxidant on membrane integrity. Live filamentous bacteria fluoresce green, while the dead ones fluoresce red (Ramirez et al., 2000).

However, chlorination is sometimes unsuccessful in bulking control. Indeed, chlorine-resistant filamentous bacteria (e.g., chlorine-resistant 021N) have been reported (Séka et al., 2001b).

Hydrogen peroxide is generally added to the RAS at concentrations of 100–200 mg/L. However, as shown for chlorine, excessive levels of hydrogen peroxide can be deleterious to floc-forming bacteria (Cole et al., 1973). In addition to its role as an oxidizing agent, hydrogen peroxide may also act as a source of oxygen in the aeration tank. Ozone was also proposed for curing filamentous bulking (Colignon et al., 1986).

9.5.2 Treatment with Flocculants and Coagulants

Synthetic organic polymers, lime, and iron salts may be added to the mixed liquor to improve bridging between the flocs and thus promote sludge settling. However, the

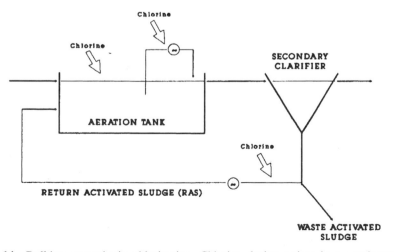

Figure 9.14 Bulking control via chlorination. Chlorine dosing points in an activated sludge system. From Jenkins et al. (1984).

addition of lime and iron salts increases the solids load, and the use of polymers is costly. The addition of cationic polymers at concentrations of 15–20 mg/L resulted in a successful control of bulking in wastewater from the brewing industry. Although treatment with polymers and coagulants leads to an immediate improvement in sedimentation, their effect is of short duration because they exert no adverse effect on filamentous microorganisms. Séka and collaborators (2001) formulated an additive made of a polymer (flocculant), talc (a ballasting agent), and a quaternary ammonium compound CTAB (cetyltrimethyl ammonium bromide, which acts as a biocide) to improve the settling of filamentous bulking activated sludge. This additive led to a long-lasting improvement of settling and retarded sludge rising due to denitrification.

9.5.3 Manipulation of Return Activated Sludge (RAS) Flow Rates

The clarifier in the activated sludge process has two functions: clarification (i.e., floc removal to obtain a clear effluent) and thickening of the sludge. The degree of thickening achieved in the clarifier is given by Eq. (9.3):

$$\frac{X_u}{X} = \frac{(Q + Q_r)}{Q_r} \tag{9.3}$$

where X_u = RAS (return activated sludge) suspended solids concentration (w/v); X = MLSS in aeration tank (w/v); Q = influent flow rate (V/t); and Q_r = RAS flow rate (V/t).

Bulking interferes essentially with sludge thickening in the clarifier and results in a decrease of the RAS suspended solids concentration (X_u). This decrease must be compensated by an increase in RAS flow rate (Q_r). Thus, increasing RAS flow rate helps prevent failure of the clarifier. Reduction of MLSS concentration in the clarifier feed can also help control bulking. This reduction can be achieved by decreasing the mixed liquor solids inventory, which is obtained by increasing the sludge wasting rate (Jenkins et al., 1984).

9.5.4 Biological Selectors

Biological selectors are alternative process configurations that favor the growth of floc-forming bacteria over filamentous bacteria and, thus, help control bulking. A selector is a tank or compartment in which certain parameters (e.g., F/M ratio, electron acceptor) can be manipulated to discourage the overgrowth of undesirable filamentous micro-organisms. The incoming wastewater and the RAS are mixed in the selector tank under the desired conditions prior to entering the aeration basin (Sykes, 1989).

There are three categories of biological selectors: aerobic, anoxic, and anaerobic.

9.5.4.1 Aerobic Selectors.   The concept of *kinetic* selection was introduced during the 1970s (Chudoba et al., 1973). The kinetics are based on the Monod's equation (see Chapter 2). At relatively high substrate concentrations ($S > K_s$), the specific growth rate is controlled by μ_{max}. However, at low substrate concentrations ($S < K_s$), the specific growth rate is controlled mainly by K_s. Filamentous bacteria are slow-growing organisms (*K*-strategists) with lower μ_{max} and K_s than floc-forming microorganisms (r-strategists) (Chudoba et al., 1973) (Fig. 9.15). In contrast, floc-forming bacteria predominate under high substrate concentrations and are also called μ_{max}-strategists.

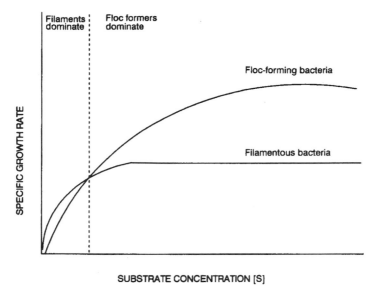

Figure 9.15 Kinetic selection of filamentous microorganisms based on Monod's equation. From Chudoba et al. (1973).

Thus, an aerobic selector consists of creating a substrate concentration gradient (F/M gradient) across the bioreactor. This gradient can be created by using, for example, several reactors in series or establishing an F/M gradient within the same tank (Fig. 9.16a; Chudoba et al., 1973; Sykes, 1989). This configuration gives a selective advantage to floc-forming bacteria, which take up most of the soluble substrate at the head of the reactor.

9.5.4.2 Anoxic Selectors.
An anoxic condition is defined as the absence of oxygen and the presence of nitrate as the electron acceptor. Pilot plant studies have demonstrated the positive effect of anoxic conditions on sludge settling (Chambers, 1982; Price, 1982). This approach consists of setting up an anoxic reactor followed by an aerobic one. In the anoxic reactor, the floc-forming bacteria predominate over the filamentous bacteria because they can take up organic substrates, using nitrate as an electron acceptor. Some filamentous microorganisms (e.g., *S. natans*, 021N, type 0092) cannot use nitrate or nitrite as electron acceptors (Horan et al., 1988; Wanner et al., 1987a) although some can reduce nitrate to nitrite but at much lower rates than floc-forming bacteria (Martins et al., 2004). Nitrate is provided via the recycling of RAS as well as mixed liquor (Barnard, 1973; Sykes, 1989) (Fig. 9.16b). The subsequent low organic substrate concentration in the aerobic reactor is not sufficient enough to sustain the growth of filamentous bacteria. The use of an anoxic selector was successful in controlling the growth of *Nostocoida lumicola* (Nowak and Brown, 1990).

9.5.4.3 Anaerobic Selectors.
Anaerobiosis is defined as the lack of oxygen as well as nitrate as electron acceptors. Anaerobic conditions suppress the growth of filamentous bacteria such as *Sphaerotilus natans* and type 021N. An anaerobic selector is based on the ability of floc-forming bacteria to accumulate polyphosphates (i.e., luxury uptake of phosphorus by poly-P bacteria; see Chapter 3) under aerobic conditions and to use them as a source of energy for uptake of soluble organic substrates under anaerobic conditions.

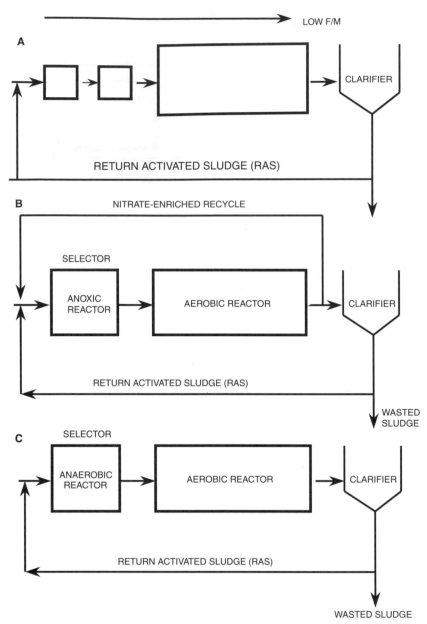

Figure 9.16   Biological selectors for bulking control: (a) Aerobic selector; (b) anoxic selector; (c) anaerobic selector. From Chudoba et al. (1973).

Under anaerobic conditions, filamentous bacteria are unable to uptake organic substrates at a rate comparable to that of poly-P bacteria (Wanner et al., 1987b). Thus, processes designed for biological phosphorus removal help select for floc-forming bacteria in the anaerobic reactor (Sykes, 1989) (Fig. 9.16c). The accumulated carbohydrates will be metabolized in the aerobic reactor. However, an excessive growth of *Thiothrix* was observed under anaerobic conditions under which this filamentous microorganism is able to utilize organic substrates by dissimilatory sulfate reduction and subsequently cause bulking (Wanner et al., 1987b).

9.5.5 Biological Control

Microorganisms, mainly bacteria and actinomycetes, isolated from various sources (e.g., activated sludge, compost, soil) are capable of lysing filamentous bacteria. An active lytic microorganism against type 021N was isolated from soil (Yagushi et al., 1991). Predacious ciliated protozoa (e.g., *Trithigmostoma cucullulus*) are also able to ingest filamentous microorganisms, and their growth in aeration tanks is followed by a decrease of the sludge volume index, an indication of their controlling effect on sludge bulking. Inoculation of bulking activated sludge with these protozoa also results in a decrease of SVI (Fig. 9.17; Inamori et al., 1991). This approach needs further exploration under field conditions.

9.5.6 Other Specific Methods

Preaeration of wastewater to remove sulfides helps control the growth of *Thiothrix*, but higher dissolved oxygen concentrations did not reduce type 0961 growth (Farquhar and Boyle, 1972; Nowak et al., 1986). Filamentous sulfur bacteria (*Thiothrix*, *Beggiatoa*, type 021N) are unable to grow well under low pH conditions (Williams and Unz, 1989). Thus, adjustment and maintenance of wastewater at low pH (e.g., pH 5.5) may help control the growth of filamentous sulfur bacteria, although the possibility exists that the low pH might promote the growth of fungal filaments (Unz and Williams, 1988).

Iron compounds (e.g., ferrous sulfate, potassium ferrate, Fe-cystein) strongly inhibit the respiration of filamentous bacteria such as *Sphaerotilus*, *Thiothrix*, or type 021N (Chang et al., 1979; Kato and Kazama, 1991; Lee, Koopman, and Bitton, unpublished results). These chemicals deserve further exploration.

Some investigators recommend a two-phase approach to solving a bulking episode: (1) chlorination to kill the extended filamentous microorganisms; and (2) identification of the causative microorganism(s) in order to make specific design or operational changes (Jenkins et al., 1984).

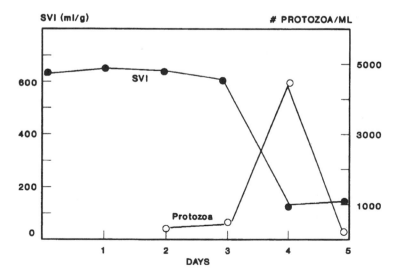

Figure 9.17 Control of filamentous bacteria by predacious protozoa. From Inamori et al. (1991). (Courtesy of Pergamon Press.)

9.6 FOAMING IN ACTIVATED SLUDGE

9.6.1 Introduction

Foaming is a common problem encountered in many wastewater treatment plants around the world. A recent survey indicated that two-thirds of the 114 U.S. plants surveyed experienced foaming at one time or another (Pitt and Jenkins, 1990). A 40 percent incidence was observed in South Africa (Blackbeard et al., 1986). Half to more than 92 percent of the plants surveyed in Australia experienced a foaming problem during the sampling period (Blackall et al., 1988; 1991; Seviour et al., 1990) and 20 percent of 6000 activated sludge plants surveyed in France were affected by foaming (Pujol et al., 1991).

The following are types of foams and scums that are found in activated sludge (Jenkins et al., 1984):

1. Undegraded surface-active organic compounds
2. Poorly biodegradable detergents, which produce white foams
3. Scums due to rising sludge resulting from denitrification in the clarifier
4. Brown scums due to excessive growth of actinomycetes. Figure 9.18 shows the excessive scum growth in the aeration basins at a wastewater treatment plant in Atlanta, GA. This type of scum appears to be the most troublesome in activated sludge and will be covered in some detail.

9.6.2 Problems Caused by Foams

Scum formation in activated sludge plants causes several problems:

1. Excess scum can overflow into walkways and cause slippery conditions, leading to hazardous situations for plant workers.
2. Excess scum may pass into activated sludge effluent, resulting in increased BOD and suspended solids in the effluent.
3. Foaming causes problems in anaerobic digesters.

Figure 9.18 Scum covering the aeration basins at a wastewater treatment plant in Atlanta. (Courtesy of Mesut Sezgin.)

4. Foaming can produce nuisance odors, especially in warm climates.

5. Potential infection of wastewater workers with opportunistic pathogenic actinomycetes such as *Nocardia asteroides.*

9.6.3 Foam Microbiology

Actinomycetes, particularly the mycolic acid-containing bacteria (mycolata), are the most important agents responsible for the brown viscous scum in activated sludge. These organisms generally proliferate in aeration basins operated at high mean cell residence time, but they may also be found in the clarifier. They can now be identified in situ, using fluorescent rRNA-targeted oligonucleotide probes after alcohol preservation and enzyme treatment of the activated sludge samples (Schuppler et al., 1998). Immunomagnetic separation (IMS) using an anti-mycolic acid polyclonal antibody helped target *Gordonia amarae* and other mycolata (Morisada et al., 2002).

Foaming is mainly associated with the mycolic acid-containing actinomycetes group, which belongs to the family Nocardiaceae. Genera from this group that have been observed in foams are *Gordonia* (*G. amarae*, formerly *Nocardia amarae*), *Nocardia* (*N. asteroides, N. caviae, N. pinensis*), *Skermania piniformis* (formerly known as *Nocardia pinensis*), *Tsukamurella*, and *Rhodococcus*. There is a relationship between foam-causing organisms and the initiation and stability of foam in activated sludge. Using FISH to detect mycolic acid-containing actinomycetes (mycolata), Davenport et al. (2000) established a relationship between their numbers and foam production in activated sludge. The predominant (79 percent) mycolata found in foaming activated sludge were the rod- and coccoidal-shaped mycolata (Davenport and Curtis, 2002). The threshold foaming concentration of mycolata was estimated at about 2×10^6 cells mL^{-1}. Using rRNA-targeted hybridization probes, a cause–effect relationship between *G. amarae* and foaming was established. Threshold *Gordonia* levels for foam formation and stability were approximately 2×10^8 μm mL^{-1} and 1×10^9 μm mL^{-1}, respectively (de los Reyes and Raskin, 2002).

Other filamentous organisms observed in foams are *Microthrix parvicella*, *Streptomyces* spp., *Micromonospora*, and type 0675 (Goddard and Foster, 1987a; Klatte et al., 1994; Lechevalier and LeChevalier, 1974; 1975; Lemmer and Kroppenstedt, 1984; Madoni et al., 2000; Pujol et al., 1991; de los Reyes et al., 1997; Seviour et al., 1990; Sezgin and Karr, 1986; Sezgin et al., 1988). Figure 9.19 shows micrographs of *Nocardia* foams in mixed liquor. Other microorganisms causing foaming problems are *Nostocoida limicola* and type 0041. Like *Gordonia amarae*, they have the ability to produce biosurfactants (Goddard and Forster, 1987b; Madoni et al., 2000; Sutton, 1992; Wanner and Grau, 1989). Several other filamentous microorganisms have been reported in foams, but their association with foam formation is not clear.

Gordonia amarae and, to a lesser extent, *Nocardia pinensis*, are the major organisms found in foams examined by U.S. and Australian investigators. *Nocardia pinensis* displays a characteristic "pine tree" branching with acute branching angles and can be distinguished microscopically from other *Gordonia amarae*-like organisms. Nuisance foams are produced when *Nocardia* occurs in mixed liquor at levels exceeding 26 mg *Nocardia* per g VSS (Jenkins, 1992). *Rhodococcus* sp. was the dominant organism in foams examined in Europe. *Microthrix parvicella*, a nonbranching filamentous organism implicated in bulking, is sometimes a dominant organism in foams, as observed in surveys conducted in France and South Africa (Blackall et al., 1988; 1991; Blackbeard et al., 1986; Pujol et al., 1991).

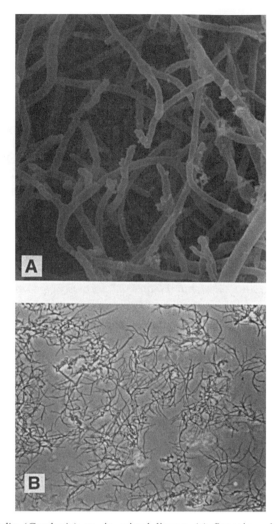

Figure 9.19 *Nocardia* (*Gordonia*) sp. in mixed liquor: (a) Scanning electron micrograph of *Nocardia* foam at the University of Florida activated sludge process. (Courtesy of J. Awong and G. Bitton.) (b) Phase contrast micrograph of *Nocardia* sp. (Courtesy of Mesut Sezgin.)

Foam microorganisms use several growth substrates varying from sugars to high molecular weight polysaccharides, proteins, and aromatic compounds (Lemner, 1986; Lemner and Kroppenstedt, 1984). It has been suggested that scum actinomycetes survive in the aeration tank because they are able to switch from *K*-strategy (i.e., ability to grow at low substrate concentrations because of low K_s) to μ_{max}-strategy (i.e., ability to produce high biomass when sufficient nutrients are present) (Lemner, 1986.). However, study of the growth kinetics of *G. amarae* showed that this actinomycete has a relatively low μ_{max} (0.087 h^{-1}) and a relatively high K_s (675 mg/L). These values lead to the conclusion that excessive growth of this actinomycete in activated sludge cannot be associated with favorable growth kinetics (Baumann et al., 1988), but can possibly be explained by the production of biosurfactants and selective utilization of hydrophobic compounds such as hydrocarbons by *G. amarae*, which has a hydrophobic surface (Lemmer and Baumann, 1988a). These characteristics are essential for foam production and transport of the cells to the bubble phase (Blackall and Marshall, 1989; Blackall et al., 1988). Anionic surfactants and their biodegradation products can significantly enhance foaming in *Nocardia*-containing activated sludge (Ho and Jenkins, 1991). Other advantages of actinomycetes over other wastewater bacteria are their higher resistance to desiccation and ultraviolet irradiation and their ability to store polyphosphates and poly-β-hydroxybutyric acid (Lemmer and Baumann, 1988b).

9.6.4 Mechanisms of Foam Production

The causes and mechanism(s) of foam production are not well understood. Some possible mechanisms are the following (Soddell and Seviour, 1990):

1. Gas bubbles produced by aeration or metabolism (e.g., N_2) may assist in flotation of foam microorganisms.
2. The hydrophobic nature of cell walls of foam microorganisms help their transport to the air-water interface.
3. Biosurfactants produced by foam microorganisms assist in foam formation (Pagilla et al., 2002).
4. Foams are associated with relatively long retention times (>9 days), warm temperatures ($>18°C$) (Pipes, 1978) and with wastewaters rich in fats (Eikelboom, 1975).

9.6.5 Foam Control

Numerous measures for controlling foams in activated sludge have been proposed. Several of these cures are not always successful and have not been rigorously tested under field conditions. The control measures proposed are the following (Jenkins et al., 1984; Soddell and Seviour, 1990):

1. *Chlorination of foams* (chlorine sprays) or return activated sludge (RAS). Some operators reported success after chlorination of RAS. However, excessive chlorine levels may cause floc dispersion and effluent deterioration. This practice was not successful in controlling scums in an Atlanta plant (Sezgin and Karr, 1986).

   In Phoenix, Arizona, *Nocardia* foaming was controlled by spraying the foam with a high concentration ($2000-3000$ mg/L) of chlorine (Albertson and Hendricks, 1992).

2. *Increase in sludge wasting.* *Gordonia amarae* takes $5-7$ days to develop colonies on agar media while *N. pinensis* requires up to 21 days to form colonies and occurs in foaming plants with sludge ages of $17-30$ days (Sodell and Seviour, 1990; 1994). Thus, foaming may be due to relatively long mean cell retention time (MCRT) and can be controlled by increasing sludge wasting (i.e., reducing sludge age), causing *Nocardia* to be washed out. Actinomycete numbers declined to undetectable levels at an MCRT of 2.2 days at $16°C$ and 1.5 days at $24°C$ (Fig. 9.20; Cha et al., 1992). A survey indicated that MCRT reduction was the most common strategy used by U.S. wastewater treatment plants, with a success rate of 73 percent (Pitt and Jenkins, 1990). However, this control measure is not always successful in eliminating foaming in full-scale wastewater treatment plants (Mori et al., 1992) and is not desirable when long retention times are required (e.g., nitrification process). At the R.M. Clayton plant in

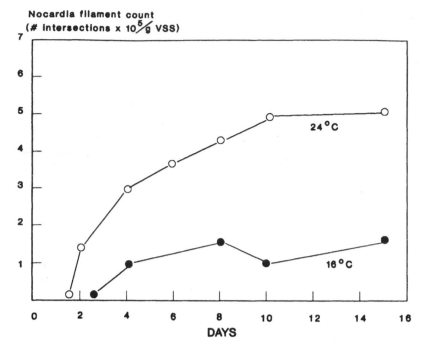

Figure 9.20 Effect of MCRT and temperature on *Nocardia* populations in beach-scale activated sludge units. From Cha et al. (1992). (Courtesy of Water Environment Federation.)

Atlanta, scum occurrence was eliminated by reducing sludge age from 10 days to less than 3 days for a period of approximately 25 days (Sezgin and Karr, 1986).

3. *Use of biological selectors.* The use of an anoxic selector was sometimes successful in controlling the establishment of scum in a wastewater treatment plant in Georgia (Sezgin and Karr, 1986). Bench-scale studies showed than an aerobic selector can control *Nocardia* populations at an MCRT of 5 days (Cha et al., 1992).

4. *Reducing air flow in the aeration basin.* This helps control scum accumulation because filamentous microorganisms are strict aerobes (Sezgin and Karr, 1986).

5. *Reduction in pH, and oil and grease levels.* This approach also resulted in a decrease in scum accumulation (Sezgin and Karr, 1986).

6. *Addition of anaerobic digester supernatant to wastewater.* This supernatant is toxic to pure cultures of *Nocardia* (Lechevalier et al., 1977; Lemmer and Kroppenstedt, 1984). However, under field conditions, the addition of this supernatant to wastewater is not always successful in controlling actinomycetes foams (Blackall et al., 1991). The toxic agent found in anaerobic digester supernatant needs to be fully characterized.

7. *Water sprays* to control foam buildup. This approach does not result in a complete mechanical collapse of the foam.

8. *Antifoam agents, iron salts, and polymers.* Their use has produced mixed results. Under laboratory conditions, it was shown that a suspension of montmorillonite (a three-layer clay mineral), at a concentration of 100 μg/mL, prevents the formation of stable foam by *G. amarae* (Blackall and Marshall, 1989). The addition of a cationic polymer (0.5 mg/L) to the mixed liquor of an activated sludge unit in California helped control foaming due to *Nocardia* (Shao et al., 1997).

9. Some *foam accumulation in low-turbulence zones.* This was observed in the aeration tank. This problem can be avoided through the proper location of the aerators (Blackall et al., 1991).

10. *Physical removal of the foam.* The skimmed off scum should not be recycled into the primary clarifier or the aeration tank, but should be wasted (Lemmer and Baumann, 1988a, b).

11. *Use of antagonistic microflora.* The use of bacteria and predatory protozoa to control foam actinomycetes has not been successful (Soddell and Seviour, 1990).

12. *Potential use of actinophages (i.e., actinomycete-lysing phage).* Several actinophages were isolated from foaming activated sludge and may potentially serve as potential biological tools for foam control (Thomas et al., 2002).

9.7 WEB RESOURCES

http://www.dec.state.ny.us/website/dow/bwcp/ta_sludbulk.html (bulking in activated sludge)
http://www.wrc.org.za/publications/watersa/2003/october/1583.pdf (Water Research Commission, South Africa)
http://aem.asm.org/cgi/content/full/67/11/5303
http://www.wrc.org.za/publications/watersa/1999/October/oct99_p397.pdf (Water Research Commission, South Africa)
http://www.scitrav.com/wwater/asp1/foams.htm (presentation by Dr Jiri Wanner)
http://www.rpi.edu/dept/chem-eng/Biotech-Environ/Biocontrol/waterlnk.html (wide world of activated sludge)
http://www.scitrav.com/wwater/waterlnk.htm (Wastewater world wide)

9.8 QUESTIONS AND PROBLEMS

1. Discuss the various separation problems encountered in activated sludge.

2. How do we assess bulking in activated sludge?

3. What are some physiological differences between floc-forming bacteria and filamentous bacteria?

4. Cite the important factors involved in filamentous bulking.

5. Explain the kinetics selection theory of Chudoba concerning the relationship between floc-forming bacteria and filamentous bacteria.

6. What is the cause of rising sludge, and how do we control it?

7. Explain the competition between floc-forming and filamentous bacteria in terms of *r-strategy* and *K-strategy*.

8. What is the basis for the use of biological selectors for controlling filamentous bacteria in activate sludge?

9. Give three types of biological selectors and explain the mechanisms involved.

10. What do we try to accomplish when using oxidants to control bulking?

11. Discuss the main microorganisms involved in foaming in activated sludge.

12. What are the mechanisms involved in foaming?

13. What is the most used approach for controlling foaming? Explain.

9.9 FURTHER READING

Eikelboom, D.H. 1975. Filamentous organisms observed in bulking activated sludge. Water Res. 9: 365–388.

Eikelboom, D.H., and H.J.J. van Buijsen. 1981. *Microscopic Sludge Investigation Manual*, Report # A94a, TNO Research Institute, The Netherlands.

Jenkins, D., M.G. Richard, and G.T. Daigger. 1984. *Manual on the Causes and Control of Activated Sludge Bulking and Foaming*, Water Research Commission, Pretoria, South Africa.

Kämpfer, P., and M. Wagner. 2002. Filamentous bacteria in activated sludge: Current taxonomic status and ecology, pp. 1287–1306, In: *Encyclopedia of Environmental Microbiology*, Gabriel Bitton, editor-in-chief, Wiley-Interscience, N.Y.

Martins, A.M.P., K. Pagilla, J. J. Heijnena, and M.C.M. van Loosdrecht. 2004. Filamentous bulking sludge – a critical review. Water Res. 38: 793–817.

Soddell, J.A., and R.J. Seviour. 1990. Microbiology of foaming in activated sludge foams. J. Appl. Bacteriol. 69: 145–176.

Wanner, J. 1994. Activated Sludge Bulking And Foaming Control. Technomic Publishing Co, 327 pp.

Wanner, J. 2002. Filamentous bulking in activated sludge, control of, pp. 1306–1315, In: *Encyclopedia of Environmental Microbiology*, Gabriel Bitton, editor-in-chief, Wiley-Interscience, N.Y.

10

PROCESSES BASED ON ATTACHED MICROBIAL GROWTH

10.1 INTRODUCTION

In a fixed-film biological process, microorganisms are attached to a solid substratum where they reach relatively high concentrations. The support materials include gravels, stones, plastic, sand, or activated carbon particles. Two important factors that influence microbial growth on the support material are the flow rate of wastewater as well as the size and geometric configuration of particles.

Biofilm reactors comprise trickling filters, rotating biological contactors (RBC), and submerged filters (downflow and upflow filters). These reactors are used for oxidation

Wastewater Microbiology, Third Edition, by Gabriel Bitton

Copyright © 2005 John Wiley & Sons, Inc.

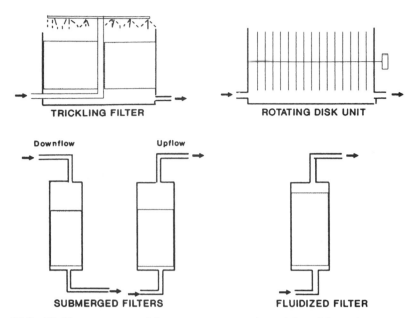

Figure 10.1 Biofilm reactors used for wastewater treatment. Adapted from Harremoes (1978).

of organic matter, nitrification, denitrification, or anaerobic digestion of wastewater (Harremoes, 1978) (Fig. 10.1). Several rate-limiting phenomena are involved in biofilms. Electron donors and acceptors must diffuse inside the biofilms and reaction products must be transported out of the biofilm (Harremoes, 1978) (Fig. 10.2).

Fixed-biofilm processes offer the following advantages (Rittmann et al., 1988):

1. They allow the development of microorganisms with relatively low specific growth rates (e.g., methanogens).
2. They are less subject to variable or intermittent loadings.
3. They are suitable for small reactor size.
4. For fixed-film processes such as trickling filters, the operational costs are lower than for activated sludge.

However, in industrial systems, fixed-biofilm processes lead to biofouling, the undesirable overgrowth of microorganisms on surfaces. Biofouling adversely affects heat exchangers, water distribution pipes (see Chapter 16), ship hulls, and medical devices (e.g., catheters, implants). Extracellular polymeric substances (EPS) in biofilms provide the mechanical stability necessary to withstand shear forces and protection against biocides (Flemming and Wingender, 2002). Biofouling is controlled by using physical or chemical methods or a combination of both. Physical methods include physical removal of biofilms or application of low-intensity electric fields or ultrasound energy across the biofilm. Chemical control involves the use of oxidizing agents (e.g., peroxides, halogens, ozone), and nonoxidizing biocides (e.g., surface active agents, aldehyde-based chemicals, or phenol derivatives) (Cloete and Brözel, 2002).

Two types of fixed-film reactors (trickling filters and rotating biological contactors) are described in this chapter. Anaerobic fixed-film bioreactors are described in Chapter 13.

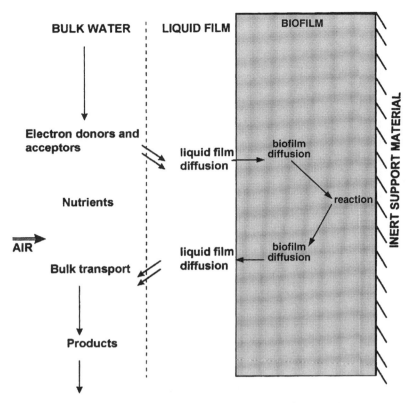

Figure 10.2 Potentially rate-limiting phenomena involved in biofilm reactions. From Harremoes (1978).

10.2 TRICKLING FILTERS: PROCESS DESCRIPTION

The *trickling* or *percolating filter* was introduced in 1890 and is one of the earliest systems for biological waste treatment. It has four major components (Fig. 10.3):

1. A *circular or rectangular tank containing the filter medium* with a bed depth of approximately 1.0–2.5 m. The filter medium provides a large surface area for microbial growth. The ideal filter medium should provide a large surface area to maximize microbial attachment and growth. It should also provide sufficient void space for air diffusion as well as allowing sloughed microbial biofilm to pass through. It should not be toxic to microorganisms and should be chemically and mechanically stable (Grady and Lim, 1980).

   The filter media used in trickling filters are stones (crushed limestone and granite), ceramic material, treated wood, hard coal, or plastic media. Selection of filter media is based on factors such as specific surface area, void space, unit weight, media configuration and size, and cost (U.S. EPA, 1977). The smaller the size, the higher the surface area for microbial attachment and growth, but the lower the percentage of void space. Plastic media, introduced in the 1970s, are made of PVC or polypropylene and are mainly used in high-rate trickling filters. They have a low bulk density and offer optimum surface area ($85–140\ \text{m}^2/\text{m}^3$) and much higher void space (up to 95 percent) than that of other filter media. Thus, filter clogging is considerably minimized

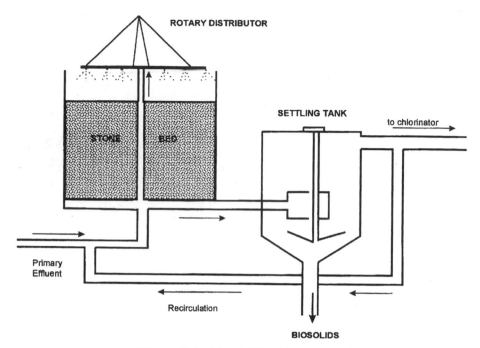

Figure 10.3 The trickling filter process.

when these media are used. Plastic is also a light material that requires less reinforced concrete tanks than do stone media. Therefore, biological tower reactors containing these materials can be as high as 6–10 m.

2. A *wastewater distributor*, which allows a uniform hydraulic load over the filter material. It has one to four arms and its configuration and speed depends on the filter media used. Hydraulic load varies from less than 5 m^3/m^2/day for low-rate filters to more than 25 m^3/m^2/day for high-rate filters (Nathanson, 1986). Wastewater is percolated or trickled over the filter and provides nutrients for the growth of microorganisms on the filter surface.

3. An *underdrain system* for collection of liquid and introduction of air. The underdrain collects treated wastewater as well as biological solids (i.e., microbial biomass) that have been sloughed off the biofilm material.

4. A *final clarifier*. This is also called the humus tank, for separation of solids from the treated wastewater.

Trickling filters can operate in different modes (Forster and Johnston, 1987):

- *Single pass mode.* The single pass mode has an organic loading rate of 0.06–0.12 kg BOD m^{-3} day^{-1}.
- *Alternating double filtration (ADF) mode.* The ADF mode involves the alternate use (1–2 weeks interval) of two sets of filters and humus tanks and allows higher organic loading rates with no problems of filter clogging.
- *Recirculation mode.* Trickling filter effluents are partially recirculated through the filter to increase the treatment efficiency of the filter media (Metcalf and Eddy, 1991). A portion of the treated effluent is returned to the filter. The *recirculation ratio, R,* is the ratio of the flow rate of recirculated effluent to the flow rate of the

wastewater influent (Nathanson, 1986):

$$R = Q_R/Q \qquad (10.1)$$

where Q_R = the flow rate of recirculated trickling filter effluent; and Q = the flow rate of wastewater influent.

Changes in the quality and quantity of wastewater can be handled by adjusting the rate of recirculation. Recirculation improves contact between wastewater and the filter material, helps dilute high-BOD or toxic wastewater, increases dissolved oxygen for biodegradation of organics and for tackling odor problems, improves distribution of the influent on the filter surface, prevents the filter from drying out during the night when the wastewater flow is low, and avoids ponding (i.e., puddles at the surface of the filter) (Davis and Cornwell, 1985; U.S. EPA, 1977).

10.3 BIOLOGY OF TRICKLING FILTERS

A trickling filter essentially converts soluble organic matter to biomass, which is further removed via settling in the final clarifier. Organic loading is the rate at which BOD is applied to the trickling filter, and is expressed in kg BOD applied per m^3 of filter per day. A typical organic loading is $0.5 \, kg/m^3/day$ and may vary from 0.1–0.4 for low-rate filters to $0.5–1 \, kg \, BOD/m^3/day$ for high-rate filters. This parameter is important to the performance of the trickling filter and may dictate the hydraulic loading onto the filter. Removal of BOD by trickling filters is approximately 85 percent for low-rate filters and 65–75 percent for high-rate filters (U.S. EPA, 1977).

10.3.1 Biofilm Formation

The biofilm forming on the surface of the filter media in trickling filters is called the *zoogleal film*. It is composed of bacteria, fungi, algae, protozoa, and other life forms (Fig. 10.4). Scanning electron micrographs of biofilms forming on surfaces are displayed in Figure 10.5. The processes involved in biofilm formation in wastewater are similar to those occurring in natural aquatic environments. After conditioning of the substratum with organic materials, the surface is colonized by bacteria, followed by other life forms. Bacterial adsorption to the substratum requires the formation of a polymer-containing matrix, named *glycocalyx*. These extracellular polymers help anchor the biofilm microorganisms to the surface of the filter material (Bitton and Marshall, 1980). The sequence of events that lead to biofilm formation is described in more detail in Chapter 16. The glycocalyx also provides a surface that is rich in polyanionic compounds that complex metal ions. Biofilm microorganisms degrade the organic matter present in wastewater. The increase in biofilm thickness leads, however, to limited oxygen diffusion to the deeper layers of the biofilm, creating an anaerobic environment near the filter media surface. Microorganisms in the deeper layers face a reduced supply of organic substrates and enter into the endogenous phase of growth. They are subsequently sloughed off the surface. Sloughing is followed by the formation of a new biofilm (Metcalf and Eddy, 1991).

10.3.2 Organisms Present in Trickling Filter Biofilms

Trickling filters are notable for the diversity of life forms that participate in wastewater treatment, making this process relatively more complex than the activated sludge

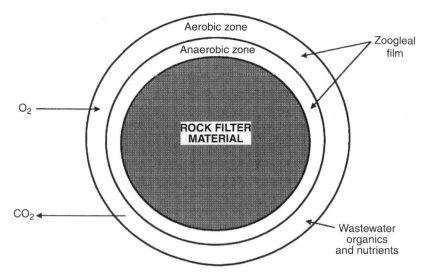

Figure 10.4 Zoogleal film formation on packing media in trickling filters. Adapted from Zajik (1971).

process. In addition to prokaryotic and eukaryotic microorganisms, trickling filters contain higher life forms such as rotifers, nematodes, annelid worms, snails, and many insect larvae. The reader should consult a color atlas of the organisms most frequently observed in wastewater (Berk and Gunderson, 1993). The major groups of organisms encountered in trickling filters are the following:

10.3.2.1 *Bacteria.*
These are active in the uptake and degradation of soluble organic matter. Colloidal organic matter is also trapped in the filter and degraded by extracellular enzymes. Some bacterial genera that are active in trickling filters are *Zooglea*, *Pseudomonas*, *Flavobacterium*, *Achromobacter*, *Alcaligenes*, filamentous bacteria (e.g., *Sphaerotilus*), and nitrifying bacteria (*Nitrosomonas* and *Nitrobacter*). Nitrifying bacteria convert ammonia into nitrate via nitrite and are outcompeted by heterotrophs for oxygen and ammonia in the presence of organic materials (see Chapter 3).

We have witnessed major improvements in the methodology for studying biofilm microbiology (this topic is further discussed in Chapter 16). Microsensors with micron-size tips are now available for measuring chemical concentration gradients (e.g., methane, oxygen, carbon dioxide, hydrogen sulfide, nitrate and nitrite levels, pH; oxido-reduction potential). Figure 10.6 (Yu and Bishop, 2001) illustrates the use of micro-electrodes to measure the profiles of pH, oxido-reduction potential, O_2, and sulfide in a biofilm (Yu and Bishop, 2001). Microelectrodes, in combination with other methods such as fluorescent in situ hybridization (FISH), can be used to study the community structure and activity of nitrifiers and other microorganisms in biofilms and activated sludge flocs (Biesterfeld et al., 2003; Okabe et al., 2002; Satoh et al., 2003; Yu et al., 2004). Confocal laser-scanning microscopy (based on the use of a laser beam to excite a fluorescent dye bound to a target structure inside the cell), in conjunction with 16S rRNA-targeted oligonucleotide probes, has shown that nitrifiers are mostly found in the oxic layers of the biofilm, but some can be detected in the anoxic layers. *Nitrosomonas* (green cells) occurs in clusters in close proximity to less dense aggregates of *Nitrobacter* (red cells)

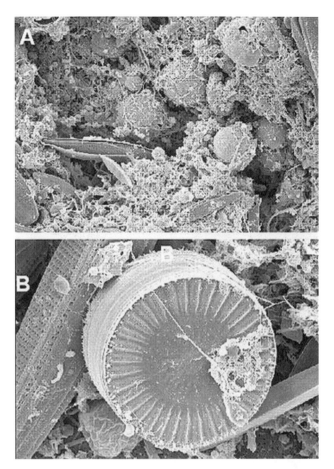

Figure 10.5 Scanning electron micrographs forming on surfaces: (a) General view of a biofilm; (b) diatoms in a biofilm.

(Schramm et al., 1996) (Fig. 10.7). A similar study was undertaken on a biofilm that carried out nitrification–denitrification and enhanced phosphorus removal (Gieseke et al., 2001). It was shown that nitrification was oxygen-limited and was confined to the first 200 μm of the biofilm. As regards the ammonia-oxidizing bacteria (AOB), the first 100 μm of the biofilm harbored a mixture of *Nitrosomonas europaea*, *Nitrosomonas oligotropha*, and *Nitrosomonas communis*, whereas at deeper layers only *N. oligotropha* was found. In a nitrifying trickling filter, the amount of biofilm and AOB (predominance of *Nitrosomonas*) numbers and activity (Fig. 10.8; Persson et al., 2002) decreased with increasing filter depth (Persson et al., 2002). The *amoA* gene present in a nitrifying biofilm was amplified using in situ PCR (*AmoA* gene encodes ammonium monooxygenase, which catalyses the oxidation of ammonium to hydroxylamine). Biofilm cells that have the *AmoA* genes were detected on the surface of the biofilm (Hoshino et al., 2001).

As found in other studies of engineered systems, *Nitrospira* sp. was the predominant nitrite-oxidizing bacteria (NOB) found in the nitrifying biofilm (Nogueira et al., 2002; Okabe et al., 2002). It is, however, desirable to detect the expression of the *AmoA* gene (i.e., detection of *amoA* mRNA) in order to obtain information about the viability/activity of ammonium oxidizers. This approach was used in the real-time monitoring of the activity of ammonia-oxidizing bacteria in nitrifying biofilm. The *amoA* mRNA transcription

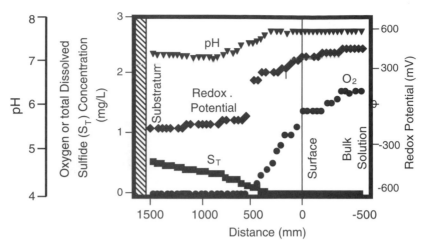

Figure 10.6 Profiles of oxygen, sulfide, oxido-reduction potential (ORP) levels, and pH as a function of distance from the surface of a biofilm. From Yu and Bishop (2001).

responded to the inhibiting effect of low pH and was induced following addition of ammonia (Aoi et al., 2002). Fluorescent in situ hybridization (FISH) (see Chapter 11 for more details on this technique) was used to quantify AOB in a full-scale nitrifying trickling filter. A relationship was found between AOB populations and ammonia removal rates or nitrate + nitrite generation rates (Biesterfeld et al., 2001; 2003).

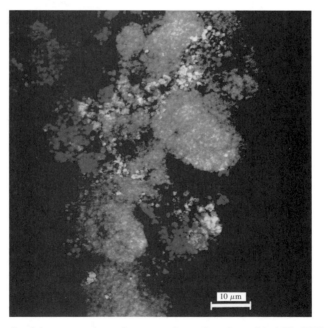

Figure 10.7 Confocal laser-scanning microscopy in conjunction with 16S rRNA-targeted oligo-nucleotide probes showing clusters of nitrifiers (red cells) in close proximity to more dense clusters of denitrifiers (green cells) in a 20 μm thick section of biofilm. From Schramm et al. (1996). (Courtesy of R. Amann and with permission of the publisher.) Figure also appears in color figure section.

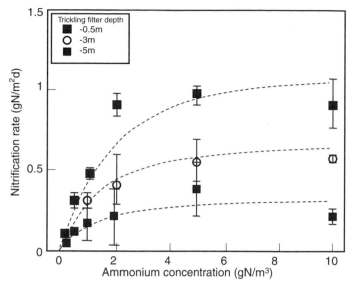

Figure 10.8   Ammonia-oxidizing bacteria activity as a function of depth in a nitrifying trickling filter. From Persson et al. (2002).

Nitrification efficiency is operationally defined as the percentage of ammonia removed during treatment. This definition does not take into account other nitrogen transformations, such as nitrogen assimilation and denitrification (Parker and Richards, 1986). As discussed for activated sludge, the extent of nitrification in trickling filters depends on a variety of factors, including temperature, dissolved oxygen, pH, presence of inhibitors, filter depth and media type, loading rate, and wastewater BOD (Balakrishnan and Eckenfelder, 1969; Parker and Richards, 1986; U.S. EPA, 1977). Low-rate trickling filters allow the development of a high nitrifying population. Conversely, high-rate filters, due to higher loading rates and continuous sloughing of the biofilm, do not allow nitrification to proceed (U.S. EPA, 1977). For rock media filters, organic loading should not exceed 0.16 kg BOD_5/m^3/day (U.S. EPA, 1975). The decrease of nitrification at higher loading rates is due to the predominance of heterotrophs in the biofilm (Parker and Richards, 1986). Higher loading rates (0.35 kg BOD_5/m^3/day) are allowable in plastic-media trickling filters because of the higher surface area of the plastic media (Stenquist et al., 1974). Figure 10.9 (Parker and Richards, 1986) shows the relationship of BOD_5 level and nitrification in a 16 ft plastic media tower (combined carbon oxidation–nitrification system) in Garland, Texas, and Atlanta, Georgia. Nitrification was initiated when the BOD_5 level was less than 20 mg/L and occurred only at the bottom of the tower where the BOD_5 level was relatively low and where the competition between heterotrophs and nitrifiers was low (Wanner and Gujer, 1984). If two filters are used, heterotrophic growth occurs in the first filter and nitrification in the second filter.

Methane-producing bacteria and sulfate-reducing bacteria (SRB) were also detected in trickling filter biofilms that provide anoxic microsites for the development of these bacteria. The oxidation of sulfides within biofilms may contribute to the maintenance of SRB in that environment (Lens et al., 1995).

10.3.2.2   *Fungi.*   These organisms are also active in the biofilm in connection with waste stabilization. They predominate, however, only under low pH conditions, a situation

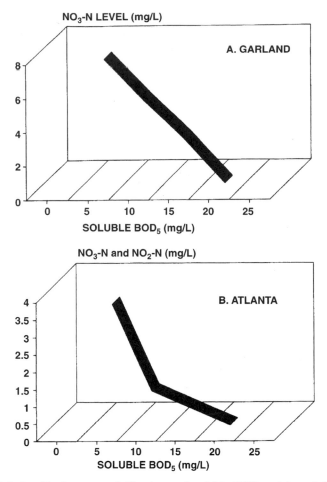

Figure 10.9 Relationship between nitrification and soluble BOD₅. Adapted from Parker and Richards (1986).

that can be created by the introduction of some acidifying industrial wastes. Some examples of fungi found in trickling filters are *Fusarium*, *Penicillium*, *Aspergillus*, *Mucor*, *Geotrichum*, and yeasts. The hyphae growth is helpful for the transfer of oxygen to the lower depths of the biofilm.

10.3.2.3 Algae. Many types of algae also grow on the biofilm surface (e.g., *Ulothrix*, *Phormidium*, *Anacystis*, *Euglena*, *Chlorella*). They produce oxygen during daytime following photosynthesis, and some species of blue-green algae are also able to fix nitrogen. In contrast to activated sludge, algae and fungi are important components of biofilms in trickling filters.

10.3.2.4 Protozoa. These unicellular eukaryotic organisms feed on biofilm bacteria. Continuous removal of bacteria by protozoa helps maintain a high decomposition rate (Uhlmann, 1979). The protozoa occurring in biofilms are flagellates (e.g., *Bodo*, *Monas*), ciliates (e.g., *Colpidium*, *Vorticella*) and ameba (e.g., *Amoeba*, *Arcella*).

10.3.2.5 Rotifers (e.g., Rotaria). These are also encountered in biofilms.

10.3.2.6 *Macroinvertebrates.*

Several groups of macroinvertebrates (nematodes, lumbricidae, collembola, and diptera) are found in trickling filters. Insect larvae (e.g., chironomids, *Psychoda alternata*, *P. severini*, *Sylvicola fenestralis*) feed on the biofilm and help control its thickness, thus avoiding clogging of the filter by microbial exopolymers. These larvae develop into adult insects ("filter flies") in 2–3 weeks and may become a nuisance, particularly to wastewater treatment plant operators. Flies in numbers as high as 30,000 flies/m$^2 \cdot$ day have been reported (Edeline, 1988). Insects are generally controlled by increasing the wetting of the filter surface since *Psychoda* larvae emerge only in dry filters, and by chemical control via insecticides (Forster and Johnston, 1987). Commercial preparations of *Bacillus thuringiensis* var. *israelensis* can also be applied to filter beds to control insect larvae. The spores of this entomogenous pathogen contain a protoxin that is activated in the insect gut, causing its death (Houston et al., 1989a, b).

Cold temperatures and toxicants slow down predator (protozoa and macroinvertebrates) activity and thus increase the chances of filter clogging with subsequent ponding. This may adversely affect the performance of the humus tank in the spring when predator activity resumes. Excess biofilm ("spring sloughing") may overload the settling tank with subsequent increase of solids in the final effluent (Forster and Johnston, 1987). Nitrification is also affected by low winter temperatures.

More work is needed for a more thorough understanding of the microbial ecology of trickling filters.

10.3.3 Biofilm Kinetics

The proper functioning of a trickling filter depends on the growth kinetics within the biofilm that develops on the filter material. Biofilm growth is described by the following equation (La Motta, 1976; Uhlmann, 1979):

$$dX/dt = \mu X - kX \tag{10.2}$$

where $\mu X =$ growth; $kX =$ loss; $X =$ number of microorganisms (microbial biomass); $\mu =$ growth rate constant (h^{-1}); $k =$ decay rate constant (h^{-1}).

The growth rate μ depends on the wastewater loading rate while k depends on factors such as removal by grazing, decay of bacterial and fungal biomass, and hydraulic load.

At steady state:

$$dX/dt = 0 \tag{10.3}$$

which implies that

$$\mu X = kX \tag{10.4}$$

This situation is desirable for the proper functioning of the trickling filter. Biofilm thickness depends on the strength (i.e., BOD$_5$) of the incoming wastewater and controls substrate removal by the filter. Above a critical value, biofilm thickness no longer controls substrate removal (Uhlmann, 1979) (Fig. 10.10). Since oxygen diffusion within the biofilm is limited to approximately ≤ 0.3 mm, the thickness of the active portion of the biofilm is controlled by the extent of oxygen diffusion within the film.

Removal of BOD by the active portion of the biofilm proceeds according to the Monod's equation and thus depends on substrate concentration (see Chapter 2).

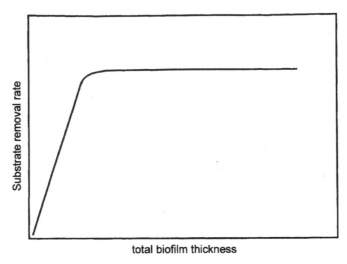

Figure 10.10 Effect of biofilm thickness on substrate removal rate. Adapted from Uhlmann (1979).

10.3.4 Some Advantages and Disadvantages of Trickling Filters

Some advantages and problems related to trickling filters are the following:

10.3.4.1 Advantages. Trickling filters are attractive to small communities because of easy operation, low maintenance costs, and reliability. They are used to treat toxic industrial effluents and are able to withstand shock loads of toxic inputs. The sloughed biofilms can also be easily removed by sedimentation.

10.3.4.2 Disadvantages. High organic loading may lead to filter clogging as a result of excessive growth of slime bacteria in biofilms. Excessive biofilm growth can also cause odor problems in trickling filters. Clogging restricts air circulation, resulting in low availability of oxygen to biofilm microorganisms. However, modifications have helped improve the BOD removal of trickling filters (Best et al., 1985; Grady and Lim, 1980). The following are some of these improvements:

1. Alternating double filtration (ADF), which consists of alternating two filters for receiving the waste
2. Slowing down wastewater distribution
3. Use of plastic materials in the filter for increased surface area and improvement of air circulation
4. Management of odors problems by increasing air flow by means of forced ventilation.

10.4 REMOVAL OF PATHOGENS AND PARASITES BY TRICKLING FILTERS

The removal of pathogens and parasites by trickling filters is generally low and erratic. Bacterial removal is inconsistent and varies from 20 to >90 percent, depending on the operation of the trickling filter. Removal of *Salmonella* by trickling filters is lower than the activated sludge process and may vary from 75 to 95 percent (Feachem et al., 1983).

The removal of viruses by trickling filters is also generally low and erratic. Filtration rate affects the removal of viruses, and probably other pathogenic microorganisms. Laboratory experiments showed that at medium filtration rate (10 MGD/acre), the removal of viruses, coliforms, fecal streptococci, and BOD and COD, was greater than at a higher rate (23 MGD/acre) (Clarke and Chang, 1975). At the Kerrville, Texas, wastewater treatment plant, the mean removal of total and fecal coliforms by a trickling filter was 92 and 95 percent, respectively. However, the removal of enteroviruses was erratic and varied from 59 to 95 percent (Moore et al., 1981). Similarly, removal of bacterial phage is inconsistent and varies between 40 and 90 percent, depending on the season of the year (Kott et al., 1974). In New Zealand, two trickling filter systems removed between 0 and 20 percent of viruses despite a significant reduction of fecal coliforms (>90 percent removal) (Lewis et al., 1986). Similar observations were made in Japan; the removal efficiency of trickling filters is lower for viruses than for indicator bacteria. The mechanism of removal of viruses by trickling filters is poorly understood. Some investigators have suggested that viruses are removed by adsorption to the biofilm material (Omura et al., 1989).

Cysts and oocysts removal by trickling filters is also generally low and erratic, with removals varying from 10 to 99 percent (Schwartzbrod et al., 2002). Trickling filter plants in India removed from 74 to 91 percent of *Entamoeba histolytica*, and the removal of *Giardia* cysts is similar to that of *Entamoeba* (Panicker and Krishnamoorthi, 1978). The removal efficiency of trickling filters is generally lower than that of activated sludge as cysts are frequently detected in effluents (Casson et al., 1990). However, no significant difference was found between activated sludge and trickling filters as regards the removal of *Giardia* cysts and *Cryptosporidium* oocysts (Robertson et al., 2000).

10.5   ROTATING BIOLOGICAL CONTACTORS

10.5.1   Process Description

A rotating biological contactor (RBC) is another example of fixed film bioreactor. This process was conceived in Germany at the beginning of the last century, introduced in the United States during the 1920s, but was marketed only in the 1960s (Huang and Bates, 1980; U.S. EPA, 1977).

An RBC consists of a series of disks mounted on a horizontal shaft that rotate slowly in the wastewater. The disks are approximately 40 percent submerged in wastewater (Fig. 10.11). At any time, the submerged portion of the disk removes BOD as well as dissolved oxygen. The rotation of the disks provides aeration as well as the shear force that causes sloughing of the biofilm from the disk surface. Increased rotation improves oxygen transfer and enhances the contact between attached biomass and wastewater (Antonie, 1976; Hitdlebaugh and Miller, 1981; March et al., 1981). The advantages offered by RBC are short residence time, low operation and maintenance costs, and production of a readily dewatered sludge that settles rapidly (Weng and Molof, 1974).

10.5.2   RBC Biofilms

As discussed for trickling filters, there is an initial adsorption of microorganisms to the disk surface to form a 1–4 mm-thick biofilm that is responsible for BOD removal in RBCs. The rotating disks provide a large surface area for the attached biomass.

Biofilms developing on RBCs comprise a complex and diverse microbial community made of eubacteria, filamentous bacteria, protozoa, and metazoa. Commonly observed filamentous organisms include *Sphaerotilus*, *Beggiatoa*, *Nocardia*, and filamentous algae such as *Oscillatoria* (Hitdlebaugh and Miller, 1981; Kinner et al., 1983; Pescod and Nair, 1972; Pretorius, 1971; Torpey et al., 1971). Biofilm examination by transmission

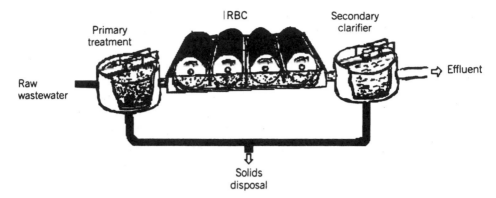

Figure 10.11 Rotating biological contactor. From the U.S. EPA (1975).

electron microscopy shows that *Sphaerotilus* contains many poly-β-hydroxybutyrate inclusions, an indication of storage of excess carbon by the bacteria. These inclusions may account for 11 to more than 20 percent of the dry weight of this bacterium (Rouf and Stokes, 1962). In one study, scanning electron microscopy showed that the RBC biofilm was composed of two layers: an outer whitish layer containing *Beggiatoa* filaments and an inner black layer (due to ferrous sulfide precipitation) containing *Desulfovibrio*, a sulfate-reducing bacterium (Alleman et al., 1982).

- *Anaerobic zone.* In this layer, fermentative bacteria provide the endproducts (organic acids, alcohols) used by sulfate-reducing bacteria.
- *Aerobic zone.* Hydrogen sulfide produced by sulfate-reducing bacteria in the anaerobic zone diffuses into the aerobic zone, and is readily used by *Beggiatoa* as an electron donor. H_2S is oxidized to elemental sulfur.

The organism succession on RBC surfaces is similar to that observed in activated sludge and is as follows (Kinner and Curds, 1989). Bacterial colonization is followed by protozoan flagellates and small amoebae $\rightarrow$ free-swimming bacteriovorous ciliates (e.g., *Colpidium*) $\rightarrow$ nematodes $\rightarrow$ stalked ciliates (e.g., *Vorticella*) $\rightarrow$ rotifers.

After reaching a certain thickness, the biofilm sloughs off and the sloughed material ultimately reaches the final clarifier.

The first stages of an RBC mostly remove organic materials (i.e., BOD_5 removal) while subsequent stages remove NH_4 as a result of nitrification, when the BOD_5 is low enough. Ammonia oxidizers cannot effectively compete with the faster growing heterotrophs that oxidize organic matter. Nitrification occurs only when the BOD is reduced to approximately 14 mg/L and increases with rotational speed (Weng and Molof, 1974). Performance of RBCs is negatively affected by low dissolved oxygen in the first stages and by low pH in the later stages, when nitrification occurs (Hitdlebaugh and Miller, 1981).

Rotating biological contactors have some of the advantages of trickling filters (e.g., low cost, low maintenance, resistance to shock loads) but lack some of the disadvantages (e.g., filter clogging, filter flies).

10.5.3 Pathogen Removal in RBCs

Little is known concerning the removal of indicator and pathogenic microorganisms in RBCs. They appear to be fairly efficient in removing indicator bacteria. A total of 1 log or more of fecal coliforms is removed by this process (Sagy and Kott, 1990). It was found that adsorption to the biofilm and the grazing activity of protozoa and nematodes (in the second and third stage of the RBC system) were the most important factors controlling *E. coli* removal (Tawfik et al., 2002; 2004). Between 79 and 99 percent of Campylobacters were removed by this treatment process (Jones, 2001).

In conclusion, trickling filters appear to be less popular than activated sludge for the treatment of domestic wastewater. Hence, little research has been undertaken since the 1980s on the microbiology of this process. Conversely, environmental microbiologists are now focusing on the microbiology and molecular ecology of biofilms, and this will help in a deeper understanding of the microbial ecology of trickling filters and other biofilm-based processes. Biofilms will be covered in more detail in Chapter 16.

10.6 WEB RESOURCES

http://www.scitrav.com/wwater/waterlnk.htm (collection of web resources concerning wastewater treatment)
http://www.sequencertech.com/biotechnology/trikling_randommedia/random_media.htm (picture of plastic packing material in trickling filter)
http://projects.andassoc.com/tomscreek/alternative/trickling_filters.htm (diagram and picture of trickling filter)
http://www.swbic.org/education/env-engr/secondary/intro/secondary.html#tf (principles of trickling filtration; pictures)

10.7 QUESTIONS AND PROBLEMS

1. What is the rationale for using fixed biofilm processes for treating wastewater?
2. What types of microorganisms would be most benefited by fixed-film processes?
3. Discuss the consequences of biofouling and the approaches used to control it.
4. Describe the sequence of events involved in biofilm formation.
5. Describe the most recent methodology that is available for studying microbial activity in biofilms.
6. What is the role of protozoa, rotifers, and macroinvertebrates in trickling filters?
7. What are the most important factors controlling biofilm thickness?
8. Which of the two processes (activated sludge or trickling filter) would you recommend for removing pathogens and parasites?
9. Are there any biosolids separation problems in trickling filters?

10.8 FURTHER READING

Edeline, F. 1988. *L'Epuration biologique des eaux residuaires: Theorie et Technologie*, Editions CEBEDOC, Liege, Belgium, 304 pp.
Forster, C.F., and D.W.M. Johnston. 1987. Aerobic processes, pp. 15–56, In: *Environmental Biotechnology*, C.F. Forster and D.A.J. Wase, Eds., Ellis Horwood Ltd., Chichester, U.K.

Grady, C.P.L., Jr., and H.C. Lim. 1980. *Biological Waste Treatment.* Marcel Dekker, New York, 963 pp.

Harremoes, P. 1978. Biofilm kinetics, pp. 71–109, In: *Water Pollution Microbiology*, Vol. 2, R. Mitchell, Ed., John Wiley & Sons, New York

USEPA. 1977. *Wastewater Treatment Facilities for Sewered Small Communities*, EPA Report No EPA-625/1-77-009, Environmental Protection Agency, Washington, D.C.

11

WASTE STABILIZATION PONDS

11.1 INTRODUCTION

Treatment of wastewater in ponds is probably the most ancient means of waste treatment known to humans. Oxidation ponds are also called *stabilization ponds* or *lagoons* and serve mostly small rural areas, where land is readily available at relatively low cost. They are used for secondary treatment of wastewater or as polishing ponds. It is estimated that there are over 7000 waste stabilization ponds in the United States (Mara, 2002). The various types of stabilization ponds will be discussed in the following sections.

Waste stabilization ponds are classified as facultative, aerobic, anaerobic, aerated, high-rate aerated, and maturation ponds (Hammer, 1986; Hawkes, 1983b; Nathanson, 1986; Reed et al., 1988).

11.2 FACULTATIVE PONDS

Facultative ponds are the most common type of lagoons for domestic wastewater treatment. Waste treatment is provided by both aerobic and anaerobic processes. These ponds range from 1 to 2.5 m in depth and are subdivided into three layers: an upper aerated zone, a middle facultative zone, and a lower anaerobic zone. The detention time varies between 5 and 30 days (Hammer, 1986; Hawkes, 1983b; Negulescu, 1985). Among the advantages of these ponds are low initial cost and ease of operation. Some disadvantages include odor problems associated mainly with algal growth and H_2S production, and mosquitos, which are of public health concern.

11.2.1 Biology of Facultative Ponds

Waste treatment in oxidation ponds is the result of natural biological processes carried out mainly by bacteria and algae. Waste treatment is carried out by a mixture of aerobic, anaerobic, as well as facultative microorganisms. These ponds allow the accumulation of solids, which are degraded anaerobically at the bottom of the pond. Many categories of organisms play a role in the treatment process. These include mainly algae, heterotrophic and autotrophic bacteria, and zooplankton. The microbiological processes in facultative ponds are summarized in Figure 11.1 (U.S. EPA, 1977).

11.2.1.1 Activity in the Photic Zone. In the photic zone, photosynthesis is carried out by a wide range of algal species (mostly green algae, euglenophyta, and diatoms), producing from 10 to 66 g algae/m^2/day (Edeline, 1988). Chlorophyll *a* concentration in facultative ponds ranges between 500 and 2000 mg/L (Mara, 2002). The most common species encountered in oxidation ponds are *Chlamydomonas*, *Euglena*, *Chlorella*, *Scenedesmus*, *Microactinium*, *Oscillatoria*, and *Microcystis* (Hawkes, 1983b). The type of predominant algae is determined by a variety of factors. Motile algae tend to predominate in turbid waters because they can control their position within the water column to

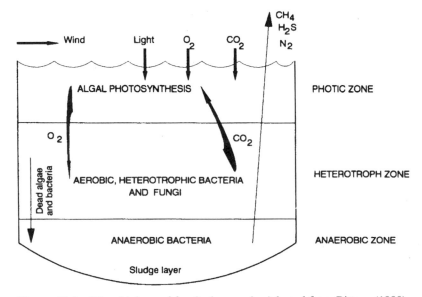

Figure 11.1 Microbiology of facultative ponds. Adapted from Bittons (1980).

optimally use the incident light for photosynthesis (Mara, 2002). Diatoms prevail at lower temperatures than blue-green algae (Edeline, 1988). Algal photosynthesis depends on available temperature and light. In the presence of high algal numbers, light penetration is limited to the first 2 ft of the water column. Wind-induced mixing is important for the maintenance of aerobic conditions within the pond and for providing the exchange of nutrients and gases between phototrophs and heterotrophs. This exchange is impeded when the pond becomes stratified, a phenomenon that occurs under warm conditions, in the absence of natural circulation. Stratification is caused by the establishment of a temperature difference between the warm upper layer or *epilimnion* and the lower and colder layer, the *hypolimnion*. The zone between the epilimnion and hypolimnion is called the *thermocline* and is characterized by a sharp decrease in temperature (Curds and Hawkes, 1983) (Fig. 11.2).

Algae are also involved in nutrient uptake, mainly nitrogen and phosphorus. Some algae are able to fix nitrogen (e.g., blue-green algae), while most others utilize ammonium or nitrate. Photosynthesis leads to an increase in pH, particularly if the treated wastewater has a low alkalinity; this may create conditions for removal of nutrients. At high pH, phosphorus precipitates as calcium phosphate, and ammonium ion may be lost as ammonia. Furthermore, algal photosynthesis produces oxygen, which is used by heterotrophic microorganisms. Oxygen concentration reaches a peak at mid-afternoon and then decreases to a minimum during the night. Other photosynthetic microorganisms in oxidation ponds are the photosynthetic bacteria that use H_2S as electron donor instead of

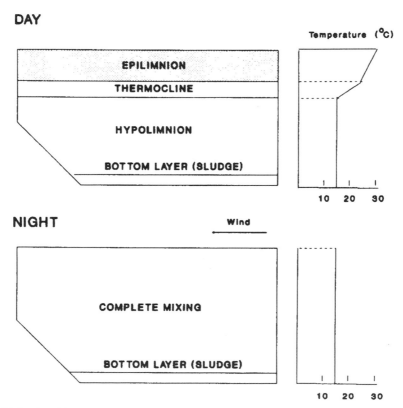

Figure 11.2 Stratification in oxidation ponds. From Curds and Hawkes (1983). (Courtesy of Academic Press.)

H_2O. Therefore, their main role is the removal of H_2S, which causes odor problems (see Chapter 2). The numbers of both algae and photosynthetic bacteria decrease with increased organic loading (Houghton and Mara, 1992).

Algae are inhibited by ammonia and hydrogen sulfide, which are produced in waste stabilization ponds, and both inhibit photosynthesis.

11.2.1.2 Heterotrophic Activity.

Bacterial heterotrophs are the principal microbial agents responsible for organic matter degradation in facultative ponds. The role of fungi and of protozoa appears to be less significant. Heterotrophic activity results in the production of CO_2 and micronutrients necessary for algal growth. In return, algae provide oxygen that is necessary for aerobic heterotrophs to oxidize organic matter. Surface reaeration is another source of oxygen for heterotrophs.

Dead bacterial and algal cells and other solids settle to the bottom of the pond where they undergo anaerobic decomposition (Fig. 11.1). Anaerobic microbial activity results in the production of gases such as methane, hydrogen sulfide, carbon dioxide, and nitrogen. Hydrogen sulfide production by sulfate-reducing bacteria may encourage the growth of photosynthetic bacteria, namely the purple sulfur bacteria (*Chromatium*, *Thiocapsa*, *Thiopedia*) (Edeline, 1988; Holm and Vennes, 1970; Houghton and Mara, 1992). Blooms of Rhodospirillaceae (non-sulfur purple bacteria) in sewage lagoons have also been documented (Jones, 1956). Photosynthetic bacteria are found below the algal layers in facultative ponds (see Chapter 2 for more details on their ecology). They help protect pond algae from the deleterious effect of H_2S. Although some of the carbon is lost to the atmosphere as CO_2 or CH_4, most of it is converted to microbial biomass. Unless the microbial cells are removed by sedimentation or some other solids removal process (e.g., intermittent sand filtration), little carbon reduction is obtained in oxidation ponds. Furthermore, the microbial cells in pond effluents may exert an oxygen demand in receiving waters (U.S. EPA, 1977).

11.2.1.3 Zooplankton Activity.

Zooplankton (rotifera, cladocera, and copepoda) prey on algal and bacterial cells and can play a significant role in controlling these populations. Their activity is thus important to the operation of the pond. Cladocera (e.g., *Daphnia*) are filter feeders that feed mostly on bacterial cells and detrital particles, but less on filamentous algae. They are therefore helpful in reducing the turbidity of pond effluents. Zooplankton biomass can, however, be adversely affected in stratified ponds. This is due to the high concentrations of unionized ammonia resulting from high pHs associated with algal photosynthesis (Arauzo, 2003).

11.2.2 Effect of Temperature on Pond Operation

Temperature plays an important role with regard to the activity of phototrophs and heterotrophs in wastewater ponds. It also significantly affects anaerobic waste degradation in the pond sediments. No methanogenic activity and subsequent reduction of sludge volume occurs at temperatures below 15°C. Loading of BOD varies from 2.2 g/m^2.day in cold climates to 5.6 g/m^2.day in warmer climates (Hammer, 1986). During the colder months, the ponds become anaerobic because of the lack of solar radiation and, hence, photosynthesis. One of the empirical equations developed for expressing the loading rate of a pond is the Gloyna equation, which gives the pond volume as a function of the prevailing temperature (Gloyna, 1971):

$$V = (3.5 \times 10^{-5})NqL_{a} \times 1.085^{(35-T_{m})} \qquad (11.1)$$

where V = pond volume (m^3); N = number of people contributing the waste; q = per capita waste contribution (L d^{-1}); L_{a} = BOD (mg L^{-1}); and T_{m} = average water temperature of coldest month (°C).

In warm climates, the effluent has a BOD of <30 mg/L. However, the suspended solid concentration may be high due to the presence of algal cells in the pond effluents.

11.2.3 Removal of Suspended Solids, Nitrogen, and Phosphorus by Ponds

Oxidation pond effluents often have a high level of suspended solids composed mostly of algal cells and wastewater solids (Reed et al., 1988). The algae often exert an oxygen demand in the receiving stream. Thus, the lagoon effluents need to be treated by intermittent sand filters, microstrainers, or constructed wetlands (Steinmann et al., 2003). To obtain a high-quality effluent with low turbidity (about 1 NTU), the pond effluent can be treated by chemical coagulation, clarification, and passage through a continuous backwash sand filter (e.g., Dynasand filter). This sand filter can handle high algae concentrations because of its countercurrent flow pattern and continuous backwashing of the sand particles. The clean sand is deposited afterwards on top of the filter (Fraser and Pan, 1998).

Nitrogen is removed by ponds by a number of mechanisms, including nitrification/denitrification, volatilization as ammonia, and algal uptake (see Chapter 3). Ponds remove approximately 40–80 percent of nitrogen. Phosphorus removal by ponds is low. Only 26 percent removal was obtained in two experimental lagoons in series under the arid climate of Marrakech, Morocco (Mandi et al., 1994). Phosphorus removal can be increased by in-pond treatment with iron and aluminum salts or with lime.

11.3 OTHER TYPES OF PONDS

Aerobic ponds are shallow ponds (0.3–0.5 m deep) and are generally mixed to allow the penetration of light necessary for algal growth and subsequent oxygen generation. The detention time of wastewater is generally 3–5 days.

Aerated lagoons are 2–6 m deep with a detention time of <10 days. They are used to treat high-strength domestic wastewater. They are mechanically aerated with air diffusers or mechanical aerators. Treatment (i.e., BOD removal) depends on aeration time, temperature, and type of wastewater. At 20°C, there is an 85 percent BOD removal with an aeration period of 5 days. Faulty operation of the aerated lagoon may result in foul odors.

Anaerobic ponds (Hammer, 1986) have a depth of 2.5–9 m and a relatively long detention time of 20–50 days (Metcalf and Eddy, 1991; Reed et al., 1988). These ponds serve as a pretreatment step for high-BOD organic wastes with high protein and fat content (e.g., meat wastes) and with high concentration of suspended solids. Organic matter is biodegraded under anaerobic conditions to CH_4, CO_2, and other gases such as H_2S (see Chapter 13 for more details on the microbiology of anaerobic digestion of wastes). These ponds do not require expensive mechanical aeration and generate small amounts of sludge. Some problems associated with these ponds are the production of odorous compounds (e.g., H_2S), sensitivity to toxicants, and the requirement of relatively high temperatures. Anaerobic digestion of wastewater is virtually halted at $<10°C$. These lagoons are not suitable for domestic wastewaters, which have a characteristic low BOD.

Maturation or tertiary ponds are $1-2$ m deep and serve as tertiary treatment for wastewater effluents from activated sludge or trickling filters. The detention time is approximately 20 days. Oxygen provided by surface reaeration and algal photosynthesis is used for nitrification. Their role is to further reduce BOD, suspended solids, and nutrients (nitrogen and phosphorus), and to further inactivate pathogens.

11.4 PATHOGEN REMOVAL BY OXIDATION PONDS

Removal or inactivation of pathogens in oxidation ponds is controlled by a variety of factors, among which are temperature, sunlight, pH, lytic action of bacteriophages, predation by macroorganisms, and attachment to settleable solids.

11.4.1 Factors Controlling Removal of Bacterial Pathogens

Oxidation ponds remove a significant percentage (90–99 percent) of indicator and pathogenic bacteria. Table 11.1 shows inactivation of fecal coliforms, *Vibrio cholerae*, and *Pseudomonas aeruginosa* in a series of two experimental ponds with a total detention time of 16 days in Marrakech, Morocco. The reduction of fecal coliforms and *P aeruginosa* was 98 and 92.2 percent, respectively, while no reduction was observed for *V. cholerae*, which displayed a peak during the summer season (Oufdou, 1994). The removal of fecal coliforms increased to 99.6 percent when the retention time in the experimental ponds was raised to 50 days (Mandi et al., 1994). Coliform die-off in ponds generally increases with an increase in temperature, retention time, and pH, but decreases with an increase in BOD_5 and pond depth (Saqqar and Pescod, 1992). Other factors include aeration, antibacterial extracellular algal compounds, nutrient depletion, and sunlight intensity (Fernandez et al., 1992; Qin et al., 1991). There are several possible reasons for the efficient removal of bacteria:

1. Long detention times used in oxidation ponds.
2. High pH generated as a result of photosynthesis. Fecal coliform decline is higher in waste stabilization ponds where pH exceeds 9 (Parhad and Rao, 1974;

TABLE 11.1. Decline of Pathogenic and Indicator Microorganisms in a Series of Two Experimental Facultative Ponds in Marrakech, Morocco[a]

Organism	% Reduction
Fecal coliforms	97.9
Pseudomonas aeruginosa	92.2
Vibrio cholerae	No reduction observed (this pathogen peaks during summer)

[a]The two ponds had a total detention time of 16 days. During the 18-month study, the temperature varied between 11 and 32°C. Adapted from Oufdou (1994).

Pearson et al., 1987). A study of the interaction between *Chlorella* and *E. coli* showed that the decline of the latter was due to the high pH (10–10.5) generated as a result of algal photosynthesis and growth. However, *E. coli* and *Chlorella* are able to grow together when the wastewater is buffered to pH 7.5. This phenomenon is illustrated in Figure 11.3 (Parhad and Rao, 1974). Oufdou (1994) also showed that a high pH in oxidation ponds was deleterious to *E. coli* but not to *V. cholerae*.

3. Predation by zooplankton. We have seen that these organisms play a significant role in the control of bacterial populations in ponds.

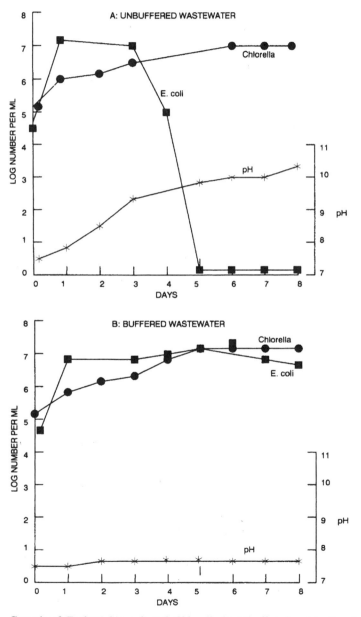

Figure 11.3 Growth of *Escherichia coli* and *Chlorella* in unbuffered and buffered wastewater. From Parhad and Rao (1974). (Courtesy of the Water Environment Federation.)

4. Inactivating effect of sunlight. Although the UV-B (wavelength between 280 and 320 nm) portion of sunlight plays a significant role in the destruction of coliforms and other microorganisms in aquatic environments (Herndl et al., 1993; Moeller and Calkins, 1980; Weinbauer et al., 1997), longer wavelengths may also adversely affect biological activity. Since UV light does not penetrate well into the water column, visible light could affect bacterial die-off. Its impact is increased at high dissolved oxygen concentrations as well as high pH levels (Curtis et al., 1992a, b). Table 11.2 shows the effect of solar radiation on the decay rates of *E. coli* and *V. cholerae* at the entrance and exit of a series of two lagoons. The decay rates are higher for *E. coli* than for *V. cholerae* after exposure to light. The impact of solar radiation is increased at high dissolved oxygen concentrations as well as high pH levels (Curtis et al., 1992a, b; Oufdou, 1994).

11.4.2 Bacterial Die-Off Kinetics

Several kinetic models have been proposed for the prediction of bacterial die-off in waste stabilization ponds. As shown previously, several factors affect bacterial decay rates in ponds: mainly temperature, solar radiation, predation, and antibiosis resulting from algal growth in the pond. Several investigators have proposed die-off models based mainly on the effects of temperature and solar radiation.

A simple model was proposed by Marais (1974):

$$N_e/N_i = 1 + K\theta \tag{11.2}$$

where N_e = coliform number in the pond effluent (number/100 mL); N_i = coliform number in pond influent (number/mL); K = first-order die-off coefficient (day^{-1}); and θ = theoretical retention time (days).

Temperature is a primary factor in bacterial die-off. There is a relationship between the decay constant K and temperature:

$$K_T = K_{20}C^{T-20} \tag{11.3}$$

where K_{20} = decay constant at 20°C; C = constant; and T = mean water temperature in the pond (°C).

TABLE 11.2. Effect of Solar Radiation on the Decay Rates (h^{-1}) of *E. coli* and *V. cholerae* in Wastewater Lagoons[a]

Sample Location	Exposure Conditions	*E. coli*	*V. cholerae*
Lagoon influent	Light	0.051	0.042
	Dark	0.021	0.038
Effluent first lagoon	Light	0.070	0.022
	Dark	0.020	0.019
Effluent second lagoon	Light	0.093	0.012
	Dark	0.024	0.010

[a]The two lagoons had a total detention time of 16 days.
Adapted from Oufdou (1994).

In a completely mixed pond within a temperature range of 5–21°C, K_T is given by the following equation:

$$K_T = 2.6(1.19)^{T-20} \tag{11.4}$$

However, decay rates are sometimes lower than those predicted by Eq. (11.4). For example, the relationship between decay rates and temperature for a number of waste stabilization ponds in Kenya gave the following equation (Mills et al., 1992):

$$K_T = 0.712(1.166)^{T-20} \tag{11.5}$$

In a series of n ponds, N_e, the bacterial number in the pond effluent, is given by Eq. (11.6):

$$N_e = \frac{N_i}{(K\theta_1 + 1)(K\theta_2 + 1)\ldots(K\theta_n + 1)} \tag{11.6}$$

where n = number of ponds in the series. It is assumed that all ponds have the same size. θ_1, θ_2, and θ_n are the retention times in pond 1, 2, and n, respectively. For a fixed total retention time, the removal efficiency increases as the number of ponds in the series increases.

Solar radiation greatly influences the rate of coliform die-off in oxidation ponds (Sarikaya and Saarci, 1987). The relationship between the decay constant K and light intensity I is given by Eq. (11.7):

$$K = K_d + K_s(I) \tag{11.7}$$

where K_d = decay rate constant in the dark for $I = 0$ (h^{-1}); K_d is temperature-dependent. K_s = decay rate constant due to the effect of light (h^{-1}); and I = light intensity (cal/cm^2/h).

The relationship between K and I at temperatures ranging from 25 to 30°C is given by Eq. (11.8) (Sarikaya and Saaci, 1987).

$$K = 0.018 + 0.012(I) \tag{11.8}$$

Therefore, the change in bacterial concentration N with time is given by Eq. (11.9):

$$dN/dt = -KN = -(K_d + K_sI)N \tag{11.9}$$

Pond depth also plays a role in bacterial decay in waste stabilization ponds. The effect of pond depth on coliform decay rate constant is given by Eq. (11.10) (Sarikaya et al., 1987).

$$K = 1.156 + 5.244 \times 10^{-3} \frac{S_0}{kH}(1 - e^{-kH}) \tag{11.10}$$

where K = decay rate constant (day^{-1}); S_0 = daily solar radiation [cal/(cm^2.day)]; k = light attenuation coefficient (m^{-1}); and H = pond depth (m). Equation (11.10) shows that bacterial decay rates in shallow ponds are higher than in deeper ponds.

Concentration of algae can also be incorporated in die-off kinetic models. A multiple linear regression equation gives the bacterial die-off rate K as a function of temperature, as well as algal concentration and influent COD loading rate (Polprasert et al., 1983):

$$K = f(T, C_s, OL) \tag{11.11}$$

where T = temperature (°C); C_s = algae concentration (mg/L); and OL = COD loading rate (kg COD/ha/d). K is given by the following equation (Polprasert et al., 1983):

$$e^K = 0.6351(1.0281)^T (1.0016)^{C_s} (0.9994)^{OL} \tag{11.12}$$

A model was developed for a waste stabilization pond system in Jordan. The model shows that the coliform reduction rate increases with increasing temperature and pH and with decreasing soluble BOD_5 (Saqqar and Pescod, 1992).

$$K_b = 0.50(1.02)^{T-20} (1.15)^{(pH-6)} (0.99784)^{SBOD-100} \tag{11.13}$$

where K_b = fecal coliform reduction rate (day^{-1}); T = water temperature (°C); and SBOD = soluble BOD_5 (mg/L).

11.4.3 Virus Removal and/or Inactivation in Oxidation Ponds

Temperature and solar radiation are important factors that control virus persistence in oxidation ponds (Bitton, 1980a). Virus removal is expected to be high under a hot and sunny climate in the upper layer of the pond. The survival of poliovirus type 1 in a model pond is displayed in Figure 11.4 (Funderburg et al., 1978). During the summer season, it takes 5 days to obtain a 2–log reduction of viruses; in winter, 25 days are required to achieve a similar reduction. Biological factors are also involved in virus inactivation in ponds. As shown for bacteria, high pH resulting from heavy growth of algae may increase inactivation of virus in oxidation ponds. Virus adsorb to suspended solids that settle in pond sediments where they survive for longer periods than in the water column. Disturbance of the contaminated sediments, however, may help in the resuspension of viral and bacterial pathogens.

11.4.4 Protozoa Cysts and Helminth Eggs

The factors influencing cyst and oocyst removal in stabilization ponds are retention time, temperature, pH, and solar radiation. A review of the literature showed close to 100 percent removal of *E. histolytica*, *Giardia* and *Cryptosporidium* reported in several countries (Schwartzbrod et al., 2002). The reported removal of *Giardia* cysts by waste stabilization ponds in Kenya and France was 99.1 percent (25 days retention time) and 99.7 percent (40 days retention time), respectively. However, the cysts were always found in the ponds' final effluents (Grimason et al., 1996). The percentage removal of *Giardia* and *Entamoeba histolytica* in an aerated lagoon varies from 67 to 100 percent (Panicker and Krishnamoorthi, 1981). More than 97 percent of *C. parvum* oocysts were inactivated when placed inside cellulose semi-permeable bags suspended in a high-rate algal pond. It was suggested that the high inactivation was due to the high pH generated by the high algal biomass, solar radiation, and ammonia (Araki et al., 2001).

Sedimentation is the main factor responsible for the removal of helminth parasite eggs (Bouhoum et al., 2000). Physicochemical factors (e.g., ionic conditions) may also adversely affect the viability of helminth eggs, as shown by a study on eggs in a

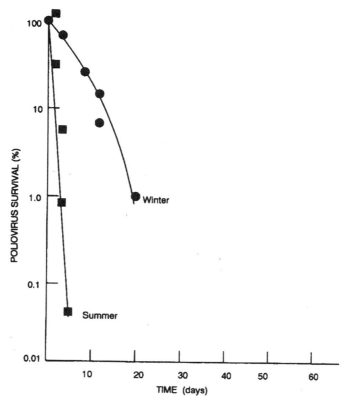

Figure 11.4 Effect of season of the year on the survival of poliovirus in a mode oxidation pond. Adapted from Funderburg et al. (1978).

high-rate algal pond in semi-permeable cellulose bags (Araki et al., 2000). The percent removal of parasites (*Ascaris lumbricoides*, hookworm) in an aerated lagoon varied between 50 and 100 percent (Panicker and Krishnamoorthi, 1981). Under the arid climate of Marrakech, Morocco, 100 percent removal of helminth eggs (nematodes such as *Ascaris*, *Trichuris*; cestodes such as *Taenia*, *Hymenolepis*) was achieved after raw wastewater treatment by two lagoons in series with a total detention time of 10–50 days. Most of the eggs were transferred to the lagoon sediments (Mandi et al., 1994; Schwartzbrod et al., 1989). Parasite egg removal in a series of five ponds (anaerobic pond → facultative pond → three maturation ponds in series) in Brazil was 100 percent when the combined hydraulic retention time was about 25 days. Moreover, most of the eggs (99.9 percent) were removed after the secondary facultative pond (Stott et al., 2003a). However, eggs of *Necator americanus* were not well removed in a stabilization pond system with a total retention period of 23 days (Ellis and Korth, 1993). Helminth ova can survive for years in the sludge accumulated at the bottom of the pond and may pose a potential hazard when the pond is dredged (Nelson et al., 2004).

The percentage removal of nematode eggs follows Eq. (11.14) (Ayres et al., 1992):

$$R = 100[1 - 0.14e^{-(0.38\theta)}] \tag{11.14}$$

where R = percent nematode removal; and θ = retention time (days).

For design purposes, it was recommended to use the lower 95 percent confidence limit of Eq. (11.14) to meet the WHO (1989) guideline of less than one nematode egg/L for irrigation purposes.

Biosolids at the bottom of a pond contain high concentrations of protozoan cysts and parasite eggs (Bouhoum et al., 2000). In a study in central Mexico, it was found that biosolids accumulate in wastewater stabilization ponds at a rate of approximately 0.04 m^3/person/year. Batch and core tests showed that the first-order decay rates of bacterial indicators (fecal coliforms, fecal enterococci), F+ coliphage, and *Ascaris* eggs in pond biosolids were approximately 0.1, 0.01, and 0.001 d^{-1}, respectively. This means that the inactivation of bacterial pathogens may take several months, while the helminth egg inactivation may take several years. This points to the need to remove ponds out of operation prior to biosolids removal or to treat biosolids properly before disposal (Nelson et al., 2004).

11.5 WEB RESOURCES

http://sorrel.humboldt.edu/~ere_dept/marsh/flow3a.html#oxpond
http://www.rpi.edu/dept/chem-eng/Biotech-Environ/FUNDAMNT/streem/methods.htm
http://216.239.57.104/search?g=cache: L8Hi39ck454J:www.epa.gov/ORD/NRMRL/Pubs/625R00008/tfs7.pdf+stabilization+ponds&hl=en&start=8
http://www.irc.nl/content/view/full/8237

11.6 QUESTIONS AND PROBLEMS

1. How is nitrogen removed in oxidation ponds?
2. What would be the role of photosynthetic bacteria in oxidation ponds?
3. What is the main use of anaerobic ponds?
4. How would you calculate the decay constant of coliforms in an oxidation pond at 31°C?
5. Explain stratification in an oxidation pond.
6. How does a Dynasand filter work? Give a diagram and applications of this filter.
7. Why do we have to treat lagoon effluents?
8. What is the main factor contributing to the removal of helminth eggs in oxidation ponds?
9. What are the main factors controlling the inactivation of microbial pathogens in oxidation ponds?
10. Explain the beneficial relationship between bacteria and algae in oxidation ponds.
11. Why is the pH sometimes high in oxidation ponds? At what time of the day does pH reach its peak under a sunny sky?
12. Why are pond sediments of concern from a public health viewpoint?

11.7 FURTHER READING

Edeline, F. 1988. *L'Epuration biologique des eaux residuaires: theorie et technologie*, Editions CEBEDOC, Liege, Belgium, 304 pp.

Hawkes, H.A. 1983. *Stabilization ponds, In: Ecological Aspects of Used-Water Treatment*, Vol. 2, C.R. Curds and H.A. Hawkes, Eds., Academic Press, London.

Mara, D.D. 2002. Waste stabilization ponds, pp. 3330–3337 In: *Encyclopedia of Environmental Microbiology*, Gabriel Bitton, editor-in-chief, Wiley-Interscience, N.Y.

Mara, D., and N.J. Horan, Eds. 2003. *The Handbook of Water and Wastewater Microbiology*, Academic.

U.S. EPA. 1977. Wastewater Treatment Facilities for Sewered Small Communities. Report No EPA-625/1-77-009, U.S. Environmental Protection Agency, Washington, D.C.

12

SLUDGE MICROBIOLOGY

Wastewater Microbiology, Third Edition, by Gabriel Bitton
Copyright © 2005 John Wiley & Sons, Inc.

12.1   INTRODUCTION

Sludge (biosolids) is composed mostly of solids generated during wastewater treatment processes. Sludge treatment and disposal are probably the most costly operations in wastewater treatment plants. The U.S. Environmental protection Agency (U.S. EPA) has reported that approximately 7 million dry tons of sludge are produced annually in the United States, and sludge production is expected to increase in the future. Member States of the European Community produce approximately 5.5 million dry tons of biosolids per year (Bowden, 1987). France produces a little less than 900,000 dry tons of sludge/year (Sachon et al., 1997).

The types of sludges that must be treated or disposed of are the following (Davis and Cornwell, 1985):

- *Grit.*   Grit is a mixture of coarse and dense materials such as sand, bone chips, and glass. These materials are collected in the grit chamber and are directly disposed into a landfill.
- *Primary sludge.*   This is the sludge generated by primary treatment of wastewater and accumulated in the primary clarifier. It contains 3–8 percent solids.
- *Secondary sludge.*   Secondary sludge is generated by biological treatment processes (e.g., activated sludge, trickling filter). The solids are mostly organic and range from 0.5 to 2 percent for activated sludge and up to 5 percent for trickling filter sludge.
- *Tertiary sludge.*   This sludge results from tertiary (chemical) treatment of wastewater.

12.2   SLUDGE PROCESSING

We will now review the various physical, chemical, and biological treatments available for processing sludges generated by biological treatment processes (Davis and Cornwell, 1985; Forster and Senior, 1987; Metcalf and Eddy, 1991) (Fig. 12.1).

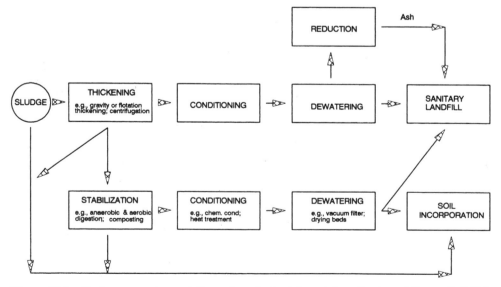

Figure 12.1   Sludge processing and disposal methods. Adapted from Davis and Cornwell (1985) and Metcalf and Eddy (1991).

12.2.1 Screening

Raw sludge is sometimes screened to remove coarse materials that may cause problems such as blockage of pipes and pumps.

12.2.2 Thickening

The goal of thickening is to increase the solid concentration of sludge. Thickening can be accomplished in tanks where the solids are allowed to settle to the bottom or centrifuged in the presence or absence of chemical conditioners. This process can increase the solids concentration of primary sludge to approximately 12 percent. This operation reduces the transportation cost of sludge to the disposal site.

12.2.3 Dewatering

Sludges are dewatered by filtration or application onto drying beds. Filtration helps achieve a higher solids concentration than thickening. The most common filtration device is the filter press made of a series of plates fitted with a filter cloth or, more recently, membrane filters. Filtration is carried out by applying positive pressure or vacuum. Vacuum filtration may be applied to raw sludge before incineration or to stabilize sludge prior to disposal on land. In the presence of conditioners, filtration helps achieve a solids content of 20–40 percent, depending on sludge type and filtration mode. The filtrate has a high BOD, however, and must be recycled through the treatment plant.

Sludge may also be applied on drying beds to reach a solids content of approximately 40 percent after a drying period ranging from 10 to 60 days, depending on weather conditions. Sand drying beds are commonly used, especially in small plants. The sludge is removed mechanically or manually after the drying period.

12.2.4 Conditioning

Conditioning facilitates the separation of solids from the liquid phase. This is accomplished by chemical or heat treatment. Chemical treatment consists of adding conditioners to help improve sludge filterability. These chemicals may be inorganic salts (alum, ferrous and ferric salts, lime) or the increasingly popular synthetic organic polymers called *polyelectrolytes*. Sludge can also be heated under pressure (1000–2000 kPa) at temperatures ranging from 175 to 230°C (Davis and Cornwell, 1985). This treatment also reduces the affinity of sludge for water.

12.2.5 Stabilization

The purpose of stabilization is to break down the organic fraction of the sludge in order to reduce its mass and to obtain a product that is less odorous as well as safer from a public health standpoint. Sludge stabilization includes the following technologies: anaerobic digestion, aerobic digestion, composting, lime stabilization, and heat treatment (Metcalf and Eddy, 1991). Lime and heat stabilization are covered in Section 12.3.9.

12.2.5.1 Anaerobic Digestion. This topic will be covered in detail in Chapter 13.

12.2.5.2 Aerobic Digestion. Aerobic digestion consists of adding air or oxygen to sludge contained in a 10–20 ft deep open tank. A typical circular aerobic digester is shown in Figure 12.2 (U.S. EPA, 1977). Oxygen concentration must be maintained

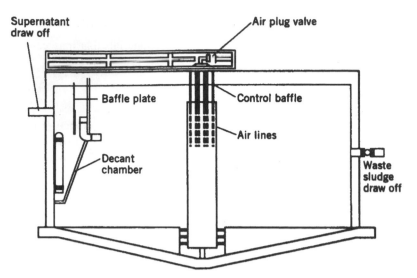

Figure 12.2   Aerobic sludge digester. From U.S. EPA (1977).

above 1 mg/L to avoid production of foul odors. The mean cell residence time in the digester is 12–60 days, depending on the prevailing temperature. Microorganisms degrade aerobically available organic substrates and reduce the volatile solids of the sludge by 40–50 percent (U.S. EPA, 1992a). They enter the endogenous phase of growth when the biodegradable organic matter is depleted. The prevailing conditions in the tank also promote nitrification, which converts NH_4 to NO_3. The end result is a reduction in sludge solids. The stabilized sludge is allowed to settle and the supernatant is generally recycled to the head of the plant because it has a high BOD and high N and P levels (BOD = 100–500 mg/L; nitrate concentration = 200–500 mg/L; total P concentration = 50–200 mg/L) (U.S. EPA, 1977).

An important innovation in aerobic digestion of sludges has been the development and implementation of *autoheated thermophilic aerobic digestion*. This process generates heat as a result of free energy released by the aerobic degradation of organic matter by sludge microorganisms. Much of the energy resulting from the oxidation of organic compounds is released as heat. The heat energy produced by the aerobic digestion of primary and secondary sludges is around 25 kcal/L (Metcalf and Eddy, 1991). The rise in temperature is a function of the level of organic matter and is given by Eq. (12.1) (Forster and Senior, 1987; Jewell and Kabrick, 1980):

$$\Delta T = 2.4\Delta(\text{COD}) \tag{12.1}$$

where ΔT = increase in temperature in °C; ΔCOD = COD (mg/L) oxidized.

Most of the heat generated during biological oxidation is lost during sludge aeration. Therefore, to achieve *autoheating*, the heat resulting from organic matter oxidation can be conserved by providing enough biodegradable organic matter, by insulating the bioreactor, and by increasing oxygen transfer efficiency to levels exceeding 10.

The advantages of aerobic digestion are low capital cost, easy operation, and production of odorless stabilized sludge. Some disadvantages include high consumption of energy necessary for supplying oxygen, dependence upon weather conditions, and an endproduct with relatively low dewatering capacity.

12.2.6 Composting

The main objectives of composting are the production of stabilized organic matter accompanied by odor reduction and the destruction of pathogens and parasites.

12.2.6.1 Process Description. Composting consists essentially of mixing sludge with a bulking agent, stabilizing the mixture in the presence of air, curing, screening to recover the bulking agent, and storing the resulting compost material. There are three main types of composting systems (Benedict et al., 1988; Pedersen, 1983; Reed et al., 1988):

1. *Aerated static pile process* (Beltsville system) (Benedict et al., 1988) (Fig. 12.3). This process is the most widely used in the United States (Goldstein, 1988) and consists of mixing dewatered sludge (raw, anaerobically or aerobically digested sludge) with a bulking agent (new or recycled) such as wood chips, leaves, corncobs, bark, peanut and rice hulls, or dried sludge. Wood chips are the most commonly used bulking agent in composting (they have a high carbon content and a high C/N ratio). Bulking materials offer structural support and favor aeration during composting. The pile is covered with screened compost to reduce or remove odor and to maintain high temperatures inside the pile. Aeration is provided by blowers and air diffusers during a 21-day active composting period in the pile. Afterwards, the compost is cured for at least 30 days, dried, and screened to recycle the bulking agent.

2. *Windrow Process* (Benedict et al., 1988) (Fig. 12.4). Dewatered sludge is mixed with the bulking agent and stacked in 1 to 2 m-high rows called *windrows*. The composting period lasts approximately 30–60 days. Aeration is provided by turning the windrows two or three times per week. Additional induced aeration may be provided in the *aerated windrow process*.

3. *Enclosed systems.* These systems are enclosed to ensure a better control of temperature, oxygen concentration, and odors during composting. They require little space and minimize odor problems. Their cost is, however, higher than open systems.

12.2.6.2 Process Microbiology. The composting process is dominated by the degradation of organic matter by microorganisms under aerobic, moist, and warm conditions. Composting results in the production of a stable product that is used as a soil conditioner, or a fertilizer, or as feed for fish in aquaculture. Furthermore, this process results in pathogen inactivation, as well as transformation of organic forms of nitrogen and phosphorus into inorganic forms that are more bioavailable for uptake by agricultural crops (Polprasert, 1989; Walker and Wilson, 1973).

Composting is a self-heating aerobic biodegradation process that results in temperature elevation and consists of several temperature-controlled phases, each driven by specific groups of organisms (Beffa et al., 1996a; Finstein, 1992; Fogarty and Tuovinen, 1991; Polprasert, 1989):

1. *The latent phase* for the acclimatization of microbial populations to the compost environment.

2. *The mesophilic phase*, dominated by bacteria that raise the temperature following decomposition of organic matter.

3. *The thermophilic phase*, characterized by the growth of thermophilic bacteria, fungi, and actinomycetes. Organic matter is degraded at high rates during this phase.

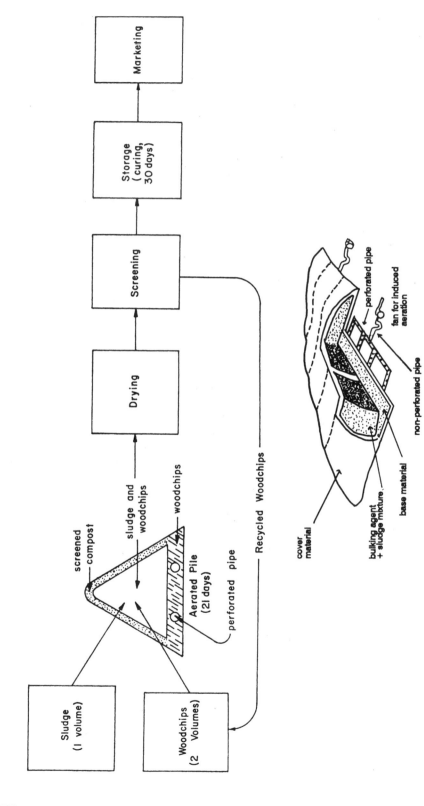

Figure 12.3 Composting. Aerated static pile process. From Benedict et al. (1988). (Courtesy of Noyes Data Corp.)

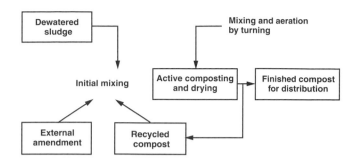

Figure 12.4 Composting. Conventional windrow process. From Benedict et al. (1988). (Courtesy of Noyes Data Corp. and W. D. Burge.)

4. *The cooling phase*, during which temperature decreases again to the mesophilic range (thermophilic microorganisms are replaced by mesophilic microorganisms).

5. *The maturation phase*, in which the temperature drops to ambient levels, allowing the establishment of biota from higher trophic levels (e.g., protozoa, rotifers, beetles, mites, nematodes) as well as other important processes such as nitrification, which is sensitive to high temperatures.

The composting mass may be regarded as a microbial ecosystem. Microbial succession during composting is relatively rapid, since it may take only a few days to reach 55°C. The thermophilic stage continues until depletion of available substrates, leading to a decrease in heat output.

Plate-count methods have traditionally been used to study microbial community composition and succession in composts. Presently, lipid biomarkers (e.g., phospholipids fatty acids (PLFA) analysis) are used (Klamer and E. Bååth,1998; Steger et al., 2003).

Microbial succession during composting is as follows (Beffa et al., 1996a) (Fig. 12.5):

- *Mesophilic phase* (20–40°C). Active degradation of organic matter occurs during the mesophilic phase, which is dominated mostly by bacteria and thermotolerant fungi. Fermenting bacteria (*Leuconostoc*, *pediococcus*) were also found in a laboratory-scale compost, using molecular-based methods (Ishii et al., 2000).

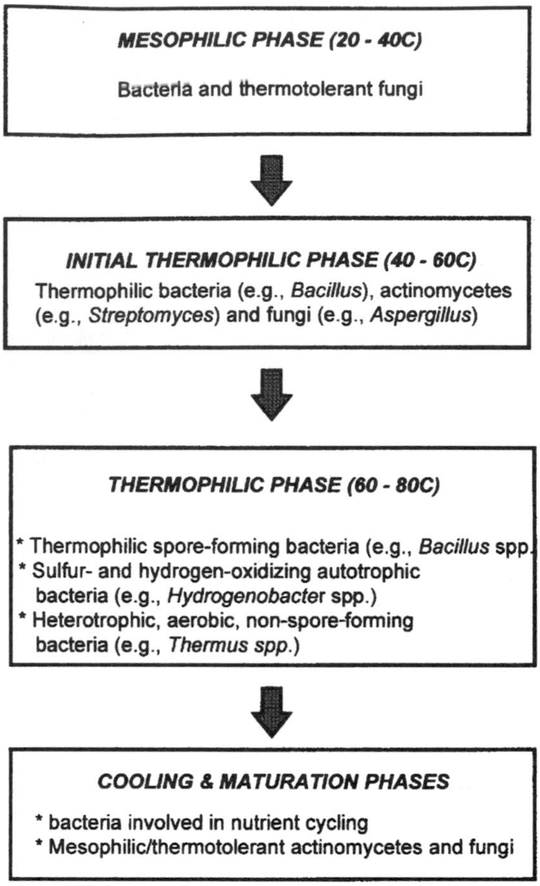

```
┌─────────────────────────────────────────────┐
│        MESOPHILIC PHASE (20 - 40C)          │
│                                             │
│      Bacteria and thermotolerant fungi      │
└─────────────────────────────────────────────┘
                      ⬇
┌─────────────────────────────────────────────┐
│    INITIAL THERMOPHILIC PHASE (40 - 60C)    │
│   Thermophilic bacteria (e.g., Bacillus),   │
│           actinomycetes                     │
│   (e.g., Streptomyces) and fungi (e.g.,     │
│           Aspergillus)                      │
└─────────────────────────────────────────────┘
                      ⬇
┌─────────────────────────────────────────────┐
│       THERMOPHILIC PHASE (60 - 80C)         │
│                                             │
│ * Thermophilic spore-forming bacteria       │
│   (e.g., Bacillus spp.)                     │
│ * Sulfur- and hydrogen-oxidizing autotrophic│
│   bacteria (e.g., Hydrogenobacter spp.)     │
│ * Heterotrophic, aerobic, non-spore-forming │
│   bacteria (e.g., Thermus spp.)             │
└─────────────────────────────────────────────┘
                      ⬇
┌─────────────────────────────────────────────┐
│       COOLING & MATURATION PHASES           │
│                                             │
│  * bacteria involved in nutrient cycling    │
│  * Mesophilic/thermotolerant actinomycetes  │
│    and fungi                                │
└─────────────────────────────────────────────┘
```

Figure 12.5   Microbial succession during composting. Adapted from Beffa et al. (1996a).

- *Initial thermophilic phase* (40–60°C).   This phase is characterized by an increase in the number of thermophilic bacteria (e.g., *Bacillus*), actinomycetes (e.g., *Streptomyces*), and fungi (e.g., *Aspergillus fumigatus*). The optimum growth temperature of thermophilic actinomycetes (50–55°C) is generally higher than that of thermophilic fungi (40–55°C). Thermophilic actinomycetes (e.g., *Streptomyces* sp. and *Thermoactinomyces* sp.) and fungi (e.g., *Aspergilus fumigatus*) play a role in the degradation of complex organics such as cellulose and lignin (Polprasert, 1989).

- *Thermophilic phase* (60–80°C).   A diverse bacterial community has been reported in composts at temperatures of 60–80°C. Composts harbor obligately thermophilic, spore-forming bacteria such as *Bacillus stearothermophilus*, *B. thermoglucosidasius*, *B. pallidus*, or *B. thermodenitrificans* (Beffa et al., 1997; Duvoort-van Engers and Coppola, 1986; Strom, 1985a, b). The presence of thermophilic *Bacillus* species (*B. coagulans*, *B. badius*) was confirmed in a laboratory compost reactor, using denaturing gradient gel electrophoresis (DGGE) analysis of polymerase chain reaction (PCR)-amplified small subunit (ssu) rRNA genes (Ishii et al., 2000). Other thermophilic bacteria have been isolated in composts at temperatures above 60°C and they include obligate (e.g., *Hydrogenobacter* spp.; optimum 70–75°C) and facultative (e.g., *Bacillus schlegelii*; optimum 65–70°C) sulfur- and hydrogen-oxidizing autotrophic bacteria, as well as heterotrophic, aerobic, nonspore-forming *Thermus* strains (e.g., *T. thermophilus*, *T. aquaticus*; optimum 65–75°C) (Beffa et al., 1996b, c). Figure 12.6 displays a scanning electron micrograph of *Thermus thermophilus* isolated from a hot compost (Beffa et al., 1996c). Thermophilic bacteria play a role in the degradation of carbohydrates and proteins.

- *Cooling and maturation phases.* These phases are characterized by the development of bacteria implicated in C, N, and S cycling, as well as a diverse assemblage of mesophilic/thermotolerant actinomycetes and fungi. These microorganisms play a role in compost mineralization and maturation. Nutrient cycling involves nitrogen fixation, nitrification, hydrogen oxidation, and sulfur oxidation.

12.2.6.3 *Factors Affecting Composting.*

Several physical and chemical parameters control the activity of microorganisms during composting: temperature, moisture, pH, aeration, carbon and nitrogen content, as well as the type of material being composted and the composting system used.

- *Temperature*

This is probably the most important factor affecting microbial activity in composts. It can be controlled by adjusting the moisture and aeration levels (McKinley and Vestal, 1985; Polprasert, 1989). It was thought that the optimum temperature for composting sewage sludge, as measured by microbial activity, is between 55 and 60°C and should not exceed 60°C (Bach et al., 1984; Fogarty and Tuovinen, 1991; McKinley et al., 1985; Nakasaki et al., 1985a; Strom, 1985a). We have seen, however, that a diverse bacterial community develops at temperatures higher than 60°C. We will see that the high temperatures achieved during composting help in pathogen inactivation (see Section 12.3.7). In the static pile process, a temperature of 55°C or above must be maintained for at least three days to ensure effective pathogen inactivation (Benedict et al., 1988).

- *Aeration*

Aeration provides oxygen to the aerobic compost microorganisms, helps control the compost temperature, and removes excess moisture as well as gases (Fogarty and

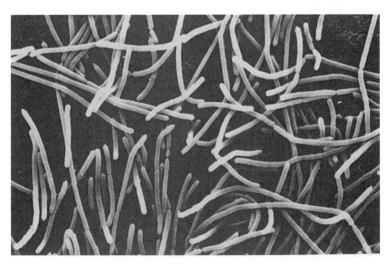

Figure 12.6 Scanning electron micrograph of *Thermus thermophilus* isolated from a hot compost. From Beffa et al. (1996c). (Courtesy of T. Beffa and permission from the publisher.)

Tuovinen, 1991). Turning frequency of the compost controls the homogeneous distribution of substrates, water, and air. Frequent turning (daily to weekly) of the compost mass leads to a significant increase in temperature and extends the thermophilic phase in the compost (Beffa et al., 1997).

• *Moisture Control*

This is also an important factor during composting. The optimum moisture content should be 50–60 percent and should not exceed 65 percent in enclosed composting systems.

• *Other Factors*

Composts should be nutritionally balanced to guarantee an optimum degradation of substrates. The optimum C/N ratio should be 25:1. Levels of toxic compounds (heavy metals, organic toxicants, salts) should be monitored to avoid inhibition of compost microorganisms and to comply with disposal regulations (Metcalf and Eddy, 1991).

12.2.6.4 Advantages and Disadvantages of Composting. Composting produces a stable endproduct that is a good source of nutrients and serves as a useful soil amendment in agriculture and horticulture. A further advantage is the ability of composting to inactivate pathogenic microorganisms. However, compost is labor-intensive and is sometimes the source of foul odors.

12.3 PATHOGEN AND PARASITE REMOVAL DURING SLUDGE TREATMENT

We have previously shown that biological wastewater treatment processes do not completely remove or inactivate pathogenic and parasitic organisms. Many of these organisms become bound to solids after wastewater treatment and are merely transferred to wastewater sludge. Further treatment is necessary to eliminate them, or at least reduce their numbers significantly. We will now review the removal/inactivation capacity of some of the sludge treatment processes.

12.3.1 Pathogens and Parasites Found in Sludge

Bacterial pathogens of primary concern in sludge include *Salmonella*, *Shigella*, *Campylobacter*, *Yersinia*, *Leptospira*, and pathogenic *E. coli* strains. As regards viral pathogens, primary sludge collected in Nancy, France, was found to harbor 97-1230 PFU of enteroviruses/L sludge. Five out of six samples contained hepatitis A antigen as determined by radio-immunoassay (Albert et al., 1990). Other virus types detected in anaerobically digested sludges are polioviruses, coxsackie A and B viruses, echoviruses, and reoviruses. The virus load reported was as high as 15 TCID50/mL (TCID = tissue culture infective dose) (Lydholm and Nielsen, 1981). Similar virus types are generally detected in aerobically digested sludge, but in smaller numbers than in anaerobically digested sludge (Bitton et al., 1985; Goddard et al., 1981; Goyal et al., 1984; Scheuerman, 1984). Astroviruses, a leading cause of viral gastroenteritis, were detected in both untreated and treated sludge samples, using cell culture in combination with molecular-based methods (Chapron et al., 2000). Although enteroviruses in biosolids have been studied, there is a lack of information concerning the presence in biosolids of other viruses (e.g., caliviruses, rotaviruses).

The helminth parasites most found in sludge are *Ascaris* species such as *A. lumbricoides* (human intestinal roundworm) and *A. suum* (pig roundworm), *Taenia saginata*, *Toxocara* (e.g., *T. cati*, the roundworm of cats, and *T. canis*, the roundworm of dogs), *Trichuris* (e.g., *T. trichiura*, the human whipworms) (Little, 1986). The numbers of parasitic eggs detected per kilogram of anaerobically digested sludge (35°C; 42 day detention time; 25 percent solids) from the Caen-Mondeville agglomeration in France were 2200–2400 for *Taniidae*, 410–1200 for *Ascaris* spp., 350–410 for *Toxocara* spp., and 4910–7250 for the family Trichuridae. A total of 8 percent of the parasitic eggs were viable (Barbier et al., 1990). The sludge concentrations of total and viable nematode eggs were 2–53 eggs/10 g and 2–45 eggs/10 g of biosolids, respectively (Gantzer et al., 2001).

In the United States, the use and disposal of wastewater sludge are regulated under 40 CFR Part 503 regulation, promulgated in 1993 (Federal Register, 1993). This regulation established requirements (1) for controlling and reducing pathogens in wastewater sludge, and (2) for reducing the ability of sludge to attract vectors (e.g., insects, rodents, birds) that may be involved in pathogen transmission (U.S. EPA, 1992b). Sludge designated as "Class A" is considered safe and can be used without any restrictions and should have the following criteria: $<1 \times 10^3$ MPN fecal coliforms per gram dry weight; <3 MPN *Salmonella* sp. per 4 g dry weight (d.w.); <1 PFU viruses per 4 g d.w.; and <1 helminth ova per 4 g dry weight. As a comparison, France defines "sanitized" sludge (equivalent to Class A sludge in the United States) as material that has been treated so that pathogens and parasites are no longer detectable (*Salmonella* [<8 MPN/10 g DW], enteroviruses [<3 MPN/10 g DW], viable nematode eggs [<3 MPN/10 g DW]) (Gantzer et al., 2001). Public access to land treated with Class B sludge is restricted for up to a year following sludge application.

We will discuss the requirements for pathogen control in sewage sludge. There are two categories of treatments to reduce pathogen levels in sludges (Appleton et al., 1986; Pedersen, 1981; 1983; U.S. EPA, 1992b; Venosa, 1986; Yeager and O'Brien, 1983).

12.3.1.1 *Processes to Significantly Reduce Pathogens (PSRP).* This category includes aerobic digestion (60 days at 15°C to 40 days at 20°C), anaerobic digestion (60 days at 20°C to 15 days at 35–55°C), lime stabilization (pH 12 after 2 h contact time), mesophilic composting, air drying, and low-temperature composting (minimum: 40°C for at least 5 days). Most facilities are equipped with processes that qualify as PSRP. The use of sludge in the PSRP category for land application is allowed. There are, however, restrictions with regard to crop production (no food crops should be grown within 18 months), animal grazing (prevention of grazing for at least one month by animals that provide products that are consumed by humans), and public access to the treated site (public access must be controlled for at least 12 months).

12.3.1.2 *Processes to Further Reduce Pathogens (PFRP).* This category includes heat treatment (at 180°C for 30 min), irradiation, high-temperature composting, thermophilic aerobic digestion (10 days at 55–60°C), and heat drying. Some PFRP (e.g., high-temperature composting, thermophilic aerobic digestion, heat drying) do not require prior PSRP. Such PFRP processes are required if edible crops are exposed to sludge. There are no restrictions on the use of PFRP sludge on land.

We will now examine the effect of some of the sludge treatment processes on inactivation of pathogens and parasites.

12.3.2 Anaerobic Digestion

Detention time and temperature are the major factors affecting pathogen survival during anaerobic digestion of sludge. The sludge feeding protocol also influences the reduction of microbial pathogens. The draw/fill mode (the digested sludge is withdrawn before feeding the digester) achieves greater pathogen reduction than the fill/draw mode (feed is added before withdrawing product). The difference in pathogen reduction between the two modes is smaller for viruses than for bacteria (Farrel et al., 1988; Pedersen, 1983; Venosa, 1986).

Anaerobic digestion achieves 1– to 3–log reduction in pathogenic and indicator bacteria. For example, a study of six wastewater treatment plants showed that anaerobic digestion led to 2.3–log reduction of fecal coliforms at a hydraulic retention time (HRT) of 22 days and 2.9–log reduction at an HRT of 32 days (Dahab and Surampalli, 2002). However, at an SRT as low as 10 days, mesophilic two-phase anaerobic digestion achieves a greater reduction of bacterial indicators (TC, FC, FS) than does conventional anaerobic digestion. Enterovirus reduction is similar in both processes (Lee et al., 1989). As regards *Salmonella*, anaerobic digestion, in 5 out of 6 plants, produced sludge that did not fulfil the Class A requirement of <3 MPN/4 g sludge (dry weight) (see Section 12.3.1) (Dahab and Surampalli, 2002).

Anaerobic digestion appears to be efficient in protozoan cyst inactivation, which essentially depends on digestion temperature. It achieves a 3–log reduction in viability of cysts of *Giardia muris* (as measured by *in vivo* cyst excystation) in 7.9 days, in 19 h, and in 13.6 min at 21.5, 37, and 50°C, respectively. Protozoan cysts (*Giardia muris*, *Giardia lamblia*, *Cryptosporidium parvum*) are eliminated within 24 h at 37°C (Gavaghan et al., 1992; van Praagh et al., 1993).

As regards the helminth ova, their levels in digested and lagooned sludges from Chicago are shown in Table 12.1 (Arther et al., 1981). Embryonation tests showed that 55 percent of the ova in anaerobically digested sludge are viable (Arther et al., 1981). Thus, parasites (e.g., *Ascaris*, *Trichuris*, *Toxocara*, *Capillaria*) generally survive well the mesophilic anaerobic digestion process (Gantzer et al., 2001). *Ascaris* ova are destroyed only by thermophilic digestion or by heating at 55°C for 15 min (Pike et al., 1988). Thermophilic anaerobic digestion at 53°C reduces this parasite to undetectable levels (Lee et al., 1989).

Varying levels of viruses are detected in anaerobically digested sludge. In France, they were found at levels varying from 50 to 130 PFU/L, and 44 percent of the digested sludge samples were positive for viruses (Schwartzbrod and Mattieu, 1986). Thus, anaerobic digestion does not completely eliminate viruses. Enteric viruses survive in sludge that has been subjected to mesophilic (30–32°C) and sometimes following thermophilic

TABLE 12.1. Levels of Parasitic Nematode Ova in Digested and Lagooned Sludge[a]

Date Collected	No. of Ova per 100 g Dry Sludge			
	Ascaris spp.	*Toxocara* spp.	*Toxascaris leonina*	*Trichuris* spp.
May 1976	192	64	64	32
June 1976	171	150	43	43
June 1977	231	331	66	33
Oct. 1977	218	146	18	36
Mean	203	172.7	47.7	36
SD	26.8	112.7	22.4	4.9

[a]From Arther et al. (1981).
SD, standard deviation.

anaerobic digestion (50°C). Poliovirus 1, embedded within sludge flocs, survives more than 30 days of anaerobic digestion (Moore et al., 1976). Temperature plays an important role in virus inactivation in sludge (Berg and Berman, 1980; Eisenhardt et al., 1977; Ward et al., 1976). Another known virucidal agent of viruses in digesting sludge is the uncharged form of ammonia that predominates at higher pH values. Nucleic acids are the main target of inactivation by ammonia (Ward and Ashley, 1977a).

12.3.3 Aerobic Digestion

Although much of the sludge produced in the United States is treated by anaerobic digestion, many small communities use the less complex aerobic digestion, a process considered by the U.S. EPA to significantly reduce pathogen numbers in sludge before land application (U.S. EPA, 1981). *Salmonella* spp. (mostly *S. enteriditis*) was detected at levels of 0.8–33 MPN/g in aerobically digested sludges from three wastewater treatment plants in Florida (Farrah and Bitton, 1984) (Table 12.2). Reduction in enteric bacteria and viruses during aerobic digestion of sludge depends on both detention time and temperature (Martin et al., 1990; Scheuerman et al., 1991). Bacterial indicator (fecal coliforms, fecal streptococci) die-off increases as temperature is raised from 20 to 40°C (Kuchenrither and Benefield, 1983) (Fig. 12.7). Temperature was also found to be the single most important factor influencing the survival of enteric viruses (poliovirus 1, coxsackie B3, echovirus 1, rotavirus SA-11) during aerobic digestion of sludge (Scheuerman et al., 1991). Mesophilic aerobic digestion appears to be ineffective, however, in reducing the number of parasite eggs such as *Ascaris* (Pedersen, 1983). Figure 12.8 (Kabrick and Jewell, 1982) shows that thermophilic aerobic digestion where temperature is ≥45°C achieves a much higher pathogen (*Salmonella*, *P. aeruginosa*, viruses) destruction than traditional mesophilic anaerobic digestion. A similar trend was observed for parasites (Kabrick et al., 1979; Kuchenrither and Benefield, 1983). In general, no viable eggs are detected at temperatures between 48 and 55°C (Gantzer et al., 2001; Plachy et al., 1995). Thus, autoheated aerobic digestion can produce virtually pathogen- and parasite-free sludge.

12.3.4 Dual Digestion System

The dual digestion system (DDS) consists of two steps (Appleton and Venosa, 1986; Appleton et al., 1986): Step 1 includes a covered aerobic digester using pure oxygen, with a 1 day detention time. Heat (temperature around 55°C) is generated during oxidation of organic matter in the aerobic digester. Step 2 involves treatment in an anaerobic digester with an 8 day detention time.

TABLE 12.2. Detection of *Salmonella* spp. in Aerobically Digested Sludge[a]

Treatment Plant	Sludge Source	% of Samples Positive for *Salmonella* spp.	Mean MPN per g
Gainesville (Kanapaha)	Undigested	27	4.6
	Digested	36–40	0.9–2.3
Gainesville (Main street)	Undigested	29	7.5
	Digested	14–25	0.4–0.8
Tallahasse	Undigested	40	30
	Digested	100	33

[a]Adapted from Farrah and Bitton (1984).

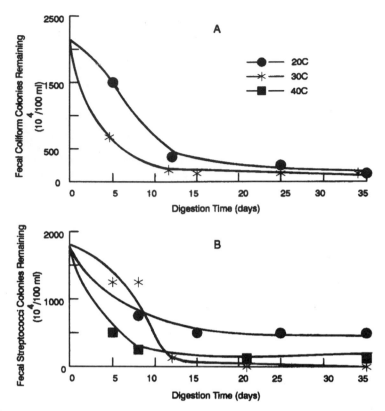

Figure 12.7   Effect of temperature on inactivation of indicator bacteria during aerobic digestion of sludge: (a) Fecal coliforms; (b) fecal streptococci. From Kuchenrither and Benefield (1983). (Courtesy of Water Environment Federation.)

The destruction of pathogenic and indicator organisms is higher in step 1 than in step 2. Reduction of microbial indicators by DDS is greater than that achieved with other stabilization processes such as high-rate anaerobic digestion, lime stabilization, or mesophilic composting. Pathogen reductions are greater than or equal to those achieved by thermophilic aerobic digestion (Appleton et al., 1986).

12.3.5   Sludge Lagooning

Pathogen inactivation in sludge lagoons depends on weather conditions and type of pathogen. At temperatures exceeding 20°C, it appears that the storage time for 1–log reduction of pathogens, should be 1, 2, and 6 months for bacteria, viruses, and parasitic eggs, respectively (Pedersen, 1983). Virus monitoring in a sludge lagoon in Jay, Florida, showed that the virus concentration dropped to undetectable levels in approximately 1–2 months following addition of fresh sludge to the lagoon (Farrah et al., 1981) (Fig. 12.9). Longer storage periods are necessary for pathogen and parasite inactivation in cold climates.

12.3.6   Pasteurization

The process of pasteurization consists of subjecting the sludge to heat treatment at 70°C for 30 min. Although energy-intensive, pasteurization destroys effectively helminth eggs

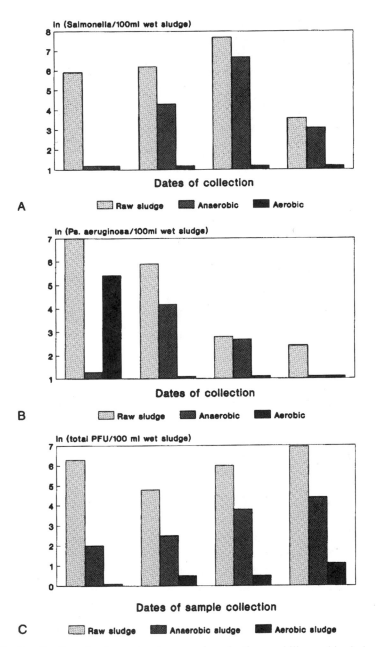

Figure 12.8 Inactivation of pathogenic microorganisms by thermophilic aerobic sludge digestion: (a) *Salmonella*; (b) *Pseudomonas aeruginosa*; (c) viruses. From Kabrick and Jewell (1982). (Courtesy of Pergamon Press.)

and most bacterial and viral pathogens. Some have proposed to heat sludge more economically at a lower temperature for a longer period of time. *Ascaris* ova are destroyed when pasteurization is carried out at temperatures above 55°C for approximately 2 h (Carrington, 1985). Pasteurization at 70°C for 30 min also ensures the destruction of more that 99 percent of *Taenia saginata* ova. This process achieves a complete inactivation of *Salmonella* and enteroviruses (Pike et al., 1988; Saier et al., 1985).

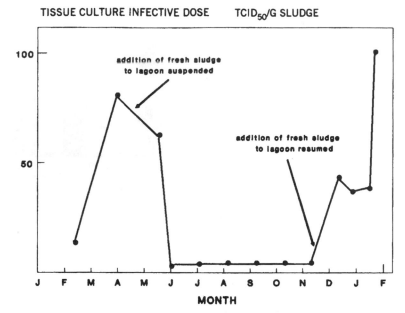

Figure 12.9   Survival of enteroviruses in a sludge lagoon in Florida. From Farrah et al. (1981). (Courtesy of American Society of Microbiology.)

The mechanism of heat inactivation of viruses in sludge has been investigated. A heat-stable ionic detergent, associated with sludge solids, appears to protect enteroviruses but not reoviruses from heat inactivation. This protective effect can be overcome by ammonia, a virucidal agent associated with the liquid portion of digested sludge at pH >8.5. It was concluded that heat treatment of sludge under alkaline conditions would be effective against viruses (Ward and Ashley, 1977a, b; Ward et al., 1976).

12.3.7   Composting

From a public health standpoint, there are two concerns over the production and use of compost material: (1) the effect of pathogenic microorganisms in the waste being composted, and (2) the health hazards of secondary pathogens (fungi, actinomycetes) to compost workers.

Temperature is the key factor controlling pathogen survival in sludge undergoing composting. However, since the compost material is quite heterogeneous, it is difficult to maintain a uniform temperature within the compost biomass. This may explain the difficulties in achieving a complete inactivation of pathogens. It was suggested that compost must be maintained at 55°C for at least 3 days at static pile facilities and for at least 15 days at windrow facilities to ensure safety of this material (Benedict et al., 1988; Epstein, 1979). This process is effective in reducing bacterial indicators (except possibly for fecal streptococci) and bacterial pathogens such as *Salmonella* and *Shigella*. However, pathogen regrowth (e.g., *Salmonella*) has been observed in composted sludge (Brandon et al., 1977; Hussong et al., 1985; Russ and Yanko, 1981). A total of 12 percent of sewage sludge compost from 30 municipalities in the United States was found to harbor *Salmonella* at densities of $\leq 10^4$ cells/g. *Salmonella* regrowth in composted sludge occurs when the moisture content is above 20 percent, at temperatures in the mesophilic range (20–40°C) and requires a C/N ratio in excess of 15:1 (Russ and Yanko, 1981).

In addition to the effect of moisture, pathogen survival in composts is also adversely affected by the indigenous compost microflora. Although the level of nutrients is sufficient for supporting growth, the indigenous microflora limits pathogen regrowth (Hussong et al., 1985). The antagonistic effect of indigenous microorganisms on *Salmonella* regrowth in composted sludge was demonstrated. This effect decreased as the storage time of the biosolids increased (Pietronave et al., 2004; Sidhu et al., 2001).

Some investigators argue that temperature should be maintained below 60°C to achieve high microbial activity during composting. Laboratory studies show evidence of effective inactivation of enteroviruses and parasites (e.g., *Giardia*, *Cryptosporidium*) during composting and curing periods (Gerba et al., 1995; Pedersen, 1983). Temperatures of 55°C or above readily inactivate protozoan cysts and helminth eggs in composting sludge. Figure 12.10 (Brandon, 1978) shows that *Ascaris* eggs are readily inactivated at 55°C, but not at 45°C. A 2–log reduction of somatic phages was obtained at the 50–55°C temperature range while 5–log reduction was obtained at 70°C (Mignotte-Cadiergues et al., 2002).

Aspergillus fumigatus is an opportunistic pathogenic fungus that optimally grows in the outer layers of composts at temperatures of 30–45°C, and decreases significantly at temperatures above 60°C. It causes allergies when inhaled during composting operations (Clark et al., 1984). This pathogen has been isolated from compost workers and can cause serious lung damage. It has been suggested that *A. fumigatus* can serve as an indicator of the presence of other pathogens (Beffa et al., 1997). *Thermoactinomyces* has also been implicated in farmer's lung disease (Blyth, 1973). Thus, proper preventive measures (e.g., wearing protective masks) should be taken to avoid spore inhalation by compost workers (see Chapter 13).

12.3.8 Air Drying

Digested sludge is dewatered on sand beds in small wastewater treatment plants. Liquid sludge is placed on a 1 ft layer of sand and allowed to dry. Water is removed via drainage and evaporation and most of it is collected within the first few days in an underdrain

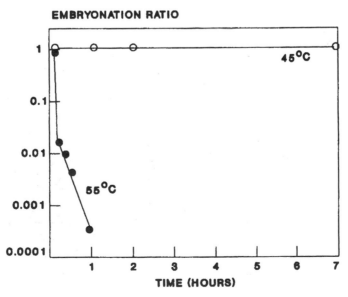

Figure 12.10 Heating inactivation of *Ascaris* eggs in composted sludge. Adapted from Brandon (1978).

system and usually returned to the plant. Afterward, water is removed mostly by evaporation. Both processes remove approximately 45–75 percent of the water (U.S. EPA, 1977). Sludge drying to 95 percent solids brings about a reduction in bacterial pathogens ranging from 0.5 to almost 4–log (Ward et al., 1981; Yeager and O'Brien, 1983).

Pathogen inactivation during storage of dewatered sludge is dependent on the storage temperature, and helminth ova are much more resistant than bacterial pathogens (Ahmed and Sorensen, 1995). Viruses and helminth ova can be detected in dried sludge and, thus, sludge drying is not a reliable process for their complete inactivation.

12.3.9 Chemical Inactivation

12.3.9.1 Lime Stabilization. Lime, added as $Ca(OH)_2$ or CaO, is an inexpensive chemical that is used as a flocculating agent and for odor control in wastewater treatment plants. The lime stabilization process consists of adding a lime slurry to liquid sludge to achieve a pH higher than 12. Modifications of this process include (Westphal and Christensen, 1983): (1) the addition of lime slurry to iron-conditioned sludge prior to vacuum filtration, and (2) addition of dry lime to dewatered conditioned sludge cake.

In order to be effective, lime stabilization must achieve a pH of ≥ 12 for at least 2 h. This treatment leads to 3– to 6–log reduction of bacterial indicators (Venosa, 1986). Figure 12.11 (Farrel et al., 1974; Pedersen, 1983) shows that pH should exceed 10.5 to achieve effective bacterial inactivation. In Germany, it was shown that lime treatment of raw sludge at pH 12.8 completely inactivated *Salmonella senftenberg* within 3 h (Pfuderer, 1985). Viruses are not detected in limed sludge after 12 h of contact time (Sattar et al., 1976). Only 1 of 10 sludge samples conditioned with lime and ferric chloride, and subsequently dehydrated, was positive for viruses (Schwartzbrod and Mattieu, 1986). However, liming does not seem to be completely effective for parasite reduction. Liming at pH >12 for 20–60 days is necessary for the elimination of viable nematode eggs (Amer, 1997; Gaspard et al., 1997).

A process, the N-Viro soil, has been accepted as a PFRP by the U.S. EPA. It consists of treating dewatered sludge with lime and then mixing it with cement kiln dust. The sludge

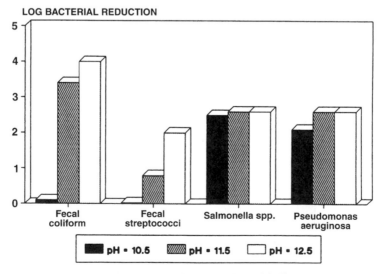

Figure 12.11 Effect of pH on inactivation of pathogenic and indicator bacteria in sludge (0.5 h detention time). From Farrel et al. (1974). (Courtesy of Water Environment Federation.)

is then allowed to cure for 3–7 days. The liming and the resulting high alkalinity of the product may be responsible for efficient inactivation of pathogenic microorganisms.

12.3.9.2 Ammonification. Addition of ammonia or ammonium sulfate to digested sludge effectively reduce parasites in sludge, but the treatment is expensive (Reimers et al., 1986).

12.3.9.3 Ozonics Process. This ozonics process consists of treating sludge for 30–90 min with a high ozone concentration (200 ppm) at low pH (pH 2.5–3.5) and under pressure (60 psi). This process is effective against bacterial pathogens and viruses but not against parasites (Reimers et al., 1986).

12.3.10 Irradiation

This treatment consists of subjecting sludge to isotopes emitting gamma radiation (cesium-137 or cobalt-60) or to high-energy electron beams. This process essentially destroys the microbial genetic material. The energy level of irradiation is relatively low and thus does not result in the production of radioactive sludge (Ahlstrom and Lessel, 1986).

The first sludge irradiator was built in Germany in the early 1970s. Other irradiators were built subsequently in the United States. In 1976, an experimental electron beam irradiator was installed in Deer Island, Massachusetts. The electrons beams are generated by a 750 kV electron accelerator. This unit, which treats a thin layer (approx. 2 mm) of liquid sludge spread on a rotating drum, is illustrated in Figure 12.12 (Ahlstrom and Lessel, 1986). This system was designed to treat approximately 100,000 gallons of sludge per day. In 1978, a sludge irradiator was built at Sandia National Laboratory in Albuquerque, New Mexico. This unit is driven by cesium-137 and treats dried or dewatered sludge (Fig. 12.13). Later on, in 1984, a larger electron beam irradiator was constructed near Miami, Florida. This unit handles thin layers of liquid sludge (2 percent solids) at an applied dose of 350–400 krad and at a rate of 120 GPM (Ahlstrom and Lessel, 1986). Afterward, the sludge is conditioned with polymers, dewatered, and later stored in a sludge lagoon to be marketed as "Daorganite" for agricultural use.

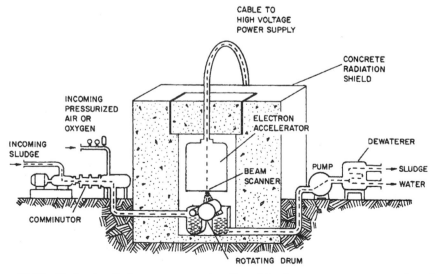

Figure 12.12 Sludge irradiation process used at Deer Island wastewater treatment plant. From Ahlstrom and Lessel (1986). (Courtesy of the Water Environment Federation.)

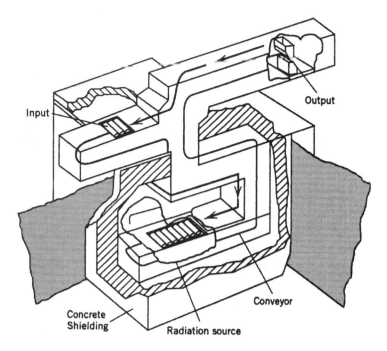

Figure 12.13   Sandia irradiator for dried sludge (Albuquerque, NM). From Ahlstrom and Lessel (1986). (Courtesy of the Water Environment Federation.)

Ionizing radiation inactivates microorganisms either directly or indirectly via production of free radicals. Nucleic acids are the main target of ionizing radiation. The sensitivity of microorganisms to sludge irradiation varies with the type of pathogen or parasite. D_{10}, the radiation dose necessary for inactivation of 90 percent of parasites/pathogens, increases as sludge solids increase. Bacteria (e.g., *Salmonella typhimurium*, *Klebsiella* sp.) are protected from ionizing radiation at reduced sludge moisture levels (Ward et al., 1981). Bacterial pathogens are efficiently reduced (6 to 8–log) at radiation doses ranging from 200 to 300 krad. Pathogens such as *Salmonella* can regrow in irradiaded sludge, but this problem is, however, not encountered in sludge irradiated at doses exceeding 400 krad. Using an electron beam accelerator, the D_{10} values obtained for *Ascaris* ova were 788 Gy (78.8 krad) and 1125 Gy (112.5 krad), depending on the suspending medium of the ova (Capizzi-Banas and Schwartzbrod, 2001). The values of D_{10}, expressed in krad, for inactivation of selected bacterial pathogens and parasites have been compiled (Ahlstrom and Lessel, 1986; Forster and Senior, 1987), and are shown in Table 12.3. It appears that viruses are the pathogens most resistant to irradiation (Yeager and O'Brien, 1983). It was recommended to use 0.5 Mrad for liquid sludges and 1.0 Mrad for dry sludges. A 1 Mrad dose is sufficient for elimination of bacterial, fungal, and viral pathogens as well as parasites (Yeager and O'Brien, 1983).

With regard to pathogen/parasite inactivation, the following processes are more efficient than irradiation alone:

• *Thermoradiation*

Combination of irradiation with heat treatment at 47°C results in higher destruction of pathogens. For example, thermoradiation, with a radiation dose of 150–200 krad and a temperature of 51°C leads to a 5–log reduction of poliovirus in five minutes (Bitton, 1980). This process also improves sludge settling and filterability.

TABLE 12.3. Sludge Irradiation: D values of Selected Pathogens and Parasites[a]

Organism	D-value (krad)[b]
Bacteria	
E. coli	$<22-36$
Micrococcus spp.	14
Klebsiella spp.	36–92
Enterobacter spp.	34–62
Salmonella typhimurium	$<50-140$
Proteus mirabilis	$<22-50$
Streptococcus faecalis	110–250
Viruses	
Poliovirus	350
Coxsackievirus	200
Echovirus	170
Reovirus	165
Adenovirus	150
Parasites	
Ascaris spp.	<66
Fungi	
Aspergillus fumigatus	50–60

[a]From Ahlstrom and Lessel (1986).
[b]D-value is the dose required to reduce the pathogen or parasite by one log.

• *Oxiradiation*

Irradiation in the presence of air or oxygen results in higher pathogen inactivation.

12.3.11 Heat Treatment

This process involves heating sludge, under pressure, at temperatures up to 260°C for approximately 30 min. This treatment stabilizes sludge and improves its dewatering. Microbial pathogens and parasites should be effectively destroyed following heat treatment.

12.3.12 Microwave Treatment

Microwave treatment has been proposed as an alternative pasteurization method for the destruction of pathogens and parasites. *Salmonella senftenberg* is effectively inactivated (in drinking water and in liquid manure) by microwave treatment at temperatures of 67–69°C with an average holding time of about 7 s (Niederwohrmeier et al., 1985). Microwave irradiation led to complete inactivation of total and fecal coliforms in sludge and was much faster than conventional heating (Hong et al., 2004).

As regards nematode eggs, Table 12.4 shows the effect of various sludge treatment processes on their survival in sludge. Aerobic thermophilic digestion, heat treatment, composting, and storage of limed sludge for 180 days produce sludge with no detectable viable nematode eggs (Gantzer et al., 2001).

12.4 EPIDEMIOLOGICAL SIGNIFICANCE OF PATHOGENS IN SLUDGE

Humans and animals can be exposed to pathogens via direct or indirect contact with sewage sludge. Direct contact with biosolids may result from walking in

TABLE 12.4. Effect of Sludge Treatment on the Viability of Nematode Eggs[a]

Sludge Treatment	Viable Nematode Eggs (Mean Number/10 g Sludge (dw))	
	Before	After
Mesophilic stabilization	9.5 ± 4.4	11.3 ± 4.3
Anaerobic mesophilic digestion	7.3 ± 5.7	3.5 ± 1.5
Aerobic thermophilic digestion	4.5 ± 3.2	<1
Composting	1.5 ± 2.0	<1
Quick lime (25%)	23.2 ± 14.6	10.5 ± 12.3
Slaked lime (62%)	6.2 ± 3.7	3.2 ± 0.8
Heat treatment	9.5 ± 4.0	<1
Storage of dehydrated sludge (180 days)	6	2
Storage of limed sludge (180 days)	13	<1

[a]Adapted from Gantzer et al. (2001).

sludge-treated land, handling of soil/produce grown on sludge-amended soil, or inhaling airborne pathogens generated in sludge application sites. Indirect contact may result from consumption of contaminated crops, products from animals grazing in pastures treated with sludge, or contaminated drinking water (U.S. EPA, 1992b).

Prospective epidemiological studies have been conducted to assess the potential health risks associated with sludge treatment and disposal. An epidemiological study of workers at composting facilities showed that workers did not experience any ill effects in comparison with a control group. Relatively minor effects were observed and consisted of skin irritation and inflammation of nose and eyes, resulting from exposure to dust and to the thermophilic fungus *Aspergillus fumigatus* (Clark et al., 1984). Another prospective epidemiological study was conducted in Ohio to determine the health risks of sludge handling to farm workers. Sludge, soil, forage, and fecal samples were analyzed for bacterial, viral, and parasitic pathogens. Serum samples were tested for antibodies to coxsackieviruses, echoviruses, and hepatitis A virus. This study has essentially shown that sludge handling did not result in any adverse health effects on the study population as compared to a control group (Ottolenghi and Hamparian, 1987). Some argue that the combination of pathogens and chemicals in biosolids facilitates infection by pathogens. Chemicals lower the host resistance to infection (Lewis and Gattie, 2002).

The presence of *Salmonella* spp., *Shigella* spp., and *Campylobacter* spp. in sludge applied to land was monitored in Ohio. No *Shigella* or *Campylobacter* were detected in sludge, but 21 serotypes of *Salmonella* were isolated, with the dominant serotype being *Salmonella infantis*. From an examination of antibodies to salmonellae, it was concluded that the risk of infection to farm populations was minimal.

The prospective epidemiological approach may not be the most suitable one for assessing the health risks associated with sludge production and disposal. There are several confounding factors associated with this approach. Sludge harbors a myriad of pathogenic microorganisms and parasites that can cause a myriad of symptoms and diseases. Moreover, the transmission of the pathogenic agents may not be limited to sludge, since other transmission routes (e.g., person-to-person, water, food) may be implicated (Jakubowski, 1986). There are several other problems associated with the assessment of

TABLE 12.5. Probability of Infection from Ingesting Biosolids or Biosolids-Amended Soil[a]

Virus[b]	Days of Exposure	
	1	10
Biosolids ingestion (50 mg)		
Rotavirus	3.67×10^{-2}	2.15×10^{-1}
Echovirus 12	1.30×10^{-4}	1.30×10^{-3}
Ingestion of biosolids-amended soil (50 mg)		
Rotavirus	8.17×10^{-5}	8.17×10^{-4}
Echovirus 12	2.64×10^{-7}	2.64×10^{-6}

[a]Adapted from Gerba et al. (2002).
[b]Virus concentration is assumed to be 1.30×10^{-3} virus/mg biosolids or soil.

the epidemiological significance of pathogenic microorganisms in wastewater and sludge. Some of these problems are (Block, 1983):

1. Methodological problems in determining the number of pathogens in sludge.
2. Pathogen densities in sludges are highly variable.
3. Lack or fragmentary nature of information about the minimum infective doses for pathogens, which may vary from <10 to 10^{10} organisms (Block, 1983; Bryan, 1977).

12.5 RISK ASSESSMENT

Haas and collaborators (1999) have developed models for the assessment of risk of infection from exposure to enteric viruses in drinking water. These models were used to assess the risk associated with activity on land treated with biosolids (Gerba et al., 2002). The risk from infection with rotavirus or echovirus 12 was estimated, assuming a minimum ingestion of 50 mg biosolids/day and a maximum of 480 mg/day. The probability of infection from ingesting biosolids was higher than from ingesting biosolid-amended soil. When the biosolids were injected into the soil, the risk was generally lower than the risk of 10^{-4} recommended by the U.S. EPA (Table 12.5; Gerba et al., 2002).

12.6 WEB RESOURCES

http://www.swbic.org/education/env-engr/sludge/sludge.html (sludge management)
http://biosolids.policy.net/ (updates on biosolids from the National Biosolids Partnership)
http://www.epa.gov/owm/mtb/biosolids/ (biosolids programs, U.S. EPA)
http://www.biosolids.state.va.us/ (the biosolids lifecycle, Virginia Department of Health)
http://ehp.niehs.nih.gov/qa/105-1focus/focusbeauty.html (NIEHS page on biosolids)
http://www.ecy.wa.gov/programs/swfa/biosolids/ (biosolids page from the Washington State Department of Ecology)
http://www.cdc.gov/niosh/docs/2002-149/2002-149.html (guidance for controlling potential risks to workers exposed to class B biosolids; from the Center for Disease Control and Prevention, Atlanta, GA)

http://www.wsscwater.com/BIOSOLIDS/faq.html (from Washington Suburban Sanitary Commission)

www.extension.usu.edu/publica/agpubs/agwm02.pdf (land application of biosolids; Utah State University)

http://muextension.missouri.edu/xplor/envqual/wq0420.htm (publications on biosolids)

12.7 QUESTIONS AND PROBLEMS

1. As regards biosolids disposal, compare the advantages and disadvantages of land disposal, incineration, and landfilling

2. How does sludge lagooning help in pathogen inactivation?

3. What is grit? How is it disposed of?

4. What does sludge pasteurization consist of?

5. What are the main factors affecting composting?

6. What is the difference between Class A and Class B sludges?

7. What is the efficiency of anaerobic digestion for removing bacterial pathogens? How about enteric viruses?

8. What are the two main requirements of U.S. regulations concerning wastewater sludges?

9. What is an autoheated aerobic sludge?

10. Give the main factor involved in pathogen inactivation during composting. What does it take to inactivate protozoan cysts and parasites?

11. What is the opportunistic pathogen of concern to compost workers?

12. What are the two main requirements of U.S. regulations concerning wastewater sludges?

13. How does lime stabilization of sludge inactivate pathogens and parasites?

14. Explain the microbial succession during composting.

15. Should we rely on sludge drying to inactivate viruses in biosolids?

16. What are the confounding factors that make it difficult to assess the epidemiological significance of pathogens and parasites in biosolids?

12.8 FURTHER READING

Bitton, G., B.L. Damron, G.T. Edds, and J.M. Davidson, Eds. 1980. *Sludge-Health Risks of Land Application.* Ann Arbor Scientific Publications., Ann Arbor, MI, 366 pp.

Block, J.C., A.H. Havelaar, and P. l'Hermite, Eds. 1986. *Epidemiological Studies of Risks Associated with the Agricultural Use of Sewage Sludge: Knowledge and Needs.* Elsevier, London, 168 pp.

Davis, M.L., and D.A. Cornwell. 1985. *Introduction to Environmental Engineering*, PWS Engineering, Boston, Ma.

Strauch, D., A.H. Havelaar, and P. L'Hermite, Eds. *Inactivation of Microorganisms in Sewage Sludge by Stabilization Processes*, Elsevier, London.

U.S. EPA. 1992. *Control of Pathogens and Vector Attraction in Sewage Skudge (including Domestic Septage)* Under 40 CFR Part 503, EPA/625R-92/013.

Venosa, A.D. 1986. Detection and significance of pathogens in sludge, IN: *Control of Sludge Pathogens*, C.A. Sorber, Ed., Water Pollution Control Federation., Washington, D.C.

ANAEROBIC DIGESTION OF WASTEWATER AND BIOSOLIDS

Wastewater Microbiology, *Third Edition*, by Gabriel Bitton
Copyright © 2005 John Wiley & Sons, Inc.

13.1   INTRODUCTION

Methane (CH_4) production is a common phenomenon in several diverse natural environments ranging from glacier ice, marine and freshwater sediments, marshes, swamps, termites, gastrointestinal tracts of ruminants, municipal solid waste landfills, oil fields, and hydrothermal vents. The total estimated CH_4 emission to the atmosphere is approximately 525×10^6 tons CH_4-C/year and this rate is rising by about 1 percent per year. Methane is a greenhouse gas and is about 20 times more effective than carbon dioxide. The major sources of CH_4 emission are wetlands, rice paddies, ruminants, and fossil fuels, and contribute about 70 percent of the total emission by the major generators of CH_4 to the atmosphere (Ritchie et al., 1997; Schimel, 2002). Chemical oxidation removes approximately 85 percent of CH_4 from the atmosphere as compared to 10 percent by biological oxidation (Born et al., 1990).

   Anaerobic digestion consists of a series of microbiological processes that convert organic compounds to methane and carbon dioxide, and reduce the volatile solids by 35 percent to 60 percent, depending on the operating conditions (U.S. EPA, 1992a). The microbiological nature of methanogenesis was discovered more than a century ago (Koster, 1988). While several types of microorganisms are implicated in aerobic processes, anaerobic processes are driven mostly by bacteria and methanogens.

   Anaerobic digestion has long been used for the stabilization of wastewater sludges. Later on, however, it was successfully used for the treatment of industrial and domestic wastewaters. This was made possible through a better understanding of the microbiology of this process and through improved reactor designs.

   Anaerobic digestion has several advantages over aerobic processes (Lettinga, 1995; Lettinga et al., 1997; Sahm, 1984; Speece, 1983; Switzenbaum, 1983):

1. Anaerobic digestion uses readily available CO_2 as an electron acceptor. It requires no oxygen, the supply of which adds substantially to the cost of wastewater treatment.

2. Anaerobic digestion produces lower amounts of stabilized sludge (3–20 times less than aerobic processes) since the energy yields of anaerobic microorganisms are relatively low. Most of the energy derived from substrate breakdown is found in the final product, CH_4. With regard to cell yields, 50 percent of organic carbon is converted to biomass under aerobic conditions, whereas only 5 percent is converted into biomass under anaerobic conditions. The net amount of cells produced per metric ton of COD destroyed is 20–150 kg, as compared with 400–600 kg for aerobic digestion. However, the amount of biomass produced can be theoretically reduced by uncoupling metabolism in aerobic activated sludge. Under laboratory conditions, the addition of *para*-nitrophenol, a known uncoupler of oxidative phosphorylation (see Chapter 2), led to a 30 percent decrease in biomass production in activated sludge (Low et al., 2000). Heavy metals, such as Cu, Zn, and Cd, can also act as uncouplers of microbial metabolism. For example, following the addition of cadmium to activated sludge, the observed growth yield Y_{obs} decreases as the C_{Cd}/X_0 ratio increases (C_{Cd} is the cadmium concentration; X_0 is the initial biomass) (Fig. 13.1; Liu, 2000).

3. Anaerobic digestion produces a useful gas, methane. This biogas contains about 90 percent of the energy, has a calorific value of approximately 9000 kcal/m^3, and can be burned on site to provide heat for digesters or to generate electricity. Little energy (3–5 percent) is wasted as heat. The biogas has been exploited for centuries as an inexpensive source of energy. Today, millions of small-scale plants operate worldwide and produce heat and light (Cowan and Burton, 2002). The biogas yield can be improved by lysing biosolids microorganisms. This is achieved by re-treating the

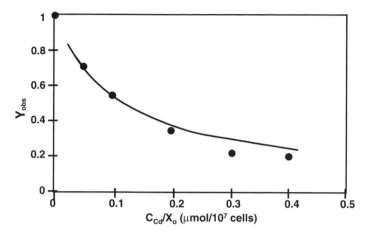

Figure 13.1 Effect of C_{Cd}/X_0 ratio on the observed growth yield of microbial biomass in activated sludge. From Liu (2000).

biosolids by mechanical (high pressure, sonication), thermal (biosolids exposure for short periods to high temperatures ranging from 60 to 225°C), or chemical (alkaline treatment with NaOH or lime) means (Sanders et al., 2002). Methane production contributes to the BOD reduction in digested sludge.

4. There is a reduction of energy required for wastewater treatment.
5. Anaerobic digestion is suitable for high-strength industrial wastes.
6. There is a possibility of applying high loading rates to the digester.
7. There is preservation of the activity of anaerobic microorganisms, even if the digester has not been fed for long periods of time.
8. Anaerobic systems can biodegrade xenobiotic compounds such as chlorinated aliphatic hydrocarbons (e.g., trichloroethylene, trihalomethanes) and recalcitrant natural compounds such as lignin (see Chapter 19).
9. Some investigators predict that anaerobic processes will be used increasingly in the future.

Some disadvantages of anaerobic digestion are:

1. It is a slower process than aerobic digestion.
2. It is more sensitive to upsets by toxicants.
3. Start-up of the process requires long periods, although the use of high-quality seed material (e.g., granular sludge) can speed up the process.
4. As regards biodegradation of xenobiotic compounds via co-metabolism, anaerobic processes require relatively high concentrations of primary substrates (Rittmann et al., 1988).

13.2 PROCESS DESCRIPTION

13.2.1 Single-Stage Digestion

Anaerobic digesters are large fermentation tanks provided with mechanical mixing, heating, gas collection, sludge addition and withdrawal, and supernatant outlets (Metcalf and Eddy, 1991) (Fig. 13.2). Sludge digestion and settling occur simultaneously

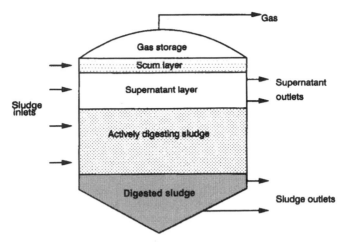

Figure 13.2   Conventional single-stage anaerobic digester. Adapted from Metcalf and Eddy (1991).

in the tank. Sludge stratifies and forms several layers from the bottom to the top of the tank: digested sludge, actively digesting sludge, supernatant, a scum layer, and gas. Higher sludge loading rates are achieved in the high-rate version where sludge is continuously mixed and heated.

13.2.2   Two-Stage Digestion

This process consists of two digesters (Metcalf and Eddy, 1991) (Fig. 13.3); one tank is continuously mixed and heated for sludge stabilization and the other one for thickening and storage prior to withdrawal and ultimate disposal. Although conventional high-rate anaerobic digestion and two-stage anaerobic digestion achieve comparable methane yield and COD stabilization efficiency, the latter process allows operation at much higher loading rates and shorter hydraulic retention times (Ghosh et al., 1985).

13.3   PROCESS MICROBIOLOGY

Consortia of microorganisms, mostly bacteria and methanogens, are involved in the transformation of complex high-molecular-weight organic compounds to methane. Furthermore,

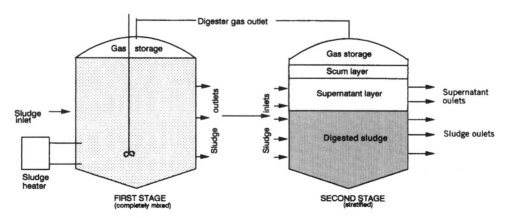

Figure 13.3   Two-stage anaerobic digester. Adapted from Metcalf and Eddy (1991).

synergistic interactions between the various groups of microorganisms are implicated in anaerobic digestion of wastes. The overall reaction is shown in Eq. (13.1) (Polprasert, 1989):

$$\text{Organic matter} \longrightarrow CH_4 + CO_2 + H_2 + NH_3 + H_2S \tag{13.1}$$

Although some fungi and protozoa (anaerobic protozoa were found in landfill material; Finlay and Fenchel, 1991) may be found in anaerobic digesters, bacteria and methanogens are undoubtedly the dominant microorganisms. Large numbers of strict and facultative anaerobic bacteria (e.g., *Bacteroides*, *Bifidobacterium*, *Clostridium*, *Lactobacillus*, *Streptococcus*) are implicated in the hydrolysis and fermentation of organic compounds.

Four categories of microorganisms are involved in the transformation of complex materials into simple molecules such as methane and carbon dioxide. These microbial groups operate in a synergistic relationship (Archer and Kirsop, 1991; Barnes and Fitzgerald, 1987; Koster, 1988; Sahm, 1984; Sterritt and Lester, 1988; Zeikus, 1980): (Fig. 13.4).

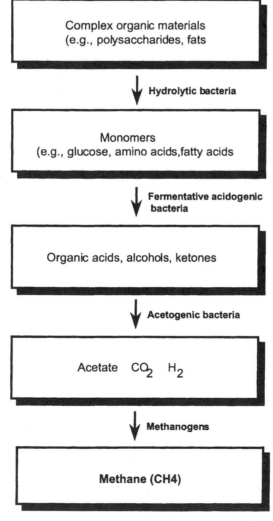

Figure 13.4 Metabolic bacterial groups involved in anaerobic digestion of wastes.

13.3.1   Group 1: Hydrolytic Bacteria

Consortia of anaerobic bacteria break down complex organic molecules (e.g., proteins, cellulose, lignin, lipids) into soluble monomer molecules such as amino acids, glucose, fatty acids, and glycerol. The monomers are directly available to the next group of bacteria. Hydrolysis of the complex molecules is catalyzed by extracellular enzymes such as cellulases, proteases, and lipases. However, the hydrolytic phase is relatively slow and can be limiting in anaerobic digestion of wastes such as raw cellulolytic wastes that contain lignin (Polprasert, 1989; Speece, 1983).

13.3.2   Group 2: Fermentative Acidogenic Bacteria

Acidogenic (i.e., acid-forming) bacteria (e.g., *Clostridium*) convert sugars, amino acids, and fatty acids to organic acids (e.g., acetic, propionic, formic, lactic, butyric, or succinic acids), alcohols and ketones (e.g., ethanol, methanol, glycerol, acetone), acetate, CO_2, and H_2. Acetate is the main product of carbohydrate fermentation. The products formed vary with the bacterial type as well as with culture conditions (temperature, pH, redox potential).

13.3.3   Group 3: Acetogenic Bacteria

Acetogenic bacteria (acetate and H_2-producing bacteria) such as *Syntrobacter wolinii* and *Syntrophomonas wolfei* (McInernay et al., 1981) convert fatty acids (e.g., propionic acid, butyric acid) and alcohols into acetate, hydrogen, and carbon dioxide, which are used by the methanogens. This group requires low hydrogen tensions for fatty acid conversion, necessitating a close monitoring of hydrogen concentration. Under relatively high H_2 partial pressure, acetate formation is reduced and the substrate is converted to propionic acid, butyric acid, and ethanol rather than methane. There is a symbiotic relationship between acetogenic bacteria and methanogens. Methanogens help achieve the low hydrogen tension required by acetogenic bacteria. Online methods for measuring the concentration of total VFAs (volatile fatty acids) and dissolved hydrogen are available and can serve as good monitoring tools for the functioning of anaerobic digestion processes (Björnsson et al., 2001).

Ethanol, propionic acid, and butyric acid are converted to acetic acid by acetogenic bacteria according to the following reactions:

$$\underset{\text{ethanol}}{CH_3CH_2OH} + H_2O \longrightarrow \underset{\text{acetic acid}}{CH_3COOH} + 2H_2 \qquad (13.2)$$

$$\underset{\text{propionic acid}}{CH_3CH_2COOH} + 2H_2O \longrightarrow \underset{\text{acetic acid}}{CH_3COOH} + CO_2 + 3H_2 \qquad (13.3)$$

$$\underset{\text{butyric acid}}{CH_3CH_2CH_2COOH} + 2H_2O \longrightarrow \underset{\text{acetic acid}}{2CH_3COOH} + 2H_2 \qquad (13.4)$$

Acetogenic bacteria grow much faster than methanogens. The former group has a μ_{max} of approximately $1\,h^{-1}$, whereas the μ_{max} of the latter is around $0.04\,h^{-1}$ (Hammer, 1986).

13.3.4   Group 4: Methanogens

As mentioned before, anaerobic digestion of organic matter in the environment releases approximately 500 million tons of methane/year into the atmosphere, representing about 0.5 percent of the organic matter derived from photosynthesis (Kirsop, 1984; Sahm, 1984). The fastidious methanogens occur naturally in deep sediments or in the rumen of herbivores. Methanogenic microorganisms grow slowly in wastewater and their generation times range from 3 days at 35°C to as high as 50 days at 10°C.

Methanogens use a limited number of substrates that include acetate, H_2, CO_2, formate, methanol, and methylamines. All of these substrates are reduced to methylCoM ($CH_3-S-CoM$), which is converted to CH_4 by methylCoM reductase (Ritchie et al., 1997).

Methanogens are subdivided into two subcategories:

- *Hydrogenotrophic methanogens* (i.e., hydrogen-using chemolithotrophs) convert hydrogen and carbon dioxide into methane:

$$CO_2 + 4\,H_2 \longrightarrow CH_4 + 2\,H_2O \qquad (13.5)$$

Most of the methanococcales and methanobacteriales use H_2 and CO_2 (Ritchie et al., 1997)

- *Acetotrophic methanogens*, also called *acetoclastic* or acetate-splitting methanogens, convert acetate into methane and CO_2:

$$CH_3COOH \longrightarrow CH_4 + CO_2 \qquad (13.6)$$

This group comprises two main genera: *Methanosarcina* (Smith and Mah, 1978) and *Methanothrix* (Huser et al., 1982) and *Methanosaeta* (Ritchie et al., 1997). During thermophilic (58°C) digestion of lignocellulosic waste, *Methanosarcina* was the dominant acetotrophic methanogen encountered in the bioreactor. After 4 months, *Methanosarcina* ($\mu_{max} = 0.3$ day^{-1}; $K_s = 200$ mg/L) was displaced by *Methanothrix* ($\mu_{max} = 0.1$ day^{-1}; $K_s = 30$ mg/L). It was postulated that the competition in favor of *Methanothrix* was due to the lower acetate K_s value of this organism (Gujer and Zehnder, 1983; Koster, 1988; Zinder et al., 1984).

About two-thirds of methane is derived from acetate conversion by acetotrophic methanogens. The other third is the result of carbon dioxide reduction by hydrogen (Mackie and Bryant, 1981).

Methanogens belong to a separate domain, the *archaea*, and differ from bacteria in the following characteristics (Sahm, 1984):

1. They differ in cell wall composition; for example, the cell wall of methanogens lacks peptidoglycan.
2. They also differ in the composition of the cell membranes, which are made of branched hydrocarbon chains attached to glycerol by ether linkages.
3. Methanogens have a specific coenzyme F_{420}, a 5-deazaflavin analog, which acts as an electron carrier in metabolism. Its oxidized form absorbs light at 420 nm (Cheeseman et al., 1972). This blue-green fluorescent coenzyme has been proposed for quantifying methanogens in mixed cultures (van Beelen et al., 1983). F_{420} determination in cell extracts is carried out by extraction followed by fluorescence measurement or by high-performance liquid chromatography (HPLC) with fluorimetric detection (Peck, 1989). Methanogenic colonies can be distinguished from nonmethanogenic ones by using fluorescence microscopy (Edwards and

McBride, 1975; Kataoka et al., 1991). However, it was found that the use of F_{420} for methanogenic consortia can be misleading as regards the determination of acetoclastic methanogenic activity (Dolfing and Mulder, 1985). Another nickel-containing coenzyme, F_{430}, is also unique to methanogens.

4. Methanogens belong to the kingdom of euryarchaeota within the archaea domain. They are strictly anaerobic and thrive in oxygen-free environments such as freshwater and marine sediments, swamps, landfills, the rumen of cows, or in anaerobic digesters. A key coenzyme involved in methane production is methyl coenzyme M (see Section 1.4.2, Chapter 1).

5. Methanogens have ribosomal RNA sequences that differ from those of bacteria and eukaryotes.

An example of a tentative general classification of methanogens is displayed in Table 13.1 (Balch et al., 1979; Simon and Goodman, 2002). Methanogens were grouped into four orders: Methanobacteriales (e.g., *Methanobacterium, Methanobrevibacter, Methanothermus*), Methanomicrobiales (e.g., *Methanomicrobium, Methanogenium, Methanospirillum, Methanococcoides*), Methanococcales (e.g., *Methanococcus*), and Methanosarcinales. At least 49 species of methanogens have been described (Vogels et al., 1988) and more are being discovered. Table 13.2 (Koster, 1988) is a compilation of some of the methanogens isolated and their respective substrates.

13.4 METHODS FOR DETECTION OF METHANOGENS

During the past decade, progress has been made in the development of procedures for determining the numbers and activity of methanogens in anaerobic digesters. Isolation of methanogens from environmental samples was a difficult task but the development of relatively recent sophisticated techniques has facilitated the search and subsequent isolation of new species of methanogens (Nichols and Nichols, 2002).

Standard microbiological enumeration techniques are not suitable for methanogens. Methanogens are fastidious and occur as microbial consortia. They are difficult to

TABLE 13.1.   Classification of Methanogens[a]

Order	Family	Genus	Species
Methanobacteriales	Methanobacteriaceae	Methanobacterium	*M. formicicum*
			M. bryanti
			M. thermoautotrophicum
			M. ruminantium
		Methanobrevibacter	*M. arboriphilus*
			M. smithii
			M. vannielli
Methanococcales	Methanococcaceae	Methanococcus	*M. voltae*
		Methanomicrobium	*M. mobile*
Methanomicrobilaes	Methanomicrobiaceae	Methanogenium	*M. cariaci*
			M. marisnigri
		Methanospirillum	*M. hungatei*
			M. barkeri
	Methanosarcinaceae	Methanosarcina	*M. mazei*

[a]From Balch et al. (1979).

TABLE 13.2. Some Isolated Methanogens and Their Substrates[a]

Bacteria	Substrate	Bacteria	Substrate
Methanobacterium bryantii	H_2	*Methanoplanus limicola*	H_2 and HCOOH
M. formicicum	H_2 and HCOOH	*M. endosymbiosus*	H_2
M. thermoautotrophicum	H_2	*Methanogenium cariaci*	H_2 and HCOOH
M. alcaliphilum	H_2	*M. marisnigri*	H_2 and HCOOH
		M. tatii	H_2 and HCOOH
Methanobrevibacter arboriphilus	H_2	*M. olentangyi*	H_2
M. ruminantium	H_2 and HCOOH	*M. thermophilicum*	H_2 and HCOOH
M. smithii	H_2 and HCOOH	*M. bourgense*	H_2 and HCOOH
Methanococcus vannielii	H_2 and HCOOH	*M. aggregans*	H_2 and HCOOH
M. voltae	H_2 and HCOOH	*Methanococcoides methylutens*	CH_3NH_2 and CH_3OH
M. deltae	H_2 and HCOOH	*Methanothrix soehngenii*	CH_3COOH
M. maripaludis	H_2 and HCOOH	*M. concilii*	CH_3COOH
M. jannaschii	H_2	*Methanothermus fervidus*	H_2
M. thermolithoautotrophicus	H_2 and HCOOH	*Methanolobus tindarius*	CH_3OH, CH_3NH_2, $(CH_3)_2NH$, and $(CH_3)_3N$
M. frisius	H_2, CH_3OH		
	CH_3NH_2, and $(CH_3)_3N$	*Methanosarcina barkeri*	CH_3OH, CH_3COOH, H_2,
			CH_3NH_2, $(CH_3)_2NH$, and $(CH_3)_3N$
Methanomicrobium mobile	H_2 and HCOOH	*Methanosarcina thermophila*	CH_3OH, CH_3COOH, H_2
M. paynteri	H_2		CH_3NH_2, $(CH_3)_2NH$, and $(CH_3)_3N$
Methanospirillum hungatei	H_2 and HCOOH		

[a]From Koster (1988), with permission from publisher.

culture in the laboratory. Another difficulty is the fact that culture techniques detect only a small fraction of methanogens.

Immunological analyses, using polyclonal (Archer, 1984; Macario and deMacario, 1988) or monoclonal (Kemp et al., 1988) antibodies, have been used as a tool for determining the numbers and identity of methanogens in anaerobic digesters. Indirect immunofluorescence (IIF) and slide immunoenzymatic assays (SIA) have shown that the methanogenic microflora of anaerobic digesters was more diverse than previously thought. The predominant species detected were *Methanobacterium formicum* and *Methanobrevibacter arboriphilus* (Macario and Macario, 1988).

Today, molecular techniques are available for identification of Archaea in general, and methanogens in particular. These techniques include, among others, analysis of 16S rRNA gene sequences (Burggraf et al., 1994; Huang et al., 2002). There are 16S rRNA-targeted oligonucleotide probes for the Euryarchaeota (methanogens belong to this kingdom) and Crenarchaeota as well as for individual species of methanogens. Such probes have been utilized for identifying methanogens in anaerobic digesters (Raskin et al., 1994). Another approach is the amplification of the gene responsible for methyl-coenzyme M reductase, which catalyses the terminal step of methanogenesis and is found only in methanogens (Nercessian et al., 1999).

Microbial activity in anaerobic digesters is usually determined by measuring volatile fatty acids (VFA) or methane. Lipid analysis has been used to determine the biomass, community structure, and metabolic status in experimental digesters. Microbial biomass, community structure, and metabolic status are indicated by determining the total lipid phosphate, phospholipid fatty acids, and poly-β-hydroxybutyric acid, respectively (Henson et al., 1989; Martz et al., 1983; White et al., 1979). Microbial activity in anaerobic sludge can also be determined by measuring ATP and INT-dehydrogenase activity. These parameters correlate well with other traditional parameters such as gas production rates (Chung and Neethling, 1989). Determination of ATP responds to pulse feeding of the digester and to the addition of toxicants (Chung and Neethling, 1988). Tests are available for the estimation of the amount of acetotrophic methanogens in sludge (Valcke and Verstraete, 1983; van den Berg et al., 1974). One of these tests measures the capacity of the sludge to convert acetate into methane. The test gives information on the percentage of acetotrophic methanogens in anaerobically digested sludge.

Phosphatase activity has also been proposed as a biochemical tool to predict digester upset or failure. An increase in acid and alkaline phosphatases can predict instability of the digestion process well in advance of conventional tests (pH, VFA, gas production) (Ashley and Hurst, 1981).

13.5 FACTORS CONTROLLING ANAEROBIC DIGESTION

Anaerobic digestion is affected by temperature, pH, retention time, chemical composition of wastewater, competition of methanogens with sulfate-reducing bacteria, and the presence of toxicants.

13.5.1 Temperature

Methane production has been documented under a wide range of temperatures ranging between 0 and 97°C. Although psychrophilic methanogens have not been isolated, thermophilic strains operating at an optimum range of 50–75°C occur in hot springs.

Methanothermus fervidus has been found in a hot spring in Iceland and grows at 63–97°C (Sahm, 1984).

In municipal wastewater treatment plants, anaerobic digestion is carried out in the mesophilic range at temperatures from 25°C to up to 40°C,with an optimum at approximately 35°C. Thermophilic digestion operates at temperature ranges of 50–65°C. It allows higher loading rates and is also conducive to greater destruction of pathogens. One drawback is its higher sensitivity to toxicants (Koster, 1988).

Because of their slower growth as compared to acidogenic bacteria, methanogens are very sensitive to small changes in temperature. As to utilization of volatile acids by methanogens, a decrease in temperature leads to a decrease of the maximum specific growth rate (μ_{max}), while the half-saturation constant K_s increases (Lawrence and McCarty, 1969). Thus, mesophilic digesters must be designed to operate at temperature of 30–35°C for their optimal functioning.

13.5.2 Retention Time

The hydraulic retention time (HRT), which depends on wastewater characteristics and environmental conditions, must be long enough to allow metabolism by anaerobic microorganisms in digesters. Digesters based on attached growth have a lower HRT (1–10 days) than those based on dispersed growth (10–60 days) (Polprasert, 1989). The retention times of mesophilic and thermophilic digesters range between 25 and 35 days, but can be lower (Sterritt and Lester, 1988).

13.5.3 pH

Most methanogens function optimally at a pH range of 6.7–7.4, but optimally at pH 7.0–7.2, and the process may fail if the pH is close to 6.0. Acidogenic bacteria produce organic acids that tend to lower the pH of the bioreactor. Under normal conditions, this pH reduction is buffered by bicarbonate produced by methanogens. Under adverse environmental conditions, the buffering capacity of the system can be upset, eventually stopping methane production. Acidity is more inhibitory to methanogens than to acidogenic bacteria. An increase in volatile acids level thus serves as an early indicator of system upset. Monitoring the ratio of total volatile acids (as acetic acid) to total alkalinity (as calcium carbonate) has been suggested to ensure that it remains below 0.1 (Sahm, 1984). One method for restoring the pH balance is to increase alkalinity by adding chemicals such as lime, anhydrous ammonia, sodium hydroxide, or sodium bicarbonate.

13.5.4 Chemical Composition of Wastewater

Methanogens can produce methane from carbohydrates, proteins, and lipids, as well as from complex aromatic compounds (e.g., ferulic, vanillic, and syringic acids). However, a few compounds such as lignin and n-paraffins are hardly degraded by anaerobic microorganisms.

Wastewater must be nutritionally balanced (nitrogen, phosphorus, sulfur, etc.) to maintain an adequate anaerobic digestion. Phosphorus limitation results in a reversible decrease in methanogenic activity. The C : N : P ratio for anaerobic bacteria is 700 : 5 : 1 (Lettinga, 1995; Sahm, 1984). However, some investigators argue that the C/N ratio for optimal gas production should be 25–30 : 1 (Polprasert, 1989). Methanogens use ammonia and sulfide as nitrogen and sulfur sources, respectively. Although un-ionized sulfide is toxic to methanogens at levels exceeding 150–200 mg/L, it is required by methanogens as a major source of sulfur (Speece, 1983). Moreover, trace elements such as iron, cobalt, molybdenum, and nickel are also necessary. Nickel, at concentrations as low as 10 μM, significantly increases methane production in laboratory digesters (Williams et al., 1986). Nickel addition can increase the acetate utilization rate of methanogens from two to as high as 10 g acetate g^{-1} VSS day^{-1} (Speece et al., 1983). Nickel enters in the composition of the co-factor F_{430}, which is involved in biogas formation (Diekert et al., 1981; Whitman and Wolfe, 1980).

13.5.5 Competition of Methanogens with Sulfate-Reducing Bacteria

Methanogens and sulfate-reducing bacteria may compete for the same electron donors, acetate and H_2. The study of the growth kinetics of these two groups shows that sulfate-reducing bacteria have a higher affinity ($K_s = 9.5$ mg/L) than methanogens ($K_s = 32.8$ mg/L) for acetate substrate. This means that sulfate-reducing bacteria can outcompete methanogens under low acetate concentrations (Oremland, 1988; Shonheit et al., 1982; Yoda et al., 1987). This competitive inhibition results in the shunting of electrons from methane generation to sulfate reduction (Lawrence et al., 1966; McFarland and Jewell, 1990). This competition was also demonstrated in biofilms, using a 30 μm diameter methane microsensor based on the activity of immobilized methane-oxidizing bacteria. Using hydrogen as electron donor, methanogenesis was inhibited in the biofilm in the presence of 2 mmol/L sulfate (Damgaard et al., 2001).

Sulfate reducers and methanogens are very competitive at COD/SO_4 ratios of 1.7–2.7. An increase of this ratio is favorable to methanogens, whereas a decrease in the ratio is favorable to sulfate reducers (Choi and Rim, 1991).

13.5.6 Toxicants

A wide range of toxicants are responsible for the occasional failure of anaerobic digesters. Toxicity becomes less severe, however, once the anaerobic microorganisms become adapted to the toxic wastewater (Lettinga, 1995). Inhibition of methanogenesis is generally indicated by reduced methane production and increased concentration of volatile acids. Some of the inhibitors of methanogenesis are discussed below.

13.5.6.1 Oxygen. Methanogens are obligate anaerobes and may be adversely affected by trace levels of oxygen (Oremland, 1988; Roberton and Wolfe, 1970). However, new findings show that methanogens can withstand oxygen, especially in granular sludge (see Section 13.6.2), where they are protected from the detrimental effect of oxygen inside the sludge aggregates (Kato et al., 1993). Oxygen can also cause the deterioration of granular sludge (see Section 13.6.2) due to the growth and attachment of filamentous microorganisms on the granules (Lettinga, 1995).

13.5.6.2 Ammonia. Un-ionized ammonia is quite toxic to methanogens. However, since the production of unionized ammonia is pH-dependent (i.e., more un-ionized form at high pH), little toxicity is observed at neutral pH. Ammonia is inhibitory to methanogens at levels of 1500–3000 mg/L. Toxicity caused by continuous addition of ammonia decreases as the solid retention time is increased (Bhattacharya and Parkin, 1989).

13.5.6.3 Chlorinated Hydrocarbons. Chlorinated aliphatics are much more toxic to methanogens than to aerobic heterotrophic microorganisms (Blum and Speece, 1992).

Chloroform is very toxic to methanogens and leads to their complete inhibition, as measured by methane production and hydrogen accumulation, at concentrations above 1 mg/L (Hickey et al., 1987) (Fig. 13.5). Acclimation to this compound increases the tolerance of methanogens to up to 15 mg/L of chloroform. Methanogen recovery depends on biomass concentration, solid retention time, and temperature (Yang and Speece, 1986).

13.5.6.4 Benzene Ring Compounds.

Pure cultures of methanogens (e.g., *Methanothix concilii*, *Methanobacterium espanolae*, *Methanobacterium bryantii*) are inhibited by benzene ring compounds (e.g., benzene, toluene, phenol, pentachlorophenol). Pentachlorophenol is the most toxic of all the benzene ring compounds tested (Patel et al., 1991). Among the phenolic compounds, the order of inhibition of methanogenesis is nitrophenols > chlorophenols > hydroxyphenols (Wang et al., 1991). Methanogenic activity in granules from an upflow anaerobic sludge blanket (UASB) was inhibited by aromatic compounds in the following order: cresol > phenol > hydroxyphenol > phathalates (Fang et al., 1997). Among N-substituted aromatic compounds, nitrobenzene was one of the most toxic to acetoclastic methanogens (Donlon et al., 1995).

13.5.6.5 Formaldehyde.

Methanogenesis is severely inhibited at formaldehyde concentration of 100 mg/L, but appears to recover at lower formaldehyde concentrations (Hickey et al., 1987; Parkin and Speece, 1982) (Fig. 13.6).

13.5.6.6 Volatile Acids.

If the pH is maintained near neutrality, volatile acids such as acetic or butyric acid appear to exert little toxicity toward methanogens. Propionic acid, however, displays toxicity to both acid-forming bacteria and methanogens.

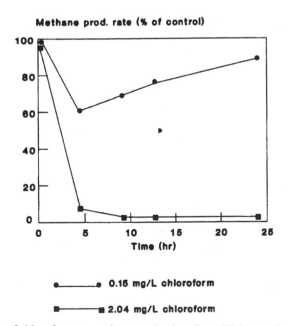

Figure 13.5 Effect of chloroform on methane production. From Hickey et al. (1987). (Courtesy of Pergamon Press.)

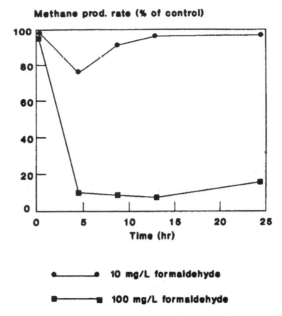

Figure 13.6 Effect of formaldehyde on methanogenesis. From Hickey et al. (1987). (Courtesy of Pergamon Press.)

13.5.6.7 Long-Chain Fatty Acids. The long-chain fatty acids (e.g., caprylic, capric, lauric, myristic, and oleic acids) inhibit the activity of acetoclastic methanogens (e.g., *Methanothrix* spp.) in acetate-fed sludge (Koster and Cramer, 1987).

13.5.6.8 Heavy Metals. Heavy metals (e.g., Cu^{2+}, Pb^{2+}, Cd^{2+}, Ni^{2+}, Zn^{2+}, Cr^{+6}) found in wastewaters and sludges from industrial sources are inhibitory to anaerobic digestion (Lin, 1992; Mueller and Steiner, 1992). The sequence of inhibition of anaerobic digestion of municipal sludge was found to be Ni > Cu > Cd > Cr > Pb. Toxicity increased as the heavy metal affinity for sludge decreased. Thus, lead is the least toxic heavy metal, due to its high affinity for sludge (Mueller and Steiner, 1992). As regards wastewater from the photoprocessing industry, the anaerobic digestion process can tolerate relatively high concentrations of silver (at least up to 100 mg Ag/L) (Pavlostathis and Maeng, 2000). Metal toxicity is reduced after reaction with hydrogen sulfide, which leads to the formation of insoluble heavy metal precipitates.

However, as discussed previously, some metals (e.g., nickel, cobalt, and molybdenum), at trace concentrations, may stimulate methanogens (Murray and van den Berg, 1981; Shonheit et al., 1979; Whitman and Wolfe, 1980).

13.5.6.9 Cyanide. This chemical is used in industrial processes such as metal cleaning and electroplating. While cyanide toxicity is both concentration- and time-dependent, methanogen recovery depends on biomass concentration, solid retention time, and temperature (Fedorak et al., 1986; Yang and Speece, 1985)

13.5.6.10 Sulfide. Sulfide is one of the most potent inhibitors of anaerobic digestion (Anderson et al., 1982). Since un-ionized hydrogen sulfide diffuses through the cell membrane more rapidly than ionized species, sulfide toxicity is highly dependent on pH (Koster et al., 1986). For example, at the neutral pH that is typical of methanogenesis,

20–50 percent of the dissolved sulfide is in the form of H_2S (Colleran et al., 1995). Sulfides are toxic to methanogens when their levels exceed 150–200 mg/L. Acid-forming bacteria are less sensitive to hydrogen sulfide than methanogens. Within the latter group, hydrogen-oxidizing methanogens seem to be more sensitive than acetoclastic methanogens (Rinzema and Lettinga, 1988).

The inhibitory effect of sulfides may be due to sulfide per se, but it may also be caused by sulfate, which serves as a terminal electron acceptor for sulfate-reducing bacteria, possibly resulting in competition between this bacterial group and methanogens since both groups utilize the same substrates (see Section 13.5.5). No inhibition was observed when the TOC/sulfate molar ratio was 1.3 (Karhadkar et al., 1987). In a thermophilic anaerobic fermenter, sulfide inhibition could be minimized by controlling the pH and by biomass recycling to select for sulfide-tolerant microflora (McFarland and Jewell, 1990).

13.5.6.11 *Tannins.*

Tannins are phenolic compounds originating from grapes, bananas, apples, coffee, beans, and cereals. These compounds are generally toxic to methanogens. While gallotannic acid is highly toxic to methanogens, its monomeric derivatives, gallic acid and pyrogallol, are much less toxic (Field and Lettinga, 1987). Tannins inhibit methanogens by possibly reacting with accessible enzyme sites.

13.5.6.12 *Salinity.*

Salinity is another cause of toxicity encountered in anaerobic digestion of wastes. Since potassium antagonizes sodium toxicity, this type of toxicity can be countered by adding K salts to wastewater.

13.5.6.13 *Feedback Inhibition.*

Anaerobic systems may also be inhibited by several of the intermediates produced during the process. High concentrations of these intermediates (H_2, volatile fatty acids) are toxic via feedback inhibition (Barnes and Fitzgerald, 1987)

In order to avoid some of the toxicity problems discussed above, it was suggested that two-phase anaerobic digestion systems be used to separate acidogenic bacteria spatially from methanogens (Cohen et al., 1980; Ghosh and Klass 1978; Pipyn et al., 1979). Some of the advantages of the two-phase system are enhanced stability and increased resistance to toxicants (toxicants are removed or reduced in the first stage). A long solid retention time (SRT) also allows methanogens to acclimate to toxicants such as ammonia, sulfides, and formaldehyde. Thus, anaerobic digestion of industrial wastes containing toxic chemicals should be undertaken in reactors (e.g., anaerobic filter, anaerobic fluidized bed, anaerobic upflow sludge blanket) that allow a long SRT at relatively low hydraulic retention times (Bhattacharya and Parkin, 1988; Parkin et al., 1983).

13.6 ANAEROBIC TREATMENT OF WASTEWATER

Anaerobic digestion processes have been traditionally used for the treatment of wastewater sludges. There is growing interest in anaerobic treatment of wastewater as an energy-efficient approach to waste treatment. For processing of low-strength wastes, the SRT must be controlled independent of the HRT. A strategy for a successful treatment of wastewater via anaerobic systems is to find means to concentrate the slow-growing microorganisms.

We will now discuss the microbiological aspects of the main types of anaerobic systems used in wastewater treatment. The reactor configurations are displayed in Fig. 13.7 (Speece, 1983).

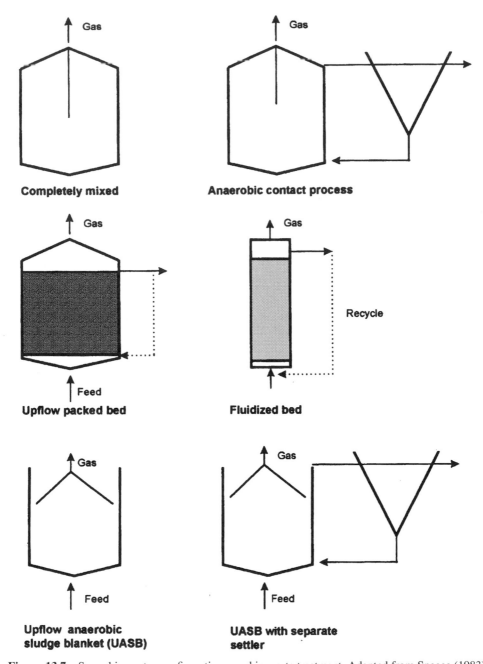

Figure 13.7 Some bioreactor configurations used in waste treatment. Adapted from Speece (1983).

13.6.1 Septic Tanks

The septic tank system, the oldest and most widely used anaerobic treatment system, was introduced at the end of the 19th century. Approximately 25 percent of the U.S. population is served by septic tanks. Septic tanks are the most important contributors of contaminated effluents to groundwater.

A septic tank system is composed of a tank and an absorption field (Fig. 13.8).

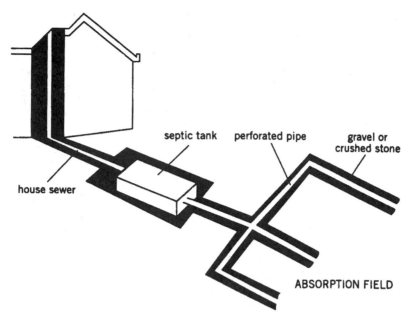

Figure 13.8   Septic tank system. Adapted from U.S. Department of Health, Education and Welfare (1969).

13.6.1.1   Tank.   The primary function of the tank is the biological digestion of waste organic matter under anaerobic conditions, followed by solid separation from the incoming wastewater (Scalf et al., 1977). The tank is made of concrete, metal, or fiberglass, and is designed to remove the wastewater solids to avoid clogging of the absorption field. Wastewater undergoes anaerobic digestion, resulting in the production of sludge called *septage* and a floating layer of light solids (fats) called *scum.* The detention time of wastewater within a septic tank varies from 24 to 72 h. A septic tank system generates a relatively small amount of septage (1000–2000 gallons/tank/2–5 years). Septage is mainly disposed of by land application or combined with municipal wastewater to be treated in wastewater treatment plants. Septage is approximately 50 times as concentrated as municipal wastewater. For a plant loaded with municipal wastewater to 75 percent of its design capacity, the recommended septage loading is approximately 1 percent (Segall and Ott, 1980). The tank should be inspected regularly and should be cleaned every 3–5 years to remove the accumulated septage. In hot climates, septic tanks provide efficient removal of suspended solids (approximately 80 percent) and BOD (>90 percent). Otherwise, septic tanks achieve 65–80 percent and 70–80 percent removal of BOD and suspended solids, respectively. However, anaerobic digestion in the tank provides limited inactivation of pathogens (Hagedorn, 1984). Viruses are not significantly inactivated following wastewater digestion in septic tanks and may therefore be transported into groundwater (Hain and O'Brien, 1979; Sinton, 1986; Vaughn et al., 1983)

13.6.1.2   Absorption Field.   The effluents from septic tanks reach an absorption field through a system of perforated pipes, which are surrounded by gravel or crushed stones. The septic tank effluent is treated by the soil as it percolates downward to the groundwater. There are several designs of soil absorption systems, including trenches, beds, mounds, and seepage pits (U.S. EPA, 1980). Proper functioning of the soil absorption field depends on many factors, including among them the wastewater characteristics, rate of wastewater loading, geology, and soil characteristics. Under normal conditions, the absorption field should operate under unsaturated flow conditions (Scalf et al., 1977).

Septic tank effluents are major contributors to groundwater pollution with pathogenic microorganisms (see Chapter 21). Wastewater volumes as high as $820–1460 \times 10^9$ gallons/year are contributed by septic tanks in the United States (Office of Technology Assessment,1984). These relatively large wastewater inputs to the subsurface environment are responsible for a significant percentage of disease outbreaks resulting from the consumption of untreated groundwater. Septic tanks are probably the major contributors of enteric viruses found in the subsurface environment. The U.S. Census Bureau (1993) estimated that 15 million households in the United States use private wells as their source of drinking water. Monitoring of private household wells located near septage application sites or in subdivisions served by septic tanks in Wisconsin showed the presence of enteric viruses in 8 percent of the samples. The viruses detected in the well water were hepatitis A virus, Norwalk-like viruses, rotaviruses, and enteroviruses (Borchardt et al., 2003).

Drinking water wells should therefore be properly situated to avoid groundwater contamination and to allow for sufficient safe distance to bring about an efficient inactivation of microbial pathogens. Geostatistical techniques have been considered to estimate virus inactivation rates in groundwater in order to predict safe septic tank setback distances for installation of drinking water wells (Yates, 1985; Yates et al., 1986; Yates and Yates, 1987).

13.6.2 Upflow Anaerobic Sludge Blanket

The upflow anaerobic sludge blanket (UASB) uses immobilized biomass to allow the retention of the sludge in the treatment system. It was introduced at the beginning of the century and, after numerous modifications, it was put into commercial use in the Netherlands for the treatment of industrial wastewater generated by the food industry (e.g., beet-sugar, corn, and potato starch).

The UASB-type digester consists of a bottom layer of packed sludge, a sludge blanket and an upper liquid layer (Lettinga, 1995) (Fig. 13.9). Wastewater flows upward through a sludge bed, which is covered with a floating blanket of active bacterial flocs. Settler screens separate the sludge flocs from the treated water and gas is collected at the top of the reactor (Schink, 1988). This process results in the formation of a compact *granular sludge*, which settles well and which withstands the shear force caused by the upflow of wastewater. The sludge is immobilized by the formation of highly settleable microbial

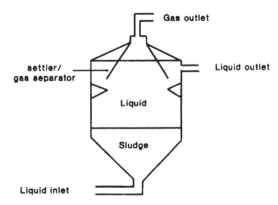

Figure 13.9 Upflow anaerobic sludge blanket (UASB) process. Adapted from Lettinga et al. (1980).

aggregates that grow into distinct *granules* ($<1-5$ mm) that have a high VSS content and specific activity. Immunological techniques, scanning electron microscopic (SEM) examination, and energy-dispersive X-ray analysis of granular sludge have shown that the granules are composed of methanogens such as *Methanothrix soehngenii* (Brummeler et al., 1985; Hulshoff Pol et al., 1982; 1983), *Methanobacterium*, *Methanobrevibacter*, *Methanosarcina*, as well as Ca precipitates (Vissier et al., 1991; Wu et al., 1987). Scanning electron microscopy and transmission electron microscopy (TEM) have shown that the granules are three-layered structures (MacLeod et al., 1990) (Fig. 13.10). The inner layer consists of *Methanothrix*-like cells, which may act as nucleation centers necessary for the initiation of granule development. The middle layer consists of bacterial rods that include both H_2-producing acetogens and H_2-consuming organisms. The outermost layer consists of a mixture of rods, cocci, and filamentous microorganisms. This layer compiles a mixture of fermentative and H_2-producing organisms. Thus, a granule appears to harbor the necessary physiological groups to convert organic compounds to methane. The layered structure of anaerobic sludge granules was confirmed using 16S rRNA-targeted probes (Harmsen et al., 1996; Liu et al., 2002a; Sekiguchi et al., 1999). Both mesophilic and thermophilic granules displayed layered structures, with the outer layer containing mostly bacterial cells while the inner layer contained mostly archaeal cells (*Methanosaeta-*, *Methanobacterium-*, *Methanospirillum-*, and *Methanosarcina*-like cells) (Sekiguchi et al., 1999).

Micro-electrodes, in combination with genetic probes, are useful in showing pH, methane, and sulfide profiles within the granules. This approach provides useful information on the interaction between methanogens and sulfate-reducing bacteria (SRB) within the granules. For example, in granules fed glucose as a substrate, it was shown that pH increased from the outer to the inner portion of the granules (i.e., conversion of acids to methane in the inner portion). This technology also proved that sulfate reduction occurred at the outer portion of the granule while FeS precipitation occurred in the inner portion (Yamaguchi et al., 2001). In methanogenic–sulfidogenic bioreactors, the SRB were located in the surface layer down to a depth of 100 μm, while the methanogens were located in the core of the aggregates (Santegoeds et al., 1999).

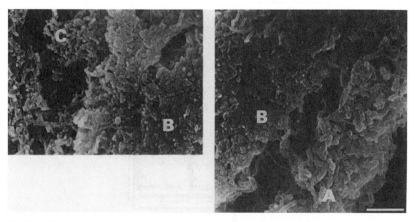

Figure 13.10 Three-layered structure of a sludge anaerobic granule (bar = 5 μm): A, exterior heterogenous layer with rods, filaments, and cocci; B, second layer containing predominantly rods and cocci; C, homogenous core containing a large number of cavities, surrounded by one bacterial morphology. From MacLeod et al. (1990). (Courtesy of American Society of Microbiology and S. Guiot.)

The microbiological composition of granules depends on the type of growth substrate (Grotenhuis et al., 1991). Factors affecting the rate of granulation include wastewater characteristics (higher rate when wastewater is composed of soluble carbohydrates), conditions of operation (e.g., sludge loading rate), temperature, pH, and availability of essential nutrients (Hulshoff Pol et al., 1983; Wu et al., 1987). Problems associated with granulation include deterioration of sludge granules, attachment of fast-growing bacteria, flotation, and calcium carbonate scaling (Lettinga et al., 1997). The treatment of distillery wastewater by the UASB process resulted in 92 percent BOD removal (Pipyn et al., 1979).

Novel versions of UASB reactors include, among others, the expanded granular sludge bed (EGSB) reactor, which functions at high flow velocities and allows the treatment of very low-strength wastewaters even at temperatures lower than 10°C (Kato et al., 1994; Rebac et al., 1995).

13.6.3 Anaerobic Filters

Anaerobic filters (Fig. 13.11) were first introduced at the beginning of the last century and further developed in 1969 by Young and McCarty. These filters are the anaerobic equivalent of trickling filters. They contain support media (rock, gravel, plastic) with a void space of approximately 50 percent or more (Frostell, 1981; Jewell, 1987). The bulk of anaerobic microorganisms grow attached to the filter medium, but some form flocs that become trapped inside the filter medium. The upflow of wastewater through the reactor helps retain suspended solids in the column. This process is particularly efficient for wastewaters rich in carbohydrates (Sahm, 1984). The loading rate varies with the type of waste and with the type of support medium. It generally falls within the range of 5–$20 \, \text{kg COD m}^{-3} \, \text{d}^{-1}$

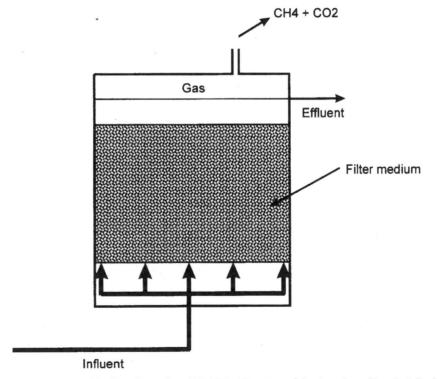

Figure 13.11   Anaerobic filter. From Jewell (1987). (Courtesy of the American Chemical Society.)

(Barnes and Fitzgerald, 1987). This system achieves modest BOD removal but higher removal of solids. Approximately 20 percent of the BOD is converted to methane.

Another version of the anaerobic filter is the *thin film reactor* developed by van den Berg and collaborators (1981). This reactor contains several clay tubes 5–10 cm in diameter. Incoming wastewater flows downward and is treated by the 1 to 3 mm thick anaerobic biofilm that develops on the surfaces of the clay tubes (Fig. 13.12).

13.6.4 Anaerobic Attached-Film Expanded-Bed and Fluidized-Bed Reactors

These reactors were introduced during the 1970s. Some features distinguish the expanded beds from the fluidized ones (Fig. 13.13). Bed expansion caused by the upflow of wastewater is much greater in expanded than in fluidized beds. Wastewater flows upward through a sand bed (diameter <1 mm), which provides a surface area for biofilm growth. The flow rate is high enough to obtain an expanded or fluidized bed. This in turn necessitates the recirculation of the wastewater through the bed. This process is effective for the treatment of low-strength organic substrates (COD less than 600 mg/L) at short hydraulic retention times (several hours) and allowing high solid retention times (SRT) (Speece, 1983; Switzenbaum, 1983). This process offers several advantages:

1. Good contact is achieved between wastewater and microorganisms.
2. Clogging and channeling are avoided.
3. High biomass concentrations can be achieved, and this is associated with reduced reactor volume.
4. Biofilm thickness can be controlled.
5. Successful treatment of low-strength wastewater (≤600 mg/L COD) is attained at low temperature and at relatively short hydraulic retention times (<6 h).

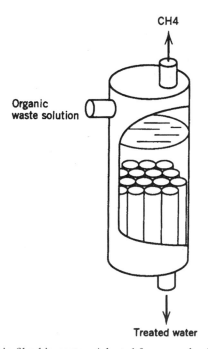

Figure 13.12 Thin film bioreactor. Adapted from van den Berg et al. (1981).

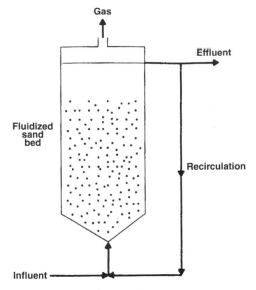

Figure 13.13   Fluidized-bed reactor.

Some investigators have argued that fluidized-bed reactors can be applicable to aerobic treatment. An advantage of such an application would be lower detention times, and thus, higher loading rates. Furthermore, nitrification would be favored as a result of the retention of the slow-growing nitrifiers in the biofilm (Rittmann, 1987).

Some of the disadvantages of this process are the high energy necessary for sufficient upflow velocity for bed expansion and the relatively large volume (30–40 percent) occupied by the support material. Some have proposed an *anaerobic expanded micro-carrier bed* (MCB) to solve these problems. Powdered zeolite is used as support material in the MCB process, which promotes the formation of granular sludge as in the UASB process (Yoda et al., 1989).

13.6.5   Anaerobic Rotating Biological Contactor

An anaerobic rotating biological contactor is similar to its aerobic counterpart (see Chapter 10) except that the reactor is sealed to create anaerobic conditions (Laquidara et al., 1986)

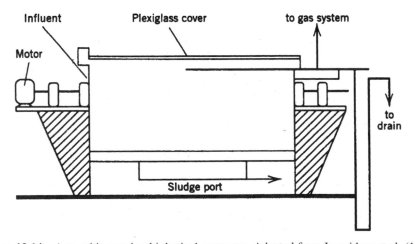

Figure 13.14   Anaerobic rotating biological contactor. Adapted from Laquidara et al. (1986).

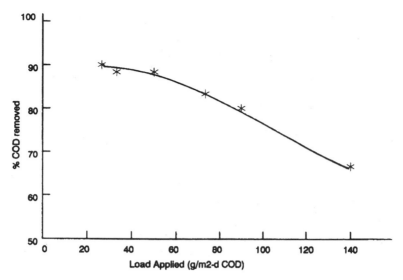

Figure 13.15 Efficiency of an anaerobic biological contactor as a function of organic load. From Laquidara et al. (1986). (Courtesy of the Water Environment Federation.)

(Fig. 13.14). This process allows greater disk submergence because oxygen transfer is not considered. Development of the attached anaerobic biofilm is a function of applied organic loading and time. Approximately 85 percent COD removal is achievable even at loading rates as high as $90 \text{ g/m}^2/\text{day}$ COD (Laquidara et al., 1986) (Fig. 13.15). At this organic loading, methane is produced at a rate of $20 \text{ L/m}^2/\text{day}$. Some of the advantages of this fixed-film anaerobic system are (Laquidara et al., 1986).

1. Potential for higher organic loadings;
2. Cell residence time independent of hydraulic detention time;
3. Low production of waste solids;
4. Ability to withstand toxic shock loads;
5. Methane production.

Close examination of the anaerobic processes for the treatment of wastewater shows that more research is needed concerning the microbiological aspects of these processes. A thorough examination of attached biofilms is also warranted.

13.7 WEB RESOURCES

http://www.montana.edu/wwwpb/pubs/mt9401.html (septic tanks: Montana State University)

http://www.inspect-ny.com/septbook.htm (general reference on septic tanks)

http://www.italocorotondo.it/tequila/ (septic tanks)

http://202.114.65.51/fzjx/wsw/newindex/tuku/MYPER/A02.htm (SEM and TEM pictures of methanogens)

http://www.labmed.umn.edu/meth/meth_map.html (methanogenesis, University of Minnesota)

http://toxics.usgs.gov/definitions/methanogenesis.html (information on methanogenesis from USGS) http://faculty.washington.edu/leighj/mmmethanogenesis.html (methanogenesis, University of Washington)

http://sos6448.ifas.ufl.edu/protected/Unit-3/unit%203%20paragraphs/methane-production-and-oxidation.htm (methane production and oxidation in wetlands; University of Florida)

http://www.sidwell.edu/us/science/vlb6/labs/Classification_Lab/Archaea/(Archaea domain)

http://www.sidwell.edu/us/science/vlb6/labs/Classification_Lab/Archaea/Methanogens/ (methanogens, University of Nijmegu, Netherlands)

13.8 QUESTIONS AND PROBLEMS

1. Anaerobic processes are driven mostly by what types of microorganisms?
2. Why does anaerobic digestion produce a low amount of biomass?
3. How could we decrease the amount of produced biomass in aerobic processes?
4. Give the two types of methanogens and explain how they produce methane.
5. What is the problem with septic tank effluents?
6. Methanogens belong to the Archaea domain. How do they differ from bacteria?
7. How does a UASB work?
8. What is the respective position of sulfate-reducing bacteria and methanogens in a UASB granule?
9. What are the approaches used to detect methanogens?
10. Give the main factors controlling methanogenesis.
11. Explain the competition between methanogens and sulfate-reducing bacteria during anaerobic digestion.
12. Describe a septic tank.
13. What are the four categories of microorganisms involved in anaerobic digestion?
14. What would happen if septic tanks were operated in the aerobic mode?
15. How is septage handled?
16. What are the advantages of anaerobic digestion?

13.9 FURTHER READING

Barnes, D. and P.A. Fitzgerald. 1987. Anaerobic wastewater treatment processes, pp. 57–113, In: *Environmental Biotechnology*, C.F. Forster and D.A.J. Wase, Eds., Ellis Horwood Ltd., Chichester, U.K.

Jewell, W.J. 1987. Anaerobic sewage treatment. Environ. Sci. Technol. 21: 14–20.

Kirsop, B.H. 1984. Methanogenesis. Crit. Rev. Biotechnol. 1: 109–159.

Koster, I.W. 1988. Microbial, chemical and technological aspects of the anaerobic degradation of organic pollutants. pp. 285–316, In: *Biotreatment Systems*, Vol. 1, D.L. Wise, Ed., CRC Press, Boca Raton, FL.

Lettinga, G. 1995. Anaerobic digestion and wastewater treatment systems. Antonie van Leeuwenhoek 67: 3–28.

Lettinga, G., J. Field, J. van Lier, G. Zeeman, and L.W. Hulshoff Pol. 1997. Advanced anaerobic wastewater treatment in the near future. Water Sci. Technol. 35 (10): 5–12.

Nichols, P.D., and C.A. Mancuso Nichols. 2002. Archaea: Detection methods, pp. 246–259, In: *Encyclopedia of Environmental Microbiology*, G. Bitton, Editor-in-Chief, Wiley Interscience, Hoboken, N.J.

Sahm, H. 1984. Anaerobic wastewater treatment. Adv. Biochem. Eng. Biotechnol. 29: 84–115.

Speece, R.E. 1983. Anaerobic biotechnology for industrial waste treatment. Environ. Sci. Technol. 17: 416A–427A.

Zehnder, A.J.B., Ed. 1988. *Biology of Anaerobic Microorganisms*, John Wiley & Sons, New York.

14

BIOLOGICAL AEROSOLS AND BIOODORS FROM WASTEWATER TREATMENT PLANTS

Wastewater Microbiology, *Third Edition*, by Gabriel Bitton
Copyright © 2005 John Wiley & Sons, Inc.

14.1   INTRODUCTION

Aerobiology deals with the study of transport, fate, and public health aspects of biological aerosols. The latter are defined as biological contaminants occurring as suspended particles in the air. The increased interest in aerobiology in recent years is probably due to concerns over the health aspects of indoor air and the threat of bioterrorism and biological warfare.

The size of aerosolized biological particles varies widely from viruses to airborne algae or protozoa. Microorganisms may also attach to airborne dust particles, which may enter into and cause damage to the respiratory system. Airborne microorganisms may adversely affect humans by causing allergies, respiratory problems, infectious diseases, and hypersensitivity reactions. They may also increase the endotoxin level in the lungs (Hensel and Petzoldt, 1995). Infections and allergic reactions caused by biological aerosols depend on the type of microorganism, its pathogenicity, and the susceptibility of the host. A total of 200 to 250 million episodes of respiratory infections occur each year in the United States, with subsequent loss of billions of dollars in medical care costs and lost work (U.S. DHEW, 1992).

The sources of biological aerosols are of two kinds: natural and production sources. *Natural sources* include coughing, sneezing, shedding from human skin or animal hide, and disturbance of soil and aquatic (e.g., bursting bubbles that have a high concentration of microorganisms) environments by wind action. In humans, a cough produces 3000 droplets, while a sneeze generates up to 40,000 droplets. Indoor sources also contain mite fragments, insect feces, as well as microbial pathogens such as streptococci, staphylococci, *Legionella*, Pseudomonas, and *E. coli*, species as well as respiratory viruses (Stites, 2002),

Production sources include agricultural practices (e.g., cleaning of silos, chicken houses), textile mills (e.g., inhalation of *Bacillus anthracis*), processing of diseased animals in abattoirs and rendering plants, medical and dental facilities (e.g., breathing equipment, humidifying devices in nurseries, surgical pumps, rotary instruments used by dentists), aerosols from laboratory operations (e.g., blenders, sonication devices), and finally wastewater treatment operations. Several infections can be acquired in dental clinics by dentists, dental assistants, and patients, particularly in those who are immunosuppressed (Grenier, 1995). The American Dental Association recommended that professionals should always wear protective masks, gloves, and eye glasses with lateral protective shields (Council on Dental Materials, 1988).

Aerosols in indoor environments are produced by human and animal sources (droplets resulting from sneezing, coughing, or talking, desquamation of skin or hide), plants, ventilation and air conditioning, home humidifiers, household toilets, floor materials and draperies (Spendlove and Fannin, 1983).

Early laboratory studies, using simulated wastewaters contaminated with micro-organisms, have suggested that aeration of wastewater may be responsible for the generation of biological aerosols, some of which could infect humans and cause diseases (Dart and Stretton, 1980).

This chapter also covers the microbiological aspects of bioodors production in wastewater treatment plants as well as microbiological approaches for treating the odors.

14.2 DEFENSE MECHANISMS OF THE RESPIRATORY SYSTEM AGAINST AIRBORNE PARTICLES

In the respiratory system, air is conveyed from the nose to the pharynx, larynx (voice box), trachea (windpipe), bronchi, bronchioles, terminal bronchioles, and finally to the alveoli (Fig. 14.1) (Vander et al., 1985). Airborne particles are retained according to their size at the various levels of the respiratory tract. Particles of $10-20$ μm are trapped in the nose. Particles ranging in size between 2 and 10 μm are retained in nasal passages and/or in the bronchial tree (trachea, bronchi, and bronchioles), whereas those <2 μm in size deposit in terminal bronchioles and alveoli.

Particle size determines the degree of penetration and retention in the respiratory system. Particles with a size range of $1-2$ μm escape trapping by the upper respiratory tract and display the greatest deposition at the level of the alveoli.

The respiratory system, however, provides defense mechanisms against inhaled airborne particles. In the tracheobronchial region, inhaled particles are transported by the *mucociliary escalator* up to the mouth, where they are expelled or swallowed. In the pulmonary region, foreign particles are ingested by alveolar macrophages.

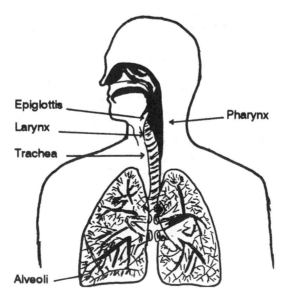

Figure 14.1   Human respiratory system.

14.3 SAMPLING OF BIOLOGICAL AEROSOLS

Several samplers have been proposed for the collection of biological aerosols in various fluids or directly on solid growth media. An important consideration is the ability to gain information on the particle size distribution of biological aerosols. This information is important from a public health viewpoint since, as discussed above, the depth of penetration of biological aerosols into the respiratory system depends on the aerosol size. Selective growth media are needed for the recovery of specific categories of airborne microorganisms. Furthermore, airborne microorganisms are subjected to desiccation, sunlight, and other environmental stresses. Thus, the collection fluid must be suitable for recovering stressed microorganisms and consists of buffered salt solution, gelatin phosphate solution, skim milk, or fluids supplemented with protective agents such as amino acids (e.g., cystein), sugars (e.g., trahalose), betain, pyruvate, catalase, or glycerol (Hensel and Peyzoldt, 1995). Growth media may also have to be modified to recover the stressed airborne microorganisms. For example, the recovery of airborne bacteria is greatly enhanced when resuscitation agents such as betain and catalase are incorporated into the growth medium (Marthi and Lighthart, 1990; Marthi et al., 1991). Flow cytometry, fluorescent in situ hybridization (FISH), and polymerase chain reaction (PCR) and real-time PCR (see Chapters 1 and 2) can also be used to detect airborne microorganisms (Alvarez et al., 1995; Lange et al., 1997; Makino and Cheun, 2003).

The various approaches for collecting biological aerosols are described below.

14.3.1 Sedimentation

Airborne microorganisms are allowed to deposit on a surface coated with a nutrient medium or on open Petri dishes containing a suitable growth medium. The composition of the growth medium depends on the type of airborne microorganism sought. However, this approach only provides qualitative information.

14.3.2 Membrane Filters

Air is passed through a membrane filter preferentially made of polycarbonate, which can then be placed on a suitable growth medium. The filter may also be eluted, and the eluates may be plated on appropriate solid growth media to give an indication of viable airborne microorganisms, or may be stained with acridine orange or DAPI and examined via epifluoresence microscopy to obtain a total (viable plus nonviable cells) microbial count (see Chapter 2) (Palmgren et al., 1986; Roane and Pepper, 2000; Thorne et al., 1992).

14.3.3 Reuter Centrifugal Air Sampler

Air is collected by centrifugation followed by impaction on an agar medium surface. This portable sampler collects microorganisms on an agar medium-coated plastic strip lining the sampler drum. The maximum air sampling capacity is 11.3 ft^3 (Placencia et al., 1982).

14.3.4 Liquid Impingement Device (All-Glass Impinger)

This is one of the most popular approaches for collecting biological aerosols (Fig. 14.2). A vacuum pump is attached to the impinger and draws the air sample, resulting in the collection of biological aerosols in a suitable fluid (e.g., phosphate buffer, peptone water). After aerosol collection, the fluid can be plated directly on a suitable growth medium or

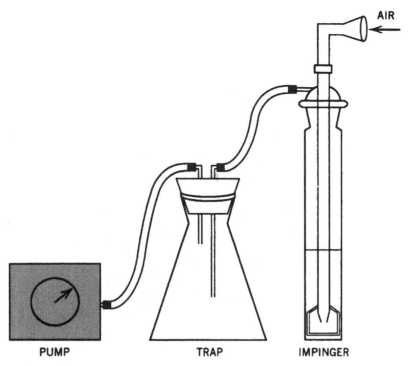

Figure 14.2 Liquid impingement device for collecting biological aerosols.

can be concentrated by membrane filtration and the filter placed on a solid nutrient medium. The collection fluid can also be plated on various media for the detection of various nutritional groups of airborne microorganisms, or can be subjected to biochemical, immunological, or molecular analyses (Buttner et al., 1997). For viruses and bacterio-phages, the collecting fluid should be preferably concentrated prior to assay on a suitable host cell. A drawback of this device is the loss of microbial viability due to impingement, sudden hydration, and osmotic shock (Thorne et al., 1992).

14.3.5 Sieve-Type Sampler

The most used sieve-type air sampler (Fig. 14.3) is the *Andersen multistage sieve sampler.* A measured volume of air is drawn and passed through a series of sieves of decreasing pore size. Growth media plates are included at each stage. This sampler, in contrast to the liquid impinger, gives information, useful from a public health viewpoint, about the particle size distribution of airborne microorganisms. Some drawbacks of the Andersen sampler include the possible overloading of the growth media plates, microbial stress due to impaction on agar surface, aggregation of microorganisms, desiccation of the collected microorganisms after long sampling times, and the detection of only viable airborne microorganisms (Andersen, 1958; Lembke et al., 1981; Thorne et al., 1992).

14.3.6 May Three-Stage Glass Impinger

The multistage May sampler is a liquid impingement device with the advantage of particle size distribution. This three-stage glass impinger helps collect particle sizes of 6.0, 3.3, and 0.7 μm, which, respectively, simulate approximately the three major portions

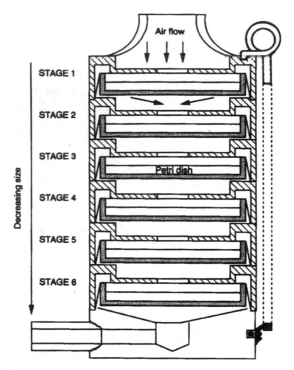

Figure 14.3   Sieve-type sampler (Andersen sampler) for biological aerosols.

of the respiratory tract: nasopharyngeal, tracheobronchial, and pulmonary regions (May, 1966; Zimmmerman et al., 1987) (Fig. 14.4). The results obtained with this sampler correlate well with those obtained with the Andersen sampler (Zimmerman et al., 1987).

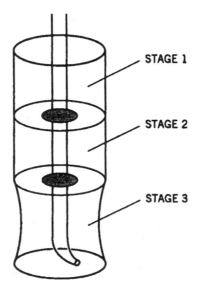

Figure 14.4   May three-stage glass impinger. Adapted from Zimmerman et al. (1987). (Courtesy of the American Society of Microbiology.)

14.3.7   Large-Volume Electrostatic Precipitators

Since enteric viruses are found in relatively low concentrations in the air, it is necessary to use large-volume electrostatic precipitators for their detection. These costly air samplers can be operated at flow rates of approximately 1000 L/min. The aerosols are collected in a liquid medium, and thousands of liters must be sampled (20,000–50,000 L) for virus detection in air samples (Shuval et al., 1989; Sorber et al., 1984).

14.4   FACTORS CONTROLLING THE SURVIVAL OF BIOLOGICAL AEROSOLS

Microorganisms are subjected to an initial shock, immediately following their aerosolization. A rapid die-off of microorganisms is observed following this initial shock due mainly to desiccation. Thereafter, microbial decay will continue at a slower rate as the aerosols are transported downwind from the source. The persistence of airborne microorganisms depends upon several environmental factors, the most important of which are relative humidity (RH), desiccation, solar radiation, oxygen concentration, and temperature (Cox, 1987; Mohr, 1991).

14.4.1   Relative Humidity

Relative humidity (RH) is probably the most crucial factor controlling aerosol stability (Cox, 1987). An RH below 50 percent is generally deleterious to most vegetative bacteria (Hensel and Petzoldt, 1995). Conflicting results have been obtained regarding the effect of relative humidity on microbial stability and are probably due to variations in experimental procedures (Mohr, 1991). Bacteria and viruses generally survive best at high relative humidity. Nonenveloped viruses (picornaviruses and adenoviruses) survive best at high relative humidity, whereas enveloped viruses (e.g., myxoviruses) survive best at low relative humidity. Figure 14.5 illustrates the effect of relative humidity on an enveloped virus (influenza virus) and a nonenveloped virus (poliovirus) (Hemmes et al., 1960). However, rotaviruses are inactivated at high RH and survive best at 50 percent relative humidity (Ijaz and Sattar, 1985; Sattar et al., 1984). Controlled field experiments with

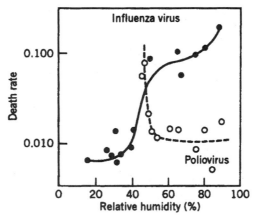

Figure 14.5   Influence of relative humidity on the survival of influenza virus and poliovirus. From Hemmes et al. (1960).

a nalidixic acid-resistant *Escherichia coli* showed that survival was correlated with relative humidity (Teltsch and Katzenelson, 1978) (Fig. 14.6). However, in atmospheres containing inert gases such as nitrogen, argon, or helium, the survival of *E. coli* was best at low RH (Cox, 1987).

14.4.2   Temperature

Stability of microbial aerosols generally decreases as temperature increases (Marthi et al., 1990).

14.4.3   Solar Radiation

Aerosols survive longer at night, in the absence of sunlight (Shuval et al., 1989). Controlled field experiments with aerosolized *E. coli* showed that bacterial survival was negatively correlated with solar radiation intensity (Teltsch and Katzenelson, 1978) (Fig. 14.7).

14.4.4   Type of Microorganism

Aerosol persistence depends on the type of microorganisms. In aerosols, enteroviruses are generally hardier than bacteria or coliphages. Encapsulated bacteria (e.g., *Klebsiella*) appear to survive better than their nonencapsulated counterparts.

14.4.5   Other Factors

Other factors affecting the viability of airborne microorganisms include fluctuations in air pressure, free radicals of oxygen, atmospheric pollutants (e.g., NO_2, SO_2, O_3), sampling methodology, and collection medium (Mohr, 1991; 1997).

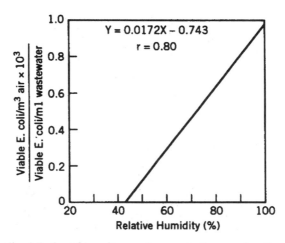

Figure 14.6   Normalized *Escherichia coli* aerosol concentration as a function of relative humidity. Adapted from Teltsch and Katzenelson (1978).

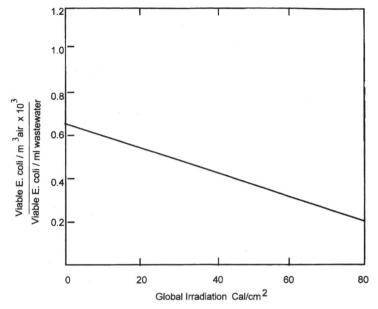

Figure 14.7 Normalized *Escherichia coli* aerosol concentration as a function of solar irradiation. From Teltsch and Katzenelson (1978). (Courtesy of the American Society of Microbiology.)

14.5 PREDICTIVE MODELS FOR ESTIMATING DOWNWIND LEVELS OF AIRBORNE MICROORGANISMS

Biological aerosol monitoring is costly and labor-intensive, particularly with regard to airborne viruses that occur at relatively very low concentrations in the air we breathe. Dispersion models, borrowed from the field of aerosol mechanics and requiring the integration of biological and meteorological data, have been developed to predict the downwind concentrations of aerosolized microorganisms from known sources (e.g., activated sludge units, spray irrigation sites) and to assess the possible health significance of microbial aerosols. These models assume a Gaussian (normal) distribution of pollutants at any specific downwind location (Mohr, 1991; Reed et al., 1988; U.S. EPA, 1982). Early dispersion models dealt with the dispersion and deposition of inert particles (Pasquill, 1961). Later models took into account the viability of microorganisms in the aerosolized state (Camann et al., 1978; Lighthart and Frisch, 1976; Lighthart and Mohr, 1987; U.S. EPA, 1982).

The downwind concentration C_d of aerosolized microorganisms at distance d from the source is given by the following equation (U.S. EPA, 1982):

$$C_d = QD_d e^{xa} + B \qquad (14.1)$$

where Q = microbial concentration at the source (number/m^3); D_d = atmospheric dispersion factor (s/m^3); x = decay rate (s^{-1}); a = aerosol age (downwind distance/wind speed) (s); and B = background concentration of biological aerosol (number/m^3).

The microbial concentration at the source is given by:

$$Q = WFEI \qquad (14.2)$$

where W = microbial concentration in wastewater (number/L); F = flow rate of wastewater (L/s); E = aerosolization efficiency = fraction or percentage of sprayed wastewater that becomes aerosolized ($E = 0.3$ percent for wastewater; 0.04 percent for sludge applied with a spray gun); and I = survival factor of aerosolized microorganisms, which is the fractional reduction of microorganisms during the aerosol formation process ($I > 0$). I varies with the pathogen under consideration.

Certain model components are site-specific and need to be determined at the site under study. These include meteorological data (wind direction and speed, atmospheric stability), microbial concentration in wastewater, flow rate, and background aerosol concentration. Other model components are non-site-specific and include aerosolization efficiency E, microbial decay rate (depends on the microorganism type and is generally obtained from laboratory studies), and survival factor I, which represents the initial shock during the aerosolization process.

A similar dispersion model has been used to predict the fate of airborne microorganisms in a spray irrigation site in Pleasanton, California. This model also incorporates essential microbiological parameters such as decay rates (Camann et al., 1978; Johnson et al., 1980).

14.6 PRODUCTION OF BIOLOGICAL AEROSOLS BY WASTEWATER TREATMENT OPERATIONS

The first study on bioaerosols and wastewater was possibly the one carried out by Horrocks (1907) on aerosols generated in sewers. The link between bioaerosol generation and wastewater treatment operation was first examined in the 1930s (Fair and Wells, 1934). However, interest in wastewater aerosols was seriously expressed only in the 1970s and 1980s when relatively better sampling techniques were developed (Dowd, 2002).

Biological aerosols are produced whenever aeration is used in treatment processes (e.g., activated sludge, aerated lagoons, aerobic digestion of sludge). Mechanical devices that assist in the aerosolization of raw wastewater or wastewater effluents may lead to the production of biological aerosols. For example, airborne bacterial densities above the aeration tank of an activated sludge system appear to be directly related to the aeration rate. Figure 14.8 shows that bacterial densities are related to aeration rate

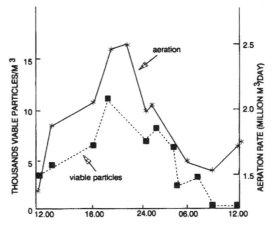

Figure 14.8   Effect of time of the day and aeration rate on bacterial densities above an activated sludge unit. From Fedorak and Westlake (1980). (Courtesy of the Water Environment Federation.)

and peak during night time (Fedorak and Westlake, 1980). In a wastewater treatment plant in Italy, pathogenic enteric bacteria (*Salmonella enteritidis* and *S. boydii*) and reoviruses were isolated in 2 and 46 percent of the samples, respectively (Carducci et al., 2000). Biological aerosols are also generated by routine activities indoors (e.g., sludge dispensing or floor mopping) (Fedorak and Westlake, 1980). Bioaerosols may be transported for relatively great distances from the wastewater treatment plant.

Spray irrigation with wastewater effluents also leads to the production of biological aerosols. Indicators (total and fecal coliforms, *Clostridium*) and pathogenic bacteria (e.g., *Mycobacterium*, *Klebsiella pneumonae*, *Salmonella*) as well as enteroviruses have been isolated at spray irrigation sites, sometimes hundreds of meters from the production source (Applebaum et al., 1984; Camann et al., 1988; Fannin et al., 1985; Moore et al., 1979; Shuval et al., 1989; Teltsch and Katzenelson, 1978; Teltsch et al., 1980). In Israel, spray irrigation with oxidation pond effluents produced bacterial (total bacteria, total and fecal coliforms, fecal streptococci), viral, and algal aerosols up to a distance of 730 m from the source. At this distance, 79 percent of the air samples were positive for indicator bacteria during daytime and 100 percent were positive during night time. Approximately 10 percent of the samples were positive for viruses. Virus concentration in positive samples ranged from 0.03 to 2 PFU/m^3 (Shuval et al., 1989). Other studies in Israel have demonstrated the presence of enteroviruses (poliovirus 2, echovirus type 1, 7, 17, and 25, coxsackievirus B1, and other unidentified viruses), *Salmonella*, and total coliforms in biological aerosols (Teltsch et al., 1980).

In Lubbock, Texas, a poor-quality trickling filter effluent was used for spray irrigation of agricultural crops. The fecal coliform levels in the wastewater effluent exceeded 10^6/100 mL, and the enterovirus level during the summer season sometimes exceeded 1000 PFU/L. However, impoundment of the wastewater effluent in a reservoir reduced the fecal coliform level by 99 percent and the enterovirus level to less than 10 PFU/L (Moore et al., 1988). Spray irrigation with the effluent before reservoir impoundment significantly increased the downwind levels of indicator bacteria (fecal coliform, fecal streptococci, *Clostridium perfringens*, mycobacteria) and bacterial phage relative to background levels (Camman et al., 1988) (Table 14.1). Microbial levels decreased as the distance from the spray irrigation site increased. Enteroviruses, mostly polioviruses, were consistently detected in air samples at 44 to 60 m downwind from the spray irrigation site (Table 14.2; Camman et al., 1988). Thus, storage impoundment of wastewater should be an integral part of the irrigation scheme because this practice significantly reduces indicator and pathogen levels in wastewater and aerosol samples (Camman et al., 1988; Moore et al., 1988).

Air samples were taken at the Muskegon County (Michigan) Wastewater Management System No. 1, with an Army prototype XM2 Biological Sampler, which collects 1050 L of air/min (Brenner et al., 1988). Bacteria (86–7143 CFU/m^3) and coliphage (0–9 PFU/m^3) were isolated but no animal virus was recovered, even in air sample volumes of >400 m^3.

In sludge application sites, microbial indicators (total coliforms, fecal coliforms, fecal streptococci, bacteriophages) and mycobacteria were detected downwind from the spray irrigation site. Enteroviruses were not detected in a 1470 m^3 pooled air sample. It was concluded that sludge application to land via spray irrigation is not a serious threat to human health (Sorber, et al., 1984).

Molecular methods, such as PCR, can help trace the source of the aerosolized microorganisms. For example, during application of biosolids, 30 *Clostridium* isolates were recovered from air samples. Ten of these isolates displayed banding patterns similar to those of Clostridia isolated from the biosolids (Dowd and Pillai, 1999). The use of these methods should be expanded to screen pathogens, instead of indicators, in air samples.

TABLE 14.1. Biological Aerosols Downwind from a Site Spray-Irrigated with Wastewater Effluents[a]

	Microorganism Concn. in Air (No./m^3)				
	Ambient Background		Downwind of Irrigation Line		
Microorganism Group/ Wastewater Source	Homes	Fields	30–89 m	90–149 m	250–409 m
Fecal coliforms, CFU	0.01	<0.006			
Pipeline			180	1.8	0.3
Reservoir			1.4	0.2	<0.08
Fecal streptococci, CFU	0.5	0.07			
Pipeline			140	16	0.5
Reservoir			0.5	0.2	0.3
Mycobacteria, CFU	0.05	0.1			
Pipeline			2.5	0.8	0.2
Reservoir			0.05	0.1	<0.03
Clostridium perfringens, CFU		0.08			
Pipeline			9	1.2	0.6
Coliphage, PFU	<0.005	<0.003			
Pipeline			9.9	1.8	0.1
Reservoir			0.06	0.07	0.07
Enteroviruses, PFU					
Pipeline			0.05		

[a]Adapted from Camann et al. (1988).

14.7 HEALTH HAZARDS OF BIOLOGICAL AEROSOLS GENERATED BY WASTE TREATMENT OPERATIONS

An evaluation of the potential hazard due to biological aerosols must take into account several critical factors, particularly the pathogen concentration of the wastewater effluent

TABLE 14.2. Viral Aerosols Downwind from Two Spray Irrigation Sites[a]

	Enterovirus Density		
Distance from Spray Line (m)	Host Cell Line	In Wastewater (PFU/mL)	In Air (PFU/m^3)
Hancock farm, Texas			
60	HeLa	0.16	0.0029
	RD		0.0015
46	HeLa	0.10	0.011
	RD		0.018
44	HeLa	2.2	16.2
	RD		18.3
49	HeLa	0.066	0.010
	RD		0.013
Pleasanton, California			
63	HeLa	0.036	0.0047
63	HeLa	0.18	0.0074

[a]Adapted from Camann et al. (1988).

being sprayed, the degree of aerosolization (which depends on meteorological conditions and type of equipment), and meteorological conditions (wind velocity, temperature, solar radiation, and relative humidity).

14.7.1 Spray Irrigation with Wastewater Effluents and Liquid Sludge

We have shown that microbial aerosols are produced following spray irrigation of agricultural land with wastewater effluents and, sometimes, liquid sludges. Studies have been undertaken in the United States, Canada, Israel, and other countries on the health aspects of aerosols generated by spray irrigation of wastewater (Bausum et al., 1983; Camann et al., 1986; 1988; Fattal et al., 1986; Sekla et al., 1980). High-pressure sprinklers can aerosolize from 0.1 to 1 percent of the wastewater effluent, leading to the spreading of aerosols hundreds of meters from the production source. Furthermore, 66–78 percent of particles generated as a result of spray irrigation are within the 1 to 5 μm range and are efficiently deposited into the lungs (Bausum et al., 1983). Others have reported that 20 percent of the bacterial aerosols collected with an Andersen sampler at a spray irrigation site at 730 m from the source, were in the respirable range (0.65–2.0 μm) (Shuval et al., 1989). However, in the United States, no significant increase of incidence of human disease was observed as a result of exposure to biological aerosols generated by sewage treatment plants or by spray irrigation with wastewater effluents (U.S. EPA, 1982). Most epidemiological studies have shown little evidence of increased risk of infection or increased disease occurrence related to spray irrigation with wastewater. The Lubbock Infection Surveillance Study (LISS), conducted to monitor microbial infections in a rural community living near a wastewater spray irrigation site, indicated that spray irrigation with wastewater effluents did not result in any significant increase in rotavirus infections at spray irrigation sites (rotavirus infection was defined as a greater than two-fold increase in rotavirus serum antibody between two blood collection periods) (Camann et al., 1986; Ward et al., 1989).

In Israel, a retrospective epidemiological study was conducted to record disease incidence resulting from irrigation with oxidation pond effluents. Clinical records for enteric diseases were examined for kibbutzim (i.e., agricultural settlements) practicing spray irrigation with wastewater and compared with those from kibbutzim that did not irrigate with wastewater. Although preliminary data showed an increase in the incidence of some diseases (e.g., shigellosis, salmonellosis, hepatitis) in kibbutzim practicing spray irrigation (Katzenelson et al., 1976), there was no conclusive evidence of significant increased risk of disease resulting from exposure to aerosolized wastewater effluents. Except for echovirus type 4 (probably due to an echovirus 4 epidemic prior to the collection of blood samples for the study), no significant increase of antibodies to seven enteroviruses was observed in communities exposed to biological aerosols from sprinkler irrigation with partially treated wastewater (Fattal et al., 1987). However, within 11 kibbutzim, the practise of spray irrigation with wastewater effluents resulted in a small excess risk of enteric diseases during the irrigation season, particularly for the 0 to 4 year age group. No such excess was observed however on a year-round basis (Fattal et al., 1984; 1986).

Among the possible reasons for difficulties in drawing definite conclusions with regard to human health hazards due to biological aerosols are the following.

1. Airborne microbial pathogens are not unique to wastewater.
2. There are confounding factors, such as other sources and routes of infection.

3. Induced immunity occurs with low aerosol levels.

4. Sensitivity of monitoring techniques is insufficient.

5. Minimum infectious doses for many aerosolized microorganisms have not been established.

6. Difficulties are encountered in epidemiological studies. These include the lack of sensitive exposed populations.

14.7.2 Health Risks Associated with Composting of Wastewater Sludge

Some composting operations may result in worker exposure to pathogens from aerosols or from direct contact with the composted material. The high temperatures ($55-65°C$) generated during composting are efficient for inactivation of pathogenic microorganisms normally found in wastewater sludge. However, these temperatures allow the proliferation of thermophilic actinomycetes as well as pathogenic fungi such as *Aspergillus fumigatus*. A study of workers in four composting facilities in the United States has found effects on compost workers (Clark et al., 1984) that include nasal, ear, and skin infections, burning eyes and skin irritation, and throat and anterior nares cultures positive for *A. fumigatus*. Infection is caused by inhalation of spores, which may reach the alveoli. Its course is determined by the immunological status of the host, as infection can be very severe in immunodeficient patients. Allergic bronchopulmonary aspergillosis also develops in people with asthma (Clark et al., 1981; 1984; Rippon, 1974). Workers also display high levels of antibodies to endotoxin (lipopolysaccharides) produced by gram-negative bacteria. Endotoxins may cause headache, fever, and nose and eye irritation. They may play a role in the "sick building syndrome" and in the severity of asthma (Michel et al., 1991; Teeuw et al., 1994).

14.7.3 Health Hazards to Wastewater Treatment Plant Workers

Opportunities exist for wastewater treatment operators to become exposed to pathogenic bacteria, viruses, parasites, and endotoxins from biological aerosols or from contact with contaminated materials and surfaces in the plant. Sewage workers are known to be stricken by the so-called "sewage worker's syndrome," which consists of acute rhinitis, fever, and general malaise (Rylander et al., 1976). There are reports of association of exposure to wastewater aerosols and the occurrence of viral diseases (Heng, 1994). In a wastewater treatment plant in Oslo, Norway, *Salmonella* spp. was isolated from wastewater, sludge, and from floor surfaces in an eating area, but there was a poor correlation between the *Salmonella* serotypes isolated from the plant and those isolated directly from the population (Langeland, 1982). Similarly, no correlation was detected between bacterial isolates from air samples in a plant treating hospital wastewater and those from the oral cavity of plant workers (Orsini et al., 2002).

Airborne endotoxins exert a deleterious effect on the structure and function of the respiratory system (Olenchock, 1997). Endotoxin concentrations in a wastewater treatment plant in Finland varied between 0.6 and $310 \, ng/m^3$. A proposed occupational exposure limit (8 h time-weighted average) for airborne endotoxin is $30 \, ng/m^3$ (Palchak et al., 1988). Operators may display an increased level of antibodies to endotoxins from gram-negative bacteria, and to some enteroviruses, as shown for hepatitis A and coxsackie B3 viruses (Clark et al., 1977; Rylander et al., 1976; Skinhoj et al., 1981). In Copenhagen, Denmark, higher levels of antibody against HAV were observed among sewage workers (81 percent) than among gardeners (61 percent) or city clerks

(48 percent) (Skinhoj et al., 1981). An outbreak of Pontiac fever was reported among sewage workers repairing a sludge decanter. They tested positive for *Legionella*, which was also detected in the biosolids (Gregersen et al., 1999).

However, the slightly increased risk of infection resulting from exposure to wastewater does not appear to translate into increased overt disease incidence among wastewater treatment plant workers.

Preventive measures must be taken to reduce the potential for disease incidence among plant workers and surrounding populations in the vicinity of wastewater treatment plants and spray irrigation sites. For spray irrigation, the wastewater effluents should be disinfected prior to application, and buffer zones and physical barriers must be established between the bioaerosol source and the public. In wastewater treatment plants aeration tanks should be covered, and workers should be encouraged to wear protective gear (Dowd, 2002).

14.8 MICROBIOLOGICAL ASPECTS OF BIOODORS GENERATED BY WASTEWATER TREATMENT PLANTS

During wastewater treatment operations, odor-producing compounds are produced by the anaerobic decomposition of organic matter containing sulfur and nitrogen. Odors from treatment facilities are the subject of many complaints from the public, especially from populations in the vicinity of the plant. The design of new wastewater treatment plants must take into consideration the control of odors generated by the facility (Gostelow et al., 2001). Bioodor sources can be both anaerobic (e.g., generation of H_2S) as well as aerobic processes (e.g., stripping of odorous compounds).

14.8.1 Odor Measurement

These volatile compounds are of relatively low molecular weight and are measured and identified using analytical or sensory means. There is, however, a poor relationship between the two approaches (Gostelow et al., 2001).

14.8.1.1 *Analytical Measurements.* They provide information on the physical and chemical properties of the odorous compound and its concentration. The air sample is subjected to a separation step, which is followed by analysis using gas chromatography–mass spectrometry (GC–MS).

14.8.1.2 *Sensory Measurements.* These are subjective measurements that use the human nose as a tool, and an odor panel consisting of several people who rank the odor on a scale of 0 (no odor) to 5 (strong odor). Objective sensory measurements consist of using the nose to smell a series of dilutions of the odorous sample to obtain a threshold concentration. The odorous sample is assigned a *threshold odor number* (TON) defined as the concentration below which the odor is no longer detectable by the human nose.

14.8.2 Types of Odor-Causing Compounds

Volatile odorous compounds are transferred from wastewater to the air. Their release into the air depends on three main factors, which include the compound concentration in wastewater, turbulence, and rate of ventilation of the headspace above the water (Quigley and Corsi, 1995).

The main bioodor compounds causing problems in wastewater treatment plants are sulfur and nitrogen compounds as well as organic acids, aldehydes, and ketones (Table 14.3), and are discussed below (Devinny and Chitwood, 2002; Gostelow et al., 2001; Henry and Gehr, 1980; Jenkins et al., 1980, U.S. EPA, 1985; van Langenhove et al., 1985).

14.8.2.1 *Inorganic Gases.*

Hydrogen sulfide is produced via sulfate reduction by anaerobic sulfate-reducing bacteria such as *Desulfovibrio desulfuricans*, which use sulfate as an electron acceptor. It can also be produced via the anaerobic decomposition of sulfur-containing amino acids such as methionine, cysteine, and cystine (see Chapter 3). However, sulfate reduction is the most important contributor of H_2S in wastewater. Sulfate reduction to sulfide is controlled by several factors, including BOD, sulfate concentration, and dissolved oxygen concentration (Devinny and Chitwood, 2002). The produced H_2S is a corrosive, colorless toxic gas with a "rotten-egg" odor and has a low olfactory threshold (<0.2 ppm) (National Research Council, 1979). It is quite toxic to humans and is fatal at concentrations exceeding 500 ppm (National Research Council, 1979). This gas is also corrosive to materials (concrete, copper, lead, iron) commonly found in wastewater treatment plants.

Hydrogen sulfide is responsible for odor problems. The partitioning of H_2S between the liquid and gas phases depends mainly on pH (Fig. 14.9; Gostelow et al., 2001), initial dissolved hydrogen sulfide concentration, and temperature. At pH 7.0, H_2S represents 50 percent of the dissolved sulfides in wastewater. Its concentration increases as the pH level decreases (Pisarczyk and Rossi, 1982; Sawyer and McCarty, 1967) and its solubility in wastewater also decreases as temperature increases. Figure 14.10 shows that, at one atmosphere pressure, H_2S level in air increases as temperature and dissolved hydrogen sulfide increase (Bowker et al., 1989).

14.8.2.2 *Mercaptans.*

The most important volatile organic compounds contributing to odor problems in wastewater treatment plants are the mercaptans (e.g., ethyl mercaptan,

TABLE 14.3. Some Odorous Sulfur Compounds in Wastewater[a]

Substance	Formula	Characteristic Odor	Odor Threshold	Molecular Weight
Allyl mercaptan	$CH_2{=}CH{-}CH_2{-}SH$	Strong garlic–coffee	0.00005	74.15
Amyl mercaptan	$CH_3{-}(CH_2)_3$ $-CH_2{-}SH$	Unpleasant, putrid	0.0003	104.22
Benzyl mercaptan	$C_6H_5CH_2{-}SH$	Unpleasant, strong	0.00019	124.21
Crotyl mercaptan	$CH_3{-}CH{=}CH$ $-CH_2{-}SH$	Skunk-like	0.000029	90.19
Dimethyl sulfide	$CH_3{-}S{-}CH_3$	Decayed vegetables	0.0001	62.13
Ethyl mercaptan	$CH_3CH_2{-}SH$	Decayed cabbage	0.00019	62.10
Hydrogen sulfide	H_2S	Rotten eggs	0.00047	34.10
Methyl mercaptan	CH_3SH	Decayed cabbage	0.0011	48.10
Propyl mercaptan	$CH_3{-}CH_2$ $-CH_2{-}SH$	Unpleasant	0.000075	76.16
Sulfur dioxide	SO_2	Pungent, irritating	0.009	64.07
Tert-butyl mercaptan	$(CH_3)_3C{-}SH$	Skunk, unpleasant	0.00008	90.10
Thiocresol	$CH_3{-}C_6H_4{-}SH$	Skunk, rancid	0.000062	124.21
Thiophenol	C_6H_5SH	Putrid, garlic-like	0.000062	110.18

[a]From U.S. EPA (1985).

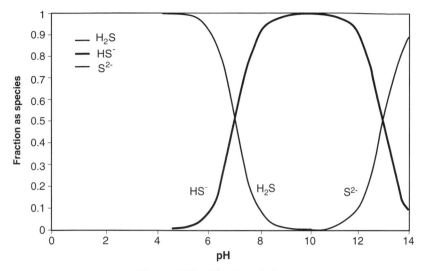

Figure 14.9 H_2S dissociation.

methyl mercaptan, isopropyl mercaptan), organic sulfides (e.g., diallyl sulfide, dimethyl sulfide, methyl isopropyl sulfide, diethyl sulfide, methyl pentyl sulfide), polysulfides (e.g., dimethyl disulfide, methyl ethyl disulfide), and thiophenes (e.g., alkylthiophenes) (Forster and Wase, 1987; van Langenhove et al., 1985). Some characteristics of odorous sulfur compounds are displayed in Table 14.3 (U.S. EPA, 1985).

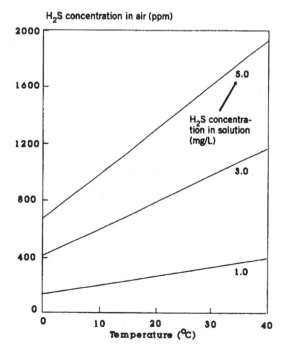

Figure 14.10   Effect of temperature and dissolved hydrogen sulfide on H_2S concentration in air. Adapted from Bowker et al. (1989).

14.8.2.3   Other Compounds.   Other compounds that contribute to odor problems
include nitrogenous compounds (e.g., ammonia, methylamine, ethylamine; see Table 14.4),
organic acids, phenol, and *p*-cresol (van Langenhove et al., 1985).

14.8.3   Preventive and Treatment Methods for Odor Control in Wastewater Treatment Plants

14.8.3.1   Preventive Measures.   Several wastewater process units or operations
(e.g., preliminary treatment, primary treatment, trickling filters, rotating biological contactors, activated sludge) can be the source of odors if improperly designed and maintained
(e.g., poor housekeeping, insufficient aeration). Several operating practices can help
prevent odors in wastewater treatment plants. Preventive measures include mechanical
cleaning of collection systems, frequent washing to remove grit and organic debris,
frequent scraping to remove scums and grease, keeping vents clear in tricking filters to
keep biofilms under aerobic conditions, and reduce foam spray in aeration tanks. In activated sludge systems, tank walls, air pipes, and diffusers should be kept clean (Bowker
et al., 1989; Harshman, 2003; U.S. EPA, 1985).

Malodorous gas production can also be prevented by addition of chlorine, hydrogen
peroxide, potassium permanganate, sodium nitrate (e.g., the BIOXIDE process, which
uses nitrate to convert dissolved sulfide to sulfate), anthraquinone (which disrupts the
sulfate reduction process), and iron salts (Fe^{2+} or Fe^{3+}) to the wastewater. Another
preventive measure is the covering of odor-producing units and treatment of odorous
air prior to release (Bowker et al., 1989).

14.8.3.2   Treatment Methods.   Wastewater odors can be treated before their release
into the atmosphere. The various treatment methods are the following.

• *Wet scrubbers*

The odorous gases are contacted with a solution, which may contain an oxidizing agent
such as chlorine, potassium permanganate, hydrogen peroxide, or ozone. The treated air
is then discharged to the atmosphere. Wet scrubbers remove more than 90 percent of
gases. Both H_2S and organic odorous compounds are targeted by wet scrubbers.

**TABLE 14.4.   Some Nitrogenous Odorous Compounds Associated with
Wastewater Treatment**[a]

Compound	Formula	Odor
Ammonia	NH_3	Sharp, pungent
Methylamine	CH_3NH_2	Fishy
Dimethylamine	$(CH_3)_2NH$	Fishy
Trimethylamine	$(CH_3)_3N$	Fishy, ammoniacal
Ethylamine	$C_2H_5NH_2$	Ammoniacal
Diamine (i.e., cadaverine)	$NH_2(CH_2)_5NH_2$	Decomposing meat
Pyridine	C_6H_5N	Disagreeable, irritating
Indole	C_8H_6NH	Fecal, nauseating
Skatole	C_9H_8NH	Fecal, nauseating

[a]Adapted from Gostelow et al. (2001).

• *Combustion of Odors*

This is done by a direct flame process at temperatures varying from approximately 500°C to more than 800°C or by catalytic oxidation (the catalyst may be platinum or palladium) at lower temperatures within the range 300–500°C (Bowker et al., 1989; Henry and Gehr, 1980; Pope and Lauria, 1989).

• *Adsorption to Activated Carbon and Other Media*

Activated carbon offers a large surface area (≤ 950 m^2/g of carbon) for the nonselective adsorption of malodorous gases. It is regenerated by thermal or chemical treatment. A 1 kg amount of activated carbon can treat 276–735 m^3 of air (Huang et al., 1979). Activated carbon is specifically used in the rendering and food industries (Henry and Gehr, 1980). Activated carbon is particularly effective in removing sulfur-containing odorous compounds, which include methyl mercaptan, dimethyl sulfide, and dimethyl disulfide (Hwang et al., 1994), but is not as effective for ammonia and other nitrogen-based compounds (Table 14.4). Other adsorbents are peat biofilters, activated alumina impregnated with potassium permanganate, wood bark, mixtures of wood chips and iron oxide, or a chelated iron adsorbent. The efficiency of the latter was demonstrated under field conditions in a wastewater treatment plant in Hawaii (Mansfield et al., 1992).

• *Ozone Contactors*

Odor-causing compounds are readily oxidized with ozone. Hydrogen sulfide, methyl mercaptan, and amines are oxidized to sulfur, methyl sulfonic acid, and amine oxides, respectively.

• *Biofiltration*

Biofiltration consists of passing contaminated air through a moist porous medium (e.g., soil, compost) that contains attached microorganisms that drive the biooxidation of bioodors. The odorous compounds are transferred from the air phase to the water phase and are subsequently absorbed by biofilm microorganisms on the surface of the porous medium.

Soil/compost filters ("Bodenfilters") consist of a network of perforated PVC pipes buried 1–3 m below the soil or compost surface (Fig. 14.11; U.S. EPA, 1985; Pomeroy, 1982). As gases flow upward through soil macropores, they are adsorbed by the soil and compost matrix and rapidly oxidized by microbiological or chemical means (Bohn and Bohn, 1988; Rands et al., 1981). Odorous gas oxidation via microbial action appears to be the primary means for odor reduction in biofilters. Adsorption of gases to soil varies with soil type (sandy loams are typically used), and gas type. Gas retention is relatively high in soils with high content of clay minerals and humic substances. A biofilter medium must be kept moist and warm to maintain microbial activity and thus achieve a good performance. The biofilter moisture is controlled by periodic sprinkling of water on the filter surface (Fig. 14.12; Devinny and Chitwood, 2002). A humidification chamber is also used to moisten the air, thus avoiding drying of the filter medium. Anaerobic conditions developing as a result of excess water inside the bed should be avoided. This may be accomplished by covering the bed or by draining the excess water (U.S. EPA, 1985). There are also closed biofilters, which allow better process control (Fig. 14.13; Devinny and Chitwood, 2002).

In the biofilter, volatile organic compounds (VOC) are degraded by heterotrophic microorganisms whereas inorganic compounds such as H_2S and NH_3 are oxidized by autotrophic microorganisms (see Chapter 3). Since microbial activity leads to the

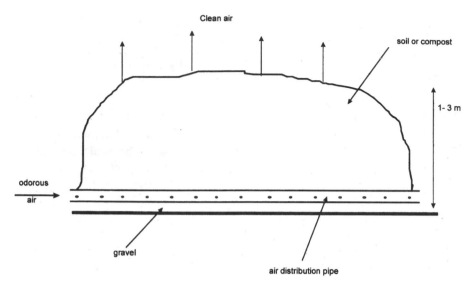

Figure 14.11   Soil/compost filter for odor control. From U.S. EPA (1985).

production of acids, it is sometimes necessary to add buffering materials such as oyster shells or dolomite to the biofilter (Devinny and Chitwood, 2002).

Some data are available regarding the identity of the microorganisms responsible for the degradation of methyl sulfides and hydrogen sulfide. Methyl sulfide, dimethyl sulfide, and dimethyl disulfide can be oxidized by bacteria belonging to the genera *Thiobacillus* or the methylotroph *Hyphomicrobium* (de Bont et al., 1981; Kanagawa and Kelly, 1986; Sivela and Sundman, 1975; Suylen and Kuenen, 1986; Zhang et al., 1991) or by fungi (Phae and Shoda, 1991). These compounds, along with methylmercaptans, are efficiently removed from contaminated air by oxidation to sulfate by a culture of *Thiobacillus thioparus* (Kanagawa and Mikami, 1989).

Installation and operating costs for soil/compost beds are considerably lower than for incineration or wet scrubbing. The practice of using soil beds for the treatment of gases is environmentally safe because soils do not adsorb more gas than they can handle, thus avoiding groundwater pollution.

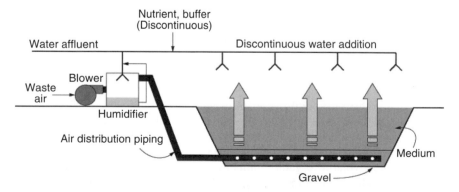

Figure 14.12   Open bed biofilter. From Devinny and Chitwood (2002).

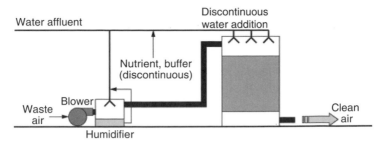

Figure 14.13 Closed bed biofilter. From Devinny and Chitwood (2002).

· Biotrickling Filters for the Treatment of Contaminated Air

Biotrickling filters act as biological scrubbers of contaminated air. They consist of a bed packed with rocks or plastic modules. The contaminated air is blown over the bed, which contains attached microorganisms (see Chapter 10) that biodegrade volatile organic compounds (VOC) or oxidize bioodor-causing chemicals such as H_2S. A feed solution is circulated through the packed bed and provides water and nutrients to the biofilm microorganisms (Fig. 14.14; Deshusses and Cox, 2002). The removal efficiency of odors and VOCs by this system varies between 60 and 99.9 percent.

· Other Microbiological Methods

Other methods have been suggested to remove H_2S, a foul-smelling and dangerous gas. Bioreactor columns loaded with attached photosynthetic bacteria (e.g., *Chlorobium*) remove up to 95 percent of H_2S from anaerobic digester effluent (Kobayashi et al., 1983). Photosynthetic bacteria, especially members of chromatiaceae (purple sulfur bacteria) and chlorobiaceae (green sulfur bacteria) are known to use H_2S as an electron donor in photosynthesis and oxidize it to elemental sulfur and sulfate (Pfenning, 1978). Aerated biofilters containing acclimated microorganisms can remove more than 80 percent of sulfur- and nitrogen-containing malodorous compounds (Hwang et al., 1994).

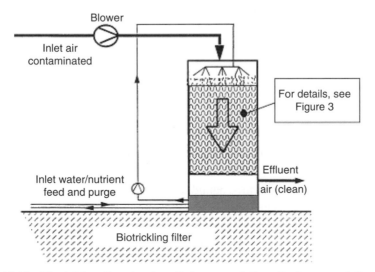

Figure 14.14 Biotrickling filter for air pollution control. From Deshusses and Cox (2002).

• *Electron Beam Treatment*

Recently, the feasibility of using electron beam treatment for the removal of odorous compounds was explored. An odorous compound, thioanisole (methyl phenyl sulfide), was efficiently removed using this technology. The process is based on the reaction of odor-causing compounds with reactive intermediates (e.g., hydrated electrons, ˙OH radicals) (Tobien et al., 2000).

14.9 WEB RESOURCES

General

http://www.arche.psu.edu/iec/abe/ (aerobiological engineering)

Indoor Bioaerosols

http://www.indoor-air.org/index.php/50 (indoor air quality)
http://www.osp.state.nc.us/divinfo/frames/divisions/rcs/hygiene/microbes.pdf
http://www.epa.gov/iedweb00/pubs/bio_1.html (biological pollutants in the home)

Bioaerosol Sampling

http://www.skcinc.com/bioaerosol.asp
http://www.cdc.gov/niosh/nmam/pdfs/chapter-j.pdf (sampling and characterization of bioaerosols)
http://www.arche.psu.edu/iec/abe/biodet.html

14.10 QUESTIONS AND PROBLEMS

1. Discuss the health impact of bioaerosols on wastewater treatment operators.
2. What bioaerosol sampling method should we use to obtain information on how deep the aerosols penetrate inside the respiratory system?
3. What are the reasons for using predictive models for estimating the levels of viral aerosols?
4. What measures should be taken to reduce the health risks of spray irrigation with wastewater effluents?
5. What are the defenses of the respiratory system against bioaerosols?
6. Give some natural sources of bioaerosols.
7. What type of aerosol collector would you use to sample airborne enteric viruses?
8. What is the effect of relative humidity on bioaerosols?
9. What are the difficulties encountered in attempts to draw definite conclusions concerning the health hazards of bioaerosols?
10. How is solar radiation affecting bioaerosols?
11. Explain the biofiltration method for treating bioodors in wastewater treatment plants.

14.11 FURTHER READING

Applebaum, J., N. Gutman-Bass, M. Lugten, B. Teltsch, B. Fattal, and H.I. Shuval. 1984. Dispersion of aerosolized enteric viruses and bacteria by sprinkler irrigation with wastewater. Monog. Virol. 15: 193–210.

Cox, C.S. 1987. *The Aerobiological Pathway of Microorganisms*, John Wiley & Sons, Chichester, U.K.

Cox, C.S., and C.M. Mathes. Eds. 1995. *Bioaerosols Handbook*, Lewis, Boca Raton, FL.

Devinny, J.S., and D.E. Chitwood. 2002. Biofiltration and bioodors, pp. 593–601, In: *Encyclopedia of Environmental Microbiology*, Gabriel Bitton, editor-in-chief, Wiley-Interscience, N.Y.

Dowd, S.E. 2002. Wastewater and biosolids as sources of airborne microorganisms, pp. 3320–3330, In: *Encyclopedia of Environmental Microbiology*, Gabriel Bitton, editor-in-chief, Wiley-Interscience, N.Y.

Gostelow, P., S.A. Parsons, and R.M. Stuetz. 2001. Odour measurements for sewage treatment works. Water Res. 35: 579–597.

Mohr, A.J. 1991. Development of models to explain the survival of viruses and bacteria in aerosols, pp. 160–190, In: *Modeling the Environmental Fate of Microorganisms*, C.J. Hurst, Ed., American Society Microbiology, Washington, DC.

Stetzenbach, L.D. 1997. Introduction to aerobiology, pp. 619–628, In: *Manual of Environmental Microbiology*, C.J. Hurst, G.R. Knudsen, M.J. McInerney, L.D. Stetzenbach, and M.V. Walter, Eds., ASM Press, Washington, D.C.

PART D

MICROBIOLOGY OF DRINKING WATER TREATMENT

15

MICROBIOLOGICAL ASPECTS OF DRINKING WATER TREATMENT

15.1 INTRODUCTION

Drinking water safety is a worldwide concern. Contaminated drinking water has the greatest impact on human health worldwide, especially in developing countries. The distribution, by continent, of the global population not served with improved water supply is illustrated in Figure 15.1 (WHO, 2003). It is estimated that 1.1 billion of the world's population does not have access to safe clean water (WHO/UNICEF, 2000; WHO, 2003).

Wastewater Microbiology, *Third Edition*, by Gabriel Bitton
Copyright © 2005 John Wiley & Sons, Inc.

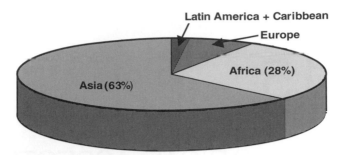

Figure 15.1 Distribution of the global population not served (1.1 billion) with improved water supply. From WHO (2003).

The World Health Organization (WHO) reported that 80 percent of diseases and one-third of deaths in developing countries are due to consumption of contaminated water (WHO, 1996a). Drinking nontreated or improperly treated water is a major cause of illness in developing countries. For example, in 1980, 25,000 persons per day (approximately 9 million/year) died worldwide as a result of consumption of contaminated water. A more recent WHO estimate shows that water-related diseases are responsible for more than 5 million deaths per year (WHO, 1996a). In fact, 25 percent of hospital beds were occupied by people who became ill after consuming contaminated water (U.S. EPA, 1989c; WHO, 1979). According to UNICEF, 3.8 million children under the age of 5 years died in 1993 from diarrheal diseases worldwide. In India, more than 300,000 children die each year from these diseases (Reuther, 1996). A retrospective epidemiological study conducted in the United States showed that waterborne disease outbreaks are due to the consumption of untreated and inadequately treated water (e.g., inadequate disinfection or filtration, temporary interruption of disinfection) or to subsequent contamination (e.g., cross-connection, back siphonage) in the distribution network (Lippy and Waltrip, 1984) (Fig. 15.2). In 1991 and 1992, 34 drinking water associated outbreaks were reported in the United States. The etiologic agent was unknown in 23 of the outbreaks, while the remaining outbreaks were attributed to *Giardia*, *Cryptosporidium*, *Shigella sonnei*, hepatitis A virus (HAV), or chemicals (Moore et al., 1994) (Fig. 15.3).

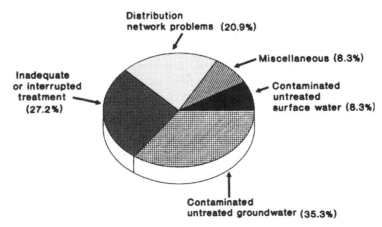

Figure 15.2 Waterborne disease outbreaks caused by deficiency in public water systems. From Lippy and Waltrip (1984). (Courtesy of the American Water Works Association.)

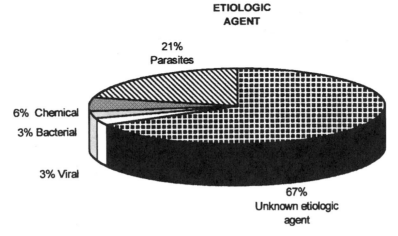

ETIOLOGIC AGENT

21% Parasites

6% Chemical

3% Bacterial

3% Viral

67% Unknown etiologic agent

Figure 15.3 Etiologic agents responsible for outbreaks associated with drinking water in 1991 and 1992. From Moore et al. (1994).

Ingestion of even low levels of pathogens, particularly viruses, may pose some risk of infection, clinical illness or even mortality to susceptible populations (Gerba and Haas, 1986) (see Chapter 4). An example of risk associated with viruses in drinking water is shown in Table 15.1 (Mena et al., 2003). An exponential risk assessment model (described in Chapter 22) was used to assess the risk associated with the presence of coxsackie viruses in drinking water. The model for the general population assumes a consumption of 2 L of drinking water per day and a virus concentration of 0.13 PFU/L (highest concentration) or 10^{-3} MPN CPU (most probable number of cytopathic units)/L (lowest concentration). The daily risk of infection from drinking treated surface water was 10^{-5}, whereas the annual risk was 10^{-2}. A higher risk was associated with drinking nondisinfected groundwater (Mena et al., 2003).

Advances in drinking water research followed by the establishment of multiple barriers against microbial pathogens and parasites have significantly increased the safety of the water we drink daily, particularly in industrialized nations. This multiple-barrier system includes source water protection, reliable water treatment (pretreatment, coagulation, flocculation, sedimentation, filtration, disinfection), and protection of the water distribution network (Berger and Oshiro, 2002; LeChevallier, 2002).

In this chapter, we will cover the microbiological quality of source waters as well as the public health aspects of water treatment processes. Distribution of potable water, with particular emphasis on biofilm formation in distribution lines, will be covered in Chapter 16.

TABLE 15.1. Risks Associated with Coxsackievirus in Drinking Water[a]

Type of Risk[b]	5×10^{-3} MPN CPU/L Surface Water		0.13 PFU/L Groundwater	
	Day	Year	Day	Year
Risk of infection	7.75×10^{-5}	2.79×10^{-2}	2.01×10^{-3}	5.21×10^{-1}
Risk of illness	5.81×10^{-5}	2.09×10^{-2}	1.51×10^{-3}	3.91×10^{-1}
Risk of death	3.43×10^{-7}	1.23×10^{-4}	8.91×10^{-6}	2.30×10^{-3}

[a]Adapted from Mena et al. (2003).

[b]General population: 2 L drinking water/day/person.

15.2 MICROBIOLOGICAL QUALITY OF SOURCE WATER

The desirable approach is to withdraw water from the best available and safest source. Sometimes, due to the scarcity of quality source water, communities must use reclaimed water to augment their water supplies (see Chapter 23).

Communities obtain their potable water from surface or underground sources. However, both types of water can become contaminated by biological and chemical pollutants originating from point and nonpoint sources. Point sources, for which the source can be identified, include discharges from wastewater treatment plants, and land disposal of wastewater effluents and biosolids. The wastes are from both human and animal sources (e.g., animal feeding operations). Nonpoint sources are diffuse sources of pollution and include, for example, urban and agricultural runoffs, water recreation facilities (e.g., swimming, boating), and wildlife (many human pathogens and parasites use animals as reservoirs). Source water can also become contaminated by opportunistic pathogens (e.g., *Ps. Aeruginosa, Legionella pneumophila, Aeromonas, Mycobacterium, Flavobacterium, Naegleria fowleri, Acanthamoeba* spp.) that are part of the indigenous microorganisms in soil and aquatic environments (Berger and Oshiro, 2002).

Surface waters include lakes, rivers, and streams, and their quality rapidly changes as a response to changes in the surrounding watershed. For example, relatively high concentrations of nutrients (N, P) result in eutrophication of surface waters with excessive growth of algae, leading to excessive levels of microorganisms and turbidity in the source water. In the United States, surface waters are used as source water by an estimated 6000 community water systems and serve a population of approximately 155 million people (Craun, 1988; Federal Register, 1987). Surface waters are often contaminated by domestic wastewater, stormwater runoff, cattle feedlot runoff, discharges from food processing plants, and by resuspension of microbial pathogens and parasitic cysts and ova that have accumulated in bottom sediments. Lake destratification leads to a resuspension of microorganisms, humic acids, and turbidity in the water column (Geldreich, 1990). This leads to increased levels of suspended solids, nutrients, BOD, and microbial pathogens and parasites in surface waters. In some areas, these surface waters are practically "diluted wastewaters." Problems arise when upstream communities discharge pathogen-laden wastewater effluents into surface waters that become drinking water supplies to downstream communities.

The pathogen load downstream from a pollution source depends on the extent of the natural self-purification processes, which are controlled by temperature, solar radiation, dilution, available nutrients, and biological factors, which include competition with indigenous microorganisms, protozoan grazing of pathogenic bacteria, or ability to enter the viable but nonculturable (VBNC) state as a survival strategy.

Contamination of groundwaters is also well documented (see Chapter 21). Concern over these subsurface waters stems from the fact that they supply the drinking water needs of more than 100 million people in the United States, and they are often consumed without any treatment. Unconfined aquifers are of most concern as regards fecal contamination from surface and underground sources. Viruses are of most concern to public health professionals because they can be readily transported over a greater distance through soils and aquifers.

Source water protection can be achieved by identifying the contamination sources and by taking protective measures. These measures may include the use of physical barriers to exclude humans and animals, the reduction or elimination of certain activities such as cattle grazing and sewage discharges, and the establishment of land-use restrictions (Robertson and Edberg, 1997).

15.3 OVERVIEW OF PROCESSES INVOLVED IN DRINKING WATER TREATMENT PLANTS

Water contains several chemical and biological contaminants that must be removed efficiently in order to produce drinking water that is safe and aesthetically pleasing to the consumer. The chemical contaminants include nitrate, heavy metals, radionuclides, pesticides, and other xenobiotics. The finished product must also be free of microbial pathogens and parasites, turbidity, color, taste, and odor. To achieve this goal, raw water (surface water or groundwater) is subjected to a series of physicochemical processes that will be described in detail. Disinfection alone is sufficient if the raw water originates from a protected source. More commonly, several processes are used to treat water. Disinfection may be combined with coagulation, flocculation, and filtration. Additional treatments to remove specific compounds include preaeration and activated carbon treatment. The treatment train depends on the quality of the source water under consideration.

There are two main categories of water treatment plants:

- *Conventional filter plants.* Conventional plants serve an estimated population of 108 million consumers in the United States (U.S. EPA, 1989c) (Fig. 15.4). The leading processes in this type of plant are coagulation and filtration. The raw water is rapidly mixed with a coagulant (aluminum sulfate, ferric chloride, ferric sulfate). After coagulation, the produced flocs are allowed to settle in a clarifier. Clarified effluents are then passed through sand or diatomaceous earth filters. Water is finally disinfected before distribution.
- *Softening plants.* The leading process in these plants is water softening, which helps remove hardness due to the presence of Ca and Mg in water, and results in the formation of Ca and Mg precipitates. After settling of the precipitates, the water is filtered and disinfected.

15.4 PROCESS MICROBIOLOGY AND FATE OF PATHOGENS AND PARASITES IN WATER TREATMENT PLANTS

In water treatment plants, microbial pathogens and parasites can be physically removed by processes such as coagulation, precipitation, filtration, and adsorption, or they can be inactivated by disinfection or by the high pH resulting from water softening.

There are several types of pathogens and parasites of most concern in drinking water (AWWA, 1987). (Chapter 4 should be consulted for more details on these pathogens and parasites.)

1. *Viruses.* These are occasionally detected in drinking water from conventional water treatment plants that meet the standards currently used to judge treatment efficiency. For example, enteroviruses have been detected at levels ranging from 3 to 20 viruses per 1000 L in finished drinking water from water treatment plants that include prechlorination, flocculation, sedimentation, sand filtration, ozonation, and final chlorination (Payment, 1989a). However, at a water filtration plant in Laval, Canada, no virus was detected in 162 finished water samples (1000–2000 L per sample), although coliphage and Clostridia were detected in finished water (Payment, 1991). In South Korea, tap water processed via flocculation/sedimentation, filtration, and chlorination contained infectious enteroviruses

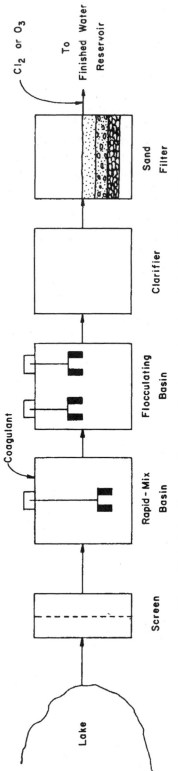

Figure 15.4 Flow diagram of a conventional water filtration plant. From Bitton (1980).

(poliovirus type 1, echovirus type 6, coxsackie B viruses) and adenoviruses (e.g., adenoviruses type 40 and 41) in 39–48 percent of the samples examined by cell culture assays followed by PCR amplification (Lee and Kim, 2002). Virus removal was also investigated at a full-scale water treatment plant in Mery sur Oise, France. The treatment train in this plant included the following: preozonation step followed by coagulation–flocculation, sand filtration, ozonation, activated carbon, and final ozonation step. No virus was detected in the final product (Joret et al., 1986) (Table 15.2). Similar results were reported at three water treatment plants in Spain (Jofre et al., 1995).

Isolation of enteric viruses has been documented worldwide in partially treated drinking water, in marginally operated and even in apparently "well-operated" water treatment plants (Bitton et al., 1986). These findings may be due to relatively drastic changes in source water quality or to equipment and process failure.

2. *Cryptosporidium* and *Giardia lamblia*. The methodology for detecting these parasites is under development, but the skill to routinely monitor their presence in drinking water is not yet available in water treatment plants. These parasites have been detected in drinking water, and outbreaks have been described in Chapter 4.

3. *Opportunistic pathogens*. These are waterborne pathogens (e.g., *Pseudomonas putida*, *Alcaligenes*, *Acinetobacter*, and *Flavobacterium* species), which cause secondary infections in hospitals and, potentially, among consumers.

4. *Legionella*. This pathogen is an example of a nonenteric microorganism that can be transmitted by inhalation of drinking water aerosols from shower heads or humidifiers. Nosocomial Legionnaire's disease may be contracted by exposure to *Legionella* from the water distribution system in hospitals (Best et al., 1984).

5. The U.S. EPA is now examining a list of microbial pathogens in drinking water for potential regulation. These pathogens are transmitted via the waterborne route but other routes may be involved. This list includes *Aeromonas hydrophila*, *Helicobacter pylori*, *Mycobacterium avium intracellulare* complex (MAIC), adenoviruses, caliciviruses, coxsackie viruses, echoviruses, and *Microsporidia* protozoan parasites) (Hilborn et al., 2002; U.S. EPA, 1998). Several unit processes and operations are used in water treatment plants to produce microbiologically and chemically safe drinking water. The extent of treatment depends on the source of raw water, with surface waters generally requiring more treatment than is needed for groundwaters. The unit processes designed for water treatment, with the exception of the disinfection step, do not address specifically the destruction or removal of parasites or bacterial and viral pathogens.

TABLE 15.2. Virus Concentrations (PFU/1000 L) at Various Stages of a Water Treatment Plant[a]

Sampling Event	Stored Water	Sedimentation	Sand Filtration	Ozonation
1	10.4	<25	9.1	<1
2	6.1	132	<1	<1
3	100	75	<2	<2
4	90	5	<1	<1
5	10	20	3	<1
6	30.7	10	5	<1

[a]Adapted from Joret et al. (1986).

15.4.1 Pretreatment of Source Water

Pretreatment is a range of steps that are designed to improve the quality of the source water prior to entry in the water treatment plant (LeChevallier, 2002).

15.4.1.1 *Storage of Raw Water (Off-Stream Reservoirs).* Raw water can be stored in reservoirs to minimize fluctuations in water quality. Storage can affect the microbiological quality of the water. Water quality in the reservoirs is affected by physical (e.g., settling of solids, evaporation, gas exchange with atmosphere), chemical (e.g., oxidation–reduction, hydrolysis, photolysis), and biological processes (e.g., nutrient cycling, biodegradation, pathogen decay) (Oskam, 1995). The reduction of pathogens, parasites, and indicator microorganisms during storage is variable and is influenced by a number of factors, such as temperature, sunlight, sedimentation, and biological adverse phenomena such as predation, antagonism, and lytic action of bacterial phages. Temperature is a significant factor controlling pathogen survival in reservoirs.

It appears that, under optimal conditions, storage in reservoirs can lead to approximately 1 to 2–log reduction of bacterial and viral pathogens although higher reductions have been observed. Protozoan cysts are removed by entrapment into suspended solids followed by settling into the sediments.

15.4.1.2 *Roughing Filters.* Roughing filters contain coarse media (gravel, rocks), which help to reduce water turbidity and bacterial concentrations (approximately one–log redution).

15.4.1.3 *Microstrainers.* Microstrainers are made of woven stainless steel or polyester wires with a pore size ranging from 15 to 45 μm. They retain mostly algae and relatively large protozoa. Filamentous or colonial algae are more efficiently removed than unicellular algae.

15.4.1.4 *River Bank Filtration (RBF).* This pretreatment has been practiced in Europe since the 1870s. In Germany, approximately 16 percent of the drinking water is produced from water processed by RBF or infiltration (Kuehn and Mueller, 2000).

River bank filtration is the seepage of water through the bank of a river or lake to the production well of the water treatment plant. This practise provides certain advantages such as removal of pathogens and parasites, removal of algal cells, reduction in turbidity and natural organic matter, and dilution with groundwater. Particle removal is due to the combined effect of adsorption, straining, and biodegradation. River bank filtration decreases the concentration of assimilable organic carbon (AOC) (see Chapter 16 for more details on AOC), leading to a reduction of the biological growth potential of the water (Fig. 15.5; Kuehn and Mueller, 2000). A study of five RBFs in the United States showed an efficient removal of *Cryptosporidium* and *Giardia*. In the Netherlands, RBF treatment provided a 4–log removal of viruses and 5 to 6–log removal of F-specific coliphages (Havelaar et al., 1995; Ray et al., 2002).

15.4.2 Prechlorination

A prechlorination step is sometimes included to improve unit process performance (e.g., filtration, coagulation–flocculation) or to oxidize color-producing substances such as humic acids. Although prechlorination reduces somewhat the levels of pathogenic

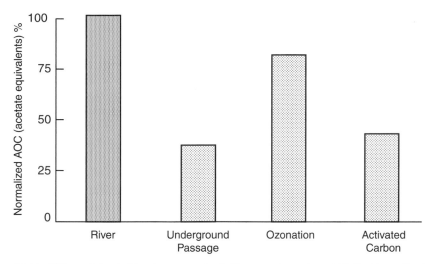

Figure 15.5 Effect of bank filtration on assimilable organic carbon (AOC). From Kuehn and Mueller (2000).

microorganisms, its use is being questioned due to increased chances of forming trihalomethanes (see Chapter 6).

15.4.3 Coagulation–Flocculation–Sedimentation

Coagulation involves the destabilization and interparticle collisions of colloidal particles (e.g., mineral colloids, microbial cells, virus particles) by coagulants (Al and Fe salts) and sometimes by coagulant aids (e.g., activated silica, bentonite, polyelectrolytes, starch). The most common coagulants are alum, ferric chloride, and ferric sulfate. The process of interparticle contacts and formation of larger particles is called flocculation. After mixing, the colloidal particles form flocs that are large enough to allow rapid settling (Letterman et al., 1999; Williams and Culp, 1986) (Fig. 15.6). The pH of the water is probably the most significant factor affecting coagulation. Other factors include turbidity, alkalinity, temperature, and mixing regimen. Coagulation is the most important process used in water treatment plants for the clarification of colored and turbid waters.

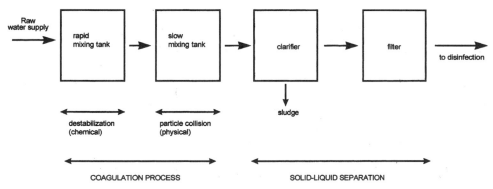

Figure 15.6 Schematic diagram of a coagulation process. From Williams and Culp (1986). (Courtesy of van Nostrand Reinhold Co., NY.)

Jar tests have shown that the removal of bacteria and protozoan (*Cryptosporidium*, *Giardia*) cysts may range from 1 to 2–logs. The removal of viruses is variable and ranges from 1 to more than 3–logs (Bell et al., 2000). Under laboratory conditions, coagulation–flocculation was effective in removing 90–99 percent of viruses from water (Bitton, 1980a) (Table 15.3). In three Spanish water treatment plants, the prechlorination–flocculation–sedimentation step was most efficient in phage removal (somatic, male-specific and *Bacteroides fragilis*-specific phages) (Jofre et al., 1995). Coagulation removes 74–99.4 percent of *Escherichia coli* and coliforms (Sobsey and Olson, 1983). Although algal cells are well removed by coagulation–flocculation, no efficient removal has been reported for algal toxins.

However, pathogen and parasite removal under field conditions may be lower than under controlled laboratory conditions. In a pilot water treatment plant treating water from the Seine River in France, indigenous virus removal by coagulation–flocculation varied from 31 to 90 percent with an average removal of 61 percent. This shows that virus removal obtained with laboratory strains is much higher than with indigenous viruses (Joret et al., 1986).

Endotoxins or lipopolysaccharides (LPS) are components of the outer membrane of most gram-negative bacteria and some cyanobacteria. They have been associated with acute respiratory illnesses, water fever, gastrointestinal disorders, and allergic reactions. The different processes involved in water treatment were shown to remove 59 to 97 percent of the endotoxin activity. The highest removal was observed following coagulation, clarification, and rapid sand filtration (Rapala et al., 2002).

Coagulation is sometimes improved by using *coagulant aids*, such as polyelectrolytes, bentonite, or activated silica. Polyelectrolytes help form large flocs that settle out rapidly. Their concentration is an important parameter, since excessive dosage can inhibit flocculation.

Coagulation merely transfers pathogenic microorganisms from water to the flocculated material, which is incorporated in a sludge that must be disposed of properly.

15.4.4 Water Softening

Hardness is caused by the presence of calcium and magnesium in the water. There are two categories of hardness: *carbonate hardness*, which is due to bicarbonates of Ca and Mg, and *noncarbonate hardness*, which is due to Ca and Mg chlorides. Hardness is responsible

TABLE 15.3. Removal of Viruses by Coagulation–Sedimentation[a]

Coagulant	Concn. of Coagulant (ml/L)	Clay (mg/L)	Virus	Virus (%)	Turbidity (%)
Alum	10	50	Poliovirus 1	86	96
	25.7	120	Phage T4	98	99
	25.7	120	Phage MS2	99.8	98
Ferric sulfate	40	–	Poliovirus 1	99.8	–
	40	–	Phage T4	99.8	–
Ferric chloride	60	50	Poliovirus 1	97.8	97.5
	60	100	Poliovirus 1	93.3	97.8
	60	500	Poliovirus 1	99.7	99.7

[a]From Bitton (1980a).

for increased soap consumption and scale formation in pipes. *Water softening* is the removal of Ca and Mg hardness by the lime-soda process or by ion-exchange resins (Bitton, 1980a). The lime-soda process consists of adding lime (calcium hydroxide) and soda ash (sodium carbonate) to the water. The carbonate hardness is removed by the following:

$$Ca(HCO_3)_2 + Ca(OH)_2 \longrightarrow 2 CaCO_3 + 2 H_2O \tag{15.1}$$

Ca-bicarbonate lime Ca-carbonate

$$Mg(HCO_3)_2 + 2 Ca(OH)_2 \longrightarrow Mg(OH)_2 + CaCO_3 + 2 H_2O \tag{15.2}$$

Mg-bicarbonate lime Mg-hydroxide

The high pH (>11) generated by water softening with lime leads to an effective inactivation of bacterial and viral pathogens. Poliovirus type 1, rotavirus and HAV are effectively removed (>95 percent removal) during water softening (pH 11) (Rao et al., 1988) (Table 15.4). Bacterial pathogens are also efficiently reduced following liming to reach a pH that exceeds 11. The inactivation rate is temperature-dependent (Riehl et al., 1952; Wattie and Chambers, 1943).

Microbial removal during water softening is due (1) to microbial inactivation at detrimental high pH values (pH ≥ 11) by the loss of structural integrity or inactivation of essential enzymes, and (2) to physical removal of microorganisms by adsorption to positively charged magnesium hydroxide flocs ($CaCO_3$ precipitates are negatively charged and do not adsorb microorganisms).

As regards ion-exchange resins, Ca and Mg are removed from water by exchange with Na present on the exchange sites. Viruses are removed by anion-exchange resins, but not as much by cation-exchange resins. However, ion-exchange resins cannot be relied upon to remove microbial pathogens.

15.4.5 Filtration

Filtration is defined as the passage of fluids through porous media to remove turbidity (suspended solids, such as clays, silt particles, microbial cells) and flocculated particles. This process depends on the filter medium, concentration and type of solids to be filtered out, and the operation of the filter.

Filtration is one of the oldest processes used for water treatment. In 1685, an Italian physician, Luc Antonio Porzio, conceived a filtration system for protecting soldiers' health in military installations. This process has been in use since and has contributed greatly to the reduction of waterborne diseases such as typhoid fever and cholera. An examination of waterborne outbreaks around the world clearly shows that filtration has

TABLE 15.4. Removal of Virus by Water Softening[a]

Virus	Virus Input (PFU/500 mL)	Percentage Removal		
		Virus	Total Hardness	Turbidity
PV	3.9×10^6	96	54	70
RV	7.8×10^5	>99	47	60
HAV	6.8×10^8	>97	76	62

PV, poliovirus; RV, rotavirus; HAV, hepatitis A virus.
[a]pH adjusted to 11.0.
Adapted from Rao et al. (1988).

been historically instrumental as a barrier against pathogenic microorganisms and has largely contributed to the reduction of waterborne diseases. Figure 15.7 shows the dramatic reduction of typhoid fever death rate in Albany, New York, after the adoption of sand filtration and chlorination about a decade later (Logsdon and Lippy, 1982; Willcomb, 1923).

15.4.5.1 Slow Sand Filtration. Although more popular in Europe than in the United States, slow sand filters serve mostly communities of less than 10,000 people because capital and operating costs are lower than for rapid sand filters (Slezak and Sims, 1984). In the United States, the first slow sand filter was installed in Lawrence, Massachusetts, to remove *Salmonella typhi* from water.

Slow sand filters (Fig. 15.8) contain a layer of sand (60–120 cm depth) supported by a graded gravel layer (30–50 cm depth). The sand grain size varies between 0.15 and 0.35 mm, and the hydraulic loading range is between 0.04 and 0.4 m/h (Bellamy et al., 1985a).

Biological growth inside the filter comprises a wide variety of organisms, including bacteria, algae, protozoa, rotifers, microtubellaria (flatworms), nematodes (round worms), annelids (segmented worms), and arthropods (Duncan, 1988). The buildup of a biologically active layer, the *schmutzdecke*, occurs during the normal operation of a slow sand filter. The top layer is composed of biological growth and filtered particulate matter. This leads to a head loss across the filter, a problem corrected by removing or scraping the top layer of sand. The length of time between scrapings depends on the turbidity of the raw water and varies between 1 to 2 weeks and several months (Logsdon and Hoff, 1986). Scraping is followed by replenishing of the filter bed with clean sand, an operation called *resanding*. There is sometimes a deterioration of water quality for some days after scraping, but it later improves during the ripening period (Cullen and Letterman, 1985). However, some investigators (Logsdon and Lippy, 1982) have not observed a deterioration after scraping. Bacterial activity in the schmutzdecke helps

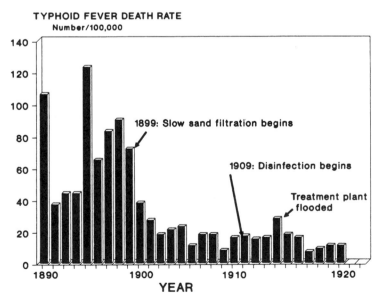

Figure 15.7   The impact of water filtration and chlorination on typhoid fever death rate in Albany, NY. From Logsdon and Lippy (1982). (Courtesy of the American Water Works Association.)

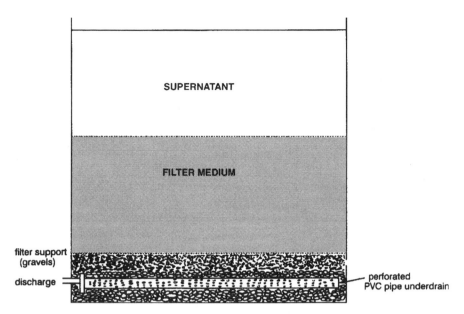

Figure 15.8 Slow sand filter. From Slezak and Sims (1984). (Courtesy of the American Water Works Association.)

remove assimilable organic compounds (AOC), some of which are precursors of chlorinated organics such as trihalomethanes (Collins et al., 1992; Eighmy et al., 1992; Fox et al., 1984).

The removal efficiency for coliform bacteria is influenced by temperature, sand size, and filter depth (Bellamy et al., 1985a, b). Figure 15.9 (Bellamy et al., 1985b) shows the effect of filter depth and sand size on total coliform removal. Removal of *Giardia* and total coliforms exceed 99 percent, even at the highest loading rate of 0.4 m/h. Pilot plant studies showed that slow sand filtration achieves 4 to 5–log reductions of coliforms (Fox et al., 1984). A survey of slow sand filters in the United States showed that most the plants reported coliform levels of 1 per 100 mL or less (Slezak and Sims, 1984). In water treatment plants in Paris, France, slow sand filters were found to be more efficient in myco-bacteria removal than rapid sand filters (Le Dantec et al., 2002). No *Cryptosporidium* oocysts were detected in treated water from two water treatment plants using slow sand filters, despite the frequent isolation of this parasite in the raw waters entering the

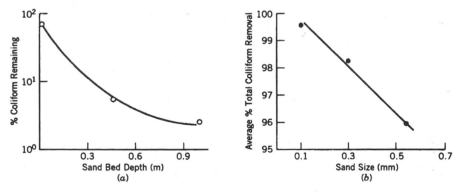

Figure 15.9 Effect of sand bed depth and sand size on the removal of total coliforms by slow sand filtration. From Bellamy et al. (1985). (Courtesy of the American Water Works Association.)

plants (Chauret et al., 1995). The microbial community developing on the sand filter greatly influences the removal of bacteria, protozoan cysts, and turbidity by slow sand filters (Bellamy et al., 1985a; Cleasby et al., 1984; Huisman and Wood, 1974).

With regard to viruses, a removal exceeding 99.999 percent was observed in a slow sand filter operated at 11°C at a rate of 0.2 m/h (Poynter and Slade, 1977). An established filter appears to perform better than a new filter with regard to virus removal (Wheeler et al., 1988). The factors controlling virus removal by slow sand filters include filter depth, flow rate, temperature, and the presence of a well-developed biofilm on the filter.

15.4.5.2 Rapid Sand Filtration.
A rapid sand filter consists of a layer of sand supported by a layer of anthracite, gravel, or calcite (Bitton, 1980a). The rapid sand effluent is collected by an underdrain system. Rapid sand filters are operated at filtration rates of 5–24 $m^3/h/m^2$ as compared with 0.1–1 $m^3/h/m^2$ for slow sand filters (Huisman and Wood, 1974; Logsdon and Hoff, 1986). These filters are periodically cleaned by backwashing (i.e., reversing the flow) at a sufficient flow rate to allow a thorough cleaning of the sand.

This process appears to be less effective for the removal of bacteria, viruses, and protozoan cysts unless it is preceded by coagulation–flocculation. The removal of *Salmonella* and *Shigella* by filtration and coagulation is similar to that of coliforms. *Giardia* cysts pass through water treatment plants that use direct filtration without any chemical pretreatment. The breakthrough is more pronounced during winter months when temperature is below 5°C (Hibler and Hancock, 1990). Significant changes in flow rates may result in deterioration of water quality due to the release of retained particles, including protozoan cysts.

The combination of filtration with coagulation–flocculation is particularly important in respect to *Giardia* cysts and *Cryptosporidium* oocysts removal from water (Logsdon et al., 1981). For low-turbidity water, proper chemical coagulation of the water before rapid sand filtration is necessary to achieve a good removal of turbidity and protozoan cysts and oocysts. In a plant in Japan, coagulation–flocculation–sedimentation followed by rapid sand filtration removed approximately 2.5–$\log_{10}$ of *Cryptosporidium* and *Giardia* (Hashimoto et al., 2002). It was suggested that the removal of turbidity could serve as a surrogate indicator of *Giardia* cysts removal in water with low turbidity (Al-Ani et al., 1986). However, *Giardia* and *Cryptosporidium* monitoring in filtered drinking water samples indicated that turbidity removal was a good predictor of *Cryptosporidium* oocysts removal but not *Giardia* cysts (LeChevallier et al., 1991a). Similarly, as reported in Chapter 4, oocyst-sized polystyrene microspheres appear to be reliable surrogates for *C. parvum* oocyst removal by filtration (Emelko and Huck, 2004). Filtration, without the prior addition of coagulants, removes approximately 90 percent of *Cryptosporidium* oocysts. Surface coating of sand particles with a hydrous iron aluminum oxide was found to improve the removal of *Cryptosporidium* (Shaw et al., 2000). However, the relatively long-term effect of the coating on oocyst removal is not known at the present time. If the source water is a surface water, then a dual barrier is necessary to prevent contamination with *Giardia lamblia* cysts. This dual barrier consists of coagulation–sand filtration to retain cysts and turbidity, and disinfection (AWWA, 1987).

Entamoeba histolytica cysts are also effectively removed when rapid sand filtration is preceded by coagulation (Baylis et al., 1936). Because sand particles are essentially poor adsorbents toward viruses, pathogen removal by sand filtration is variable and often low. However, coagulation prior to sand filtration removes more than 99 percent of viruses (Bitton, 1980a) (Table 15.5). Coagulation, settling, and filtration removes more than one–log of hepatitis virus, simian rotavirus (SA-11), and poliovirus (Table 15.6; Rao et al., 1988). A pilot plant, using the Seine River water, which contained 190–1420 PFU/1000 L, removed 1 to 2–logs of viruses after coagulation–flocculation

TABLE 15.5. Removal of Poliovirus by Rapid Sand Filters[a]

Treatment	% Removal
A. Sand filtration	1–50
B. Sand filtration + coagulation with alum	
1. Without settling	90–99
2. With settling	>99.7

[a]The flow rate was 2–6 gpm/ft^2.
Adapted from Robeck et al. (1962).

followed by sand filtration (Joret et al., 1986). A preozonation step (0.8 ppm ozone level) greatly improved virus removal, which was increased to 2 to 3–logs. Helminths form relatively large eggs, which are effectively removed by sedimentation, coagulation, and filtration.

Thus, optimum coagulation followed by filtration in water treatment plants is essential for parasites and pathogens control. Other parameters that adversely affect filter operation are sudden changes in water flow rates, interruptions in chemical feed, inadequate filter backwashing, and the use of clean sand, which was not allowed to undergo a ripening period (Logsdon and Hoff, 1986; Ongerth, 1990).

Because of their limited capacity to retain solids, rapid sand filters must be *backwashed* to remove the trapped solids from the filter matrix. The water treatment plant finished water is generally used as backwash water. However, the spent filter backwash water (SFBW) must be adequately treated prior to reuse within the plant or disposal into a receiving stream (Arora and LeChevallier, 2002). The recycling of SFBW, mostly by treatment plants that use surface water as the source water, is regulated by the U.S. EPA, due to the increased levels of biological (protozoan cysts, bacterial, and viral pathogens) and chemical (disinfection byproducts, metals) contaminants in SFBW. The SFBW must be treated to reduce the risk of breakthrough of pathogens and parasites into finished water. The treatments include sedimentation (preferably in the presence of coagulants), dissolved air flotation, filtration through granular media or membrane filters, and disinfection, preferably after clarification.

Dual-stage filtration (DSF) is an alternative to rapid sand filtration for small water treatment plants. This process consists of chemical coagulation followed by a filter assembly consisting of two tanks, a depth clarifier and a depth filter. At a flow rate of 10 gpm/ft^2, this treatment removed more than 99 percent of *Giardia* cysts from water with an effluent turbidity less than 1 NTU (Horn et al., 1988).

TABLE 15.6. Removal of Virus by Coagulation–Settling–Sand Filtration[a]

	Viral Assays, Total PFU/200 L (Percentage Removed)		
Virus	Input	Settled Water	Filtered Water
PV	5.2×10^7	1.0×10^6 (98)	8.7×10^4 (99.84)
RV	9.3×10^7	4.6×10^6 (95)	1.3×10^4 (99.987)
HAV	4.9×10^{10}	1.6×10^9 (97)	7.0×10^8 (98.6)

[a]Adapted from Rao et al. (1988).
PV, poliovirus; RV, rotavirus; HAV, hepatitis A virus.

15.4.5.3 *Diatomaceous Earth Filtration.* Diatomaceous earth (DE) or diatomite is made of the remains of siliceous shells of diatoms. Diatomaceous earth filtration includes two steps (Lange et al., 1986):

- *Precoating.* A precoat of DE is applied on a porous filter septum that serves as a support for the buildup of a $\frac{1}{8}$ in. precoat of filter medium (Bitton, 1980a) (Fig. 15.10) to form a filter cake.
- *Filtration.* Raw water, treated with additional DE (body feed), is passed though the filter cake to keep the filter running for longer periods of time.

At the end of a run, the filter cake is removed and replaced by fresh diatomite.

The performance of a DE filter is affected by DE grade, thickness of DE cake, size of microorganisms, and chemical conditioning of DE. Chemical coating of DE with Al and Fe salts or cationic polyelectrolytes improves the microbe-removal efficiency of this material. A cationic polyelectrolyte, at a concentration of 0.15 ppm, brings about 100 percent removal of poliovirus, as compared with 62 percent without polyelectrolyte (Bitton, 1980b). Virus removal by DE is also enhanced by in situ precipitation of metallic salts (Al, Fe) (Farrah et al., 1991). The size of the DE used for precoating and body feed also influences the efficiency of this material for removing microorganisms (Logsdon and Lippy, 1982; Logsdon and Hoff, 1986). Removal of bacteria is influenced by DE grade, hydraulic loading rate, influent bacterial concentration, and filtration time. Chemical coating with alum and coagulants greatly improves their removal (Table 15.7) (Hunter et al., 1966; Lange et al., 1986; Schuler and Ghosh, 1990).

The DE is generally effective in removing cysts of protozoan parasites. The removal of *Giardia lamblia* cysts and *Cryptosporidium* oocysts by DE exceeds 99 percent (Logsdon and Lippy, 1982; Schuler and Ghosh, 1990). Furthermore, removal of *Cryptosporidium* is greatly improved upon addition of alum (Schuler and Ghosh, 1990). A 6–log removal of *Cryptosporidium* was obtained when using a DE precoat of 1 kg/m^3 (20 lb/100 ft^2), a body feed of 5 mg/L, and a filtration rate of 1 gpm/ft^2 (2.5 m/h) (Ongerth and Hutton, 2001).

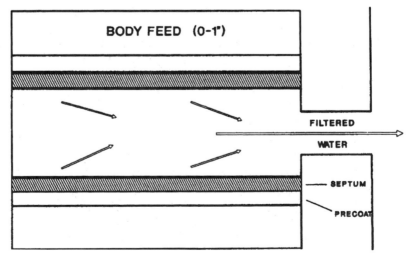

Figure 15.10 Diatomite filter. Adapted from Baumann (1971).

TABLE 15.7. Alum Enhancement of Removal of Bacteria by Diatomaceous Earth

Grade	Alum-to-DE Ratio[a] $\left(\dfrac{\text{g Alum}}{\text{g DE}}\right)$	Body Feed Concn (mg/L)	Average Percentage Removals		
			Total Coliform	Standard Plate Count	% Turbidity Removal
B[b]	0.02	25	–	–	66.11
	0.04	25	99.02	95.02	86.41
	0.04	25	99.02	95.02	86.41
	0.05	25	99.86	98.56	98.38
D[c]	0.05	25[d]	98.01	79.31	79.06
	0.05	25	99.56	93.25	94.41
	0.05	50	96.33	99.57	98.61
	0.08	25	99.83	99.52	98.80

[a]Alum coating ratio is grams of alum as $Al_2SO_4 \cdot 14.3 \, H_2O$ slurry per gram of diatomaceous earth in slurry.
[b]C-545.
[c]C-503.
[d]Body feed concentration was increased to 50 mg/L after 3 h of testing at 25 mg/L.
Adapted from Lange et al. (1986).

15.4.6 Activated Carbon

Activated carbon is an adsorbent derived from wood, bituminous coal, lignite, or other carbon-containing materials, and is the most widely utilized adsorbent for the treatment of water and wastewater. It is activated by a combustion process to increase its internal surface area. It offers a large internal surface area ($500-600 \, m^2/g$) for the adsorption of taste, odor, and color compounds, excess chlorine, toxic and mutagenic substances (e.g., chlorinated organic compounds, including trihalomethanes), trihalomethane precursors, pesticides, phenolic compounds, dyes, toxic metals, and substances that cause biological aftergrowth (Allen, 1996; Najm et al., 1991; van Puffelen, 1983).

Activated carbon may be used in the form of granular activated carbon (GAC) applied after sand filtration and before chlorination, and in the form of powdered activated carbon (PAC), which has a smaller particle size than GAC and can be applied at various points in water and wastewater treatment plants, mainly prior to filtration. Powdered activated carbon treatment (PACT) is less costly than GAC treatment because the powdered carbon is applied only when needed, leading to lower amounts of carbon used. The PACT is used in both aerobic and anaerobic wastewater treatment systems (McKay, 1996).

Activated carbon has been known for its purification properties since antiquity (i.e., early Egyptians), but its microbiological aspects have only recently been investigated (LeChevallier and McFeters, 1990; Weber et al., 1978). The functional groups on the carbon surface help in the adsorption of microorganisms including bacteria (rods, cocci, and filamentous bacteria), fungi, and protozoa. Scanning electron microscopic observations have shown colonization of the carbon surface by microorganisms (polysaccharide-producing bacteria, stalked protozoa). The dominant bacterial genera identified on GAC particles or in interstitial water were *Pseudomonas, Alcaligenes, Aeromonas, Acinetobacter, Arthrobacter, Flavobacterium, Chromobacterium, Bacillus, Corynebacterium, Micrococcus, Paracoccus,* and *Moraxella* (Camper et al., 1986; Wilcox et al., 1983).

Activated carbon has the ability to remove organic materials that, in turn, serve as carbon substrates that support bacterial growth in the filter matrix (De Laat et al., 1985;

Servais et al., 1991). The magnitude of biological assimilation depends on various factors, including temperature, running time and intensity, and frequency of filter backwashing (van der Kooij, 1983). Removal of organic compounds appears to result from microbial proliferation in the activated carbon column. Some of these bacteria may, however, produce endotoxins, which may enter the treated water. Treatment to enhance bacterial growth on GAC produces biological activated carbon (BAC), which has been recommended for increasing the removal of organics, as well as extending the lifetime of GAC columns. Bacterial growth in activated carbon columns can be enhanced by ozone, which makes organic matter more biodegradable. Micropollutants, such as phenol, can be degraded by biofilms grown on ozonated natural organic matter (DeWaters and DiGiano, 1990). The bacterial community in BAC is made mostly of aerobic, gram-negative, oxidase-negative, catalase-positive, motile rods. Other bacterial genera are *Pseudomonas*, *Acinetobacter*, *Enterobacter*, and *Moraxella*-like bacteria (Rollinger and Dott, 1987).

Pathogenic bacteria may be successful in colonizing mature GAC filters (Grabow and Kfir, 1990; LeChevallier and McFeters, 1985a). However, pathogenic and indicator bacteria can be inhibited by biofilm microorganisms on activated carbon. This phenomenon is attributable to nutritional competitive inhibition or to the production of bacteriocin-like substances by microbial community in the filter (Camper et al., 1985; LeChevallier and McFeters, 1985a; Rollinger and Dott, 1987).

Problems arise when bacteria or bacterial microcolonies attached to carbon filters are sloughed off the filter or when bacteria-coated carbon particles penetrate the distribution system (Camper et al., 1986). Some of these carbon fines can become associated with biofilms and may serve as inoculants that induce the regrowth of potential pathogens in distribution systems (Morin et al., 1996). An examination of 201 samples showed that heterotrophic bacteria and coliform bacteria are associated with the carbon particles. The attached bacteria display increased resistance to chlorination (Camper et al., 1986; LeChevallier et al., 1984a; Stewart et al., 1990). Figure 15.11 shows that *E. coli* and naturally occurring heterotrophic bacteria attached to carbon particles are more resistant to chlorine than are unattached bacteria (LeChevallier et al., 1984a). Monochloramine appears to control biofilm bacteria more efficiently than does free chlorine or chlorine dioxide (LeChevallier et al., 1988).

Operational variables contributing to the release of particles from activated carbon beds include filter backwashing and increase in bed depth, applied water turbidity, and filtration rate (Camper et al., 1987). Thus, activated carbon particles entering the water distribution system can carry potential pathogens that are resistant to chlorination as well as nutrients that support microbial growth within the distribution system.

Viruses are adsorbed to activated carbon by electrostatic forces and the interaction is controlled by pH, ionic strength, and organic matter content of the water. The competition of organics with viruses for the attachment sites on the carbon surface makes this material an unreliable sorbent for removing viruses from water (Bitton, 1980a).

The following are some possible disadvantages of carbon filters:

1. Production and release of endotoxins;
2. Creation of anaerobic conditions inside filters with subsequent production of odorous compounds (e.g., H_2S);
3. Production of effluents with high colony counts;
4. Occasional growth of zooplankton in carbon filters and their release in the filter effluents.

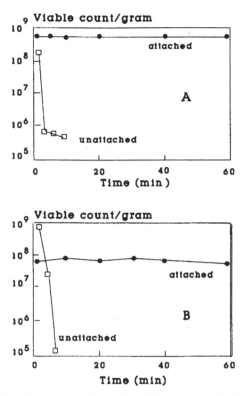

Figure 15.11 Increased resistance to chlorination of bacteria attached to GAC: (a) *Escherichia coli*; (b) heterotrophic plate count. Adapted from LeChevallier et al. (1984).

15.4.7 Membrane Systems

Membranes can be used to treat water to remove a wide range of contaminants. Membrane processes include reverse osmosis (removal of cations, anions, metals, organics, and microorganisms), nanofiltration (mostly removing calcium and magnesium), ultrafiltration (pore size below 0.1 μm; removes colloidal particles, organics, and microorganisms), and microfiltration (Pizzi, 2002).

Microfiltration, using microporous filters, is used in water and wastewater treatment or as a post treatment to remove suspended solids, algae, bacterial pathogens, as well as cysts and oocysts of protozoan parasites. Marketed membranes are made of cellulose acetate, polypropylene, polyvinylidene fluoride, polyethersulfone, and other proprietary materials, and have various nominal pore sizes ranging from 0.01 to 0.2 μm. Membrane challenge pilot studies showed that membranes can remove from 4.9 to 5.8–$\log_{10}$ units for *Giardia* cysts, and from 5.8 to 6.8–$\log_{10}$ units for *Cryptosporidium* oocysts (States et al., 2000). The removal mechanism is probably physical straining (Jacongelo et al., 1995). Microfiltration was able to achieve a 4–log removal of *B. subtilis* spores from water and 3.4–log removal of coliphage from a wastewater effluent (Farahbakhsh and Smith, 2004; Huertas et al., 2003).

However, membranes are subject to fouling by colloidal particles and natural organic matter (NOM), leading to the formation of a cake on the membrane surface. The fouling was shown to enhance coliphage removal by microfiltration (Farahbakhsh and Smith, 2004). The NOM fraction contributing to fouling comprises small, neutral, hydrophilic compounds (Carroll et al., 2000). Coagulation prior to microfiltration reduces membrane fouling.

15.4.8 Biological Treatment of Water

Water containing organic compounds as well as ammonia nitrogen is biologically instable. The presence of organic and inorganic electron donors in water causes significant problems such as trihalomethane formation following disinfection with chlorine, taste and odor problems, regrowth of bacteria in distribution systems, and reduced bed life of GAC columns (Hozalski et al., 1992). Several European countries (e.g., France, Germany, Netherlands) and Japan include biological processes in the treatment train in an effort to remove nutrient levels in the treated water and thus obtain biologically stable water, which limits the growth of microorganisms in distribution pipes and reservoirs (van der Kooij, 1995). To prevent or reduce biofilm formation in water distribution systems, it was recommended that the assimilable organic carbon (AOC; see Chapter 16) concentration of drinking water should be <10 µg of acetate-C equivalents/L (van der Kooij, 1990; 1992). However, certain bacteria such as *Aeromonas* grow in biofilms even at AOC concentrations below 10 µg of acetate-C equivalents/L. Hence, a biofilm formation rate (BFR) measurement was proposed to better assess the biostability of drinking water in distribution systems. A BFR value below 10 pg $ATP/cm^2 \cdot day$ indicates biostable water (van der Kooij, 2002). Furthermore, biostability can also be affected by the materials in contact with drinking water. This led to the determination of the biofilm formation potential, which varies from less than 10 to 3000 pg ATP/cm^2, depending on the material being used (van der Kooij and Veenendaal, 1994).

Biological treatment (e.g., biofiltration involving, for example, biologically active GAC filters, anthracite, or sand) is based on aerobic biofilm processes that offer several advantages over physicochemical processes in regard to drinking water treatment. Granular activated carbon provides a large surface area for the accumulation of microorganisms as a biofilm. Three important factors affecting biodegradable organic matter removal by biofilters are the presence of chlorine in the backwash water, media type, and temperature (Liu and Slawson, 2001).

The advantages provided by biofiltration are the following (Andersson et al., 2001; Bouwer and Crowe, 1992; Hozalski and Bouwer, 1998; Hozalski et al., 1992; LeChevallier et al., 1992; Manem and Rittmann, 1992; Nerenberg et al., 2000; Rittmann, 1987; 1989; 1995a; Rittmann and Snoeyink, 1984):

1. Biotreatment removes organic compounds (total and assimilable organic carbon), thus reducing bacterial growth in water distribution systems and producing biostable water.

2. Taste and odor compounds can be removed (e.g., geosmin and 2-methyl isoborneol). Ozonation followed by biofiltration provide an effective treatment for these compounds.

3. Biotreatment, using GAC filtration, can also remove ammonia from water and reduces the formation of chlorine demand (theoretical consumption of 7.6 mg of Cl_2 per mg of $N-NH_4$).

4. Trihalomethane precursors and other disinfection byproducts may be removed. A comparison of chemical and biological treatments shows that the latter produces drinking water with lower mutagenic activity.

5. Iron and manganese may be removed.

6. Xenobiotics (e.g., petroleum hydrocarbons, halogenated hydrocarbons, pesticides, chlorinated phenols, and benzenes) may be biodegraded and removed.

Backwash of biological filters with water partially removes the attached biomass but maintains the ability of the filter to remove biodegradable organic carbon. Preozonation enhances the biodegradability of organic compounds by biofilms by oxidizing large-molecular-weight compounds to smaller ones that are more easily biodegraded. Thus, ozonation increases the AOC in plant effluents and distribution systems (Escobar and Randall, 2001). Ozonation followed by biofiltration should be able to remove AOC and thus reduce bacterial regrowth in distribution systems.

15.4.9 Disinfection

Disinfection is the last barrier against the entry of pathogens and parasites in our drinking water. Disinfection addresses specifically the *inactivation* of disease-causing organisms. For a detailed discussion of this topic, the reader is referred to Chapter 6.

15.5 WEB RESOURCES

http://www.awwa.org (American Water Works Association)
http://www.epa.gov/ogwdw000/sdwa/sdwa.html (Safe Drinking Water Act)
http://www.epa.gov/ogwdw000/protect.html (source water protection)
http://www.epa.gov/fedrgstr/ (Federal Register Environmental documents)
http://www.epa.gov/OGWDW/ (U.S. EPA groundwater and drinking water)
http://www.epa.gov/ogwdw000/dwhealth.html (drinking water and health)
http://www.cee.vt.edu/program_areas/environmental/teach/wtprimer/rapid/rapid.html (rapid sand filtration from Virginia Tech)
http://www.cee.vt.edu/program_areas/environmental/teach/wtprimer/slowsand/slow sand.html (slow sand filtration page from Virginia Tech)
http://www.cee.vt.edu/program_areas/environmental/teach/wtprimer/wtprimer.html (water treatment primer from Virginia Tech)
http://www.nesc.wvu.edu/ndwc/pdf/OT/TB/TB14_slowsand.pdf (a National Drinking Water Clearinghouse fact sheet)
http://www.swbic.org/education/env-engr/index.html (online course on water and wastewater treatment)

15.6 QUESTIONS AND PROBLEMS

1. What are the pathogens and parasites of most concern in drinking water?
2. Can we rely on ion-exchange resins to remove/inactivate pathogens?
3. What are endotoxins? Are they a problem in drinking water?
4. Why does water softening inactivate pathogens?
5. What is the purpose of biological treatment of drinking water? What are the advantages of the practise?
6. What is *schmutzdecke*?
7. Is clean sand a good adsorbent for viruses?

8. What process most likely removes *Giardia* cysts and *Cryptosporidium* oocysts in a water treatment plant?
9. What is river bank filtration?
10. What could be used as a surrogate for cyst and oocyst removal by sand filtration?
11. How do we remove helminth eggs in water treatment plants?
12. Outside the waterborne route, are there any other routes by which pathogens in drinking water can infect consumers?
13. How do we remove pathogens and parasites in swimming pools?
14. What are GAC and PAC? What are they used for?
15. What is the effect of bacterial colonization of activated carbon on the sensitivity of pathogens to chlorination?
16. Could we rely on activated carbon to remove viruses?
17. Give the different membrane processes and the contaminants they remove
18. What is reverse osmosis? How are contaminants removed by this process?

15.7 FURTHER READING

AWWA "Organisms in Water" Committee. 1987. Committee report: Microbiological considerations for drinking water regulation revisions. J. Amer. Water Works Assoc. 79: 81–88.

Bitton, G., S.R. Farrah, C. Montague, and E.W. Akin. 1986. Global survey of virus isolations from drinking water. Env. Sci. Technol. 20: 216–222.

Geldreich, E.E. 1990. Microbiological quality of source waters for water supply. pp. 3–31, In: *Drinking Water Microbiology*, G.A. McFeters, Ed., Springer-Verlag, New York.

Gray, N.F. 1994. *Drinking Water Quality: Problems and Solutions*. John Wiley & Sons, Chichester, U.K., 315 pp.

LeChevallier, M.W. 2002. Microbial removal by pretreatment, coagulation and ion exchange, pp. 2012–2019, In: *Encyclopedia of Environmental Microbiology*, Gabriel Bitton, editor-in-chief, Wiley-Interscience, N.Y.

LeChevallier, M.W., and G.A. McFeters. 1990. Microbiology of activated carbon, pp. 104–119, In: *Drinking Water Microbiology*, G.A. McFeters, Ed., Springer Verlag, New York.

McFeters, G.A., Ed. 1990. *Drinking Water Microbiology*, Springer-Verlag, New York.

Olson, B.H., and L.A. Nagy. 1984. Microbiology of potable water. Adv. Appl. Microbiol. 30: 73–132.

Pizzi, N.G. 2002. *Water Treatment Operator Handbook*. Amer. Water Assoc., Denver, CO.

U.S. EPA. Drinking Water Health Effects Task Force. 1989. *Health Effects of Drinking Water Treatment Technologies*, Lewis, Chelsea, MI., 146 pp.

16

MICROBIOLOGICAL ASPECTS OF DRINKING WATER DISTRIBUTION

Wastewater Microbiology, Third Edition, by Gabriel Bitton
Copyright © 2005 John Wiley & Sons, Inc.

16.1   INTRODUCTION

Drinking water quality can deteriorate during storage and transport through water distribution pipes before reaching the consumer. In the United States, approximately 18 percent of the outbreaks reported for public water systems are due to contamination of the water distribution system by microbial pathogens and toxic chemicals (Craun, 2001). This deterioration has several causes (Smith, 2002; Sobsey and Olson, 1983):

- improperly built and operated storage reservoirs, which should be covered to prevent airborne contamination and to exclude animals;
- loss of disinfectant residual.

Of the deficiencies reported in water distribution systems, back siphonage and cross-contamination are major causes of contamination (Craun, 2001).

Some of the consequences are:

- excessive growth and colonization of water distribution pipes by bacteria and other organisms, some of which are pathogenic;
- microbial regrowth in storage reservoirs;
- taste and odor problems due to the growth of algae, actinomycetes and fungi;
- corrosion problems.

Drinking water may contain humic and fulvic acids, as well as easily biodegradable natural organics such as carbohydrates, proteins, and lipids. The presence of dissolved organic compounds in finished drinking water is responsible for several problems, among which are taste and odors, enhanced chlorine demand, trihalomethane formation, and bacterial colonization of water distribution lines (Allen et al., 1980; Bourbigot et al., 1984; Hoehn et al., 1980; Rook, 1974; Servais et al., 1991). Water utilities should maintain the distribution system by periodically flushing the lines to remove sediments, excessive bacterial growth, and encrustations due to corrosion.

Biofilms develop at solid–water interfaces and are widespread in natural environments as well as in engineered systems. They are ubiquitous and are commonly found in trickling filters, rotating biological contactors, activated carbon beds, pipe surfaces, groundwater aquifers, aquatic weeds, tooth surfaces (i.e., dental plaque), and medical prostheses such as catheters, pacemakers, and artificial joints (Anwar and Costerton, 1992; Bryers and Characklis, 1981; Characklis, 1988; Hamilton, 1987; Schachter, 2003; Trulear and Characklis, 1982; van der Wende and Characklis, 1990). Some of the concerns over biofilms in various fields are shown in Table 16.1 (Characklis et al., 1982; Trulear and Characklis, 1982).

16.2 BIOFILM DEVELOPMENT IN DISTRIBUTION SYSTEMS

Biofilms are relatively thin layers (up to a few hundred microns thick) of microorganisms that form microbial aggregates and also attach and grow on surfaces. Biofilms formed by prokaryotic microorganisms resemble in some ways tissues formed by eukaryotic cells (Costerton et al., 1995). Observations of living biofilms by scanning confocal laser microscopy have helped in the understanding of biofilm structure. Biofilm microorganisms form microcolonies separated by open water channels (Donlan, 2002; Stoodley et al., 2002). Liquid flow through these channels allow the diffusion of nutrients, oxygen, and antimicrobial agents into the cells (Fig. 16.1; Donlan, 2002). Biofilms also include corrosion byproducts, organic detritus, and inorganic particles such as silt and clay minerals. They contain heterogeneous assemblages of microorganisms, depending on the chemical composition of the pipe surface, the chemistry of finished water, and oxido-reduction potential in the biofilm. They take days to weeks to develop, depending on nutrient availability and environmental conditions. Biofilm growth proceeds up to a critical thickness (approximately $100–200\ \mu m$) when nutrient diffusion across the biofilm becomes limiting. The decreased diffusion of oxygen is conducive to the development of facultative and anaerobic microorganisms in the deeper layers of the biofilm.

A number of processes contribute to biofilm development on surfaces exposed to water flow. The processes involved are the following (Bitton and Marshall, 1980; Gomez-Suarez et al., 2002; Marshall, 1976; Olson et al., 1991; Trulear and Characklis, 1982). The sequence of events is described in Figure 16.2 (Gomez-Suarez et al., 2002).

16.2.1 Surface Conditioning

Surface conditioning is the first step in biofilm formation. Minutes to hours after exposure of a substratum to water flow, a surface conditioning layer, made of proteins, glycoproteins, humic-like substances, and other dissolved or colloidal organic matter initially adsorb to the surface. This results in a modified substratum that is different from the original one (Schneider and Leis, 2002). The conditioning film is a source of nutrients for bacteria, particularly in oligotrophic environments such as drinking water or groundwater.

16.2.2 Transport of Microorganisms to Conditioned Surfaces

Diffusion, convection, and turbulent eddy transport are involved in turbulent flow regime. Chemotaxis may also enhance the rate of bacterial adsorption to surfaces under more quiescent flow.

TABLE 16.1. Some Concerns About Accumulation of Biofilms[a]

Effects	Specific Process	Concerns
Heat transfer reduction	Biofilm formation on condenser tubes and cooling tower fill material. Energy losses.	Power industry Chemical process industry U.S. Navy Solar energy systems
Increase in fluid frictional resistance	Biofilm formation in water and waste water conduits as well as condenser and heat exchange tubes. Causes increased power consumption for pumped systems or reduced capacity in gravity sytems. Energy losses.	Municiple utilities Power industry Chemical process industry
	Biofilm formation on ship hulls causing increased fuel consumption. Energy losses.	U.S. Navy Shipping industry
Mass transfer and chemical transformations	Accelerated corrosion caused by processes in the lower layers of the biofilm. Results in material deterioration in metal condenser tubes, wastewater conduits, and cooling tower fill.	Power industry U.S. Navy Municiple utilities Chemical process industry
	Biofilm formation on remote sensors, submarine periscopes, sight glasses, and so on, causing reduced effectiveness.	U.S. Navy Water quality data collection
	Detachment of microorganisms from biofilms in cooling towers. Releases pathogenic organisms (e.g., *Legionella* in aerosols).	Public health
	Biofilm formation and detachment in drinking water distribution systems. Changes water quality in distribution systems.	Municiple utilities Public health
	Biofilm formation on teeth. Causes dental plaque and caries.	Dental health
	Attachment of microbial cells to animal tissue. Causes disease of lungs, intestinal tract, and urinary tract.	Human health
	Extraction and oxidation of organic and inorganic compounds from water and wastewater (e.g., rotating biological contactors, biologically aided carbon absorption, and benthal stream activity). Reduced pollutant load.	Wastewater treatment Water treatment Stream analysis
	Biofilm formation in industrial production processes reduces product quality.	Pulp and paper industry
	Immobilized organisms or community of organisms for conducting specific chemical transformation.	Chemical process industry

[a]Adapted from Trulear and Characklis (1982).

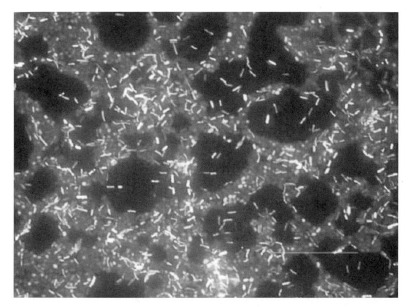

Figure 16.1 Polymicrobic biofilm grown on a stainless steel surface in a laboratory potable water biofilm reactor for 14 days, then stained with 4,6-diamidino-2-phenylindole (DAPI) and examined by epifluorescence microscopy (bar, 20 μm). From Donlan (2002).

16.2.3 Adhesion of Microorganisms to Surfaces

According to the thermodynamic theory, adhesion of a microorganism to a substratum is favored when the free energy of adhesion is negative ($\Delta G_{adh} < 0$) (Gomez-Suarez et al., 2002). Furthermore, according to the DLVO theory (Derjaguin, Landau, Verwey and Overbeek theory), adhesion is a balance between Lifshitz–Van der Waals forces and repulsive or attractive electrostatic forces due to electrical surface charges on both microbial and substratum surfaces. Hydrophobic interactions are also involved in microbial adhesion to surfaces.

16.2.4 Cell Anchoring to Surfaces

Following adhesion to a given surfaces, microbial cells are anchored to the surface, using extracellular polymeric substances (EPS). These EPS, made of polysaccharides such as mannans, glucans, and uronic acids, proteins, nucleic acids, and lipids help support biofilm structure. They display hydrophilic and hydrophobic properties. They help in the adhesion of microorganisms to surfaces and their cohesion within biofilms, and play a structural role in biofilms (Characklis and Cooksey, 1983; Costerton and Geesey, 1979; Flemming and Wingender, 2002). Extracellular polymeric substances also help protect microorganisms from protozoan predation, chemical insult, and antimicrobial agents.

Other attachment organelles include flagella, pili (fimbriae), stalks, or holdfasts (e.g., *Caulobacter*) (Bitton and Marshall, 1980; Olson et al., 1991). Bacterial growth on stainless steel and PVC surfaces, as measured via epifluorescence microscopy, was shown to proceed with a doubling time of 11 days to 4 months. Afterward, growth slowed down and the doubling time was 47 days. The mean number of bacteria on the surfaces was 4.9×10^6 cells/cm^2 (Pedersen, 1990).

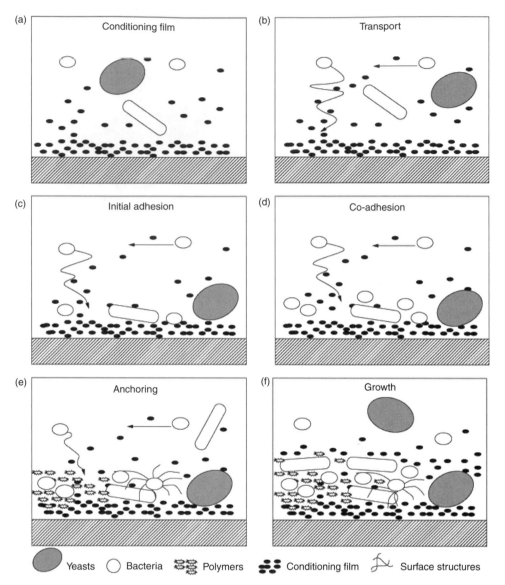

Figure 16.2   Sequence of events in biofilm formation. From Gomez-Suarez et al. (2002).

16.2.5   Cell Growth and Biofilm Accumulation

Biofilms essentially consist of microbial aggregates embedded in extracellular polymeric substances (EPS) attached to a solid surface (Rittmann, 1995b). The study of single-species biofilms has helped in the understanding of the steps involved in biofilm accumulation, which depends on several factors.

1. *Concentration of assimilable organic carbon* (AOC).   The presence of even microgram levels of organic matter in distribution lines allows the growth and accumulation of biofilm microorganisms (Block et al., 1993; Ollos et al., 2003; Servais, 1996).

2. *Limiting nutrients.* Microbial growth in drinking water can also be highly regulated by the availability of phosphorus, which can act as a limiting nutrient for microbial growth in drinking water (Charnock and Kjønnø, 2000; Keinänen et al., 2002; Miettinen et al., 1997; Sathasivan and Ohgaki, 1999). In most Finnish drinking waters, microbial growth was well correlated with the concentration of microbially available phosphorus (Lehtola et al., 2002).

3. *Disinfectant concentration.* Biofilm thickness and activity are inhibited by disinfection with free chlorine. For example, the biofilm thickness ($\sim 10^3$ pg ATP/cm^2) in the presence of a chlorine concentration <0.05 mg/L was two orders of magnitude greater than the thickness (~ 10 pg ATP/cm^2) in the presence of a chlorine level of 0.30 mg/L (Hallam et al., 2001). Attached microorganisms are better controlled by monochloramine than by free chlorine (see also Chapter 6). In a bench-scale distribution system, it was found that a free chlorine residual of 0.5 mg/L or chloramine at 2 mg/L reduced biofilm microorganisms by several orders of magnitude (Ollos et al., 2003). Bismuth thiols present another approach for controlling biofilms. Although their action is slow, their efficiency increases with temperature. They inhibit polysaccharide production, therefore slowing down biofilm formation (Codony et al., 2003).

4. *Type of pipe material.* Biofilm thickness also depends on the type of pipe material. Plastic-based materials (polyethylene or polyvinyl chloride) support less attached biomass than iron (e.g., gray iron) or cement-based materials (e.g., asbestos-cement, cemented cast iron) (Fig. 16.3; Niquette et al., 2000). Corrosion and surface roughness both contribute to increased attached biomass. Material type can also influence the community composition in the biofilms (Kalmbach et al., 2000).

5. *Other factors.* A number of other factors (pH, redox potential, TOC, temperature, hardness) control the growth of microorganisms on pipe surfaces (Olson and Nagy, 1984).

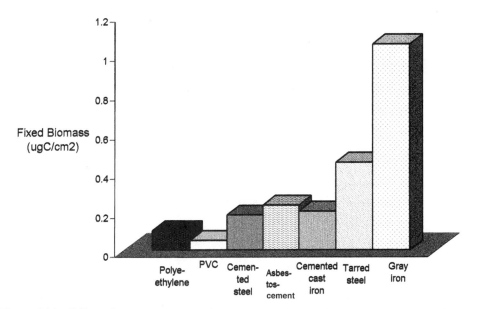

Figure 16.3 Effect of pipe materials on fixed bacterial biomass in a drinking water distribution system. Adapted from Niquette et al. (2000).

6. Flow velocity or shear.
7. Temperature.

16.2.6 Biofilm Ecology

Biofilms display a high ecological diversity. The proximity of microbial cells in biofilms offers opportunities for communication and exchange of genetic material between microorganisms (Wuertz, 2002). The three major means of horizontal transfer of genes are transformation, conjugation, and transduction (see Chapter 1 for more details). The use of modern techniques such as confocal laser scanning microscopy and fluorescently labeled (e.g., tagging with green fluorescent protein) plasmids has facilitated the study of gene transfer in biofilms and in the environment in general. These methods have generally shown that gene transfer rates are much higher than those determined by plating on selective media.

Much work is presently being done on the molecular mechanisms of biofilm development. For their survival and interactions in biofilms, bacteria are able to communicate with cells of the same species or with other species. *Quorum-sensing* diffusible signal molecules (e.g., N-acylhomoserine lactone), called autoinducers, play an important role in density-dependent gene expression and biofilm differentiation. Bacteria undergo changes when the signal molecule has reached a threshold concentration and when the bacterial population has reached a certain density. N-acylhomoserine lactone activity was demonstrated in single-species biofilms (e.g., *Pseudomonas aeruginosa*) and in natural occurring biofilms (McLean et al., 1997). Autoinducer-2 is another cell-to-cell signal molecule that may function in cell communication in some bacteria (Winzer et al., 2002). In addition to biofilm formation, quorum sensing in both gram-negative and gram-positive bacteria is also known to be involved in the regulation of several activities, such as virulence, motility, sporulation, production of secondary metabolites such as antibiotics, conjugation, symbiosis, and individual survival strategies (Miller and Bassler, 2001; Withers et al., 2001).

16.2.7 Biofilm Detachment from Surfaces

Biofilm detachment occurs following *erosion* (detachment of small pieces of biofilm), *sloughing* (detachment of larger pieces of biofilm, sometimes extending all the way to the substratum), and *scouring* following abrasion or scraping (Rittmann and Laspidou, 2002).

Biological processes also involved in biofilm detachment include, among other factors, a decrease in EPS production and predation on biofilm microorganisms. Detachment events are genetically regulated (Stoodley et al., 2002).

The net accumulation of biofilms on surfaces is described by the following equation (Rittmann and Laspidou, 2002):

$$X_f L_f = \frac{YJ}{b + b_{det}} \tag{16.1}$$

where X_f = biomass density (mg biomass/cm^3); L_f = biofilm thickness (cm); Y = true yield of biomass (mg biomass/mg substrate); J = substrate flux (mg substrate/cm^2-day); b = specific decay rate of biofilm microorganisms (d^{-1}); and b_{det} = specific detachment rate (d^{-1}).

Biofilm accumulation increases when the substrate is increased or when detachment is decreased. Under steady-state conditions, the average specific growth rate (μ_{ave}) is equal to the specific detachment rate (b_{det}). Slow-growing microorganisms (e.g., nitrifiers, methanogens) are protected from detachment because they live in the deeper layers of the biofilm.

There are several factors affecting biofilm detachment (Pedersen, 1990; Rittmann and Laspidou, 2002). Detachment is:

- increased by tangential flow shear stress, which is related to the detachment rate (at least for smooth surfaces);
- increased by axial force on the biofilm, due to abrasion and to pressure changes caused by turbulent changes and expressed by the Reynolds number Re;
- decreased by surface roughness;
- affected by physiological parameters such as EPS content (higher detachment at low EPS), or biofilm density (lower detachment for dense biofilms).

16.2.8 Methodology for Biofilm Study

Several direct and indirect methods are available for the characterization of biofilms. Some of these methods, including molecular techniques, are summarized in Table 16.2 (Characklis et al., 1982; Lazarova and Manem, 1995; Neu and Lawrence, 2002; Schmidt et al., 2004).

Total cell count in biofilms is difficult due to the presence of aggregated cells. This task is made easier by using modern techniques such as confocal scanning laser microscopy, which gives information about the biofilm structure. The nondestructive confocal laser-scanning microscopy, in conjunction with 16S rRNA-targeted oligonucleotide probes, helps to show the heterogeneous structure of biofilms and their complex biodiversity by giving information on the distribution and activity of specific groups of microorganisms in biofilms (Costerton et al., 1995; Schramm et al., 1996). rRNA-targeted, fluorescently labeled oligonucleotide probes are useful in identifying biofilm microorganisms and community composition. For example, the predominant bacteria isolated from four water distribution systems in Germany and Sweden belonged to the β proteobacteria, with *Aquabacterium* accounting for 19–53 percent of the bacteria (Kalmbach et al., 2000).

Microelectrode probes give information on the physicochemical properties (pH, temperature, O_2, NH_4, NO_3, NO_2, H_2S, CH_4) of the biofilm microenvironment (Burlage, 1997). These microelectrodes have helped shed light on the distribution of nitrifiers and sulfate reducers along the depth of biofilms.

Microbial activity within biofilms can also be assessed by using dyes that measure membrane integrity, enzyme activity (e.g., esterase activity), or the reduction of tetrazolium salts by biofilm microorganisms (see Chapter 2 for more details).

16.3 SOME ADVANTAGES AND DISADVANTAGES OF BIOFILMS IN DRINKING WATER TREATMENT AND DISTRIBUTION

The development of biofilms on surfaces can be beneficial or detrimental to processes in water and wastewater treatment plants. Trickling filters and rotating biological contactors are examples of processes that rely on microbial activity in biofilms in wastewater treatment plants.

TABLE 16.2.   Methodology for the Characterization of Biofilms[a]

Type	Analytical Method
Direct measurement of biofilm quantity	Biofilm thickness (using microscopy, image analysis, thermal resistance)
	Biofilm mass
	Total cell count (staining with acridine orange (AO) or DAPI
Indirect measurement of biofilm quantity	Extracellular polymeric substances (EPS)
	Specific biofilm constituents (polysaccharides, proteins, nucleic acids)
	Total organic carbon (TOC)
	Total proteins
	Peptidoglycan
	Lipid biomarkers
Microbial activity within biofilms	Viable cell count (plate counts quantity: Live/Dead BacLight™)
	Active bacteria (direct viable count)
	ATP
	Lipopolysaccharides
	Substrate removal rate
	Intracellular and extracellular enzymes
	Dehydrogenase activity using INT or CTC
	Oxygen uptake rate (OUR)
	DNA
	rRNA
	mRNA
Microbial identification	Immunofluorescence (monoclonal and polyclonal antibodies)
	FISH (fluorescent in situ hybridization)
	mRNA amplified by polymerase chain reaction
	Green fluorescent protein (GFP)
	Commercial fluors (e.g., FUN-1 for fungi)
Microenvironment	Fluor conjugates for determining diffusion and permeability in biofilms
pH	Microelectrodes for determining pH in biofilms
Indirect measurement of biofilm quantity: effects of biofilm on transport properties	Frictional resistance
	Heat transfer resistance

Some advantages offered by biofilms are (Rittmann, 1995b):

- The fact that any time a solid surface is exposed to water, biofilm formation follows demonstrating well the benefits of biofilms to microorganisms.

- Biofilms are useful in natural systems for the self-purification of surface waters and groundwater, and in engineered fixed-film processes (e.g., trickling filters, rotating biological contactors) for the treatment of water and wastewater. They are desirable whenever microorganisms must be retained (i.e., need for a high mean cell residence time) in a system with a short hydraulic retention time, as is the case for biological treatment of drinking water, which involves the processing of large volumes of water with very low concentrations of biodegradable compounds.

- Many of the processes used in drinking water treatment allow biofilm formation and include, among others, granular activated carbon, slow and rapid sand filters, and fluidized beds.

However, biofilms can cause problems in water treatment and distribution systems (Rittmann, 1995b; Smith, 2002; van der Wende and Characklis, 1990):

- Because of mass-transport resistance, bacteria are exposed to lower concentrations of substrates than in the bulk liquid.
- Biofilm accumulation increases fluid frictional resistance in distribution pipelines (Bryers and Characklis, 1981) (Fig. 16.4), leading to an increase in pressure drop or to reduced water flow if the pressure drop is held constant. The community structure of the biofilm also affects frictional resistance. A predominantly filamentous biofilm appears to increase frictional resistance (Trulear and Characklis, 1982).
- Anaerobic conditions lead to the production of H_2S, a toxic gas characterized by a rotten egg odor.
- Biofilms have an impact on public health (e.g., accumulation of pathogens and parasites in pipe biofilms, which are habitats for opportunistic pathogens such as *Legionella* and *Mycobacterium avium*, infections related to implants, dental plaques). Biofilms developing on implanted medical devices harbor microbial pathogens such as *Candida albicans*, *Staphylococcus aureus*, *Pseudomonas aeruginosa*, *Kelbsiella pneumoniae*, and *Enterococcus* spp. Most of the circulatory and urinary tract infections are due to biofouled implanted devices. These infections increase patient mortality and the cost of medical care. Biofilms associated with medical implants are often resistant to antibacterial and antifungal drugs (Anwar and Costerton, 1992; Donlan, 2002). Great efforts are being made to prevent the formation of biofilms on medical devices and to develop drugs to treat existing biofilms. One original control method is to block quorum-sensing pathways in order to prevent the formation of biofilms on medical devices. A wide range of chemicals are being screened by biotechnological companies to find a way to control biofilms. Some of

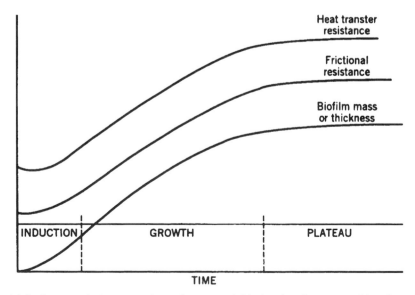

Figure 16.4   Increase in heat transfer resistance and frictional resistance resulting from biofilm growth. From Bryers and Characklis (1981). (Courtesy of Pergamon Press.)

these chemicals are the halogenated furanones produced by marine algae (Schachter, 2003).

- Corrosion of pipes is associated with biofilm growth.
- Aesthetic problems include complaints about red and black waters, which result from the activity of iron- and manganese-oxidizing bacteria (e.g., *Galionella*, *Hyphomicrobium*). Accumulation of biofilms causes taste and odor problems.
- Resistance of biofilm microorganisms to disinfection by chlorine increases, contributing to the regrowth of indicator and pathogenic bacteria in distribution systems. It has been suggested that this increased resistance may be due to diffusional resistance of the biofilm matrix as measured with a chlorine microelectrode (de Beer et al., 1994), protective effect by extracellular polymeric materials, selection of biofilm microorganisms with enhanced resistance to the disinfectant, or to attachment of bacteria to biological (e.g., macroinvertebrates and algae) and nonbiological surfaces (particles, activated carbon). As shown for activated carbon, biofilm bacteria appear to be more resistant to residual chlorine than suspended bacteria (LeChevallier et al., 1988). Particulates, including particulate organic matter, protect pathogenic microorganisms from disinfectant action (Hoff, 1978). It was found that *Enterobacter cloacae*, in the presence of particulates from water distribution systems, is protected from chlorine action as a result of its attachment to the particles (Herson et al., 1987). Chloramines appear to be more efficient in biofilm control than free chlorine (HOCl or OCl$^-$). This may be due to the lower affinity of chloramines for bacterial polysaccharides (LeChevallier et al., 1990; van der Wende and Characklis, 1990). It has been proposed that, for biofilm control, free chlorine should be used as a primary disinfectant but the residual should be converted to chloramine (LeChevallier et al., 1990). The increased resistance of biofilm microorganisms to chlorine applies to other antibacterial agents such as antibiotics. The free chlorine residual decreases as the water flows through the distribution system.
- Influence of the pipe material: Laboratory and field studies generally show that iron pipes display the highest biofilm growth and heterotrophic plate counts in water, as compared to cement, epoxy, and PVC pipes. This effect is even more pronounced when high TOC water flows through cast iron pipes (Camper et al., 2003). The pipe material has a strong influence on chlorine decay in distribution systems. Some pipes are quite reactive (e.g., unlined iron) while others are unreactive towards chlorine (e.g., PVC, cement-lined iron) (Hallam et al., 2002). This, as well as the influence of biofilms, lead to an increase in heterotrophic plate count (HPC) in the distribution system.

The drinking water industry has three major options for controlling biofilm growth in water distribution systems (Camper et al., 2003):

- Maintain a proper level of disinfectant. The general strategy is to convert free chlorine to chloramines.
- Reduce the level of organic carbon in finished water. This can be accomplished by using membrane filtration, enhanced coagulation, or biological filtration (see Chapter 15).
- Consider the effect of the pipe material. Replace deteriorating iron pipes with other materials (e.g., PVC pipes).

The choice of the proper option depends on cost consideration and conditions prevailing at the plant.

16.4 GROWTH OF PATHOGENS AND OTHER MICROORGANISMS IN WATER DISTRIBUTION SYSTEMS

16.4.1 Public Health Implications of Pathogen Growth in Water Distribution Systems

Pathogen accumulation in biofilms of water distribution systems may contribute to the spread of waterborne diseases. Bacteria grow in water distribution systems and colonize pipe connections, tubercles, and dead ends. Early on, the presence of a wide range of microorganisms (eubacteria, filamentous bacteria, actinomycetes, diatoms) was demonstrated in distribution pipes by scanning electron microscopy (SEM). Some of the bacteria were seen attached to the surfaces by means of extracellular fribrillar materials (Ridgway and Olson, 1981). Turbercles provide a high surface area for microbial growth and protect microorganisms from the lethal action of disinfectants. Turberculated cast iron pipe sections from the water distribution system in Columbus, Ohio, metropolitan area harbored high numbers of aerobic and anaerobic (e.g., sulfate-reducing bacteria) bacteria. Some samples contain up to 3.1×10^7 bacteria/g of tubercle material (Allen et al., 1980; Geldreich, 1980; Tuovinen and Hsu, 1982) (Fig. 16.5; Liu et al., 2002b). Several investigators have also reported the proliferation of iron and manganese bacteria that grow attached to pipe surfaces and, subsequently, cause a deterioration of water quality (e.g., pipe clogging and color problems). Attached iron bacteria such as *Gallionella* have stalks that are partially covered with iron hydroxide. This was confirmed by X-ray energy-dispersive microanalysis (Ridgway et al., 1981). Several materials used in reservoirs and water distribution systems were found to support bacterial growth (e.g., *Pseudomonas, Aeromonas, Mycobacterium, Legionella, Flavobacterium, Acinetobacter, Klebsiella*). In Germany and France, mycobacterial species were found in 90 percent of biofilm samples taken from water treatment plants (Schulze-Robbecke et al., 1992).

Figure 16.5 Pipe tubercles biofilm as observed with SEM. Reprinted from Liu et al. (2002b). (With permission from Elsevier.)

As mentioned in a preceding section, biofilm growth depends on the plumbing material. For example, ethylene–propylene and latex surfaces support more bacterial growth than PVC or stainless steel (Table 16.3). The development of abundant biofilm growth is related to the ability of the plumbing material to supply nutrients to bacteria (Rogers et al., 1994). Bacterial growth is limited when the assimilable organic carbon (AOC) ranges between 10 and 15 μg of acetate equivalent per liter (van der Kooij and Hijnen, 1985b). Coliform (e.g., *K. pneumonae, Escherichia coli, E. aerogenes, E. cloacae*) growth has been demonstrated under very low nutrient conditions in water distribution systems. Environmental isolates appear to grow better (i.e., they have higher growth rates, higher yields, and higher affinity for substrates) than clinical isolates and are thus better indicators of the conditions prevailing in distribution lines (Camper et al., 1991). In the United States, the presence of coliform bacteria in distribution systems is regulated by the Total Coliform Rule (TCR) (U.S. EPA, 1989g). Compliance with TCR is based on coliform testing each month. The sampling frequency depends on the size of the population served by the water utility.

Bacterial growth in water distribution systems is important from a public health viewpoint because isolation of pathogenic bacteria (e.g., *Pseudomonas* spp., *Klebsiella pneumonae, Yersinia, Legionella, Mycobacterium* spp., *Campylobacter, Helicobacter*) and the routine occurrence of opportunistic pathogens have been documented in the literature (Keevil, 2002). Pathogenic viruses and parasites can also be trapped within biofilms (Quignon et al., 1997; Rodgers and Keevil, 1995). Standard culture-based methods do not give an indication of the presence of viable but nonculturable (VBNC) pathogens that may remain infectious to consumers. Nonculture-based methods, such a lipid biomarker analysis, can provide information on the biofilm microbial community structure and metabolic state. The detection of oxirane fatty acids following exposure of the biofilm to chlorine was proposed as a biosensor to assess the effectiveness of this disinfectant (Smith et al., 2000).

The heterotrophic plate count (HPC) indicates bacterial levels in water distribution systems and should not exceed 500 organisms/mL. A high HPC is indicative of deterioration of water quality in distribution pipes. The HPC includes gram-negative bacteria belonging to the following genera: *Pseudomonas, Aeromonas, Klebsiella, Flavobacterium, Enterobacter, Citrobacter, Serratia, Acinetobacter, Proteus, Alcaligenes, Enterobacter*, and *Moraxella*. Both *Aeromonas* and *Pseudomonas* have been proposed as indicators of potential bacterial regrowth in water distribution systems due to their ability to grow at very low substrate concentrations (van der Kooij, 1988; Ribas et al., 2000).

TABLE 16.3. Comparison of Different Materials for Their Ability to Support Biofilm Growth[a]

Material	Colonization (mean CFU/cm^2)	
	Total Bacteria	*Legionella pneumophilia*
Glass	1.90×10^5	1.70×10^3
Stainless steel	2.13×10^5	1.03×10^4
Mild steel	1.69×10^6	2.06×10^4
PVC	6.23×10^6	7.75×10^3
Ethylene–propylene	1.8×10^7	1.44×10^5
Latex	5.50×10^7	2.20×10^5

[a]Adapted from Rogers et al. (1994).

Sphingomonas spp. can also be found in water distribution lines. They are often overlooked due to their slower growth than other members of the heterotrophic plate count (Koskinen et al., 2000).

Gram-positive bacteria found are *Bacillus* and *Micrococcus* (Agoustinos et al., 1992). The growth of these bacteria can also be promoted by trace organics in the distribution lines (van der Kooij and Hijnen, 1988). The HPC in chlorinated distribution water serving an Oregon coastal community have been characterized (LeChevallier et al., 1980). Actinomycetes and *Aeromonas* spp. were shown to be the microorganisms most frequently detected in the distribution system. At the Pont-Viau water filtration plant in Canada, the HPC (20 or 35°C) increased as the treated water traveled through the distribution system. *Pseudomonas aeruginosa*, *Aeromonas hydrophila*, and *Clostridium perfringens* were detected in the distribution system (Payment et al., 1989) (Table 16.4). Many of the bacteria isolated in water distribution systems are opportunistic pathogens. For example, most *Aeromonas* isolates from 13 Swedish drinking water distribution systems produced virulence factors such as cytotoxins, suggesting potential public health problems (Kühn et al., 1997). The presence of high numbers of opportunistic pathogens in drinking water is of concern because these microorganisms can lead to infection of certain segments of the population (newborn babies, the sick, and the elderly). They can also cause secondary infections among patients in hospitals (see more details in Chapter 4).

16.4.2 *Legionella* Growth in Water Distribution Systems

Owing to their airborne transmission through showering, much work has been carried out on the survival and growth of Legionellae in potable water distribution systems and plumbing in hospitals and homes (Colbourne et al., 1988; Muraca et al., 1988; Stout et al., 1985). Furthermore, legionellae appear to be more resistant to chlorine than *E. coli* (States et al., 1989), and small numbers may survive in distribution systems that

TABLE 16.4. Bacteria Isolated on R2A or M-Endo Media from Water Distribution System Samples[a]

Colonies Detected on R2A Medium (20 or 35°C)		Colonies from M-Endo Medium	
o	*Acinetobacter* spp.	o	*Aeromonas hydrophila*
	Alcaligenes spp.	o	*Citrobacter freundii*
	Arthrobacter spp.	o	*Enterobacter aerogenes*
o	*Bacillus* spp. (mainly *cereus* and *sphaericus*)	o	*Enterobacter agglomerans*
	Corynebacterium spp.	o	*Enterobacter cloacae*
	Empedobacter spp.	o	*Hafnia alvei*
o	*Flavobacterium* spp.		*Klebsiella oxytoca*
	Flexibacter spp.		*Klebsiella ozaenae*
	Micrococcus spp.	o	*Klebsiella pneumoniae*
o	*Moraxella* spp.	o	*Serratia fonticala*
o	*Pseudomonas* (non-*aeruginosa*)	o	*Serratia liquefaciens*
o	*Serratia marcescens*	p	*Vibrio fluvialis*
	Spirillum spp.		
	Sporosarcina spp.		
	Staphylococcus spp.		
p	*Vibrio fluvialis*		

o, Opportunistic pathogen; p, primary pathogen.
[a]From Payment et al. (1989).

have been judged to be microbiologically safe. These organisms may also grow to detectable levels inside hot water tanks in hospitals and homes and thus pose a health threat (Stout et al., 1985; Witherell et al., 1988). *Legionella* survives well at 50°C (Dennis et al., 1984), and environmental isolates are able to grow and multiply in tap water at 32, 37, and 42°C (Yee and Wadowsky, 1982). The enhanced survival and growth in these systems has also been linked to stagnation (Ciesielski et al., 1988), stimulation by rubber fittings in the plumbing system (Colbourne et al., 1984), and trace concentrations of metals such as Fe, Zn, and K (States et al., 1985; 1989). It has also been found that sediments found in water distribution systems and tanks (i.e., scale and organic particulates) and the natural microflora significantly improved the survival of *Legionella pneumophila*. Sediments indirectly stimulate *Legionella* growth by promoting the growth of commensalistic microorganisms (Stout et al., 1985; Wadowsky and Yee, 1985).

Another condition that is favorable for the multiplication of *Legionella* in hot water tanks is the stagnation of the water within the tank (Ciesielski et al., 1984). *Legionella pneumophila* was detected in higher numbers in tanks that were not in use than in online tanks. Thus, the prevention of stagnation of water in these tanks may help control the multiplication of *Legionella*. Several methods have been considered for controlling *Legionella* in water: superchlorination (chlorine concentration of 2–6 mg/L), chloramination, maintaining the hot water tanks at temperatures above 50°C, ultraviolet irradiation (Antopol and Ellner, 1979; Knudson, 1985; Kool et al., 2000; Muraca et al., 1987; Stout et al., 1986), biocides (Fliermans and Harvey, 1984; Grace et al., 1981; Skaliy et al., 1980; Soracco et al., 1983), and alkaline treatment (States et al., 1987). The control of legionellae in potable water systems may also be achieved via the control of protozoa, which may harbor these pathogens (States et al., 1990).

16.4.3 Growth of Nitrifiers in Water Distribution Systems

Bacterial proliferation may also result from nitrification in water distribution systems. Many water treatment plants around the world now use inorganic chloramines as disinfectants to reduce the formation of disinfection byproducts, which are of health concern (see Chapter 6). However, surveys have shown that a large percentage (about two-thirds) of the plants using chloramination experience nitrification problems. Ammonia oxidation by ammonia-oxidizing bacteria (AOB) results in an increase in the concentration of nitrite, which has a significant chlorine demand and increases the breakdown of chloramine residuals. This results in the proliferation of heterotrophic bacteria in the water distribution system (Wolfe et al., 1988). Indicators of nitrification include decrease in chloramine residual, elevated bacterial count, decrease in dissolved oxygen concentration, and increase in nitrite (NO_2–N = 15–100 μg/L during nitrification episodes) or nitrate. The occurrence of nitrification is influenced by several factors, including increased water temperature, chloramine residual, possible synergistic relationship between AOB and heterotrophic bacteria, as well as operational practises such long detention times leading to the growth of ammonia-oxidizing bacteria (Wolfe and Lieu, 2002).

The main preventive/control measures for nitrification in water distribution systems include the following (Harrington et al., 2002; Pintar and Slawson, 2003; Wolfe and Lieu, 2002):

- Maintenance of an adequate chloramine residual (≥ 2 mg/L).
- Increasing the chlorine-to-ammonia ratio (traditionally, this ratio is 3:1 to optimize monochloramine formation) (Haas, 1999). This is considered by some as a long-term preventive measure.

- Improving the removal of natural organic matter (NOM), which is responsible for the depletion of chloramines with subsequent release of ammonia. This can be accomplished by enhanced coagulation (i.e., coagulation at a pH of approximately 6).
- Periodic flushing of the distribution system to remove biofilms harboring nitrifying bacteria.
- Switching to free chlorine to control biofilms.
- Breakpoint chlorination: this is not a long-term strategy.

Drinking water suppliers may also consider biological denitrification to remove nitrate from drinking water. Other commonly used methods include ion exchange, electrodialysis, reverse osmosis, and membrane bioreactors (Mateju et al., 1992; Nuhoglu et al., 2002).

16.5 DETERMINATION OF ASSIMILABLE ORGANIC CARBON IN DRINKING WATER

Bacterial regrowth in distribution systems is influenced by several factors, among them the level of trace biodegradable organic materials found in finished water, water temperature, nature of pipe surfaces, disinfectant residual concentration, and detention time within the distribution system (Joret et al., 1988). Drinking water harbors bacteria (e.g., *Flavobacterium*, *Pseudomonas*) that can grow at extremely low substrate concentrations (van der Kooij and Hijnen, 1981). *Pseudomonas aeruginosa* (strain P1525) and *P. fluorescens* (strain P17) are able to grow in treated water at relatively low concentrations ($\mu g/L$) of low-molecular-weight organic substrates such as acetate, lactate, succinate, and amino acids. The K_s values for bacteria isolated from drinking water may vary from 2 to 100 $\mu g/L$ in the presence of substrates such as glucose, acetate, oleate, or starch (van der Kooij, 1995). The growth kinetic parameters of some bacteria isolated from drinking water are shown in Table 16.5 (van der Kooij, 2002). A strain of *Klebsiella pneumonae* is

TABLE 16.5. Growth Kinetics of Bacteria Isolated from Drinking Water[a]

Microorganism	Substrate	Temperature ($^\circ$C)	K_s (μg C/L)	V_{max} (h^{-1})
A. hydrophila	Acetate	15	11	0.15
A. hydrophila	Glucose	15	16	0.28
A. hydrophila	Amylase	15	93	0.26
A. hydrophila	Oleate	15	2.1	0.23
Flavobacterium sp.	Glucose	15	3.3–109	0.15–0.21
Flavobacterium sp.	Maltose	15	23.7	0.37
Flavobacterium sp.	Amylose	15	26	0.50
Flavobacterium sp.	Amylopectin	15	11	0.48
Klebsiella pneumoniae	Maltose	15	51	0.49
Klebsiella pneumoniae	Maltopentaose	15	92	0.41
Pseudomonas fluorescens	Acetate	15	4	0.18
Pseudomonas fluorescens	Glucose	15	57	0.22
Spirillum sp.	Oxalate	15	15.2	0.24
Citrobacter freundii	Glucose	15	95	0.17
Enterobacter sp.	Glucose	15	60	0.21
E. coli	Lactose	25	142	0.37

[a]Adapted from van der Kooij (2002).

able to use organic compounds at concentrations of a few micrograms per liter. This bacterium grows on maltose with a yield $Y = 4.1 \times 10^6$ CFU per 1 μg C (van der Kooij and Hijnen, 1988). *Klebsiella pneumonae* is capable of regrowth in distribution systems as well as in drinking water stored in redwood reservoirs (Clark et al., 1982; Geldreich and Rice, 1987; Seidler et al., 1977; Talbot et al., 1979).

Natural organic matter (NOM) is made of two fractions: biodegradable organic matter (BOM) and nonbiodegradable fraction (Fig. 16.6; Volk and LeChevallier, 2001). Bioassays have been developed for BOM to assess the biological stability of drinking water by determining the potential of drinking water to support microbial growth, to exert a chlorine demand, or to lead to the formation of disinfection byproducts. These bioassays include (Huck, 1990; Volk and LeChevallier, 2000): (1) biomass-based methods, the goal of which is to measure the production of bacterial biomass following consumption of assimilable organic carbon (AOC); and (2) dissolved organic carbon (DOC)-based methods, the goal of which is to assess chlorine demand and disinfection byproducts formation by measuring biodegradable dissolved organic carbon (BDOC).

16.5.1 Biomass-Based Methods

These methods are based on the measurement of the growth potential of pure bacterial cultures or indigenous microorganisms following incubation in the water samples for several days. Colony forming units or ATP are determined to assess bacterial numbers or biomass.

A pasteurized or filtered water sample is seeded with pure bacterial cultures of *Pseudomonas fluorescens* (strain P17), *Spirillum* (strain NOX), or *Flavobacterium* (strain S12). *Pseudomonas fluorescens* (strain P17) utilizes a wide range of low-molecular-weight substrates, whereas *Spirillum* (strain NOX) utilizes only carboxylic acids. *Flavobacterium* sp. (strain S12) specializes in the utilization of carbohydrates. The growth of these reference bacteria in a given water sample is compared with that obtained in a

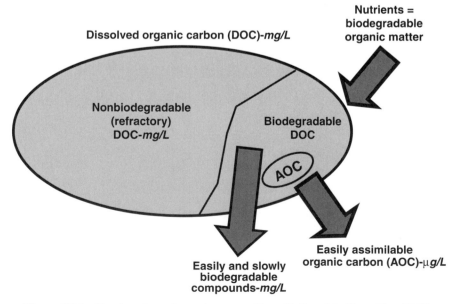

Figure 16.6 Fractionation of organic matter. From Volk and LeChevallier (2000).

sample spiked with substrates such as acetate or oxalate. The assimilable organic carbon (AOC) is calculated using the following formula (van der Kooij, 1990; 2002):

$$AOC \ (\mu g\,C/L) = \frac{N_{max} \times 1000}{Y} \tag{16.2}$$

where N_{max} = maximum colony count (CFU/mL); Y = yield coefficient (CFU/mg of carbon); and AOC concentration is expressed as μg of acetate-C equivalents/L.

When using *P. fluorescens strain* P17, AOC is calculated by using the yield factor Y for acetate ($Y = 4.1 \times 10^6$ CFU/μg C) (van der Kooij and Hijnen, 1985b; van der Kooij et al., 1982a). For *Spirillum* sp., strain NOX, $Y_{acetate} = 1.2 \times 10^7$ CFU/μg C. *Flavobacterium* has a yield of $2–2.3 \times 10^7$ CFU/μg C (van der Kooij and Hijnen, 1984). The concentration of AOC in the sample is expressed as μg acetate C equivalents per liter. This approach necessitates a relatively long incubation period (van der Kooij, 2002; van der Kooij et al., 1982b; van der Kooij and Hijnen, 1984; 1985b). P17 and NOX strains can also be used in combination, sequentially or simultaneously, to measure AOC in environmental samples. Efforts have been made to simplify and standardize this method (Kaplan et al., 1993).

Other bacterial mixtures (e.g., mixture of *P. fluorescens*, *Curtobacterium* sp., *Corynebacterium* sp.) have also been proposed to estimate AOC. Calibration curves help estimate the AOC from the bacterial colony count (Kemmy et al., 1989) (Fig. 16.7). Some investigators have criticized the use of pure bacterial cultures for measuring AOC in environmental samples and prefer the use of natural assemblages of bacteria derived from the sample being examined. Biomass production can also be determined via ATP measurement (Stanfield and Jago, 1987) or turbidimetry (Werner, 1985). Rapid determination of AOC can be achieved by increasing the incubation temperature and

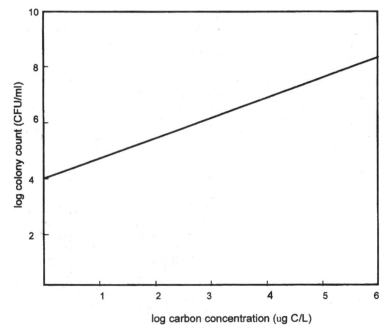

Figure 16.7 Calibration curves for estimation of assimilable organic carbon (AOC) from bacterial colony count. From Kemmy et al. (1989). (Courtesy of Pergamon Press.)

bacterial inoculum density, and enumerating the test organisms via ATP measurement (LeChevallier et al., 1993). Assimilable organic carbon can also be conveniently determined with a bioluminescent biosensor consisting of *P. fluorescens* P17 with the luciferase operon (Noble et al., 1994).

It is estimated that the assimilable organic carbon (AOC) in tap water is 0.1–9 percent of the total organic carbon (TOC) (van der Kooij and Hijnen, 1985; van der Kooij et al., 1982b; LeChevallier et al., 1991c).

The AOC levels in various environmental samples, using the method based on the growth potential of *P. fluorescens* P17, are displayed in Table 16.6 (van der Kooij et al., 1982b). Levels of AOC vary from less than 1 μg C/L to >3 mg/L and account for 0.03–27 percent of the total dissolved organic carbon. The average AOC values of drinking water in distribution systems of five plants in China varied between 177 and 235 μg acetate-C eq/L (Liu et al., 2002b). The AOC values in several types of drinking water in the United States ranged between 18 and 322 μg/L (Kaplan et al., 1992). Another U.S. survey of 95 treatment plants revealed that AOC levels varied from 18 to 214 μg/L (Volk and LeChevallier, 2000). The AOC of drinking water from 20 water treatment plants in the Netherlands varied between 1 and 57 μg acetate-C eq/L. Most of the AOC was utilized by *Spirillum*, indicating that AOC was composed mostly of carboxylic acids. The AOC represented less than 1.7 percent of dissolved organic carbon. A significant correlation was observed between AOC of water and counts of heterotrophic bacteria in distribution systems (Escobar et al., 2001; van der Kooij, 1992). This technique is useful for determining the effect of water treatment processes on AOC and the release of biodegradable materials from pipe surfaces (van der Kooij et al., 1982b). Coagulation, sedimentation, sand filtration, and activated carbon (GAC) treatment generally reduce the AOC concentration whereas ozonation, and to a lesser extent chlorination, increase the AOC levels as a result of production of low-molecular-weight compounds (van der Kooij and Hijnen, 1985; LeChevallier et al., 1992; Servais et al., 1987; 1991). It was recommended that the AOC concentration of drinking water in distribution systems should be <10 μg of acetate-C equivalents per liter to limit regrowth potential of heterotrophic bacteria (van der Kooij, 1990; 1992). Concentration of AOC is also an important parameter to consider in regard to the biological clogging of sand beds, which occurs as a result of infiltration of pretreated surface water in recharge wells. The AOC level should be <10 μg acetate-C equivalent/L to prevent biological clogging (Hijnen and van der Kooij, 1992).

Determination of AOC level, based on the growth potential of *P. fluorescens*, does not appear to be a good indicator of the growth potential of coliforms in water distribution

TABLE 16.6. Concentration of Assimilable Organic Carbon (AOC) in Various Water Samples[a]

Source of Water	Dissolved Organic Carbon (DOC) (mg C/L)	Assimilable Organic Carbon (AOC) (mg C/L)
Biologically treated wastewater	13.5	3.0–4.3
River Lek	6.8	0.062–0.085
River Meuse	4.7	0.118–0.128
Brabantse Diesbosch	4.0	0.08–0.103
Lake Yssel, after open storage	5.6	0.48–0.53
River Lek, after bank filtration	1.6	0.7–1.2
Aerobic groundwater	0.3	<0.15

[a]Adapted from van der Kooij et al. (1982b).

systems; coliform monitoring appears to be more suitable (McFeters and Camper, 1988). In a water distribution system in New Jersey, coliform regrowth was associated with rainfall, water temperature higher than 15°C, and AOC levels higher than 50 μg acetate-C equivalent/L (LeChevallier et al., 1991b). Coliform regrowth is limited at AOC levels lower than 100 μg/L (LeChevallier et al., 1992) (Fig. 16.8).

A *Coliform Growth Response* test, using *Enterobacter cloacae* as the test organism, was developed by the U.S. EPA to measure specifically the growth potential of coliform bacteria in water. Some investigators recommend the use of environmental coliform isolates found in drinking water (e.g., *Enterobacter aerogenes*, *Klebsiella pneumonae*, *E. coli*) to determine AOC in distribution systems (Camper et al., 1991). These isolates appear to be more adapted to oligotrophic conditions than their clinical counterparts (McFeters and Camper, 1988; Rice et al., 1988).

16.5.2 DOC-Based Methods

These assays are based on the measurement of dissolved organic carbon before and after incubation of the sample in the presence of an indigenous bacterial inoculum (e.g., river water or sand filter bacteria). It is argued that indigenous bacterial populations are more suitable than pure cultures for testing the biodegradation of natural organic compounds (Block et al., 1992; Neilson et al., 1985). The biodegradable dissolved organic carbon, *BDOC*, is given by the following formula:

$$BDOC \ (mg/L) = initial \ DOC - final \ DOC \qquad (16.3)$$

The general approach is as follows. A water sample is sterilized by filtration through a 0.2 μm pore size filter, inoculated with indigenous microorganisms and incubated in the dark at 20°C for 10–30 days, until DOC reaches a constant level. The BDOC is the difference between the initial and final DOC values (Servais et al., 1987). This method showed that the BDOC in the Seine River in France was approximately 0.7 mg/L (Servais et al., 1989).

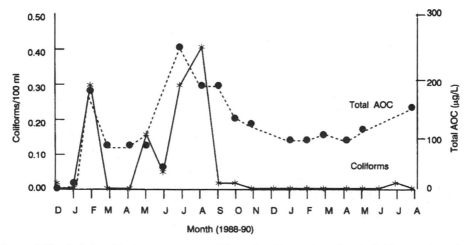

Figure 16.8 Relationship between mean coliform densities and total assimilable organic carbon (AOC) levels (LeChevallier et al., 1992. J. Amer. Water Works Assoc. 84: 136–146). Courtesy of Amer. Water Works Association.

Another approach consists of seeding the water sample (300 mL) with prewashed biologically active sand with mixed populations of attached indigenous bacteria. A European collaborative study has demonstrated that the source of the indigenous bacterial inoculum is not a major source of variance in the BDOC procedure (Block et al., 1992). The BDOC is estimated by monitoring the decrease in dissolved organic carbon (Joret et al., 1988). The experimental set-up for this method is displayed in Figure 16.9 (Joret and Levi, 1986; Joret et al., 1988). The incubation period at 20°C is 3–5 days. One advantage of this method is the use of biofilm microorganisms as inoculum, thus simulating situations occurring in water distribution systems. The monitoring of water treatment plants in suburbs of Paris, France, showed that treated drinking water contains enough BDOC to sustain bacterial growth in distribution systems (Joret et al., 1988). Substrate availability, as measured by this technique, appears to correlate well with the regrowth potential of bacteria (Joret et al., 1990). A survey of treatment plants in the United States showed that BDOC levels varied between 0.03 and 1.03 mg/L [geometric mean of 0.32 mg/L for all sites (Volk and LeChevallier, 2000).

Another approach consists of passing water continuously through one or two glass columns filled with sand or sintered porous glass and conditioned to obtain the development of a biofilm on the supports provided. The BDOC is the difference between the inlet of the first column and the outlet of the second column (Frias et al., 1992; Ribas et al., 1991) (Fig. 16.10). The BDOC is obtained in hours to days, depending of the type of water being assayed. Immobilized cells bioreactors have also been considered for determining BDOC in drinking water. The pre-aerated drinking water is passed through the bioreactor with a hydraulic retention time of 3 h. This rapid procedure does not require a start-up period (Khan et al., 2003).

The AOC fraction represents the easily degradable compounds, whereas the BODC fraction represents a wider range of compounds, some of which are slowly biodegradable. It was reported that AOC values in drinking water represent 15–22 percent of the BDOC concentrations. This may be mainly due to the fact that, in the BDOC method, several drinking water microorganisms are involved in the biodegradation of the wide range of compounds found in drinking water. The BDOC approach also necessitates a longer incubation (up to 4 weeks) than the AOC method (1 week) (van der Kooij, 1995). A weak but significant relationship was found between the AOC and BDOC values from 359 effluent samples from 31 sites in the United States (Volk and LeChevallier, 2000).

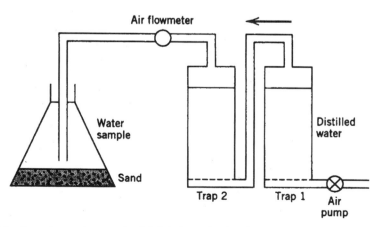

Figure 16.9 Experimental set-up for BDOC determination. From Joret et al. (1988). (Courtesy of the American Water Works Association.)

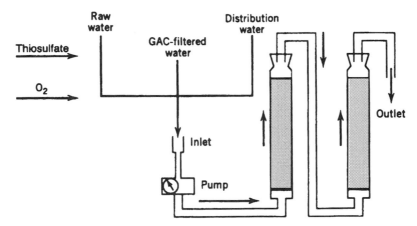

Figure 16.10 Apparatus for BDOC determination. Adapted from Ribas et al. (1991).

16.6 OTHER BIOLOGICAL PROBLEMS ASSOCIATED WITH WATER TREATMENT AND DISTRIBUTION

16.6.1 Taste and Odor Problems

Several surveys have shown that water treatment plants, especially those using surface waters, have taste and odor problems. One such survey in the United States and Canada showed that 43 percent of the 800 plants surveyed had experienced taste and odor problems (Suffet et al., 1996). This problem preoccupies consumers, who often base their perception of safe water on objectionable tastes and odors in the drinking water.

The sources of the taste and odor in water may be anthropogenic or natural.

Anthropogenic sources include phenol, chlorinated phenols (2-chlorophenol, 2,4-dichlorophenol, 2,6-dichlorophenol (Burttschell et al., 1959), hydrocarbons, and halogenated compounds such as chloroform (Zoeteman et al., 1980).

Among *natural* sources, the major taste- and odor-causing compounds are geosmin and 2-methyl isoborneol (MIB), which are products of actinomycete metabolism (e.g., *Streptomyces* and *Nocardia* species), and cyanobacteria (e.g., *Oscillatoria*, *Anabaena*, *Lyngbia*) (Gerber, 1979; Izaguirre et al., 1982; Lalezary et al., 1986; Markovic and Kroeger, 1989; Medsker et al., 1968; Safferman et al., 1967). Geosmin, 2-methyl isoborneol, sesquiterpenes, β-cyclocitral, 3-methyl-1-butanol, and others have been isolated from water supplies and implicated as the cause of earthy-musty odor of drinking water (Hayes et al., 1989; Izaguirre et al., 1982). Geosmin and 2-methyl isoborneol have very low threshold odor concentrations in the nanograms per liter (Cees et al., 1974; Persson, 1979). Other microbial metabolites are 2-isopropyl-3-methoxypyrazine and 2-isobutyl-3 methoxypyrazine (Lalezary et al., 1986) (Fig. 16.11).

Several approaches are used to control taste- and odor-causing compounds in water treatment plants (Namkung and Rittmann, 1987):

1. *Adsorption to solids.* Adsorption to activated carbon is the most popular treatment for removing these compounds. Concentrations as low as 5 mg/L can successfully reduce geosmin and MIB to acceptable levels (Lalezary-Craig et al., 1988). The adsorption of these odorous compounds on powdered activated carbon (PAC) depends on the type of carbon used, the nature and concentration of natural

Molecular structure	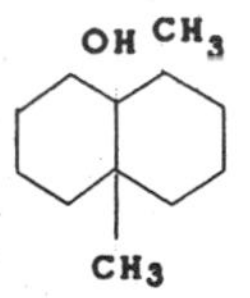	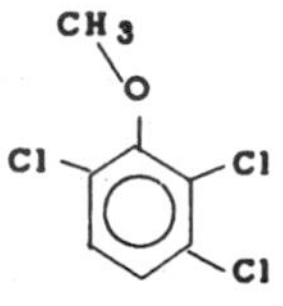	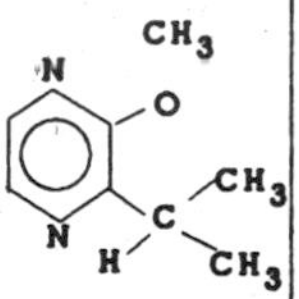	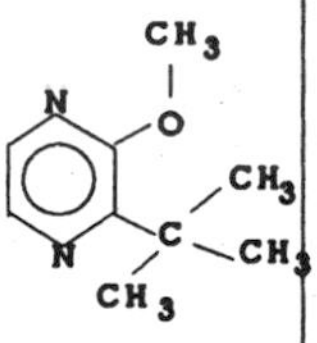	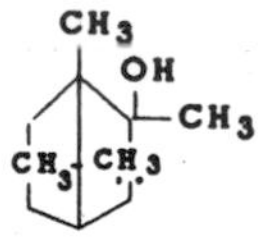
Symbol	Geosmin	TCA	IPMP	IBMP	MIB
Molecular weight	182	212	152	166	168
Molecular formula	$C_{12}H_{22}O$	$C_7H_5OCl_3$	$C_8H_{12}ON_2$	$C_9H_{14}ON_2$	$C_{11}H_{20}O$
Name	trans-1,10-dimethyl trans-9 decalol	2,3,6-trichloro anisole	2-isopropyl-3-methoxy pyrazine	2-isobutyl-3-methoxy pyrazine	2-methyl-isoborneol

Figure 16.11   Characteristics of five taste and odor compounds. Adapted from Lalezary et al. (1986).

organic matter (NOM), which competes for the same sites on the activated carbon, PAC concentration, and contact time (Cook et al., 2001). Among many carbon types tested, bituminous activated carbon was found to display the highest adsorption capacity towards MIB (Chen et al., 1997a). Zeolites have also been suggested for taste and odor removal (Ellis and Korth, 1993).

2. *Oxidation processes (chlorination, ozonation, K permanganate).*   Chlorine dioxide is one of the most effective oxidants for removing taste- and odor-causing compounds, but is not effective in eliminating odors caused by hydrocarbons (Walker et al., 1986). Chlorine dioxide is also effective against geosmin-producing actinomycetes such as *Streptomyces griseus* (Whitmore and Denny, 1992). Peroxone, a mixture of ozone and hydrogen peroxide, appears to be more effective for the oxidation of MIB and geosmin than ozone alone. The optimum $H_2O_2:O_3$ ratio for oxidation of taste and odor compounds is 0.1 to >0.3, depending on the water being treated (Ferguson et al., 1990).

3. Ozonation followed by biofiltration is effective in removing MIB and geosmin (Nerenberg et al., 2000).

4. *Biodegradation.*   Various laboratory bacterial cultures have been considered for the removal of these compounds. *Bacillus cereus* can degrade geosmin in water (Hoehn, 1965). While MIB was found to be degraded by *Bacillus fusiformis* and *Bacillus sphaericus* as shown by microscopy and 16S rRNA probes (Lauderdale et al., 2004). The taste- and odor-causing compounds (e.g., geosmin and MIB) are also used co-metabolically by biofilms grown on fulvic acid, which serves as a primary substrate (Namkung and Rittmann, 1987).

5. *Artificial groundwater recharge*. This practise (i.e., percolation of surface water through sand and gravel ridges) leads to higher removal of odorous compounds, including geosmin and 2-methyl isoborneol, than is achieved with alum coagulation (Savenhed et al., 1987).

16.6.2 Algae

Algae may cause the following problems in water treatment plants.

1. Sand filters become clogged, particularly during the warm season.

2. Taste and odor problems: Cyanobacteria (*Oscillatoria, Anabaena, Microcystis*) are responsible for taste and odor problems in drinking water (Medsker et al., 1968). The chrysophite, *Synura petersenii*, is responsible for the cucumber odor of some drinking waters. This odor is caused by a chemical identified as 2,6-nonadienal (Hayes et al., 1989). Another diatom species, *Sinura uvella* inparts a cod-liver oil odor to water (Juttner, 1981).

3. Trihalomethane precursor levels increase, with a subsequent increase in trihalomethanes upon chlorination of drinking water (Karimi and Singer, 1991). Several blue-green algae (*Anabaena, Anacystis*), green algae (*Selenastrum, Scenedesmus*), and diatoms (*Navicula*) have been implicated in chloroform production upon chlorination (Oliver and Schindler, 1980).

4. Raw surface waters may contain high numbers of cyanobacteria (e.g., *Microcystis aeruginosa, Anabaena flos-aquae, Oscillatoria*), many of which produce allergenic, hepatoxic, neurotoxic, and possibly tumor-promoting toxins (Carmichael, 1989; Falconer and Buckley, 1989). Based on their chemical structure, these toxins are divided into cyclic peptides, alkaloids and lipopolysaccharides (Sivonen, 1999). *Microcystins* (e.g., microcystin-LR) are cyanobacterial hepatoxins that have been isolated from freshwater environments and cause liver tumors in humans and animals. Failure of hemodialysis treatment systems to remove these toxins has led to several patient deaths in Brazil. Indeed, high liver cancer rates have been correlated with high toxin concentrations in drinking water sources in China. These toxins are released in the water following cell death. A waterborne outbreak of diarrheal illness (watery diarrhea) was associated with cyanobacteria-like organisms in Chicago, ILL (Centers for Disease Control, 1991). Some of the long-term chronic effects of these algal byproducts on human health are not well known (Falconer, 1989).

There are several treatments for removing algal cells from water (Drikas et al., 2001). Algal cells are effectively removed by coagulation/flocculation using Al- and Fe-based compounds. Owing to the low density of algal flocs, dissolved air flotation (DAF) is effective in removing algal cells, particularly cyanobacteria, which have gas vacuoles that serve as flotation devices. Filtration through coarse filters does not effectively remove algal cells, while fine filters are prone to clogging. These treatment processes remove intact algal cells. However, lysed algal cells release toxins that are not well removed by the conventional coagulation/flocculation–filtration–chlorination treatment train, although the addition of activated carbon filtration or ozonation to the treatment sequence leads to their removal (Himberg et al., 1989). One to three–log removal of microcystins was obtained following conventional water treatment, and in all drinking water samples the toxin concentration was below the WHO microcystin guideline of 1000 ng/L (Karner et al., 2001).

In finished waters, the algae found include green algae (e.g., *Chlorella*, *Scenedesmus*, *Ankistrodesmus*), cyanobacteria (e.g., *Schizothrix*), diatoms (e.g., *Achnanthes*), and flagellated pyrrophytes (e.g., *Glenodinium*). Algae have a tendency to grow in water reservoirs, especially during the warm season. Algal blooms in reservoirs can be prevented by covering the reservoir to block sunlight and are generally effectively controlled by copper or chlorine treatment. Chlorine, at a level of approximately 1 ppm can effectively control algae such as *Chlorella* (Kay et al., 1980).

16.6.3 Fungi

Fungi are often isolated from water distribution systems (Hinzelin and Block, 1985; O'Connor et al., 1975; Rosenszweig and Pipes, 1989). Filamentous fungi occur in drinking water at levels up to approximately 10^2 CFU/100 mL (Olson and Nagy, 1984), while yeast cell levels may reach numbers of $\leq 10^3$/mL (O'Connor et al., 1975). Approximately 50 percent of samples taken from chlorinated groundwater or surface water contained fungi. Filamentous fungi account for most the isolates (89 percent), whereas yeasts account for 11 percent of the isolates. Genera such as *Penicillium*, *Verticillium*, *Fusarium*, *Altenaria*, *Trichoderma*, *Mucor*, *Cephalosporium*, *Cladosporium*, *Aspergillus*, *Aureobasidium*, *Phoma*, *Rhizoctonia*, *Stachibotrys*, *Cladosprium*, *Mucor*, *Epicoccum*, *Phialophora*, *Candida*, and *Rhodotorula* are encountered in drinking water and biofilms because of their ability to survive chlorination and to grow on the surface of drinking water reservoirs and distribution pipes (Doggett, 2000; Nagy and Olson, 1982; Rosenszweig et al., 1989). Fungi were found in biofilms of water distribution pipes at concentrations varying from 4.0 to 25.2 CFU/cm^2, with *Aspergillus* and *Penicillium* being the dominant fungi. Yeasts were found at lower concentrations (Table 16.7; Doggett, 2000).

Monitoring of mesophilic fungi in raw and tap waters from several municipalities in Finland showed that they occur at concentrations up to approximately 100/L of tap water (Niemi et al., 1982). Approximately 50 percent of potable water samples taken from small municipal water distribution systems contain fungi at levels of 1–6 fungal propagules/50 mL. The predominant genera found were *Aspergillus*, *Alternaria*, *Cladosporium*, and *Penicillium* (Rosenszweig et al., 1986). Water treatment processes can remove up to 2–logs of fungal numbers. However, these microorganisms appear to be more resistant

TABLE 16.7. Fungal Concentration on Water Distribution Pipe Surfaces[a]

Site No.	Surface Type	Biofilm Thickness (mm)	CFU/cm^2 Yeasts	CFU/cm^2 Filamentous Fungi
1	PVC	<1	8.9	4.0
2	Iron	4	0	8.9
3	Iron	3	5.8	5.5
4	Iron	3	0	20.0
5	Iron	3	1.3	23.2
6	Iron	2	7.0	24.7
7	Iron	2	6.6	25.2
8	Iron	3	5.9	14.7

[a]Adapted from Doggett (2000).

to chlorine and ozone than bacteria (Haufele and Sprockhoff, 1973; Rosenszweig et al., 1983).

Fungi may cause the following problems in water distribution systems:

1. They exert a chlorine demand and may protect bacterial pathogens from inactivation by chlorine (Rosenszweig and Pipes, 1989; Rosenszweig et al., 1983; Seidler et al., 1977).

2. They may degrade some of the jointing compounds used in distribution systems.

3. Some form humic-like substances that may act as precursors of trihalomethanes (Day and Felbeck, 1974).

4. They may cause taste and odor problems.

5. Some fungi may be pathogenic (e.g., *A. flavus*, *A. fumigatus*) and some may cause allergic reactions in sensitive persons (Rosenszweig et al., 1986). The public health significance of their presence in water is practically unknown.

16.6.4 Actinomycetes

Soil runoff is the most likely source of actinomycetes in water. Actinomycetes are found in drinking water following water treatment (Niemi et al., 1982), and the most widely found actinomycetes in water distribution systems belong to the genus *Streptomyces*. Other genera found are *Nocardia* and *Micromonospora*. These organisms are found at levels up to 10^3 CFU/100 mL. Actinomycetes were the most frequently detected microorganisms in chlorinated distribution water from a coastal community in Oregon (LeChevallier et al., 1980).

Chemical coagulation followed by slow sand filtration and disinfection removes actinomycetes from water. However, routine chlorination or chloramination has little effect on actinomycetes and is not very effective in taste and odor reduction (Jensen et al., 1994; Sykes and Skinner, 1973).

16.6.5 Protozoa

Protozoa may be found in treated drinking water (e.g., *Hartmanella*) and some are part of the microflora of biofilms developing on the surface of water reservoirs (e.g., *Bodo*, *Vorticella*, *Euplotes*). Although much effort has been focused on pathogenic protozoa (e.g., *Giardia lamblia*, *Cryptosporidium*, *Entamoeba histolytica*, *Acanthamoeba*, *Balantidium coli*), little is known about the ecology of these microorganisms in water treatment plants and distribution networks. Protozoa were found at average concentrations of 10^5 cells/L in distribution systems and 10^3 cells/cm^2 in biofilms. Flagellates were predominant in water whereas biofilms harbored mostly ciliates and thecamoebae. Protozoa graze on biofilm bacteria, but the grazing rates are lower than those found in other aquatic environments (Sibille et al., 1998). They may reach bacterial microcolonies deep in the biofilm by swimming through the water channels in the biofilms. Some of the predatory amoeba are human parasites (*Acanthamoeba*, *Hartmanella*) (Keevil, 2002).

Amoebae harbor pathogenic microorganisms (e.g., *Legionella*, *Mycobacterium* spp., viruses) and may protect them from disinfection. Protozoa (e.g., ciliates, amebas) graze also on *Cryptosporidium* oocysts, which gain protection from environmental stresses and, probably, from chlorination (Stott et al., 2003b).

To address protozoan cyst removal from water, amendments of the Safe Drinking Water Act (PL 93-523) mandate EPA to require filtration for all surface water supplies.

16.6.6 Invertebrates

This group includes nematodes, oligochetes (annelids), amphipods, crustaceans, flat-worms, water mites and insect larvae (e.g., chiromonid larvae). Figure 16.12 (van Lieverloo et al., 2002; 2004) displays examples of invertebrates and protozoa commonly found in distribution lines. Their presence in water treatment plants and in distribution lines triggers complaints from consumers. They may be of public health significance because these organisms may harbor potential pathogens and protect them from disinfectant action. For example, nematodes may harbor pathogenic bacteria, shielding them from disinfectants (Caldwell et al., 2003).

Invertebrates can be controlled in water distribution lines by chemical (copper sulfate, chlorine) and physical methods (air scouring, pipe and hydrant flushing, filtration of source waters) or a combination of both approaches. Invertebrate growth in distribution line sediments can be prevented by reducing the food supply. This can be achieved by controlling the formation of biofilms, which, through sloughing, are the main contributors to sediment accumulation inside the pipes (Gray, 1994; Levy, 1990; Levy et al., 1986; van Lieverloo et al., 2002).

Figure 16.13 (van Lieverloo et al., 2002) illustrates the various organisms living inside distribution pipes.

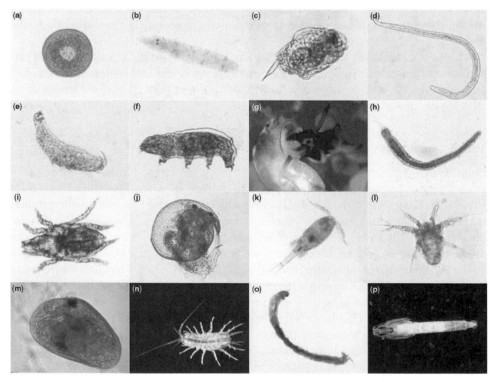

Figure 16.12 Examples of protozoan and invertebrates commonly found in water distribution systems: (a) Testacea (shelled amoebae); (b) Turbellaria (flat worms); (c) Rotifers; (d) Nematodes (round worms); (e) Gastroticha; (f) Tardigrada; (g) Gastropoda (snails); (h) Oligochetes (common worms); (i) Hydrachnellae (water mites); (j) Cladocera (water fleas); (k) Copepoda; (l) Larvae of copepoda; (m) Ostracoda; (n) Asellidae (aquatic sow bugs); (o) Larvae of Chironomidae; (p) Adult of Chironomidae. From van Liverloo et al. (2002).

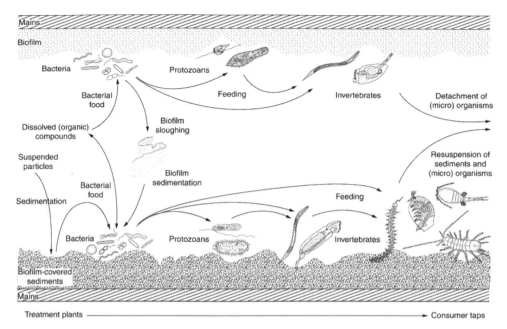

Figure 16.13   Organisms in drinking water distribution systems. From van Liverloo et al. (2002).

16.6.7   Endotoxins

These heat-stable lipopolysaccharides are structural components of the outer membrane of gram-negative bacteria and some cyanobacteria, and may be detected in surface and groundwater and in drinking water produced from reclaimed wastewater (Burger et al., 1989; Carmichael, 1981a,b; Goyal and Gerba, 1982b). High concentrations of endotoxin-producing cyanobacteria (e.g., *Schizothrix calcicola*) have been found in uncovered reservoirs containing finished water (Sykora et al., 1980). Raw (untreated) water may contain endotoxin concentrations ranging from 1 to 1049 ng/mL (approximately 10–10,000 endotoxin units (EU)/mL) (Anderson et al., 2002).

Water treatment processes (e.g., coagulation, flocculation, sedimentation and filtration) may remove from 78 to 98 percent of endotoxins (Rapala et al., 2002). Endotoxins can also be removed by oxidants such as chlorine, monochloramine, or potassium permanganate (Anderson et al., 2003). Water distribution systems may contain from 2 to 32 EU/mL (Korsholm and Søgaard, 1988; Rapala et al., 2002).

Although endotoxins have been associated with human health problems (fever, diarrhea, vomiting, and allergies), the full implications of their presence in drinking water are still unclear.

16.6.8   Iron and Manganese Bacteria

Iron and manganese bacteria attach to and grow on pipe surfaces and subsequently cause a deterioration of water quality (e.g., pipe clogging and color problems leading to laundry staining).

16.7   DRINKING WATER QUALITY AT THE CONSUMER'S TAP

Drinking water quality at the consumer's tap is affected by both the service lines and the home devices installed by the consumer to improve taste and odor problems.

16.7.1   Effect of Service Lines on Drinking Water Quality

In addition to water distribution pipes, service lines (i.e., house plumbing systems) can also influence the microbiological and chemical quality of drinking water. Conditions prevailing in service lines and conducive to bacterial proliferation include long residence time, absence of chlorine residual, higher temperature, and high surface-to-volume ratio (ratio of pipe surface to volume of water). Total bacterial numbers, as measured by epifluorescence microscopy, are generally much higher in the first flush than in the distribution system (Prévost et al., 1997). Depending on the distance of the home to the treatment plant, a flushing time of 2–10 min is necessary to lower bacterial numbers to background levels. Similarly, lead and copper concentrations were high in the first flush and decreased to background levels within 2 min of flushing.

16.7.2   Point-of-Use (POU) Home Devices for Drinking Water Treatment

The public at large is interested in *point-of-use home devices* to remove toxic and carcinogenic chemicals and improve the aesthetic quality of drinking water (removal of taste and odor, turbidity, color). The home devices may be connected to a third faucet separate from the traditional cold and hot water faucets. Treatment is accomplished by filtration, adsorption, ion exchange, reverse osmosis, distillation or UV irradiation (Geldreich and Reasoner, 1990; Reasoner et al., 1987). The most frequently used process is filtration through activated carbon.

The contaminants of concern are pathogenic bacteria, viruses, protozoan cysts (e.g., *Giardia*), toxic metals (e.g., cadmium, mercury, lead, arsenic), iron, manganese, organic substances of potential health significance (e.g., trichloroethylene, hydrocarbons, benzene), particulates, color, odor, and chlorine taste (Geldreich and Reasoner, 1990). As to microbial contaminants, the U.S. EPA requires a minimum removal capacity of 99.9, 99.99, and 99.9999 percent for *Giardia* cysts, viruses, and bacteria, respectively.

Such POU devices may be installed in a home as faucet add-on units consisting of small activated carbon cartridges, in-line devices (filters, reverse osmosis units) that are installed under the kitchen sink, or pour-through pitchers (Fig. 16.14; Reasoner, 2002). There are also point-of-entry (POE) devices for treating the entire home water supply. Table 16.8 summarizes the basic processes involved in water treatment by home units (Reasoner, 2002).

However, there are some problems associated with the use of these devices (Reasoner et al., 1987; Reasoner, 2002):

1. Heterotrophic bacteria and, possibly, pathogenic microorganisms, may colonize the activated carbon surface, leading to the occurrence of high levels of bacteria in the product water. An example of HPC bacteria found in the product water is shown in Figure 16.15 (Reasoner et al., 1987; Geldreich et al., 1985; Taylor et al., 1979; Wallis et al., 1974). Static conditions overnight or following vacation periods as well as favorable temperatures and nutritional conditions provide opportunity for bacterial growth in the treatment device. Figure 16.16 shows an increase in HPC in the first-draw filtered water (Snyder et al., 1995). Pathogenic bacteria such as *Pseudomonas aeruginosa* or *Klebsiella pneumonae* are also able to colonize the filter surface (Geldreich et al., 1985; Reasoner et al., 1987; Tobin et al., 1981) and may pass in the product water. However, the epidemiological significance of this phenomenon is unknown. It is thus advisable to flush the unit for 1–3 min prior to use in the morning or after returning from vacation.

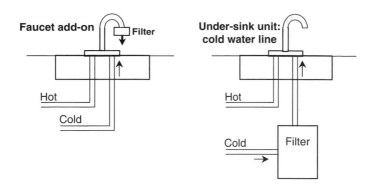

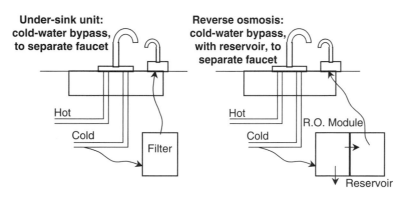

Figure 16.14   Point-of-use devices. From Reasoner (2002).

2. Silver-impregnated filters are sold by some manufacturers to control bacterial growth inside the filter. The antibacterial effect of silver is rather slow and significant bacterial reductions are obtained only after hours of contact (Bell, 1991). Furthermore, at the concentrations used in point-of-use devices, silver does not exert any significant detrimental effect on heterotrophic bacterial growth (possibly due to the selection of silver-resistant bacteria) and has no significant antiviral effect (Gerba and Thurman, 1986; Tobin et al., 1981). Despite the presence of silver, granulated activated carbon filters are prone to colonization by bacteria. Furthermore, because of concern over

TABLE 16.8.   Processes Involved in Point-of-Use (POU) and in Point-of-Entry (POE) Devices[a]

Process	Contaminant(s) Removed
Adsorption (activated carbon)	Chlorine, taste and odors, organics
Mechanical filtration	Particulates, color, turbidity, asbestos fibers, cysts, and oocysts
Reverse osmosis	Total dissolved solids, metals, nitrate, bacteria, viruses, cysts, and oocysts
Water softening (cationic)	Ca, Mg, Fe, Mn, Ba, Ra
Water softening (anionic)	Sulfate, nitrate, bicarbonate, chloride, arsenic
Distillation	Inorganics, dissolved solids, organics
Disinfection (chemical or UV)	Bacteria, viruses, cysts, and oocysts

[a]Adapted from Reasoner (2002).

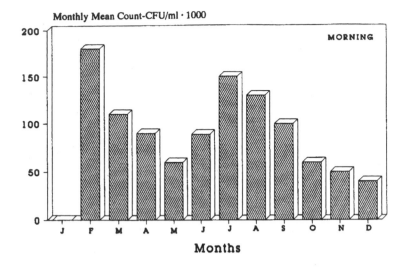

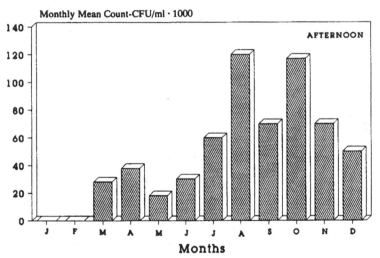

Figure 16.15   Mean heterotrophic plate count found in product water from home point-of-use devices. From Reasoner et al. (1987). (Courtesy of the American Water Works Association.)

consumers' health, the silver released in the treated water cannot exceed 50 μg/L (Geldreich et al., 1985). Electrochemical disinfection, using an applied potential of 0.8 V, has also been proposed for bacterial inactivation on activated carbon (Matsunaga et al., 1994).

Water filtration units based on reverse osmosis may produce water with bacterial counts of approximately $10^3 - 10^4$ CFU/mL. Bacterial isolates were identified as *Pseudomonas*, *Flavobacterium*, *Alcaligenes*, *Acinetobacter*, *Chromobacterium*, and *Moraxella* (Payment, 1989b). A prospective epidemiological study undertaken in a suburban area in Montreal, Canada, showed a correlation between bacterial counts in drinking water at 35°C and the incidence of gastrointestinal symptoms in 600 households consuming water treated by reverse osmosis (RO). The level of gastrointestinal illnesses in the group consuming regular tap water was 30 percent higher that in the group consuming water treated by reverse osmosis (Payment et al., 1991; 1993). A follow-up study showed that 14–40 percent of gastrointestinal

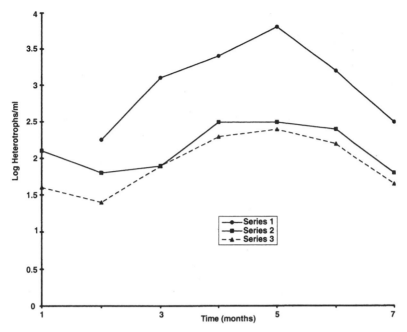

Figure 16.16   Mean heterotrophic plate counts of spring water systems (series 1: first-draw filtered water sample; series 2: unfiltered water sample; series 3: postflush filtered water sample. From Snyder et al. (1995). (With permission of the publisher.)

illnesses were due to consumption of tap water meeting microbiological standards. Children (2–5 years old) were the most affected by gastrointestinal illnesses (Payment et al., 1997).

Thus, although reverse osmosis appears to reduce the incidence of gastrointestinal illnesses, treatment of drinking water by some home devices may have an adverse effect on the microbiological quality of drinking water. A novel POU device, based on inactivation of pathogens by pulsed broad spectrum white light, achieved inactivation exceeding $4-\log_{10}$ for viruses (poliovirus and rotavirus) and *Cryptospodium parvum*, and $7-\log_{10}$ for *Klebsiella terrigena*. The device was operated at a flow rate of 4 gallons/min (Huffman et al., 2000).

The use of point-of-use devices by consumers is considered by some experts as a potential health hazard, and some strongly believe that water treatment should be carried out by trained professionals and not by poorly trained consumers (Geldreich et al., 1985).

16.8   BOTTLED WATER

People consume bottled water for a variety of reasons, ranging from complaints about bad taste of tap water to concerns about chemicals and pathogens and parasites. In 1999, bottled water sales were approximately 35 billion dollars worldwide, and 5 billion dollars for the United States alone. In 1990, Californians consumed more than 700 million gallons of bottled water. In the United Kingdom, consumption increased from >5 million liters in 1975 to 1.38 billion liters in 2000 (www.globaldrinks.com).

In Europe, the bottled water industry has traditionally projected an image of health. Drinking and bathing in mineral waters has been claimed to cure many human ailments (e.g., migraine headaches, arthritis, kidney infections, excess cholesterol, gout, obesity, and the improvement of metabolism in general). Spas across Europe have enjoyed a reputation for offering relaxation and fitness to their customers. Bottled water is perceived by consumers as being of better quality and better taste than tap water. However, the higher quality of bottled water is sometimes a misconception.

16.8.1 Sources and Categories of Bottled Water

Bottled water comprises any one of the following categories (Gray, 1994; Rosenberg, 2002):

- *Mineral water.* This is water that contains dissolved cations and anions (e.g., Ca^{2+}, Mg^{2+}, Na^+, HCO_3^-). The mineral content varies from one source to another and may sometimes be lower than community water supplies. *Natural mineral water* is water that is drawn from a spring. It becomes enriched with minerals as it flows through underground formations. Manufacturers are not allowed to alter the mineral content of this type of water. In France, "natural mineral water" must also have a demonstrated therapeutic value.
- *Sparkling water.* This is water that is "naturally" carbonated with carbon dioxide. It can also be made bubbly by adding dissolved carbon dioxide. Some brands of bottled water (e.g., Perrier) are made bubbly by harvesting the "natural" carbon dioxide at the source and reinjecting it into the water during the bottling operation.
- *Flat or still water.* This is mineral or tap water without gas bubbles.
- *Spring water.* This is water drawn from a spring, which is groundwater emerging at the Earth surface. Approximately 25 percent of the bulk water sold in the United States is spring water. This water may be disinfected with ozone.
- *Distilled or purified water.* This is water from which minerals have been removed by processes such as distillation (i.e., water vaporized and then condensed), deionization (i.e., passage of water through special resins), or reverse osmosis (i.e., passage of water through special membranes, which retain the minerals).
- *Tap water* is also bottled with, sometimes, additional treatment. In the United States, 25–40 percent of bottled water is bottled tap water.

16.8.2 Microorganisms Found in Source Water

This topic has been covered in Chapter 15.

16.8.3 Water Treatment Before Bottling

Depending on the source, water may be treated via activated carbon filtration, ultraviolet radiation, ozonation, or reverse osmosis (see Chapter 6 for further details on disinfection). In Europe, source water is not subjected to any of these treatments.

16.8.4 Microorganisms Found in Bottled Water

As discussed in Section 16.5, bacteria present in bottled water may attach to the container surface and form biofilms, as they use the small amounts of organic compounds in the water to reach levels as high as 10^4-10^5 bacteria/mL. Owing to the nutritional stress in bottled water, some bacteria may enter into the viable but nonculturable (VBNC) state. Several factors control bacterial growth in bottled water (Rosenberg, 2002). Bottled water stored at room temperature allows higher bacterial growth than when stored under refrigeration. Moreover, plastic containers allow higher bacterial growth than glass bottles. Bacteria survive and grow less in carbonated than in noncarbonated water and this is due to the lower pH in carbonated water. Bottled water may also harbor microorganisms (e.g., *Pseudomonas* sp., *Acantamoeba*, caliciviruses) that are harmful to the consumers. Two of 13 brands of mineral waters from Brazil harbored *Cryptosporidium* at concentrations of 0.2–0.5 oocysts/L (Franco and Neto, 2002).

Mineral waters are derived from subsurface environments, which have very low concentrations of available organic matter. The oligotrophic and starved bacteria living under these conditions develop various survival strategies. One important strategy is the viable but nonculturable (VBNC) state. Several pathogenic bacteria have been shown to enter into the VBNC state and retain infectivity. Following bottling of mineral water, the heterotrophic bacteria experience the "bottle effect" and their number increases but is fairly constant after a few weeks. This phenomenon was demonstrated using both culture methods and molecular techniques (e.g., FISH) (Leclerc and Moreau, 2002). Although epidemiological data did not show any public health hazard associated with bottled mineral water, it is recommended to monitor routinely this resource for the presence of indicators (e.g., *E. coli*, HPC), and regularly for the presence of viruses and protozoan parasites.

16.8.5 Regulations Concerning Bottled Water

Regulations concerning bottled water are generally much stricter in Europe than in the United States. From a regulatory viewpoint, bottled water sold in interstate commerce is considered in the United States as a food item and, as such, is regulated by the federal Food and Drug Administration (FDA). Tap water produced by municipal water treatment plants is regulated by the Environmental Protection Agency (EPA). In Europe, the production of natural mineral water is tightly regulated by the Ministry of Health or an equivalent agency. For example, in France, natural mineral water must originate from groundwater free from pollution sources. It is subjected to chemical and microbiological analyses by health authorities. No chemical treatment, including disinfection (except for CO_2 addition), is allowed. The water is generally bottled at the source and must be free of pathogens and parasites. If contamination occurs, the operation must be shut down until the source of contamination is identified and corrective measures taken by the responsible authorities.

16.9 WEB RESOURCES

http://www.dartmouth.edu/~gotoole/reviews.html (biofilms)
http://www.personal.psu.edu/faculty/j/e/jel5/biofilms/ (biofilm online manual)
http://www.personal.psu.edu/faculty/j/e/jel5/biofilms/primer.html (biofilm primer)
http://www.erc.montana.edu/ (Center for Biofilm Engineering, Montana State University, Bozeman)
http://www.erc.montana.edu/Res-Lib99-SW/Image_Library/ (pictures and drawings of biofilm microorganisms: Center for Biofilm Engineering, Montana State University, Bozeman)
http://www.biofilmclub.co.uk/index.html (Biofilm Club, UK)

16.10 QUESTIONS AND PROBLEMS

1. Discuss biofilm structure.
2. Give examples of treatment processes discussed in previous chapters that are based on biofilm formation.
3. What are the steps involved in biofilm formation in aquatic environments?
4. What is the role of extracellular polymeric substances in biofilm formation?
5. What molecular methods have been used to study biofilms?
6. What type of microorganisms would be benefited by growth in biofilms?
7. Explain the AOC and BDOC methods for determining organic carbon in drinking water.
8. Discuss the impact of biofilms on public health?
9. How do biofilms interfere with disinfection?
10. What are the mechanisms involved in the adsorption of bacteria to surfaces?
11. What is the chlorine species that best controls biofilms?
12. Give the methods used by the water industry to control biofilms.
13. Why is *Legionella* a problem in potable water systems?
14. Why do chloramines affect nitrification in water distribution systems?
15. From a kinetics and nutritional viewpoint, how would you characterize drinking water bacteria?
16. How do we determine microbial biomass in the AOC assay?
17. How can we control taste- and odor-causing compounds in water treatment plants?
18. Why is there a concern about the presence of algae and cyanobacteria in drinking water?
19. How can we remove endotoxins in drinking water?
20. Would you recommend the use of faucet add-on units for treating your drinking water? Explain your answer.

16.11 FURTHER READING

Costerton, J.W., Z. Lewandowski, D.E. Caldwell, D.R. Korber, and H.M. Lappin-Scott. 1995. Microbial biofilms. Annu. Rev. Microbiol. 49: 711–745.

Characklis, W.G., and K.E. Cooksey. 1983. Biofilms and microbial fouling. Advances Appl. Microbiol. 29: 93–138.

Gray, N.F. 1994. *Drinking Water Quality: Problems and Solutions*. John Wiley & Sons, Chichester, U.K., 315 pp.

Keevil, C.W. 2002. Pathogens in environmental biofilms, pp. 2339–2356, In: *Encyclopedia of Environmental Microbiology*, Gabriel Bitton, editor-in-chief, Wiley-Interscience, N.Y.

van der Kooij, D. 2002. Assimilable organic carbon (AOC) in treated water: Determination and significance, pp. 312–327, In: *Encyclopedia of Environmental Microbiology*, Gabriel Bitton, editor-in-chief, Wiley-Interscience, N.Y.

Leclerc H., and A. Moreau. 2002. Microbiological safety of natural mineral water. FEMS Mirobiol. Rev. 26: 207–222.

McFeters, G.A., Ed. 1990. *Drinking Water Microbiology*, Springer-Verlag, N.Y

Olson, B.H., and L.A. Nagy. 1984. Microbiology of potable water. Adv. Appl. Microbiol. 30: 73–132.

Rittmann, B.E. 1995. Fundamentals and application of biofilm processes in drinking-water treatment, pp. 61–87, In: *Water Pollution: Quality and Treatment of Drinking Water*, Springer, New York.

Rittmann, B.E. and C.S. Laspidou. 2002. Biofilm detachment, pp. 544–550, In: *Encyclopedia of Environmental Microbiology*, Gabriel Bitton, editor-in-chief, Wiley-Interscience, N.Y.

Rosenberg, F.A. 2002. Bottled water, microbiology of, pp. 795–802, In: *Encyclopedia of Environmental Microbiology*, Gabriel Bitton, editor-in-chief, Wiley-Interscience, N.Y.

Wolfe, R.L., and N.I. Lieu. 2002. Nitrifying bacteria in drinking water, pp. 2167–2176, In: *Encyclopedia of Environmental Microbiology*, Gabriel Bitton, editor-in-chief, Wiley-Interscience, N.Y.

17

BIOTERRORISM AND DRINKING WATER SAFETY

17.1 INTRODUCTION

Bioterrorism involves the deliberate use of microbial pathogens or microbe-derived products to cause harm to humans, livestock or agricultural crops.

A biological weapon is a four-component system made of the payload (biological agent), munition (container ensuring the maintenance of the potency of the biological agent), delivery system (e.g., missile, vehicle) and a dispersion system (e.g., sprayer) (Hawley and Eitzen, 2001). The production of biological weapons and their delivery are relatively inexpensive as compared to conventional weapons. The cost to inflict civilian casualties is much lower for biological weapons than for conventional weapons.

In this chapter, we will introduce the main bioterrorism (BT) agents and biotoxins and we will discuss their potential impact on the safety of drinking water supplies.

17.2   EARLY HISTORY OF BIOLOGICAL WARFARE

Harmful biological agents have been used for hundreds of years to cause harm to humans, animals, and agricultural crops (Christopher et al., 1997; Hawley and Eitzen, 2001; Sheehan, 2002). Some notorious examples illustrating the use of biological agents to cause harm to the enemy in war times include:

- *Fourteenth century.*   The Tartar army catapulted diseased bodies of dead soldiers into territory held by the enemy.
- *1754–1767.*   Smallpox-contaminated blankets were sent to native American Indians, who experienced an outbreak of smallpox.
- *1914–1918.*   During World War I, the Germans used *B. anthracis* and *Burkholderia mallei* to infect livestock, horses, and mules of the Allied Forces.
- *1932–1945.*   Japan subjected prisoners to infectious agents such as *B. anthracis*, *Shigella*, *Salmonella*, *Yersinia pestis*, and smallpox virus.
- *World War II (1939–1945).*   The Germans experimented on prisoners in concentration camps by exposing them to infectious agents.
- *1943.*   The United States started a biological weapons program at Fort Detrick, MD, involving *B. anthracis* and other BT agents. Studies were carried out on weaponization systems as well as the survival of BT agents in the environment. The Allies dropped anthrax bombs on Gruinard Island off the coast of Scotland. The island was decontaminated some fifty years later. The United States terminated its biological program in 1969.
- *1962–1968 Vietnam War.*   The Viet Kong used spear traps and spikes contaminated with human and animal excreta.
- *1970s.*   The former Soviet Union also maintained an extensive biological warfare program. An outbreak of inhalation anthrax accidentally occurred in a production facility in 1979, resulting in the deaths of 66 persons. The program was terminated following the fall of the Soviet Union.
- *1991 Gulf War.*   Iraq admitted to having engaged in the development of a biological warfare program. After the war, Iraq claimed that it had destroyed its biological agents.

17.3   BIOTERRORISM MICROBIAL AGENTS AND BIOTOXINS

17.3.1   Some Important Features of BT Agents

Some characteristics of the BT agents must be taken into account when assessing the threat posed by the pathogens and when comparing chemical and biological weapons.

- Compared to chemical agents, microbial agents are infectious and can multiply within the host. Bioterrorism agents can be transmitted via person-to-person and via the waterborne, foodborne, airborne, and fomite routes (see Chapter 4). Person-to-person transmission may be quite easy for certain agents such as the smallpox virus.
- The microbial dose necessary for causing infection (i.e., minimum infectious dose) may vary from a few to thousands of microorganisms (see Chapter 4). The minimum infectious doses of some inhaled or ingested BT agents are given in Table 17.1 (Burrows and Renner, 1999; Sheehan, 2002).

TABLE 17.1. Characteristics of Some Biological Warfare Agents[a]

Disease Agent	Potential for Weaponization	Estimated Infectious Dose[b]	Equivalent Drinking Water Concentration (number/L)[c]		Water Threat
			5 L/day	15 L/day	
Anthrax (*Bacillus anthracis*)	Yes	6000 spores (inh)	171	57	Yes
Brucellosis (*Brucella*)	Yes	10–100 cells	300	100	Probable
Cholera (*Vibrio cholerae*)	Unknown	10^3 (ing)	30	10	Yes
Plague (*Yersinia pestis*)	Probable	10^2–10^3 (inh) 70 (ing)	2	<1	Yes
Glanders (*Burkholderia mallei*)	Probable	3.2×10^6	9×10^4	3×10^4	Unlikely
Tularemia (*Francisella tularensis*)	Yes	10^8 cells (ing) 10–50 cells (inh)	3×10^6	0^6	Yes
Q fever (*Coxiella burnetii*)	Yes	25 (unsp)	<1	<1	Possible
Smallpox virus	Possible	10–100 particles (inh)			?
Cryptosporidiosis (*Cryptosporidium*)	Unknown	132 oocysts (ing)	3	1	Yes

[a]Adapted from Burrows and Renner (1999); Bitton (1999).
[b]inh = inhalation route; ing = ingestion route; unsp = unspecified route.
[c]The equivalent drinking water concentration is obtained by dividing by 7 (the maximum number of days for accumulation of the infectious organism without any clearance) and by 5 or 15 to account for water consumption of 5 L/day and 15 L/day, respectively.

- The production of biological agents may be less complex and less costly than that for chemical poisons.
- Stability of the BT agents in the environment (water, soil, groundwater, sediments, air, surfaces, etc.).
- Impact of water treatment processes, including disinfection, on their removal and/or inactivation.
- Biological agents are colorless and odorless, and the onset of symptoms varies from days to weeks for these agents, as compared to minutes or hours for chemical agents.
- In the United States, the baseline incidence of diseases reported to the National Notifiable Disease Surveillance System is generally low, except possibly for tularemia and brucellosis.

17.3.2 Categories of BT Agents

The Center for Disease Control and Prevention (CDC) in the United States has categorized BT agents as follows (Table 17.2; Atlas, 2002; Rotz et al., 2002).

17.3.2.1 *Category A.* These agents pose the highest threat and are of most concern to public health. They include variola virus (smallpox), *B. anthracis*, *Yersinia pestis*,

TABLE 17.2. Bioterrorism: Categories of Biological Agents of Concern to Public Health[a]

Biological Agent	Disease
Category A[b]	
Variola major	Smallpox
Bacillus anthracis	Anthrax
Yersinia pestis	Plague
Clostridium botulinum	Botulism
Francisella tularensis	Tularemia
Filoviruses and Arenaviruses	Viral hemorrhagic
(e.g., Ebola virus, Lassa virus)	fevers
Category B	
Coxiella burnetii	Q fever
Brucella spp.	Brucellosis
Burkholderia mallei	Glanders
Alphaviruses	Encephalitis
Rickettsia prowazekii	Typhus fever
Biotoxins (e.g., ricin,	Toxic syndromes
Staphylococcal enterotoxin B)	
Chlamydia psittaci	Psittacosis
Foodborne agents (e.g., *Salmonella*)	
Waterborne agents (e.g., *vibrio cholera*,	
caliciviruses, *Cryptosporidium*)	
Category C	
Emerging agents (e.g., Nipah virus, hantavirus)	

[a]Adapted from Rotz et al. (2002).
[b]From a public health threat viewpoint: A > B > C.

Clostridium botulinum toxin, *Francisella tularensis*, and viruses causing hemorrhagic fevers (e.g., Ebola, Marburg, Lassa).

17.3.2.2 *Category B.*   These agents cause lower morbidity and mortality than category A agents. They include *Brucella* (brucellosis), *Coxiella burnetti* (Q fever), *Burkholderia mallei* (glanders), alphaviruses, Venezuelan encephalomyelitis, eastern and western equine encephalomyelitis, ricin toxin from *Ricinus communis* (castor beans), and *Staphylococcus* enterotoxin B. A subset of category B agents includes pathogens that are foodborne or waterborne. These pathogens include *Salmonella* spp., *Shigella* spp., *Escherichia coli* O157:H7, *Vibrio cholerae*, *Cryptosporidium parvum*, *Giardia lamblia*, hepatitis A and E viruses, and caliciviruses.

17.3.2.3 *Category C.*   This includes emerging pathogens such as hantaviruses, tick-borne hemorrhagic fever viruses, or tick-borne encephalitis viruses.

17.3.3 Major BT Agents

Table 17.1 gives a list of the main BT agents, their estimated infectious dose, and their potential for weaponization.

17.3.3.1 *Bacillus anthracis.*   *Bacillus anthracis* is a gram-positive, spore-forming, and rod-shaped bacterium responsible for anthrax. The spores (Fig. 17.1) are extremely

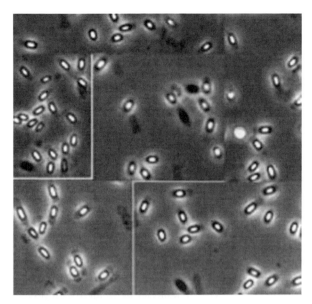

Figure 17.1 *Bacillus anthracis* endospores. Adapted from Nicholson and Galeano (2003).

resistant to environmental stresses and survive for long periods in soils (survival for up to about 50 years on Guinard Island). This makes the resistant spores highly desirable for use as biological warfare agents. Although anthrax is primarily a disease of animals, it can also infect humans who contract the disease through contact with infected animals or through occupational exposure to infected animal products. Although person-to-person transmission is rare, *B. anthracis* can be transmitted via inhalation, gastrointestinal, and cutaneous routes. The inhalation route is fatal in most cases with a mortality rate exceeding 80 percent (Baillie and Read, 2001; Dixon et al., 1999; Franz et al., 1997), while the mortality rate of cutaneous anthrax is up to 25 percent. The infectious dose for *B. anthracis* is approximately 6000 spores via the inhalation route. If we assume that the infective dose via inhalation and ingestion routes are equivalent, the corresponding spore concentration in drinking water should be 171 spores/L following consumption of 5 L/day over 7 days (Burrows and Renner, 1999).

Anthrax is characterized by flu-like symptoms including headache, fever, myalgia, cough, and mild chest discomfort. This is followed by chest wall edema and hemorrhagic meningitis. Death occurs if treatment is delayed beyond 48 h. Treatment consists of administering the antibiotic ciprofloxacin to patients.

The methodology for detecting airborne *B. anthracis* consists of membrane filtration followed by plating the concentrate on blood agar supplemented with polymyxin or on *B. cereus* agar, by detecting components of spore glycoproteins, or via real-time PCR (Fox et al., 2003; Makino and Cheun, 2003).

17.3.3.2 *Yersinia pestis*.   Plague is caused by the gram-negative bacterium *Yersinia pestis*, which is amenable to aerosolization. There are three clinical forms of plague: bubonic, septicemic, and pneumonic. The disease may result in respiratory failure and shock. Death occurs if the patient is not treated within 24 h. In addition to its person-to person transmission, it may cause massive casualties when spread via the airborne route. It is transmitted to humans from infected rodents via flea bites and causes bubonic plague. The infectious dose for humans via inhalation is around 100 to 1000 organisms. If ingested, the infectious dose is around 70 organisms (Cooper et al., 1986).

17.3.3.3 Brucella. *Brucella* is the etiologic agent for brucellosis, which causes fever, sweating, and pain, but has a low fatality rate. *Brucella suis* has been weaponized as an aerosolized BT agent. Although *Brucella* (*B. suis*) is mostly pathogenic to livestock, some species are pathogenic to humans. Although less resistant to environmental stresses than *B. anthacis*, *Brucella* is a desirable biological warfare agent due to its low infectious dose (10–100 cells). The mortality rate of brucellosis is around 5 percent (Eitzen et al., 1998). Since brucellosis can be acquired through consumption of contaminated milk, drinking water should also be considered for the transmission of this pathogen.

17.3.3.4 Francisella tularensis. *Francisella tularensis* is a gram-negative nonspore forming bacterium, which causes tularemia (fever, weight loss, pneumonia). This pathogen attacks mostly animals, and humans may become infected following contact with diseased animals. The fatality rate is around 35 percent owing to the nonspecific symptoms. This pathogen has a low infectious dose (10–50 cells) (Eitzen et al., 1998; Franz et al., 1997). The patient can be treated with doxycycline.

17.3.3.5 Burkholderia mallei. This gram-negative bacterium is the causative agent of glanders, a disease affecting mostly horses, donkeys, and mules, but which may also infect humans via the airborne route. It was used during World Wars I and II to infect animals and sometimes humans. An important symptom is the inflammation of mucus membranes of the nose. The septicemic form of glanders, if left untreated, has a 100 percent mortality rate. The minimum infectious dose is not known.

17.3.3.6 Vibrio cholerae. This bacterial pathogen was described in Chapter 4. Because it is strictly a waterborne pathogen, it has not been considered for use in the airborne state.

17.3.3.7 Clostridium botulinum. Botulism is caused by a neurotoxin produced from the anaerobic, spore-forming bacterium *Clostridium botulinum*. This neurotoxin has a toxic dose of 0.001 µg/kg body weight, and is 15,000 times more toxic than the nerve agent VX and 100,000 times more toxic than sarin (Arnon et al., 2001). It causes nerve and muscle paralysis and may culminate in respiratory failure. The route of exposure to this toxin could be via aerosolization or water contamination. Treatment consists of rapid administration of botulism antitoxin.

17.3.3.8 Smallpox Virus. Although it was considered eradicated in the 1970s, the smallpox virus is still a serious threat to public health if deliberately released by terrorist groups (Henderson et al., 1999). This virus seems to have a low infectious dose and can be contracted through aerosolization and through person-to-person contact. The fatality rate of this virus is around 30 percent. It may possibly be weaponized, but its water threat is unknown at the present time.

17.3.3.9 Hemorrhagic Fever Viruses. These viruses (e.g., Marburg, Ebola, Lassa) could also be used in biological warfare. They have a low infectious dose and may cause a high mortality rate among the victims. However, their transmission via the waterborne route is unknown.

17.3.4 Biotoxins

Table 17.3 shows the major bacterial, fungal, and algal biotoxins, their potential for weaponization and threat to drinking water supplies.

17.3.4.1 *Clostridium botulinum Toxins.*

Clostridium botulinum, an anaerobic spore-forming bacterium, produces the most potent and deadliest neurotoxins known to humans. The LC_{50} of the botulism toxin is less than $0.01~\mu g/kg$. Exposure to this toxin generally occurs through ingestion of contaminated canned food or through the respiratory route. Exposure to this neurotoxin results in muscle paralysis and death by respiratory failure. An antitoxin is available and is effective against type E toxin.

17.3.4.2 *Mycotoxins.*

Mycotoxins are secondary metabolites produced by fungi such as *Penicillium*, *Fusarium*, or *Aspergillus*. Exposure to these toxins occurs via inhalation, ingestion, or skin contact. They are extremely stable under environmental conditions. *Penicillium* species produce the carcinogens islandotoxin and patulin and the hepatoxicant luteoskyrin. *Fusarium moniliforme* is one of the most common fungi colonizing corn. It produces fusarin, a potent mutagen, and fumonisin, a cancer promoter (Chu and Li, 1994). *Fusarium* also produces the T2 toxin, which affects mostly livestock at micrograms or nanograms per gram of food. Signs of exposure of animals to T2 toxin range from feed refusal to cardiovascular shock (Albertson and Oehme, 1994). A dipstick enzyme immunoassay is available for detecting ng of T2 toxin (De Saeger and van Peteghem, 1996). This toxin is probably amenable to weaponization and would pose a water threat.

Aflatoxins are produced by *Aspergillus flavus* or *A. parasiticus* and are potent carcinogens, mutagens, and teratogens. Humans and animals come into contact with these chemicals through consumption of contaminated corn, peanuts, and other agricultural crops. Humid local storage conditions are responsible for high aflatoxin concentrations in food crops. In the United States the action level for total aflatoxin is $20~\mu g/kg$ of food. Aflatoxins have been weaponized and pose a threat to the safety of drinking water (Zilinskas, 1997).

TABLE 17.3. Water Threat from Biotoxins[a]

Biotoxin	Amenability to Weaponization	NOAEL[b] or LD_{50}	Water Threat
Aflatoxins	Yes	$LD_{50} = 10-100$ mg/person	Yes
Anatoxin A	Unknown	$LD_{50} = 200~\mu g/kg$ (mice)	Probable
Botulinum toxins	Yes	$0.003~\mu g/kg$ (mice) $0.006~\mu g/kg$ (humans)	Yes
Microcystins	Possible	ID50 $= 1-10$ mg/person NOAEL $= 10~\mu g/L$ WHO standard $= 1~\mu g/L$ (lifetime exposure)	Yes
Ricin	Yes	$LD_{50} = 20$ mg/kg (mice) NOAEL $= 2~\mu g/L$	Yes
Saxitoxin	Possible	ID50 $= 0.3-1$ mg/person NOAEL $= 0.1~\mu g/L$	Yes
T2 mycotoxin	Probable		Yes
Tetrodotoxin	Possible	$LD_{50} = 30~\mu g/kg$ NOAEL $= 0.1-30~\mu g/L$	Yes

[a]Adapted from Burrows and Renner (1999).
[b]LD_{50} = lethal dose affecting 50 percent of the population; NOAEL = no observed adverse effect level.

17.3.4.3 Ricin Toxin. This toxin is produced by the castor plant *Ricinus communis*. It is extremely toxic ($LD_{50} = 20$ mg/kg for mice) via the oral and respiratory routes. Ricin production is relatively inexpensive but its weaponization for mass destruction is not well known. No vaccine is available against ricin.

17.3.4.4 Saxitoxin. Dinoflagellates produce an alkaloid, saxitoxin, which acts as a neurotoxin. It is the cause of paralytic shellfish poisoning. The main exposure routes are ingestion or exposure to aerosols. Ingestion of this biotoxin may lead to paralysis while breathing it may lead to respiratory failure and death. The LD_{50} for this toxin is 0.3 to 1 mg/person (Burrows and Renner, 1999).

17.3.4.5 Microcystins. These toxins are produced by cyanobacteria (*Microcystis aeruginosa*) and are highly toxic to humans and animals via the ingestion or respiratory routes. Microcystin-LR is a prime candidate for weaponization.

17.3.4.6 Tetrodotoxin. Tetrodotoxin is produced by certain species of puffer fish (fugu fish in Japan). This potent ($LD_{50} = 30$ μg/kg) neurotoxin accumulates in the fish ovaries and liver and has a 60 percent mortality rate.

17.4 DELIBERATE CONTAMINATION OF WATER SUPPLIES

Historically, during World War II, the Japanese contaminated Chinese water supplies with *B. anthacis*, *Shigella* spp., *Salmonella* spp., *Vibrio cholera*, and *Yersinia pestis*. Thus, the possibility exists that terrorists could strike the water supply of cities around the world. Different actions could be taken to adversely disrupt the system (Denileon, 2001):

- *Physical destruction of the system*, leading to the disruption of water supply to customers due to destruction of equipment (pumps, chlorinators, power source, computers for data acquisition) and structures. The release of chlorine gas could affect the plant personnel as well as the adjacent neighborhoods.
- *Deliberate chemical contamination of the water supply.* The water distribution system is particularly vulnerable to attack because of its accessibility (through distribution reservoirs and fire hydrants).
- *Cyber attack* to disrupt the operation of the plant.
- *Bioterrorist attack.* Drinking water treatment plants have been identified as possible targets of bioterrorism (Brosnan, 1999). Although most biological warfare agents are intended to be delivered via the more efficient aerosol route, the waterborne route would be of no less concern. Deliberate water contamination may occur at the source water, on site at the water treatment plant, or in the distribution networks.

There are, however, safeguards to the deliberate contamination of drinking water (Khan et al., 2001):

- Effect of dilution on the introduced BT agents.
- Physical, chemical, and biological factors contribute to the inactivation of any introduced microorganisms at the source.
- Water treatment plants offer multiple barriers that help in the physical removal (e.g., coagulation/flocculation, sand filtration, activated carbon) and inactivation of

pathogens and parasites (e.g., disinfection, water softening) (see Chapters 6 and 15). The U.S. Army mobile water treatment units use reverse osmosis (RO) in addition to the above treatment processes. Reverse osmosis (RO) is assumed to remove protozoan cysts, bacteria, viruses, and biotoxins. However, RO treatment may sometimes fail, resulting in lower removal of pathogens.

Some work has been undertaken to determine the fate of *B. anthracis* following water disinfection. A wide range of physical and chemical methods have been used to inactivate spores of *B. anthracis* and other bacilli (*B. cereus, B. globigii, B. subtilis*) in the environment, which includes water, air, and contaminated surfaces (Spotts et al., 2003). The chemicals used include hypochlorite, hydrogen peroxide, peracetic acid, formaldehyde, glutaraldehyde, sodium hydroxide, ethylene oxide, chlorine dioxide, or ozone. Chemical disinfection achieves from one to 6–log inactivation of the spores. *Bacillus* spores can also be inactivated by UV (Nicholson and Galeano, 2003), gamma irradiation, and microwave radiation (Bitton, Park, and Melker, unpublished data).

Unfortunately, there is a lack of information concerning the removal of biological warfare agents in conventional water treatment plants as well as in hand-held devices based on filtration through activated carbon or reverse osmosis (Burrows and Renner, 1999). In the home environment, point of use (POU) devices should also be investigated for their ability to remove BT agents and biotoxins (see Chapter 16 for more information on these devices).

Despite these safeguards, some pathogens or parasites (e.g., *Cryptosporidium*) break though the multiple barrier system. An example of the ability of drinking water to cause mass casualties in urban populations is the *Cryptosporidium* outbreak in Milwaukee, United States, in 1993. This outbreak affected 403,000 consumers, of whom 4400 were hospitalized and 54 died of the disease (Kaminski, 1994; MacKenzie et al., 1994). *Cryptosporidium* oocycts are not effectively inactivated by disinfectants and are more effectively removed by proper sand filtration. Many waterborne outbreaks involving several pathogens or parasites have been documented worldwide. Similarly to municipal tap water, bottled water can also be subject to accidental or intentional contamination with microbial pathogens.

There is also little information concerning the fate of biotoxins in drinking water and their removal by water treatment processes. The little information we have concerns some algal biotoxins (e.g., microcystins) that cause problems in water treatment plants using surface waters as the source water. Activated carbon and reverse osmosis units can potentially help in their removal from drinking water.

Most of the risk assessments performed by government and other agencies concerned mostly the military (Rotz et al., 2002). However, civilian populations include a wide range of age and health status groups and this makes the assessment more complex. In addition to the airborne route, civilian populations may also be exposed via the foodborne and waterborne routes. An example of intentional contamination of the food supply has been documented in Oregon where a salad bar was deliberately contaminated with *Salmonella* sp. (Torok et al., 1997).

17.5 EARLY WARNING SYSTEMS FOR ASSESSING THE CONTAMINATION OF SOURCE WATERS

Source waters are prone to natural, accidental, or intentional (e.g., bioterrorist activity) contamination. Early warning systems (EWS) that monitor the quality of source waters or water distribution systems are a necessity in order to protect the consumers. Such EWS must be based on rapid detection technologies to allow implementation of an effective response by the local or national public health authorities and emergency management officials (Foran, 2000).

In drinking water treatment plants with intakes from rivers, the most important contaminants are hydrocarbons, chemicals released as a result of industrial and transportation accidents, pesticides from soil runoffs, and microbial pathogens and parasites from untreated or inadequately treated effluents from wastewater treatment plants. The main components of an EWS include pollutant detection, characterization for confirmation, communication to the proper authorities and the public, and response to the pollution event (Gullick et al., 2003).

If contamination is suspected, the U.S. EPA recommends that the utilities conduct field-screening tests, which include radioactivity measurement, pH, chlorine, cyanide, and other tests if deemed necessary. Several other analytical tests are available for the preliminary assessment of water safety. A monitoring program may comprise several analytical procedures and approaches (Gullick et al., 2003; States et al., 2004) (Fig. 17.2):

- *Physical analyses.* Online probes measure pH, conductivity, dissolved oxygen, or turbidity. The latter has been correlated with the presence of oocysts of hardy parasites such as *Cryptosporidium* (see Chapter 4).

- *Chemical analyses.* Various online analytical probes are available for the measurement of anions (e.g., chloride, nitrate, nitrite), cations, certain metals (e.g., copper, lead, cadmium), chlorophyll, and organic compounds (e.g., analyses for total organic carbon, surfactants, hydrocarbons, pesticides, or various organics by using gas or liquid chromatography). Portable GC/MS devices can be used for onsite detection of chemical threats.

- *Microbiological tests.* Some examples are microbial biosensors to detect metals, nucleic acid probes, immunoassays, portable fluorometers to measure algal chlorophyll, and microchip array technology.

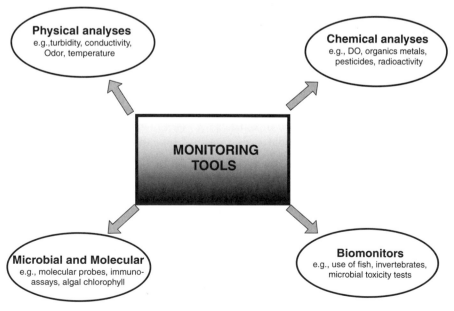

Figure 17.2   Monitoring tools for biological and chemical threats.

TABLE 17.4. Rapid Methodology for Detecting Biological and Chemical Threats to Our Drinking Water Supply[a]

Test	Product Name and Manufacturer	Contaminants Detected	Detection Limits	Assay Time (min)
Immunoassays	BTA test strips (tetracore)	Bioterrorism agents (BT) (e.g., *B. anthracis*)	10^5 CFU/mL	15
	SMART tickets (New Horizon Diagnostic)	Biotoxins	2–50 μg/L	
Enzymatic assays	Severn Trent Services	Organophosphates, carbamates	0.1–5 mg/L	5
Polymerase chain reaction (PCR)	R.A.P.I.D (Idaho Technology)	BT agents and other pathogens	10^3 CFU/mL	90
Portable GC/MS device	HAPSITE (Inficon, Inc.)	Volatile organic compounds	Low μg/L to mg/L	90
Rapid acute toxicity tests	See Chapter 19 of this book			

[a]Adapted from States et al. (2004).

- *Rapid immunoassays.*   As an example, some test strips allow the detection of *Bacillus anthracis*, *Yersinia pestis*, *Francisella tularensis*, *Clostridium botulinum* toxin, ricin, and staphylococcal enterotoxin B (see www.tetracore.com). However, these test strips have a relatively low sensitivity (detection limit is around 10^5 bacteria/mL) and sometimes low specifity.
- *Rapid enzymatic tests.*   Enzymatic tests are quite useful in detecting the presence of certain toxic chemicals in environmental samples. An example is a test based on the specific inhibition of acetyl cholinesterase by organophosphates and carbamates.
- *Molecular-based methods.*   A wide range of methods (e.g., polymerase chain reaction; PCR) are now available for the detection of human pathogens and parasites. Such PCR-based detection methods are more sensitive than immunoassays (see Chapters 1 and 4 for more details).
- *Online toxicity testing.*   Use of online biomonitors based on the activity of bacteria, algae, mussels, *Daphnia*, or fish (see Chapter 19 for more details on microbial tests).

A summary of the rapid methods for detecting biological and chemical threats is displayed in Table 17.4 (States et al., 2004).

17.6 PROTECTION OF DRINKING WATER SUPPLIES

The Biological and Toxin Weapons Convention (BTWC) of 1972 (http://www.fas.org/nuke/control/bwc/text/bwc.htm) deals with biological warfare agents (including bacteria, protozoa, and viruses) and biotoxins (e.g., botulinum toxin) produced by microorganisms. It forbids their development, production, and stockpiling. Limited information is available on the safe levels of these agents in drinking water, and no current regulation or guideline covers specifically their presence in drinking water supplies (Pontius, 2002). However, the 9/11/01 terrorist attack on American soil has led to a closer partnership between the U.S. EPA and the drinking water community to safeguard the safety of the drinking water supplies. They were joined in this effort by the American Water Works Association (AWWA), which helps in the dissemination of information concerning the security of water utilities. In 2002, the U.S. Congress passed the Bioterrorism Prevention and Response Act (also called the Bioterrorism Act) (PL 107-188), which mandates security requirements (vulnerability assessment and emergency response plans) for approximately 8000 water utilities in the United States (Anonymous, 2003; Pontius, 2003). Vulnerability assessment includes water collection, pretreatment and treatment, distribution pipes, storage facilities, safety issues related to the use and storage of chemicals used in the plant, and computerized data acquisition systems. Following the vulnerability assessment, the utilities must submit emergency response plans to respond to a potential terrorist attack. These plans may be different from those addressing natural disasters such as an earthquake or a tornado.

Information Sharing and Analysis Centers (ISACs) were established to promote the exchange of information between water utilities and security agencies. WaterISAC is a clearinghouse for the flow of information concerning biological, chemical, and cyber threats to water and wastewater treatment plants. Some WaterISAC resources include databases on contaminants and emergency contacts, and new search engines. An example of a useful database is the United Kingdom Water Industry Research (UKWIR) concerning toxicity and microbiology data (Sullivan, 2004).

The deliberate contamination of our drinking water should be prevented by protecting water reservoirs and water distribution systems. The government should stockpile, if available, vaccines against microbial pathogens and antidotes against chemical agents.

17.7 WEB RESOURCES

www.tetracore.com (test strips for the detection of bioterrorism agents and biotoxins)
http://www.waterisac.org/ (web page for WaterISAC concerning the flow of information between utilities and security agencies)
http://www.amwa.net/security (water security information from the Association of Metropolitan Water Agencies)
http://www.epa.gov/safewater/security/ (U.S. EPA page on water security)
http://www.fas.org/nuke/control/bwc/text/bwc.htm (Biological and Toxin Weapons Convention (BTWC) of 1972, which deals with biological warfare agents and biotoxins)
http://www.epa.gov/ogwdw000/security/index.html (EPA water infrastructure security)

17.8 FURTHER READING

Atlas, R.M. 2002. Bioterrorism: From threat to reality. Ann. l Rev. Microbiol. 56: 167–185.
Burrows, I.D., and S.E. Renner. 1999. Biological warfare agents as threats to potable water. Environ. Health Perspectives 107: 975–984.
Christopher, G.W, T.J. Cieslak, J.A. Pavlin, and E.M. Eitzen, Jr. 1997. Biological warfare: a historical perspective. J. Amer. Med. Assoc. 278(5): 412–417.
Denileon, G.P. 2001. The who, what, why and how of counter terrorism issues. J. Amer. Water Wks. Assoc. 93(5): 78–85.
Gullick, R.W., W.M. Grayman, R.A. Deininger, and R.M. Males. 2003. Design of early warning systems for source waters. J. Amer. Water Wks. Assoc. 95(11): 58–72.

Hawley, R.L., and E.M. Eitzen Jr. 2001. Biological weapons – A primer for microbiologists. Ann. Rev. Microbiol. 55: 235–253.

Khan, A.L., D.L. Swerdlow, and D.D. Juranek. 2001. Precautions against biological and chemical terrorism directed at food and water supplies. Pub. Health Rep. 116: 3–14.

Rotz, L.D., A.S. Khan, S.R. Lillibridge, S.M. Ostroff, and J. M. Hughes. 2002. Public health assessment of potential biological terrorism agents. Emerg. Infect. Dis. 8(2): 225–229.

Sheehan, T. 2002. Bioterrorism, pp. 771–782, In: *Encyclopedia of Environmental Microbiology*, Gabriel Bitton, editor-in-chief, Wiley-Interscience, N.Y.

PART E

BIOTECHNOLOGY IN
WASTEWATER TREATMENT

18

POLLUTION CONTROL BIOTECHNOLOGY

18.1 INTRODUCTION

Advances in microbial genetics and genetic engineering have given great impetus to the field of pollution control biotechnology. In some countries, efforts are being focused on the application of biotechnology to wastewater treatment. There is a general interest in the application of genetic engineering methods, immobilization techniques for waste treatment, as well as the development and improvement of bioreactors for wastewater treatment (Matsui et al., 1991). Owing to the lack of information on the fate of genetically engineered microorganisms in the environment, only microorganisms isolated by traditional enrichment techniques are being marketed.

In this chapter, we will discuss the use of commercial enzyme and microbial blends, immobilized microorganisms, and molecular techniques for enhancing biological wastewater treatment. We will also explore the role of microorganisms in metal removal in wastewater treatment plants.

18.2 USE OF COMMERCIAL BLENDS OF MICROORGANISMS AND ENZYMES IN WASTEWATER TREATMENT

18.2.1 Use of Microorganisms in Pollution Control

The early attempts to use microorganisms in the pollution control field focused on anaerobic digestion. Later, microorganisms capable of degrading pesticides and other chemicals in industrial wastes were isolated and used in commercial preparations designed for pollution control. During the 1970s, microbial preparations were marketed for enhancing the operation efficiency of waste treatment processes. This approach is referred to as *bioaugmentation* (Johnson et al., 1985).

18.2.2 Production of Microbial Seeds

Microbial strains for enhancing biodegradation of specific chemicals are generally isolated from environmental samples (wastewater, biosolids, compost, soil) and selected by conventional enrichment techniques (Walter, 1997). They are grown in nutrient media that contain a specific organic chemical as the sole source of carbon and energy or as a sole source of nitrogen. Furthermore, strains that can handle relatively high concentrations of the target chemical are selected. Some of the microbial strains may be subsequently irradiated to obtain a desirable mutation (Johnson et al., 1985) (Fig. 18.1). Many of the microbial seeds available on the market are commercialized for bioremediation of oil pollution (Dott et al., 1989; Venosa et al., 1992).

Before using commercial preparations of microorganisms for pollution control, one must first gain information about the biodegradability of the chemical pollutant under investigation. This can be done by searching the literature or by undertaking biodegradation studies under laboratory conditions. Bioassays are also used for assessing the toxicity of the wastewater under consideration and the commercial preparation of microbial seeds (see Chapter 20). Enrichment culture techniques have been used to isolate commercial bacterial strains that degrade petroleum hydrocarbons. For example, mixtures of bacterial isolates belonging to the genus *Pseudomonas* are marketed for the in situ biorestoration of aquifers contaminated with aliphatic or aromatic hydrocarbons (von Wedel et al., 1988). The selected strains are grown in large fermenters and then concentrated via centrifugation or filtration. They are then preserved via lyophilization, drying, or freezing. Spore-forming bacteria that are able to degrade xenobiotics can also be stored in dry form as hardy spores (Sembries and Crawford, 1997). For successful results, the applied microorganism must withstand the desired environmental conditions. These conditions include temperature, pH, dissolved oxygen, nutrient availability, and ability to withstand potential toxicity of the wastewater.

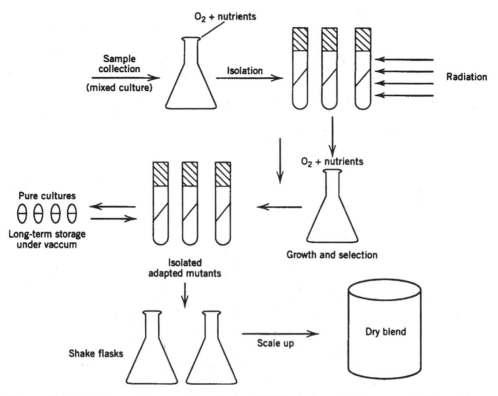

Figure 18.1 Isolation and preparation of microbial blends for pollution control. Adapted from Johnson et al. (1985).

18.2.3 Use of Bioaugmentation in Waste Treatment

Bioaugmentation, a technology known for decades, is the use of selected microbial strains isolated from the environment to improve some of the processes involved in traditional waste treatment or in bioremediation sites. Commercial bioaugmentation products are single or consortia of microorganisms with desirable degradative properties and with desirable characteristics (e.g., the ability to grow under low nutrient conditions, to withstand high toxicant concentrations, and to survive in the contaminated environment). Industrial wastewater treatment plants are presently the most important users of bioaugmentation products. The selected microorganism is generally added to a bioreactor to maintain or enhance the biodegradation potential in the bioreactor. Unfortunately, information on the formulation of mixtures of microbial cultures is scanty because of trade secrets (Forsyth et al., 1995; Grubbs, 1984; Rittmann et al., 1990). Bioaugmentation can help in increasing the resistance of a conventional activated sludge against shock loadings of toxic compounds (Quan et al., 2003). Applications of bioaugmentation include the following.

- *Increased BOD removal.* Microbial strains may be used to *increase BOD removal* in wastewater treatment plants. For example, anoxygenic phototrophic bacteria have been examined for their capacity to reduce BOD/COD in waste treatment (Sasikala and Ramana, 1995).
- *Reduction of sludge volume.* Production of large amounts of biosolids is a serious problem associated with aerobic waste treatment and, thus, reduction of biosolids

volumes is highly desirable (see Chapter 12). The reduction is the result of increased organic removal after the addition of a mixed culture of selected microorganisms. Reductions in volumes of generated sludge of 17 to nearly 30 percent have been documented.

- *Use of mixed cultures in the digestion of biosolids.* In aerobic digesters, the use of mixed cultures has led to significant savings in energy requirements. In anaerobic digesters, bioaugmentation has resulted in enhanced methane production.

- *Biotreatment of hydrocarbon wastes.* Commercial bacterial formulations have been traditionally used for the treatment of hydrocarbon wastes. For example, the addition of cultures of mutant bacteria improved the effluent quality of a petrochemical wastewater treatment plant (Christiansen and Spraker, 1983).

- *Biotreatment of hazardous wastes.* The use of added microorganisms for treating hazardous wastes (e.g., phenols, ethylene glycol, formaldehyde) has been attempted and has a promising future. Bioaugmentation with parachlorophenol-degrading bacteria achieved a 96 percent removal in 9 h as compared with a control that exhibited 57 percent removal after 58 h (Kennedy et al., 1990) (Fig. 18.2). *Candida tropicalis* cells have been also used to remove high concentrations of phenol in wastewater (Kumaran and Shivaraman, 1988). *Delsulfomonile tiedjei,* when added to a methanogenic upflow anaerobic granular-sludge blanket (UASB), increased the ability of the bioreactor to dechlorinate 3-chlorobenzoate (Ahring et al., 1992). Anoxygenic phototrophic bacteria have also been considered for the degradation of toxic compounds in wastes (reviewed by Sasikala and Ramana, 1995). In wastewater treatment, bioaugmentation was shown to increase the biodegradation of 2,4-dichlorophenol (Quan et al., 2004), phenolic compounds (Hajji et al., 2000; Selvaratnam et al., 1997), and chloroaniline (Boon et al., 2000).

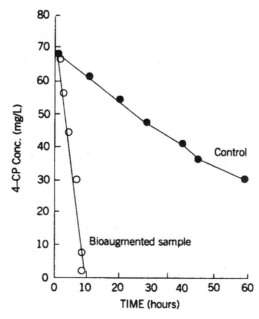

Figure 18.2 Biodegradation of parachlorophenol via bioaugmentation. Adapted from Kennedy et al. (1990).

Some major drawbacks of bioaugmentation are the need for an acclimation period prior to onset of biodegradation, and short survival or lack of growth of microbial inocula in the seeded bioreactors. Evaluations of some of these commercial products are sometimes negative or inconclusive. Furthermore, addition of cell-free extracts does not significantly improve the performance of biological waste treatment (Jones and Schroeder, 1989). Some novel bioaugmentation schemes for the degradation of hazardous wastes in wastewater treatment plants are being investigated. For example, the *enricher– reactor process*, which involves the maintenance of a separate offline acclimated culture for addition to an activated sludge system or a fluidized-bed reactor, does improve the biodegradation and removal of xenobiotics (Babcock et al., 1992; Ro et al., 1997). The enrichment cultures can be maintained in the enricher–reactor under optimum growth conditions by adding a less hazardous inducing compound of similar structure as the target compound (Babcock and Stenstrom, 1993). Periodic addition of biomass from an enricher–reactor to a fluidized fixed film reactor led to more than 90 percent removal of 1-naphthylamine (Ro et al., 1997). This bioaugmentation process is suitable for treating low-strength wastewaters.

18.2.4   Use of Enzymes in Waste Treatment

Enzymes play a key role in the hydrolysis and biotransformation of organics in wastewater treatment plants. Several enzymes may be detected in wastewater samples, including catalase, phosphatases, aminopeptidases, and esterases (Boczar et al., 1992; Hosetti and Frost, 1994). It has also been suggested that enzymes can be added to wastewaters to improve the treatability of xenobiotic compounds.

Earlier studies showed that microbial exoenzymes are used to detoxify pesticides in soils (Munnecke, 1981). For example, carbaryl is transformed to 1-naphtol which is 920 times less toxic than the parent compound. Parathion hydrolases, produced by *Pseudomonas* sp. and *Flavobacterium* species, degrade parathion to diethylthiophosphoric acid and *p*-nitrophenol (Mulbry and Karns, 1989). These enzymes have been used for the cleanup of containers for parathion, detoxification of wastes containing high concentrations of organophosphates, and in soil cleanup operations (Karns et al., 1987).

Less is known about the use of enzymes in wastewater treatment plants. To be useful, the added enzymes must be stable under conditions (e.g., temperature, pH, toxicant levels) prevailing in waste treatment processes. An application of this technology is illustrated by the use of specific enzymes to reduce the production of excessive amounts of extracellular polysaccharides during wastewater treatment. An overproduction of polysaccharides may lead to increased water retention, resulting in reduced sludge dewatering. A number of enzymes can degrade these exopolymers (Sutherland, 1977). Phage-induced depolymerases have been described in several phage-bacterial sytems (Adams and Park, 1956; Bartel et al., 1968; Bessler et al., 1973; Nelson et al., 1988; Vandenbergh et al., 1985). A phage-induced depolymerase has been isolated from a sludge sample and found to readily increase the degradation, as shown by viscosity reduction, of exopolysaccharides of sludge bacteria (Nelson et al., 1988) (Fig. 18.3). The addition of this enzyme or mixtures of enzymes could eventually be used to improve sludge dewatering.

The polymerization and precipitation of aromatic compounds (e.g., substituted phenols and anilines) in drinking water and wastewater can be catalyzed by specific enzymes such as horseradish peroxidase (HRP). This enzyme catalyzes the oxidation of phenol and chlorophenols by hydrogen peroxide (Kilbanov et al., 1983; Maloney et al., 1986; Nakamoto and Machida, 1992; Nicell et al., 1992).

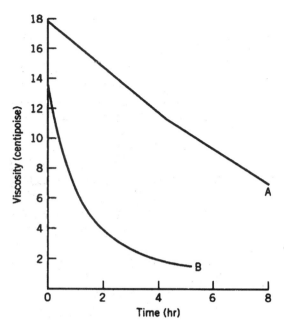

Figure 18.3 Viscosity reduction of polysaccharides by depolymerase. A, 1.8×10^9 PFU/mL; B, 18×10^9 PFU/mL. Adapted from Nelson et al. (1988).

Extracellular fungal laccases from *Trametes versicolor* or *Botrytis cinerea* can function under adverse conditions (e.g., in presence of organic solvents, low pH, high temperature) and can be useful for the dechlorination of chlorinated phenolic compounds or oxidation of aromatic compounds. It was suggested that these enzymes can be used for the treatment of effluents generated by the pulp and paper industry (Milstein et al., 1989; Roy-Arcand and Archibald, 1991; Slomczynski et al., 1995).

To respond to the need to find enzymes that can function under adverse environmental conditions, attention is being focused on the "extremozymes" of microorganisms that live in ocean vents, hot springs, and other adverse environments (Brown, 1996). The usefulness of these enzymes is best exemplified by *Thermus aquaticus*, a thermophilic bacterium that has been isolated from Yellowstone hot springs in the United States. This bacterium is the source of the heat-stable *Taq* polymerase enzyme used in polymerase chain reaction (PCR). Other commercial applications of extremozymes include their use in gas drilling and detergents, and potentially in detoxification of chemicals in wastewater treatment plants.

18.3 USE OF IMMOBILIZED CELLS IN WASTE TREATMENT

Cell immobilization technology offers several advantages such as cell protection from biotic and abiotic stresses, protection from toxic chemicals, and increase in cell survival and metabolic activity.

Some wastewater treatment processes are based on naturally immobilized micro-organisms. Aggregated cells in activated sludge systems as well as cells attached to rock or plastic surfaces in trickling filters or rotating biological contactors, are good examples of exploitation of immobilized cells in waste treatment (Webb, 1987).

Fluidized-bed reactors, as used in water and wastewater treatment, are other examples of cell immobilization on surfaces such as sand or other particles. These fluidized-bed reactors are used in the anaerobic treatment of wastewater (see Chapter 13). The use of immobilized microorganisms in wastewater treatment is now under consideration in several laboratories.

18.3.1 Immobilization Techniques

Various approaches have been taken for the immobilization of microorganisms, animal and plant cells, and organelles. The various immobilization techniques used are summarized in Table 18.1 (Brodelius and Mosbach, 1987). The most popular approach to immobilization is entrapment of cells in polymeric materials such as alginate, carrageenan, polyurethane foam, polyacrylamide, or silica gel. Owing to the toxicity and detrimental effect of polyacrylamide on cell viability, natural algal polysaccharides such as alginate and carrageenan have been the polymers of choice for microbial cell immobilization (Cheetham and Bucke, 1984). Carrageenan is an algal polysaccharide extracted from algae of the class Rhodophyceae, while alginate is extracted from algae belonging

TABLE 18.1. Cell Immobilization Techniques[a]

Cell/Cell Component	Immobilization Technique/Support
Microorganisms	Entrapment
	Alginate
	Carrageenan
	Polyacrylamide
	Polyacrylamide-hydrazide
	Agarose
	Photo-cross-linkable resin, prepolymers, and urethane
	Epoxy carrier
	Chitosan
	Cellulose; cellulose acetate
	Gelatin
	Adsorption
	Sand beads
	Porous brick, porous silica
	Celite
	Wood chips
	Covalent binding
	Hydroxyethyl acrylate
Animal cells	Hollow fiber
	Entrapment (agarose and fibrin)
Plant cells	Entrapment
	Polyurethane
	Agarose, carrageenan, alginate
Plant protoplasts	Microcarriers
	Entrapment
	Alginate, agarose, and carrageenan
Organelles	Entrapment
	Alginate
	In cross-linked protein

[a]Adapted from Brodelius and Mosbach (1987).

to the class Phaeophyceae (e.g., *Laminaria hyper borea* or *Macrocystis pyrifera*). Alginate reacts with most divalent cations, particularly Ca^{2+}, to form gels. Briefly, the cells are mixed with a solution of sodium alginate and the mixture is poured dropwise over a solution of $CaCl_2$. Beads form instantaneously and are left in the $CaCl_2$ solution for approximately 1 h for complete gel formation (Bucke, 1987). However, calcium alginate beads can be destroyed if the surrounding medium contains phosphates or other calcium-chelating agents. Loss of bead integrity may also result from excess cell growth or from gas production by the immobilized microorganisms. Microorganisms can also be immobilized in porous silica gels. *Spirulina platensis*, immobilized in silica gel, was found to be an efficient adsorbent for cadmium (Rangsayatorn et al., 2004).

Other techniques include the immobilization of activated sludge microorganisms in polyvinyl alcohol (PVA) dropped into a saturated boric acid solution or refrigerated for gel formation (Lin and Chen, 1993; Matsui et al., 1991). Photo-cross-linkable PVA can be used as a mild method of gel entrapment (Ichimura and Watanabe, 1982). Cells can also be co-immobilized with magnetic particles in polyacrylamide gels. Batch and continuous flow experiments showed almost 100 percent phenol removal for at least 40 days by sludge microorganisms co-immobilized with ferromagnetic particles (Ozaki et al., 1991). A distinct advantage of this system is the ability to recover the immobilized microorganisms with a magnet.

Encapsulation of microorganisms within polymers generally produces relatively large beads (2–3 mm in diameter), in which immobilized microorganisms may be subjected to oxygen limitations. Additionally, these beads would not be suitable for certain environmental applications (e.g., groundwater restoration). An improved procedure for bacterial encapsulation produces microspheres with a 2 to 50 μm diameter. Entrapped *Flavobacterium* cells are as active as free cells as regards pentachlorophenol degradation (Stormo and Crawford, 1992).

18.3.2   Examples of the Use of Immobilized Cells in Waste Treatment

The use of immobilized enzymes for the degradation of pesticides has been widely investigated. However, immobilized cells have also been considered for the treatment of various wastes, for the decontamination of water or wastewater containing natural or xenobiotic compounds, and for decontamination of soils and aquifers (Cassidy et al., 1996; Crawford and O'Reilly, 1989). The following are some examples of the use of this technology in pollution control.

18.3.2.1 Removal of Brown Lignin Compounds.   Brown lignin compounds found in paper mill effluents can be removed by immobilized white-rot fungus (*Coriolus versicolor*) (Livernoche et al., 1983).

18.3.2.2 Biodegradation of Phenolic Compounds.   Several studies addressed the degradation of phenolic compounds by using immobilized bacteria (Bettmann and Rehm, 1984; Bisping and Rehm, 1988; Hackel et al., 1975; Heitkamp et al., 1990). Although lower rates of phenol degradation were achieved by an immobilized consortium of methanogens, the immobilized microorganisms were able to tolerate higher phenol concentrations (Dwyer et al., 1986). Chlorinated phenols are degraded by bacteria (*Pseudomonas*, *Arthrobacter*) immobilized on the chitin surface by covalent bonding (Portier, 1986), by *Alcaligenes* sp. entrapped in polyacrylamide-hydrazide (Bisping and Rhem, 1988), or by *Rhodococcus* immobilized on a polyurethane carrier (Valo et al., 1990). The biodegradation of 2-chlorophenol by activated sludge microorganisms immobilized

in calcium alginate was also demonstrated (Sofer et al., 1990). Bioreactors containing *Flavobacterium* immobilized in calcium alginate can degrade pentachlorophenol at a maximum degradation rate of 0.85 mg PCP/g beads/h (O'Reilly et al., 1988). Tyrosinase immobilized on magnetite, cation exchange resins or siliceous supports can remove rapidly, via oxidation, phenol, chlorophenols, methoxyphenols and cresols from waste-water. The colored products formed may be efficiently removed by treatment with chitosan, a polycation derived from chitin (Seetharam and Saville, 2003; Wada et al., 1992; 1993). A bioreactor packed with a highly porous nylon biocarrier achieved a high removal of *p*-nitrophenol (Heitkamp and Stewart, 1996).

18.3.2.3 *CH₄ Production by Immobilized Methanogens.* (Karube et al., 1980). Anaerobic waste treatment may consist of two-stage bioreactors containing immobilized microorganisms. The first stage contains acid formers while the second stage contains methanogens (Messing, 1988).

18.3.2.4 *Dehalogenation of Chloroaromatics.* Chloroaromatics (e.g., mono-chlorobenzoates, 2,4-dichlorophenoxyacetic acid) may be dehalogenated by immobilized *Pseudomonas* sp. cells (Sahasrabudhe et al., 1991).

18.3.2.5 *Use of Immobilized Nitrifiers and Denitrifiers.* These may be used to tackle nitrogen pollution problems (Nilsson and Ohlson, 1982; Nitisoravut and Yang, 1992; van Ginkel et al., 1983). Nitrifying bacteria, immobilized in polyethylene glycol resin and added to activated sludge as suspended beads, enhance nitrification in activated sludge, thereby reducing the time necessary for complete nitrification (Tanaka et al., 1991; 2003). The enhancement of nitrification by the immobilized bacteria is illustrated in Figure 18.4 (Tanaka et al., 1991). Cell immobilization in polyvinyl alcohol (PVA), photo-cross-linkable PVA (PVA-SbQ), or polyethylene glycol has also been considered

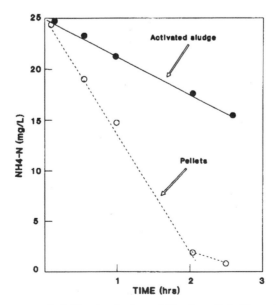

Figure 18.4 Enhancement of nitrification by immobilization of nitrifiers. Adapted from Tanaka et al. (1991).

for enhancing nitrification in waste treatment (Ariga et al., 1987; Geenens et al., 2002; Myoga et al., 1991; Vogelsang et al., 1997). Activated sludge nitrifiers, immobilized in PVA-SbQ/alginate beads, maintained their nitrifying activity for a period of 10 months (Vogelsang et al., 1997). Pilot tests with "pellet reactors" containing immobilized nitrifiers showed enhanced nitrification in activated sludge (Geenens et al., 2002). Nitrifying granules forming in an aerobic upflow fluidized-bed reactor were able to carry out nitrification, resulting in the removal of ammonia at a rate of 1.5 kg-N/m^3/day (Tsuneda et al., 2003). It appears that the immobilized nitrifiers are more stable than free cells (Asano et al., 1992).

Bioreactors containing denitrifiers immobilized in cellulose triacetate or PVA carry out denitrification in wastewater with up to 99 percent efficiency (Lin and Chen, 1993; Nitisoravut and Yang, 1992). As shown in Chapter 3, denitrifiers need a carbon source (e.g., methanol) to efficiently reduce NO_3 to N_2. To avoid residual methanol in the effluents, denitrifiers and methanogens can be co-immobilized in PVA beads where the methanogens turn methanol into methane gas (Chen et al., 1997b). Because methanol is toxic in drinking water, entrapped denitrifiers can also use ethanol as a carbon source to remove nitrate from groundwater (Qian et al., 2001).

18.3.2.6 Immobilized Activated Sludge Microorganisms.
A two-step process, consisting of a reactor containing immobilized activated sludge microorganisms followed by a biofilm reactor, achieved a high treatment efficiency when the BOD load was 1.4 kg/m^3/day. Moreover, the operation of the first bioreactor under anaerobic conditions helped prevent the biofouling of the beads by filamentous fungi (Inamori et al., 1989). Cellulose triacetate was used as the immobilizing agent to entrap activated sludge microorganisms for the simultaneous removal of carbon and nitrogen (Yang et al., 2002). Some advantages of the entrapped mixed microbial cells process are (1) high density of microbial cells; (2) ability to co-immobilize a mixture of microorganisms; (3) long-term performance (i.e., years) of the immobilized cells.

18.3.2.7 Use of Immobilized Algae to Remove Micronutrients from Wastewater Effluents.
Scenedesmus entrapped in carrageenan beads or *Phormidium* immobilized on the surface of chitosan flakes were shown to remove nitrogen and phosphorus from wastewater effluents (Chevalier and de la Noue, 1985; Proulx and de la Noue, 1988).

18.3.3 Use of Immobilized Cells and Enzymes in Biosensor Technology

A biosensor is composed of a biological sensing element (e.g., immobilized microorganism, enzyme, nucleic acid, or antibody), which, upon interaction with an analyte, produces a signal that is transmitted to a transducer that converts it into an electrical signal (Burlage, 1997) (Fig. 18.5). Biosensors are made of a wide range of biological elements and transducers (Turner et al., 1987) (Table 18.2). Several biosensors have been developed and applied to microbiological and biochemical processes in food, clinical, pharmaceutical, and food industries as well as wastewater treatment (Bryers and Hamer, 1988; Karube, 1987; Karube and Tamiya, 1987). A breakthrough in medical technology was the development of biosensors capable of rapidly measuring glucose and urea in body fluids. Examples of biosensors that can be applied to wastewater treatment are those designed for the detection of BOD, ammonia, organic acids, and methane (Karube and Tamiya, 1987; Matsui et al., 1991).

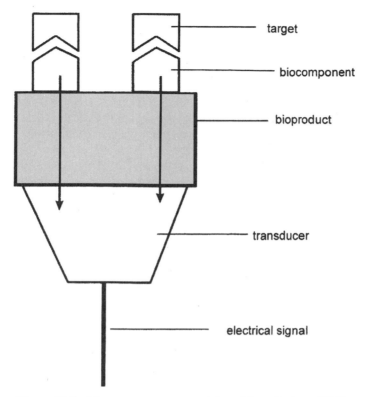

Figure 18.5 Biosensor components. Adapted from Burlage (1997).

18.3.3.1 BOD Sensors. More rapid methods for determination of BOD are being sought. Biosensors using pure microbial cultures (e.g., *Bacillus subtilis*, *Klebsiella oxytoca*, *Clostridium butyricum*, *Pseudomonas putida*, *Trichosporon cutaneum*) or mixtures of activated sludge microorganisms have been considered and some are commercially available. Biofilm-based biosensors consist of immobilized microorganisms trapped between a porous membrane and a gas-permeable membrane (Fig. 18.6) (Liu and Mattiasson, 2002). The response is a change in dissolved oxygen concentration.

A biosensor consisting of immobilized yeast, *Trichosporon cutaneum*, and an oxygen probe was developed for BOD estimation (Karube, 1987; Karube et al., 1977; Karube and

TABLE 18.2. Biosensor Components[a]

Biological Elements	Transducers
Organisms	Potentiometric
Tissues	Amperometric
Cells	Conductimetric
Organelles	Impedimetric
Membranes	Optical
Enzymes	Calorimetric
Receptors	Acoustic
Antibodies	Mechanical
Nucleic acids	"Molecular" electronic
Organic molecules	

[a]Adapted from Turner et al. (1987).

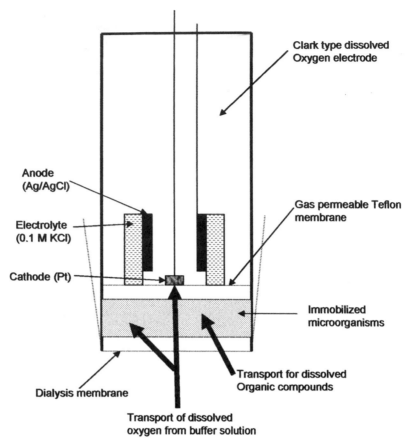

Figure 18.6   BOD biosensor. Combination of immobilized microorganisms with a Clark-type oxygen electrode. From Liu and Mattiasson (2002).

Tamiya, 1987). The BOD biosensor includes an oxygen electrode, which consists of a platinum cathode and an aluminum anode bathing in saturated KCl solution, and a Teflon® membrane. Yeast cells are immobilized on a porous membrane and are trapped between the porous and the Teflon membranes. Oxygen consumption by the immobilized microorganisms causes a decrease in current until a steady state is reached. A good correlation was observed between the current drop and BOD values as obtained by standard methods. The BOD biosensor measures BOD at 3–60 mg/L (Karube and Tamiya, 1987). A biosensor using *T. cutaneum* immobilized in polyvinyl alcohol displays a very short response time (<30 s) and is stable for 48 days. This sensor showed a good correlation with the 5-day BOD test (Riedel et al., 1990) (Fig. 18.7). A biosensor based on activity of bioluminescent bacteria was also proposed to measure BOD (Hyun et al., 1993).

A biosensor can also incorporate a combination of microorganisms to increase the range of substrates assimilated by the microorganisms. Hence, a BOD sensor made of co-immobilized *Trichosporon cutaneum* and *Bacillus licheniformis* was proposed (Suriyawattanakul et al., 2002). This justifies the use of activated sludge microorganisms in biosensor construction (Lui et al., 2000).

18.3.3.2 Methane Biosensor.

This biosensor consists of immobilized methanotrophic bacteria (*Methylomonas flagellata*) in contact with an oxygen electrode.

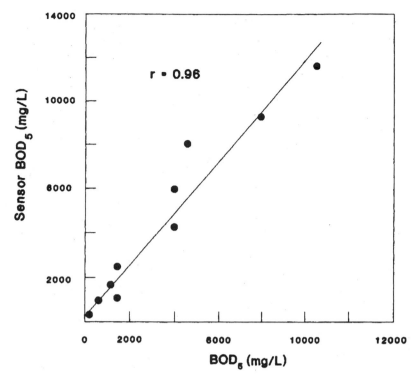

Figure 18.7 Correlation between biosensor BOD value with the 5-day BOD value of wastewater. From Riedel et al. (1990). (Courtesy of Pergamon Press.)

The immobilized bacteria use methane as well as oxygen according to the following reaction (Matsunaga et al., 1980):

$$CH_4 + NADH_2 + O_2 \longrightarrow CH_3OH + NAD + H_2O \qquad (18.1)$$

Oxygen consumption leads to a decrease in current, which is proportional to methane concentration in the sample. A methane amperometric microsensor (30 μm diameter) with a response time of 30 s was used to measure methane profiles within biofilms (Damgaard and Revsbech, 1997; Damgaard et al., 2001). This technology makes it possible to study methane dynamics within short time periods and biofilm depths of a few millimeters. A methane profile of a natural biofilm is displayed in Figure 18.8 (Damgaard et al., 2001).

18.3.3.3 *Ammonia and Nitrate Biosensors.*

This ammonia biosensor, based on amperometry, consists of immobilized nitrifying bacteria (e.g., *Nitrosomonas europaea*) and a modified oxygen electrode. The biosensor, with a lifetime of approximately two weeks, was used for ammonia determination in wastewaters. Based on the conversion of nitrate to N_2O by an immobilized denitrifying bacterium *Agrobacterium* sp. (Larsen et al., 1996), the nitrate biosensor has been used to measure nitrate profiles in biofilms (Schramm et al., 1996). A recombinant bioluminescent cyanobacterium *Synechocystis* was used as a whole-cell biosensor for monitoring nitrate (NO_3^-) bioavailability in aquatic environments (Mbeunkui et al., 2002).

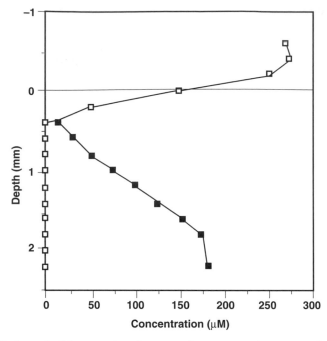

Figure 18.8 Methane (solid squares) and oxygen (open squares) concentration profiles in an undistributed natural biofilm incubated in tap water for 24 h with no added electron donors. From Damgaard et al. (2001).

18.3.4 Advantages and Disadvantages of Immobilized Cells

Some advantages of cell immobilization are the following (Brodelius and Mosbach, 1987; Bryers and Hamer, 1988; Cassidy et al., 1986; Webb, 1987):

1. Continuous reactor operation without risk of cell washout
2. Ease of cell separation from reaction mixture
3. High cell density
4. Ability to reuse cells
5. Ability of different microbial species spatially separated to perform different functions
6. Enhanced overall productivity due to increased cell concentration in a given volume
7. Enhanced stability of immobilized microorganisms or enzymes
8. Increased plasmid stability
9. Decrease in volume of the bioreactors
10. Protection of immobilized microorganisms from toxicity and environmental stresses
11. Easy and low cost operation (e.g., rotating biological contactors; trickling filters)

However, cell immobilization has limitations:

1. Diffusion problems due to high cell density and to low solubility of oxygen in water
2. Changes in cell physiology that might affect productivity

3. Changes in composition of the microbial population, which can be a problem in wastewater, which harbors mixed populations of microorganisms

4. Cost of immobilization when using artificially captured cell systems

18.4 ROLE OF MICROORGANISMS IN METAL REMOVAL IN WASTEWATER TREATMENT PLANTS

Physical/chemical processes are generally employed for heavy metal removal from wastewater. These include ion exchange, oxidation/reduction, precipitation, ultra-filtration, and many others. Much of the particulate-associated metals is removed by sedimentation during primary treatment of wastewater. The removal of soluble forms in the activated sludge system depends on the type of metal. It is around 50–60 percent for Cd, Hg, Cu, and Zn, but it may be lower for other metals (e.g., Ni, Co) (Sterrit and Lester, 1986).

Microorganisms offer an alternative to physical/chemical methods for metal removal/ recovery (Forster and Wase, 1987). Metals are removed (i.e., immobilized) by the activated sludge biomass. Their removal increases with sludge age and is partially due to increase in mixed liquor solid concentration.

18.4.1 Metal Removal by Wastewater Microorganisms

Microorganisms (bacteria, cyanobacteria, algae, fungi) have been used as biosorbents for a wide range of metals because they offer a high surface area for metal binding and their production is relatively inexpensive. Another advantage is the ability to recycle the microbial biomass.

It is well known that extracellular polymeric substances produced by some microorganisms in marine waters, freshwaters, and wastewater can scavenge metals following interaction of carboxyl groups on acidic polysaccharides with metal ions (Weiner, 1997). Extracellular polymers produced by microorganisms commonly found in activated sludge (see Chapter 8) display a great affinity for metals. Several bacterial types (e.g., *Zooglea ramigera*, *Bacillus licheniformis*), some of which have been isolated from activated sludge, produce extracellular polymers that are able to complex and subsequently accumulate metals such as iron, copper, cadmium, nickel, or uranium. The accumulated metals can be easily released from the biomass by treatment with acids. For example, *Zooglea ramigera* can accumulate up to 0.17 g of Cu per g of biomass (Norberg and Persson, 1984; Norberg and Rydin, 1984). This bacterium, when immobilized in alginate beads, is also able to accumulate Cd from solutions containing Cd concentrations as high as 250 mg/L (the alginate beads adsorb some of the cadmium) (Kuhn and Pfister, 1990). Some microorganisms may also synthesize siderophores that chelate iron and facilitate its transport into the cell (Lundgren and Dean, 1979).

Nonliving immobilized microbial systems are also able to remove metals from wastewaters. Patented processes involving immobilized bacteria, fungi, and algae, have been developed to remove heavy metals from wastewater (Brierley et al., 1989). A proprietary product, called *Algasorb*, consists of algae cells embedded in a silica gel polymeric material, and can remove heavy metals, including uranium (Anonymous, 1991). Biosorption is the sequestration of metals and radionuclides by dead or live fungal, bacterial, or algal cells and can be utilized to adsorb metals in wastewater. The mechanisms involved in biosorption include adsorption, ion exchange, as well as electrostatic and hydrophobic interactions (Barkay and Schaefer, 2001; Schiewer and Volesky, 2000). Fungal mycelia (e.g., *Aspergillus*, *Penicilium*) have also been considered for metal removal from wastewater (Galun et al., 1982) and thus may offer a good alternative for detoxification of effluents. Waste fungal mycelia (*Aspergillus niger*, *Penicillium chrysogenum*, *Claviceps paspali*) from industrial fermentation plants, have been successfully used as biosorbents of zinc ions (Luef et al., 1994). Fungi remove metals from solutions by sorption to the fungal surface or by a much slower energy-dependent intracellular uptake. Biosorption column studies have shown that immobilized *Aspergillus oryzae* remove Cd efficiently from solution. It was shown that detergent treatment of fungal biomass considerably improves metal removal (Ross and Townsley, 1986). Cadmium appears to adsorb to the fungal biomass, but active uptake of the metal by the fungus does not appear to be significant (Kiff and Little, 1986). Therefore, most of the removal of metals by fungi does not appear to be linked to a metabolic process. Table 18.3 (Eccles and Hunt, 1986) lists some microorganisms used for removal and/or recovery of metals from industrial wastewaters.

18.4.2 Mechanisms of Metal Removal by Microorganisms

In the environment, including wastewater treatment plants, metals are removed by microorganisms by the following mechanisms (Brierley et al., 1989; Sterrit and Lester, 1986; Trevors, 1989; Trevors et al., 1985):

18.4.2.1 Adsorption to Cell Surfaces. Microorganisms bind metals as a result of interactions between metal ions and the negatively charged microbial surfaces. Gram-positive bacteria are particularly suitable for metal binding. Fungal and algal cells also display a high affinity for heavy metals (Darnall et al., 1986; Ross and Townsley, 1986). Sorption of metals to activated sludge solids has been found to conform to the Langmuir and Freundlich isotherms. Figure 18.9 (Mullen et al., 1989) displays the Freundlich isotherms for the adsorption of cadmium and copper to *Bacillus cereus* and *Pseudomonas aeruginosa*.

18.4.2.2 Complexation and Solubilization of Metals. Microorganisms may produce organic acids (e.g., acetic, citric, oxalic, gluconic, fumaric, lactic, and malic

TABLE 18.3. Some Microorganisms Involved in Metal Removal/Recovery from Industrial Wastewaters[a]

Microorganism Removed/Recovered	Metal
Zooglea ramigera	Copper
Saccharomyces cereviseae	Uranium and other metals
Rhizopus arrhizus	Uranium
Chlorella vulgaris	Gold, zinc, copper, mercury
Aspergillus orhizae	Cadmium
Aspergillus niger	Copper, cadmium, zinc
Penicillum spinulosum	Copper, cadmium, zinc
Trichoderma viride	Copper
AMT-Bioclaim[TM]	Biotechnology-based use of granulated product derived from biomass

[a]Adapted from Eccles and Hunt (1986).

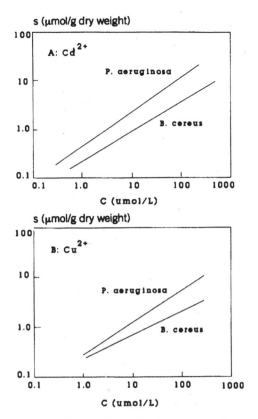

Figure 18.9 Freundlich isotherms for sorption of cadmium (a) and copper (b) by *B. Cereus* and *P. aeruginosa*. Adapted from Mullen et al. (1989).

acids), which may chelate toxic metals, resulting in the formation of metallorganic molecules. These organic acids help in the solubilization of metal compounds and in their leaching from surfaces. Metals may also be biosorbed or complexed by carboxyl groups found in microbial polysaccharides and other polymers. This phenomenon is of great importance in wastewater treatment plants, particularly those using the activated sludge process, where industrial wastes are treated (Bitton and Freihoffer, 1978; Brown and Lester, 1979; 1982; Loaec et al., 1997; McLean et al., 1990; Sterritt and Lester, 1986; Rudd et al., 1984). Under iron-limiting conditions, microorganisms produce compounds called *siderophores* that are involved in iron solubilization, but may also play a role in the solubilization of other metals.

Microorganisms increase the solubility of metals by decreasing the pH of their milieu. Metabolic activity of heterotrophic microorganisms produces protons and CO_2, (production of carbonic acid), which help metal solubilization. Acidification is also the result of the activity of chemolithotrophs such as the iron- and sulfur-oxidizing bacteria (e.g., *Thiobacillus ferrooxidans, Thiobacillus thiooxidans*; see Chapter 3). Chemolithotrophs are used in the bioleaching of metals from contaminated sediments and biosolids (Crowley and Dungan, 2002).

18.4.2.3 Precipitation.
Some bacteria promote metal precipitation by producing ammonia, organic bases, or hydrogen sulfide, which precipitate metals as hydroxides or sulfides. Sulfate-reducing bacteria (e.g., *Desulfovibrio, Desulfotomaculum*) transform

SO_4 to H_2S, which promotes the extracellular precipitation of highly insoluble metal sulfides (Webb et al., 1998; White et al., 1997). *Klebsiella aerogenes* is able to detoxify cadmium to a cadmium sulfide (CdS) form, which precipitates as electron-dense granules at the cell surface. This process is induced by cadmium (Aiking et al., 1982). *Desulfotomaculum auripigmentum* reduces As^{5+} to As^{3+} and S^6 to S^{2-}, which leads to the intracellular and extracellular precipitation of arsenic trisulfide (As_2S_3) (Newman et al., 1997).

18.4.2.4. Volatilization.

Some metals are transformed to volatile species as a result of microbial action. For example, bacterially mediated methylation converts Hg^{2+} to dimethyl mercury, a volatile compound. Some bacteria have the ability to detoxify mercury by transforming Hg^{2+} into Hg^0, a volatile species. This detoxification process is plasmid-encoded and is regulated by the *mer* operon, which consists of several genes. The most important gene is the *merA* gene, which is responsible for the production of mercuric reductase, the enzyme that catalyzes the transformation of Hg^{2+} to Hg^0. Bacteria and fungi can also methylate metalloids (e.g., As, Se), producing volatile compounds.

18.4.2.5 Intracellular Accumulation of Metals.

Microbial cells may accumulate metals that gain entry into the cell via specific transport systems.

18.4.2.6 Redox Transformations of Metals.

Bacteria have the ability to reduce metals to ions with reduced solubility. They use the metals as electron acceptors in respiration (dissimilatory metal reduction). Examples are the reduction of Fe^{3+} to Fe^{2+} by *Geobacter metallireducens*, and Cr^{6+} to Cr^{3+} by *Pseudomonas fluorescens* or *Bacillus* sp. Other redox transformations are displayed in Table 18.4 (Crowley and Dungan, 2002).

18.4.2.7 Use of Recombinant Bacteria for Metal Removal.

Metal removal by adsorbents from water and wastewater is strongly influenced by physicochemical parameters such as ionic strength, pH, and the concentration of competing inorganic and inorganic compounds. Recombinant bacteria and cyanobacteria are being investigated

TABLE 18.4. Biologically Driven Redox Transformations of Some Metal Ions[a]

Metal	Reaction	Microorganisms
As	$As^{5+} \rightarrow As^{3+}$	Gram-positive bacteria
Cr	$Cr^{6+} \rightarrow Cr^{3+}$	*Pseudomonas, Bacillus, Enterobacter Streptomycetes*
Cu	$Cu^{2+} \rightarrow Cu^+$	*Thiobacillus*
Fe	$Fe^{3+} \rightarrow Fe^{2+}$	*Geobacter, Shewanella*
Hg	$Hg^{2+} \rightarrow Hg^0$	Diverse bacteria
Mn	$Mn^{6+} \rightarrow Mn^{4+}$	Diverse bacteria
	$Mn^{4+} \rightarrow Mn^{2+}$	Diverse bacteria
Se	$Se^{6+} \rightarrow Se^0$	Diverse bacteria
	$Se^{6+} \rightarrow Se^{4+}$	
	$Se^{4+} \rightarrow Se^0$	
U	$U^{6+} \rightarrow U^{4+}$	*Geobacter, Desulfovibrio, Micrococcus*

[a]Adapted from Crowley and Dungan (2002).

for enhanced removal of specific metals from contaminated water. For example, a genetically engineered *E. coli*, which expresses metal transport systems and metallothioneins (metal binding intracellular protein widespread in animals and in certain fungi), was able to selectively bioaccumulate Hg^{2+} and Ni^{2+} (Chen and Wilson, 1997; Krishnaswamy and Wilson, 2000). Recombinant *E. coli* with synthetic phytochelatins bound to outer membrane proteins, was able to bioaccumulate up to 60 nmoles of Cd^{2+}/mg dry cells (Bae et al., 2000). Similarly, insertion of mouse metallothionein into the outer membrane of *Ralstonia eutropha* enhanced cadmium sorption (Valls et al., 2000).

18.5   POTENTIAL APPLICATIONS OF MOLECULAR TECHNIQUES IN WASTE TREATMENT

The application of molecular techniques to domestic and industrial waste treatment is in its infancy. The relatively slow use of this technology in full-scale waste treatment is partly the result of our lack of knowledge concerning the release of engineered microorganisms into the environment. In addition, the proposed new technology does not appear to be more economical than existing technologies.

Major applications of molecular techniques include the enhancement of biodegradation of xenobiotics in wastewater treatment plants and the use of nucleic acid probes to detect pathogens and parasites in wastewater effluents and other environmental samples.

Some potential improvements of biological treatment are displayed in Table 18.5 (Rittmann, 1984). Some of these improvements can be accomplished by modifying existing technologies, selecting for novel microorganisms via traditional methods, or by using molecular techniques. Successful use of these technologies may result, for example, in

TABLE 18.5.   Some Improvements in Biological Treatment of Municipal Wastewaters[a]

Improvement	Problem Type[b]	Likely Solution Type[c]
Eliminate activated sludge bulking	2	a, d
Improve biofilm attachment	2	a, d
Stable nitrification	2	a
	3	d
Prevent sloughing in trickling filters	2	a
Reduce O_2 limitation in aerobic processes	3	d
Reduce energy consumption	2	a, b
Reduce sludge quantities produced	2	a, b
Enhance removal of phosphorus	2, 3	a, b, c, d
Biodegrade xenobiotic organics	2	a, b, c, d
Resist toxic upsets	2	a, d
Prevent generation of odors	2, 3	a, b, c
Make simple efficient processes for small communities	2	a, b

[a]From Rittman (1984).
[b]Problem types: 1, not feasible; 2, not reliable or efficient; 3, not economic.
[c]Solution types: a, improve exiting process; b, use of new process; c, use a novel microorganism; d, apply genetic manipulation.

enhancement of denitrification in sewage treatment plants, increased growth rates of nitrifying bacteria, speeding up nitrification and reducing the sludge age in wastewater treatment plants, improved bacterial flocculation in activated sludge, improved performance of biofilm processes, enhanced biological phosphorus removal (e.g, enhanced P removal by a recombinant strain of *E. coli* that displayed a phosphorus content of 16 percent on a dry weight basis; Hardoyo et al., 1994; Kato et al., 1993), improved performance of methanogens, means for controlling the acid buildup ("pickling") of anaerobic digesters, resistance to toxic hazardous wastes, better control of activated sludge bulking, and a better understanding of the microbial ecology of wastewater treatment by using tools such as rRNA, rDNA, mRNA, or reporter proteins to draw conclusions on the phylogenic identity and phenotypic activity of wastewater microorganisms (Johnston and Robinson, 1982; Rittmann, 1984; 2002).

Genetically engineered microbes can be useful in several areas of waste treatment, some of which are biomass production, biodegradation of recalcitrant molecules, removal of toxic metals, fermentation (methane and organic acid production), enhancement of enzyme activity, and increased resistance to toxic inhibition. Waste treatment processes can be improved by selection of novel microorganisms that can thrive in domestic wastewater and perform a desirable function (Patterson, 1984; Rittmann, 1984). Two steps are involved in the application of recombinant DNA technology to waste treatment (Rittmann, 1984). The first step consists of finding a microorganism that has the desirable function (e.g., ability to degrade a pesticide). The second step consists of transferring this desirable function to a suitable host, preferably a microorganism with some relevance from an environmental viewpoint. Efforts are now being focused on understanding the genetic basis of biodegradation of xenobiotics in the environment in general and in biological treatment processes in particular (Timmis et al., 1994). Advances in molecular technology will help future development of engineered microbial strains with enhanced and broader biodegradative ability (Chaudhury and Chapalamadugu, 1991; Rittmann et al., 1990; Sayler and Blackburn, 1989). A useful application of genetic engineering would be in the treatment of industrial wastes. These wastes offer a harsh environment for the maintenance and growth of genetically engineered microorganisms (GEMs). Extremes in temperatures, pH, salinity, ionic composition, oxygen, and redox potential are often encountered in industrial wastes. Successful treatment of these wastes by conventional biological treatment technologies often necessitates several adjustments (temperature, pH, salinity) of wastewater. Some (Kobayashi, 1984; Shilo, 1979) have proposed the use of GEMs originating from extreme environments (e.g., hypersaline waters, acid hot springs, alkaline lakes). These microorganisms would be able to degrade industrial wastes in special biological reactors. Biofilm reactors would constitute an acceptable option because they help minimize the potential release of GEMs into the environment.

Other potential applications include the use of genetically engineered multi-plasmid *Pseudomonas* strains for degrading several components found in crude oils (Friello et al., 1976). The enhancement of the level of several enzymes has been achieved by molecular biologists using recombinant DNA technology. These enzymes include DNA ligase, tryptophan synthetase, α-amylase, benzylpenicillin acylase, and others (Demain, 1984). The new molecular techniques can help alter enzymes to improve their stability and catalytic efficiency, broaden their substrate range, or help to create multifunctional hybrid enzymes with improved substrate flux through catabolic pathways (Timmis et al., 1994). Enhancing the production of enzymes implicated in the degradation of recalcitrant organic molecules would also be a useful undertaking.

However, the use of DNA technology in pollution control is offset by the following limitations (Hardman, 1987; Timmis et al., 1994):

1. *Multistep pathways in the degradation of xenobiotics.* This raises the possibility that an engineered organism may not be able to completely mineralize the target xenobiotic.

2. *Limited degradation.* The engineered microorganism may be capable of degrading only one or two compounds.

3. *Limited or absence of knowledge about the degradative pathways of interest.* This knowledge would be useful for the identification of the responsible genes to be cloned.

4. *Substrate concentration.* Degradative enzymes may not be induced below a threshold substrate concentration (this is discussed in more detail in Chapter 19). Toxicity may be a problem at high substrate concentration.

5. *Substrate bioavailability.* For example, hydrophobic compounds tend to adsorb to surfaces, leading to low availability to microorganisms.

6. *Instability of the recombinant strain of interest in the natural environment.* Little is known about the competitive ability of the recombinant strains with the more adapted indigenous microbial populations. Microcosms simulating activated sludge systems have been useful for studying the fate of GEMs degrading substituted aromatic compounds. Recent studies showed that the GEMs survive well and degrade aromatic compounds in the complex activated sludge environment. There was no demonstrable negative effect of the GEMs on the indigenous microflora (Dwyer et al., 1988).

7. *Public concern about the accidental or deliberate release of GEMs into the environment.* Such concerns can limit their application. Issues of concern include the persistence and reproduction of recombinant microorganisms in the natural environment, their interaction with indigenous organisms, and their impact on ecosystem function. Of specific concern is the persistence of added plasmids in wastewater treatment plants (Phillips et al., 1989). The use of suicide genes has been suggested for controlling the spread of GEMs in the environment. The GEMs would carry an inducible suicide gene that, once induced, would kill the GEM and thus halt its spread (Fox, 1989).

The benefits and risks of GEM use in the environment, and in wastewater treatment plants in particular, need to be addressed (Tiedje et al., 1989).

Today, we have the technologies to remove organic matter, suspended solids, nutrients, metals, pathogens, and parasites, and even recalcitrant compounds from wastewater. However, these so-called "high-tech" solutions are expensive, and energy intensive. Sustainable wastewater treatment will be based on processes with low resource consumption and little impact on the environment (Henze, 1997).

However, the relevance of information provided by modern analytical techniques to the practicing engineer is yet to be evaluated. Some of the modern molecular techniques have confirmed what we already know from traditional methodology; others are helping in establishing a relationship between a microbiological parameter (e.g., species composition of the community) and a process parameter, and are pointing to new directions for improving the water and wastewater treatment processes. A collaborative study (France–Netherlands–United States) of a nitrifying activated sludge has been undertaken to compare traditional measurement parameters such as TKN, BOD, and SRT (solids retention time) with community structure information derived from the use of rRNA probes. rRNA probes have been used to obtain the ratio of ammonia-oxidizing bacteria (probe Nso1225) to total bacteria (probe Eub338). This ratio correlated well with the model-derived ratio $X_{ao}/X_h + X_{ao}$ (where X_{ao} = concentration of active ammonia oxidizers; X_h = concentration of active heterotrophs) obtained by considering traditional measurements parameters. The probe analysis also showed that at low solid retention times (<4 days), the ammonia oxidizers concentration was <1 percent, and this could serve as a signal of washout of nitrifiers from the bioreactor (Rittmann, 2002; Rittmann et al., 1999). Other studies have shown a correlation between FISH assay and activity tests results regarding ammonia oxidizers in biofilms (Biesterfeld and Figueroa, 2002).

These preliminary efforts point to the need to use molecular tools to complement traditional measurement parameters. In the future, we will witness a bridging of the gap between the scientist in the laboratory and the practicing engineer in the field (Wilderer et al., 2002).

18.6   MEMBRANE BIOREACTORS FOR WASTEWATER TREATMENT

Membrane bioreactors (MBR) combine biological with physical processes to treat wastewater. We have discussed in Chapter 9 the solid separation problems encountered in conventional activated sludge plants. Membranes can thus be used to separate the biomass and other solids from the treated effluent, thus bypassing the need for a settling tank. They are used as external modules or as modules immersed in the aeration tank (Fig. 18.10) (Ben Aim and Semmens, 2002).

Some advantages of MBRs are (Ben Aim and Semmens, 2002; DiGiano et al., 2004; Jefferson et al., 2000):

- Wastewater biodegradation in activated sludge is more efficient in MBRs.
- MBRs can be operated with high concentrations of biomass.

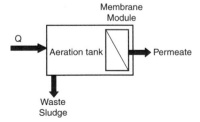

Figure 18.10   External and submerged membrane modules. From Ben Aim and Semmens (2002).

- Settling does not depend on the formation of biological flocs in the aeration tank. The membrane retains the settling as well as the planktonic microorganisms.
- Microorganisms have the same residence time in the system. Slow-growing bacteria (e.g., nitrifiers) are maintained on the membrane in contrast to conventional systems.
- Removal of pathogens and parasites is efficient.
- The residence time of colloidal particles and macromolecules does not depend on the hydraulic retention time.
- MBRs require less space than traditional systems as a result of lower hydraulic retention time.
- Production of biosolids is lower in MBRs that in conventional activated sludge.
- Postdisinfection of the particle-free effluents is more effective.
- Owing to their simplicity, they would be ideal in a decentralized wastewater treatment system.

However, membranes are amenable to fouling (Judd, 2004; Lim et al., 2004). The main contributors to fouling are the biomass, extracellular polymeric substances (EPS), colloidal particles, and biofilm formation on the membrane. Fouling is alleviated by periodic backwashing of the membranes with oxidizing agents or acids.

Membrane bioreactor technology has been used worldwide in several industrial applications. It has also been successful in treating gray water (Jefferson et al., 2000). It is an essential component of global water sustainability because it could help in promoting water reuse and decentralization of wastewater treatment plants (DiGiano et al., 2004). The technology meets the environmental, technical, economic, and socio-cultural criteria of water sustainability (Table 18.6; Balkema et al., 2002; DiGiano et al., 2004). It would be of great help to developing countries where 1.1 billion people do not have access to safe drinking water.

TABLE 18.6. **Sustainability Criteria for MBR Technology**[a]

Criteria	Indicators	Improvement Needed	Good Now
Economic	Cost and affordability	X	
Environmental	Effluent water quality		
	Microorganisms		X
	Suspended solids		X
	Biodegradable organics		X
	Nutrient removal		X
	Chemical usage	X	
	Energy	X	
	Land usage		X
Technical	Reliability		X
	Ease of use	X	
	Flexible and adaptable		X
	Small-scale systems		X
Socio-cultural	Institutional requirements	X	
	Acceptance	X	
	Expertise	X	

[a]From Balkema et al. (2002); DiGiano et al. (2004).

18.7 WEB RESOURCES

http://www.ncbi.nlm.nih.gov (National Center for Biotechnology Information)
http://www.iwahq.org.uk/template.cfm?name=home (International Water Association)
http://www.ceb.utk.edu/ (Center for Environmental Biotechnology, University of Tennessee, Knoxville)
http://www.ornl.gov/sci/microbialgenomes/ (Microbial Genome Program, U.S. Department of Energy)
http://www.nal.usda.gov/bic/bio21/ (environmental biotechnology)
http://wwtp.org/index.php (Activated Sludge Biomolecular Database; Syracuse University)
http://evolution.genetics.washington.edu/phylip/software.html (phylogeny software packages)
http://www.ncbi.nlm.nih.gov/About/primer/microarrays.html (microarray primer)
http://www.tigr.org (Institute for Genomic Research)
http://www.filtsepbuyersguide.com/WZ/FiltSep/lat_feat/feat_art/000015/show/ (MBR technology)
http://www.werf.org/press/spring02/02sp_MBR.cfm (document on MBR from the Water Environment Federation)
http://www.desline.com/articoli/4546.pdf (removal of endocrine disruptors by MBR technology)

18.8 QUESTIONS AND PROBLEMS

1. Define bioaugmentation.
2. What are some applications of immobilized enzymes?
3. What are some applications of biosensors based on enzyme or microbial immobilization?
4. What could be some applications of bioaugmentation in wastewater treatment plants?
5. What would be the use of the enzymes laccases in wastewater treatment?
6. What are some applications of immobilized enzymes?
7. What are some applications of biosensors based on enzyme or microbial immobilization?
8. Why would extremozymes be of potential importance in waste treatment?
9. Give three cell immobilization techniques.
10. What are some mechanisms of removal of metals by microorganisms in wastewater?
11. How do microorganisms complex metals?
12. What would be the potential use of genetically modified microorganisms in wastewater treatment plants?
13. What are some concerns over the release of genetically modified microorganisms in the environment?
14. Explain the concept of enricher–reactors in wastewater treatment.
15. What are some advantages of membrane bioreactors?
16. What are some disadvantages of cell immobilization?

18.9 FURTHER READING

Amann, R., and W. Ludwig. 2000. Ribosomal RNA-targeted nucleic acid probes for studies in microbial ecology. FEMS Microbiol. Rev. 24: 555–565.

Atlas, R.M. 1991. Environmental applications of the polymerase chain reaction. ASM News 57: 630–632.

Brierley, C.L., J.A. Brierley, and M.S. Davidson. 1989. Applied microbial for metal recovery and removal from wastewater. p. 359–381, In: *Metal Ions and Bacteria*, T.J. Beveridge, and R.J. Doyle, Eds., John Wiley and Sons, N.Y.

Crowley, D.E., and R.S. Dungan. 2002. Metals: Microbial processes affecting metals, pp.1878–1893, In: *Encyclopedia of Environmental Microbiology*, Gabriel Bitton, editor-in-chief, Wiley-Interscience, N.Y.

Cheetham, P.S.J., and C. Bucke. 1984. Immobilization of microbial cells and their use in waste water treatment, pp. 219–235, In: *Microbiological Methods for Environmental Biotechnolog*, J.M. Grainger, and J.M. Lynch, Eds., Academic Press, London.

Hardman, D.J. 1987. Microbial control of environmental pollution: the use of genetic techniques to engineer organisms with novel catabolic capabilities. pp. 295–317, In: *Environmental Biotechnology*, C.F. Forster, and D.A.J. Wase, Eds., Ellis Horwood Ltd., Chichester, U.K.

Johnson, J.B., and S.G. Robinson. 1982. Opportunities for development of new detoxification processes through genetic engineering. pp. 301–314, In: *Detoxification of Hazardous Wastes*, J.H. Exner, Ed., Ann Arbor Scientific Publications, Ann Arbor, MI.

Kobayashi, H.A. 1984. Application of genetic engineering to industrial waste/wastewater treatment, pp. 195–214, In: *Genetic Control of Environmental Pollutants*, G.S. Omenn, and A. Hollanender, Eds., Plenum Press, New York.

Liu, J., and B. Mattiasson. 2002. Microbial BOD sensors for wastewater analysis. Water Res. 36: 3786–3802.

Patterson, J.W. 1984. Perspectives on opportunities for genetic engineering applications in industrial pollution control, pp. 187–193, In: *Genetic Control of Environmental Pollutants*, G.S. Ommen, and A. Hollaender, Plenum Press, New York.

Rittmann, B.E. 2002. The role of molecular methods in evaluating biological treatment processes. Water Environ. Res. 74: 421–427.

Rittmann, B., and P.L. McCarty. 2001. *Environmental Biotechnology: Principles and Applications*, McGraw-Hill, New York.

Sayler, G.S., and A.C. Layton. 1990. Environmental application of nucleic acid hybridization. Ann. Rev. Microbiol. 44: 625–648.

Tiedje, J.M., R. Colwell, Y.L. Grossman, R.E. Hodson, R.E. Lenski, R.N. Mack, and P.J. Regal. 1989. The planned introduction of genetically engineered organisms: Ecological considerations and recommendations. Soc. Ind. Microbiol. News 39: 149–165.

Yang, P.Y., K. Cao, and S.J. Kim. 2002. Entrapped mixed microbial cell process for combined secondary and tertiary wastewater treatment. Water Environ. Res. 74: 226–234.

PART F

FATE AND TOXICITY OF CHEMICALS IN WASTEWATER TREATMENT PLANTS

19

FATE OF XENOBIOTICS AND TOXIC METALS IN WASTEWATER TREATMENT PLANTS

Wastewater Microbiology, Third Edition, by Gabriel Bitton
Copyright © 2005 John Wiley & Sons, Inc.

19.1 INTRODUCTION

During the last few decades, an array of foreign compounds called xenobiotics (i.e., foreign to biological systems) have been introduced into the environment. These compounds are generally resistant or recalcitrant to biodegradation. Some chemicals such as the halogenated organic compounds (halogenated hydrocarbons, halogenated aromatics, pesticides, PCBs) are very resistant to microbial action. However, there are also refractory compounds of natural origin. These compounds include humic substances and lignin, as well as halogenated natural compounds generally found in the marine environment.

Industrial wastewaters contain relatively high concentrations of recalcitrant organic compounds that may be toxic, mutagenic, carcinogenic, or estrogenic (see Chapter 20) and may be bioaccumulated or biomagnified (i.e., their concentration increases at higher trophic levels) by the biota. These wastewaters, sometimes treated on site mostly via physicochemical processes, are discharged to surface waters or to conventional municipal wastewater treatment plants.

Microorganisms play a key role in biogeochemical cycles, particularly the carbon cycle. They degrade natural and anthropogenic compounds and release CO_2, CO, or CH_4. It is widely assumed that microorganisms can degrade any natural organic compounds although some (e.g., humic compounds, lignin) are quite resistant to biodegradation and have formed as a result of conditions unfavorable to their biodegradation.

Biotransformation is the alteration of organic compounds by microbial action, sometimes by *consortia* of microorganisms. The biological transformation of a xenobiotic may result in mineralization, accumulation, or polymerization of the compound with naturally occurring compounds (e.g., humic compounds) (Bollag, 1979; Hardman, 1987). (Fig. 19.1). Mineralization refers to the transformation of organic compounds to halide (e.g., chlorine, bromine), CO_2, and/or CH_4 (Rochkind-Dubinsky et al., 1987). Despite the abundant literature on biodegradability of toxic pollutants in wastewater treatment plants, work is being done to identify the microorganisms responsible for their biodegradation in these engineered systems.

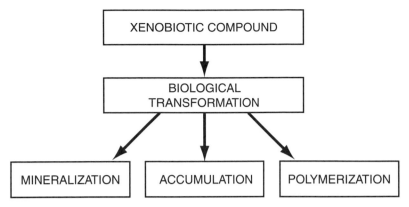

Figure 19.1 The biological fate of xenobiotic compounds in the environment. From Hardman (1987). (Courtesy of Ellis Horwood, Ltd.)

19.2 BIODEGRADATION IN AQUATIC ENVIRONMENTS

Biological, chemical, and environmental factors affect the fate of chemicals in the environment (Giger and Roberts, 1978; Hardman, 1987; Johnston and Robinson, 1984; Leisinger et al., 1981; Sayler and Blackburn, 1989; Sayler et al., 1984; Young and Cerniglia, 1995). Table 19.1 provides more details concerning these factors.

TABLE 19.1. Factors Controlling the Fate of Xenobiotics in the Environment[a]

Factors	Consequences
Chemical factors	
Molecular weight or size	Limited active transport
Polymeric nature	Extracellular metabolism required
Aromaticity	Oxygen-requiring enzymes (in aerobic environment)
Halogen substitution	Lack of dehalogenating enzymes
Solubility	Competitive partitioning
Toxicity	Enzyme inhibition, cell damage
Environmental factors	
Dissolved oxygen	O_2-sensitive and O_2-requiring enzymes
Temperature	Mesophilic temperature optimum
pH	Narrow pH optimum
Dissolved carbon	Concentration dependence or organic/pollutant complexes for growth
Particulates, surfaces	Sorptive competition for substrate
Light	Photochemical enhancement
Nutrient and trace elements	Limitations on growth and enzyme synthesis
Biological factors	
Enzyme ubiquity	Low frequency of degradative species
Enzyme specificity	Analogous substrates not metabolized
Plasma-encoded enzymes	Low frequency of degradative species
Enzyme regulation	Repression of catabolic enzyme synthesis
	Required acclimation or induction
Competition	Extinction of low-density populations
Habitat selection	Lack of establishment of degradative populations

[a]Adapted from Sayler et al. (1984).

Several investigators have isolated and identified microorganisms or microbial consortia that are capable of partially or completely degrading several classes of trace organics. More recently, much effort has been devoted to understanding the genetics of biodegradation, with emphasis on the study of catabolic plasmids, which code for the biodegradation of xenobiotics. The ultimate goal is the construction of microbial strains that can be used for the bioremediation of hazardous waste sites. Several halogenated xenobiotics are, however, quite resistant to microbial action and several causes have been suggested to explain their recalcitrance.

19.2.1 Recalcitrance of Organic Compounds

Resistance to biodegradation can result from the following (Alexander, 1985; Anderson, 1989; Hanstveit et al., 1988):

1. Molecular structure (e.g., substitutions with chlorine and other halogens).
2. Failure of compound to enter the cell due to absence of suitable permeases.
3. Unavailability of the compound as a result of insolubility or adsorption may make it inaccessible for microbial action.
4. Unavailability of the proper electron acceptor.
5. Unfavorable environmental factors such as temperature, light, pH, O_2, moisture, or redox potential.
6. Unavailability of other nutrients (e.g., N, P) and growth factors necessary for microorganisms.
7. Compound toxicity can affect the biodegradation potential by microorganisms, which have numerous ways to detoxify chemicals (e.g., catabolic plasmids). Some metabolites formed as a result of biodegradation may be more toxic than the parent compound.
8. Low substrate concentration can also affect biodegradation by microorganisms. Organisms growing at very low substrate concentration have a high affinity for substrates (i.e., very low half-saturation constant K_s). Some microorganisms in the environment may not be able to assimilate and grow on organic substrates below a threshold concentration. Most biodegradation studies have been carried out using relatively high substrate concentrations. This is not a realistic approach as environmental concentrations are much lower (ppm or ppb level) than those that are sometimes used under laboratory conditions (Alexander, 1985).

On the basis of substrate concentration, there are two categories of microorganisms: (1) *copiotrophs*, which grow rapidly in the presence of high substrate concentrations; and (2) *oligotrophs*, which grow slowly in the presence of low substrate concentrations. The latter have a high affinity for substrates (i.e., low K_s) and, thus, they grow slowly but efficiently in the presence of low substrate concentrations. Microorganisms respond to low nutrient concentrations by adapting morphologically, leading to cells with high surface-to-volume ratios.

19.2.2 Dehalogenation of Organic Compounds

Research efforts have focused on the mechanisms of dehalogenation of synthetic halogenated organics because these chemicals are resistant to degradation (i.e., recalcitrant), are lipophilic, and tend to bioaccumulate in the food chain. Microorganisms must be capable of cleaving the carbon–halogen bond of halogenated compounds.

However, this necessitates a lag period for the induction of dehalogenase enzymes, called hydrohalidases, which hydrolyse the carbon–halogen bond (Johnston and Robinson, 1984). Some mechanisms of dehalogenation are the following (Colwell and Sayler, 1978; Linkfield et al., 1989; Sims et al., 1991):

- *Reductive dehalogenation* is the replacement of one Cl atom with one H atom. Reductive dehalogenation of halogenated organic compounds occurs under anaerobic conditions, particularly under methanogenic conditions, and requires the induction of dehalogenating enzymes.
- Hydrolytic dehalogenation is the replacement of a halogen by a hydroxyl group.

19.2.3 Co-metabolism

In co-metabolism, an organic compound is converted to metabolic products but does not serve as a source of energy or carbon to microorganisms (Alexander, 1981; 1985). The organisms derive their energy and carbon from a primary substrate but none from the xenobiotic, which acts as a secondary substrate. The reactions involved in co-metabolism include dehalogenation, introduction of hydroxyl groups, ring cleavage, or oxidation of methyl groups (Rittmann et al., 1988).

Co-metabolism has been demonstrated in pure and mixed cultures for a wide range of compounds (insecticides, herbicides, surfactants, aliphatic and aromatic hydrocarbons, and other industrial chemicals), and it has been suggested that it occurs in nature (Alexander, 1979; Horvath, 1972). There is some evidence for the occurrence of co-metabolism in wastewater (Jacobson et al., 1980). Metabolic products of ^{14}C-labeled herbicides have been demonstrated in nonsterile but not in sterile primary sewage effluent after a relatively long period of incubation. However, the carbonaceous substrates were not incorporated into wastewater microorganisms, an indication that they did not serve as a source of carbon to microorganisms.

19.2.4 Genetic Regulation of Xenobiotic Biodegradation

The genes regulating the catabolism of many xenobiotics are plasmid-borne. Catabolic plasmids (or degradative plasmids) are extrachromosomal DNA elements that control the transformation of xenobiotic compounds. They have been identified in a limited number of bacterial types, mostly in the genus *Pseudomonas* (Sayler and Blackburn, 1989) (Table 19.2). These catabolic plasmids may be lost when the microorganisms are

TABLE 19.2. Some Degradative Plasmids[a]

Plasmid	Substrate	Conjugative Ability
TOL	Toluene, *meta*-xylene, *para*-xylene	+
CAM	Camphor	+
OCT	Octane, hexane, decane	−
SAL	Salicylate	+
NAH	Naphthalene	+
NIC	Nicotine/nicotinate	+
pJP1	2,4-Dichlorophenoxyacetic acid	+
pAC21	4-Chlorobinphenyl	+
pAC25	3-Chlorobenzoate	+
pAC27	3- and 4-chlorobenzoate	+

[a]Adapted from Hooper (1987) and Sayler and Blackburn (1989).

not maintained on the substrate specific for the enzyme encoded by the plasmid. Catabolic plasmids may complement chromosome-encoded pathways. A well-described plasmid is the TOL plasmid (pWWO). This 117 kb size plasmid encodes for the degradation of toluene and xylenes, and for its own replication and transfer (Saunders and Saunders, 1987; Sayler and Blackburn, 1989). Multiplasmid microbial strains have been constructed for the biodegradation of hydrocarbons in crude oil (Friello et al., 1976). These strains are capable of degrading toluene, xylenes, camphor, octane, and naphthalene (Saunders and Saunders, 1987). Degradative plasmids have also been constructed for the biodegradation of highly persistent and toxic xenobiotics, namely the chlorinated compounds. However, the use of these novel plasmid-bearing microorganisms under field conditions remains to be fully investigated.

Genetic ecologists work towards enhancing the capabilities of natural microbial communities at the gene level through gene amplification and/or increased expression (Olson and Goldstein, 1988; Olson, 1991). The increased activity of the natural microbial communities would lead to an enhanced degradation of toxic organic pollutants and to biotransformation of toxic metals to species that are less toxic as well as less available to the biota. This approach appears to be less troublesome than the one based on the introduction of genetically engineered microorganisms into the environment. There are certain advantages to a genetic ecological approach. First, genes control specific functions, whereas whole cells carry out a multitude of functions, many of which are useless to a particular goal. In addition, genes controlling a specific function (e.g., degradation of a xenobiotic compound) can be amplified through genetic manipulations (e.g., increasing the copy number of a desirable plasmid). Because several catabolic plasmids have been reported, the biodegradation of a given xenobiotic can be increased by encouraging gene transfer via manipulation of the environment (Olson, 1991). Using culture-independent methods, more than 300 catabolic genes have identified to date (Widada et al., 2002).

19.3   FATE OF XENOBIOTICS IN WASTEWATER TREATMENT PLANTS

There is great concern over the impact of toxic industrial pollutants on wastewater treatment processes and upon receiving waters such as lakes and rivers. Hazardous organic and inorganic pollutants are removed by biological and advanced wastewater treatment processes by the following mechanisms: sorption to sludge biomass and to added activated carbon, volatilization, chemical oxidation, chemical flocculation, and biodegradation (Grady, 1986; Metcalf and Eddy, 1991).

19.3.1   Physico-Chemical Processes

19.3.1.1   Sorption to Sludge Biomass and Activated Carbon.   Nonpolar trace organics tend to adsorb to wastewater solids and some are removed following sedimentation, resulting in their transfer into sludge solids. It is well known that bacterial, algal, and fungal cells are capable of adsorbing and accumulating organic pollutants (Baughman and Paris, 1981; Lal and Saxena, 1982). The activated sludge biomass is able to adsorb organic pollutants such as lindane, diazinon, pentachlorophenol, and PCBs. Adsorption of these compounds generally fits the Freundlich isotherm. There is a good correlation between compound adsorption and the octanol/water partition coefficient. Because the adsorption of organic compounds is generally reversible, concern has arisen over their leaching following land application of sludges (Bell and Tsezos, 1987).

Activated carbon is suitable for removing residual refractory organic compounds and heavy metals. The carbon columns can be operated in the downflow or upflow modes. A process, called PACT, that combines the use of activated carbon and activated sludge, removes ammonia, toxic metals, and trace organics.

19.3.1.2 Volatilization.

Air stripping removes volatile organic compounds (VOC) and ammonia. Because air stripping transfers VOC from the water phase to the air phase, concern has arisen over the release of potential toxicants into the air. This has mainly to do with chlorinated compounds, 31–71 percent of which air stripping transfers to the atmosphere (Parker et al., 1993).

19.3.1.3 Chemical Oxidation.

In addition to their disinfecting property, disinfectants such as chlorine dioxide or ozone are capable of oxidizing trace organics.

19.3.1.4 Photocatalytic Degradation of Organics in Wastewater.

This treatment, which consists of exposing wastewater to sunlight in the presence of a suitable catalyst (e.g., titanium dioxide), achieves complete mineralization of toxic organics (e.g., phenol, chlorinated compounds, surfactants) (Koramann et al., 1991; Okamoto et al., 1985; Ollis, 1985). The large-scale application of this technology to waste treatment is yet to be demonstrated.

19.3.1.5 Chemical Coagulation.

Chemical coagulation is suitable for removing toxic metals and trace organics.

19.3.2 Microbial Processes: Biodegradation

Biodegradation of organics by wastewater microorganisms is variable and generally low for chlorinated compounds. Exposure of wastewater microbial communities to toxic xenobiotics results in the selection of resistant microorganisms that have the appropriate enzymes to use the xenobiotic as the sole source of carbon and energy. This process is called acclimation or adaptation. Most microorganisms need a period of acclimation before the onset of metabolism. Prior exposure to the xenobiotic helps reduce the acclimation period (Hallas and Heitkamp, 1995). Microbiologists have isolated a vast array of microorganisms that have the ability to degrade organic toxicants. Table 19.3 lists some of these microbial species (Kumaran and Shivaraman, 1988). Since several xenobiotics are aromatic molecules, the microorganisms must be capable of cleaving the aromatic ring of these compounds. The biochemical pathways have been elucidated for many of the compounds. The fission of the aromatic ring is mediated by enzymes called monooxygenases, which introduce molecular oxygen into the ring before cleavage.

The biodegradability (i.e., susceptibility of organic substrates to microbial attack) of organic compounds is often assessed via substrate disappearance or by BOD or COD removal. More accurate approaches include recovery of radiolabeled parent substrate and/or metabolic products, and mineralization products (measuring CO_2 or CH_4 production) (Rochkind-Dubinsky et al., 1987).

Some assay systems for biodegradability tests include the following (Anderson, 1989) (Fig. 19.2):

1. Measurement of oxygen consumption via manometric and electrolytic systems (Fig. 19.2a).
2. Measurement of CO_2 evolution, using infrared or chemical methods (Fig. 19.2b).

TABLE 19.3.   Some Microorganisms Involved in the Biodegradation of Xenobiotics[a]

Organic Pollutants	Organism
Phenolic compounds	*Achromobacter, Alcaligenes, Acinetobacter, Arthrobacter, Azotobacter, Bacillus cereus, Flavobacterium, Pseudomonas putida, P. aeruginosa,* and *Nocardia* *Candida tropicalis, Debaromyces subglobosus,* and *Trichosporon cutaneoum*
Benzoates and related compounds	*Arthrobacter, Bacillus* spp., *Micrococcus, Moraxella, Mycobacterium, P. putida,* and *P. fluorescens*
Hydrocarbons	*Escherichia coli, P. putida, P. aeruginosa,* and *Candida*
Surfactants	*Alcaligenes, Achromobacter, Bacillus, Citrobacter, Clorstidium resinae, Corynebacterium, Flavobacterium, Nocardia, Pseudomonas, Candida,* and *Cladosporium*
Pesticides	
DDT	*P. aeruginosa*
Linurin	*B. sphaericus*
2,4-D	*Arthrobacter* and *P. cepacia*
2,4,5-T	*P. cepacia*
Parathion	*Pseudomonas* spp. and *E. coli*; *P. stutzeri,* and *P. aeruginosa*

[a]From Kumaram and Shivaraman (1988).

3. Use of radiolabeled substrates (^{14}C-labeled substrates).

4. Measurement of the disappearance of the chemical by gas chromatography or HPLC.

5. Determination of the decrease of dissolved organic carbon (DOC). For example, in Europe, the Association Francaise de Normalisation (AFNOR) test measures the reduction of DOC in aquatic samples. The pass level is 70 percent DOC reduction. In Japan, the Ministry of International Trade and Industry (MITI) test measures BOD reduction. The pass level is 60 percent BOD reduction.

6. Biodegradability under anaerobic conditions (Fig. 19.2c): The test essentially consists of spiking a diluted sludge sample with the test chemical in a sealed bottle and measuring gas production ($CH_4 + CO_2$) by gas chromatography or by other means (Owen et al., 1979; Shelton and Tiedje, 1984; Young, 1997).

19.4   REMOVAL OF TOXIC ORGANIC POLLUTANTS BY AEROBIC BIOLOGICAL PROCESSES

Several organic compounds have been detected in municipal wastewaters and are contributed by households, industry, commercial establishments, schools, universities, hospitals, and urban stormwaters. These compounds belong to several categories, which include phenols, aliphatic, polycyclic, and chlorinated aromatic hydrocarbons, phthalates, phosphate esters, ethers, terpenes, sterols, aldehydes, acids, and their esters. Hydrophobic organic compounds are generally effectively removed by wastewater treatment (Paxeus et al., 1992). The removal of organic compounds by biological waste treatment depends

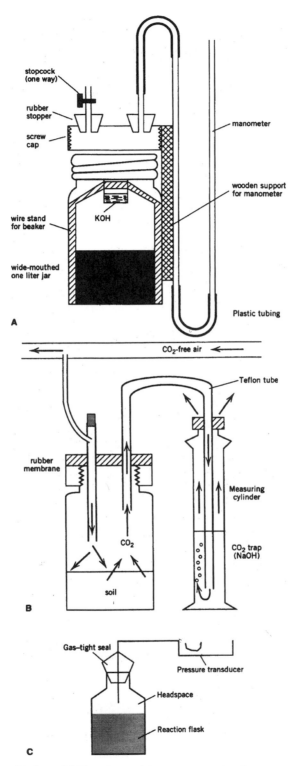

Figure 19.2 Some biodegradability tests: (a) measurement of oxygen consumption; (b) measurement of CO_2 evolution; (c) measurement of gas CH_4/CO_2 production under anaerobic conditions. From Anderson (1989). (Courtesy of Gulf Publishing Co.)

mainly on waste composition, type of treatment, and solid retention time (SRT). Conventional aerobic biological treatment removes up to 85 percent of dissolved organic carbon. Table 19.4 shows that about one-half of the dissolved organic carbon remaining after biological treatment (activated sludge, trickling filter, or stabilization ponds) is comprised of humic, fulvic, and hymathomelanic acids. The easily degradable compounds such as carbohydrates and proteins represent approximately 25 percent of the dissolved organic carbon (DOC) in biologically treated wastewater (Giger and Roberts, 1978; Manka et al., 1974).

The removal of several organic toxicants by activated and trickling filter processes has been investigated (Hannah et al., 1986; 1988). In general, the activated sludge process is quite efficient in decreasing the concentration of many priority pollutants and other xenobiotics to concentrations below detection limits (Grady, 1986). The average removal for seven volatile organic pollutants exceeded 91 percent for both activated sludge and trickling filter processes. However, removal is variable for semivolatile compounds and varies from 41 to 91 percent for trickling filters, and from 57 to 96 percent for activated sludge (Hannah et al., 1988; Table 19.5). In general, the activated sludge process provides the best removal for both volatile and semivolatile organic pollutants. Facultative lagoons with a long detention time (25.6 days) are the best alternative to activated sludge for the removal of organics (Hannah et al., 1986). A rotating biological contactor removed 75–96 percent of chlordane from contaminated wastewater. It was suggested that biodegradation is the main mechanism responsible for chlordane removal (Sabatini et al., 1990). Anionic surfactants such as the linear alkylbenzenesulfonates (LAS) are well removed (average removal above 99 percent) after activated sludge treatment (Brunner et al., 1988). Cationic surfactants are removed in the activated sludge process via adsorption/precipitation as well as biodegradation. Phthalate esters, commercially used as plasticizers and some listed as priority pollutants by the U.S. EPA, are rapidly biodegraded by activated sludge microorganisms. Their biodegradability decreases as their molecular weight increases (O'Grady et al., 1985; Shugatt et al., 1984). Most *para*-dichlorobenzene (i.e., >76 percent) is biodegraded during activated sludge treatment. Some of it is volatilized, and the extent of volatilization increases with increasing aeration rate (Topping, 1987).

TABLE 19.4.   Distribution of Categories of Organic Compounds in Biologically Treated Effluents[a]

Chemical Class	Total Chemical Oxygen Demand (%)		
	Trickling Filter	Stabilization Pond	Activated Sludge (Extended Aeration)
Proteins	21.6	21.1	23.1
Carbohydrates	5.9	7.8	4.6
Tannins and lignins	1.3	2.1	1.0
Anionic detergents	16.6	12.2	16.0
Ether extractables	13.4	11.9	16.3
Fulvic acid	25.4	26.6	24.0
Humic acid	12.5	14.7	6.1
Hymathomelanic acid	7.7	6.7	4.8

[a]From Manka et al. (1974).

TABLE 19.5. Removal of Volatile and Semivolatile Compounds by Biological Treatment[a]

Pollutants	% Removal	
	Trickling Filter	Activated Sludge
A. Semivolatile compounds		
Bis(2-ethylhexyl)phthalate	76	71
Dibutylphthalate	81	71
Naphthalene	89	95
Phenanthrene	90	93
Pyrene	83	91
Fluoranthene	85	92
Isophorone	65	96
Bis(2-chloroethyl)ether	31	64
p-Dichlorobenzene	86	94
Phenol	91	91
2-4-Dichlorophenol	55	86
Pentachlorophenol	41	60
Lindane	46	57
Heptachlor	64	67
B. Volatile compounds		
Carbon tetrachloride	90	81
1,1-Dichloroethane	92	97
1,1-Dichloroethylene	>97	>97
Chloroform	89	98
1,2-Dichloroethylene	87	95
Bromoform	82	68
Ethylbenzene	>98	>98

[a]Adapted from Hannah et al. (1988).

19.4.1 Pentachlorophenol

Pentachlorophenol (PCP) is a biocide widely used as a wood preservative. It is also used as a fungicide/bactericide in many other products (Guthrie et al., 1984). Strains of *Flavobacterium* and *Pseudomonas* are known to degrade PCP under aerobic conditions (Saber and Crawford, 1985; Watanabe, 1973). *Flavobacterium* and *Pseudomonas* strains, isolated from PCP-contaminated soils, are capable of mineralizing up to 200 ppm and 160 ppm of PCP, respectively (Radehaus and Schmidt, 1992; Saber and Crawford, 1989). *Phanerochaete* spp., the white-rot fungus, is also capable of degrading PCP as well as other xenobiotics such as polychlorinated biphenyls, chlorinated anilines, polycyclic aromatic hydrocarbons, dioxins, polychlorinated dibenzofurans, pesticides (e.g., chlordane, lindane, toxaphene, or munitions (e.g., trinitrotoluene) (Barr and Aust, 1994; Bennett and Faison, 1997; Brodkorb and Legge, 1992; Field et al., 1992; Lamar and Dietrich, 1990; Lamar et al., 1990; Shim and Kawamoto, 2002; Takada et al., 1996). A select list of environmental pollutants degraded by *Phanerochaete* is given in Table 19.6 (Reading et al., 2002). A U.S EPA survey has shown, however, that PCP removal by activated sludge is generally low and erratic. Much of the removal is due to PCP adsorption to sludge solids (Feiler, 1980). In laboratory-scale activated sludge units, PCP biodegradation increases with increasing solids retention time (SRT). Sorption is an important removal mechanism at low SRTs (Jacobsen et al., 1991), particularly for nonacclimated biomass (Blackburn et al., 1984; Edgehill and Finn, 1983a; Moos et al., 1983;

TABLE 19.6. Example of Environmental Pollutants Degraded by *Phanerochaete Chrysosporium*[a]

Chemical Category	Examples
Biopolymers	Cellulose lignin
Synthetic polymers	Polyacrylamide, nylon
Chlorinated aromatic compounds	Dichlorophenols, trichlorophenols, pentachlorophenol
Pesticides	DDT, chlordane, lindane, toxaphen
Polycyclic aromatic compounds	Benzo(a)pyrene, anthracene, naphthalene
Polycyclic chlorinated aromatic compounds	PCBs, dioxin
Munitions	RDX (hexahydro-1,3,5-trinitro-1,3,5-triazine), HMX (octahydro-1,3,5,7-tetranitro-1,3,5,7-tetraazocine), nitroglycerin
Dyes	Azo dyes, crystal violet

[a]Adapted from Reading et al. (2002).

Rochkind-Dubinsky et al., 1987). The PCP is biotransformed (and sometimes completely mineralized) in reactors that contain acclimated biomass or are amended with PCP-degrading bacteria.

19.4.2   Polychlorinated Biphenyls (PCBs)

Because they are thermally stable and have good dielectric properties, polychlorinated biphenyls (PCB) are used as dielectric fluids and as fire retardants. Bacteria (e.g., *Pseudomonas*, *Achromobacter*) and fungi (e.g., *Aspergillus*, *Phanerochaete chrysosporium*) can degrade PCBs in the environment (Ahmed and Focht, 1973; Dmochewitz and Ballschmiter, 1988; Takase et al., 1986; Yadav et al., 1995) and the biodegradation occurs under both aerobic and anaerobic conditions (Abramowicz, 1990). The genes that regulate PCB biodegradation are either plasmid- or chromosome-encoded (Chaudhury and Chapalamadugu, 1991). The few studies dealing with the fate of PCBs in the activated sludge system have shown the involvement of sorption and stripping in the removal of these compounds, but there is little definite proof of biotransformation or mineralization (Herbst et al., 1977; Kaneko et al., 1976; Rochkind-Dubinsky et al., 1987; Tucker et al., 1975).

19.4.3   Chloroaromatics

Chlorobenzenes or chlorophenoxy herbicides are metabolized to chlorophenols, which in turn can be metabolized to chlorocatechols or chloroanisoles (Rochkind-Dubinsky et al., 1987). Chloroaromatics that contain nitrogen are metabolized to chloroanilines, which form chlorocatechols, among other products. Chlorocatechols can be further metabolized to nonchlorinated byproducts such as pyruvate, succinate, and acetyl-CoA. There are few data demonstrating the biotransformation or mineralization of chloroaromatics in full-scale treatment plants (Rochkind-Dubinsky et al., 1987). Table 19.7 (Jordan, 1982) shows the removal of four chloroaromatics by an activated sludge unit. Their removal varies from 30 percent to 80 percent.

TABLE 19.7. Removal of Chloroaromatic Compounds by an Activated Sludge Unit[a]

Compound	Percentage Removal
2,4-Dichlorophenol	46
1,3-Dichlorobenzene	30
1,4-Dichlorobenzene	80
1,2,4-Trichlorobenzene	79

[a]Adapted from Jordan (1982).

19.4.4 Aromatic and Polynuclear Aromatic Hydrocarbons (PAHs)

Polynuclear aromatic hydrocarbons include compounds such as benzene, xylene, toluene, naphthalene, anthracene, styrene, phenantrene, and benzo(a)pyrene. Several of these chemicals or their oxidation products are carcinogenic. These chemicals are degraded under aerobic conditions (Crawford and O'Reilly, 1989). Under these conditions, benzene is converted to catechol, followed by ring fission to give ultimately CO_2 and H_2O (Gibson and Subramanian, 1988) (Fig. 19.3). With regard to hydrocarbons with side chains, biodegradation may proceed either through ring fission or by oxidation of the side chain (Gibson and Subramanian, 1988).

19.4.5 Trichloroethylene (TCE)

The aerobic degradation of TCE by mixed microbial consortia was demonstrated in experimental expanded-bed bioreactors. More than 90 percent of TCE is degraded when propane or methane plus propane are used as primary substrates. The microbial consortia use propane more efficiently than methane, and no TCE intermediates (e.g., dichloroethylene, vinyl chloride) are detected in the bioreactors (Phelps et al., 1990) (Fig. 19.4).

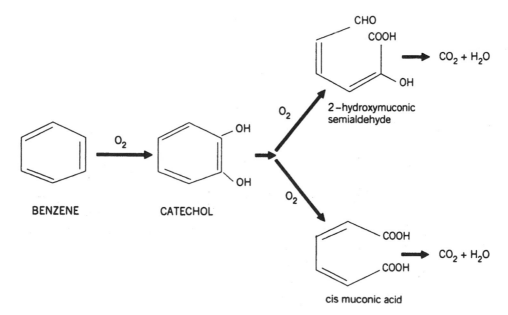

Figure 19.3 Bacterial catabolism of benzene. Adapted from Gibson and Subramanian (1988).

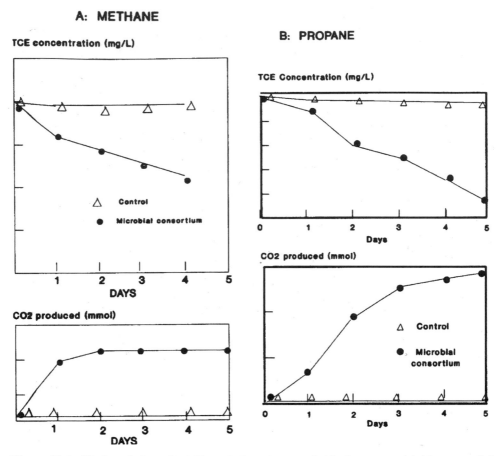

Figure 19.4 Biodegradation of trichloroethylene is expanded-bed reactors: (a) bioreactor fed methane daily; (b) bioreactor fed propane daily. From Phelps et al. (1990). (Courtesy of the American Society of Microbiology.)

Fixed-film bioreactors, using methane or natural gas as the primary substrate, are able to remove TCE at levels of <1,000 μg/L (TCE is toxic to methanotrophic bacteria at concentrations higher than 1000 μg/L). Methanotrophic biofilms are also capable of removing more than 90 percent of TCE from vapor streams generated by air stripping of polluted groundwater and soil venting (Canter et al., 1990). In expanded-bed bioreactors, microbial consortia fed methane or propane can degrade mixed organic wastes containing benzene, toluene, xylene, TCE, vinyl chloride, and nine other chlorinated hydrocarbons. These consortia are able to degrade 80–95 percent TCE and 99 percent of the other organics (Phelps et al., 1991). Removal of nitrogen and phosphorus from wastewater can also be achieved by using methanotrophic attached-film bioreactors. Phosphorus is removed to <0.1 mg P/L and the ammonia removal efficiency varied between 94 and 99 percent (Jewell et al., 1992).

19.4.6   Endocrine Disrupters

Several natural and anthropogenic compounds found in the environment may interfere with the endocrine system of aquatic and terrestrial organisms. They have been called endocrine disruptors or hormonally active agents (HAA). Sources of HAA include agricultural runoffs and effluents from industrial and domestic wastewater treatment plants. The major categories of HAAs are the phytoestrogens (natural plant HAAs), phthalates (plasticizers), alkylphenols (e.g., nonylphenol, octylphenol), as well as natural and synthetic estrogens. The natural estrogens include 17β-estradiol (E2), estrone (E1), and the degradation product estriol (E3). The major synthetic hormones utilized in birth control pills are 17α-ethynylestradiol (EE2) and mestranol.

Several studies have shown that caged male fish exposed to sewage effluents have high plasma vitellogenin concentrations. Vitellogenin is found only in female fish, and its presence in male fish signals an estrogenic response to HAAs present in the effluents. This has stimulated research on the fate of HAAs in wastewater treatment plants. Worldwide surveys have shown that these pollutants are indeed detected in raw wastewater as well as treatment plants effluents. Table 19.8 (Ward, 2003) shows a brief summary of the literature on the concentrations of E1, E2, E3, and EE2 in both influents and effluents. The HAAs are removed mainly by adsorption to solids and biodegradation. In general, the activated sludge process removes over 85 percent of E2, E3, and EE2, but E1 removal is lower (Johnson and Sumpter, 2001). Nonylphenol, an alkyl phenol, is lipophilic and tends to adsorb to biosolids, where it can reach concentrations from <1 to more than $1200\,\mu g/g$ d.w. (Bennie et al., 1998; Lee and Paert, 1995). On the basis of in vivo potency, it was suggested that EE2 is the most important estrogenic compound in sewage effluents and could potentially exert the biggest impact on wildlife (Johnson and Sumpter, 2001).

19.4.7 Pharmaceuticals

A wide range of pharmaceuticals and feed additives for animals find their way into wastewater (Halling-Sorensen et al., 1998). The sources include the dumping of pharmaceuticals through the kitchen sink or following excretion of unchanged chemicals or

TABLE 19.8. Concentrations (ng/L) of Natural and Synthetic Hormones in Wastewater Treatment Plants[a]

Plant	17β-estradiol (E2)	Estrone (E1)	Estriol (E2)	17α-estradiol (EE2)
Plant no. 1				
Influent	11.6	51.8	80.4	3.0
Effluent	1.4	18.4	3.0	0.4
Plant no. 2				
Influent	$<0.5-20$	$<0.5-75$	$2-120$	$<0.5-6$
Effluent	$<0.5-7$	$<0.5-52$	$<0.5-28$	$<0.5-2.2$
Plant no. 3				
Influent	11	44	73	NM
Effluent	1.6	17	2.3	NM
Plant no. 4 effluent				
Winter	$7-88$	$15-220$	NM	NM
Summer	$4-8.8$	$27-56$	NM	NM
Plant no. 5 effluent	0.9	4.5	NM	<0.3
Plant no. 6 effluent	$2.7-48$	$1.4-76$	NM	$0.2-7$

[a]Adapted from a compilation by Ward (2003).
NM = not measured.

metabolites though urine and feces from humans and animals. Some are quite resistant to biodegradation in wastewater treatment plants and end up in surface water, groundwater, and, potentially, in potable water supplies. There is now ongoing research into the fate of pharmaceuticals, drugs residuals, and personal care products in the environment and, particularly, in wastewater treatment plants. Little is known about the long-term health risks of their presence in our environment (Ongerth and Khan, 2004).

19.5 REMOVAL OF TOXIC ORGANIC POLLUTANTS BY ANAEROBIC AND ANOXIC BIOLOGICAL PROCESSES

In anaerobic respiration the oxidation of organic matter is coupled with the reduction of alternate electron acceptors such as nitrate (denitrification), sulfate (sulfate reduction), ferric iron (iron reduction), and CO_2 (methanogenesis). During the past two decades, significant advances have been made in our understanding of anaerobic processes (Suflita and Sewell, 1991). New high-rate anaerobic treatment processes have been developed and show great potential for the treatment of industrial wastewaters. These processes include suspended biomass and attached growth systems (Sterrit and Lester, 1988). Combined anaerobic–aerobic processes also show great promise for the treatment of complex wastewaters that contain refractory and toxic compounds (Lettinga, 1995).

The advantages of anaerobic treatment of wastewaters and sludges have been discussed in Chapter 13. We will now discuss the anaerobic biodegradation of selected categories of xenobiotics.

19.5.1 Phenolic Compounds

Phenolic compounds (e.g., phenol, catechol, resorcinol, *p*-cresol) make up 60–80 percent of coal conversion wastewaters. They are degraded to methane and carbon dioxide under anaerobic conditions (Blum et al., 1986; Fedorak and Hrudey, 1984; Khan et al., 1981; Suidan et al., 1981; Young and Haggblom, 1989). Culture acclimation improves the metabolic and gas production rates for phenolic compounds. These compounds may also serve as a carbon source and electron donors to denitrifiers and thus are degraded under anoxic conditions (Fedorak and Hrudey, 1988).

Chlorinated phenols are used as biocides (e.g., wood preservation) and their formation under environmental conditions results from the use of chlorine as an oxidant in the paper industry or as a disinfectant in water and wastewater containing phenols (Ahlborg and Thunberg, 1980). Concern has arisen over the release of chlorophenols into the environment because some of these compounds are toxic, some are carcinogenic, some may act as precursors of dioxins (Boyd and Shelton, 1984), and others may act as endocrine disrupters. Some chlorophenols (e.g., pentachlorophenol, 2,4-dichlorophenol, 2,4,5-trichlorophenol) have been placed on the U.S. EPA list of priority pollutants. Chlorophenols are biodegraded under both aerobic and anaerobic conditions (Häggblom and Valo, 1995; Tokuz, 1989). They are degraded in wastewater sludge under methanogenic conditions via reductive dechlorination (chlorine is removed from the aromatic ring) as the initial step followed by mineralization to CO_2 and CH_4 (Krumme and Boyd, 1988; Nicholson et al., 1992; Tiedje et al., 1987; Woods et al., 1989b). In fresh sludge, the relative rate of biodegradation is *ortho > meta > para* (Boyd and Shelton, 1984; Boyd et al., 1983). *o*-Chlorophenol is degraded to phenol, thus indicating reductive dechlorination of the molecule (Boyd et al., 1983). In acclimated sludge, more than 90 percent of 2-chlorophenol, 4-chlorophenol, and 2,4-dichlorophenol were mineralized to CO_2 and CH_4 (Boyd and Shelton, 1984). Chlorophenol, used as the sole source of carbon and energy in an anaerobic upflow bioreactor, was dechlorinated by reductive dechlorination and mineralized to CO_2 and CH_4. Most of the methanogenic activity was located at the bottom of the bioreactor (Krumme and Boyd, 1988). There is evidence that chlorophenol biodegradation may also occur under sulfate-reduction conditions (Häggblom and Young, 1990; Kohring et al., 1989).

Pentachlorophenol (PCP) is also biodegraded during anaerobic digestion of sludge and is transformed by reductive dechlorination to 3,5-dichlorophenol and other chlorinated phenols. Two-thirds of labeled PCP were mineralized to CO_2 and CH_4 (Guthrie et al., 1984; Mikesell and Boyd, 1985; 1986).

The biodegradation of phenolic compounds under anaerobic conditions has been summarized by Fedorak and Hrudey (1988). In general, alkyl phenolics (e.g., cresols, dimethyl phenols, ethyl phenols) are the most resistant to biodegradation by methanogenic cultures.

19.5.2 Benzene and Toluene

Benzene and toluene are degraded under methanogenic conditions to CO_2 and CH_4 by a methanogenic consortium originally isolated from sewage sludge. Most of the CO_2 is derived from the methyl group of toluene (Grbic-Galic and Vogel, 1987). Benzene can also be completely oxidized to CO_2 by sulfate-reducing bacteria and this phenomenon has been demonstrated in groundwater under field conditions (Anderson and Lovley, 2000; Lovley et al., 1995). Chlorinated benzenes are widely used as fungicides, industrial solvents, and in the manufacturing of various chemicals. They are widely detected in environmental samples, including wastewater, surface waters, groundwater, and sediments (Fathepure et al., 1988). Chlorobenzene and dichlorobenzene are used by bacteria (e.g., *Alcaligenes* spp. and *Pseudomonas* spp.) as the sole source of carbon (Chaudhury and Chapalamadugu, 1991). Available evidence shows that chlorobenzenes are degraded to chlorophenols and chlorocatechols (Rochkind-Dubinsky et al., 1987). Hexachlorobenzene is biodegraded to tri- and dichlorobenzene in anaerobically digested sludge by reductive dechlorination at a rate of 13.6 mm/L/day (Fathepure et al., 1988).

19.5.3 Chlorinated Hydrocarbons

Chlorinated hydrocarbons (pentachloroethylene, trichloroethylene, and carbon tetrachloride) are also degraded under anaerobic conditions. Pentachloroethylene (PCE) is transformed by reductive dehalogenation to trichloroethylene (TCE), dichloroethylene (DCE), and vinyl chloride (VC) (Fathepure et al., 1987; Vogel and McCarty, 1985). Almost complete dechlorination of PCE to ethene is possible under anaerobic conditions in the presence of methanol. The TCE may be degraded under both aerobic or anaerobic conditions. Under methanogenic conditions, TCE biodegradation results in the formation of vinyl chloride, which is more genotoxic than the parent compound (Vogel and McCarty, 1985) (Fig. 19.5). Furthermore, vinyl chloride can be mineralized to CO_2, under aerobic conditions, by a wide range of microorganisms (Davis and Carpenter, 1990; Hartmans et al., 1985). Carbon tetrachloride is transformed to chloroform, dichloromethane, and CO_2 (Egli et al., 1988), but it can be mineralized directly to CO_2, under denitrification conditions, without simultaneous production of chloroform, a dangerous byproduct (Criddle et al., 1990).

19.5.4 Other Compounds

Several other compounds are biodegraded under anaerobic conditions. These include phthalic acid esters, anionic surfactants, or polyethylene glycol (PEG). Phthalic acid esters can be

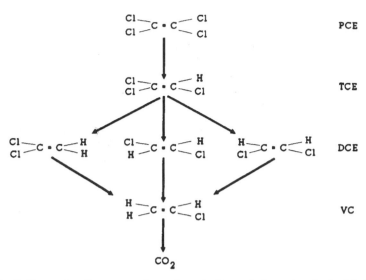

Figure 19.5 Pathway for the conversion of pentachloroethylene to vinyl chloride through reductive dehalogenation. From Vogel and McCarty (1985). (Courtesy of the American Society of Microbiology.)

biodegraded aerobically in activated sludge (O'Grady et al., 1985; Hannah et al., 1986) as well as anaerobically during anaerobic sludge digestion. However, some of these compounds (e.g., di-2-ethylhexyl and di-*n*-octyl phthalates) are resistant to biodegradation (Ziogou et al., 1989). Anionic surfactants such as the linear alkylbenzenesulfonates (LAS) are partially transferred to sludge. They are slightly affected by anaerobic digestion and thus accumulate at levels in the range of approximately 4 g/kg of sludge (Brunner et al., 1988). Polyethyleneglycol is relatively resistant to microbial degradation in aerobic environments (Cox, 1978), whereas, under anaerobic conditions, methanogenic consortia biodegrade PEG to ethanol, acetate, methane, and ethylene glycol (Dwyer and Tiedje, 1983).

Several anthropogenic compounds can be metabolized by denitrifying organisms. Substrates metabolized by denitrifiers include benzoate (Garcia et al., 1981), phthalates (Aftring and Taylor, 1981), nonionic detergents (Dodgson et al., 1984), toluene, *m*-xylene (Zeyer et al., 1989), and chlorinated compounds (Bouwer and McCarty, 1983; Criddle et al., 1990).

19.6 BIODEGRADATION IN BIOFILMS

Adsorption of microorganisms to surfaces leads to the formation of biofilms in diverse environments such as trickling filters, water distribution pipes, and aquifer materials. We have discussed the mechanisms involved in biofilm formation as well as the microbial ecology of biofilms in Chapter 16.

Trace organics can be removed from water and wastewater by physical, chemical, and biological processes. Fixed-film or biofilm processes are simple and inexpensive means for biological removal of trace organics from water and wastewater (Namkung et al., 1983; Rittmann and McCarty, 1980a,b; Williamson and McCarty, 1976). Models have been proposed to describe the kinetics of trace organics uptake and subsequent growth and decay of biofilm microorganisms. The parameters of biofilm kinetic are generally estimated through conventional batch and continuous culture techniques, but can also be determined in situ by using biofilm reactors (Rittmann et al., 1986).

19.6.1   Steady-State Model

The steady-state model (microbial biomass growth balances biomass loss) describes the growth of biofilm bacteria in the presence of a single growth-limiting substrate (Rittmann and McCarty, 1980a, b; McCarty et al., 1984). The model considers the diffusion of the substrate from the bulk solution into the biofilm and its subsequent utilization by biofilm bacteria, as well as bacterial growth and decay (Rittmann and McCarty, 1980a) (Fig. 19.6).

$$L_f = \frac{YJ}{bX_f} \tag{19.1}$$

where L_f = steady-state biofilm thickness (cm); Y = true yield of cell biomass per substrate used (mg VSS/mg substrate); X_f = biofilm cell density (mg VSS/cm^3); b = bacterial decay rate (day^{-1}); and J = substrate flux (mg/cm^2/day). J is the rate at which a substrate is transported from the bulk solution into the biofilm (Rittmann, 1987).

J is given by Fick's first law:

$$J = D\frac{S - S_s}{L} \tag{19.2}$$

where S = substrate concentration in the bulk solution (mg/L); S_s = substrate concentration at the biofilm surface (mg/L); L = effective diffusion layer (cm); and D = molecular diffusivity of substrate in water (cm^3/day).

There is a minimum substrate flux to support a deep biofilm. For heterotrophic microorganisms, this minimum was reported to be 0.1 mg BOD$_L$/cm^2/day (BOD$_L$ is the ultimate BOD as measured by oxygen demand; see Chapter 7). Substantial growth of the biofilm is obtained at J values three times higher than the minimum substrate flux. Nitrifiers are repressed at these high J values and are able to compete favorably with

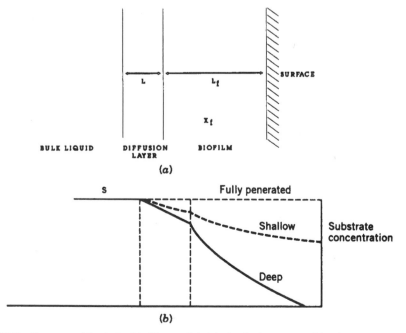

Figure 19.6   Conceptual basis for biofilm model: (a) physical concepts; (b) substrate concentration profiles. Adapted from Rittmann and McCarty (1980a).

heterotrophs only at J values lower than the minimum of 0.1 mg $BOD_L/cm^2/day$ (Rittmann, 1987).

There is a minimum steady-state substrate concentration, S_{min}, below which there is no biofilm growth. Figure 19.7 (Rittmann and McCarty, 1980b) shows that no further substrate degradation occurs when the substrate concentration approaches S_{min}. S_{min}, is given by the following formula (Rittmann, 1987; Rittmann and McCarty, 1980a, b; 1981; Rittmann et al., 1986, Rittmann et al., 1988):

$$S_{min} = \frac{K_s b}{Yk - b} = \frac{K_s b}{\mu_{max} - b} \qquad (19.3)$$

where K_s = half-saturation constant (mg/L); b = bacterial decay rate (day^{-1}); Y = yield of cell biomass per substrate used (mg VSS/mg substrate); k = maximum specific substrate utilization rate (mg/mg VSS); and μ_{max} = maximum specific growth rate (day^{-1}).

In oligotrophic biofilms, bacteria have a high affinity for substrates (i.e., low K_s) and therefore a low S_{min} (Rittmann et al., 1986). Oligotrophic microorganisms (or K-strategists) also have a low μ_{max}, and their K_s values vary between 0.0005 and 0.33 mg/L for suspended and biofilm oligotrophs (Rittmann et al., 1986) (Table 19.9), whereas copiotrophs display K_s values of 1–20 mg/L (van der Kooij et al., 1982a; van der Kooij and Hijnen, 1984; 1985a; Rittmann et al., 1986; Schmidt and Alexander, 1985). Acetate-fed biofilm oligotrophs growing on activated carbon have a K_s value of <10 µg/L. The μ_{max} values for these oligotrophs vary between 0.14 and 0.20 h^{-1} (Nakamura et al., 1989), whereas their maximum substrate uptake rate varies between 0.15 and 1.7 g COD/g biomass-day (Rittmann et al., 1986).

For simple substrates, under aerobic conditions, S_{min} varies between 0.1 and 1 mg/L (Stratton et al., 1983). The S_{min} under methanogenic conditions is much higher and ranges from 3.3 to 67 mg/L (Rittmann et al., 1988).

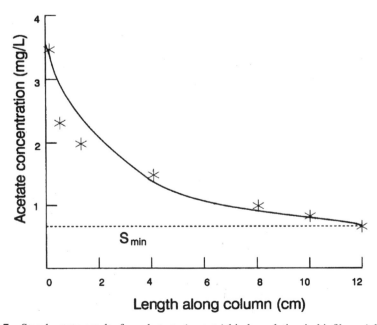

Figure 19.7 Steady-state results for substrate (acetate) biodegradation in biofilms. Adapted from Rittmann and McCarty (1980b).

TABLE 19.9. K_s Values of Suspended Biofilm Oligotrophs[a]

Microorganisms and Substrate	K_s (mg/L)
Pseudomonas aeruginosa	
Acetate	0.075
Aspartate	0.23
Flavobacterium spp.	
Maltose	0.065
Glucose	0.33
Starch	0.037
Amylose	0.083
Amylopectin	0.029
Spirillum spp.	
Oxalate	0.005
Salmonella typhimurium	
Glucose	0.25
Biofilm cultures	
Acetate	0.018
Phenol	0.012
Salicylate	0.0005–0.11

[a]Adapted from Rittmann *et al.* (1986).

19.6.2   Nonsteady-State Model

Substrate removal by biofilms at substrate concentrations below S_{min} can be achieved by a nonsteady-state biofilm (Rittmann and McCarty, 1981). Under laboratory conditions, this type of biofilm sustained more than 85 percent removal of galactose for a year, at substrate concentrations below S_{min} (galactose $S_{min} = 0.41$ mg/L) (Rittmann and Brunner, 1984). Co-metabolism can also explain the utilization of trace organics by biofilms at concentrations below S_{min}.

19.7   METAL BIOTRANSFORMATIONS

Heavy metals sources in wastewater treatment plants include mainly industrial discharges and urban stormwater runoff. Biological treatment processes (activated sludge, trickling filter, oxidation ponds) remove from 24 percent (e.g., Cd) to 82 percent (e.g., Cu, Cr) of metals (Hannah et al., 1986) (Table 19.10). Toxic metals may adversely affect biological treatment processes as well as the quality of receiving waters. They are inhibitory to

TABLE 19.10.   Percentage Metal Removal by Wastewater Treatment Plants[a]

Metal	Primary Clarification	Trickling Filter	Activated Sludge	Aerated Lagoon	Faculative Lagoon
Cr	7	52	82	71	79
Cu	19	60	82	74	79
Ni	4	30	43	35	43
Pb	30	48	65	58	50
Cd	12	28	24	—	32

[a]From Hannah et al. (1986).

both anaerobic and aerobic processes in wastewater treatment (Barth et al., 1965). The impact of metals on wastewater microorganisms as well as bioassays for assessing their toxic action will be discussed in Chapter 20.

19.7.1 Metabolic Activity and Metal Biotransformations

The metabolic activity of microorganisms may result in the solubilization, precipitation, chelation, biomethylation, or volatilization of heavy metals (Bremer and Geesey, 1991; Iverson and Brinckman, 1978). Microbial activity may result in the following results:

1. Production of strong acids such as H_2SO_4 by chemoautotrophic bacteria (e.g., *Acidibacterium*, (formerly *Thiobacillus*), which dissolve minerals (see Chapter 3 for further details).

2. Production of organic acids (e.g., citric acid), that dissolve but that also chelate metals to form metallorganic molecules.

3. Production of ammonia or organic bases, which precipitate heavy metals as hydroxides.

4. Extracellular metal precipitation: Sulfate-reducing bacteria (see Chapter 3) produce H_2S, which precipitates heavy metals as insoluble sulfides. Another example is *Klebsiella planticola* Strain Cd-1, which can precipitate Cd in the presence of thiosulfate (Sharma et al., 2000).

5. Production of extracellular polysaccharides, which can chelate heavy metals and thus reduce their toxicity (Bitton and Freihofer, 1978).

6. Ability of certain bacteria (e.g., sheathed filamentous bacteria) to fix Fe and Mn on their surface in the form of hydroxides or some other insoluble metal salts.

7. Biotransformation by bacteria that have the ability to biomethlylate or volatilize (e.g., Hg), oxidize (e.g., As), or reduce (e.g., Cr) heavy metals (see Section 19.7.2).

19.7.2 Biotransformations of Specific Metals and Metalloids

Biomethylation reactions transform inorganic forms of the metal to organic forms and may lead to the production of toxic gases. Several heavy metals (As, Se, Hg, Sn, Pb, Te) undergo methylation (see Table 18.4 in Chapter 18; Crowley and Dungan, 2002).

Biomethylation of mercury was first reported in 1969. Since then, many bacterial (e.g., *Pseudomonas fluorescens*, *E. coli*, *Clostridium* sp.) and fungal (*Aspergillus niger*, *Saccharomyces cerevisiae*) isolates have been shown to methylate Hg(II) to methyl mercury under mostly anaerobic conditions. Methylcobalamin (derivative of Vitamin B_{12}) is a coenzyme that donates methyl groups and stimulates biomethylation.

$$\underset{\text{mercury}}{Hg^{2+}} \longrightarrow \underset{\substack{\text{methyl} \\ \text{mercury}}}{Hg^{+}-CH_3} \longrightarrow \underset{\text{dimethyl mercury}}{Hg(CH_3)_2} \tag{19.4}$$

Methyl mercury accumulates under anaerobic conditions and is degraded under aerobic conditions. Sulfate-reducing bacteria are the predominant group capable of biomethylating mercury (Compeau and Bartha, 1985). Molybdate, an inhibitor of sulfate reducers, decreases mercury methylation in anoxic environments by more than 95 percent (Compeau and Bartha, 1987).

Microorganisms can also transform Hg^{2+} or organic mercury compounds (methyl mercury, phenylmercuric acetate, ethyl mercuric phosphate) to the volatile form Hg^0. Mercury volatilization is a detoxification mechanism, the genetic control of which has been elucidated. Plasmid genes code for resistance to metals such as mercury or cadmium. Resistance to mercury is carried out by the *mer* operon, which consists of a series of genes (*merA*, *merC*, *merD*, *merR*, *merT*). The *merA* gene is responsible for the production of mercuric reductase enzyme, which transforms Hg^{2+} to Hg^0 (Silver and Misra, 1988).

Cadmium (Cd^{2+}) can be accumulated by bacteria (e.g., *E. coli*, *B. cereus*) and fungi (e.g., *Aspergillus niger*). This metal can also be volatilized in the presence of vitamin B_{12}. As for mercury, lead can be methylated by bacteria (e.g., *Pseudomonas*, *Alcaligenes*, *Flavobacterium*) to tetramethyl lead $(CH_3)_4Pb$.

Fungi (e.g., *Aspergillus*, *Fusarium*) are capable of transforming arsenic to trimethyl arsine, a volatile form with a garlic-like odor. Methanogens convert inorganic arsenic to dimethylarsine under anaerobic conditions.

Selenium is methylated to volatile alkylselenides through the metabolic activity of bacteria (*Aeromonas*, *Flavobacterium*) and fungi (*Penicillium*, *Aspergillus*). As seen for mercury, methylcobalamin is the methyl donor.

Chromium-rich wastewaters are generated by industrial processes such as leather tanning, metal plating, and cleaning. An *Enterobacter cloacae* strain, isolated from municipal wastewater, was able to reduce hexavalent chromium (Cr^{6+}) to trivalent chromium (Cr^{3+}), which precipitates as a metal hydroxide at neutral pH, reducing the bioavailability and toxicity of this metal (Ohtake and Hardoyo, 1992).

19.8 MOBILE WASTEWATER PROCESSING SYSTEMS FOR BIODEGRADATION OF HAZARDOUS WASTES

Mobile waste treatment systems consist of modular equipment units that can be transported to a contaminated site by truck or railcar. They may also consist of larger units that require assembly on the site. Mobile waste treatment technology is not new, as it has been used for many years by the military. The U.S. EPA has used this technology for emergency response and remedial action for hazardous waste sites and contaminated aquifers. Mobile waste treatment systems are being investigated as possible alternatives to land treatment. Moreover, the use of the transportable units avoids the risks to humans and the environment that are associated with the transport of hazardous wastes to conventional treatment facilities. However, because of the limited data available on the performance and reliability of these systems, communities are sometimes reluctant to adopt them to tackle their pollution problems.

Mobile systems treat wastes via physical, chemical, and biological processes. These processes are listed in Table 19.11. We will focus on the biological processes used in mobile waste treatment systems (Glynn et al., 1987).

19.8.1 Aerobic Treatment Processes

Processes such as activated sludge or fixed-film bioreactors are used in mobile treatment units. In some instances, powdered activated carbon is added to wastewater to help remove halogenated organics. Removal efficiency for organics is sometimes improved by the addition of commercial acclimated bacterial strains. The removal efficiency for some organics can be as high as 99 percent. These systems can also be used for on-site treatment of groundwater from contaminated aquifers. Sludges generated by the mobile treatment units must be dewatered and sent off to a treatment and disposal facility.

TABLE 19.11.　Processes Involved in Mobile Waste Treatment Systems[a]

Category	Process
Physical processes	Thermal treatments (e.g., incineration, wet air oxidation)
	Air stripping
	Adsorption to activated carbon
	Evaporation/dewatering
	Filtration
	Ion exchange
Chemical processes	Reduction/oxidation (redox)
	Precipitation
	Neutralization
	Dechlorination
Biological processes	Aerobic digestion
	Anaerobic digestion
	In situ biodegradation

[a]From Glynn et al. (1987).

19.8.2　Anaerobic Digestion Processes

The processes involved in anaerobic digestion have been described in Chapter 13. Anaerobic digestion units are useful for the handling of high-strength wastewaters and leachates from hazardous waste sites. However, manufacturers rarely recommend them over aerobic systems because the anaerobic biodegradation process can be inhibited by several factors, including sensitivity to heavy metal toxicity, variable temperature and pH, and sudden changes in wastewater characteristics.

19.9　WEB RESOURCES

http://www.labmed.umn.edu/ (biodegradation database from the University of Minnesota)

http://ei.cornell.edu/biodeg/ (biodegradation from Cornell University)

http://www.ce.utexas.edu/prof/mckinney/papers/UTCHEM_BIO/UTCHEM_BIO. html (biodegradation model, University of Texas, Austin)

http://toxics.usgs.gov/bib/bib-Biodegradation.html (Bibliography of Publications on Natural Attenuation and Biodegradation from USGS)

http://www.cbs.umn.edu/cbri/ (Center for Biodegradation Research, University of Minnesota)

http://www.lbl.gov/NABIR/researchprogram/researchtopics/biotransformation.html (research programs in metal biotransformations)

http://www.status-umwelthormone.de/report_final_web.pdf (seminar on endocrine disruptors, Berlin, April 2001)

http://www.epa.gov/oscpmont/oscpendo/history/endo2_2.htm (endocrine disruptors, from U.S. EPA)

http://www.epa.gov/scipoly/oscpendo/index.htm (endocrine disruptor screening programs, U.S. EPA)

http://www.nap.edu/books/0309064163/html/index.html (Issues in Potable Reuse: The Viability of Augmenting Drinking Water Supplies with Reclaimed Water (1998); Commission on Geosciences, Environment and Resources)

http://www.envirpharma.org/presentations.php?p_type=speeche&p_lang=en (series of slide presentations on pharmaceuticals in the environment (conference in Lyon, France, April 2003)

19.10 QUESTIONS AND PROBLEMS

1. What are the physico-chemical processes operating in the removal of xenobiotics from wastewater?
2. What is *photocatalytic degradation*?
3. What is the fate of phthalates in activated sludge?
4. How do methanotrophs remove trichloroethylene?
5. What are the possible causes of xenobiotic recalcitrance?
6. How do microorganisms biotransform metals?
7. Which metals can be methylated by microorganisms?
8. Why should we be concerned about the presence of endocrine disrupters in wastewater effluents?
9. Why is vitellogenin used as an indicator for the presence of endocrine disrupters in aquatic environments?
10. What are the different approaches for measuring the biodegradation of an organic toxicant in wastewater?
11. Explain co-metabolism.
12. Compare oligotrophic and copiotrophic microorganisms.
13. What is a nonsteady-state biofilm model?

19.11 FURTHER READING

Alexander, M. 1985. Biodegradation of organic chemicals. Environ. Sci. Tech. 18: 106–111.

Chaudhury, G.R., and S. Chapalamadugu. 1991. Biodegradation of halogenated organic compounds. Microbiol. Rev. 55: 59–79.

Grady, C.P.L. 1986. Biodegradation of hazardous wastes by conventional biological treatments. Haz. Wastes and Haz. Material. 3: 333–365.

Hardman, D.J. 1987. Microbial control of environmental pollution: the use of genetic techniques to engineer organisms with novel catabolic capabilities. pp. 295–317, In: *Environmental Biotechnology*, C.F. Forster, and D.A.J. Wase, Eds., Ellis Horwood, Chichester, U.K.

Leisinger, T., R. Hutter, A.M. Cook, and J. Nuesch, Eds. 1981. *Microbial Degradation of Xenobiotics and Recalcitrant Compounds*, Academic Press, San Diego.

Maier, R.M. 2000. Microorganisms and organic pollutants, pp. 363–402, In: *Environmental Microbiology*, R.M. Maier, I.L. Pepper, and C.P. Gerba, Eds., Academic Press, San Diego.

Rittmann, B.E., D.E. Jackson, and S.L. Storck. 1988. Potential for treatment of hazardous organic chemicals with biological processes, pp. 15–64, In: *Biotreatment Systems*, Vol. 3, D.L. Wise, Ed. CRC Press, Boca Raton, FL.

Rochkind-Dubinsky, M.L., G.S. Sayler, and J.W. Blackburn. 1987. *Microbiological Decomposition of Chlorinated Aromatic Compounds*, Marcel Dekker, New York, 315 pp.

Sayler, G.S., and J.W. Blackburn. 1989. Modern biological methods: The role of biotechnology, pp. 53–71, In: *Biotreatment of Agricultural Wastewater*, M.E. Huntley, Ed. CRC Press, Boca Raton, FL.

Young, L.Y., and C.E. Cerniglia, Eds. 1995. *Microbial Transformation and Degradation of Toxic Organic Chemicals*, Wiley-Liss, New York.

20

TOXICITY TESTING IN WASTEWATER TREATMENT PLANTS USING MICROORGANISMS

20.1 INTRODUCTION

Serious concern has arisen over the release of more than 50,000 xenobiotics into the environment. Their impact on aquatic environments, including wastewaters, is generally

Wastewater Microbiology, Third Edition, by Gabriel Bitton
Copyright © 2005 John Wiley & Sons, Inc.

determined by acute and chronic toxicity tests, using mostly fish and invertebrate bioassays (Peltier and Weber, 1985). However, because of the large inventory of chemicals, short-term bioassays are now being considered for handling this task. These tests are based chiefly on inhibition of the activity of enzymes, bacteria, fungi, algae, and protozoa (Bitton and Dutka, 1986; Blaise, 1991; Dutka and Bitton, 1986, Liu and Dutka, 1984; Wells et al., 1998). These enzymatic and microbial assays, also called *microbiotests*, are simple, rapid, cost-effective, and can be miniaturized. The advantages of microbiotests are summarized in Table 20.1 (Blaise, 1991).

20.2 IMPACT OF TOXICANTS ON WASTEWATER TREATMENT

Toxic inhibition by organic (e.g., chlorinated organics, phenolic compounds, surfactants, pesticides) and inorganic (e.g., heavy metals, sulfides, ammonia) chemicals is a major problem encountered during the biological treatment of industrial and domestic wastewaters. Some of the chemicals that enter wastewater treatment plants, particularly the volatile compounds, may pose a potentially health threat to plant operators. Many of the toxic chemicals or their metabolites are, however, transferred to wastewater sludges. The application of these sludges to agricultural soils may result in the uptake and accumulation of toxic and genotoxic chemicals by crops and grazing animals, eventually posing a threat to humans (see Chapter 21).

TABLE 20.1.   Attractive Features of Microbiotests[a]

Feature	Explanatory Remark
Inexpensive or cost-efficient	Cost is test-dependent and can vary from a few dollars to several hundred dollars.
Generally not labor-intensive	As opposed to steps involved in undertaking fish bioassays, for example.
High sample throughput potential	When automation technology can be applied.
Cultures easily maintained or maintenance-free	Freeze-drying technology can be applied.
Modest laboratory and incubation space requirement	As opposed to a specialized laboratory essential for fish bioassays, for example.
Insignificant postexperimental chores	Owing to disposable plasticware, which is recycled instead of having to be washed for reuse, as in the case of large experimental vessels.
Low sample volume requirements	Often, a few milliliters suffice to initiate tests instead of liters.
Sensitive/rapid responses to toxicants	Short life cycles of (micro)organisms enable endpoint measurements after just minutes or several hours of exposure to toxic chemicals.
Precise/reproducible responses	High number of assayed organisms, increased number of replicates, and error-free robotic technology are contributors to this feature.
Surrogate testing potential	Microbiotests are adequate substitutes for macrobiotests in some cases.
Portability	For cases in which microbiotests are amenable to being applied in the field.

[a]From Blaise (1991).

Chemical toxicants may also adversely affect biological treatment processes (Koopman and Bitton, 1986). Toxic inhibition is sometimes a serious problem in plants treating industrial effluents. Activated sludge is the aerobic process that has been studied primarily with regard to toxic inhibition (see the impact of toxicants on anaerobic digestion in Chapter 13). The major effects of toxicants on activated sludge are reduced BOD and COD removal, reduced efficiency in solids separation, and modification of sludge compaction properties.

Chemical toxicants can also diminish the quality of receiving waters. Toxic wastewater effluents may threaten aquatic organisms in receiving waters, the use of which may be restricted. Guidelines are available for the levels of several heavy metals in receiving waters, but less is known as regard the levels of organic toxicants.

Some of the human-made chemicals may disrupt the endocrine system of aquatic organisms and humans. These chemicals can mimic the natural estrogens (i.e., female hormones) and can compete for the estrogen receptor sites in cells (Folmar et al., 1996; Janssens et al., 1997; Purdom et al., 1994). Endocrine disrupters (ED) of varying potencies, include natural estrogens (e.g., 17β-estradiol), synthetic steroids (e.g., 17α-ethynylestradiol entering in the composition of contraceptive pills), phytoestrogens, pesticides, and alkylphenols. The latter are the biodegradation products of the nonionic surfactants alkylphenol-polyethoxylates (NPE), which are extensively used in industrial processes (Giger et al., 1987). Metabolites of NPE (e.g., nonylphenol) have been detected in wastewater effluents and biosolids (Ahel et al., 1994; Brunner et al., 1988). Nonylphenol, being more lipophilic and more persistent than NPE, tends to be well adsorbed by biosolids, where it can reach concentrations from <1 to more than 1200 $\mu g/g$ d.w. (Bennie et al., 1998; Lee and Paert, 1995; Lee et al., 1997). An analysis of seven effluents in the United Kingdom showed the presence of natural and synthetic steroids consisting of 17β-estradiol (concentration range of $4-48$ ng/L of effluent), estrone (concentration range of $1-76$ ng/L effluent), and 17α-ethynylestradiol (concentrations range from nondetectable to 7 ng/L) (Desbrow et al., 1998). Other investigators found estrogen concentrations in tens of ng/L in sewage effluents, or river water impacted by wastewater effluents. Very low concentrations of endocrine disrupting compounds were also found in finished drinking water, but the public health significance of these findings is unknown at the present time.

Purdom and colleagues (1994) demonstrated that domestic wastewater effluents were estrogenic to fish. The effluents stimulated the production of vitellogenin (VTG) in male fish exposed to wastewater effluents. Stimulation of VTG in male trout has also been shown for receiving waters in the United Kingdom (Harries et al., 1996; 1997) and in the United States (Folmar et al., 1996). Following measurement of vitellogenin blood levels in fish in Belgium, estrogenic activity was observed in a canal predominantly impacted by domestic wastewater discharges (De Coen et al., 1999).

There is not yet a systematic monitoring program for endocrine disrupters in water and wastewater treatment plants.

20.2.1 Heavy Metals

Heavy metals are major toxicants found in industrial wastewaters and may adversely affect the biological treatment of wastewater. The sources of heavy metals in wastewater treatment plants are mainly industrial discharges and urban stormwater runoffs. Some of the heavy metals are priority pollutants (e.g., Cd, Cr, Pb, Hg, Ag). Heavy metal toxicity is mainly due to soluble metals. Toxicity is controlled by various factors such as pH, type and concentration of complexing agents in wastewater, antagonistic effects by toxicant mixtures, oxidation state of the metal, and redox potential (Babich and Stotzky, 1986; Jenkins et al., 1964; Kao et al., 1982; Sujarittanonta and Sherrard, 1981). Metals may be complexed with natural (e.g., humic substances) or anthropogenic substances (e.g., nitrilotriacetic acid, which is used as a builder in detergents). Microorganisms can also affect the complexation of metals and modify their solubility (see Chapter 19).

Anaerobic wastewater treatment processes appear to be more sensitive to heavy metals than are aerobic processes (Barth, 1975). As regards the aerobic treatment of wastewater, heavy metals inhibit two important processes, namely COD removal and nitrification. Shock loads of heavy metals (e.g., Hg, Cd, Zn, Cr, Cu) may lead to deflocculation of activated sludge (Henney et al., 1980; Lamb and Tollefson, 1973; Neufield, 1976). The compaction behavior of activated sludge may be affected as a result of differential effects of heavy metals on filamentous and floc-forming microorganisms. Activated sludge bulking appears to occur less in systems receiving heavy metals (Barth et al., 1965; Neufield, 1976). This phenomenon is also observed in the presence of organic toxicants (Monsen and Davis, 1984).

Methanogenesis is affected by heavy metals, although some metals (e.g., Ni, Co), at trace levels, may be stimulatory to methanogens (see Chapter 13). Heavy metal toxicity depends on sulfide and phosphate levels, as well as other ligands present in wastewater. Other inhibitors of methanogenesis include ammonia, phenols, chlorinated hydrocarbons, benzene ring compounds, formaldehyde, detergents, and sulfides (this topic is covered in more detail in Chapter 13).

Activated sludge and facultative lagoons provide the best removal for toxic metals such as Cd, Cr, Cu, Zn, Ni, and Pb. Metals are generally concentrated in sludges (Hannah et al., 1986). Metal removal by activated sludge is due to sorption of the metals to flocs. Removal by biological solids depends on pH, solubility, and concentration of metals, concentration of organic matter, amount of biomass, and biological solid retention time (Cheng et al., 1975; Nelson et al., 1981; Sujarittanonta and Sherrard, 1981). The affinity of biological solids for heavy metals was found to follow the order: $Pb > Cd > Hg > Cr^{3+} > Cr^{6+} > Zn > Ni$ (Neufield and Hermann, 1975). Other means of removal of heavy metals by microorganisms are complexation by carboxyl groups of microbial polysaccharides and other polymers, precipitation (e.g., precipitation of Cd by *Klebsiella aerogenes*), volatilization (e.g., mercury), and intracellular accumulation (Fig. 20.1) (Valls and de Lorenzo, 2002).

Biofilms protect microorganisms from the deleterious effects of metals. It was reported that biofilm microorganisms were 2 to 600 times more resistant to metals than planktonic (i.e., free-floating) microorganisms. This protective effect is due to the metal-complexing ability of extracellular polymeric substances, which are abundantly found in biofilms (Teitzel and Parsek, 2003). The scavenging of metals by biofilm processes can be exploited for metal removal from wastewater.

The role of microorganisms in metal removal in waste treatment was explored in Chapter 18.

20.2.2 Organic Toxicants

There are two concerns over the fate of organic toxicants in wastewater treatment plants:

1. *Biodegradation of organic toxicants in wastewater treatment plants.* It is desirable that xenobiotics be mineralized to CO_2 or, at least, to less toxic metabolites (see Chapter 19).

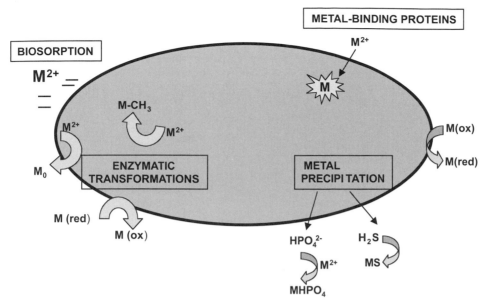

Figure 20.1   Bacterial response to heavy metals. From Valls and de Lorenzo (2002).

2. *Toxicity of xenobiotics to waste treatment organisms.* There can be subsequent reduction in the removal of biogenic organic compounds (i.e., lower BOD reduction) or inhibition of nitrification (Boethling, 1984) and methane production (Fedorak and Hrudey, 1984).

The toxicity of phenolic compounds has been widely studied in wastewater treatment plants. Phenol and *p*-cresol inhibit methane production when their concentrations are above 2000 and 1000 mg/L, respectively (Fedorak and Hrudey, 1984). Phenol toxicity is also observed in fixed-film biological reactors at concentrations of 1000–3000 mg/L (Sayama and Itokawa, 1980). Pentachlorophenol, a widely used pesticide in the United States, is inhibitory to ATP synthesis and methane production in methanogenic bacteria (Roberton and Wolfe, 1970). It is also inhibitory to unacclimated methanogens at a threshold concentration of 200 μg/L. Acclimation raises the threshold to 600 μg/L (Guthrie et al., 1984). Nitrophenols (e.g., 2,4-dinitrophenol, 4-nitrophenol, 2-nitrophenol) are toxic to anaerobic systems, with 2,4-dinitrophenol exerting the highest toxicity (Uberol and Bhattacharya, 1997). The toxic effect of organic and inorganic toxicants on methanogenesis was also discussed in Chapter 13.

Inhibitors affect substrate removal by microorganisms in a manner similar to that affecting enzymatic activity (Hartmann and Laubenberger, 1968; Volskay et al., 1988). There are four major types of reversible inhibitors (competitive, noncompetitive, uncompetitive, and mixed-type) affecting biodegradation, based on their effect on the kinetic parameters μ_{max} and K_s (see Chapter 2 for more details). Respiration-based toxicity studies in pilot plant bioreactors have shown that 13 of 33 RCRA-listed (RCRA, Resource Conservation and Recovery Act) organic compounds displayed an EC_{50} of >100 mg/L; therefore, they will not have a significant impact on activated sludge operation at the lower concentrations actually found. Characterization of the inhibition caused by RCRA compounds showed that none of the compounds displayed competitive inhibition (Volskay et al., 1990) (Table 20.2).

TABLE 20.2. Characterization of Inhibition Caused
by RCRA-Listed Organic Compounds

Inhibitor	Type of Inhibition
Carbon tetrachloride	Mixed
Chlorobenzene	Mixed
Chloroform	Mixed
1,2-Dichloroethane	Mixed
1,2-Dichloropropane	Mixed
2,4-Dimethylphenol	Uncompetitive
Methylene chloride	Mixed
Nitrobenzene	Uncompetitive
Phenol	Uncompetitive
Tetrachloroethylene	Noncompetitive
Toluene	Mixed
1,1,1-Trichloroethane	Mixed
1,1,2-Trichloroethane	Mixed
Trichloroethylene	Mixed

20.3 TOXICITY ASSAYS USING ENZYMES AND MICROORGANISMS

Applications of microbial toxicity assays in wastewater treatment plants fall into four categories:

1. The first category involves the use of these assays to monitor the toxicity of wastewaters at various points in the collection system, the major goal being the protection of biological treatment processes from toxicant action. These screening tests should be useful for pinpointing the source of the toxicants entering the wastewater treatment plant.

2. The second category involves the use of these toxicity assays in process control to evaluate pretreatment options for detoxifying incoming industrial wastes.

3. The third category concerns the application of short-term microbial and enzymatic assays to detect toxic inhibition of biological processes used in the treatment of wastewaters and sludges.

4. The last category deals with the use of these rapid assays in Toxicity Reduction Evaluation (TRE)/Toxicity Identification Evaluation (TIE) to characterize the problem toxic chemical(s) (see Section 20.4).

We will now review the most used enzymatic and microbial assays for toxicity assessment in wastewater treatment plants.

20.3.1 Enzymatic Assays

Enzymes are proteins that serve as catalysts of biological reactions in animal, plant, and microbial cells. The use of enzymes for indicating the adverse effect of toxic chemicals on microbial populations in soil is well known (Burns, 1978). In aquatic environments, some enzymes (e.g., dehydrogenases) are well correlated with microbial activity (Dermer et al., 1980). Chemical toxicity may be conveniently and rapidly determined in water and wastewater by using simple and low-cost enzymatic assays that can be miniaturized and automated. Several enzymes (dehydrogenases, ATPase, phosphatase, esterase, urease, luciferase, β-galactosidase, α-glucosidase) were explored for assessing toxicity in aquatic environments, including wastewater treatment plants. Table 20.3 lists some short-term toxicity assays based on enzymatic activity or biosynthesis (Bitton and Koopman, 1986; 1992; Christensen et al., 1982, Obst et al., 1988).

Dehydrogenases are the enzymes most used in toxicity testing in wastewater treatment plants. Dehydrogenase activity is assayed by measuring the reduction of oxidoreduction dyes such as triphenyl tetrazolium chloride (TTC), nitroblue tetrazolium(NBT), 2-(p-iodophenyl)-3-(p-nitrophenyl)-5-phenyltetrazolium chloride (INT), or resazurin. Several toxicity tests, based on inhibition of dehydrogenase activity in wastewater, have been developed (Bitton and Koopman, 1986).

There are tests based on the inhibitory effect of chemicals on enzyme biosynthesis, instead of enzymatic activity. For example, the *de novo* biosynthesis of β-galactosidase in *Escherichia coli* is more sensitive to toxicants than enzymatic activity (Dutton et al., 1988). Toxi-Chromotest is a commercial toxicity assay based on the inhibitory effect of chemicals on β-galactosidase biosynthesis in *E. coli* (Reinhartz et al., 1987). However, enzyme biosynthesis is less sensitive than other assays (e.g., Microtox, *Ceriodaphnia dubia*) used for assessing the toxicity of wastewater effluents or sediment extracts (Koopman et al., 1988, 1989; Kwan and Dutka, 1990). A higher sensitivity is obtained with a toxicity test based on the inhibitory effect of chemicals on the biosynthesis of α-glucosidase in *Bacillus licheniformis* (Campbell et al., 1993; Dutton et al., 1990).

TABLE 20.3. Some Short-Term Toxicity Assays Based on Enzymatic Activity or Biosynthesis[a]

Enzyme	Endpoint Measured	Comments
Dehydrogenases	Measure reduction of oxido-reduction dyes such as INT or TTC	Widely tested in water, wastewater, soils, sediments
ATPase	Measure phosphate concentration using ATP as a substrate	In vivo and in vitro have been used
Esterases	Nonfluorescent substrates degraded to fluorescent products	Acetylcholinesterase sensitive to organophosphates and carbamates
Phosphatases	Measure organic portion of substrate (e.g., phenols) or inorganic phosphate	Sensitive to heavy metals in soils
Urease	Measure ammonia production from urea	Studied mostly in soils
Luciferase	Measure light production using ATP as a substrate	Used in ATP–TOX bioassay in conjunction with inhibition of ATP levels in a bacterial culture
β-galactosidase	Measure hydrolysis of o-nitrophenyl-D-galactoside	Toxicant effect on both enzyme activity and biosynthesis was tested
α-glucosidase and β-glucosidase	p-nitrophenyl-α-D-glucoside p-nitrophenyl-β-D-glucoside	Toxicant effect on enzyme activity and biosynthesis have been tested
Tryptophanase	Add Ehrlich's reagent coside and measure absorbance at 568 nm	Toxicant effect on enzyme biosynthesis has been tested

[a]From Bitton and Koopman (1992).

20.3.2 Microbial Bioassays

Several bacterial assays are available for determining the toxicity of environmental samples (Bitton and Dutka, 1986; Dutka and Bitton, 1986; Janssen, 1997; Liu and Dutka, 1984; Wells et al., 1998). A selection of bacterial assays is displayed in Table 20.4.

Toxic chemicals may adversely affect the light output of bioluminescent bacteria. A toxicity assay, commercialized under the name of Microtox, uses freeze-dried cultures of the bioluminescent marine bacterium, *Vibrio fischeri* (formerly known as *Photobacterium phosphoreum*) (Bulich et al., 1979; Bulich, 1986). (Fig. 20.2). This test is based on the inhibition of *V. fischeri* by toxic chemicals (Fig. 20.3). It has been used for determining the toxicity of wastewater effluents, complex industrial wastes (oil refineries, pulp, and paper), fossil fuel process water, sediment extracts, sanitary landfill leachates, stormwaters, surface waters, groundwaters, hazardous waste leachates, and air samples (Munkittrick et al., 1991). This assay shows a good correlation with fish, *Daphnia*, and algal bioassays (Blaise et al., 1987; Curtis et al., 1982; Giesy et al., 1988; Logue et al., 1989). Microtox is, however, not as sensitive to toxic metals. Recent developments in the Microtox test system include online monitoring, chronic toxicity testing, solid phase assays, and genotoxicity testing (Mutatox) (Johnson, 1998; Qureshi et al., 1998). Microtox has been tested for online monitoring of toxicants entering wastewater or drinking water treatment plants (Levi et al., 1989; Qureshi et al., 1998). A 22 h chronic test version of Microtox is now available and is comparatively as sensitive as *Ceriodaphnia* with regard to the toxicity of inorganic and organic toxicants (Bulich and Bailey, 1995). Genotoxicity testing is based on the effect of DNA-damaging agents on a dark mutant of *Photobacterium leiognathi* (Jarvis et al., 1996; Ulitzur, 1986). A genetically engineered *Pseudomonas* containing the *lux* operon that codes for bioluminescence has also been used for toxicity testing of wastewater. Similarly to Microtox, the test is based on inhibition of bioluminescence by

TABLE 20.4. Some Short-Term Bacterial Toxicity Assays[a]

Assay	Basis for the Test
Microtox	Inhibition of bioluminescence of *Vibrio fischerii*
Spirillum volutans	Toxicants cause loss of coordination of rotating fascicles of flagella with concomitant loss of motility
Growth inhibition	Measure growth inhibition of pure (e.g., *Aeromonas*, *Pseudomonas*) or mixed cultures via absorbance determination for microbial suspensions or via measurement of zones of inhibition on solid growth media
Viability assays	Measured effect of toxicants on the viability of bacterial cultures on agar plates
ATP assay	Inhibitory effect of toxic chemicals on ATP levels in microorganisms
ATP–TOX assay	Test based on the growth inhibition, via ATP measurement, of bacterial culture and inhibition of luciferase activity
Respirometry	Measures effect of toxicants on microbial respiration in environmental samples
Toxi-Chromotest	Based on inhibition of biosynthesis of β-galactosidase in *Escherichia coli*
α-Glucosidase biosynthesis assay	Based on inhibition of biosynthesis of α-glucosidase in *Bacillus licheniformis*
Nitrobacter bioassay	Measures inhibition of nitrite oxidation to nitrate
Microcalorimetry	Measures decreases in heat production by microbial communities

[a]From Bitton and Koopman (1992).

Figure 20.2 Microtox model 500 toxicity analyzer. (Courtesy of A. Bulich, Microbics Corp., Carlsbad, CA.)

toxicants (Ren and Frymier, 2003). Similarly, Pseudomonad species, isolated from a wastewater treatment plant treating phenolic compounds, were engineered to carry a chromosomal copy of the *lux* operon (*luxCDABE*) derived from *Photorhabdus luminescens* and used for toxicity monitoring in the plant (Wiles et al., 2003).

Toxicity testing may also be based on growth inhibition of pure or mixed bacterial cultures or wastewater microorganisms (Alsop et al., 1980; Trevors, 1986). The assays consist of determining changes in bacterial densities via measurement of the optical density of the bacterial suspensions, determination of inhibition zones on solid growth media, or by ATP measurement. The ATP–TOX assay is based on both the growth inhibition, by ATP measurement, of *E. coli*, and inhibition of luciferase activity (Xu and Dutka, 1987). In a test based on inhibition of ^{14}C glucose uptake by activated sludge microorganisms, activated sludge is spiked with labeled glucose; substrate uptake is

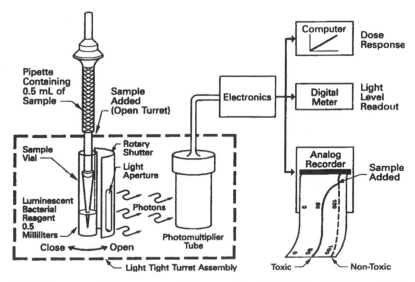

Figure 20.3 Microtox toxicity test system (Courtesy of A. Bulich, Microbics Corp., Carlsbad, CA.)

measured following a 15 min incubation (Larson and Schaeffer, 1982). This test is presently rarely utilized due to its use of radioactive compounds.

Toxic chemicals may also exert an adverse effect on nutrient cycling (C, N, P, and S cycles) by microorganisms. As to the nitrogen cycle, nitrification is probably the step most sensitive to environmental toxicants. A nitrifying process is impacted more rapidly than a carbonaceous BOD removal process. Toxicity assays based on the inhibition of both *Nitrosomonas* and *Nitrobacter* were developed for determining the toxicity of wastewater samples (Alleman, 1988; Williamson and Johnson, 1981). However, *Nitrosomonas* appears to be much more sensitive to toxicants than *Nitrobacter* (Blum and Speece, 1992). A continuous monitor based on the activity of an immobilized nitrifier (*Nitrosomonas europaea*) was developed to monitor river and wastewater toxicity (Tanaka et al., 1998). The impact of toxic chemicals on the carbon cycle is conveniently determined by measuring the inhibition of microbial respiration, which can be measured with oxygen electrodes, manometers, or electrolytic respirometers (King and Dutka, 1986).

Respiration Inhibition Kinetics Analysis (RIKA) involves the measurement of the effect of toxicants on the kinetics of a biogenic substrate (e.g., butyric acid) removal by activated sludge microorganisms. The kinetic parameters studied are q_{max}, the maximum specific substrate removal rate (determined indirectly by measuring the V_{max}, the maximum respiration rate), and K_s, the half-saturation constant. The procedure consists of measuring with a respirometer the Monod kinetic parameters, V_{max} and K_s, in the absence and in the presence of various concentrations of the inhibitory compound (Volskay and Grady, 1990; Volskay et al., 1990). Figure 20.4 (Volksay et al., 1990) illustrates the results obtained by using the RIKA procedure with 1,2-dichloropropane.

Exposure of microbial cells to toxicants also results in the induction of *stress proteins*, which can be detected by polyacrylamide gel electrophoresis. Some of these proteins overlap with heat shock (Neidhardt et al., 1984) and starvation proteins (Matin et al., 1989), but others are produced in response to specific inorganic and organic toxicants (Blom et al., 1992). The pattern of stress proteins may possibly be used as an index of exposure to toxic chemicals. Heat shock gene expression is also induced by other environmental stresses, including toxic chemicals. Biosensor bacterial strains containing heat shock gene-bioluminescence gene fusions, have been considered for detection of chemicals that induce light production. These biosensors, however, respond only to relatively

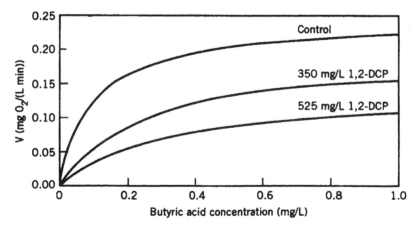

Figure 20.4   Toxicity of 1,2-dichloropropane (1,2-DCP), using the RIKA procedure (1,2-DCP is an inhibitor which increases K_s). Adapted from Volksay et al. (1990).

high concentrations of chemicals. Bioluminescence induction is generally followed by inhibition of light production at higher levels of the toxic chemical. Thus, several dilutions of a given environmental sample must be tested (van Dyk et al., 1994). Recent developments in toxicity testing using bacteria include the construction of genetically engineered bacteria that respond to *specific* categories of toxic chemicals (Belkin, 1998).

Owing to their roles as primary producers and crucial components of the food chain, micro-algae are useful indicators of environmental quality. They have been traditionally used as indicator microorganisms to assess the eutrophication potential of surface waters. Algae may also serve as test organisms in toxicity testing. The algal bottle test proposed by the U.S. EPA (U.S. EPA, 1989c) is based on growth inhibition of the green alga, *Selenastrum capricornutum* (also known as *Pseudokirchneriella subcapitata*). This toxicity assay was later miniaturized by conducting the test in 96-well microplates, which allows the use of small sample volumes. The toxicity endpoints are growth inhibition as determined via cell counts or fluorescence after 96 h, ATP content after 4 h, or cell esterase inhibition after 24 h exposure to the toxic sample. This alga was shown to be quite sensitive to heavy metals and herbicides (Blaise, 2002; Hickey et al., 1991; St. Laurent et al., 1992).

20.3.3 Commercial Rapid Microbial Assays for Toxicity Testing in Wastewater Treatment Plants

Several microbial or enzymatic assays are now marketed. Some could be useful for toxicity testing in wastewater treatment plants (Bitton and Koopman, 1992).

20.3.3.1 Microtox. This toxicity test, discussed in Section 20.3.2, is probably the most popular commercial test for assessing toxicity in wastewater treatment plants. An online model is now available.

20.3.3.2 Polytox. The assay microorganisms in Polytox are a blend of bacterial strains originally isolated from wastewater. The Polytox kit, specifically designed to assess the effect of toxic chemicals on biological waste treatment, is based on the reduction of the respiratory activity of the rehydrated cultures in the presence of toxicants.

20.3.3.3 Toxi-Chromotest. Toxi-Chromotest is based on inhibition of β-galactosidase biosynthesis in *E. coli* by toxic chemicals. This test is not as sensitive as Microtox to aquatic toxicants.

20.3.3.4 MetPAD/MetPLATE. MetPAD[TM] is the first microbial assay for the direct assessment of the toxicity of specific categories of chemicals. This test is designed for the *specific determination of heavy metal toxicity* in environmental samples and is based on the inhibition of β-galactosidase activity in a mutant strain of *E. coli* by bioavailable heavy metals. MetPAD[TM] methodology is displayed in Figure 20.5. This bioassay was useful for the determination of the toxicity of bioavailable heavy metals in wastewaters and in sediment elutriates (Bitton et al., 1992b; Campbell et al., 1993). This kit is also available in the 96-well microplate format and marketed as MetPLATE[TM] and FluoroMet-PLATE (Bitton et al., 1994; Jung et al., 1996). As regards metal toxicity testing, Met-PLATE[TM] gave results similar to those given by the standard 48 h *Ceriodaphnia dubia* test and was more sensitive than Microtox (Nelson and Roline, 1998). Other developments in MetPLATE[TM] testing include a solid-phase assay for soils, sediments, and biosolids, and the ability to test heavy metal uptake and toxicity in plants (Bitton and Morel, 1998).

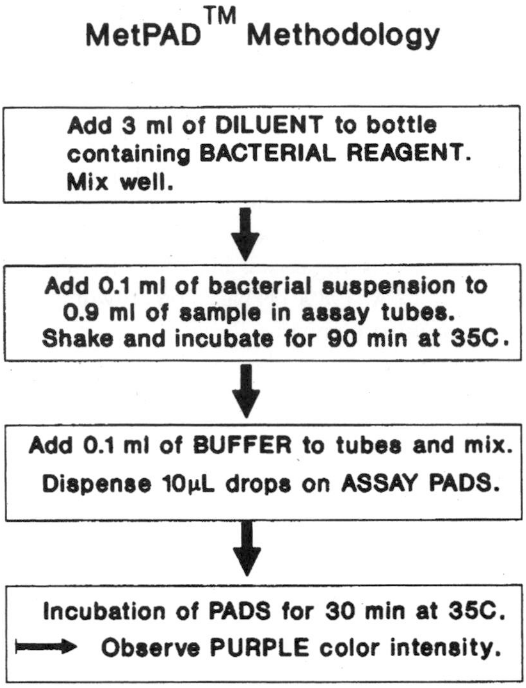

MetPAD^{TM} Methodology

```
┌─────────────────────────────────────┐
│ Add 3 ml of DILUENT to bottle       │
│ containing BACTERIAL REAGENT.       │
│ Mix well.                            │
└─────────────────────────────────────┘
                  ↓
┌─────────────────────────────────────┐
│ Add 0.1 ml of bacterial suspension to│
│ 0.9 ml of sample in assay tubes.    │
│ Shake and incubate for 90 min at 35C.│
└─────────────────────────────────────┘
                  ↓
┌─────────────────────────────────────┐
│ Add 0.1 ml of BUFFER to tubes and mix.│
│ Dispense 10µL drops on ASSAY PADS.  │
└─────────────────────────────────────┘
                  ↓
┌─────────────────────────────────────┐
│ Incubation of PADS for 30 min at 35C.│
│ ➤ Observe PURPLE color intensity.   │
└─────────────────────────────────────┘
```

Figure 20.5 MetPAD methodology for assessment of heavy metal toxicity. From Bitton et al. (1992). (Courtesy of John Wiley, NY.)

There are also specific biosensors that can detect the presence of specific metals in environmental samples. These biosensors are constructed by fusing an inducible promoter from a metal-resistance operon to a reporter gene that codes for the production of bioluminescence or a specific enzyme (e.g., β-galactosidase). Mercury and arsenite can be detected by such biosensors (Ramanathan et al., 1997).

20.4 APPLICATIONS OF MICROBIAL AND ENZYMATIC TESTS IN TOXICITY ASSESSMENT IN WASTEWATER TREATMENT PLANTS

We will discuss briefly four applications of microbial biotests.

20.4.1 Whole Effluent Toxicity (WET)

In the United States, toxicant discharges from wastewater treatment plants are regulated under the National Pollutant Discharge System (NPDS). Permits are issued to dischargers by State regulatory agencies. In addition to collecting chemical data, dischargers may also be required to carry out whole effluent toxicity tests, using standard toxicity tests with fish, invertebrates, and algae as the test organisms. Such WET tests using microorganisms such as bioluminescent bacteria or algae are more rapid and more economical than the standard assays. As an example, Figure 20.6 (Blaise, 2002) shows the response of *Selenastrum capricornutum* to pulp and paper mill effluents in Canada.

Toxicity decreased as the degree of treatment increased.

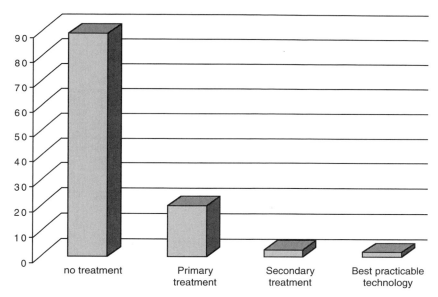

Figure 20.6 Toxicity response of *Selenastrum capricornutum* to pulp and paper mill effluents. From Blaise (2002).

20.4.2 Toxicity Identification Evaluation (TIE)

Microbial toxicity tests could also be convenient tools for conducting toxicant characterization in *Toxicity Identification Evaluation* (TIE) and *Toxicity Reduction Evaluation* (TRE), proposed by the U.S. EPA (1988; 1989a, b). These entail a series of tests that are carried out systematically to determine the sources of effluent toxicity, the specific causative toxicant(s), and the effectiveness of pollution control measures to reduce effluent toxicity.

Phase 1 of TRE consists of a series of fractionation steps that provide information about the physico-chemical properties of toxicants in a given wastewater effluent. The physico-chemical treatments are followed by acute and/or chronic toxicity assays, using daphnids, fish, or, possibly, algae, bacterial, and enzymatic assays. In acute tests, the endpoints are LC_{50} values (concentration that leads to 50 percent mortality during the testing period) or ET_{50} values (elapsed time for 50 percent mortality at one effluent concentration). These fractionation steps include the following tests (U.S. EPA, 1988) (Fig. 20.7):

1. *Baseline effluent toxicity test.* The original wastewater effluent must be tested and LC_{50} and ET_{50} determined.
2. *Degradation test.* These tests provide information concerning the biodegradation, photodegradation, and/or chemical oxidation of the effluent. This information may be useful in treatability studies during phase II of TRE.
3. *Filtration test.* This test gives an indication on whether the causative toxicant is associated with solids.
4. *Air stripping test.* This test provides information about whether the effluent toxicity is caused by volatile or oxidizable compounds. This test is carried out at ambient, acid (pH 3.0), and basic (pH 11.0) conditions.
5. *Oxidant reduction test.* This test determines the presence of oxidants (e.g., chlorine) that may be responsible for effluent toxicity.

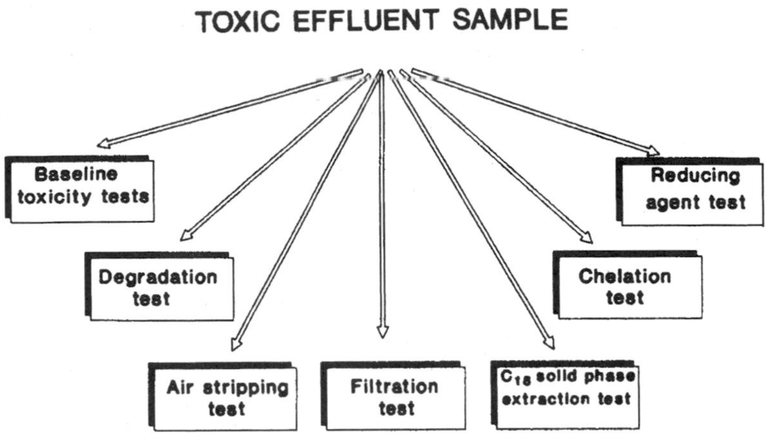

TOXIC EFFLUENT SAMPLE

Figure 20.7   Phase 1 effluent characterization tests. Adapted from U.S. EPA (1988).

6. *EDTA chelation test.*   Absence or reduction of toxicity following treatment of the sample with the sodium salt of ethylenediamine tetraacetic acid indicates the presence of cationic toxicants.

7. *Solid phase extraction (SPE) test.*   The SPE test consists of passing the filtered effluent though small columns containing a C_{18} hydrocarbon resin. Absence or reduction of toxicity following the SPE test, indicates the presence of nonpolar compounds.

8. *Other tests.*   Cationic and anionic exchange resins are also used to test for the presence of cationic and anionic toxicants, respectively. Chelating resins have also been considered for removing heavy metal toxicity. XAD resins are sometimes employed for the removal or reduction of toxicity caused by polar organic compounds (Walsh and Garnas, 1983; DiGiano et al., 1992; Mazidji et al., 1992; Mirenda and Hall, 1992).

Because these fractionation schemes generate more than 50 fractions for each sample tested, it would appear that short-term microbial assays would be useful in toxicant characterization for evaluation of toxicity reduction. Microtox, in conjunction with *Ceriodaphnia* bioassay, were used for toxicant characterization in wastewater samples from Jacksonville, Florida (Mazidji et al., 1990).

   The wastewater fractionation procedures give valuable information about the chemical properties of toxicants in wastewater, thus aiding both the evaluation of control techniques and specific identification of the problem chemicals.

20.4.3   Water Effect Ratio (WER) and Heavy Metal Binding Capacity (HMBC)

Several physico-chemical factors control metal bioavailability and, thus, toxicity in aquatic environments. These factors include pH, redox potential, alkalinity, hardness, adsorption to suspended solids, cations and anions, as well as interaction with organic compounds (Kong et al., 1995). The U.S. EPA has proposed the concept of *Water Effect Ratio* (WER) to assess the impact of water parameters on heavy metal bioavailability/toxicity in environmental samples (U.S. EPA, 1984b; 1994). This WER is the ratio of the LC_{50} derived from testing the toxicity of a metal to fish or invertebrates in metal-spiked site water to the LC_{50} of the metal-spiked laboratory water:

$$WER = LC_{50} \text{ (site water)}/LC_{50} \text{ (lab water)}$$

The *Heavy Metal Binding Capacity* (HMBC) test is based on the same concept as WER, except that MetPLATE, a bacterial assay specific for metal toxicity, is used as the toxicity test (see Section 20.3.3 for more details on MetPLATE). A ratio significantly higher than 1 means that the aquatic sample exerts an effect on the bioavailability/toxicity of the metal under investigation. This test has been used to test metal bioavailability in wastewater effluents, surface waters, and municipal solid waste landfill leachates (Huang et al., 1999; Ward, 2003). An example of HMBC values found in aquatic samples is displayed in Table 20.5 (Ward, 2003).

20.4.4 Potential Ecotoxic Effects Probe (PEEP)

No single microbial bioassay can detect all the categories of environmental toxicants with equal sensitivity. Therefore, a battery-of-tests approach has been suggested and consists of using concurrently some of the short-term assays discussed previously. The battery-of-tests approach was found to be useful in the formulation of a new index for assessing the toxic potential of wastewater effluents (Costan et al., 1993). This toxicity index, called PEEP (Potential Ecotoxic Effects Probe), is expressed as a log_{10} value, which varies from 0 to 10. This index takes into account the toxic/genotoxic response of

TABLE 20.5. HMBC[a] for Municipal Solid Waste Leachates, Lake Water, and Wastewater Effluents[b]

	HMBC[a]		
Landfill no.	Cu^{2+} (as $CuSO_4$) Mean	Zn^{2+} (as $ZnCl_2$) Mean	Hg^{2+} (as $HgCl_2$) Mean
1	54.5 (1.6–162.9)	25.7 (8.8–42.7)	30.5 ± 2.4
2	34.9 (7.2–56.2)	48.1 (6.2–115.7)	21.7 (16.7–26.7)
3	114.9 (7.9–327.9)	32.3 (8.8–101.1)	86.8 (85.2–88.3)
4	2.9 (2.7–3.1)	10.9 (10.7–11.1)	3.6 ± 0.2
5	27.5 (1.1–79.7)	6.1 (1.4–10.7)	12.9 (6.8–19.1)
6	59.7 ± 3.6	45.2 ± 9.9	100.8 ± 8.4
7	26.7 (9.9–43.5)	4.9 (3.8–5.9)	13.3 ± 1.1
Lake Alice	1.1 ± 0.1	1.8 ± 0.3	2.8 ± 0.1
Lake Beverly	<1	<1	<1
WWTP effluent	2.4 ± 0.1	2.6 ± 0.3	2.3 ± 0.1

[a]HMBC is unitless. Results are presented as mean and (range). Mean ± one standard deviation are shown for sites sampled one time. Ranges are shown for sites sampled on multiple occasions.
[b]Adapted from Ward (2003).

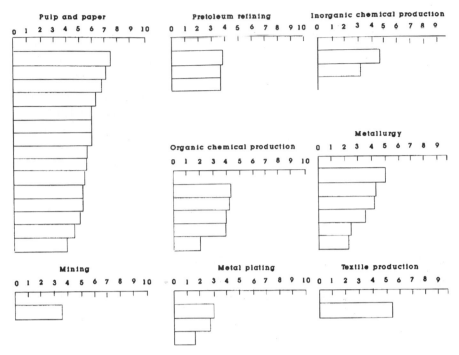

Figure 20.8 PEEP values of 37 industrial effluents from Canadian plants. From Costan et al. (1993). (Courtesy of John Wiley, NY.)

several short-term bioassays with test organisms belonging to different trophic levels (e.g., bacteria, algae, crustaceans), the persistence of the toxic/genotoxic effect, and the flow rate of the wastewater effluent. It is given by the following formula:

$$\text{PEEP} = \log_{10}\left[1 + n\frac{\sum_{i}^{N}T_i}{N}Q\right] \qquad (19.1)$$

where PEEP = Potential Ecotoxic Effects Probe (numerical value of 0–10); n = number of tests that exhibit toxic/genotoxic responses; N = maximum number of obtainable toxic/genotoxic responses; T_i = toxic response in toxicity units exhibited by a given bioassay before or after biodegradation; and Q = wastewater effluent flow rate (m³/h).

Figure 20.8 (Costan et al., 1993) shows the PEEP values of 37 industrial effluents from Canadian plants. The PEEP ranged from 0 (nontoxic effluents) to 7.5 (very toxic effluents), and was highest for pulp and paper mill effluents.

20.5 WEB RESOURCES

http://www.epa.gov/scipoly/oscpendo/(U.S. EPA endocrine disruptors screening program)
http://www.envirpharma.org/presentation/speeche/stamatelatou.pdf (effect of pharmaceuticals on wastewater treatment)
http://www.biology.pl/bakterie_sw/index_en.html (bioluminescent bacteria)
http://www.edie.net/Library/TechnologyDB/view_entry.asp?ID = 534   (whole   cell biosensors used in wastewater toxicity)

http://www.ees.ufl.edu/homepp/bitton/Metplate_frontpage/MetPAD-MetPLATE-page.html (heavy metal toxicity testing)
http://www.epa.gov/npdes/pubs/tre.pdf (EPA Toxicity Reduction Evaluation)
http://sites.huji.ac.il/applsci/es/belkin/ (bacterial biosensors for toxicity testing)
http://www.azurenv.com/mtox.htm (Microtox page)
http://www.ees.ufl.edu/homepp/bitton/Metplate_frontpage/MetPAD-MetPLATE-page.html (MetPAD/MetPLATE page)

20.6 QUESTIONS AND PROBLEMS

1. What is the impact of shock loads of heavy metals on activated sludge?
2. Compare the sensitivity of biofilm and planktonic microorganisms to metals.
3. What enzymes have been used in toxicity testing?
4. How do we assess exposure to endocrine disrupters in fish?
5. What is the basis of the following toxicity tests:
   (a) Toxi-Chromotest?
   (b) Microtox?
   (c) MetPLATE?
6. In the nitrogen cycle, what is the process most sensitive to toxic chemicals?
7. Explain the concept of *Water Effect Ratio* (WER).
8. What does Respiration Inhibition Kinetics Analysis (RIKA) consist of?
9. What are the endpoints used in toxicity testing with microscopic algae?
10. What is the toxicity of heavy metals in the presence of hydrogen sulfide?
11. What is involved in the construction of biosensors for heavy metals?
12. What is involved in phase 1 of Toxicity Reduction Evaluation (TRE)? Describe the various tests involved.
13. Explain the PEEP test for determining the toxicity of wastewater effluents.

20.7 FURTHER READING

Bitton, G., and B.J. Dutka, Eds. 1986. *Toxicity Testing Using Microorganisms*, Vol. 1. CRC Press, Boca Raton, FL.

Bitton, G., and B. Koopman. 1992. Bacterial and enzymatic bioassays for toxicity testing in the environment. Rev. Environ. Contam. Toxicol. 125: 1–22.

Blaise, C. 2002. Use of microscopic algae in toxicity testing, pp. 3219–3230, In: *Encyclopedia of Environmental Microbiology*, G. Bitton, editor-in-chief, Wiley-Interscience, N.Y.

Dutka B.J., and G. Bitton, Eds. 1986. *Toxicity Testing using Microorganisms*, Vol 2. CRC Press, Boca Raton, FL.

Janssen, C. 1997. Alternative assays for routine toxicity assessments: A review. pp. 813–839, In: *Ecotoxicology: Ecological Fundamentals, Chemical Exposure, and Biological Effects*. G. Schüürmann, and B. Markert, Eds. John Wiley & Sons, New York; Spektrum, Heidelberg, Germany.

Koopman, B., and G. Bitton. 1986. Toxicant screening in wastewater systems. p 101–132, In: *Toxicity Testing Using Microorganisms*, Vol. 2, B.J. Dutka, and G. Bitton, Eds. CRC Press, Boca Raton, FL.

Liu D., and B.J. Dutka, Eds. 1984. *Toxicity Screening Procedures using Bacterial_Systems.* Marcel Dekker, New York.

Wells, P.G., K. Lee, and C. Blaise, Eds. 1998. *Microscale Testing in Aquatic Toxicology: Advances, Techniques, and Practice.* CRC Press, Boca Raton, FL.

PART G

MICROBIOLOGY AND PUBLIC HEALTH ASPECTS OF WASTEWATER EFFLUENTS AND BIOSOLIDS DISPOSAL AND REUSE

21

PUBLIC HEALTH ASPECTS OF WASTEWATER AND BIOSOLIDS DISPOSAL ON LAND

Wastewater Microbiology, Third Edition, by Gabriel Bitton
Copyright © 2005 John Wiley & Sons, Inc.

21.1 INTRODUCTION

Wastewater treatment plants generate effluents and biosolids that must be disposed of safely and economically. In Chapters 21 and 22, we will discuss the public health aspects regarding the disposal of wastewater effluents and biosolids into the environment. We have limited our discussion to two popular and most studied approaches to waste disposal: land application (Chapter 21) and ocean outfalls (Chapter 22).

21.2 LAND TREATMENT SYSTEMS FOR WASTEWATER EFFLUENTS AND BIOSOLIDS (SLUDGES)

21.2.1 Wastewater Effluents

The main objectives of land application of wastewater are further effluent treatment, groundwater recharge, and the provision of nutrients for agricultural crops. During land treatment of wastewater effluents, biological and chemical pollutants are removed by physical (settling, filtration), chemical (adsorption, precipitation) and biological (e.g., plant uptake, microbial transformations, biological decay) processes. Land treatment systems for wastewater effluents are capable of removing microbial pathogens and parasites, BOD, suspended solids, nutrients (nitrogen and phosphorus), toxic metals, and trace organics (Fig. 21.1). Suspended solids are removed by filtration and sedimentation. Soluble organic compounds are removed by microbial action, particularly by biofilms developing on soil particles. Nitrogen is removed by sedimentation–filtration (e.g., particulate-associated organic nitrogen), adsorption to soil and volatilization (e.g., removal of NH_4 as NH_3), uptake by crops, and biological denitrification (Lance, 1972) (Fig. 21.2). Phosphorus is removed by adsorption to soil particles, chemical precipitation, and uptake by vegetation. The expected effluent chemical quality resulting from land treatment is shown in Table 21.1 (Metcalf and Eddy, 1991). The capacity of soils for metal retention is generally high, particularly in alkaline soils. The removal of trace organics by soils and aquifers is generally carried out via adsorption (organic soils have a high retention capacity), volatilization, and biodegradation (see Section 21.9). There is concern over the uptake of trace organics found in wastewaters and sludges by agricultural crops and animals (Majeti and Clark, 1981).

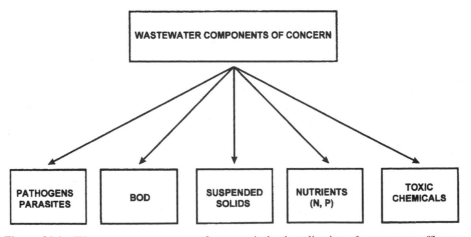

Figure 21.1 Wastewater components of concern in land application of wastewater effluents.

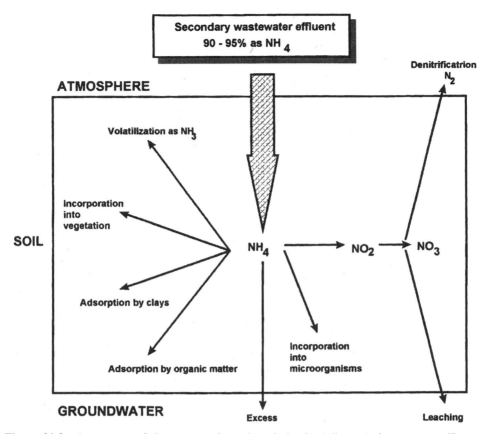

Figure 21.2 A summary of nitrogen transformations during land disposal of wastewater effluents. Adapted from Lance (1972).

There are four types of land treatment systems (Oron et al., 1991a, b; Polprasert, 1989; Reed et al., 1988; U.S. EPA, 1981), as described in the following sections.

21.2.1.1 Slow-Rate Irrigation System.

Slow-rate irrigation systems are the most frequently used land treatment systems. They provide essential nutrients to satisfy the

TABLE 21.1. Expected Chemical Quality of Wastewater Effluents After Land Application[a]

| | Value (mg/L) | | | | | |
| | Slow Rate[b] | | Rapid Infiltration[c] | | Overland Flow[d] | |
Constituent	Average	Maximum	Average	Maximum	Average	Maximum
BOD	<2	<5	2	<5	10	<15
Suspended solids	<1	<5	2	<5	15	<25
Ammonia nitrogen as N	<0.5	<2	0.5	<2	1	<3
Total nitrogen as N	3	<8	10	<20	5	<8
Total phosphorus as P	<0.1	<0.3	1	<5	4	<6

[a]From Metcalf and Eddy (1991).
[b]Percolation of primary or secondary effluent through 5 ft (1.5 m) of soil.
[c]Percolation of primary or secondary effluent through 15 ft (4.5 m) of soil.
[d]Runoff of continued municipal wastewater over about 150 ft (45 m) of slope.

growth requirements of agricultural crops. Pretreated wastewater is applied to land (soil texture ranges from sandy loams to clay loams), by sprinkler or surface distribution, at a relatively slow rate (0.5–6 m/year) (weekly loading rate of 1.3–10 cm) and serves as a source of nutrients for forage (e.g., alfalfa, bermuda grass, ryegrass), or field crops (e.g., corn, cotton, barley). (U.S. EPA, 1981) (Fig. 21.3). Physical, chemical, and biological processes contribute to the treatment of the incoming wastewater. This system provides the highest treatment potential and removes approximately 99 percent of BOD, suspended solids, and coliforms. Some limitations of the slow-rate irrigation system are land cost, high operating cost, and transport of wastewater to the treatment site.

21.2.1.2 *Overland Flow System.*

Wastewater is applied at a rate of 3–20 m/year or more and flows down a grass-covered slope (2–8 percent) with a length of 30–60 m (U.S. EPA, 1981) (Fig. 21.4). The most suitable soils are clay or clay-loamy soils with low permeability (equal to or less than 0.5 cm/h) to limit wastewater percolation through the soil profile. The treated effluent is captured in a collection channel. Nutrients (N, P, and BOD), suspended solids, and pathogens are removed as wastewater flows down the slope. This system achieves 95–99 percent removal of BOD and suspended solids. Removal of nitrogen in overland flow systems is due to nitrification followed by denitrification, and uptake by agricultural crops. Phosphorus is removed via adsorption and precipitation.

21.2.1.3 *Rapid Infiltration Systems.*

Wastewater is applied intermittently at high loading rates (6–125 m/year) onto a permeable soil (e.g., sands or loamy sands) (U.S. EPA, 1981). The hydraulic pathway displayed in Figure 21.5 shows that most of the applied wastewater flows to groundwater aquifers. The treated wastewater may be collected via recovery wells. The minimum depth to groundwater is 1 m during flooding periods and 1.5–3 m during dry periods. The treatment potential of rapid infiltration systems is lower than in slow-rate systems. Removal of nitrogen is generally low, but may be increased by encouraging denitrification. Denitrification requires adequate carbon levels (as found in primary effluents) and low oxygen levels, necessitating flooding periods as long as 9 days followed by drying periods of about 2 weeks.

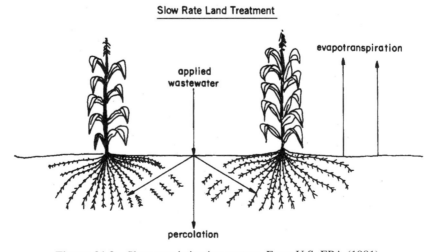

Figure 21.3   Slow-rate irrigation system. From U.S. EPA (1981).

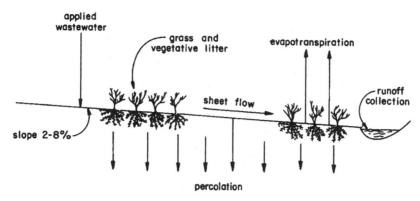

Figure 21.4 Overland flow system. From U.S. EPA (1981).

21.2.1.4 Subsurface Drip Irrigation. Subsurface drip irrigation is practiced in arid areas to save water. It consists of dripping water near the root zone (30–50 cm below the soil surface) at a rate that depends on the plant water requirements. This practice helps avoid or minimize biological aerosol production, contamination of aerial parts of crops and wastewater flow to groundwater (Oron et al., 1991a, b). Using PRD1 and MS2 phages as model viruses, it was shown that subsurface drip irrigation of turfgrass can reduce the risk of contamination by potential viral pathogens when compared to sprinkler irrigation (Enriquez et al., 2003). Similarly, as compared to sprinkler irrigation, surface and subsurface drip irrigation were found to reduce vegetable contamination with *Cryptosporidium* oocysts or *Giardia* cysts (Armon et al., 2002).

21.2.2 Biosolids (Sludges)

Approximately more than 7 million dry tons of sludge are produced annually in the United States, a number that is expected to increase in the future. Sludge is disposed of by land

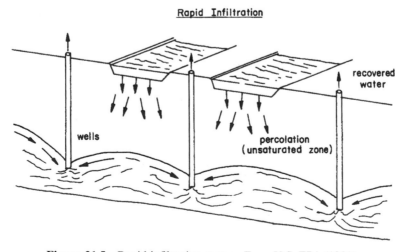

Figure 21.5 Rapid infiltration system. From U.S. EPA (1981).

application (between 21 and 39 percent, depending on the size of the plant), landfills (12–35 percent), incineration (1–32 percent, mostly by large plants), distribution and marketing (13–19 percent), and ocean dumping (1–4 percent) (U.S. EPA, 1984a). The various sludge disposal methods are summarized in Figure 21.6 (Metcalf and Eddy, 1991). Sludge is also used in horticulture and by home gardeners, generally after composting. It is applied to forests to increase productivity and to strip-mined land for reclamation. The U.S. legislation has banned sludge disposal into the ocean. Of the options available for sludge disposal, it seems that the agricultural option is the most convenient and least costly.

The European Community generates approximately 6 million dry tons of sludge, 30 percent of which is used by agriculture. In Great Britain, approximately two-thirds of the sludge produced is applied to land and one-third is dumped into the ocean. About 4 percent of the sludge is incinerated (Bruce and Davis, 1989; Forster and Senior, 1987; Kofoed, 1984; Wallis and Lehmann, 1983).

Sludge contains useful nutrients such as nitrogen (5.1 percent dry weight basis), phosphorus (1.6 percent dry weight basis), and potassium (0.4 percent dry weight basis) as well as micronutrients. Furthermore, sludge organic matter helps improve soil

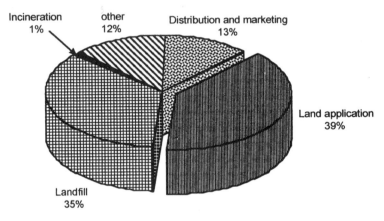

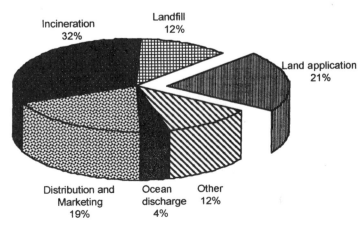

Figure 21.6   Sludge disposal options: (a) 1.0–10 Mgal/d plants; (b) >10 Mgal/d plants. Adapted from Metcalf and Eddy (1991).

structure by increasing its water-holding capacity and by aerating the soil. Design sludge-loading rates are based on nitrogen-loading rates necessary for plant growth (Metcalf and Eddy, 1991).

Wastewater sludge also contains heavy metals and trace organics that can present potential health risks to humans, animals, and agricultural crops. Much is known about phytotoxicity of heavy metals, but limited data are available concerning the impact of trace organics on soils and crops. Cadmium is the metal of most concern because it can accumulate in plants and may pose a threat to humans and grazing animals. Design sludge loading rates are based on cadmium levels in the applied sludge.

The methods of sludge application to land are the following (Bruce and Davis, 1989; Wallis and Lehmann, 1983):

1. *Surface spreading* by tanker or by rain gun. This practice suffers from uneven sludge spreading. The sludge is worked into the soil to control odors and avoid nitrogen losses.
2. *Sludge injection* into the soil. Although more expensive than surface spreading, this practice has the advantage of minimizing odors, nitrogen loss, and surface runoff.

21.3 PUBLIC HEALTH ASPECTS OF WASTEWATER AND BIOSOLIDS APPLICATION TO LAND

Problems associated with land application of wastewater and sludges are contamination of groundwater, soils, and crops with pathogens, heavy metals, nitrate, and toxic, mutagenic, carcinogenic, and endocrine disrupting organic compounds.

21.3.1 Pathogenic Microorganisms and Parasites

Pathogenic microorganisms and parasites may survive on crops, particularly leafy vegetables (e.g., spinach, lettuce), irrigated with raw wastewater or wastewater effluents. In Chile, the consumption of raw vegetables that have been irrigated with raw wastewater was found to be linked with the transmission of *Helicobacter pylori* (Hopkins et al., 1993). Market vegetables may become contaminated with pathogens and parasites (e.g., *Salmonella, Giardia lamblia, Entamoeba histolytica*) during irrigation, transportation, and subsequent handling (Ercolani, 1976; Kowal, 1982; Pude et al., 1984). Survival of pathogens on crops depends on the type of pathogen, the type of crop, and environmental conditions (e.g, sunlight, temperature, wind, rain) and may vary from days to months (Rose, 1986). The concentration of parasite eggs applied to land may be high and may reach levels of 6000–12,000 viable eggs/m^2/year. Parasite eggs, especially those of *Ascaris*, may persist in soils for years (5–7 years or more) (Little, 1986). Therefore, the irrigation of processed food crops with wastewater effluents should be stopped some weeks (4–6 weeks) before harvesting to allow a sufficient inactivation of potential pathogens. Furthermore, irrigation of food crops that are eaten raw should be restricted. Crop contamination (leaves and fruits) is limited when wastewater effluents are applied via subsurface drip irrigation. Outdoors experiments showed a limited penetration of poliovirus into tomato plants when using this irrigation method (Oron et al., 1995).

21.3.2 Chemical Contaminants

Crops and grazing animals may become contaminated with heavy metals (e.g., Cd, Zn, Cu, Ni) present in sludge. Nitrate from sludge and wastewater effluents may also contaminate groundwater used for drinking water supply (Bouchard et al., 1992). For a discussion of the health effects of nitrate in water, see Chapter 3. Lipophilic trace organic compounds are recalcitrant to biodegradation and may accumulate in the fatty tissues of grazing animals.

21.4 TRANSPORT OF PATHOGENS THROUGH SOILS

The detection of pathogenic microorganisms in groundwater has triggered research on their fate (transport and persistence) in the soil matrix.

21.4.1 Bacterial Transport Through Soils

Owing to their size, bacterial pathogens may be filtered out during their transport through the soil matrix. Bacterial removal by soils is inversely proportional to the particle size of soils. Furthermore, bacteria are charged biocolloids that can adsorb to soils, providing there are optimal conditions that favor their attachment to soil particles. These conditions are the presence of cations (for example, iron-coated sand adsorbs up to 6.9×10^8/g sand; Mills et al., 1994), presence of clay minerals that provide adsorption sites, low concentrations of soluble organics, and low pH conditions (Gerba and Bitton, 1984). However, heavy rainfall promotes bacterial transport through soils (Lamka et al., 1980; Zyman and Sorber, 1988). Laboratory experiments with sludge–soil mixtures (equivalent to an average loading rate of 0.05 tons/ha) challenged with bacterial indicators have shown that only heavy rainfall (12.3 cm/day) promotes significant downward transport of the bacteria to the bottom of an 8 in. deep column. Lower rainfall did not cause significant migration of the bacterial cells (Zyman and Sorber, 1988).

Under field conditions, indicator bacteria are efficiently retained by soils and are detected at low levels in groundwater. They are not suitable indicators of virus transport into groundwater (Alhajjar et al., 1988). After sludge application to land, most of the sludge-associated bacteria are retained at the soil surface and their transport to groundwater is unlikely (Liu, 1982).

21.4.2 Virus Transport Through Soils

The major factors controlling virus transport through soils are soil type, virus serotype, ionic strength of soil solution, pH, soluble organic compounds present in wastewater effluents, and hydraulic flow rate. Retention of virus by the soil matrix is governed primarily by adsorption to surfaces, particularly those provided by clays and other minerals such as hematite or magnetite. Both electrostatic and hydrophobic interactions are implicated in virus adsorption to soils (Bales et al., 1995; Bitton, 1980b; Bitton and Harvey, 1992; Gerba, 1984; Lipson and Stotzky, 1987; Shields and Farrah, 2002).

Adsorption of virus to soils is affected by soil texture. Clay soils generally have a greater virus-retaining capacity than sandy ones. Muck soils display a low affinity for viruses (Meschke and Sobsey, 2003; Scheuerman et al., 1979; Sobsey et al., 1980). Adsorption varies with the type and strain of virus. Some viruses (e.g., poliovirus 1) notoriously adsorb well to soils while others (e.g., echovirus 1 and 11) display a lower adsorption capacity (Gerba et al., 1980; Jansons et al., 1989b; Sobsey et al., 1986). Several enteroviruses (e.g., poliovirus, coxsackie, and echoviruses) and bacterial phages (e.g., MS2, PRD-1) have been used as models to study virus transport in soil column experiments. These model viruses do not always simulate the survival and distribution of viruses of public health importance, such as hepatitis A virus and rotaviruses (Dizer et al., 1984; Sobsey et al., 1986). Viruses adsorb poorly to soils in low ionic strength solutions. This explains why rainwater tends to desorb soil-bound viruses and redistribute them within the soil profile (Alhajjar et al., 1988; Duboise et al., 1976; Lance and Gerba, 1980). Adsorption is inhibited and, thus, virus transport is promoted, by soluble organic materials found in wastewater effluents and sludges and by humic and fulvic acids (Dizer et al., 1984; Schaub and Sorber, 1977; Scheuerman et al., 1979). Domestic wastewater contains surfactants that result from the use of laundry detergents. These surfactants inhibit virus adsorption to surfaces, thus increasing transport through soils (Chattopadhyay et al., 2002). Virus transport is also promoted by increasing hydraulic flow rate, or by increasing pH (Bales et al., 1995; Lance and Gerba, 1980; Vaughn et al., 1981).

Both column experiments and field studies have shown that sludge application to land does not result in virus transport to aquifers. Viruses have not been detected in groundwater beneath sludge application sites (Farrah et al., 1981). Indeed, sludge-associated virus particles often become trapped at the soil surface and their migration through the soil matrix is thus limited (Bitton et al., 1984; Damgaard-Larsen et al., 1977; Moore et al., 1978; Pancorbo et al., 1988). The limited leaching of viruses from sludge particles was confirmed for both viruses and bacteria (Hurst and Brashear, 1987; Liu, 1982). Table 21.2 summarizes the main factors governing the transport of microbial pathogens through soils. Several models have been proposed for predicting viral and bacterial transport through soils and aquifers; the models are based on information on microbial transport and survival in these environments that has been generated during the past 20 years (e.g., Harvey and Garabedian, 1991; Park et al., 1990; Tim and Mostaghimi, 1991; Yates and Ouyang, 1992; Yates and Yates, 1989). These models vary in complexity and require environmental (e.g., temperature), soil/hydrogeologic (e.g., texture, adsorption coefficient) and microbiological (e.g., microbial type, inactivation rate) input parameters. Most of these models simulate microbial transport under saturated flow conditions; only a few address transport under unsaturated flow conditions (e.g., VIRTUS model; Yates and Ouyang, 1992). Some of these models, after some fine-tuning, could be useful in estimating microbial pathogen numbers in groundwater following transport through soils.

TABLE 21.2. Summary of the Main Factors Governing the Transport of Microbial Pathogens Through Soils[a]

Factor	Comments
Soil type	Fine-textured soils retain microorganisms more effectively than light-textured soils. Iron oxides increase the adsorptive capacity of soils. Muck soils are generally poor virus absorbents.
Filtration	Straining of bacteria at soil surface limits their movement.
pH	Generally, adsorption increases when pH decreases.
Cations	Adsorption increases in the presence of cations (cations help reduce repulsive forces on both microorganisms and soils particles). Rainwater may desorb viruses from soil owing to its low conductivity.
Soluble organics	Generally compete with microorganisms for adsorption sites. Humic and fulvic acid reduce virus adsorption to soils.
Microbial type	Adsorption to soils varies with microbial type and strain.
Flow rate	The higher the flow rate, the lower the microbial adsorption to soils.
Saturated versus unsaturated flow	Virus movement is less under unsaturated flow conditions.

[a]Adapted from Gerba and Bitton (1984).

21.5 PERSISTENCE OF PATHOGENS IN SOILS

21.5.1 Persistence of Bacterial Pathogens

The main factors affecting the survival of pathogenic bacteria in soils are temperature, moisture content, sunlight, pH, organic matter, bacterial type, and antagonistic microflora, which include indigenous soil bacteria and predatory protozoa (Bitton and Harvey, 1992; Foster and Engelbrecht, 1973; Gerba and Bitton, 1984; Gerba et al., 1975a, b). Soil desiccation is important in the control of bacterial survival in soils. The decay rate of total and fecal coliforms in soil–sludge mixtures increases as the mixture is allowed to dry naturally (Zyman and Sorber, 1988).

21.5.2 Persistence of Viruses in Soil

The two decisive factors controlling the persistence of viruses in soils treated with wastewater effluents or sludge are soil temperature and moisture (Bitton, 1980a). Recent studies with sludge-amended desert soils confirmed the importance of these two environmental factors (Straub et al., 1992). Soil microorganisms may also produce antiviral substances that increase the rate of viral inactivation (Hurst et al., 1980; Sobsey et al., 1980). Other factors include soil and virus type (e.g., HAV has a longer persistence than other enteric viruses in soil; Sobsey et al., 1986). Viruses in sludge applied to soil persist for 23 weeks during the winter season in Denmark (Daamgaard-Larsen et al., 1977), but for only 2–4 weeks during the summer or fall in Florida (Bitton et al., 1984). Table 21.3 summarizes the main factors affecting the persistence of microbial pathogens in soils.

21.6 TRANSPORT OF PATHOGENS IN AQUIFERS

Knowledge of bacterial transport in aquifers gives useful information on (1) transport of pathogenic microorganisms, (2) facilitated transport of organic contaminants by mobile

TABLE 21.3. Summary of the Main Factors Governing the Persistence of Microbial Pathogens in Soils[a]

Factor	Comments
Physical factors	
Temperature	Longer survival at low temperatures; longer survival in winter than in summer.
Water-holding capacity	Survival is lower in sandy soils with lower water-holding capacity.
Light	Lower survival at soil surface.
Soil texture	Clays and humic materials increase water retention by soils and thus affect microbial survival.
Chemical factors	
pH	May indirectly control survival by controlling adsorption to soils, particularly for viruses.
Cations	Some (e.g., Mg^{2+}) may thermally stabilize viruses.
Organic matter	May influence bacterial survival and regrowth.
Biological factors	
Antagonism from soil microflora	Increased survival in sterile soils. No clear trend as regards the effect of soil microflora on viruses.

[a]Adapted from Bitton et al. (1987); Gerba and Bitton (1984).

bacteria, (3) movement of genetically engineered bacteria, which may be injected for aquifer bioremediation, and (4) processes involved in microbially enhanced oil recovery (MEOR) (Harvey, 1997a).

In addition to the hydrological characteristics of the aquifer (e.g., hydraulic conductivity, porosity), several physico-chemical and biological factors influence the transport of bacteria and viruses in aquifers (Gerba et al., 1991; Harvey, 1991; Yates and Yates, 1991). The physico-chemical factors comprise advection, dispersion, straining, and adsorption–desorption. Buoyant density of microorganisms is another important characteristic controlling bacterial and protozoan transport in aquifer material. Buoyant density affects both the frequency of collision of cells with aquifer material as well as their sedimentation. Culturing bacteria in rich growth media leads to the production of cells with increased size and buoyant density (Harvey et al., 1997). Biological factors include microbial survival and growth, cell size, and ability to adsorb to surfaces.

Tracers are used to understand the hydrologic properties of aquifers, the factors controlling the transport of pathogens and parasites to assess the vulnerability of aquifers to fecal contamination, and to study the transport of genetically modified microorganisms and other useful microorganisms for the bioremediation of aquifers. Several tracers, ranging from bacterial phage (e.g., PRD1) to fluorescent microspheres are under consideration and the most important ones are listed in Table 21.4 (Harvey and Harms, 2002; Harvey and Ryan, 2004).

Despite the availability of several models that simulate the transport of laboratory-grown microorganisms under controlled conditions, it is more difficult to accurately predict the transport pattern of microorganisms in aquifers.

21.7 PERSISTENCE OF PATHOGENS IN AQUIFERS

Groundwater monitoring for the past 20 years has shown that about half of all waterborne disease outbreaks are associated with contaminated groundwater. Enteric viruses are the pathogens most likely found in groundwater. Twenty-six of 34 (76 percent) drinking water-associated outbreaks in 1991–1992 in the United States, occurred in systems using untreated or inadequately treated well water (Moore et al., 1994) (Fig. 21.7). Of the 650

TABLE 21.4. Biotracers Used in the Assessment of Microbial Transport in Aquifers[a]

Phages
  MS2, f2, PRD1, T4, T7, ΦX174
  Marine phages (H4, H6, H40)
Bacteria
  Escherichia coli (antibiotic resistant)
  Pseudomonas fluorescens (antibiotic resistant)
  Bacillus sp.
  Bacillus subtilis spores
  Serratia marcescens
  Bacteria stained with DAPI and other fluorochromes
Yeasts
  Saccharomyces cerevisiae
Protozoa
  Spumella guttula (nanoflagellete)

[a]Adapted from Harvey and Harms (2002).

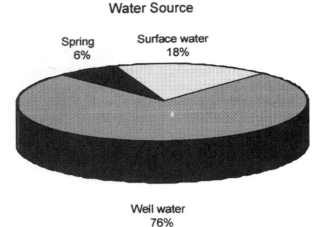

Figure 21.7   Disease outbreaks in 1991 and 1992 associated with drinking water in terms of water source. From Moore et al. (1994). (With permission of the publisher.)

outbreaks of waterborne diseases that occurred between 1971 and 1994, 58 percent were associated with groundwater, while 33 percent were associated with surface waters. Enteric viruses (hepatitis A virus, Norwalk virus and rotaviruses) were responsible for 8 percent of the outbreaks (Craun and Calderon, 1996). Gastroenteritis and hepatitis A outbreaks due to the consumption of untreated groundwater have been documented (Barwick et al., 2000; Beller et al., 1997; Bowen and McCarthy, 1983; De Serres et al., 1999; Häfliger et al., 2000). Generally, more outbreaks occur in noncommunity than in community systems using groundwater. As discussed previously, contamination may be caused by groundwater recharge, land application of wastewater effluents, or septic tank effluents. Enteric viruses (e.g., enteroviruses, hepatitis A virus, Norwalk virus, rotaviruses) have been isolated from groundwaters around the globe (Abbaszadegan et al., 2003; Bitton and Farrah, 1986; Farrah and Bitton, 1990). Recently, 4.8 percent of groundwater samples from 446 wells across the United States tested positive in infectivity assays using Buffalo green monkey kidney cells as host cells. Comparatively, RT-PCR showed that 31.5 percent of the samples were positive for enteric viruses. However, PCR data do not give information concerning the viability/infectivity of the viruses (Abbaszadegan et al., 2003).

The Groundwater Disinfection Rule was released by the U.S. EPA to address the problem of groundwater contamination by viruses and other microbial pathogens. This piece of legislation requires disinfection (e.g., chlorination, ultraviolet irradiation) for all community and noncommunity public water systems using groundwater, and it sets a maximum contaminant level goal (MCLG) of zero for viruses (Grubbs and Pontius, 1992).

Persistence of viruses is generally higher in groundwater than in surface waters. Decay rates of several enteroviruses in groundwater vary between 0.0004 and 0.0037 h^{-1} (Bitton et al., 1983; Jansons et al., 1989a). Temperature is the most decisive factor that controls virus survival in groundwater (Jansons et al., 1989a; Yates and Gerba, 1984; Yates et al., 1985). The relationship between virus inactivation rate and temperature is described by the following equation:

$$I = 0.018(T) - 0.144 \qquad (21.1)$$

where $I =$ virus inactivation rate (h^{-1}); and $T =$ groundwater temperature (°C).

Safe setback distances for drinking water wells in the vicinity of septic tanks can be predicted by geostatistical techniques that help estimate virus inactivation in groundwater, using a regression equation that describes the relationship between virus inactivation rates and groundwater temperature (Yates et al., 1985; 1986; Yates and Yates, 1987).

21.8 DISPOSAL OF SEPTIC TANK EFFLUENTS ON LAND

The description and microbiology of septic tank systems were addressed in Chapter 13. The number of septic tanks in the United States has been estimated at 22 millions units. These on-site treatment systems serve approximately one-quarter to one-third of the U.S. population (U.S. EPA, 1986). They generate approximately 800 billion gallons of sewage/year (Canter and Knox, 1984). In Florida, more than 1.3 million families (more than 27 percent of the state housing units) are served by on-site sewage disposal systems (Bicki et al., 1984). Septic tanks are major contributors to the contamination of subsurface environments. The contaminants are household chemicals (nitrate, heavy metals, organic toxicants), pathogenic microorganisms, and parasitic cysts.

Septic tank effluents contribute significantly to groundwater contamination. The use of untreated groundwater was responsible for more than one-third of disease outbreaks in the United States between 1971 and 1980 (Craun, 1986b). There are several documented instances of groundwater pollution by septic tank effluents (Bicki et al., 1984; Hagedorn, 1984). Groundwater contamination is often attributable to the use of unsuitable soils (e.g., coarse-textured soils) for receiving septic tanks effluents, high water table, saturated flow in the soil, and high loading rates. The extent of groundwater contamination depends on climatic, soil, and biological factors that influence the transport and persistence of bacterial pathogens in soils (Gerba and Bitton, 1984). Bacterial transport through the absorption field is also controlled by the degree of soil saturation with water. In unsaturated soils, bacterial movement may be restricted within 3 ft, whereas bacteria, under saturated conditions, may be transported over much greater distances. The operation of an absorption field sometimes leads to the formation of a biological clogging mat or crust, which appears to be an effective barrier to bacterial breakthrough (Bouma et al., 1972). The formation of this mat is predominantly controlled by biological factors such as the accumulation of microbial polysaccharides (Mitchell and Nevo, 1964; Nevo and Mitchell, 1967; Vandevivere and Baveye, 1992). Figure 21.8 shows scanning electron micrographs of sand grains colonized by slime-producing bacteria (Vandevivere and Baveye, 1992). Virus transport from septic tanks to groundwater or surface waters was documented in several studies. In a study by Hain and O'Brien (1979) poliovirus type 1 was not substantially inactivated in the tank and was detected in groundwater. Poliovirus or bacterial phages, when used as tracers, were detected in monitoring wells and surface waters (lake water, seawater) in the vicinity of a septic tank (Paul et al., 1995; Stramer and Cliver, 1981). Other studies have documented the breakthrough of both enteroviruses and coliform bacteria (Vaughn et al., 1983).

Septic tanks must be cleaned periodically every two to five years to remove the sludge, called *septage*, which has accumulated in the tank (Canter and Knox, 1985). Septage must be disposed of properly because it may contain high levels of inorganic nutrients (nitrogen, phosphorus), toxic heavy metals, hazardous organic compounds resulting from the use of household chemicals, pathogens, and parasites (Ridgley and Calvin, 1982; Stramer and Cliver, 1984; U.S. EPA, 1980; Ziebell et al., 1974). While much is known about the fate of bacterial pathogens, less is known with regard to viruses in septage, particularly those serving multiple housing units. Improper disposal of septage may lead to subsurface

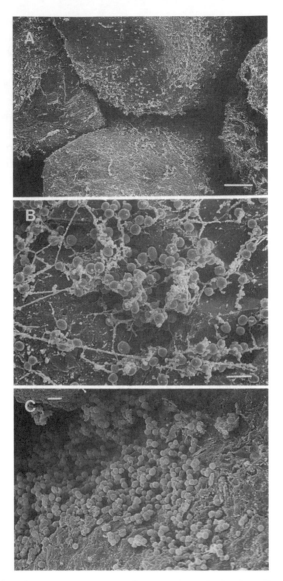

Figure 21.8   Scanning electron micrographs of sand grains colonized by slime-producing bacteria: (a) and (b) (16 mm depth), bacteria entangled in a fibrillar material (bars are 20 and 2 μm for (a) and (b) respectively); (c) (23 mm depth), no exopolymers were observed (bar, 2 μm). From Vandevivere and Baveye (1992).

pollution. Septage may be disposed of by land application (surface spreading, subsurface incorporation, trenching, and landfilling), handled in special facilities (e.g., lagooning, lime stabilization, chlorination), or treated in municipal wastewater treatment plants (U.S. EPA, 1980). Unfortunately, septage is also sometimes dumped illegally onto land and into waterways without any treatment. After a survey in Gainesville, FL, it was estimated that as much as 60 percent of septage was illegally disposed of (Gainesville Regional Utilities, 1985).

Problems with groundwater contamination with septic tank effluents prompted State and local government agencies to require minimum setback distances between septic tanks and drinking-water wells. Geostatistical techniques were proposed to predict safe setback distances between septic tanks and drinking-water wells (Yates et al., 1985; 1986; Yates and Yates, 1987; 1989; see Section 21.7).

21.9 BIODEGRADATION IN SOILS AND AQUIFERS: AN INTRODUCTION TO BIOREMEDIATION

There is a growing concern over soil and groundwater contamination with hazardous wastes, sometimes caused by effluents from industrial wastewater treatment plants. It was estimated that groundwater contamination with hazardous chemicals occurs or is suspected in 70–80 percent of land disposal facilities in the United States (Ouellette, 1991). Environmental biotechnologists are working at finding ways to reduce environmental contamination by using modern tools in microbiology, molecular ecology, chemistry, and environmental and engineering sciences (Sayler et al., 1991). The transformation of xenobiotics by microbial action has already been discussed in Chapter 19. In this section, we will concentrate solely on biotreatment strategies for soils and aquifers.

21.9.1 Bioremediation of Soils

There are four approaches to microbiological soil decontamination (Beaudette et al., 2002; Compeau et al., 1991; Hanstveit et al., 1988): (1) in situ techniques, (2) addition of microorganisms or enzymes, (3) bioreactors for treatment of excavated soils, and (4) natural attenuation. These approaches are summarized in Figure 21.9 (Beaudette et al., 2002).

21.9.1.1 In Situ Decontamination Techniques. The contaminated soil is amended with nutrients (N and P) and plowed to provide oxygen. For example, the biodegradation of diesel oil in soil can be stimulated by remediation measures consisting of liming, addition of nitrogen and phosphorus, and tilling. This approach was found to reduce total hydrocarbons by 95 percent, eliminate polycyclic aromatic hydrocarbons, resulting in complete detoxification in 20 weeks, as measured with Microtox (a microbial toxicity test) and the Ames test (a mutagenicity test) (Wang et al., 1990).

In situ biorestoration, using above-ground biodegradation cells, resulted in 95 percent removal of oil and grease in a clay soil (Vance, 1991). Some have proposed the stimulation of soil indigenous microflora by treating the soil with a structural analog of the chemical to be removed. For example, removal of 3,4-dichloroaniline can be enhanced by stimulation of aniline-degrading microorganisms (You and Bartha, 1982).

21.9.1.2 Addition of Microorganisms or Enzymes to Soils (Bioaugmentation). Addition of microorganisms and enzymes stimulates the biodegradation of xenobiotics in soils (Crawford and O'Reilly, 1989). Fungal laccases are polyphenol oxidases that catalyze the binding, by oxidative cross-coupling, of phenolic compounds to the humic fraction of soils and their subsequent immobilization and detoxification (Bollag, 1992). Fungal spores or mycelial fragments suspended in a sodium alginate hydrogel can be used to coat pellets of solid substrates for bioaugmentation of pentachlorophenol degradation in contaminated soils (Lestan and Lamar, 1996). Most efforts have been concentrated on the use of bacterial inoculants grown in large fermenters (e.g., *Arthrobacter*, *Rhodococcus chlorophenolicus*, *Flavobacterium*) for the bioremediation of pentachlorophenol-contaminated soils. Under laboratory conditions at 30°C, soil inoculated with 10^6 PCP-degrading *Arthrobacter* per gram dry soil, reduced the half-life

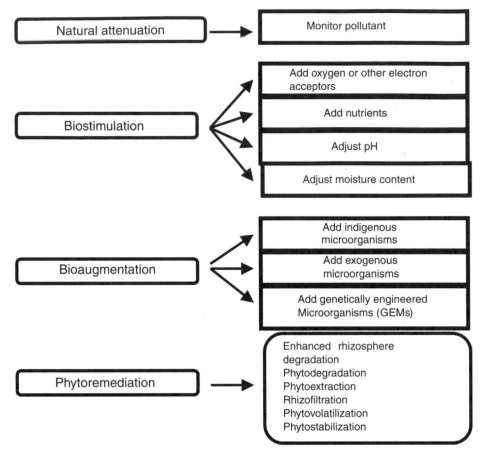

Figure 21.9   Bioremediation options. From Beaudette et al. (2002).

of PCP from two weeks to less than one day (Edgehill and Finn, 1983b). Microbial immobilization on bark chips or their encapsulation in polyurethane or alginate enhances their PCP-degrading ability, as well as their resistance to PCP toxicity (Crawford et al., 1989; Salkinoja-Salonen et al., 1989). Microcosm studies have shown that bacterial cells immobilized in gellan gum microbeads (16–53 μm diameter) can degrade gasoline, leading to 30–50 percent removal of the contaminant (Moslemy et al., 2002). These results can be applied for the in situ bioaugmentation of contaminated aquifers. The use of commercial bacterial inoculants, however, did not enhance the biodegradation of hydrocarbons in a site contaminated with bunker C fuel (Compeau et al., 1991). Successful soil bioremediation was obtained by amending a petroleum hydrocarbon-polluted soil with an enriched culture of indigenous hydrocarbon-degrading bacteria. A biological activated carbon (BAC) system, seeded with indigenous bacteria and fed hydrocarbons and nutrients, was used to enrich for the indigenous bacteria. Natural attenuation resulted in 1.7 percent removal of hydrocarbons in 32 days as compared to 42 percent for enhanced bioremediation (Fig. 21.10; Li et al., 2000). In a similar study, bioaugmentation of soils with an enriched culture of indigenous microorganisms increased the biodegradation of petroleum hydrocarbons in by 20.4 to 49.2 percent (Yerushalmi et al., 2003).

Genetically modified microorganisms (GMM) are being considered for use in environmental bioremediation for cleanup of hazardous chemicals, biomining, and as

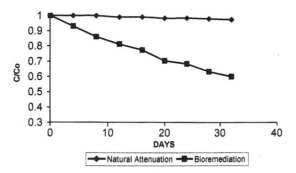

Figure 21.10 A comparison of natural attenuation and bioremediation. From Li et al. (2000).

biofertilizers and biopesticides in agriculture. Agricultural applications of GMMs include their use as biological control agents (e.g., *Bacillus thuringiensis*), frost control, bioremediation of hydrocarbons and other xenobiotics, or for increasing nitrogen fixation in soils (e.g., use of nitrogen-fixing *Rhizobium* strains) (van Overbeek and van Elsas, 2002; Ripp and Sayler, 2002). Such GMMs are applied to soil via injection, spraying, or as immobilized cells within carriers such as alginate. They can be transported vertically or laterally following rain events. Their transport from the site of application can be tracked, using cultivation-based (e.g., plating on selective growth media) or cultivation-independent (e.g., use of molecular probes) techniques.

There are three main concerns over the application of GMMs to soils:

1. *Fate of GMMs in soils (survival and transport).* The GMMs must survive in sufficient numbers after soil inoculation in order to accomplish the task at hand. Experience from soil microbiology has taught us that non-GMMs microorganisms (e.g., nitrogen-fixing bacteria such as *Rhizobium*, *Azotobacter*, or *Azospirillum*) released in soils for bioremediation or biocontrol purposes do not persist well, due to adverse conditions in soils (e.g., limiting nutrients in soils, harsh environmental conditions) and competition with indigenous soil microorganisms. The GMMs may also display a reduced level of fitness due to the additional energy necessary for carrying the recombinant genes. Culture-based and molecular-based techniques have shown that GMMs or their genetic constructs can, however, persist in soils for years after field release. Another concern is the transport of GMMs from the application site. Transport of GMMs may be due to aerosolization during application to soils, water, or animal vectors.

2. *Ecosystem alteration.* Introduction of GMMs into soils may affect the diversity of indigenous microbial populations, possibly leading to changes in primary production and nutrient cycling in soils. Owing to conflicting results obtained so far, more research is needed on this topic.

3. *Containment of the released GMMs.* Another important concern is the transfer of recombinant DNA in GMMs to indigenous microbial species in soils, possibly leading to dire ecological consequences. The possibility of gene transfer in soils mostly by processes such as conjugation and transformation (see Chapter 1 for more details on these processes) has been demonstrated (Nielsen et al., 1997; van Elsas et al., 1998). To reduce the chances of gene transfer, the foreign genes should be inserted in the chromosome rather than plasmids. Culture-based and molecular techniques can be used to assess the changes in indigenous microbial populations following the introduction of GMMs.

From a public safety viewpoint, biocontainment genes (i.e., "suicide genes") can be inserted into GMMs to ensure that they are killed when they spread out from the application site (e.g., rhizosphere). Examples of biocontainment genes are the *gef*, *hok*, *relF* genes derived from *E. coli*, and the *nuc* gene derived from *Serratia marcescens*. These genes can be placed under the control of a nutritional or environmental factor such as carbon limitation or temperature. This will ensure that the GMM will not escape from the application site (Ahrenholtz et al., 1994; Knudsen et al., 1995; van Overbeek, 1998; van Overbeek and Elsas, 2002).

Improvement of molecular techniques to monitor relatively small numbers of GMMs or their constructs in soils will help in enhancing our knowledge of their ultimate fate in soils and in predicting the risks associated with their release into the environment.

21.9.1.3 *Bioreactors for Treatment of Excavated Soils.* These bioreactors include soil slurry reactors, composting, and land treatment units. The removal of PCP and hydrocarbons, using land treatment units and soil slurries, has been demonstrated. Biodegradation of PCP in soil slurries was enhanced by the addition of a PCP microbial consortium (Compeau et al., 1991). Composting was also explored as a technology for bioremediation of soils contaminated with explosives such as trinitrotoluene (Myler and Sysk, 1991). More than 90 percent of the explosives were removed from contaminated soils within 80 days (Williams et al., 1989).

21.9.1.4 *Natural Attenuation.* *Natural attenuation* relies on natural processes to bring about the biodegradation of xenobiotics in soils. This includes physical processes (e.g., volatilization, dilution, sorption), abiotic and microbial degradation of the contaminants.

21.9.2 Treatment Strategies for Aquifers

There are three basic approaches to the treatment of contaminated aquifers (Bouwer et al., 1988): physical containment, above-ground treatment, and in situ bioremediation. A new approach that is being considered is the *natural attenuation* of contaminants in aquifers. This approach requires no intervention and consists of monitoring the change in contaminant concentration with time.

21.9.2.1 *Physical Containment.* Physical containment includes the use of temporary physical barriers to slow down or halt contaminant movement. This approach has been adopted with some success in hazardous waste sites.

21.9.2.2 *Above-Ground Treatment (Pump-and-Treat Technology).* The contaminated water is pumped out from the aquifer via extraction wells, and treated above ground, using one of several treatment processes. However, it is difficult to extract chemical contaminants adsorbed to the aquifer matrix. Since the removal of organic contaminants by this technology is relatively slow, this approach may be regarded as a means for preventing further migration of the contaminant in the aquifer (Mackay and Cherry, 1989). The major available treatment technologies are air stripping to remove volatile organic compounds, adsorption to granular activated carbon (GAC), ultrafiltration, oxidation with ozone/UV or ozone/H_2O_2, activated sludge, and fixed-film biological reactors (Bouwer et al., 1988). For example, fixed-film bioreactors, when using sand as the matrix and methane or natural gas as the primary substrate, are capable of removing up

to at least 60 percent of trichloroethylene from polluted water. They also remove more than 90 percent of TCE and TCA from vapor streams generated by air stripping of polluted groundwater (Canter et al., 1990).

21.9.2.3 *In Situ Bioremediation.*

In situ bioremediation is the enhancement of the catabolic activity of indigenous microorganisms by adding nutrients and, if necessary, oxygen (added as air, pure oxygen, or hydrogen peroxide). This approach allows the degradation of the target compound in situ, avoiding the need for excavation and treatment of the contaminated material. In situ treatment depends on aquifer characteristics (e.g., permeability as measured by hydraulic conductivity), contaminant characteristics, oxygen level, pH, availability of nutrients, redox conditions, and the presence of microorganisms able to degrade the contaminant under consideration (Alexander, 1985; Bedient and Rifai, 1992; McCarty et al., 1984; Rittmann, 1987; Thomas and Ward, 1989; Wilson et al., 1986). This approach has been used mostly for gasoline spills. Indigenous subsurface bacteria are able to grow on aromatic compounds such as naphthalene, toluene, benzene, ethylbenzene, *p*-cresol, xylene, phenol, and cresol, which are used as the sole source of carbon and energy (Brockman et al., 1989; Frederickson et al., 1991; Glynn et al., 1987). In situ bioremediation of aquifers contaminated with pentachlorophenol and creosote can be enhanced by the injection of hydrogen peroxide (100 mg/L) as well as inorganic nutrients such as nitrogen and phosphorus (Piotrowski, 1989).

Much effort has been concentrated on the fate of chlorinated aliphatic hydrocarbons (e.g., trichloroethylene, dichloroethylene) in aquifers. These chemicals undergo reductive dehalogenation under anaerobic conditions in aquifers (Vogel and McCarty,1985). Under aerobic conditions, these compounds are degraded by methane-utilizing bacteria called methanotrophs (Fogel et al., 1986; Moore et al., 1989). These bacteria use methane as the sole source of energy and as a major source of carbon (Haber et al., 1983). They can transform more than 50 percent of trichloroethene (TCE) into CO_2 and bacterial biomass (Fogel et al., 1986). High conversion rates of TCE are obtained with methanotrophs with V_{max} up to 290 nmol/min/mg of cells of *Methylosinus trichosporium* (Oldenhuis et al., 1991). A methanotrophic biofilm reactor was shown to be capable of degrading TCE and TCA in a continuous-flow operation for a period of 6 months. The maximum degradation rate for TCE was 400 μg/L.h (Strand et al., 1991). The TCE was ultimately converted to CO_2 and CO by a microbial consortium comprised of a metanotroph (*Methylocystis* sp.) and heterotrophic bacteria. The methanotroph converts TCE to glycoxylic, dichloroacetic, and trichloroacetic acids, while heterotrophic bacteria carry out the biotransformation to CO_2 and CO (Uchiyama et al., 1992). A methanotrophic bacterium, isolated from groundwater, degraded cometabolically TCE in the presence of methane or methanol used as primary substrates. TCE can also be transformed to TCE epoxide by methane monooxygenase produced by bacteria. The epoxide then breaks down spontaneously to dichloroacetic acid and glycoxilic acid (Little et al., 1988). The activity of the soluble (i.e., nonmembrane-bound) methane monooxygenase (sMMO) is associated with TCE biodegradation by pure cultures of methanotrophs (*Methylosinus trichosporium*, *Methylomonas methanica*), and by microbial isolates from groundwater (Bowman et al., 1993; Fitch et al., 1996; Koh et al., 1993; Tsien et al., 1989). The synthesis of sMMO is stimulated only under copper stress (Oldenhuis et al., 1991; Stanley et al., 1983) and is suppressed at copper concentration as low as 0.25 μM (Brusseau et al., 1990). Mutants capable of producing methane monooxygenase in the presence of high levels of copper (≤12 μM copper) have been isolated (Phelps et al., 1992). Genes encoding subunits of this enzyme have been cloned and used as probes for the detection of *Methylosinus* in bioreactors and other environmental samples (Bowman et al., 1993; Bratina and

Hanson, 1992). Biodegradation of TCE in groundwater can also be enhanced by using phenol-utilizing microorganisms (Hopkins et al., 1993).

A field demonstration of in situ biorestoration of an aquifer was undertaken at Moffett Naval Air Station (Roberts et al., 1989). Aquifer indigenous methanotrophic bacteria, stimulated by addition of oxygen and methane, were able to metabolize chlorinated aliphatic solvents such as TCE, *cis*- and *trans*-1,2-dichloroethene (DCE), and vinyl chloride (VC) under in situ conditions. The extent of biotransformation were 20 percent for TCE, 40 percent for *cis*-DCE, 85 percent for *trans*-DCE, and 95 percent for VC. Two field demonstration tests in the Savannah River site, based on the in situ enhancement of methanotrophic bacteria, showed the successful removal of TCE from aquifers (see Brigmon, 2002, for review; Kastner et al., 2000; Pfiffner et al., 1997; Travis and Rosenberg, 1997).

Another field study using methane injection was conducted in Japan to clean up groundwater contaminated with 220 μg/L of TCE. This experiment resulted in a stimulation of methanotrophs from 10 to 10^4 cells/mL and in 10–20 percent removal of TCE (Eguchi et al., 2001). Similar results were obtained by Semprini et al. (1990) as regards in situ biostimulation using methane at a site with 36–97 μg/L of TCE.

21.9.2.4 *Bioaugmentation.*

Bioaugmentation consists of adding specific microorganisms or mixed bacterial cultures to polluted aquifers to enhance the biodegradation of target contaminants (DeFlaun, 2002; Thomas and Ward, 1989). These cultures are obtained via traditional enrichment techniques, chemical mutagenesis or molecular-based techniques. However, the use of the latter approach is still being debated (see Section 21.9.1.2. of this chapter). The added bacteria should have some or all the following desirable characteristics (DeFlaun, 2002):

- *Ability to degrade the target contaminant.* Depending on the aquifer and type of contaminant, the organism should be chosen based on its ability to degrade the contaminant under aerobic or anaerobic conditions.
- *Ability to produce constitutive degradative enzymes.*
- *Ability to migrate in the aquifer material.* Adhesion-deficient bacteria should be selected because they can migrate further in the aquifer and help avoid or reduce plugging of the injection well. Other means to enhance bacterial transport in aquifers include the use of surfactant foams and liquid surfactants or the use of "ultramicro" bacteria, which travel more easily than full-size bacteria (starvation reduces bacterial size).
- *Ability of the cells to store energy-rich materials* such as poly-β-hydroxybutyric acid (PHB). This storage polymer will help in the survival of the added bacteria in the aquifer.

As an example, a bioaugmentation study was carried out in California, where commercial blends of hydrocarbon-degrading bacteria were used for the bioreclamation of soils and aquifers contaminated with petroleum hydrocarbons (von Wedel et al., 1988). The process consisted of the following:

1. Contaminated groundwater was treated above ground in chemostat bioreactors in the presence of the commercial bacterial cultures.
2. The treated groundwater was further amended with additional bacteria and was reinfiltrated into the contaminated soil for in situ treatment of both contaminated soil and aquifer.

After several months of operation, this augmented biorestoration scheme resulted in 93 percent removal of total hydrocarbons. For benzene, toluene, and xylene, the percent removal rates were 85 percent, 98 percent, and 33 percent, respectively (von Wedel et al., 1988) (Fig. 21.11).

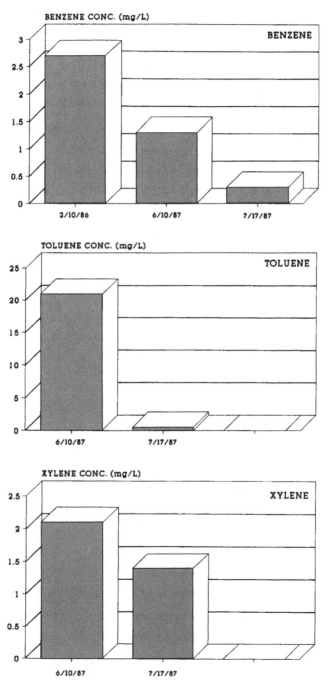

Figure 21.11 Augmented biorestoration of groundwater for reducing petroleum hydrocarbon levels. From von Wedel et al. (1988). (Courtesy of Pergamon Press, London.)

Bioaugmentation of an aquifer with a MTBE-degrading bacterial consortium and oxygen addition was successful in reducing MTBE concentrations to less than 0.001–0.01 mg/L (Salanitro et al., 2000). Other examples of field-scale demonstrations of bioaugmentation are given by DeFlaun (2002).

21.9.3   Factors Affecting Bioremediation of Soils and Aquifers

These factors include those pertaining to environmental conditions in soils or aquifers, contaminant characteristics, and the microorganisms. These factors are summarized in Figure 21.12 (Beaudette et al., 2002). Before field trials, laboratory work should be undertaken to assess the solubility, toxicity, and biodegradability of a contaminant and to determine any limiting nutrients in the aquifer. The site should also be evaluated for its suitability for bioremediation. The criteria for a suitable site are given in the following (Glynn et al., 1987).

21.9.3.1 Environmental Factors.
1. Suitable hydrogeological features must allow extraction of contaminated groundwater (i.e., recovery of the free contaminant phase) and reinjection and circulation of the treated water. Other hydrogeological characteristics that must be evaluated include depth and specific yield of the aquifer, permeability, and direction of groundwater flow.
2. Soil and aquifer parameters that affect contaminant biodegradation are the following:
   - *Electron acceptor.*   Oxygen can be added to the aquifer by injecting pure oxygen, air, or hydrogen peroxide. Hydrogen peroxide, at a concentration of up to 750 mg/L, was used as a source of oxygen for the bioremediation of a shallow sandy aquifer contaminated with fuel spills in Michigan (Wilson et al., 1989). Since H_2O_2 may be toxic to aquifer microflora, some recommend a gradual increase in its concentration to avoid toxicity. Solid peroxides (e.g., Na percarbonate, Ca peroxide) have also been considered as sources of oxygen for in situ

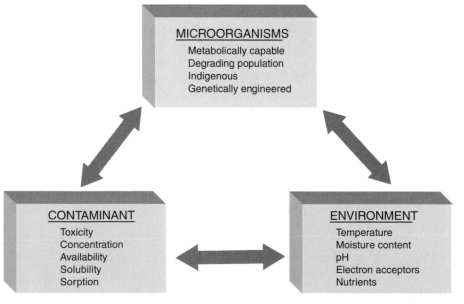

Figure 21.12   Factors considered in a bioremediation project. From Beaudette et al. (2002).

subsurface bioremediation. Toxicity is reduced when these solid peroxides are encapsulated in polyvinylidene chloride (Vesper et al., 1992). Nitrate, and perhaps nitrous oxide, can also serve as electron acceptors. Aromatic hydrocarbons (e.g., toluene, xylene, ethylbenzene) are biodegraded under denitrifying conditions in aquifer materials (Evans et al., 1991; Hutchins, 1989; 1991; Hutchins et al., 1991). Sulfate is another useful electron acceptor for the biotransformation of halogenated organic compounds (Cobb and Bouwer, 1991). Benzene bioremediation by sulfate-reducing bacteria has been demonstrated in groundwater under field conditions (Anderson and Lovley, 2000). In a petroleum-contaminated aquifer, the in situ anaerobic biodegradation of BTEX (benzene, toluene, ethylbenzene, and xylene) compounds was enhanced by the combined injection of nitrate and sulfate into the aquifer. The use of oxygen in conjunction with other electron acceptors (e.g., nitrate, sulfate) for bioremediation of aquifers could improve the biodegradation of hydrocarbons in aquifers (Cunningham et al., 2001).

- *Addition of inorganic and organic nutrients.* Sometimes, it is necessary to add nitrogen and phosphorus to enhance biodegradation in soils and aquifers (Thomas and Ward, 1989). Methane and air can enhance the biodegradation of chlorinated hydrocarbons (e.g., trichloroethylene) by methanotrophs.
- *Other factors.* Temperature, pH, soil moisture all affect biodegradation in soils and aquifers.

21.9.3.2 *Contaminant Characteristics.*
Parameters influencing the biodegradation in soils and aquifers include xenobiotic concentration, solubility, hydrophobicity, toxicity to indigenous or added microorganisms, sorption to inorganic or organic solids, and bioavailability.

21.9.3.3 *Microbiological Considerations.*
Bioremediation is affected by the presence in soils or aquifers of sufficient numbers of microorganisms capable of degrading the xenobiotic.

21.9.4 Molecular Techniques for Assessing Microbial Community Structure and Activity in Bioremediation

Several methods are available for assessing the impact of bioremediation and are as follows (Beaudette et al., 2002; Widada et al., 2002):

1. *DNA-based methods.* These methods allow the detection of genes or gene sequences in DNA extracted from environmental samples, using specific DNA probes. Greater sensitivity of these probes can be obtained by PCR amplification. These probes can be used in combination with Most Probable Number (MPN) techniques to follow the biodegradation of a given xenobiotic in soils or water.

2. *RNA-based methods.* rRNA probes are hybridized with extracted target RNA and give information on community activity and are used for the identification of potential degraders of a given contaminant. Furthermore, the detection of mRNA gives information about the expression of the gene involved in the biodegradation of a given xenobiotic.

3. *Genetic fingerprinting methods.* Following PCR amplification, products with the same length but slightly different sequences are identified. The DNA fragments

can be resolved by using denaturing gradient gel electrophoresis (DGGE). Finger-printing gives information about the community structure.

4. *Reporter gene (i.e., marker genes) systems and immunological techniques for track-ing degrading bacteria in bioaugmentation projects.* In bioaugmentation oper-ations one needs to track the persistence, transport pattern, and activity of the added bacteria. Various marker genes are used for this purpose, and some of them are listed below:

   - *Antibiotic resistance genes.* They were the first marker genes used in environ-mental microbiology. They are no longer popular due to concern over the spread of antibiotic resistance in the environment.

   - *Catabolic gene markers* such as *xylE* (encoding catechol 2,3-oxygenase), and *lacZY* (encoding β-galactosidase and lactose permease).

   - *lux reporter gene.* The *lux* operon include the *luxAB*, *luxCD*, and *luxE*, which code for various enzymes, including the luciferase enzyme responsible for the pro-duction of bioluminescence.

   - *Green fluorescent protein (GFP) gene marker.* These genes were originally iso-lated from the jellyfish *Aequorea victoria* and are responsible for the production of green fluorescence. They have been used to follow the biodegradation of xeno-biotics in contaminated soils.

The immunological approach consists of using polyclonal or monoclonal antibodies to follow the fate of introduced bacteria in soils or aquifers.

21.10 WEB RESOURCES

http://www.soil.ncsu.edu/publications/Soilfacts/AG-439-03/ (land application of sludge from North Carolina State University)
http://biology.bangor.ac.uk/~bss003/ (provides several links to biodegradation/ bioremediation sites; from the University of Wales, Bangor, UK)
http://www.brownfieldstech.org/technology/bioremediation/ (lists bioremediation technologies)
http://www.clu-in.org/remed1.cfm (EPA remediation site)
http://www.lbl.gov/NABIR/ (bioremediation research; U.S. Department of Energy)
http://commtechlab.msu.edu/sites/dLc-me/zoo/zdtmain.html (biodegradation/bio-remediation)

21.11 QUESTIONS AND PROBLEMS

1. What is the goal of land application of wastewater effluents and biosolids?
2. Compare biosolids and wastewater effluents as regards virus transport through soils.
3. How do we dispose of biosolids in the environment?
4. What are the main concerns over land application of biosolids?
5. Of the four land application systems of wastewater effluents, which one minimizes bioaerosol production?
6. Are viruses inactivated in the tanks of septic tank systems?
7. What must be done to minimize contamination of drinking water wells by septic tank effluents?

8. Give the factors affecting pathogen transport through soils.

9. Give the factors affecting pathogen survival in soils.

10. What surrogates have been used to simulate pathogen transport through soils?

11. What is the environment (aquifers or surface waters) most conducive to longer virus persistence?

12. What are the approaches used for soil bioremediation?

13. What are the main concerns over the introduction of genetically modified microorganisms in soils?

14. Briefly discuss the treatment strategies concerning aquifer restoration.

15. What desirable characteristics are required for microorganisms used in bioaugmentation of aquifers?

21.12 FURTHER READING

Abbaszadegan, M., M. Lechevallier, and C.P. Gerba. 2003. Occurrence of viruses in U.S groundwaters. J. Amer. Water Wks. Assoc. 95: 107–120.

Beaudette, L.A., M.B. Cassidy, L. England, J.L. Kirk, M. Habash, H. Lee, and J.T. Trevors. 2002. Bioremediation of soils. pp. 722–737, In: *Encyclopedia of Environmental Microbiology*, Gabriel Bitton, editor-in-chief, Wiley-Interscience, N.Y.

Bitton, G., and R.W. Harvey. 1992. Transport of pathogens through soils and aquifers. pp. 103–124, In: *Environmental Microbiology*, R. Mitchell, Ed., Wiley-Liss, New York.

Canter, L.W., and R.C. Knox. 1985. Septic Tank System: Effects on Groundwater Quality. Lewis, Chelsea, MI.

Gerba, C.P., and G. Bitton. 1984. Microbial pollutants: Their survival and transport pattern to groundwater. pp. 65–88, In: *Groundwater Pollution Microbiology*, G. Bitton and C.P. Gerba, Eds. John Wiley & Sons, New York.

Harvey, R.W. 1997. in situ and laboratory methods to study subsurface microbial transport. pp. 586–599, In: *Manual of Environmental Microbiology*, C.J. Hurst, G.R. Knudsen, M.J. McInermey, L.J. Stetzenback, and M.V. Walters, Eds., ASM Press, Washington, DC.

McCarty, P.L., B.E Rittmann, and E.J. Bouwer. 1984. Microbiological processes affecting chemical transformations in groundwater. pp. 89–115, In: *Groundwater Pollution Microbiology*, G. Bitton and C.P. Gerba, Eds., John Wiley & Sons, New York.

van Overbeek, L.S., and D. van Elsas. 2002. Genetically modified microorganisms (GMM) in soil environments. pp. 1429–1440, In: *Encyclopedia of Environmental Microbiology*, Gabriel Bitton, editor-in-chief, Wiley-Interscience, N.Y.

Prince, R. 2002. Bioremediation: An overview of how microbiological processes can be applied to the cleanup of organic and inorganic environmental pollutants. pp. 692–712, In: *Encyclopedia of Environmental Microbiology*, Gabriel Bitton, editor-in-chief, Wiley-Interscience, N.Y.

Ripp, S., and G.S. Sayler. 2002. Field release of genetically engineered microorganisms. pp. 1278–1287, In: *Encyclopedia of Environmental Microbiology*, Gabriel Bitton, editor-in-chief, Wiley-Interscience, N.Y.

Rose, J.B. 1986. Microbial aspects of wastewater reuse for irrigation. CRC Crit. Rev. Environ. Control. 16: 231–256.

U.S. EPA. 1983. *Process Design Manual for Land Application of Municipal Sludge*, Report No. EPA 625/1-83-016, U.S. Environmental Protection Agency, Cincinnati, OH.

Wallis, P.M. and D.L. Lehmann, Eds. 1983. *Biological Health Risks of Sludge Disposal to Land in Cold Climates*, Univ. of Calgary Press, Calgary, Canada.

Widada, F., H. Nojiri, and T. Omori. 2002. Recent developments in molecular techniques for identification and monitoring of xenobiotic-degrading bacteria and their catabolic genes in bioremediation. Appl. Microbiol. Biotechnol. 60: 45–59.

22

PUBLIC HEALTH ASPECTS OF WASTEWATER AND BIOSOLIDS DISPOSAL IN THE MARINE ENVIRONMENT

22.1 INTRODUCTION

Billions of gallons of wastewater effluents, sometimes receiving only primary treatment or less, are disposed of daily by the ever-increasing populations along the coastlines around the world, sometimes not far away from public bathing beaches. The effluents include domestic wastewater containing fecal materials, industrial wastes harboring metals, and xenobiotic compounds. Millions of tons of digested sludge are also disposed of at sea by pumping or transport by barges to disposal sites. Ocean outfalls are generally an inexpensive alternative to tertiary treatment, or even to biological treatment in some countries. The outfall pipes may extend from 1 to 4 miles off shore. Disposal of wastewater effluents and sludges into the ocean adversely impacts marine life and contaminates shellfish beds, and may have public health implications for swimmers at public beaches.

During the 1960s, the realization that oceans are not the "infinite sink" for dumping of human wastes triggered extensive research on the fate of pathogenic microorganisms and parasites in seawater. We will now examine the main factors controlling the fate of enteric pathogens in water and sediments in the marine environment.

22.2 GLOBAL SURVEYS OF ENTERIC PATHOGENS IN CONTAMINATED SEAWATER

Contamination of the marine environment by pathogenic microorganisms is mainly due to the disposal of wastewater or wastewater effluents into estuarine waters, to offshore disposal via sewage outfalls, and to rivers contaminated with wastewater effluents. Microbiological examination of coastal waters near sewage outfalls shows the presence of pathogenic bacteria such as *Salmonella* and *Vibrio cholerae* (Grimes et al., 1984; Morinigo et al., 1992a). Other noncholera vibrios (*Vibrio vulnificus*, *V. parahaemolyticus*, and *V. alginolyticus*) have been detected in coastal areas (Hervio_Heath et al., 2002). For example, the city of Miami, Florida, discharges domestic wastewater via a sewage outfall located 3.6 km off shore. Fecal coliforms (9000–55,000/100 mL), fecal streptococci (0.5–19/100 mL) and enteroviruses (21–59 PFU/400 L) were detected within 200 m from the outfall. Viruses were detected in the sediments at recreational bathing beaches situated up to 3.6 km from the outfall (Schaiberger et al., 1982) (Table 22.1).

Monitoring of bacterial indicators and enteroviruses in the vicinity of a sewage outfall off the Israeli coast showed that wastewater bacteria were reduced more rapidly than viruses. Only fecal streptococci displayed an inactivation rate similar to that of enteroviruses (Fattal et al., 1983) (Fig. 22.1).

TABLE 22.1. Detection of Enteroviruses in Sediments Along a Bathing Beach[a]

Station[b]	Depth (m)	Enteroviruses (PFU/L)
1	1.2	0
2	0.9	15
3	1.5	30
4	1.5	0
5	1.5	0

[a]Adapted from Schaiberger et al. (1982).
[b]Samples were collected 400 m apart on a beach 3.6 km from the outfall pipe's discharge.

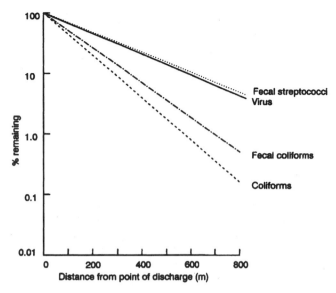

Figure 22.1 Inactivation of enteric microorganisms at various distances from a sewage outfall in Israel. From Fattal et al. (1983).

The detection of enteric viruses in seawater was also documented off the coast of several countries, including Brazil (Marques and Martins, 1983), France (Hugues et al., 1980; 1981), Italy (Petrilli et al., 1980), Spain (Bravo and de Vicente, 1992; Finance et al., 1982; Lucena et al., 1982; Morinigo et al., 1992a), and the United States (Goyal et al., 1979; Rao et al., 1984; Vaughn et al., 1979). The virus types found were identified as polioviruses, coxsackie A and B viruses, echoviruses, adenoviruses, and rotaviruses; their levels vary between 0.007 and 100 PFU/L (Bitton et al., 1985). In the vicinity of sewage outfalls, enteric viruses are also often detected in marine sediments where they may persist for long time periods (Bitton et al., 1981; Goyal et al., 1984; Rao et al., 1984; Schaiberger et al., 1982). Coliphages, *Bacteroides fragilis* phages, enteric viruses, and rotaviruses were detected in marine sediments at 300 m to 12 km off the coast of Barcelona, Spain, an area impacted by fecal pollution. There was a significant correlation between the presence of enteric viruses and *B. fragilis* phages (Jofre et al., 1989).

22.3 SURVIVAL OF PATHOGENIC AND INDICATOR MICROORGANISMS IN SEAWATER

Laboratory experiments and in situ survival studies (e.g., use of dialysis bags and flow-through systems) have shown that a number of environmental and biological factors control the fate of enteric microorganisms in the marine and other aquatic environments.

22.3.1 Temperature

Temperature is a decisive factor controlling the survival of pathogenic microorganisms in seawater. The die-off of wastewater microorganisms increases at higher temperatures (El-Sharkawi et al., 1989). Decay rates of *Escherichia coli* and enterococci in diffusion chambers were correlated with temperature in the 0–20°C range. Survival of *E. coli*

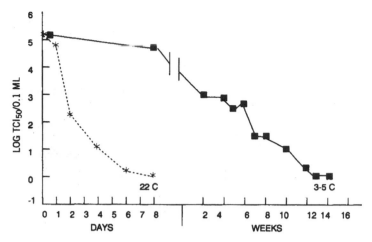

Figure 22.2 Inactivation of echovirus 6 in seawater as a function of temperature. From Won and Ross (1973). (With permission of the publisher.)

was generally lower than enterococci (Lessard and Sieburth, 1983). Figure 22.2 (Won and Ross, 1973) also shows that the survival of an enterovirus, echovirus 6, in seawater is much lower at 22°C than at 3–5°C. Similarly, hepatitis A virus (HAV) and phage indicators (F$^+$ and *B. fragilis* phages) are more persistent in seawater at 5°C than at 25°C (Chung and Sobsey, 1993b). The time to reduce virus titer by 90 percent (T_{90}) of HAV in synthetic seawater was 11 days at 25°C, while it remained stable for up to 92 days at 4°C (Crance et al., 1998).

22.3.2 Solar Radiation

Solar radiation also plays a key role in the decline of indicator and pathogenic bacteria in seawater (Bellair et al., 1977; Chamberlin and Mitchell, 1978; El-Sharkawi et al., 1989;

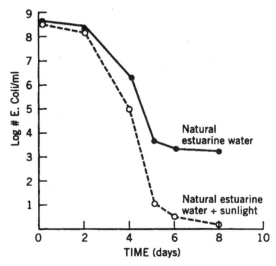

Figure 22.3 Effect of sunlight on *Escherichia coli* in seawater. Adapted from McCambridge and McMeekin (1981).

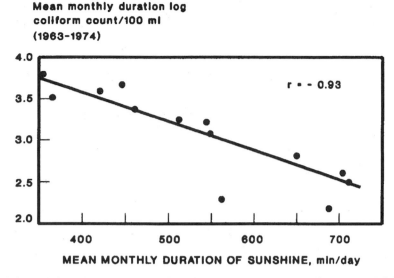

Mean monthly duration log
coliform count/100 ml
(1963-1974)

MEAN MONTHLY DURATION OF SUNSHINE, min/day

Figure 22.4   Relationship between duration of sunshine and mean coliform count (1963–1974). Fattal et al. (1983). (Courtesy of Pergamon Press.)

Fujioka et al., 1981; 2002; Gameson and Gould, 1975; Kapuschinski and Mitchell, 1982; Mascher et al., 2003). Figure 22.3 (McCambridge and McMeekin, 1981) shows that both solar radiation and biological factors have a detrimental effect on the survival of *E. coli* in seawater. Moreover, *E. coli* was more sensitive to sunlight than *Salmonella typhimurium*. Phages or fecal streptococci, particularly the enterococus subset, survive longer than fecal coliforms in seawater (Davies-Colley et al., 1994; Fujioka et al., 1981; Kapuschinski and Mitchell, 1982; Mascher et al., 2003). As regards the effect of sunlight, enterococci, rather than *E. coli*, can serve as better indicators of virus presence in marine waters (Fujioka et al., 2002). Thus, the FC/FS ratio, which was generally used to indicate the source (human vs. animal origin) of fecal pollution, would be less valid in the marine environment (Fujioka et al., 1981). Exposure to sunlight accounted for up to 83, 84, and 99 percent of the inactivation of male-specific phage, *C. perfringens*, and fecal coliforms, respectively (Burkhardt III et al., 2000). In Morecambe Bay (northwest England), *Campylobacter* numbers were lowest in May–June, the sunniest months of the year. They also exhibited diurnal variations with the highest numbers in the morning and the lowest in the afternoon (Jones et al., 1990). Analysis of hundreds of seawater samples showed a negative correlation between the mean monthly log coliform count and the mean monthly duration of sunshine ($r = 0.93$) (Fattal et al., 1983) (Fig. 22.4). Thus coliform and pathogen counts should be higher in the winter than in the summer season. Exposure to sunlight in coastal seawater can also cause sublethal injury in microorganisms and may reduce the activity of several enzymes as well as bacterial culturability, due to the production of highly reactive oxygen species (singlet oxygen, superoxide, hydrogen peroxide). Catalase may be the site of sunlight-induced damage in *E. coli*, as addition of this enzyme to minimal growth media improves the recovery of sunlight-injured cells (Arana et al., 1992; Kapuschinski and Mitchell, 1981). An enterotoxigenic *E. coli*, upon exposure to sunlight in filter-sterilized seawater, displayed a much lower plate count as compared with the dark control, but the direct counts of viable cells were slightly reduced, suggesting that the exposed bacteria have entered the viable but nonculturable state (Pommepuy et al., 1996) (Fig. 22.5) and remained pathogenic, as shown by the

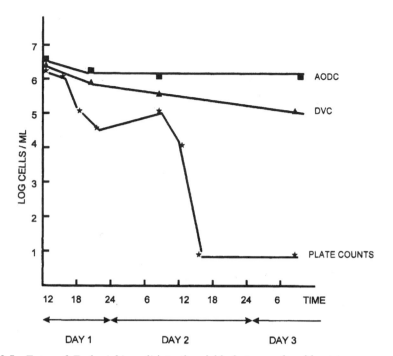

Figure 22.5 Entry of *Escherichia coli* into the viable but nonculturable state upon exposure to sunlight. (AODC, acridine orange direct count; DVC, direct counts of viable cells (cells were incubated with yeast extract and nalidixic acid and subsequently stained with acridine orange, and the elongated cells counted by epifluorescence microscopy). Adapted from Pommepuy et al. (1996).

production of enterotoxin. This clearly shows that the bacteriological methods based on culturing on growth media are not adequate for assessing the public health safety of recreational waters.

The killing effect of ultraviolet light, particularly the UV-B (280–320 nm) portion of the solar spectrum, towards viruses is well known (see Chapter 6). Solar radiation is an important contributor to the loss of infectivity of marine bacteriophages (generally used as models for animal viruses) in seawater (Noble and Fuhrman, 1997). In full sunlight, the decay rates may be from 0.4 to 0.8 h^{-1}. The decay rate was estimated at 0.033 h^{-1} when averaged over 24 h and integrated over the upper 30 m of the water column (Suttle and Chen, 1992). Exposure of poliovirus 1 for 3 h to solar radiation in Florida (light intensity was 0.646 cal/cm^2/min and the mean temperature was 26°C), resulted in approximately one–log virus inactivation (Bitton et al., 1979). A similar inactivation rate was observed for phage T_4 in seawater (Attree-Pietri and Breittmayer, 1970).

In marine waters, *Giardia* cysts are sensitive to both salinity and solar radiation. In the dark, the cysts survived up to 77 h as compared to only 3 h in the presence of light. *Cryptosporidium* oocysts displayed a higher survival than *Giardia* cysts (Johnson et al., 1997).

22.3.3 Osmotic Stress

Several investigators observed that enteric bacterial survival decreases as salinity increase. Moreover, salinity was found to increase the killing effect of sunlight. Genetically controlled osmoregulatory processes, induced by salts, help enteric bacteria survive osmotic stress in the marine environment (Munro et al., 1987). Osmoregulatory processes involve K^+ uptake as well as accumulation of compatible organic osmolytes or osmoprotectant molecules such as glycine-betaine, trehalose, and glutamate (Gauthier et al., 1991; Larsen et al., 1987; Rozen and Belkin, 2001; Strom et al., 1986). Glycine-betaine was found in marine sediments (King, 1988) and may protect enteric bacteria from osmotic stress in this environment (Munro et al., 1989).

22.3.4 Adsorption to Particulates

Adsorption of bacteria or viruses to particulates (silts, clay minerals, cell debris, or particulate organic matter) appears to provide protection to microorganisms from environmental insults (Bitton and Mitchell, 1974b; Gerba and Schaiberger, 1975). Figure 22.6 illustrates the protective effect of a clay mineral, montmorillonite, toward *E. coli* and bacteriophage T7 (Bitton and Mitchell, 1974a, b). This phenomenon was also observed in lake water (Babich and Stotzky, 1980). Solid-associated microorganisms may settle and accumulate in aquatic sediments (see Section 21.4) and may remain infective to their host cells.

22.3.5 Biological Factors

Biological factors are also implicated in the decline of enteric pathogens in the marine environment. More than 40 years ago, it was demonstrated that a heat-labile substance was implicated in virus inactivation in seawater (Plissier and Therre, 1961). Since then, this phenomenon has been repeatedly demonstrated in the marine environment around

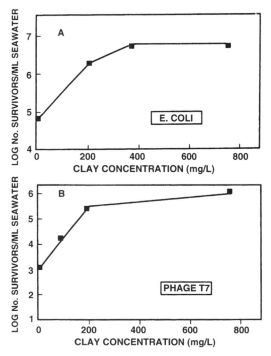

Figure 22.6   Effect of montmorillonite on the survival of *Escherichia coli* and bacteriophage T7 in seawater. Adapted from Bitton and Mitchell (1974a, b).

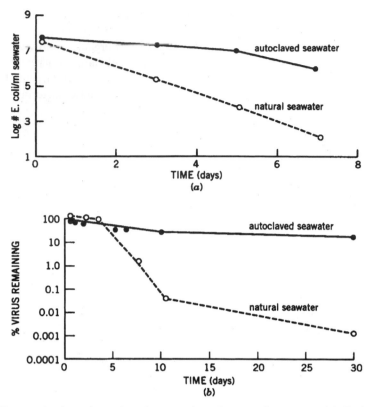

Figure 22.7 Inactivation of enteric microorganisms in natural seawater: (a) *Escherichia coli*, (b) poliovirus 1. From Bitton and Mitchell (1974b) and Shuval et al. (1971).

the globe. The contribution of small protozoan flagellates to the decay of marine bacterio-phages in seawater has been documented. These nanoflagellates can ingest approximately three viruses per flagellate per hour (Suttle and Chen, 1992). Figure 22.7 shows that both enteric bacteria and viruses are readily inactivated in natural seawater but to a lesser extent in heat-treated seawater (Bitton and Mitchell, 1974b; Mitchell, 1971; Noble and Fuhrman, 1997; Pietri and Breittmayer, 1976; Shuval et al., 1971). Thus, lytic or antagonistic micro-organisms (e.g., *Vibrio marinus*, *Bdellovibrio bacteriovorus*, predacious protozoa such as *Vexillifera*) or other unknown biological factors are implicated in inactivation of enteric microorganisms in seawater (Borrego and Romero, 1985; Magnusson et al., 1967; Patti et al., 1987).

Some investigators reported that protozoa are the most important predators of *E. coli* in the marine environment (Enzinger and Cooper, 1976). The major influence of protozoa is exerted during the first 2 days of incubation (McCambridge and McMeekin, 1980). Grazing by protozooplankton was found to be the dominant process of decay of fecal and autochthonous bacteria in both freshwater and marine aquatic environments. It was determined that grazing was responsible for more than 90 percent of the overall mortality rate of fecal and autochthonous bacteria (Menon et al., 2003).

22.3.6 Other Factors

Other factors controlling the survival of enteric pathogens in aquatic environments are microorganism type (e.g., viruses survive better than do bacteria), growth stage, nutrient availability, aggregation, pH, dissolved oxygen, or heavy metals (Bitton, 1978; Block, 1981; Gauthieŕ et al., 1992). Fecal streptococci generally persist longer in seawater than do total or fecal coliforms (Bravo and de Vicente, 1992). *Cryptosporidium* oocysts survive well in seawater at both 15°C and 30°C. Their survival was much higher that of *E. coli* or coxsackievirus A9 (Nasser et al., 2003).

At the molecular level, among the many genes playing a role in *E. coli* survival in seawater, *RpoS* genes play a dominant role, protecting this enteric bacterium from a wide range of stresses (Rozen and Belkin, 2001).

As regards the persistence of enteric pathogens in the marine environment, we can draw some general conclusions:

1. Enteric pathogens, especially viruses, are more labile in seawater than in fresh water.
2. The results of several studies show that microbial inactivation in sea water is variable and unpredictable.
3. Laboratory-grown cultures show similar inactivation trends as sewage-derived indicators as regards the effect of temperature or sunlight.
4. There are probably other, yet unkown, factors contributing to the decline of enteric pathogens in seawater.

22.4 SURVIVAL OF PATHOGENIC AND INDICATOR MICROORGANISMS IN SEDIMENTS

Wastewater and biosolids disposal off the coast from sewage outfalls or barges affect marine sediments, which are important reservoirs of pathogenic microorganisms and parasites. Pathogenic and indicator bacteria, when becoming associated with organic and inorganic particulates, settle into the bottom sediments of freshwater and marine environments, where they accumulate and reach higher concentrations than in the water column. A buildup of indicator bacteria in sediments was observed around sewage outfalls (Goyal et al., 1977) (Table 22.2). The accumulation of enteric pathogens is a result of their longer survival in sediments. Figure 22.8 shows a longer persistence of *E. coli* in marine sediments than in seawater (Gerba and McLeod, 1976). The longer survival of

TABLE 22.2. Accumulation of Indicator Bacteria in Marine Sediments[a]

	Total Coliform*	Fecal Coliform[b]
Overlying Water Station		
1	6,886	2,382
2	5,320	1,528
3	64	19
4	92	10
Sediment Station		
1	382,143	9,731
2	192,857	16,806
3	16,791	152
4	14,279	151

[a]Adapted from Goyal et al. (1977).
[b]MPN/100 mL.

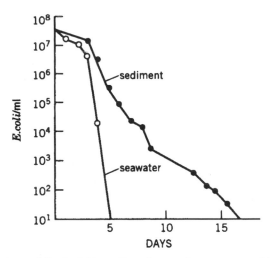

Figure 22.8 Comparison of *Escherichia coli* persistence in seawater and in sediments. Adapted from Gerba and McLeod (1976).

Salmonella typhimurium in sediments was also documented (Craig et al., 2003). Furthermore, in *E. coli*, expression of the genes responsible for osmoregulation was found to be enhanced in marine sediments containing organic matter. This may explain, at least partially, why sediments act as reservoirs for pathogens (Gauthier and Breittmayer, 1990).

Enteric viruses also find their way into sediments, where they accumulate and survive longer than in the water column (Bitton, 1980a, b). For example, hepatitis A viruses, poliovirus 1, and phage indicators (F^+ and *B. fragilis* phages) persist longer in sediments than in the water column (Chung and Sobsey, 1993a, b). As a result of sludge dumping off the Delaware–Maryland coast, bacterial indicators (TC, FC, FS) and amebas were detected as far as 40 km from the sludge dumpsite. Amebas, total coliforms, and fecal streptococci persisted longer in sediments than did fecal coliforms (O'Malley et al., 1982).

Several species of *Acanthamoeba* were isolated from the sediments of a Philadelphia–Camden sludge disposal site (Sawyer et al., 1982). Some of these ameba species are of potential health significance (e.g., *A. culbertsoni*, *A. hatchetti*).

The levels of pathogenic and indicator bacteria in intertidal sediments in Boston Harbor were followed up to 460 m from a storm and wastewater discharge point. Despite their accumulation in sediments, all bacterial indicators (total bacteria, fecal coliforms, fecal streptococci) declined, however, with increasing distance from the sewage outfall (Shiaris et al., 1987).

Thus, sediments can serve as reservoirs of enteric pathogens. Furthermore, fecal microbial indicators (fecal coliforms, fecal streptococci) and pathogens generally accumulate in the upper layer of sediments (Table 22.3; Obiri-Danso and Jones, 2000). Thus, sediment resuspension through motor boat activity, currents, swimming, and changes in water quality would increase pathogen levels in the water column.

22.5 PUBLIC HEALTH ASPECTS OF SWIMMING IN CONTAMINATED RECREATIONAL WATERS

Several diseases can be contracted after swimming in contaminated recreational waters and involve eye, ear, nose, and skin infections. The microbial pathogens and parasites

TABLE 22.3. Distribution of *Campylobacter* and Bacterial Indicators at Increasing Depths in Sediments[a]

Microorganism	Sediment Depth (cm)	Sampling Date 1 (numbers/g d.w.)[b]	Sampling Date 2 (numbers/g d.w.)[b]
Campylobacter	0–1	13	7
	4–5	0	0
	9–10	0	0
	19–20	0	0
Fecal coliforms	0–1	2273	6515
	4–5	52	52
	9–10	0	0
	19–20	0	0
Fecal streptococci	0–1	686	777
	4–5	92	53
	9–10	0	0
	19–20	0	0

[a]Adapted from Obiri-Danso and Jones (2000).
[b]Results expressed as MPN/g d.w. sediment for *Campylobacter* and fecal coliforms, and as CFUs/g d.w. sediment for fecal streptococci; d.w. = dry weight.

responsible for the diseases acquired though contact with water are summarized in Table 22.4 (Fujioka, 1997; Moe, 1997), which does not include the agents acquired through the ingestion of water.

The health risks associated with water-contact sports (swimming, water-skiing, scuba-diving, surfing) fall into the following two categories according to the source of exposure (Dufour, 1986).

22.5.1 Exposure to Water Containing Pathogenic Microorganisms of Fecal or Domestic Wastewater Origin

Several diseases have been associated with wastewater-contaminated recreational waters where swimming can be associated with the accidental ingestion of pathogens and

TABLE 22.4. Illnesses Acquired by Contact with Recreational Waters[a]

Organism	Illness/Problem
Bacteria	
Aeromonas hydrophila	Wound infection
Leptospira	Leptospirosis
Mycobacterium spp.	Skin or subcutaneous lesions
Legionellae	Legionnaire's disease or Pontiac fever
Pseudomonas spp.	Dermatitis, ear infection
Vibrio spp.	Wound infection, septicemia
Staphylococcus auerus	Wound infection
Viruses	
Adenoviruses	Conjunctivitis, pharyngitis
Protozoa	
Naegleria fowleri	Meningoencephalitis
Acanthamoeba	Subcutaneous abcesses; conjuctivitis

[a]Adapted from Fujioka (1997); Moe (1997).

TABLE 22.5.   Pathogens and Parasites Introduced by Swimmers into Recreational Waters[a]

Infectious Agent	Recreational Water
Pseudomonas aeruginosa	Bathing beach
Staphylococcus	Swimming pool
Rotaviruses and enteroviruses	Full-contact bathing in a stream
Norwalk virus	Full-contact bathing in a lake
Hepatitis A virus	Full-contact bathing in a lake
Enteroviruses	Swimming pool
Adenovirus type 4	Swimming pool
Cryptosporidium	Swimming pool, water park
Giardia	Swimming pool

[a]Adapted from Stewart et al. (2002).

parasites. These include typhoid fever, salmonellosis, shigellosis, hepatitis, and gastroenteritis. The sources of the infectious agents are wastewater treatment plant effluents, septic tank effluents, stormwater runoffs, and discharges from humans or lower animals into recreational waters (Cabelli, 1989). An important source of pathogens in recreational waters is human fecal release. It was estimated that the fecal release is 0.14 g/person and can be as high as 10 g/person (Gerba, 2000a).

Table 22.5 (Stewart et al., 2002) shows the range of pathogens introduced into recreational water by swimmers.

22.5.1.1 *Waterborne Outbreaks Associated with Recreational Swimming.*
Eleven gastroenteritis outbreaks associated with recreational waters (e.g., lakes, swimming pools, creeks) were reported in 1991–1992 in the United States (Moore et al., 1994) (Table 22.6). Hepatitis A outbreaks were associated with swimming in lakes or public pools (Mahoney et al., 1992). In a Vermont camp, 21 of 23 swimmers and counselors were infected with coxsackievirus type B4 or B5 following swimming in

TABLE 22.6.   Outbreaks of Gastroenteritis Associated with Recreational Water (1991–1992)[a]

Etiologic Agent	No. of Cases	Source (Location)
Giardia	9	Wading pool, day care center
Giardia	7	Wading pool, day care center
Cryptosporidium	26	Water slide (park)
Giardia	14	Swimming pool (park)
AGI	15	Creek (private home)
AGI	15	Dunking booth (fair)
E. coli O157:H7[b]	80	Lake (park)
Cryptosporidium	500	Wave pool (park)
Shigella sonnei	203	Lake (park)
Shigella sonnei	23	Lake (swimming area)
Giardia	4	Lake (campground)

[a]Adapted from Moore et al. (1994).
[b]Mixed outbreak of *E. coli* O157:H7 and *Shigella sonnei*.
AGI, acute gastrointestinal illness of unknown etiology.

a lake (Hawley et al., 1973). Protozoan parasites (*Giardia*, *Cryptosporidium*) were responsible for 55 percent of the outbreaks in 1991–1992, and for 71.4 percent of the 14 outbreaks reported in 1993–1994. In two of the outbreaks, the etiologic agent was not identified while the remaining outbreaks were attributed to *Shigella sonnei* or to *E. coli* O157:H7. Four of 14 recreational water-associated outbreaks reported in 1993–1994 were attributed to bacterial pathogens (Kramer et al., 1996; Moore et al., 1994). A waterborne shigellosis disease outbreak was reported in Iowa and was linked to swimming in a fecally contaminated portion of the Mississippi River (Rosenberg et al., 1976). Other outbreaks of swimming-associated shigellosis have been reported (Makintubee et al., 1987; Sorvillo et al., 1988). In a California shigellosis outbreak, about 50 percent of 68 persons with diarrheal illness had *Shigella sonnei* and *S. boydii* in their stools. Shigellosis was associated with swallowing water while swimming (Sorvillo et al., 1988). Outbreaks of hepatitis and viral gastroenteritis have been associated with swimming in contaminated aquatic environments (Bryan et al., 1974; Denis et al., 1974; Koopman et al., 1982). Prospective epidemiological studies have also shown an association between gastroenteritis and swimming (Cabelli, 1981).

22.5.1.2 *Risk Analysis.*

A prospective epidemiological study in South Africa has shown that the relative risk (i.e., incidence rate among swimmers divided by the incidence rate among nonswimmers) for developing gastrointestinal symptoms was higher at a moderately polluted beach than at a control beach (von Schirnding et al., 1993). Using a quantitative microbial risk assessment, it was found that the mean probability of becoming infected with *Salmonella derby* following swimming in a coastal water of very poor quality was 10^{-3}. For water that met regulations, the mean probability decreased to 10^{-6} (Craig et al., 2003).

A risk assessment approach was used to evaluate the probability of infection following swimming in freshwater contaminated with coxsackieviruses. The exponential risk assessment model used was (Mena et al., 2003):

$$P_i = 1 - e^{[-(1/k)N]}$$

where P_i = probability of infection; N = number of viruses ingested; $k = 129$ (k was estimated from a dose–response study on mice infection with coxsackievirus B4.

Assuming a daily exposure to 100 mL of freshwater following swimming, and a virus concentration of 0.67 MPN/L, the risk of becoming infected is 5.19×10^{-4}. If the virus concentration is 5.44 MPN/L, the risk of infection is 4.21×10^{-3} (Table 22.7; Mena et al., 2003).

22.5.1.3 *Microbial Indicators.*

The enterococci level in recreational waters appears to be a good indicator of the risk of swimming-associated gastroenteritis. The U.S. EPA

TABLE 22.7. Risks Associated with Swimming in Water Contaminated with Coxsackievirus[a]

Type of Risk[b]	0.67 MPN CPU/L		5.44 MPN CPU/L	
	1 day	10 days	1 day	10 days
Risk of infection	5.19×10^{-4}	5.18×10^{-3}	4.21×10^{-3}	4.13×10^{-2}
Risk of illness	3.89×10^{-4}	3.89×10^{-3}	3.16×10^{-3}	3.10×10^{-2}
Risk of death	2.30×10^{-6}	2.29×10^{-5}	1.86×10^{-5}	1.83×10^{-4}

[a]Adapted from Mena et al. (2003).
[b]General population at 100 mL single exposure.

guidelines for recreational waters set single-sample limits of 235 CFU/100 mL for *E. coli* and 61 CFU/100 mL for enterococci. Furthermore, the 5 day geometric mean should not exceed 126 CFU/100 mL for *E. coli* and 33 CFU/100 mL for enterococci (U.S. EPA, 2000). However, In Lake Michigan recreational water, the relationship between *E. coli* and enterococci was weak (Kinzelman et al., 2003).

22.5.2 Spas and Hot Tubs

Bathing in spas and hot tubs is a popular recreational activity, especially in industrialized countries. They offer, however, a confined environment that can be easily contaminated with skin and fecal releases from bathers. Several diseases associated with bathing in spas and hot tubs are caused by bacterial, viral, and protozoan agents such as *Pseudomonas aeruginoa, Mycobacterium, Legionella, Yersinia, Staphylococcus,* enteroviruses, *Acanthamoeba, Cryptosporidium, Giardia,* or *Naegleria fowleri* (Broadbent, 1996; Mangione et al., 2001; Stewart, 2002). Examples of pathogens and parasites transmitted via this recreational activity are illustrated in Table 22.8 (Stewart, 2002). Some of these pathogens or parasites survive the high temperatures of the spa water (e.g., *Pseudomonas aeruginosa, Legionella pneumophila, Acanthamoeba castellani*). In the United States, the Center for Disease Control and Prevention (CDC) recommends a free chlorine residual of 3 mg/L and free bromine residual of 4 mg/L (CDC, 1999).

22.5.3 Exposure to Autochthonous (Indigenous) Microorganisms

Water-associated activities may also be the source of infections caused by autochthonous microorganisms in marine and freshwater environments. The main agents are *Aeromonas, Vibrio,* and *Pseudomonas aeruginosa.* The etiologic agents in ear infections (swimmer's ear) are *Pseudomonas aeruginosa* and *Vibrio* species (e.g., *V. parahaemolyticus, V. vulnificus, V. alginolyticus*). *Pseudomonas aeruginosa* is ubiquitous in nature, but is also recovered from human stools and finds its way into wastewater. This opportunistic pathogen is responsible for ear and urinary tract infections, dermatitis, and folliculitis (Fox and Hambricks, 1984; Havelaar et al., 1983; Salmen et al., 1983), and bacteremia in patients receiving organ transplants (Brooks and Remington, 1986). This pathogen was found in 45 percent of coastal water samples in Israel and correlated well with the presence of total and fecal coliforms (Yoshpe-Purer and Golderman, 1987).

Swimming in fecally polluted waters may not cause only enteric disturbances but also ailments of the upper respiratory tract. Risk of pneumonia can be high in near drowning situations. Pneumonia cases attributed to *Pseudomonas putrefaciens, Staphylococcus aureus, Aeromonas hydrophila,* and *Legionnella pneumophila* have been documented

TABLE 22.8. Examples of Pathogens and Parasites Associated with Bathing in Spas and Hot Tubs[a]

Pathogen or Parasite	Waterborne Illness
Pseudomonas aeruginosa	Skin disease (folliculitis), external otitis
Legionella pneumophilia	Pontiac fever, Legionnaires' disease (aquired following aerosolization of the pathogens)
Mycobacterium avium complex	Respiratory illnesses
Hepatitis A virus	Hepatitis following ingestion of spa water
Acanthamoeba castellani	Keratitis
Naegleria fowleri	Primary amebic meningoencephalitis

[a]Adapted from Stewart (2002).

(Reines and Cook, 1981; Rosenthal et al., 1975; Sekla et al., 1982). Swimming in recreational waters can also be the cause of skin infections due to the presence of opportunistic microorganisms (*Aeromonas*, *Mycobacterium*, *Staphylococcus*, and *Vibrio* spp.). These infections are associated with skin abrasion and laceration. Recreational waters may serve as a vehicle for skin infections caused by *Staphylococcus aureus*. Some investigators have recommended that this organism be used as an additional indicator of the sanitary quality of recreational waters, since its presence is associated with human activity in recreational waters (Charoenca and Fujioka, 1993; Yoshpe-Purer and Golderman, 1987). A prospective epidemiological study in South Africa indicated that the relative risk of the development of skin symptoms was significantly higher among white swimmers at a moderately polluted beach than at a control beach (von Schirnding et al., 1993). Swimming in areas contaminated by urine from infected domestic or wild animals was related to outbreaks of leptospirosis, also named Weil's disease or hemorrhagic jaundice, which causes headaches, fever, chills, and nausea. The infectious agents are *Leptospira* spp. (Dufour, 1986).

Swimming in warm recreational lakes can be associated with primary meningoencephalitis (PAME) caused by a free-living ameba *Naegleria fowleri*. This protozoan gains access to the brain through the nose and can cause death within a few days (see Chapter 4).

22.6 WEB RESOURCES

http://www.epa.gov/OST/beaches/ (public health aspects of recreational waters)
http://www.nps.gov/public_health/inter/rec_water/rw.htm (National Park Service, U.S. Department of the Interior)
http://www.cdc.gov/healthyswimming/ (source: Center for Disease Control and Prevention)

22.7 QUESTIONS AND PROBLEMS

1. Why are coastal areas prone to contamination by pathogenic microorganisms and parasites?
2. How would you study the in situ survival of laboratory-grown pathogens, parasites, or indicator microorganisms in seawater or freshwater? Describe the experimental procedures.
3. What seems to be the most important factor involved in the die-off of pathogens and parasites in seawater?
4. How do pathogens reach the sediments in aquatic environments and why do they accumulate in sediments?

22.8 FURTHER READING

Dufour, A.P. 1986. Diseases caused by water contact. pp. 23–41, In: *Waterborne Diseases in the United States*, G.F. Craun, Ed., CRC Press, Boca Raton, FL.
Moe, C.L. 1997. Waterborne transmission of infectious agents. pp. 136–152, In: *Manual of Environmental Microbiology*, C.J. Hurst, G.R. Knudsen, M.J. McInerney, L.D. Stetzenbach, and M.V. Walter, Eds., ASM Press, Washington, D.C.
Moore, A.C., B.L. Herwaldt, G.F. Craun, R.L. Calderon, A.K. Highsmith, and D.D. Juranek. 1994. Waterborne diseases in the United States, 1991 and 1992. J. Amer. Water Works Assoc. 86: 87–99.
Rozen, Y., and S. Belkin. 2001. Survival of enteric bacteria in seawater. FEMS Microbiol. Rev. 25: 513–529.

23

WASTEWATER REUSE

23.1 INTRODUCTION

Indirect reuse of wastewater has been practiced for centuries around the globe. Planned reuse of this resource has been documented as early as the 16th century in Europe. In the United States, this practice was initiated around the beginning of the last century in Arizona and California for irrigation of lawns and gardens or for use as cooling water. Some investigators distinguish between "reclamation," "reuse," and "recycling." *Reclamation* is the treatment or processing of wastewater to make it reusable. *Recycling* is the internal reuse of water by a given industry before its ultimate disposal. *Wastewater reuse* is the use of water for a beneficial goal such as crop and landscape irrigation or urban applications. This practice has gained importance worldwide because of water shortages, particularly in arid areas (e.g., California, Arizona, Texas, Colorado), and wastewater disposal regulations that are becoming increasingly stringent. Essentially, water reuse is increasingly becoming an important component of sustainable management of water resources (Levine and Asano, 2004). There are two categories of health effects related to wastewater reuse, as given in the following.

23.1.1 Health Effects Due to Parasites as well as Bacterial and Viral Pathogens

The exposure routes for the infectious agents are direct contact from contaminated surfaces, accidental ingestion of contaminated water, consumption of raw vegetables irrigated with reclaimed water, and long-term exposure to biological aerosols in the vicinity of spray irrigation sites or cooling towers (see Chapters 4 and 14). The risk of infectious disease transmission is associated mainly with the use of untreated sewage or wastewater effluents of very poor quality (Cooper, 1991; Rose, 1986) and is much higher in developing nations than in industrialized countries. The risk depends on several factors, which include the type and persistence of the infectious agent, the availability of an intermediate host for certain helminth parasites, type of waste application and human exposure, human behavior (e.g., personal and food hygiene), and host immunity (Blum and Feachem, 1985; Cairncross, 1992). The risk of infection is greatly reduced when using effluents from well-operated wastewater reclamation plants. For example, at a reclamation plant in St. Petersburg, Florida, the average risk of using the effluent for nonrestricted irrigation was estimated at 10^{-6} to 10^{-8} for a single exposure to 100 mL of effluent. The treatment train in the plant included biological treatment, coagulation–sand filtration, disinfection (4 mg/L for a 45 min contact time), and storage in a reservoir for 16–24 h (Rose et al., 1996).

23.1.2 Chemicals

The chemicals of concern are pesticides, heavy metals, halogenated compounds, and other xenobiotics. The adverse effects of these chemicals, many of which are mutagenic or carcinogenic, are of special concern when the reclaimed wastewater is used for crop irrigation or groundwater recharge (Bitton and Gerba, 1984; Cooper, 1991; Nellor et al., 1985).

23.2 CATEGORIES OF WASTEWATER REUSE

The various categories of wastewater reuse are agricultural use (land application), landscape irrigation, groundwater recharge, recreational use, nonpotable urban use, potable reuse, and industrial use (Asano and Tchobanoglous, 1991) (Table 23.1).

23.2.1 Agricultural Reuse: Land Application

Reclaimed wastewater is most commonly used for irrigation of agricultural crops. Records show that the practice of sewage farming began in the 19th century in Europe (Britain, France, Germany, Austria), United States, India, and Australia (Cairncross, 1992). While the planned effluent reuse for agricultural irrigation in the United States today is less than 1 percent, other countries such as India, Israel, and South Africa are using

TABLE 23.1.   Categories of Municipal Wastewater Reuse[a]

Wastewater Reuse Categories[b]	Potential Constraints
Agricultural irrigation	Effect of water quality, particularly salts, on soils and crops
Crop irrigation	Public health concerns related to pathogens (bacteria,
Commercial nurseries	viruses, and parasites)
Landscape irrigation	Surface and groundwater pollution if not properly managed
Park	Marketability of crops and public acceptance
Schoolyard	
Freeway median	
Golf course	
Cemetery	
Greenbelt	
Residential	
Industrial reuse	Reclaimed wastewater constituents related to scaling,
Cooling	corrosion, biological growth, and fouling
Boiler feed	Public health concerns, particularly aerosol
Process water	transmission of organics, and pathogens in cooling
Heavy construction	and boiler feed water
Groundwater recharge	Organic chemicals in reclaimed wastewater and
Groundwater	their toxicological effects
replenishment	Total dissolved solids, metals, and pathogens in
Salt water intrusion	reclaimed wastewater
Subsidence control	
Recreational/	Health concerns over bacteria and viruses
environmental uses	Eutrophication due to N and P
Lakes and ponds	
Marsh enhancement	
Streamflow augmentation	
Fisheries	
Snowmaking	
Nonpotable urban uses	Public heath concerns about pathogens transmitted by
Fire protection	aerosols
Airconditioning	Effects of water quality on scaling, corrosion, biological
Toilet flushing	growth, and fouling
Potable use	Organic chemicals in reclaimed wastewater and their
Blending in water supply	toxicological effects
Pipe-to-pipe water supply	Esthetics and public acceptance
	Public health concerns on pathogen transmission
	including viruses

[a]From Asano and Tchobanoglous (1991).

[b]Arranged in descending order of volume of use.

approximately 20–25 percent of wastewater effluents for agricultural purpose (Rose, 1986). Wastewater reuse for agricultural purposes is also practiced in several areas, including North Africa (Morocco, Tunisia, Libya), the Middle East (Egypt, Israel, Jordan, Saudi Arabia), Latin America (Chile, Peru, Mexico), and Asia (India, China) (Bartone, 1991).

The advantages of land application of wastewater effluents are the supply of water and valuable nutrients to crops, and serving as an additional treatment for effluents before they reach groundwater. The main disadvantage is the potential contamination of groundwater resources and agricultural crops with parasites, bacterial and viral pathogens, toxic metals, and mutagenic/carcinogenic trace organics. Toxicity to crops is caused by excess salinity and toxic ions such as sodium, boron, chloride, cadmium, copper, zinc, nickel, beryllium, or cobalt (see Chapter 21 for more details).

There is obviously a risk of disease transmission through the use of untreated wastewater for vegetable irrigation (Lane and Lloyd, 2002). In Mexico, the irrigation of vegetable crops with domestic wastewater showed that the highest bacterial contamination was observed in leafy vegetables such as lettuce (37,000 total coliforms/100 g and 3600 fecal coliforms/100 g) and spinach (8700 total coliforms/100 g; 2400 fecal coliforms/100 g). Vegetable crops were highly contaminated with *Giardia* cysts following irrigation with raw wastewater. Coriander showed the highest cysts load (Table 23.2; Amahmid et al., 1999). Similarly, vegetables were shown to become contaminated with *Cryptosporidium* oocysts and *Giardia* cysts following irrigation with poor quality wastewater effluents (Armon et al., 2002).

Figure 23.1 shows an association between the consumption of wastewater-irrigated vegetables in Israel and the percentage of stool samples positive for *Ascaris*. After climbing to 35 percent, it decreased to less than 1 percent after the government banned the use of wastewater for vegetable irrigation (Gunnerson et al., 1984). The viral health risk associated with the consumption of lettuce irrigated with unchlorinated wastewater was assessed, using a quantitative risk analysis model, which showed that the risk of infection was less than 1 per 10,000 consumers (Petterson et al., 2001).

The common rinsing of vegetables with tap water does not reduce the indicator organisms to safe levels (Rosas et al., 1984) (Table 23.3). Outbreaks of diseases like cholera have been associated with wastewater irrigation of vegetables. Infections and disease outbreaks caused by parasites can also be linked to this practise. An epidemiological study was carried out in Beni Mellal, Morocco, an area where agricultural land is irrigated with raw wastewater. The study showed that 20.5 percent of children living in the area were infected with *Ascaris*, as compared with 3.8 percent in a control population (Habbari et al., 2000).

A coliform level of 2.2/100 mL (7 day median) is allowed for food crops in California. The U.S. Standard for irrigation of nonedible crops (e.g., seed and fiber crops) is a coliform level of 5000 per 100 mL. Access to the public is restricted by posting of warning signs, and a buffer zone is required if spray irrigation is conducted at the site. A level of 23 coliforms per 100 mL has been adopted for irrigation of pastures used for milk animals and for recreational use (e.g., golf courses). In Florida, a level of 23 total coliforms per 100 mL is allowed for most agricultural uses. Health protection measures to interrupt potential transmission routes for pathogens and parasites when applying wastes to land are shown in Figure 23.2 (Mara and Cairncross, 1989).

TABLE 23.2. Contamination Levels of *Giardia* Cysts Found During Field Tests on Various Crop Plants Irrigated with Raw Wastewater in Marrakech, Morocco[a]

Crop	% Contamination	Mean Number Cysts/kg
Coriander	44.4	2.5×10^2
Carrots	33.3	1.5×10^2
Mint	50	96
Radish	83.3	59.1

[a]Adapted from Amahmid et al. (1999).

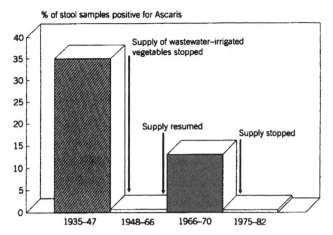

% of stool samples positive for Ascaris

Figure 23.1 Relationship between *Ascaris*-positive stool samples and supply of wastewater-irrigated vegetables in Jerusalem (1935–1982). From Gunnerson et al. (1984). (Courtesy of the American Water Works Association.)

23.2.2 Groundwater Recharge

Reclaimed water can be used for groundwater recharge to augment groundwater supplies, and to prevent saltwater intrusion in coastal areas. Recharge is carried out by surface spreading (recharge water percolates from spreading basins through the vadose zone) or by direct injection (reclaimed water is pumped directly into the groundwater zone, generally into a well-confined aquifer (Asano and Cotruvo, 2004). The reclaimed water must be of drinking water quality, requiring biological, chemical, and physical treatments. This practice raises concerns over the microbiological, chemical, and radiological quality of the water. An example is the Water Factory 21 reclamation facility in Orange County, California. The advanced treatment train for wastewater reclamation includes lime treatment, recarbonation, filtration, activated carbon adsorption, reverse osmosis, and final chlorination (Asano, 1985; Metcalf and Eddy, 1991). The treated effluent is directly injected into groundwater. A final chlorine residual is necessary to avoid bioclogging of the recharge well and aquifer (Bouwer, 1992). Inactivation and removal of pathogens in soils and groundwater are discussed in Chapter 21.

TABLE 23.3.   Effect of Vegetable Rinsing on Coliform Levels[a]

	Geometric Mean[b] of the Following Samples of the Following Types of Bacteria			
	Rinsed		Unrinsed	
Crop	TC	FC	TC	FC
Celery	300	30	1,300	300
Spinach	2,400	1,700	8,700	2,400
Lettuce	700	570	37,000	3,600
Parsley	370	300	3,100	660
Radish	650	300	2,600	360

[a]From Rosas et al. (1984).
[b]Most probable number per 100 g.
TC, total coliform; FC, fecal coliform.

MEASURES TO INTERUPT POTENTIAL TRANSMISSION
ROUTES

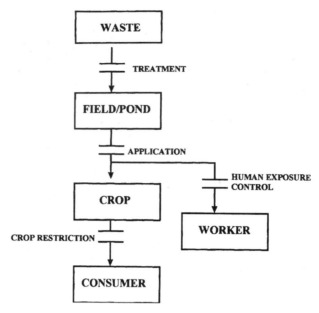

Figure 23.2   Measures to interrupt potential transmission routes for pathogens and parasites during waste application to land. From Mara and Cairncross (1989). (With permission of the publisher.)

23.2.3   Recreational Use

In arid areas, treated wastewater is used to fill recreational lakes (e.g., boating, fishing, and water sports). One of the best examples of use of reclaimed wastewater for a recreational purpose is Santee, California, where lakes receiving treated wastewater are used for boating, fishing, and even swimming. Another example of recreational use of reclaimed wastewater is South Lake Tahoe, California.

People may come into contact with potential pathogens when engaging in water sports. For a discussion of the potential public health aspects of recreational activities, see Chapter 22.

23.2.4   Urban Nonpotable Use

The category of urban nonpotable use includes the use of reclaimed wastewater for private lawns, irrigation of parks, schoolyards, golf courses, highway medians and shoulders, commercial uses (e.g., air conditioning, vehicle washing), industrial uses (e.g., process water, makeup water for evaporative cooling towers), construction projects, fire protection, and toilet flushing (Okun, 1997). Some states, such as California, have guidelines for irrigation of parks, playgrounds, schoolyards, and other public access areas. Some states (e.g., Georgia) have less stringent guidelines, while others do not allow such uses.

The city of Colorado Springs, Colorado, uses a tertiary effluent to spray-irrigate city parks. The tertiary treatment consists of activated sludge treatment followed by dual-media (sand and anthracite) gravity filtration, and chlorination to maintain a residual chlorine of 4–6 mg/L (Schwebach et al., 1988). Fecal coliform densities were 91 percent of the time below 23 per 100 mL, and 99 percent of the time below 500 per 100 mL.

A two-year prospective epidemiological study did not show any significant difference in reported gastrointestinal illnesses rates between visitors of parks irrigated with wastewater effluent and those of parks irrigated with potable water. However, a significant increase in rates was observed when the fecal coliform level of the reclaimed wastewater was above 500 per 100 mL. This study essentially showed that the city standard of 200 fecal coliforms per 100 mL was adequate for protecting public health.

Owing to the booming population in Florida and water shortages, several reuse projects are considering reclaimed water for beneficial nonpotable usage. Some of these project are described in the following sections (http://www.dep.state.fl.us/water/reuse/facts.htm).

23.2.4.1 St. Petersburg Dual Distribution System.

A *dual distribution system* consists of providing two water supply systems. One system provides water of very good microbiological and chemical quality for drinking, cooking, and washing (this category represents only 2 percent of the total amount of water used in households). The second system provides water of lower quality (reclaimed wastewater) for other household uses such as lawn irrigation or toilet flushing. The systems are color-coded in order to avoid errors. In the United States, the first dual distribution system was built in Arizona during the 1920s. A dual water supply system began operation during the 1970s in St. Petersburg, Florida. Although the major demand was initially for landscape irrigation, the system now includes industrial and commercial uses. Other dual distribution systems were installed in Irvine and San Jose, California (Okun, 1997) and other communities across the United States. Little is known about potential adverse health effects of dual distribution systems on urban populations. However, stringent water quality guidelines were adopted by the U.S. Environmental Protection Agency (EPA) for unrestricted nonpotable reuse of reclaimed water. These guidelines (Table 23.4) should not adversely affect the health of urban populations. In St. Petersburg, it was reported that sensitive ornamental plants may be adversely affected if the reclaimed wastewater contains high concentrations of chloride (>600 mg/L) (Berger, 1982; Johnson, 1991; Okun, 1997).

23.2.4.2 Conserv II Project.

The aim of this project is the use of reclaimed water in Orange County and Orlando for citrus groves, nurseries, tree farms, and golf courses.

TABLE 23.4. **Suggested U.S. EPA Treatment and Water Quality Guidelines for Unrestricted Nonpotable Reuse**[a]

Parameter	Guidelines
Treatment	Biological treatment (BOD and suspended solids ≤ 30 mg/L)
	Filtration
	Disinfection
Water quality	pH $= 6$–9
	$BOD_5 \leq 10$ mg/L
	Turbidity: 2 NTU (24 h average)
	Fecal coliforms: nondetected in 100 mL
	Chlorine residual: ≥ 1 mg/L after 30 min
	≥ 0.5 mg/L in distribution systems

[a]Adapted from U.S. EPA (1992).

BOD, biochemical oxygen demand; NTU, nephelometric turbidity unit.

23.2.4.3 *Tallahassee Spray Irrigation System.* Reclaimed water is used for irrigation of agricultural crops (corn, soybeans, coastal Bermuda grass, and other feed and fodder crops).

23.2.4.4 *Gainesville Project.* Reclaimed water from the Kanapaha treatment plant in Gainesville, Florida, is used to irrigate residential lawns, golf courses, parks, and other landscaped areas; some of it is used to recharge the Floridan aquifer.

23.2.5 Direct Potable Reuse of Reclaimed Wastewater as a Domestic Water Supply

This category includes the intentional reuse of wastewater to augment the potable water supply of a given community.

Indirect potable reuse of wastewater has been often practiced by many communities the world over. This occurs when the treated wastewater effluent of one community becomes the drinking water supply for another community downstream (e.g., Cincinnati, Ohio, which uses water from communities upstream along the Ohio River) (Dean and Lund, 1981; Donovan et al., 1980; Hammer, 1986). A planned indirect potable reuse is being considered for the city of West Palm Beach, Florida, to augment its potable water supplies. The reclaimed water is produced by tertiary treatment, followed by discharge into a constructed wetland before being allowed to recharge the groundwater aquifer. A pilot study showed that the treated water meets the primary and secondary drinking water standards (Garrigues et al., 1998).

Direct potable reuse is considered in arid zones in response to severe water shortages. Some examples of direct potable reuse of reclaimed wastewater are given in the following.

23.2.5.1 *Windhoek, Namibia.* Treated wastewater is intentionally added to the drinking water supply in Windhoek, Namibia, in southwest Africa. Biologically treated effluents are subjected to tertiary treatment, which comprises lime treatment, ammonia stripping, sand filtration, breakpoint chlorination, activated carbon, and final chlorination.

A long-term epidemiological study showed no adverse health effects associated with the consumption of directly reclaimed drinking water. Heterotrophic plate count (<100 per mL), total coliforms (0 per 100 mL) and coliphage (0 per 100 mL) were proposed for the routine monitoring of the quality of reclaimed wastewater (Grabow, 1990; Isaacson et al., 1986).

23.2.5.2 *The Denver Potable Water Reuse Demonstration Project.* The city of Denver, Colorado, situated in a semi-arid area, expects an increased demand for potable water in the next decades. The city's Water Department built a one-MGD (3.8 million L/day) water reuse plant to examine the feasibility of treating wastewater plant effluents to potable quality on a continuous basis. The plant provides multiple barriers against contaminants such as bacteria, viruses, protozoa, heavy metals, and trace organics (Lauer, 1991; Rogers and Lauer, 1992). The treatment train, illustrated in Figure 23.3 (Lauer et al., 1985), consists essentially of the following steps: lime clarification, recarbonation (use of CO_2 as a neutralizing agent), filtration, selective ion exchange for ammonia removal, two carbon adsorption steps separated by ozonation, reverse osmosis, air stripping, and final disinfection with chlorine dioxide (Lauer et al., 1985; Rogers and Lauer, 1986). The plant effectively removes total organic carbon (TOC) (Fig. 23.4) and total coliforms (Fig. 23.5) (Lauer et al., 1985). As to other indicators (fecal coliforms, fecal streptococci, heterotrophic plate count bacteria, and coliphage), the product water

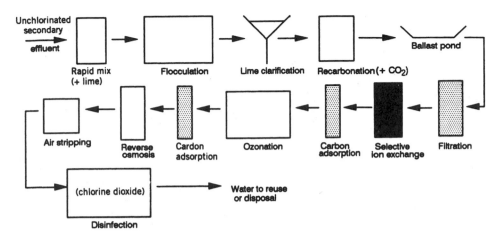

Figure 23.3 Denver's potable water reuse project. Water treatment train. Adapted from Lauer et al. (1985).

was of quality equal or better than that of the Denver drinking water. No coliphage, fecal coliform, or fecal streptococci were detected in the finished water (Arber, 1983; Rogers and Lauer, 1986). The treatment train results in an approximately 7 to 8–log decrease in total coliform counts (Fig. 23.5; Lauer et al., 1985).

23.2.5.3 San Diego Project. The city of San Diego, California, is studying the potential use of reclaimed wastewater to supplement the raw water source of potable supply (Cooper, 1991). The plan is to build a reclamation plant with a capacity of 450,000 m^3/day (120 MGD) to reclaim and beneficially reuse 86 million m^3 of reclaimed water per year by 2010 (Bayley et al., 1992). The treatment train includes primary and biological treatments followed by tertiary treatment that consists of coagulation, sand filtration, reverse osmosis, and activated carbon. The chemical and microbiological quality

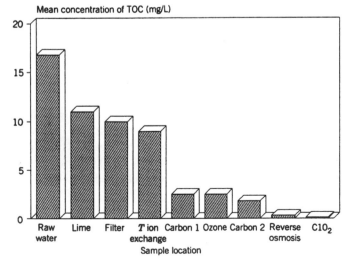

Figure 23.4 Denver's potable water reuse project. TOC removal. From Lauer et al. (1985). (Courtesy of the American Water Works Association.)

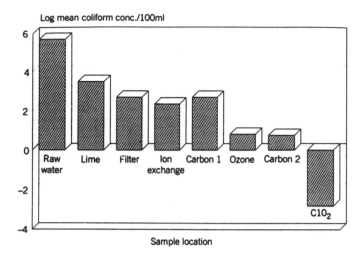

Figure 23.5 Denver's potable water reuse project. Removal of total coliforms. From Lauer et al. (1985). (Courtesy of the American Water Works Association.)

of the reclaimed wastewater was found to be equal to, or better than, that of the city's present raw water supply (Cooper, 1991).

The use of reclaimed water as a source of drinking water still raises many questions as regards the public health risks posed by some microbial pathogens as well as chemicals in the reclaimed water. Current water treatment processes can produce water that meets federal drinking water standards, but are these standards adequate regarding highly chlorine-resistant *Cryptosporidium* or the emerging pathogens such as *Cyclospora cayetanensis*, microsporidia, or *Helicobacter pylori*? Many questions remain concerning the safety of using reclaimed water to augment conventional drinking water sources, and more research is warranted in this area (Crook, 1998; NRC, 1998).

23.2.5.4 Tampa, FL. Several Florida projects involve the use of reclaimed water for nonpotable uses. However, the city of Tampa is considering an indirect potable reuse following a treatment train that includes lime–high-pH treatment, sand treatment, adsorption to activated carbon, and ozonation. Pilot-scale studies showed that the chemical treatment removed about 2–logs of *Cryptosporidium* and 6–logs of phage MS2 (Rose et al., 1999).

23.2.6 Industrial Use

Industry accounts for the highest percentage of water use. Reclaimed wastewater is used by industry mainly as cooling water for power plants. Other uses include boiler feed, pulp and paper production, washing, manufacturing, and use by mineral industries. For example, in Israel, a municipal wastewater effluent, subjected to lime treatment followed by ammonia stripping and pH adjustment, is used as a make-up water to a cooling tower serving a refinery and petrochemical complex (Rebuhn and Engel, 1988).

23.2.7 Wetlands for Wastewater Renovation

Wetlands constitute a low-cost and low-energy alternative to traditional tertiary treatment. They remove BOD and nutrients (N, P) through biological uptake by plants and microbial action. The treatment efficiency of wetlands is controlled by hydraulic loading, water depth, and extent of coverage by aquatic plants (DeBoer and Lindstedt, 1985).

Constructed wetlands (e.g., subsurface-flow wetlands) offer a means to treat wastewater using natural processes, thus improving water quality. Their construction and operation are relatively simple and they offer a promising technology for developing countries. They consist of shallow excavated basins with an inlet for wastewater input and an outlet to a receiving water. The excavated soil is seeded or planted with aquatic plants. Constructed wetlands include free water surface (FWS) and subsurface flow (SSF) systems. The FWS wetlands (Fig. 23.6; Kadlec, 2002) are continuously flooded with a water depth up to 0.5 m and are planted with emergent macrophytes (e.g., *Phragmites*, *Typha*). Later, they become colonized with floating plants and microscopic algae. The SSF wetlands consist of a saturated porous medium (e.g., sand, gravel) and emergent macrophytes (Fig. 23.7; Kadlec, 2002). Treatment is carried out by the biofilms developing on the porous medium.

23.2.7.1   *Nutrient Cycling in Wetlands.*   Wetland microorganisms are involved in the biodegradation of a wide range of organic compounds originating from internal biomass recycling and wastewater inputs. Wetland microorganisms are active in aerobic, anoxic, and anaerobic environments. In the sediments, as the oxido-reduction potential decreases with depth, the succession of microbial groups with increasing depth is as follows: denitrification $\rightarrow$ iron reduction $\rightarrow$ sulfate reduction $\rightarrow$ methanogenesis (Fig. 23.8; Kadlec, 2002). Since wetlands are major contributors of CH_4 to the atmosphere, methanogens play an important role in BOD reduction in wastewater effluents.

The microbial processes involved in this cycle have been discussed in Chapter 3. Wetlands offer favorable conditions (ample supply of carbon, presence of anaerobic zones near the sediments) for denitrification, which helps in nitrogen removal from wastewater effluents. Oxido-reduction processes drive the cycling of sulfur, iron, and manganese in wetlands (see Chapter 3).

23.2.7.2   *Pathogen and Parasite Removal in Wetlands.*   The SSF constructed wetlands allow the removal of 98 to 99 percent of indicator bacteria (total and fecal

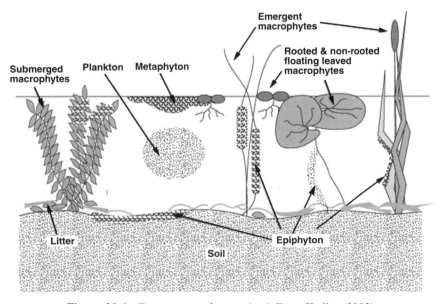

Figure 23.6   Free water surface wetland. From Kadlec (2002).

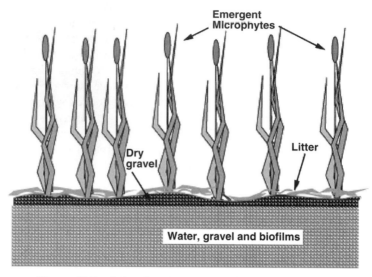

Figure 23.7   Subsurface flow wetland. From Kadlec (2002).

coliforms) and 95 percent of phage, whereas the FWS wetlands removed 95 and 93 percent of total and fecal coliforms, respectively (Gerba et al., 1999). An average removal efficiency of 85 to 94 percent was reported for fecal coliforms and fecal streptococci in a surface flow wetland planted with *Typha* (reeds) (Perkins and Hunter, 2000). Protozoa play an important role in bacterial removal in SSF wetlands. The grazing rates of protozoa on *E. coli* were estimated at 49 bacteria/ciliate/hour in planted wetlands and 9.5 bacteria/ciliate/hour in nonplanted ones (Decamp et al., 1999). Table 23.5 shows that the removal of bacterial indicators and pathogens may vary from <1–log to 3–logs (Kadlec, 2002).

Treatment wetlands perform relatively well as regards the removal of viruses. A summary of studies carried out worldwide shows that wetlands remove from 1 to 3–logs human enteric viruses or phages (Table 23.6; Kadlec, 2002).

A wide range of protozoan species are present in wetlands and play a key role in bacterial removal via grazing. As seen in Chapter 4, some of the protozoa are parasitic and form cysts and oocysts that are quite resistant to environmental stresses and to disinfection. The main mechanisms involved in cyst and oocyst removal by wetlands are sedimentation and/or adsorption to particulates followed by sedimentation. In constructed wetlands, *Giardia* cyst and *Cryptosporidium* oocyst levels were one to three orders of magnitude higher in the sediments compared to the water column (Karim et al., 2004).

Parasite eggs, due to their higher density, are effectively removed by sedimentation into wetland sediments.

23.2.8   Aquaculture for Wastewater Renovation

Aquaculture also provides a means for renovating wastewater while allowing the growth of aquatic plants such as water hyacinths (Reddy, 1984) or raising fish for human consumption (Buras, 1984).

23.2.9   Use of Attached Algae for Polishing of Wastewater Effluents

We have seen that algae are major partners in wastewater treatment in oxidation ponds, especially in the removal of nutrients such as N and P (see Chapter 11). However, a

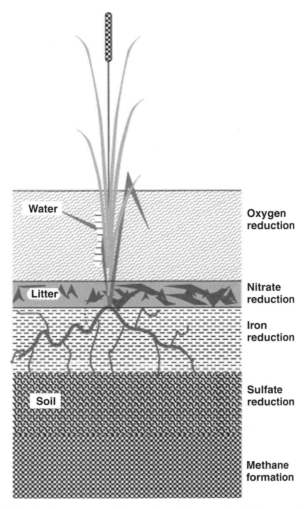

Figure 23.8 Vertical of a free water surface wetland (redox potentials range from 300 mV in the water column to −300 mV in the lower anaerobic zones). From Kadlec (2002).

TABLE 23.5.   Removal of Bacterial Pathogens and Indicators in Constructed Wetlands[a]

Pathogen/Indicator	Reduction ($\log_{10}$)
Free Water Surface Wetlands (FSW)	
E. coli	0.2–0.9
Fecal streptococci	0.7–2.2
Clostridium perfringens	2.1–3.1
Salmonella spp.	1.3–2.4
Pseudomonas aeruginosa	0.8–2.1
Yersinia enterocolitica	1.1–1.3
Subsurface Flow Wetlands (SSF)	
E. coli	0.5–0.7
Fecal streptococci	1.5
Salmonella spp.	2.1–2.4
Vibrio cholerae	3.1

[a]Adapted from Kadlec (2002).

TABLE 23.6. Virus and Bacteriophage Removal in Wetlands[a]

Location	Vegetation	Virus	% Virus Removal
Subsurface flow			
Santee, CA	Bullrush	Phage MS2 (seeded)	>99
	Bullrush	Poliovirus type 1	>99.9
	Not planted	F-specific phage	94.5
	Bullrush	F-specific phage	99
Duluth, MN	Bullrush	Somatic coliphage	90 (winter)
	Bullrush	Somatic coliphage	98.4 (summer)
Abu Atwa (Egypt)	*Phragmites*	Coliphage	99.7
Budds Farms (UK)	*Phragmites*	Coliphage	97.3
	Phragmites	Enteroviruses	99.3
Ettenbuttel (Germany)	*Phragmites*	Coliphage	96.7
Wiedersburg (Germany)	*Phragmites*	Coliphage	99.9
Free water surface			
Houghton Lake, MI	*Typha, Carex*	Reovirus, poliovirus type 1	90
Waldo, FL	Cypress new	Coliphage	85.5
Waldo, FL	Cypress old	Coliphage	99.8
Duplin County, NC	*Typha, Sparganium*	F-specific phage	81.2–89.4
		Somatic coliphage	84–92
Oxelösunf (Sweden)	*Typha*	Coliphage	94.7
Glendale, AZ	*Typha, Scirpus*	Coliphage	95

[a]Adapted from Kadlec (2002).

problem associated with the use of suspended algae is the separation of algal cells from the treated effluent. One of the solutions to this problem is the use of attached algae (i.e., periphyton) to treat wastewater. An example of algal biofilm systems is the Algal Turf Scrubber (ATS) for the removal of nutrients from wastewater (Adey and Loveland, 1998).

An experimental algal biofilm process was evaluated for the polishing of an activated sludge effluent (Schumacher and Sekoulov, 2002). The algal biofilm obtained mainly consisted of green algae (*Ulothrix, Stigeoclonium, Chlamydomonas*) and cyanobacteria (*Oscillatoria*). As seen for suspended algae in oxidation ponds (see Chapter 11) the algal biofilm activity led to an increase of dissolved oxygen and pH and the reduction of N, P, and fecal coliforms.

23.3 THE U.S. EXPERIENCE IN WASTEWATER REUSE

Most wastewater reuse projects are located in the arid and semi-arid areas of western and southwestern United States. The state of California uses approximately 1 million m^3 of reclaimed wastewater/day (Asano and Tchbanoglous, 1991; Ongerth and Jopling, 1977). This state early recognized wastewater as a valuable resource, actively promoted the use of reclaimed wastewater for irrigation and other uses (Crook, 1985; Metcalf and Eddy, 1991) (Table 23.7), and has the highest number of water reuse projects. Southern California imports water from Northern California and the Colorado River and has thus promoted several water reclamation projects (Nichols, 1988). The leading projects are Water Factory 21 and Irvine Ranch Water District. Water Factory 21 produces a highly polished tertiary effluent that meets California potable water standards. This effluent is mixed with deep-well groundwater and is used for groundwater recharge. The Irvine

TABLE 23.7. **State of California Wastewater Reclamation Criteria**[a]

Use of Reclaimed Wastewater	Description of Minimum Treatment Requirements			
	Primary	Secondary and Disinfected	Secondary Coagulated, Filtered, and Disinfected	Coliform MPN/100 mL Median (Daily Sampling)
Irrigation				
Fodder crops	X			No requirement
Fiber	X			No requirement
Seed crops	X			No requirement
Produce eaten raw, surface-irrigated		X		2.2
Produce eaten raw, spray-irrigated			X	2.2
Processed produce, surface-irrigated	X			No requirement
Processed produce, spray-irrigated		X		23
Landscapes: golf course cemeteries, freeways		X		23
Landscapes: parks, playgrounds, schoolyards			X	2.2
Recreational impoundments				
No public contact		X		23
Boating and fishing only		X		2.2
Body access (bathing)			X	2.2

[a]Adapted from Metcalf and Eddy (1991).

Ranch Water District uses reclaimed water in a dual distribution system for office buildings.

In southern California, approximately 27 percent of the effluents from reclamation plants are reused. About two-thirds of the reused water is intended for groundwater recharge by surface spreading or deep-well injection. The public health aspects of the use of reclaimed water for groundwater recharge at Whittier Narrows, California, have been evaluated. Epidemiological studies did not show evidence of any measurable health effects associated with the consumption of the water in the study area (Nellor et al., 1985).

The California Department of Health Services has established criteria to address public health concerns over wastewater reuse. These criteria specify the level of wastewater treatment and coliform levels for various types of uses of reclaimed wastewater (crop and landscape irrigation, impoundments, and groundwater recharge) (Table 23.7). Tertiary treatment (oxidation, coagulation, clarification, filtration, and disinfection) is required for water used for spray or surface irrigation of food crops that are eaten raw. The tertiary effluent must be essentially pathogen free. The guidelines establishing the treatment and bacteriological quality of reclaimed water call for 2.2 total coliforms per 100 mL (with an upper limit of 23 TC per 100 mL in 10 percent of samples) and a turbidity standard of 2 NTU. These guidelines apply to uses such as irrigation of food crops, recreational uses permitting unrestricted body contact, and irrigation of parks, playgrounds, and schoolyards.

According to some investigators, the California standards are too strict and are not based on sound epidemiological evidence (Shuval, 1991). It was argued that these standards could hardly be achieved by most of the U.S. wastewater treatment plants. More liberal guidelines were proposed by public health experts meeting in Engelberg, Switzerland, in 1985 (International Reference Centre for Waste Disposal, 1985; World Health Organization (WHO), 1989) (Table 23.8). For example, the new proposed fecal coliform guideline for wastewater used for irrigation of crops likely to be eaten uncooked is ≤ 1000 fecal coliforms/100 mL instead of ≤ 2.2 coliforms/100 mL. The wastewater effluent should also contain less than 1 nematode egg per liter. Proposed national standards in many countries (e.g., Cyprus, France, Jordan, Tunisia) are based on the WHO guidelines on fecal coliforms and nematodes levels in wastewater effluents (Hespanhol and Prost, 1994).

Water reuse in Arizona is used in 180 plants treating approximately 200 MGD (Rose, 1986; Rose et al., 1989c). The State established a compliance program for monitoring viruses, *Giardia*, and fecal coliforms in reused water (Anonymous, 1984). Arizona is the only state in the United States that adopted standards for enteric viruses. The standards specify that virus levels should not exceed 1 PFU/40 L for reclaimed water used for spray irrigation of food eaten raw or for unrestricted-access water sports. For irrigated landscape areas and golf courses with full access to the public, the virus level should not exceed 125 PFU/40 L. With regard to *Giardia*, none should be detected in 40 L of water. Virus monitoring for activated sludge and oxidation pond effluents showed that about 60 percent of the samples met the compliance standard of 1 PFU/40 L. Furthermore, 97 percent of sand-filtered activated sludge effluents met the virus standard and two-thirds of these samples met the *Giardia* standard (Arizona, 1984; Rose and Gerba, 1991; Rose et al., 1989a, b).

In Florida, reuse of reclaimed water has increased from 362 MGD to 826 MGD and the trend is expected to continue because of the expected increase of the state population to 20 million by the year 2020. Approximately 40 percent of the reclaimed water is used to irrigate public access areas, including parks, golf courses, schools, and residential areas (York and Wadsworth, 1998). The State of Florida requires that reclaimed water must receive secondary treatment followed by sand filtration and "high-level disinfection," which is achieved by meeting performance criteria regarding total suspended solids and fecal coliforms. The reclaimed water must meet a limit of 5 mg/L total suspended solids, a chlorine residual of 1 mg/L after 30 min contact time, and no detectable fecal coliforms/100 mL (no sample should exceed 25 fecal coliforms/100 mL). These criteria would probably lead to a virus-free effluent but would not guarantee the complete removal of protozoan parasites such as *Cryptosporidium* and *Giardia*, necessitating their monitoring in reclaimed water (York and Burg, 1998).

The wastewater treatment industry is constantly facing new challenges, exemplified by the presence in reclaimed wastewater of trace contaminants such as xenobiotics, pharmaceuticals, or endocrine disruptors (see Chapter 19). So far, we know that these trace contaminants are detected in effluents, but we know very little about their long-term health effects. Furthermore, the problem must be constantly reassessed as our detection methodology becomes more sophisticated. The industry must also take into account the cost of any new treatment technology (e.g., ultrafiltration, nanofiltration, or reverse osmosis).

Finally, the inclusion of life-cycle analysis in product development should result in environmentally friendly chemicals and, thus, help prevent the release of hazardous contaminants into the waste stream (Levine and Asano, 2004).

Presently, the trend is a move towards decentralized treatment systems, which would promote water reuse by offering an economic advantage (e.g., no major distribution pipe network, treated water used locally) and by posing a lower risk of waterborne pathogen infections (Crites and Tchobanoglous, 1998; Fane et al., 2002).

TABLE 23.8. WHO Recommended Microbiological Quality Guidelines for Wastewater Use in Agriculture[a]

Category	Reuse Conditions	Exposed Group	Intestinal Nematodes (Arithmetic Mean no. of Eggs per Liter)	Fecal Coliforms (Geometric Mean no. per 100 mL)	Wastewater Treatment Expected to Achieve the Required Microbiological Quality
A	Irrigation of crops likely to be eaten uncooked, sports fields, public parks	Workers, consumers, public	≤1	≤1000	A series of stabilization ponds designed to achieve the microbiological quality indicated, or equivalent treatment
B	Irrigation of cereal crops, industrial crops, fodder crops, pasture, and trees	Workers	≤1	No standard recommended	Retention in stabilization ponds for 8–10 days or equivalent helminth and fecal coliform removal
C	Localized irrigation crops in category B if exposure of workers and the public does not occur	None	Not applicable	Not applicable	Pretreatment as required by the irrigation technology but not less than primary sedimentation

[a]From World Health Organization (1989).

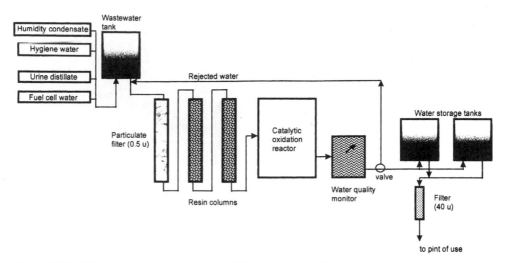

Figure 23.9 Wastewater treatment on board the International Space Station. Adapted from Hurst et al. (1998).

23.4 WATER REUSE IN SPACE

Drinking water is an expensive commodity in space exploration. It was estimated that it would cost more than US $8000 to send one gallon of water into space (Nicks, 1986). Therefore, wastewater recycling is a necessity under these conditions. NASA engineers and scientists are investigating means of recycling water from wastewater generated by astronauts working in space stations. The wastewater contains human fecal wastes, urine, wash water, and humidity condensate (Wachinski, 1988). However, this is no easy task, since processes for waste collection and wastewater and water treatment must be designed to function in a zero-gravity environment. Some processes considered for waste treatment in space include dry incineration at 600°C, wet oxidation at 230–290°C at pressures of 70–150 bars, and supercritical water oxidation at 374°C and 215 bars (Wachinski, 1988). On board the International Space Station, wastewater that will be reclaimed is composed of urine distillate, wastewater from personal hygiene activities (e.g., showering, handwashing), and moisture from perspiration, respiration, ovens, and showers. It was proposed that wastewater should be reclaimed by a treatment train that principally includes the following (Fig. 23.9): coarse and fine filters to remove particles, ion-exchange resins to remove organic and inorganic contaminants, and a high-temperature catalytic oxidation reactor to oxidize small organic contaminants. The water is iodinated to prevent microbial growth in the storage tanks. This treatment train was found to be quite efficient, as it reduces bacterial and phage levels by several orders of magnitude (Hurst et al., 1997).

23.5 WEB RESOURCES

http://www.italocorotondo.it/tequila/ (water use in the agricultural sector; European online course; go to module 5)

http://ohioline.osu.edu/b860/index.html (use of reclaimed wastewater through irrigation; Ohio State University)

http://www.onsiteconsortium.org/files/wastereuse.pdf

http://www.dep.state.fl.us/water/reuse/facts.htm (water reuse in Florida)
http://www.epa.gov/owow/wetlands/ (EPA web page for wetlands)
http://www.epa.gov/owow/wetlands/construc/content.html (constructed wetlands: EPA site)
http://www.extension.umn.edu/distribution/naturalresources/DD7671.html (constructed wetlands; University of Minnesota)
http://www.hydromentia.com/ats.html (algal turf scrubber)

23.6 QUESTIONS AND PROBLEMS

1. What are the main concerns about wastewater reuse?
2. Give the different categories of wastewater reuse. Which one of these categories is meeting some reluctance from consumers?
3. Discuss the practice of groundwater recharge with reclaimed wastewater (consult the most recent research papers).
4. Update the State and Federal regulations concerning wastewater reuse in California and Florida (consult State documents and Federal Register).
5. What is a dual distribution system?
6. Compare indirect to direct potable reuse of wastewater.
7. Discuss the role of wetlands in BOD and nutrient (N, P) removal (consult the literature).
8. What are the advantages of the Algal Turf Scrubber (ATS) over suspended algae for treating wastewater effluents?
9. Discuss the latest developments and the economical considerations for water reuse in space (consult the literature).

23.7 FURTHER READING

Asano, T., and G. Tchobanoglous. 1991. The role of wastewater reclamation and reuse in the USA. Water Sci. Technol. 23: 2049–2059.

Asano, T., and J.A. Cotruvo. 2004. Groundwater recharge with reclaimed municipal wastewater: health and regulatory considerations. Water Res. 38: 1941–1951.

AWWA-WEF. 1998. *Water Reuse*, Conference Proc. Feb. 1–4, 1998, Lake Buena Vista, FL.

Cooper, R.C. 1991. Public health concerns in wastewater reuse. Water Sci. Technol. 24: 55–65.

Grabow, W.O.K. 1990. Microbiology of drinking water treatment: Reclaimed wastewater, pp. 185–203, In: *Drinking Water Microbiology*, G.A. McFeters, Ed., Springer-Verlag, New York.

Levine, A.D., and T. Asano. 2004. Recovering sustainable water from wastewater. Environ. Sci. Technol. 38: 201A–208A.

National Research Council. 1998. *Issues in Potable Reuse: The Viability of Augmenting Drinking Water Supplies with Reclaimed Water*, National Academy Press, Washington, D.C.

Okun, D.A. 1997. Distributing reclaimed water through dual systems. J. Amer. Works Assoc. 89: 52–64.

Rose, J.B. 1986. Microbial aspects of wastewater reuse for irrigation. Crit. Rev. Env. Control 16: 231–256.

Rose, J.B., L.J. Dickson, S.R. Farrah, and R.P. Carnahan. 1996. Removal of pathogenic and indicator microorganisms by a full-scale water reclamation facility. Water Res. 30: 2785–2797.

Shuval, H.I., Ed. 1977. *Water Renovation and Reuse*, Academic Press, San Diego.

REFERENCES

Abad, F.X., R.M. Pinto, and A. Bosch. 1994. Survival of enteric viruses on environmental fomites. Appl. Environ. Microbiol. 60: 3704–3710.

Abad, F.X., R.M. Pinto, C. Villena, R. Gajardo, and A. Bosch. 1997. Astrovirus survival in drinking water. Appl. Environ. Microbiol. 63: 3119–3122.

Abbaszadegan, M., C.P. Gerba, and J.B. Rose. 1991. Detection of *Giardia* cysts with a cDNA probe and applications to water samples. Appl. Environ. Microbiol. 57: 927–931.

Abbaszadegan, M., M.S. Huber, C.P. Gerba, and I.L. Pepper. 1993. Detection of enteroviruses in groundwater with the polymerase chain reaction. Appl. Environ. Microbiol. 59: 1318–1324.

Abbaszadegan, M., M.S. Huber, C.P. Gerba, and I.L. Pepper. 1997. Detection of viable *Giardia* cysts by amplification of heat shock-induced mRNA. Appl. Environ. Microbiol. 63: 324–328.

Abbaszadegan, M., M. Lechevallier, and C.P. Gerba. 2003. Occurrence of viruses in U.S ground-waters. J. Amer. Water Works Assoc. 95: 107–120.

Abraham, J.V., R.D. Butler, and D.C. Sigee. 1997. Ciliate populations and metals in an activated sludge plant. Water Res. 31: 1103–1111.

Abram, J.W., and D.B. Nedwell. 1978. Hydrogen as a substrate for methanogenesis and sulphate reduction in anaerobic salt marsh sediments. Arch. Microbiol. 117: 93–97.

Abu-Ghararah, Z.H. and C.W. Randall. 1990. The effect of organic compounds on biological phosphorus removal. Water Sci. Technol. 23: 585–594.

Acher, A., and B.I. Juven. 1977. Destruction of fecal coliforms in sewage water by dye-sensitized photooxidation. Appl. Environ. Microbiol. 33: 1019–1023.

Acher, A., E. Fischer, R. Zellingher, and Y. Manor. 1990. Photochemical disinfection of effluents: Pilot plant studies. Water Res. 24: 837–843.

Acher, A., E. Fisher, and Y. Manor. 1994. Sunlight disinfection of domestic effluents for agricultural use. Water Res. 28: 1153–1160.

Acher, A., E. Fischer, R. Turnheim, and Y. Manor. 1997. Ecologically friendly wastewater disinfection techniques. Water Res. 31: 1398–1404.

Adal, K.A., C.R. Sterling, and R.L. Guerrant. 1995. *Cryptosporidium* and related species, pp. 1107–1128, In: *Infections of the Gastrointestinal Tract*, M.J. Blaser, P.D. Smith, J.I. Ravdin, H.B. Greenberg, and R.L. Guerrant, Eds., Raven Press, New York.

Adams, B.L., T.C. Bates, and J.D. Oliver. 2003. Survival of *Helicobacter pylori* in a natural fresh-water environment. Appl. Environ. Microbiol. 69: 7462–7466.

Adams, M.H. 1959. *Bacteriophages*. Intersciences, New York.

Adams, M.H., and B.H. Park. 1956. An enzyme produced by a phage-bacterial system. Virology 2: 719–736.

Adcock, P.W., and C.P. Saint. 2001. Development of glucosidase agar for the confirmation of water-borne Enterococcus. Water Res. 35: 4243–4246.

Adey, W.H., and K. Loveland. 1998. *Dynamic Aquaria: Building Living Ecosystems*. Acad. Press.

Aftring, R.P., and B.F. Taylor. 1981. Aerobic and anaerobic catabolism of phtalate acid by a nitrate-respiring bacterium. Arch. Microbiol. 130: 101–104.

Agoustinos, M.T., S.N. Venter, and R. Kfir. 1992. Assessment of water quality problems due to microbial growth in drinking water distribution systems. IWPRC International Symposium, Washington, D.C., May 26–29, 1992.

Ahel, M., W. Giger, and M. Koch. 1994. Behaviour of alkylphenol polyethoxylate surfactants in the aquatic environment. I. Occurrence and transformation in sewage treatment. Water Res. 28: 1131–1142.

Ahlborg, U.G., and T.M. Thunberg. 1980. Chlorinated phenols: Occurrence, toxicity, metabolism, and environmental impact. Critical Rev. Toxicol. 7: 1–35.

Ahlstrom, S.B., and T. Lessel. 1986. Irradiation of municipal sludge for pathogen control, pp. 1–19, In: C.A. Sorber, Ed., *Control of Sludge Pathogens*, Water Poll. Control Fed., Washington, D.C.

Ahmed, A.U., and D.L. Sorensen. 1995. Kinetics of pathogen destruction during storage of dewatered biosolids. Water Environ. Res. 67: 143–150.

Ahmed, M., and D.D. Focht. 1973. Degradation of polychlorinated biphenyl by two species of *Achromobacter*. Can. J. Microbiol. 19: 47–52.

Ahrenhotz, I., M.G. Lorenz, and W. Wackernagel. 1994. A conditional suicide system in Escherichia coli based on the intracellular degradation of DNA. Appl. Environ. Microbiol. 60: 3746–3751.

Ahring, B.K., N. Christiansen, I. Mathrani, H.V. Hendriksen, A.J.L. Macario, E.C. De Macario. 1992. Introduction of a de novo bioremediation ability, aryl reductive dechlorination, into anaerobic granular sludge by inoculation of sludge with Desulfomonile tiedjei. Appl. Environ. Microbiol. 3677–3682.

Aieta, E.M., and J.D. Berg. 1986. A review of chlorine dioxide in drinking water treatment. J. Am. Water Works Assoc. 78: 62–72.

Aiking, H., K. Kok, H. van Heerikhuizen, and J. van't Riet. 1982. Adaptation to cadmium by *Klebsiella aerogenes* growing in continuous culture proceeds mainly via formation of cadmium sulfate. Appl. Environ. Microbiol. 44: 938–944.

Akin, E.W., W.F. Hill, G.B. Cline, and W.H. Benton. 1976. The loss of poliovirus 1 infectivity in marine waters. Water Res. 10: 59–63.

Al-Ani, M.Y., D.W. Hendricks, G.S. Logsdon, and C.P. Hibler. 1986. Removing *Giardia* cysts from low turbidity waters by rapid rate filtration. J. Am. Water Works Assoc. 78: 66–73.

Albert, M., E. Biziagos, J.M. Crance, R. Deloince, and L. Schwartzbrod. 1990. Detection des virus enteriques cultivables in vitro et de l'antigene du virus de l'hepatite A dans les boues primaires de stations d'epuration. J. Fr. Hydrol. 2: 275–283.

Alberts, B., D. Bray, J. Lewis, M. Raff, K. Roberts, and J.D. Watson. 1989. *Molecular Biology of the Cell*. Garland, New York.

Albertson, G.E., and P. Hendricks. 1992. Bulking and foaming organism control at Phoenix, AZ WWTP. Water Sci. Technol. 26: 461–472.

Albertson, J.S., and F.W. Oehme. 1994. Animal and plant toxins, pp. 263–283, In: *Basic Environmental Toxicology*, L.G. Cockerham and B.S. Shane, Eds., CRC Press, Boca Raton, FL.

Alcaid, E., and E. Garay. 1984. R-plasmid transfer in *Salmonella spp.* isolated from wastewater and sewage-contaminated surface waters. Appl. Environ. Microbiol. 48: 435–438.

Alexander, M. 1977. *Introduction to Soil Microbiology*. Wiley, New York, 2nd Ed.

Alexander, M. 1979. Role of cometabolism, pp. 67–75, In: *Microbial Degradation of Pollutants in Marine Environments*, A.L. Bourquin, and P.H. Pritchard, Eds., U.S. Environmental Protection Agency, Gulf Breeze, Fla.

Alexander, M. 1981. Biodegradation of chemicals of environmental concern. Science 211: 132–138.

Alexander, M. 1985. Biodegradation of organic chemicals. Environ. Sci. Tech. 19: 106–111.

Alhajjar, B.J., S.L. Stramer, D.O. Cliver, and J.M. Harkin. 1988. Transport modelling of biological tracers from septic systems. Water Res. 22: 907–915.

Allard, A., B. Albinsson, and G. Wadell. 1992. Detection of adenoviruses in stools from healthy persons and patients with diarrhea by two-step polymerase chain reaction. J. Med. Virol. 37: 149–157.

Alleman, J.E. 1988. Respiration-based evaluation of nitrification inhibition using enriched *Nitrosomonas* cultures, pp. 642–650, In: *Biotechnology for Degradation of Toxic Chemicals in Hazardous Wastes Sites*, Scholze, R.J, E.D. Smith, J.T. Bandy, Y.C. Yu, and J.V. Basilico, Eds., Noyes Dat Corp., Park Ridge, NJ.

Alleman, J.E., J.A. Veil, and J.T. Canaday. 1982. Scanning electron microscope evaluation of rotating biological contactor biofilm. Water Res. 16: 543–550.

Allen, M.J., R.H. Taylor, and E.E. Geldreich. 1980. The occurrence of microorganisms in water main encrustations. J. Amer. Water Works Assoc. 72: 614–626.

Allen, S.J. 1996. Types of adsorbent materials, pp. 59–97, In: *Use of Adsorbents for the Removal of Pollutants from Wastewater*, G. McKay, Ed., CRC Press, Boca Raton, FL.

Alsop, G.M., G.T. Waggy, and R.A. Conway. 1980. Bacterial growth inhibition test. J. Water Pollut. Control Fed. 52: 2452–2456.

Altekruse, S.F. M.L. Cohen, and D.L. Swerdlow. 1997. Emerging foodborne diseases. Emerging Infect. Dis. 3: 285–293.

Altmar, R.L., T.G. Metcalf, F.H. Neill, and M.K. Estes. 1993. Detection of enteric viruses in oysters by using the polymerase chain reaction. Appl. Environ. Microbiol. 59: 631–635.

Altmar, R.L., F.H. Neill, J.L. Romalde, F. LeGuyader, C.M. Woodley, T.G. Metcalf, and M.K. Estes. 1995. Detection of Norwalk virus and hepatitis A virus in shellfish tissues with the PCR. Appl. Environ. Microbiol. 61: 3014–3018.

Alvarez, A.J., M.P. Buttner, and L. Stetzenbach. 1995. PCR for bioaerosol monitoring: Sensitivity and environmental interference. Appl. Environ. Microbiol. 61: 3639–3644.

Alvarez, M.E., and R.T. O'Brien. 1982. Mechanism of inactivation of poliovirus by chlorine dioxide and iodine. Appl. Environ. Microbiol. 44: 1064–1071.

Amahmid, O., S. Asmama, and K. Bouhoum. 1999. The effect of wastewater reuse in irrigation on the contamination level of food crops by *Giardia* cysts and *Ascaris* eggs. Int. J. Food Microbiol. 49: 19–26.

Amann, R., and W. Ludwig. 2000. Ribosomal RNA-targeted nucleic acid probes for studies in microbial ecology. FEMS Microbiol. Rev. 24: 555–565.

Amer, A.A., 1997. Destruction of sludge pathogenic bacteria using quick lime and cement dust. Egypt. J. Soil Sci. 37: 343–354.

Andersen, A.A. 1958. New sampler for the collection, sizing and enumeration of viable airborne particles. J. Bacteriol. 76: 471–484.

Anderson, G.K., T. Donnelly, and K.J. McKeown. 1982. Identification and control of inhibition in the anaerobic treatment of industrial wastewater. Process Biochem. 17: 28–32.

Anderson, J.P.E. 1989. Principles of and assay systems for biodegradation, pp. 129–145, In: *Biotechnology and Biodegradation*, D. Kamely, A. Chakrabarty, and G.S. Omenn, Eds., Gulf Pub. Co., Houston, TX.

Anderson, R.T., and D.R. Lovley. 2000. Anaerobic bioremediation of benzene under sulfate-reducing conditions in a petroleum-contaminated aquifer. Environ. Sci. Technol. 34: 2261–2266.

Anderson, W.B., P.M. Huck, D.G. Dixon, and C.I. Mayfield. 2003. Endotoxin inactivation in water by using medium-pressure UV lamps. Appl. Environ. Microbiol. 6: 3002–3004.

Anderson, W.B., R.M. Slawson, and C.I. Mayfield, 2002. A review of drinking-water-associated endotoxin, including potential routes of human exposure. Can. J. Microbiol. 48: 567–587.

Anderson, W.B., I. Colin, D. Mayfield, G. Dixon, and P.M. Huck. 2003. Endotoxin inactivation by selected drinking water treatment oxidants. Water Res. 37: 4553–4560.

Andersson, A., P. Laurent, A. Kihn, M. Prévost, and P. Servais. 2001. Impact of temperature on nitrification in biological activated carbon (BAC) filters used for drinking water treatment. Water Res. 35: 2923–2934.

Andreasen, K., and P.H. Nielsen. 2000. Growth of *Microthrix parvicella* in nutrient removal activated sludge plants: studies of in situ physiology. Water Res. 34: 1559–1569.

Andrin, C., and J. Schwartzbrod. 1992. Isolating *Campylobacter* from wastewater. IWPRC Int. Symp., Washington, D.C., May 26–29, 1992.

Anonymous. 1984. Water Quality Standards for Wastewater Reuse, Arizona Department of State, Phoenix, AZ.

Anonymous, 1991. Microbes help clean heavy metals. ASM News 57: 296.

Anonymous. 2003. Protecting our water- Drinking water security in America after 9/11. J. Amer. Water Wks. Assoc. 95 (7): 37–45.

Ansari, S.A., S.R. Farrah, and G.R. Chaudhry. 1992. Presence of human immunodeficiency virus in wastewater and their detection by the polymerase chain reaction. Appl. Environ. Microbiol. 58: 3984–3990.

Antonie, R.L. 1976. Fixed biological surfaces-wastewater treatment: The rotating biological contactor, CRC Press, Boca Raton, FL.

Antopol, S.C., and P.D. Ellner. 1979. Susceptibility of *Legionella pneumophila* to ultraviolet radiation. Appl. Environ. Microbiol. 38: 347–348.

Anwar, H., and J.W. Costerton. 1992. Effective use of antibiotics in the treatment of biofilm-associated infections. ASM News 58: 665–668.

Aoi, Y., Y. Shiramasa, S. Tsuneda, A. Hirata, A. Kitayama, and T. Nagamune. 2002. Real-time monitoring of ammonia-oxidizing activity in a nitrifying biofilm by amoA mRNA analysis. Water Sci. & Technol. 46 (1–2): 439–472.

APHA. 1989. *Standard Methods for the Examination of Water and Wastewater.* 17th Ed. American Public Health Association, Washington, D.C.

APHA. 1999. *Standard Methods for the Examination of Water and Wastewater.* 20th Ed., American Public Health Association, Washington, D.C.

Applebaum, J., N. Gutman-Bass, M. Lugten, B. Teltsch, B., and H.I. Shuval. 1984. Dispersion of aerosolized enteric viruses and bacteria by sprinkler irrigation with wastewater. Monog. Virol. 15: 193–210.

Appleton, A.R., Jr., and A.D. Venosa. 1986. Technology evaluation of the dual digestion system. J. Water Pollut. Control Fed. 58: 764–769.

Appleton, A.R., Jr., C.J. Leong, and A.D. Venosa. 1986. Pathogen and indicator organism destruction by the dual digestion system. J. Water Pollut. Contr. Fed. 58: 992–999.

Araki, S., J.M. González, E. Luis, and E. Bécares. 2000. Viability of nematode eggs in high rate algae ponds. The effect of physico-chemical conditions. Water Sci. Technol. 42 (10): 10–11.

Araki, S., S. Martín-Gomez, E. Bécares, E. De Luis-Calabuig, and F. Rojo-Vazquez. 2001. Effect of high-rate algal ponds on viability of *Cryptosporidium parvum* oocysts. Appl. Environ. Microbiol. 67: 3322–3324.

Arana, I., A. Muela, J. Iriberri, L. Egeas, and I. Barcina. 1992. Role of hydrogen peroxide in loss of culturability mediated by visible light in *Escherichia coli* in a freshwater ecosystem. Appl. Environ. Microbiol. 58: 3903–3907.

Arauzo, M. 2003. Harmful effects of un-ionised ammonia on the zooplankton community in a deep waste treatment pond. Water Res. 37: 1048–1054.

Arber, R.P. 1983. From wastewater to drinking water. Civil Eng. (ASCE) Feb. 1983: 46–49.

Archer, D.B. 1984. Detection and quantitation of methanogens by enzyme-linked immunosorbent assay. Appl. Environ. Microbiol. 48: 797–801.

Archer, D.B., and B.H. Kirsop. 1991. The microbiology and control of anaerobic digestion, pp. 43–91, In: *Anaerobic Digestion: A Waste Treatment Technology*, A. Wheatly, Ed., Elsevier Appl. Sciences, London, U.K.

Arhing, B.K., N. Christiansen, I. Mathrani, H.V. Hendriksen, A.J.L. Macario, and E.V. de Macario. 1992. Introduction of *de novo* bioremediation ability, aryl reductive dechlorination, into anaerobic granular sludge by inoculation of sludge with *Desulfomonile tiedjei*. Appl. Environ. Microbiol. 58: 3677–3682.

Ariga, O., H. Takagi, H. Nishizawa, and Y. Sano. 1987. Immobilization of microorganisms with PVA hardened by iterative freezing and thawing. J. Ferment. Technol. 65: 651–658.

Arizona. 1984. *Water Quality Standards of Wastewater Reuse.* Arizona Department of State, Phoenix, AZ.

Armon, R., D. Gold, M. Brodsky, and G. Oron. 2002. Surface and subsurface irrigation with effluents of different qualities and presence of *Cryptosporidium* oocysts in soil and on crops. Water Sci. Technol. 46 (3): 115–122.

Arnon, S.S., R. Schechter, T.V. Inglesby, D.A. Henderson, and J.G. Bartlett. (Working Group on Civilian Biodefense). 2001. Botulinum toxin as a biological weapon: medical and public health management. J. Amer. Med. Assoc. 285: 1059–1070.

Armstrong, G.L., J.J. Calomiris, and R.J. Seidler. 1982. Selection of antibiotic-resistant standard plate count bacteria during water treatment. Appl. Environ. Microbiol. 44: 308–316.

Armstrong, G.L., D.S. Shigeno, J.J. Calomiris, and R.J. Seidler. 1981. Antibiotic-resistant bacteria in drinking water. Appl. Environ. Microbiol. 42: 277–283.

Arora, H., and M.W. LeChevallier. 2002. Occurrence of protozoa in spent filter backwash water, pp. 2261–2267, In: *Encyclopedia of Environmental Microbiology*, G. Bitton, editor-in-chief, Wiley-Interscience, N.Y.

Arther, R.G., P.R. Fitzgerald, and J.C. Fox. 1981. Parasitic ova in anaerobically digested sludge. J. Water Pollut. Control Fed. 53: 1334–1338.

Arun, V., T. Mino, and T. Matsuo. 1988. Biological mechanism of acetate uptake mediated by carbohydrate consumption in excess phophorus removal systems. Water Res. 22: 565–570.

Arvin, E. 1985. Observations supporting phosphate removal by biologically mediated chemical precipitation: A review. Water Sci. Technol. 15: 43–63.

Arvin, E., and G.H. Kristensen. 1983. Phosphate precipitation in biofilms and flocs. Water Sci. Technol. 15: 65–85.

Asano, H., H. Myoga, M. Asano, and M. Toyao. 1992. A study of nitrification utilizing whole micro-organisms immobilized by the PVA-freezing method. Water Sci. Technol. 26: 1037–1046.

Asano, T., Ed. 1985. *Artificial Recharge of Groundwater*. Butterworth Pub., Boston, MA.

Asano, T., and G. Tchobanoglous. 1991. The role of wastewater reclamation and reuse in the USA. Water Sci. Technol. 23: 2049–2059.

Asano, T., and J.A. Cotruvo. 2004. Groundwater recharge with reclaimed municipal wastewater: Health and regulatory considerations. Water Res. 38: 1941–1951.

Ashgari, A., S.R. Farrah, and G. Bitton. 1992. Use of hydrogen peroxide treatment and crystal violet agar plates for selective recovery of bacteriophages from natural environments. Appl. Environ. Microbiol. 58: 1158–1162.

Ashley, N.V., and T.J. Hurst. 1981. Acid and alkaline phosphatase activity in anaerobic digested sludge: A biochemical predictor of digester failure. Water Res. 15: 633–638.

Ashokkumar, M., T. Vu, F. Grieser, A. Weerawardena, N. Anderson, N. Pilkington, and D.R. Dixon. 2003. Ultrasonic treatment of *Cryptosporidium* oocysts. Water Sci. Technol. 47 (3): 173–177.

Atlas, R.M. 1986. *Basic and Practical Microbiology*. Macmillan, New York.

Atlas, R.M. 1991. Environmental applications of the polymerase chain reaction. ASM News 57: 630–632.

Atlas, R.M. 2002. Bioterrorism: From threat to reality. Ann.l Rev. Microbiol. 56: 167–185.

Atlas, R.M., and R. Bartha. 1987. *Microbial Ecology: Fundamentals and Applications*. Addison Wesley, Reading, MA.

Atlas, R., A. Bej, R. Steffan, J. Dicesare, and L. Haff. 1989. Detection of coliforms in water by poly-merase chain reaction (PCR) and gene probe methods, In: 89th Annual Meeting, American Society Microbiology, New Orleans, LA, 14–18, 1989.

Atmar, R.L., T.G. Metcalf, F.H. Neill, and M.K. Estes. 1993. Detection of enteric viruses in oysters by using the polymerase chain reaction. Appl. Environ. Microbiol. 59: 631–635.

Attree-Pietri, C., and J.P. Breittmayer. 1970. Etude comparee de germes-tests de contamination fecale et de bacteriophages en eau de mer. J. Fr. Hydrol. 10: 103–106.

Auling, G., F. Pilz, H.-J. Busse, S. Karrasch, M. Streichan, and G. Schon. 1991. Analysis of poly-phosphate accumulating microflora in phosphorus-eliminating, anaerobic-aerobic activated sludge systems by using diaminopropane as a biomarker for rapid estimation of *Acinetobacter* spp. Appl. Environ. Microbiol. 57: 3585–3592.

AWWA. 1985a. Waterborne *Giardia*: It's enough to make you sick (Roundtable). J. Am. Water Works Assoc. 77: 14–15.

AWWA. 1985b. Trends in ozonation. Roundtable held in Washington, D.C., June 25, 1985. J. Amer. Water Works Assoc. 77: 19–30.

AWWA. 1988. *Cryptosporidium* roundtable (Nov. 18, 1987). Baltimore, MD. J. Amer. Water Works Assoc. 80: 14–27.

AWWA. 2000. Committee Report: Disinfection at large and medium-sized systems. J. Amer. Water Works Assoc. 92: 32–33.

AWWA "Organisms in Water" Committee. 1987. Committee report: Microbiological consideration for drinking water regulations revisions. J. Amer. Water Works Assoc. 79: 81–88.

AWWA-WEF. 1998. *Water Reuse*. Conference Proc. Feb. 1–4, 1998, Lake Buena Vista, FL.

Ayres, R.M., G.P. Alabaster, D.D. Mara, and D.L. Lee. 1992. A design equation for human intestinal nematode egg removal in waste stabilization ponds. Water Res. 26: 863–865.

Babcock, R.W. Jr., and M.K. Stenstrom. 1993. Use of inducer compounds in the enricher-reactor process for degradation of 1-naphthylamine wastes. Water Environ. Res. 65: 26–33.

Babcock, R.W., Jr., K.S. Ro, C.-C. Hsieh, and M.K. Stenstrom. 1992. Development of an off-line enricher-reactor process for activated sludge degradation of hazardous wastes. Water Environ. Res. 64: 782–791.

Babich, H., and G. Stotzky. 1980. Reductions in inactivation rates of bacteriophages by clay minerals in lake water. Water Res. 14: 185–187.

Babich, H., and G. Stotzky. 1986. Environmental factors that affect the utility of microbial assays for the toxicty and mutagenicity of chemical pollutants, pp. 9–42, In: *Toxicity Testing Using Micro-organisms*, Vol. 2, B.J. Dutka, and G. Bitton, Eds., CRC Press, Boca Raton, FL.

Bach, P.D., M. Shoda, and H. Kubota. 1984. Rate of composting of dewatered sewage sludge in continuously mixed isothermal reactor. J. Ferment. Technol. 62: 285–292.

Bader, T.F. 1995. *Viral Hepatitis: Practical Evaluation and Treatment*. Hogrefe & Huber Pub., Seattle, pp. 234.

Bae, W., W. Chen, A. Mulchandani, and R.K. Mehra. 2000. Enhanced bioaccumulation of heavy metals by bacterial cells displaying synthetic phytochelatins. Biotech. Bioeng. 70: 518–524.

Baillie, L., and T.D. Read. 2001. *Bacillus anthracis*, a bug with attitude! Current Opinion Microbiol. 4 (1): 78–81.

Baker, G.C., L.A. Tow, and D.A. Cowan. 2003. PCR-based detection of non-indigenous micro-organisms in 'pristine' environments. J. Microbiol. Methods 53: 157–164.

Baker, K.H. Baker, J.P. Hegarty, B. Redmond, N.A. Reed, and D. S. Herson. 2002. Effect of oxidizing disinfectants (chlorine, monochloramine, and ozone) on *Helicobacter pylori*. Appl. Environ. Microbiol. 68: 981–984.

Balakrishnan, S., and W.W. Eckenfelder. 1969. Nitrogen relationships in biological treatment processes: II. Nitrification trickling filters. Water Res. 3: 167–174.

Balch, W.E., G.E. Fox, L.J. Magnum, C.R. Woese, and R.S. Wolfe. 1979. Methanogenesis: Reevaluation of a unique biological group. Microbiol. Rev. 143: 260–296.

Bales, R.C., S. Li, K.M. Maguire, M.T. Yahya, C.P. Gerba, and R.W. Harvey. 1995. Virus and bacteria transport in a sandy aquifer, Cape Cod, MA. Ground Water 33: 653–881.

Balkema, A.J., H.A. Preisig, R. Otterpohl, and F. Lambert. 2002. Indicators for the sustainability assessment of wastewater treatment systems. Urban Water 4: 153–161.

Bancroft, K., P. Chrostowski, R.L. Wright, and I.H. Suffet. 1984. Ozonation and oxidation competition values. Relationship to disinfection and microorganisms regrowth. Water Res. 18: 473–478.

Barbaree, J.M., B.S. Fields, J.C. Feeley, G.W. Gorman, and W.T. Martin. 1986. Isolation of protozoa from water associated with a legionellosis outbreak and demonstration of intracellular multiplication of *Legionellla pneumophila*. Appl. Environ. Microbiol. 51: 422–424.

Barbier, D., D. Ferrine, C. Duhamel, R. Doublet, and P. Georges. 1990. Parasitic hazard with sewage sludge applied to land. Appl. Environ. Microbiol. 56: 1420–1422.

Barkay, T., and J. Schaefer. 2001. Metal and radionuclide bioremediation: Issues, considerations and potentials. Current Opinion Microbiol. 4 (3): 318–323.

Barker, J., M.R.W. Brown, P.J. Collier, I. Farrell, and P. Gilbert. 1992. Relationship between *Legionella pneumophila* and *Acanthamoeba polyphaga*: Physiological status and susceptibility to chemical inactivation. Appl. Environ. Microbiol. 58: 2420–2425.

Barnard, J.L. 1973. Biological denitrification. J. Water Pollut. Control Fed. 72: 705–709.

Barnard, J.L. 1975. Biological nutrient removal without the addition of chemicals. Water Res. 9: 485–490.

Barnes, D., and P.J. Bliss. 1983. *Biological Control of Nitrogen in Wastewater Treatment*, E. & F.N. Spon, London.

Barnes, D., and P.A. Fitzgerald. 1987. Anaerobic wastewater treatment processes, pp. 57–113, In: *Environmental Biotechnology*, C.F. Forster, and D.A.J. Wase, Eds., Ellis Horwood Ltd., Chichester, U.K.

Barnes, R., J.I. Curry, L.M. Elliott, C.R. Peter, B.R. Tamplin, and B.W. Wilcke, Jr. 1989. Evaluation of the 7-hr membrane filter test for quantitation of fecal coliforms in water. Appl. Environ. Microbiol. 55: 1504–1506.

Baron, J. 1997. Repair of wastewater microorganisms after ultraviolet disinfection under seminatural conditions. Water Environ. Res. 69: 992–996.

Baronti, C., R. Curini, G. D'Ascenzo, A. Di Corcia, A. Gentili, and R. Samperi. 2000. Monitoring natural and synthetic estrogens at activated sludge sewage treatment plants and in receiving river water. Environ. Sci. Technol. 34: 5059–5066.

Barr, D.P., and S.D. Aust. 1994. Mechanisms white rot fungi use to degrade pollutants. Environ. Sci. Technol. 28: 78A-87A.

Barraclough, D., and G. Puri. 1995. The use of super(15)N pool dilution and enrichment to separate the heterotrophic and autotrophic pathways of nitrification. Soil Biol. Biochem. 27: 17–22.

Barrette, W.C., Jr., D.M. Hannum, W.D. Wheeler, and J.K. Hurst. 1988. General mechanism for the bacterial toxicity of hypochlorous acid: Abolition of ATP production. Biochem. 28: 9172–9178.

Bartel, P.F., G.K. Lam, and T.E. Orr. 1968. Purification and properties of polysaccharide depolymerase associated with phage-infected *Pseudomonas aeruginosa*. J. Biol. Chem. 243: 2077–2088.

Barth, E.F. 1975. The effects and removal of heavy metals in biological treatment: Discussion, In: *Heavy Metals in Aquatic Environments*, P.A. Krenkel, Ed., Pergamon Press, Oxford.

Barth, E.F., M.B. Ettinger, B.V. Salotto, and G.N. McDermott. 1965. Summary report on the effects of heavy metals on biological treatment processes. J. Water Pollut. Control Fed. 37, 86–96.

Bartley, T.D., T.J. Quan, M.T. Collins, and S.M. Morrison. 1982. Membrane filter technique for the isolation of *Yersinia enterocolitica*. Appl. Environ. Microbiol. 43: 829–834.

Barton, L.L., and R.M. Plunkett. 2002. Sulfate reducing bacteria: Environmental and technological aspects, pp. 3087–3096, In: *Encyclopedia of Environmental Microbiology*, G. Bitton, editor-in-chief, Wiley-Interscience, N.Y.

Bartone, C.R. 1991. International perspective on water resources management and wastewater reuse: Appropriate technologies. Water Sci. Technol. 23: 2039–2047.

Barwick, R.S., D.A. Levy, G.F. Craun, M.J. Beach, and R.L. Calderon. 2000. Surveillance for waterborne-disease outbreaks – United States, 1997–1998. Morb. Mortal. Wkly. Rep. CDC Surveill. Summ. 49: 1–21.

Bates, R.C., P.T.B. Shaffer, and S.M. Sutherland. 1977. Development of poliovirus having increased resistance to chlorine inactivation. Appl. Environ. Microbiol. 33: 849–853.

Battigelli, D.A., D. Lobe, and M.D. Sobsey. 1993. Inactivation of hepatitis A virus and other enteric viruses in water by ultraviolet. Water Sci. Technol. 27: 339–342.

Baudart, J., J. Coallier, P. Laurent, and M. Prévost. 2002. Rapid and sensitive enumeration of viable diluted cells of members of the family enterobacteriaceae in freshwater and drinking water. Appl. Environ. Microbiol. 68: 5057–5063.

Baughman, G.L., and D.F. Paris. 1981. Microbial bioconcentration of organic pollutants from aquatic systems: A critical review. Crit. Rev. Microbiol. 8: 205–228.

Baumann, E.R. 1971. Diatomite filtration of potable water, pp. 280–294, In: *Water Quality and Treatment: A Handbook of Public Water Supplies*, McGraw-Hill, New York.

Baumann, M., H. Lemmer, and H. Ries. 1988. Scum actinomycetes in sewage treatment plants. Part 1. Growth kinetics of *Nocardia amarae* in chemostat culture. Water Res. 22: 755–759.

Bausum, H.T., S.A. Schaub, R.E. Bates, H.L. McKim, P.W. Schumacher, and B.E. Brockett. 1983. Microbiological aerosols from a field-source wastewater irrigation system. J. Water Poll. Control Fed. 55: 65–80.

Bavykin, S.G., J.P. Akowski, V.M. Zakhariev, V.E. Barsky, A.N. Perov, and A.D. Mirzabekov. 2001. Portable system for microbial sample preparation and oligonucleotide microarray analysis. Appl. Environ. Microbiol. 67: 922–928.

Bayley, H.E., J.G. Moutes, and F.D. Schlesinger. 1992. An analysis of changing constraints and planning for flexibility in a water reclamation program. Water Sci. Technol. 26: 1525–1535.

Baylis, J.R., O. Gullans, and B.K. Spector. 1936. The efficiency of rapid sand filters in removing the cysts of amoebic dysentery organisms from water. Public Health. Rep. 50: 1567–1571.

Beaudette, L.A., M.B. Cassidy, L. England, J.L. Kirk, M. Habash, H. Lee, and J.T. Trevors. 2002. Bioremediation of soils, pp. 722–737, In: *Encyclopedia of Environmental Microbiology*, G. Bitton, editor-in-chief, Wiley-Interscience, N.Y.

Bedient, P.B., and H.S. Rifai. 1992. Bioremediation, pp. 117–141, In: *Groundwater Remediation*, R.J. Charbeneau, P.B. Bedient, and R.C. Loehr, Eds., Technomic, Lancaster, PA.

Beech, I.B. 2002. Biocorrosion: Role of sulfate reducing bacteria, pp. 465–475, In: *Encyclopedia of Environmental Microbiology*, G. Bitton, editor-in-chief, Wiley-Interscience, N.Y.

van Beelen, P., A.C. Dijkstra, and G.D. Vogels. 1983. Quantitation of coenzyme F_{420} in methanogenic sludge by the use of reversed-phase high-performance liquid chromatography and a fluorescence detector. Eur. J. Microbiol. Biotechnol. 18: 67–69.

de Beer, D., R. Srinivasan, and P.S. Stewart. 1994. Direct measurement of chlorine penetration into biofilms during disinfection. Appl. Environ. Microbiol. 60: 4339–4344.

Beffa, T., M. Blanc, L. Marilley, J. Lott Fisher, P.-F. Lyon, and M. Aragno. 1996a. Taxonomic and metabolic microbial diversity during composting, pp. 149–161, In: *The Science of Composting*, M. de Bertoldi, P. Sequi, B. Lemmes, and T. Papi, Eds., Blackie Academic & Professional, London, U.K.

Beffa, T., M. Blanc, and M. Aragno. 1996b. Obligately and facultatively autotrophic sulfur- and hydrogen-oxidizing thermophilic bacteria isolated from hot composts. Arch. Microbiol. 165: 34–40.

Beffa, T., M. Blanc, P.-F. Lyon, G. Vogt, M. Marchiani, J.L. Fischer, and M. Aragno. 1996c. Isolation of *Thermus* strains from hot composts (60°C to 80°C). Appl. Environ. Microbiol. 62: 1723–1727.

Beffa, T., P.-F. Lyon, J. Lott Fisher, M. Blanc, and M. Aragno. 1997. Influence of compost systems and management on the thermogenic biodegradation, maturation and hygienization, pp. 1–6, In: *R'97 Recovery, Recycling, Reintegration*, Pal Expo, April 4–7, 1997, Geneva, Switzerland.

Behets, J., F. Seghi, P. Declerck, L. Verelst, L. Duvivier, A. Van Damme, and F. Ollevier. 2003. Detection of *Naegleria* spp. and *Naegleria fowleri*: A comparison of flagellation tests, ELISA and PCR. Water Sci. Technol. 47 (3): 117–122.

Bej, A.K., J. DiCesare, L. Haff, and R.M. Atlas. 1991a. Detection of *Escherichia coli* and *Shigella* spp. in water by using the polymerase chain reaction and gene probes for *uid*. Appl. Environ. Microbiol. 57: 1013–1017.

Bej, A.K., M.H. Mahbubani, and R.M. Atlas. 1991b. Detection of viable *Legionella pneumophila* in water by polymerase chain reaction and gene probe methods. Appl. Environ. Microbiol. 57: 597–600.

Bej, A.K., S.C. McCarty, and R.M. Atlas. 1991c. Detection of coliform bacteria and *Escherichia coli* by multiplex polymerase chain reaction: Comparison with defined substrate and plating methods for water quality monitoring. Appl. Environ. Microbiol. 57: 2429–2432.

Bej, A.K., M.H. Mahbubani, and R.M. Atlas. 1992. Detection of genus *Salmonella* using polymerase chain reaction (PCR) method (Abstract Q-251), In: 92nd meeting of American Society Microbiology, May 26–30, New Orleans, LA.

Bej, A.K., R. Steffan, J. DiCesare, L. Haff, and R.M. Atlas. 1990. Detection of coliform bacteria in water by polymerase chain reaction and gene probes. Appl. Environ. Microbiol. 56: 307–314.

Belfroid, A.C., A. van der Horst, A.D. Vethaak, A.J. Schafer, G.B.J. Rijs, J. Wegener, and W.P. Cofino. 1999. Analysis and occurrence of estrogenic hormones and their glucuronides in surface water and wastewater in the Netherlands. Sci. Total Environ. 225: 101–108.

Belkin, S. 1998. Stress-responsive luminous bacteria for toxicity and genotoxicity monitoring, pp. 171–183, In: *Microscale Testing in Aquatic Toxicology: Advances, Techniques, and Practice*, P.G., Wells, K. Lee, and C. Blaise, Eds., CRC Press, Boca Raton, FL.

Bell, K., M.W. LeChevallier, M. Abbaszadegan, G.L. Amy, S. Sinha, M. Benjamin, and E.A. Ibrahim. 2000. Enhanced and optimized coagulation for particulate and microbial removal. AWWA Res. Foundation and Amer. Water Wks. Assoc., Denver Colorado.

Bell, R.B. 1978. Antibiotic resistance patterns of fecal coliforms isolated from domestic sewage before and after treatment in an aerobic lagoon. Can. J. Microbiol. 24: 886–888.

Bell, F.A. 1991. Review of effects of silver-impregnated carbon filters on microbial water quality. J. Am. Water Works Assoc. 83: 7476.

Bell, J.P., and M. Tsezos. 1987. Removal of hazardous organic pollutants by biomass adsorption. J. Water Pollut. Control Fed. 59: 191–198.

Bell, R.B., W.R. Macrae, and G.E. Elliott. 1981. R factors in coliform-fecal coliform sewage flora of the prairies and northwest territories of Canada. Appl. Environ. Microbiol. 42: 204–210.

Bellair, J.T., G.A.P. Smith, and I.G. Wallis. 1977. Significance of diurnal variations in fecal coliform die-off in the design of ocean outfalls. J. Water Pollut. Control Fed. 49: 2022–2030.

Bellamy, W.D., G.P. Silverman, D.W. Hendricks, and G.S. Logsdon. 1985a. Removing *Giardia* cysts with slow sand filtration. J. Am. Water Works Assoc. 77: 52–60.

Bellamy, W.D., D.W. Hendricks, and G.S. Logsdon. 1985b. Slow sand filtration: Influences of selected process variables. J. Am. Water Works Assoc. 77: 62–66.

Beller, M., A. Ellis, S.H. Lee, M.A. Drebot, S.A. Jenkerson, E. Funk, M.D. Sobsey, O.D. Simmons III, S.S. Monroe, T.Ando, J.Noel, M. Petric, J.P. Middaugh, and J.S. Spika. 1997. Outbreak of viral gastroenteritis due to a contaminated well: international consequences. JAMA 278: 563–568.

Ben Aim, R.M., and M.J. Semmens. 2002. Membrane bioreactors for wastewater treatment and reuse: a success story. Water Sci. Technol. 47 (1): 1–5.

Benedict, A.H., E. Epstein, and J. Alpert. 1988. *Composting Municipal Sludge: A Technology Evaluation.* Noyes Data Corp., Park Ridge, N.J. 178 pp.

Bennett, J.W., and B.D. Faison. 1997. Use of fungi in bioremediation, pp. 758–765, In: *Manual of Environmental Microbiology*, C.J. Hurst, G.R. Knudsen, M.J. McInerney, L.D. Stetzenbach, and M.V. Walter, Eds., ASM Press, Washington, D.C.

Bennie, D.T., C.A. Sullivan, H.-B. Lee, and R.J. Maguire. 1998. Alkylphenol polyethoxylates metabolites in Canadian sewage treatment plant effluent streams. Water Qual. Res. J. Can. 33: 231–252.

Benson, H.J. 1973. Microbiological Applications: A Laboratory Manual in General Microbiology. W.C. Brown, Dubuque, IA.

Bercovier, H., M. Deral-Cochin, J. Sherman, B. Fattal, and H.I. Shuval. 1984. Seroepidemiological survey of irrigation workers in Israel and isolation of *Legionella spp.* in the environment. In: *Legionella*, C. Thornsberry, A. Balows, J.C. Feeley, and W. Jakubowski, Eds., American Society Microbiology, Washington, D.C.

Berendt, R.F. 1980. Influence of blue-green algae (cyanobacteria) on survival of *Legionella pneumophila* in aerosols. Infect. Immun. 32: 690–692.

Berg, G., Ed. 1978. *Indicators of Viruses in Water and Food*, Ann Arbor Scientific Publications, Ann Arbor, MI.

Berg, G., and D. Berman. 1980. Destruction by anaerobic mesophilic and thermophilic digestion of viruses and indicator bacteria indigenous to domestic sludges. Appl. Environ. Microbiol. 39: 361–368.

Berg, G., H. Sanjahsaz, and S. Wangwongwatana. 1989. Potentiation of the virucidal effectiveness of free chlorine by substances in drinking water. Appl. Environ. Microbiol. 55: 390–393.

Berg, G., H. Sanjahsaz, and S. Wangwongwatana. 1990. KCl potentiation of the virucidal effectiveness of free chlorine at pH 9. Appl. Environ. Microbiol. 56: 1571–1575.

Berg, J.D., and L. Fiksdal. 1988. Rapid detection of total and fecal coliforms in water by enzymic hydrolysis of 4- methylumbelliferone-β-D-galactoside. In: *Proceedings of the International Conference on Water and Wastewater Microbiology*, Newport Beach, CA. February 8–11, 1988.

Berg, J.D., P.V. Roberts, and A. Matin. 1986. Effect of chlorine dioxide on selected membrane functions of *Escherichia coli*. J. Appl. Bacteriol. 60: 213–220.

van den Berg, L., C.P. Lentz, R.J. Athey, and E.A. Rook. 1974. Assessment of methanogenic activity in anaerobic digestion. Apparatus and methods. Biotech. Bioeng. 16: 1459–1465.

van den Berg, L., and K.J. Kennedy. 1981. Support materials for stationary fixed films reactors for high rate methonogen fermentations. Biotech Lett. 3: 165–170.

Berge, J.J., D.P. Drennan, R.J. Jacobs, A. Jakins, A.S. Meyerhoff, W. Stubblefield, and M. Weinberg. 2000. The cost of hepatitis A infections in American adolescents and adults in 1997. Hepatology 31: 469–473.

Berger, B.B. 1982. Water and wastewater quality control and the public health. Ann. Rev. Public Health 3: 359–392.

Berger, P.S., and R.K. Oshiro. 2002. Source water protection: Microbiology of source water, pp. 2967–2978, In: *Encyclopedia of Environmental Microbiology*, G. Bitton, editor-in-chief, Wiley-Interscience, N.Y.

Berk, S.G., and J.H. Gunderson. 1993. *Wastewater Organisms: A Color Atlas*, Lewis, Boca Raton, FL., 25 pp.

Berman, D., and J.C. Hoff. 1984. Inactivation of simian rotavirus SA11 by chlorine, chlorine dioxide and monochloramine. Appl. Environ. Microbiol. 48: 317–323.

Berman, D., E.W. Rice, and J.C. Hoff. 1988. Inactivation of particle-associated coliforms by chlorine and monochloramine. Appl. Environ. Microbiol. 54: 507–512.

Bermudez, M., and T.C. Hazen. 1988. Phenotypic and genotypic comparison of *Escherichia coli* from pristine tropical waters. Appl. Environ. Microbiol. 54: 979–983.

Bernarde, M.A., N.B. Snow, V.P. Olivieri, and B. Davidson. 1967. Kinetics and mechanism of bacterial disinfection by chlorine dioxide. Appl. Microbiol. 15: 257–265.

Bessler, W., E. Freund-Molbert, H. Knufermann, R.C. Thurow, and S. Stirm. 1973. A bacteriophage induced depolymerase active on *Klebsiella KII* capsular polysaccharide. Virology 56: 134–151.

Best, D.J., J. Jones, and D. Strafford. 1985. The environment and biotechnology, p 213–256, In: *Biotechnology: Principles and Applications*, I.J. Higgins, D.J. Best, and J. Jones, Eds., Blackwell, Oxford.

Best, M.G., A. Goetz, and V.L. Yu. 1984. Heat eradication measures for control of nosocomial Legionnaires' disease: Implementation, education and cost analysis. Amer. J. Infect. Control 12: 26–30.

Bettmann, H., and H.J. Rehm. 1985. Continuous degradation of phenol(s) by *Pseudomonas putida* P8 entrapped in polyacrylamide-hydrazide. Appl. Microbiol. Biotech. 22: 389–393.

Betts, K.S. 2002. New drinking water hazard. Environ. Sci. Technol. 36: 92A-93A.

Beuret, C., A. Baumgartner, and J. Schluep. 2003. Virus-contaminated oysters: a three-month monitoring of oysters imported to Switzerland. Appl. Environ. Microbiol. 69: 2292–2297.

Bhattacharya, S.K., and G.F. Parkin. 1988. Fate and effect of methylene chloride and formaldehyde in methane fermentation systems. J. Water Pollut. Control Fed. 60: 531–536.

Bhattacharya, S.K., and G.F. Parkin. 1989. The effect of ammonia on methane fermentation processes. J. Water Pollut. Control Fed. 61: 56–59.

Bicki, T.J., R.B. Brown, M.E. Collins, R.S. Mansell, and D.F. Rothwell. 1984. Impact of on-site sewage disposal systems on surface and groundwater quality. Report to Florida Department of Health Rehabilitation Service, Tallahassee, FL.

Biesterfeld, S., L. Fugueroa, M. Hernandez, and P. Russell. 2001. Quantification of nitrifying bacterial populations in a full-scale nitrifying trickling filter using fluorescent *in situ* hybridization. Water Environ. Res. 73: 329–338.

Biesterfeld, S., and L. Fugueroa. 2002. Nitrifying biofilm development with time: activity versus phylogenetic composition. Water Environ. Res. 74: 470–461.

Biesterfeld, S., P. Russell, and L. Fugueroa. 2003. Linking nitrifying biofilm structure and function through fluorescent *in situ* hybridization and evaluation of nitrification capacity. Water Environ. Res. 75: 205–215.

Bifulco, J.M., and F.W. Schaefer III. 1993. Antibody-magnetite method for selective concentration of *Giardia lamblia* cysts from water samples. Appl. Environ. Microbiol. 59: 772–776.

Bisping, B., and H.J. Rehm. 1988. Multistep reactions with immobilized microorganisms. Biotech. Appl. Biochem. 10: 87–98.

Bissonette, G.K., J.J. Jezeski, G.A. McFeters, and D.G. Stuart. 1975. Influence of environmental stress on enumeration of indicator bacteria from natural waters. Appl. Microbiol. 29: 186–194.

Bissonette, G.K., J.J. Jezeski, G.A. McFeters, and D.G. Stuart. 1977. Evaluation of recovery methods to detect coliforms in water. Appl. Environ. Microbiol. 33: 590–595.

Bitton, G. 1978. Survival of enteric viruses, pp. 273–299, In: *Water Pollution Microbiology*, Vol. 2, R. Mitchell, Ed., John Wiley & Sons, New York.

Bitton, G. 1980a. *Introduction to Environmental Virology*. John Wiley & Sons, New York. 326 pp.

Bitton, G. 1980b. Adsorption of viruses to surfaces: Technological and ecological implications, pp. 331–374, In: *Adsorption of Microorganisms to Surfaces*, G. Bitton and K.C. Marshall, Eds., Wiley Interscience, New York.

Bitton, G. 1983. Bacterial and biochemical tests for assessing chemical toxicity in the aquatic environment: A review. Crit. Rev. Environ. Contr. 13: 51–67.

Bitton, G. 1987. Fate of bacteriophages in water and wastewater treatment plants, pp. 181–195, In: *Phage Ecology*, S.M. Goyal, C.P. Gerba and G. Bitton, Eds., Wiley Interscience, New York.

Bitton, G. 1999. *Wastewater Microbiology*. 2nd Ed., Wiley-Liss, New York.

Bitton, G., M. Campbell, and B. Koopman. 1992a. MetPAD: A bioassay kit for the specific determination of heavy metal toxicity in sediments from hazardous waste sites. Environ. Toxicol. Water Qual. 7: 323–328.

Bitton, G., L.T. Chang, S.R. Farrah, and K. Clifford. 1981a. Recovery of coliphages from wastewater and polluted lake water by magnetite-organic flocculation. Appl. Environ. Microbiol. 41: 93–97.

Bitton, G., Y.J. Chou, and S.R. Farrah. 1981b. Techniques for virus detection in aquatic sediments. J. Virol. Methods 4: 1–8.

Bitton, G., B.L. Damron, G.T. Edds, and J.M. Davidson, Eds. 1980. *Sludge-Health Risks of Land Application.* Ann Arbor Scientific Publications, Ann Arbor, MI, 366 pp.

Bitton, G., and B.J. Dutka, Eds. 1986. *Toxicity Testing Using Microorganisms*. Vol. 1, CRC Press, Boca Raton, FL.

Bitton, G., B.J. Dutka, and C.W. Hendricks. 1989. Microbial toxicity tests, p. 6–44 to p. 6–66, In: *Ecological Assessment of Hazardous Waste Sites*, W. Warren-Hicks, B.R. Parkhurst and S.S. Baker, Jr., Eds., EPA 600/3-89/013, U.S. Environmental Protection Agency, Corvallis, Oregon.

Bitton, G., and S.R. Farrah. 1986. Contamination des eaux souterraines par les virus. Rev. Int. Sci. Eau 2: 31–37.

Bitton, G., S.R. Farrah, C. Montague, M.W. Binford, P.R. Scheuerman, and A. Watson. 1985. *Survey of Virus Isolation Data From Environmental Samples*, Res. Report (contract # 68-03-3196), U.S. EPA, Health Effect Research Lab, Cincinnati, OH.

Bitton, G., S.R. Farrah, C. Montague, and E.W. Akin. 1986. Global survey of virus isolations from drinking water. Environ. Sci. Technol. 20: 216–222.

Bitton, G., S.R. Farrah, R.H. Ruskin, J. Butner, and Y.J. Chou. 1983. Survival of pathogenic and indicator organisms in groundwater. Ground Water 213: 405–410.

Bitton, G., R. Fraxedas, and G. Gifford. 1979. Effect of solar radiation on poliovirus: Preliminary experiments. Water Res. 13: 225–228.

Bitton, G., and V. Freihoffer. 1978. Influence of extracellular polysaccharides on the toxicity of copper and cadmium towards *Klebsiella aerogenes*. Microb. Ecol. 4: 119–125.

Bitton, G., and C.P. Gerba. Eds. 1984. *Groundwater Pollution Microbiology*. John Wiley & Sons, New York.

Bitton, G., and R.W. Harvey. 1992. Transport of pathogens through soils and aquifers, pp. 103–124, In: *New Concepts in Environmental Microbiology*, R. Mitchell, Ed., Wiley-Liss, New York.

Bitton, G., Y. Henis, and N. Lahav. 1972. Effect of several clay minerals and humic acid on the survival of *Klebsiella aerogenes* exposed to ultraviolet irradiation. Appl. Microbiol. 23: 870–874.

Bitton, G., K. Jung, and B. Koopman. 1994. Evaluation of a microplate assay specific for heavy metal toxicity. Arch. Environ. Contam. Toxicol. 27: 25–28.

Bitton G., and B. Koopman. 1986. Biochemical tests for toxicity screening, pp 27–55, In: *Toxicity Testing Using microorganisms*, G. Bitton & B.J. Dutka, Eds., CRC Press, Boca Raton, FL.

Bitton, G., and B. Koopman. 1992. Bacterial and enzymatic bioassays for toxicity testing in the environment. Rev. Environ. Contam. Toxicol. 125: 1–22.

Bitton, G., B. Koopman, and O. Agami. 1992b. MetPAD^{TM}: a bioassay for rapid assessment of heavy metal toxicity in wastewater. Water Environ. Res. 64: 834–836.

Bitton, G., B. Koopman, and K. Jung. 1995. An assay for the enumeration of total coliforms and *E. coli* in water and wastewater. Water Environ. Res. 67: 906–909.

Bitton, G., and K.C. Marshall, Eds. 1980. *Adsorption of Microorganisms to Surfaces.* Wiley Interscience, N.Y.

Bitton, G., J.E. Maruniak, and F.W. Zettler. 1987. Virus survival in natural ecosystems, pp. 301–332, In: *Survival and Dormancy of Microorganisms*, Y. Henis, Ed., Wiley-Interscience, New York., 355 pp.

Bitton, G., and R. Mitchell. 1974a. Effect of colloids on the survival of bacteriophages in seawater. Water Res. 8: 227–229.

Bitton, G., and R. Mitchell, 1974b. Protection of *E. coli* by montmorillonite in seawater. J. San. Engineering Div. 100: 1310–1313.

Bitton, G., and J.L. Morel. 1998. Microbial enzyme assays for the detection of heavy metal toxicity, pp.143–152, In: *Microscale Testing in Aquatic Toxicology: Advances, Techniques, and Practice*, P.G., Wells, K. Lee, and C. Blaise, Eds. CRC Press, Boca Raton, FL.

Bitton, G., O.C. Pancorbo, and S.R. Farrah. 1984. Virus transport and survival after land application of sewage sludge. Appl. Environ. Microbiol. 47: 905–909.

Björnsson, L., M. Murto, T.G. Jantsch, and B. Mattiasson. 2001. Evaluation of new methods for the monitoring of alkalinity, dissolved hydrogen and the microbial community in anaerobic digestion. Water Res. 35: 2833–2840.

Black, B.D., G.W. Harrington, and P.C. Singer. 1996. Reducing cancer risks by improving organic carbon removal. J. Am. Water Works Assoc. 88: 40–52.

Blackall, L. L., A.E. Harbers, P.F. Greenfield, and A.C. Hayward. 1988. Actinomycete scum problems in australian activated sludge plants. In: *International Conference Water and Wastewater Microbiology*, Vol. 2, Newport Beach, CA, Feb. 8–11, 1988.

Blackall, L.L., and K.C. Marshall, 1989. The mechanism of stabilization of actinomycete foams and the prevention of foaming under laboratory conditions. J. Ind. Microbiol. 4: 181–188.

Blackall, L. L., A.E. Harbers, P.F. Greenfield, and A.C. Hayward. 1991. Foaming in activated sludge plants: A survey in Queensland, Australia and an evaluation of some control strategies. Water Res. 25: 313–317.

Blackall, L.L., S. Rosseti, C. Christensson, M. Cunningham, P. Hartman, P. Hugenholtz, and V. Tandoi. 1997. The characterization and description of representatives of 'G' bacteria from activated sludge. Letters Appl. Microbiol. 25: 63–69.

Blackall, L.L., G.R. Crocetti, A.M. Saunders and P.L. Bond. 2002. Activated sludge: Microbiology of phosphorus removal, In: *Encyclopedia of Environmental Microbiology*, G. Bitton, editor-in-chief, Wiley, N.Y.

Blackbeard, J.R., G.A. Elkana, and G.R. Marais. 1986. A survey of filamentous bulking and foaming in activated sludge plants in South Africa. Water Pollut. Control 1: 90–100.

Blackburn, J.W., W.L. Troxler, and G.S. Sayler. 1984. Prediction of the fate of organic chemicals in a biological treatment process-an overview. Environ. Prog. 3: 163–176.

Blaise, C. 1991. Microbiotests in aquatic ecotoxicology: Characteristics, utility, and prospects. Environ. Toxicol. Water Qual. 6: 145–155.

Blaise, C. 2002. Use of microscopic algae in toxicity testing, pp. 3219–3230, In: *Encyclopedia of Environmental Microbiology*, G. Bitton, editor-in-chief, Wiley-Interscience, N.Y.

Blaise C., R. van Coillie, N. Bermingham, and G. Coulombe. 1987. Comparaison des reponses toxiques de trois indicateurs biologiques (bacteries, algues, poissons) exposes a des effluents de fabriques de pates et papiers. Rev. Int. Sci. Eau 3: 9–17.

Blaser, M.J., and L.B. Peller. 1981. *Campylobacter* enteritis. New Engl. J. Med. 305: 1444–1452.

Blaser, M.J., P.F. Smith, W.L.L. Wang, and J.C. Hoff. 1986. Inactivation of *Campylobacter* by chlorine and monochloramine. Appl. Environ. Microbiol. 51: 307–311.

Blaser. M.J., D.N. Taylor, and R.A. Feldman. 1983. Epidemiology of *Campylobacter jejuni* infections. Epidemiol. Rev. 5: 157–176.

Blatchley, E.R. III, B.A. Hunt, R. Duggirala, J.E. Thompson, J. Zhao, T. Halaby, R.L. Cowger, C.M. Straub, and J.E. Alleman. 1997. Effects of disinfectants on wastewater effluent toxicity. Water Res. 31: 1581–1588.

Blewett, D.A., S.E. Wright, D.P. Casemore, N.E. Booth, and C.E. Jones. 1993. Infective dose size studies on *Cryptosporidium parvum* using gnotobiotic lambs. Water Sci. Technol. 27: 61–64.

Block, J.C. 1981. Viruses in environmental waters, pp. 117–145, In: *Viral Pollution of the Environment*, G. Berg, Ed., CRC Press, Boca Raton, FL.

Block, J.-C. 1983. A review of some problems related to epidemiological studies, pp. 33–46, In: *Biological Health Risks of Sludge Disposal to Land in Cold Climates*, P.M. Wallis and D.L. Lehmann, Eds., University of Calgary Press, Calgary, Canada.

Block, J.C., K. Haudidier, J.L. Paquin, J. Miazaga, and T. Lévi. 1993. Biofilm accumulation in drinking water distribution system. Biofouling 6: 333–343.

Block, J.C., A.H. Havelaar, and P. l'Hermite, Eds. 1986. Epidemiological Studies of Risks Associated with the Agricultural Use of Sewage Sludge: Knowledge and Needs. Elsevier Appl. Science London, 168 pp.

Block, J.C., L. Mathieu, P. Servais, D. Fontvielle, and P. Werner. 1992. Indigenous bacterial inocula for measuring the biodegradable dissolved organic carbon (BDOC) in waters. Water Res. 26: 481–486.

Blom, A., W. Harder, and A. Matin. 1992. Unique and overlapping pollutant stress proteins of *Escherichia coli*. Appl. Environ. Microbiol. 58: 331–334.

Blum, D., and R.G. Feachem. 1985. Health aspects of night soil and sludge use in agriculture and aquaculture. Part III: An epidemiological perspective. Report No. 05/85, Dubendorf, Switzerland. International Ref. Center for Waste Disposal.

Blum, D.J.W., R. Hergenroeder, G.F. Parkin, and R.E. Speece. 1986. Anaerobic treatment of coal conversion wastewater constituents: biodegradability and toxicity. J. Water Pollut. Control Fed. 58: 122–131.

Blum, D.J.W., and R.E. Speece. 1992. The toxicity of organic chemicals to treatment processes. Water Sci. Technol. 25: 23–31.

Blyth, W. 1973. Farmer's lung disease, pp. 261–276, In: *Actinomycetales: Characteristics and Practical Importance*, G. Sykes, and F.A. Skinner, Eds., Academic Press, New York.

Boccia, D., A.E. Tozzi, and B. Cotter. 2002. Waterborne outbreak of Norwalk-like virus gastroenteritis at a tourist resort, Italy. Emerging Infect. Dis. 8: 563–568.

Bock, E., P.A. Wilderer, and A. Freitag. 1988. Growth of *Nitrobacter* in the absence of dissolved oxygen. Water Res. 22: 243–250.

Boczar, B.A., W.M. Begley, and R.J. Larson. 1992. Characterization of enzyme activity in activated sludge using rapid analyses for specific hydrolases. Water Environ. Res. 64: 792–797.

Boethling, R.S. 1984. Environmental fate and toxicity in wastewater treatment of quaternary ammonium surfactants. Water Res. 18: 1061–1076.

Bohn, H., and R. Bohn. 1988. Soil beds weed out air pollutants. Chem. Eng. April 25, 1988, p 73–76.

Bollag, J.M. 1979. Transformation of xenobiotics by microbial activity, pp. 19–27, In: *Microbial Degradation of Pollutants in Marine Environments*. A.W. Bourquin, and P.H. Pritchard, Eds., EPA-600/9-79-012, U.S. Environmental Protection Agency, Cincinnati, OH.

Bollag, J-M. 1992. Decontaminating soils with enzymes. Environ. Sci. Technol. 26: 1876–1881.

Bonjoch, X., E. Ballesté, and A. R. Blanch. 2004. Multiplex PCR with 16S rRNA gene-targeted Primers of *Bifidobacterium* spp. to identify sources of fecal pollution. Appl. Environ. Microbiol. 70: 3171–3175.

de Bont, J.A., J.P. van Dijken, and W. Harder. 1981. Dimethylsulphoxide and dimethyl sulphide as a carbon, sulfur and energy source for growh of *Hyphomicrobium*. J. Gen. Microbiol. 127: 315–323.

Boon, N., J. Goris, P. de Vos, W. Verstraete, and E.M. Top. 2000. Bioaugmentation of activated sludge by an indigenous 3-chloroaniline-degrading *Comamonas testosterone* strain I2gfp. Appl. Environ. Microbiol. 66: 2906–2913.

Borchardt, M.A., P.D. Bertz, S.K. Spencer, and D. A. Battigelli. 2003. Incidence of enteric viruses in groundwater from household wells in Wisconsin. Appl. Environ. Microbiol. 6: 1172–1180.

Born, M., H. Doerr, and I. Levin. 1990. Methane consumption in aerated soils of the temperate zone. Tellus 42B: 2–8.

Borrego, J.J., M.A. Morinigo, A. de Vicente, R. Cornax, and P. Romero. 1987. Coliphages as an indicator of faecal pollution in water. Its relationship with indicator and pathogenic microorganisms. Water Res. 21: 1473–1480.

Borrego, J.J., and P. Romero. 1985. Coliphage survival in seawater. Water Res. 19: 557–562.

Bosch, A., C. Tartera, R. Gajardo, J.M. Diez, and J. Jofre. 1989. Comparative resistance of bacteriophages active against *Bacteroides fragilis* to inactivation by chlorination or ultraviolet radiation. Water Sci. Technol. 21: 221–226.

Bosch, A., J.M. Diez, and F.X. Abad. 1993. Disinfection of human enteric viruses in water by copper:silver and reduced levels of chlorine. Water Sci. Technol. 27: 351–356.

Bouchard, D.C., M.K. Williams, and R.Y. Surampalli. 1992. Nitrate contamination of groundwater: Sources and potential health effects. J. Amer. Water Works Assoc. 84(9): 85–90.

Bouhoum, K., O. Amahmid, and S. Asmama. 2000. Occurrence and removal of protozoan cysts and helminth eggs in waste stabilization ponds in Marrakech. Water Sci. Technol., 42 (10–11): 159 –164.

Bouma, J., W.A. Ziebell, W.G. Walker, P.G. Olcott, E. McCoy, and F.D. Hole. 1972. Soil absorption of septic tank effluent: A field study of some major soils in Wisconsin. Info circ. #20, University of Wisconsin Extension and Geological Natural History Survey, Madison, WI.

Bourbigot, M.M., A. Dodin, and R. Lheritier. 1984. Bacteria in distribution systems. Water Res. 18: 585–591.

Bourbigot, M.M., M.C. Hascoet, Y. Levi, F. Erb, and N. Pommery. 1986. Role of ozone and granulated activated carbon in the removal of mutagenic compounds. Environ. Health Pers. 69: 159–163.

Bouwer, E.J., and P.B. Crowe. 1992. Assessment of biological processes in drinking water treatment. J. Am. Works Assoc. 80: 82–93.

Bouwer, E.J., and P.L. McCarty. 1983. Transformations of halogenated organic compounds under denitrification conditions. Appl. Environ. Microbiol. 45: 1295–1299.

Bouwer, E., J. Mercer, M. Kavanaugh, and F. DiGiano. 1988. Coping with groundwater contamination. J. Water Pollut. Control Fed. 60: 1415–1421.

Bouwer, H. 1992. Agricultural and municipal use of wastewater. Water Sci. Technol. 26: 1583–1591.

Bowden, A.V. 1987. Survey of European sludge treatment and disposal practices. Water Research Centre Report No. 1656-M.

Bowen, G.S., and M.A. McCarthy. 1983. Hepatitis A associated with a hardware store water fountain and a contaminated well in Lancaster County, Pennsylvania, 1980. Am. J. Epidemiol. 117: 695–705.

Bowker, R.P.G., J.M. Smith, and N.A. Webster. 1989. *Odor and Corrosion Control in Sanitary Sewerage Systems and Treatment Plants*. Hemisphere, New York, 132 p.

Bowman, J.P., L. Jimenez, I. Rosario, T.C. Hazen, and G.S. Sayler. 1993. Characterization of the methanotrophic bacterial community present in a trichlorethylene-contaminated subsurface groundwater site. Appl. Environ. Microbiol. 59: 2380–2387.

Boyce, T.G., D.L. Swerdlow, and P.M. Griffin. 1995. *Escherichi coli* O157:H7 and the hemolytic-uremic syndrome. N. Engl. J. Med. 333: 364–368.

Boyce, D.S., O.J. Sproul, and C.E. Buck. 1981. The effect of bentonite clay on ozone disinfection of bacteria and viruses in water. Water Res. 15: 759–767.

Boyd, R.F. 1988. *General Microbiology*. 2nd Ed., Times Mirror/C.V. Mosby, St. Louis, MO.

Boyd, S.A., and D.R. Shelton. 1984. Anaerobic biodegradation of chlorophenols in fresh and acclimated sludge. Appl. Environ. Microbiol. 47: 272–277.

Boyd, S.A., D.R. Shelton, D. Berry, and J.M. Tiedje. 1983. Anaerobic biodegradation of phenolic compounds in digested sludge. Appl. Environ. Microbiol. 46: 50–54.

Brandon, J.R. 1978. *Parasites in Soil/Sludge Systems*. SAND 77-1970 Report, Sandia Lab., Alburquerque, N.M.

Brandon, J.R., W.D. Burge, and N.K. Enkiri. 1977. Inactivation by ionizing radiation of *Salmonella enteriditis* serotype montevideo grown in composted sewage sludge. Appl. Environ. Microbiol. 33: 1011–1012.

Bratina, B.J., and R.S. Hanson. 1992. Ecology of methanotrophic bacteria. In: 204th. American Chemical Society National Meeting, p. 124–127, Aug. 23–28, 1992, Washington, D.C.

Bravo, J.M., and A. de Vicente. 1992. Bacterial die-off from sewage discharged through submarine outfalls. Water Sci. Technol. 25: 9–16.

Brayton, P.R., M.L. Tamplin, A. Huk, and R.R. Colwell. 1987. Enumeration of *Vibrio cholerae* 01 in Bangladesh waters by fluorescent-antibody direct viable count. Appl. Environ. Microbiol. 53: 2862–2865.

Bremer, P.J., and G.G. Geesey. 1991. Laboratory-based model of microbiologically induced corrosion of copper. Appl. Environ. Microbiol. 57: 1956–1962.

Brenner, K.F., C.C. Rankin, M. Sivaganesan, and P.V. Scarpino. 1996. Comparison of the recoveries of *Escherichia coli* and total coliforms from drinking water by the MI agar method and the U.S. Environmental Protection Agency-approved membrane filter method. Appl. Environ. Microbiol. 62: 203–208.

Brenner, K.F., P.V. Scarpino, and C.S. Clark. 1988. Animal viruses, coliphages, and bacteria in aerosols and wastewater at a spray irrigation site. Appl. Environ. Microbiol. 54: 409–415.

Brierley, C.L., J.A. Brierley, and M.S. Davidson. 1989. Applied microbial processes for metal recovery and removal from wastewater, p. 359–381, In: *Metal Ions and Bacteria*, T.J. Beveridge, and R.J. Doyle, Eds., John Wiley & Sons, New York.

Brigano, F.A.O., P.V. Scarpino, S. Cronier, and M.L. Zink. 1979. Effect of particulates on inactivation of enteroviruses in water by chlorine dioxide, In: *Progress in Water Disinfection Technology*, A.D. Venosa, Ed., EPA Publication # EPA-600/9-79-018, Cincinnati, OH.

Brigmon, R.L. 2002. Methanotrophic bacteria: Use in Bioremediation, pp. 1936–1944, In: *Encyclopedia of Environmental Microbiology*, G. Bitton, editor-in-chief, Wiley-Interscience, N.Y.

Brigmon, R.L., G. Bitton, S.G. Zam, and B. O'Brien. 1995. Development and applications of a monoclonal antibody against *Thiothrix* spp. Appl. Environ. Microbiol. 61: 13–20.

Broadbent, C. 1996. Guidance of water quality of heated spas. South Australian Health Coomission for the National Environ. Health Forum.

Brock, T.D., and M.T. Madigan. 1991. *Biology of Microorganisms*. 6th Ed. Prentice-Hall, Englewood Cliffs, N.J.

Brockman, F.J., B.A. Denovan, R.J. Hicks, and J.K. Frederickson. 1989. Isolation and characterization of quinoline-degrading bacteria from subsurface sediments. Appl. Environ. Microbiol. 55: 1029–1032.

Brodelius, P., and K. Mosbach, 1987. Overview. Methods Enzymol. 135: 173–175.

Brodisch, K.E.U., and S.J. Joyner. 1983. The role of microorganisms other than *Acinetobacter* in biological phosphate removal in activated sludge processes. Water Sci. Technol. 15: 117–122.

Brodkorb, T.S., and R.L. Legge. 1992. Enhanced biodegradation of phenanthrene in oil tar-contaminated soils supplemented with *Phanerochaete chrysosporium*. Appl. Environ. Microbiol. 58: 3117–3121.

Bronstein, L., B. Edwards, and J.C. Voyta. 1989. 1,2 Dioxetane: novel chemilumiscent enzyme substrates. Application to immunoassays. J. Biolumin. Chemilumin. 4: 99–111.

Brooks, R.G., and J.S. Remington. 1986. Transplant related infections, p. 583, In: *Hospital Infections*, J.V. Bennett, and P.S. Brachman, Eds., Little, Brown, Boston.

Brosnan, T.M. Ed. 1999. *Proceedings of the Conference on Early Warning Monitoring to Detect Hazardous Events in Water Supplies*, 17–19 May 1999, Reston, Virginia.

Brown, K.S. 1996. Scientists find jobs turning "extremozymes" into industrial catalysts. The Scientist, 10 (19): 1 (September 30, 1996).

Brown, M.J., and J.N. Lester. 1979. Metal removal in activated sludge: The role of bacterial extracellular polymers. Water Res. 13: 817–837.

Brown, L.M. 2000. *Helicobacter pylori*: epidemiology and routes of transmission. Epidemiol. Rev. 22: 283–297.

Brown, M.J., and J.N. Lester. 1982. Role of bacterial extracellular polymers in metal uptake in pure bacterial culture and activated sludge. I. Effect of metal concentration. Water Res. 16: 1539–1548.

Bruce, A.M., and R.D. Davis. 1989. Sewage sludge disposal: Current and future options. Water Sci. Technol. 21: 1113–1128.

Brummeler, E., L.W. Hulshoff Pol, J. Dolfing, G. Lettinga, and A.J.B. Zehnder. 1985. Methanogenesis in an upflow anaerobic sludge blanket reactor at pH 6 on an acetate-propionate mixture. Appl. Environ. Microbiol. 49: 1472–1477.

Brunner, P.H., S. Capri, A. Marcomini, and W. Giger. 1988. Occurrence and behaviour of linear alkylbenzenesulfonates, nonylphenol, nonylphenol mono- and nonylphenol diethoxylates in sewage and sewage sludge treatment. Water Res. 22: 1465–1472.

Brusseau, G.A., H.C. Tsien, R.S. Hanson, and S.P. Wackett. 1990. Optimization of trichloroethylene oxidation by methanotrophs and the use of a colorimetric assay to detect soluble methane mono-oxygenase activity. Biodegradation 1: 19–29.

Bryan, F.L. 1977. Diseases transmitted by foods contaminated by wastewater. J. Food Protect. 40: 45–56.

Bryan, J.A., J.D. Lehman, I.F. Setiady, and M.H. Hatch. 1974. An outbreak of hepatitis A associated with recreational lake water. Amer. J. Epidemiol. 99: 145–154.

Bryan, J.A., J.D. Lehmann, I.F. Setiady, and M.H. Hatch. 1974. An outbreak of hepatitis A associated with recreational lake water. Amer. J. Epidemiol. 99: 145.

Bryant, R.D., and E.J. Laishley. 1990. The role of hydrogenase in anaerobic biocorrosion. Can. J. Microbiol. 36: 259–264.

Bryant, R.D., W. Jansen, J. Boivin, E.J. Laishley, and J.W. Costerton. 1991. Effect of hydrogenase and mixed sulfate-reducing bacterial populations on the corrosion of steel. Appl. Environ. Microbiol. 57: 2804–2809.

Bryers, J., and W. Characklis. 1981. Early fouling biofilm formation in a turbulent flow system: Overall kinetics. Water Res. 15: 483–491.

Bryers, J.D., and G. Hamer. 1988. Application of immobilized captured microorganisms in water purification: An overview. Methods Enzymol. 137: 697–711.

Bucke, C. 1987. Cell immobilization on calcium alginate. Methods Enzymol. 135: 175–189.

Budnick, G.E., R.T. Howard, and D.R. Mayo. 1996. Evaluation of Enterolert for enumeration of enterococci in recreational waters. Appl. Environ. Microbiol. 62: 3881–3884.

Bukhari, Z., T.M. Hargy, J.R. Bolton, B. Dussert, and J.L. Clancy. 1999. Medium-pressure UV for oocyst inactivation. J. AWWA 91: 86–94.

Bukhari, Z., M.M. Marshall, D.G. Korich, C.R. Fricker, H.V. Smith, J. Rosen, and J.L. Clancy. 2000. Comparison of *Cryptosporidium parvum* viability and infectivity assays following ozone treatment of oocysts. Appl. Environ. Microbiol. 66: 2972–2980.

Bulich, A.A., M.W. Greene, and D.L. Isenberg. 1979. Reliability of the bacterial luminescence assay for determination of the toxicity of pure compounds and complex effluents, pp. 338–347, In: *Aquatic Toxicology and Risk Assessment*, D.R. Branson, and K.L. Dickson, Eds., American Society of Testing and Materials, Philadelphia.

Bulich A.A. 1986. Bioluminescent assays, pp. 57–74, In: Bitton G., and B.J. Dutka (Eds.), *Toxicity Testing Using Microorganisms*, Vol 1. CRC Press, Boca Raton, FL.

Bulich, A., and G. Bailey. 1995. Environmental toxicity assessment using luminescent bacteria, pp. 29–40, In: *Environmental Toxicity Assessment*, M. Richardson, Ed., Taylor and Francis, London.

Buras, N.S. 1984. Water reuse for aquaculture: Public Health aspects. In: *Water Reuse Symposium III*, San Diego, CA, Aug. 26–31, 1984, AWWA, Denver, CO.

Burger, J.S., W.O.K. Grabow, and R. Kfir. 1989. Detection of endotoxins in reclaimed and conventionally treated drinking water. Water Res. 23: 733–738.

Burggraf, S., T. Mayer, R. Amann, S. Schadhauser, C.R. Woese, and K.O. Stetter. 1994. Identifying members of the domain *Archaea* with rRNA-targeted oligonucleotide probes. Appl. Environ. Microbiol. 60: 3112–3119.

Burkhardt, W.III, W.D. Watkins, and S.R. Rippey. 1992. Survival and replication of male-specific bacteriophages in molluscan shellfish. Appl. Environ. Microbiol. 58: 1371–1373.

Burkhardt, W.III, and W.D. Watkins. 1992. *Clostridium perfringens* provides the only reliable measure of human contamination in the marine environment (Abstract Q-257). In: 92nd meeting of the American Society of Microbiology, May 26–30, New Orleans, LA.

Burkhardt, W.III, K.R. Calci, W.D. Watkins, S.R. Rippey, and S.J. Chirtel. 2000. Inactivation of indicator microorganisms in estuarine waters. Water Res. 34: 2207–2214.

Burkhardt, W.III, and K.R. Calci. 2000. Selective accumulation may account for shellfish-associated viral illness. Appl. Environ. Microbiol. 66: 1375–1378.

Burlage, R.S. 1997. Emerging technologies: Bioreporters, biosensors, and microprobes, pp. 115–123, In: *Manual of Environmental Microbiology*, C.J. Hurst, G.R. Knudsen, M.J. McInerney, L.D. Stetzenbach, and M.V. Walter, Eds., ASM Press, Washington, D.C.

Burnes, B.S. 2003. Antibiotic resistance analysis of fecal coliforms to determine fecal pollution sources in a mixed-use watershed. Environ. Monit. & Assess. 85: 87–98.

Burns, R.G., Ed. 1978. *Soil enzymes.* Academic Press, London.

Burrows, I.D., and S.E. Renner. 1999. Biological warfare agents as threats to potable water. Environ. Health Perspectives 107: 975–984.

Burttschell, R.H., A.A. Rosen, F.M. Middleton, and M.B. Ettinger. 1959. Chlorine derivatives of phenol causing taste and odor. J. Am. Water Works Assoc. 51: 205–214.

Busta, F.F. 1976. Practical implications of injured microorganisms in foods. J. Milk Food Technol. 39: 138–145.

Buttner, M.P., K. Willeke, and S.A. Grinshpun. 1997. Sampling and analysis of airborne microorganisms, pp. 629–640, In: *Manual of Environmental Microbiology*, C.J. Hurst, G.R. Knudsen, M.J. McInerney, L.D. Stetzenbach, and M.V. Walter, Eds., ASM Press, Washington, D.C.

Cabanes, P.-A., F. Wallet, E. Pringuez, and P. Pernin. 2001. Assessing the risk of primary amoebic meningoencephalitis from swimming in the presence of environmental *Naegleria fowleri*. Appl. Environ. Microbiol. 67: 2921–2937.

Cabelli, V.J. 1981. *Health Effects Criteria for Marine Recreational Waters.* EPA-600/1-80-031, U.S. Environmental Protection Agency, Cincinnati, OH.

Cabelli, V.J. 1989. Swimming-associated illness and recreational water quality criteria. Water Sci. Technol. 21: 13–21.

Cacciò, S.M., M. De Giacomo, F. A. Aulicino, and E. Pozio. 2003. *Giardia* cysts in wastewater treatment plants in Italy. Appl. Environ. Microbiol. 69: 3393–3398.

Cairncross, S. 1992. Control of enteric pathogens in developing countries, pp. 157–189, In: *Environmental Microbiology*, R. Mitchell, Ed., Wiley-Liss, New York.

Calabrese, J.P., and G.K. Bissonnette. 1989. Modification of standard recovery media for enhanced detection of chlorine-stressed coliform and heterotrophic bacteria (Abstract). In: 89th Ann. Meeting, the American Society Microbiology, New Orleans, LA, May 14–18, 1989.

Calabrese, J.P., and G.K. Bissonnette. 1990. Improved membrane filtration method incorporating catalase and sodium pyruvate for detection of chlorine-stressed coliform bacteria. Appl. Environ. Microbiol. 56: 3558–3564.

Caldwell, K.N., B.B. Adler, G.L. Anderson, P.L. Williams, and L.R. Beuchat. 2003. Ingestion of *Salmonella enterica* serotype Poona by a free-living nematode, *Caenorhabditis elegans*, and protection against inactivation by produce sanitizers. Appl. Environ. Microbiol. 69: 4103–4110.

Call, D.R., M.K. Borucki, and F.J. Loge. 2003. Detection of bacterial pathogens in environmental samples using DNA microarrays. J. Microbiol. Meth. 53: 235–243.

Callaway, W.T. 1968. The metazoa of waste treatment. J.Water Pollut. Control Fed. 40: R412–R422.

Calomiris, J.J., J.L. Armstrong, and R.J. Seidler. 1984. Association of metal tolerance with multiple antibiotic resistance of bacteria isolated from drinking water. Appl. Environ. Microbiol. 47: 1238–1242.

Calvo, L., and L.J. Garcia-Gil. 2004. Use of *amoB* as a new molecular marker for ammonia-oxidizing bacteria. J. Microbiol. Meth. 57: 69–78.

Camann, D.E., C.A. Sorber, B.P. Sagik, J.P. Glennon, and D.E. Johnson. 1978. pp. 240–266, In: *Risk Assessment and Health Effects of Land Application of Municipal Wastewater and Sludges*, B.P. Sagik, and C.A. Sorber, Eds., Center for Applied Research & Technology, Univ. of Texas, San Antonio, TX.

Camann, D.E., P.J. Graham, M.N. Guentzel, H.J. Harding, K.T. Kimball, B.E. Moore, R.L. Northrop, N.L. Altman, R.B. Harrist, A.H. Holguin, R.L. Mason, C.B. Popescu, and C.A. Sorber. 1986. The Lubbock Land Treatment System, Research and Demonstration Project, Vol. 4, Lubbock Infection Surveillance Study (LISS), Report EPA-600/S2-86/027d U.S. Environmental Protection Agency, Research Triangle Park, N.C.

Camann, D.E., B.E. Moore, H.J. Harding, and C.A. Sorber. 1988. Microorganism levels in air near spray irrigation of municipal wastewater: The Lubbock infection surveillance study. J. Water Pollut. Control Fed. 60: 1960–1970.

Campbell, A.T., and P. Wallis. 2002. The effect of UV irradiation on human-derived *Giardia lamblia* cysts. Water Res. 36: 963–969.

Campbell, A.T., L.J. Robertson, and H.V. Smith. 1992a. Viability of *Cryptosporidium parvum* oocysts: Correlation of in vitro excystation with inclusion or exclusion of fluorogenic vital dyes. Appl. Environ. Microbiol. 58: 3488–3493.

Campbell, A.T., R. Haggart, L.J. Robertson, and H.V. Smith. 1992b. Fluorescent imaging of *Cryptosporidium* using a cooled charge couple device (CCD). J. Microbiol. Methods 16: 169–174.

Campbell, I., S. Tzipori, G. Hutchinson, and K.W. Angus. 1982. Effect of disinfectants on survival of *Cryptosporidium* oocysts. Vet. Rec. 111: 414–415.

Campbell, M., G. Bitton, and B. Koopman. 1993. Toxicity testing of sediment elutriates based on inhibition of alpha-glucosidase biosynthesis in *Bacillus licheniformis*. Arch. Environ. Contam. Toxicol. 24: 469–472.

Camper, A.K., M.W. Lechevallier, S.C. Broadaway, and G.A. McFeters. 1985. Growth and persistence of pathogens on granular activated carbon filters. Appl. Environ. Microbiol. 50: 1378–1382.

Camper, A.K., M.W. Lechevallier, S.C. Broadaway, and G.A. McFeters. 1986. Bacteria associated with granular activated carbon particles in drinking water. Appl. Environ. Microbiol. 52: 434–438.

Camper, A.K., S.C. Broadaway, M.W. LeChevallier, and G.A. McFeters. 1987. Operational variables and the release of colonized granular activated carbon particles in drinking water. J. Am. Water Works Assoc. 79: 70–74.

Camper, A.K., and G.A. McFeters. 1979. Chlorine injury and the enumeration of waterborne coliform bacteria. Appl. Environ. Microbiol. 37: 633–641.

Camper, A.K., G.A. McFeters, W.G. Characklis, and W.L. Jones. 1991. Growth kinetics of coliform bacteria under conditions relevant to drinking water distribution systems. Appl. Environ. Microbiol. 57: 2233–2239.

Camper, A.K., K. Brastrup, A. Sandvig, J. Clement, C. Spencer, and A.J. Capuzzi. 2003. Efffect of distribution systems materials on bacterial regrowth. J. Amer. Water Wks. Assoc. 95: 107–121.

Canter, L.W., and R.C. Knox. 1984. Evaluation of septic tank system effects on groundwater quality. Report No. EPA-600/S2-84-107. R.S. Kerr Research Laboratory, Ada, OK.

Canter, L.W., and R.C. Knox. 1985. Septic Tank System: Effects on Groundwater Quality, Lewis, Chelsea, MI.

Canter, L.W., L.E. Streebin, M.C. Arquiaga, F.E. Carranza, D.E. Miller, and B.H. Wilson. 1990. Innovative processes for reclamation of contaminated subsurface environments. EPA/600/S2-90/017, U.S. Environmental Protection Agency, Ada, OK.

Capizzi-Banas, S. and J. Schwartzbrod. 2001. Irradiation of *Ascaris* ova in sludge using an electron beam accelerator. Water Res. 35: 2256–2260.

Carducci, A., E. Tozzi, E. Rubulotta, B. Casini, L. Cantiani, E. Rovini, M. Muscillo, and R. Pacini. 2000. Assessing airborne biological hazard from urban wastewater treatment. Water Res. 34: 1173–1178.

Carmichael, W.W. Ed. 1981a. *The Aquatic Environment: Algal Toxins and Health*. Plenum Press, New York.

Carmichael, W.W. 1981b. Freshwater blue-green algae toxins, pp. 7–13, In: *The Aquatic Environment: Algal Toxins and Health*, W.W. Carmichael, Ed., Plenum, New York.

Carmichael, W.W. 1989. Freshwwater cyanobacteria (blue-green algae) toxins, pp. 3–16, In: *Natural Toxins: characterization, pharmacology and therapeutics*, C.L. Ownby, and G.V. Odell, Eds., Pergamon Press, Oxford.

Carrington, B.G. 1985. Pasteurization: Effects on *Ascaris* eggs, pp. 121–125, In: *Inactivation of Microorganisms in Sewage Sludge by Stabilization Processes*, Strauch, D., A.H. Havelaar, and P. L'Hermite, Eds., Elsevier Appl. Sci., Pub., London.

Carroll, T., S. King, S.R. Gray, B.A. Bolto and N.A. Booker. 2000. The fouling of microfiltration membranes by NOM after coagulation treatment. Water Res. 34: 2861–2868.

Carson, L.A., and N.J. Petersen. 1975. Photoreactivation of *Pseudomonas cepacia* after ultraviolet exposure: A potential source of contamination in ultraviolet-treated waters. J. Clin. Microbiol. 1: 462–464.

Carraro, E., E.H. Bugliosi, L. Meucci, C. Baiocchi, and G. Gilli. 2000. Biological drinking water treatment processes, with special reference to mutagenicity. Water Res. 34: 3042–3054.

Carter, A.M., R.E. Pacha, G.W. Clark, and E.A. Williams. 1987. Seasonnal occurrence of *Campylobacter* spp. in surface waters and their correlation with standard indicator bacteria. Appl. Environ. Microbiol. 53: 523–526.

Carter, J.P., Y.H. Hsiao, S. Spiro, and D.J. Richardson. 1995. Soil and sediment bacteria capable of aerobic nitrate respiration. Appl. Environ. Microbiol. 61: 2852–2858.

Carter, J.T, E.W. Rice, S.G. Buchberger, and Y. Lee. 2000. Relationships between levels of heterotrophic bacteria and water quality parameters in a drinking water distribution system. Water Res. 34: 1495–1502.

Caruso, G., M. Mancuso, and E. Crisafi. 2003. Combined fluorescent antibody assay and viability staining for the assessment of the physiological states of *Escherichia coli* in seawaters. J. Appl. Microbiol. 95: 225–233.

Casey, T.G., M.C. Wentzel, R.E. Lowenthal, G.A. Ekama, and G.v.R. Marais. 1992. A hypothesis for the cause of low F/M filament bulking in nutrient removal activated sludge systems. Water Res. 26: 867–869.

Caslake, L.F., D.J. Connolly, V. Menon, C.M. Duncanson, R. Rojas, and J. Tavakoli. 2004. Disinfection of Contaminated Water by Using Solar Irradiation. Appl. Environ. Microbiol. 70: 1145–1151.

Cassells, J.M., M.T. Yahya, C.P. Gerba, and J.B. Rose. 1995. Efficacy of a combined system of copper an silver and free chlorine for inactivation of *Naegleria fowleri* amoebas in water. Water Sci. Technol. 31: 119–122.

Cassidy, M.B., H. Lee, and J.T. Trevors. 1996. Environmental applications of immobilized microbial cells: A review. J. Ind. Microbiol. 16: 79–101.

Casson, L.W., C.A. Sorber, J.L. Sykora, P.D. Cavaghan, M.A. Shapiro, and W. Jakubowski. 1990. *Giardia* in wastewater- Effect of treatment. J. Water Pollut. Control Fed. 62: 670–675.

Cech, J.S., P. Hartman, and J. Wanner. 1993. Competition between PolyP and non-polyP bacteria in an enhanced phosphate removal system. Water Environ. Res. 65: 690–697.

Cees, B., J. Zoeteman, and G.J. Piet. 1974. Cause and identification of taste and odour compounds in water. Sci. Total Environ. 3: 103–115.

Cellini, L., A. Del Vecchio, M. Di Candia, E. Di Campli, M. Favaro, and G. Donelli. 2004. Detection of free and plankton-associated *Helicobacter pylori* in seawater. J. Appl. Microbiol. 97: 285–292.

Centers for Disease Control and Prevention (CDC). 1991. Outbreaks of diarrheal illness associated with cyanobacteria (blue-green algae)-like bodies: Chicago and Nepal, 1989 and 1990. MMWR Morb. Mort. Weekly Rep. 40: 325–327.

Centers for Disease Control and Prevention (CDC). 1996. Nitrate-contaminated well water as a possible cause of spontaneous abortions. NMWR (In Press). (Cited by Kramer et al., 1996. J. AWWA 88: 66–80).

Centers for Disease Control and Prevention (CDC). 1999. Vessel Sanitation Program: Vessel Sanitation Operation Manual. Centers for Disease Control and Prevention, National Center for Infectious Diseases, Division of Parasitic Diseases, Atlanta, GA.

Cha, D.K., D. Jenkins, W.P. Lewis, and W.H. Kido. 1992. Process control factors influencing *Nocardia* populations in activated sludge. Water Environ. Res. 64: 37–43.

Chaiket, T., P.C. Singer, A. Miles, M. Moran, and C. Pallota. 2002. Effectiveness of coagulation, ozonation, and biofiltration in controlling DBPs. J. Amer. Water Works Assoc. 94: 81–95.

Chakrabarti, T., and P.H. Jones. 1983. Effect of molybdenum and selenium addition on the denitrification of wastewater. Water Res. 17: 931–936.

Chalfie, M., Y. Tu, G. Euskirchen, W.W. Ward, and D.C. Prasher. 1994. Green fluorescent protein as a marker for gene expression. Science 263: 802–805.

Chalmers, R.M., A.P. Sturdee, P. Mellors, V. Nicholson, F. Lawlor, F. Kenny, and P. Timpson. 1997. *Cryptosporidium parvum* in environmental samples in the Sligo area, Republic of Ireland: A preliminary report. Lett. Appl. Microbiol. 25: 380–384.

Chamberlin, C.E., and R. Mitchell. 1978. A decay model for enteric bacteria in natural waters, pp. 325–348, In: *Water Pollution Microbiology*, Vol 2, R. Mitchell, Ed., John Wiley and Sons, New York.

Chambers, B. 1982. Effect of longitudinal mixing and anoxic zones on settleability of activated sludge, In: *Bulking of Activated Sludge: Preventive and Remedial Methods*, B. Chambers, and E.J. Tomlinson, Eds., Ellis Horwood, Chichester, U.K. PAGES?

Chang, G.W., J. Brill, and R. Lum. 1989. Proportion of β-glucuronidase negative *Escherichia coli* in human fecal samples. Appl. Environ. Microbiol. 55: 335–339.

Chang, J.C.H., S.F. Ossof, D.C. Lobe, M.H. Dorfman, C.M. Dumais, R.G. Qualls, and J.D. Johnson. 1985. UV inactivation of pathogenic and indicator microorganisms. Appl. Environ. Microbiol. 49: 1361–1365.

Chang, M.-H., M.K. Glynn, and S.L. Groseclose. 2003. Endemic, notifiable bioterrorism-related diseases, United States, 1992–1999. Emerg. Infect. Dis. 9: 556–64.

Chang, S.D., and P.C. Singer. 1991. The impact of ozonation on particle stability and the removal of TOC and THM precursors. J. Amer. Water Works Assoc. 83: 71–79.

Chang, S-L. 1982. The safety of water disinfection. Ann. Rev. Public Health 3: 393–418.

Chang, S.L., R.L. Woodward, and P.W. Kabler. 1960. Survey of free living nematodes and amoebas in municipal supplies. J. Am. Water Works. Assoc. 52: 613–618.

Chang, Y., J.T. Pfeiffer, and E.S.K. Chian. 1979. Comparative study of different iron compounds in inhibition of *Sphaerotilus* growth. Appl. Environ. Microbiol. 38: 385–389.

Chang, Y., C.-W. Li, and M.M. Benjamin. 1997. Iron oxide-coated media for NOM sorption and particulate filtration. J. Amer. Water Works Assoc. 89: 100–113.

Chao, A.C., and T.M. Keinath. 1979. Influence of process loading intensity on sludge clarification and thickening characteristics. Water Res. 13: 1213–1217.

Chapron, C.D., N.A. Ballester, and A.B. Margolin. 2000. The detection of astrovirus in sludge bio-solids using an integrated cell culture nested PCR technique. J. Appl. Microbiol. 89: 11–15.

Characklis, W.G., Ed. 1988. *Bacterial Regrowth in Distribution Systems*. Research Report, AWWA Res. Foundation, Denver, CO.

Characklis, W.G., and K.E. Cooksey. 1983. Biofilms and microbial fouling. Adv. Appl. Microbiol. 29: 93–138.

Characklis, W.G., M.G. Trulear, J.D. Bryers, and N. Zelver. 1982. Dynamic of biofilm processes: Methods. Water Res. 16: 1207–1216.

Charnock, C., and O. Kjønnø. 2000. Assimilable organic carbon and biodegradable dissolved organic carbon in Norwegian raw and drinking water. Water Res. 34: 2629–2642.

Charoenca, N., and R.S. Fujioka. 1993. Assessment of *Staphylococcus* bacteria in Hawaii recreational waters. Water Sci. Technol. 27: 283–289.

Chattopadhyay, D., S.Chattopadhyay, W.G. Lyon, and J.T. Wilson. 2002. Effect of surfactants on the survival and sorption of viruses. Environ. Sci. Technol. 36: 4017–4024.

Chaudhury, G.R., and S. Chapalamadugu. 1991. Biodegradation of halogenated organic compounds. Microbiol. Rev. 55: 59–79.

Chauret, C., N. Armstrong, J. Fisher, R. Sharma, S. Springthorpe, and S. Sattar. 1995. Correlating *Cryptosporidium* and *Giardia* with microbial indicators. J. Am. Water Works Assoc. 87: 76–84.

Chauret, C.P., C.Z. Radziminski, M. Lepuil, R. Creason, and R.C. Andrews. 2001. Chlorine dioxide inactivation of *Cryptosporidium parvum* oocysts and bacterial spore indicators. Appl. Environ. Microbiol. 67: 2993–3001.

Cheeseman, P., A. Toms-Wood, and R.S. Wolfe. 1972. Isolation and properties of a fluorescent compound, factor F_{420}, from *Methanobacterium* strain M.O.H. J. Bacteriol. 112: 527–531.

Cheetham, P.S.J., and C. Bucke. 1984. Immobilization of microbial cells and their use in waste water treatment, p 219–235, In: *Microbiological Methods for Environmental Biotechnology*, J.M. Grainger and J.M. Lynch, Eds., Academic Press, London.

Chen, G., B.W. Dussert, and I.H. Suffet. 1997a. Evaluation of granular activated carbons for removal of methylisoborneol to below odor threshold concentration in drinking water. Water Res. 31: 1155–1163.

Chen, K.-C., Y.-F. Lin, and J.-Y Houng. 1997. Performance of a continuous stirred tank reactor with immobilized denitrifiers and methanogens. Water Environ. Res. 69: 233–239.

Chen, S., and D.B. Wilson. 1997b. Construction and characterization of *Escherichia coli* genetically engineered for bioremediation of Hg^{2+}-contaminated environments. Appl. Environ. Microbiol. 63: 2442–2445.

Chen, Y.S., O.J. Sproul, and A. Rubin. 1985. Inactivation of *Naegleria gruberi* cysts by chlorine dioxide. Water Res. 19: 783–789.

Chen, Y.S., J.M. Vaughn, and R.M. Niles. 1987. Rotavirus RNA and protein alterations resulting from ozone treatment. (abstr. Q-22), In: *Annual Meeting of the American Society of Microbiology*, 1987.

Chen, Y.-S, and J. Vaughn. 1990. Inactivation of human and simian rotaviruses by chlorine dioxide. Appl. Environ. Microbiol. 56: 1363–1366.

Cheng, M.H., J.W. Patterson, and R.A. Minear. 1975. Heavy metal uptake by activated sludge. J. Water Pollut. Control Fed. 47: 362–376.

Chet, I., and R. Mitchell. 1976. Ecological aspects of microbial chemotactic behavior. Annu. Rev. Microbiol. 30: 221–239.

Chevalier, P., and J. de la Noue. 1985. Wastewater nutrient removal with microalgae immobilized in carrageenan. Enzyme Microbiol. Technol. 7: 621–624.

Chiesa, S.C., and R.L. Irvine. 1985. Growth and control of filamentous microbes in activated sludge: An integrated hypothesis. Water Res. 19: 471–479.

Choi, E., and J.M. Rim. 1991. Competition and inhibition of sulfate reducers and methane producers in anaerobic treatment. Water Sci. Technol. 23: 1259–1264.

Choi, J., and R. L. Valentine. 2002. Formation of N-nitrosodimethylamine (NDMA) from reaction of monochloramine: a new disinfection by-product. Water Res. 36: 817–824.

Christensen, G.M., D. Olson, and B. Reidel. 1982. Chemical effects on the activity of eight enzymes: A review and a discussion relevant to environmental monitoring. Environ. Res. 29: 247–255.

Christensen, M.H., and P. Harremoes. 1977. Biological denitrification of sewage: A literature review. Prog. Water Technol. 8: 509–555.

Christensen, M.H., and P. Harremoes. 1978. Nitrification and denitrification in wastewater treatment, pp. 391–414, In: *Water Pollution Microbiology*, Vol. 2, R. Mitchell, Ed., John Wiley & Sons, New York.

Christiansen, J.A., and P.W. Spraker. 1983. Improving effluent quality of petrochemical wastewaters with mutant bacterial cultures, pp. 567–576, In: *Proceedings 37th. Industrial Waste Conference*, Purdue University, J.M. Bell, Ed., Ann Arbor Scientific Publications, Ann Arbor, MI.

Christopher, G.W, T.J. Cieslak, J.A. Pavlin, and E.M. Eitzen, Jr. 1997. Biological warfare: a historical perspective. J. Amer. Med. Assoc. 278 (5): 412–417.

Chu, F.S., and G.Y. Li. 1994. Simultaneous occurrence of fumonisin B1 and other mycotoxins in moldy corn collected from the People's Republic of China in regions with high incidence of esophageal cancer. Appl. Environ. Microbiol. 60: 847–852.

Chudoba, J. 1985. Control of activated sludge filamentous bulking. VI: Formulation of basic principles. Water Res. 19: 1017–1022.

Chudoba, J. 1989. Activated sludge-Bulking control, pp. 171–202, In: *Encyclopedia* of *Environmental Control Technology*, Vol. 3 (*Wastewater Treatment Technology*), P.N. Cheremisinoff, Ed., Gulf Pub. Co., Houston, TX.

Chudoba, J., J.S. Cesh, J. Farkac, and P. Grau. 1985. Control of activated sludge filamentous bulking. Experimental verification of kinetic selection theory. Water Res. 19: 191–196.

Chudoba, J., P. Grau, and V. Ottova. 1973. Control of activated sludge filamentous bulking II: Selection of microorganisms by means of a selector. Water Res. 7: 1389–1406.

Chung, H., and M.D. Sobsey. 1993a. Survival of F-specific coliphages, *Bacteroides fragilis* phages, hepatitis A virus (HAV) and poliovirus 1 in seawater and sediment. Water Sci. Technol. 27: 425–428.

Chung, H., and M.D. Sobsey. 1993b. Comparative survival of indicator viruses and enteric viruses in seawater and sediment. Water Sci. Technol. 27: 425–428.

Chung, Y.C., and J.B. Neethling. 1988. ATP as a measure of anaerobic sludge digester activity. J. Water Pollut. Control Fed. 60: 107–112.

Chung, Y.C., and J.B. Neethling. 1989. Microbial activity measurements for anaerobic sludge digestion. J. Water Pollut. Control Fed. 61: 343–349.

Ciesielski, C.A., M.J. Blaser, and W.L. Wang. 1984. Role of stagnation and obstruction of water flow in isolation of *Legionella pneumophila* from hospital plumbing. Appl. Environ. Microbiol. 48: 984–987.

Clancy, J.L., Z. Bukhari, T.M. Hargy, J.R. Bolton, B.W. Dussert, and M.M. Marshall. 2000. Using UV to inactivate *Cryptosporidium*. J. AWWA 92 (9): 97–104.

Clancy, J.L., M.M. Marshall, T.M. Hargy, and D.G. Korich. 2004. Susceptibility of five strains of *Cryptosporidium parvum* oocysts to UV light. J. Amer. Water Works Assoc. 96 (3): 84–93.

Clark, C.S., A.B. Bjornson, G.M. Schiff, J.P. Phair, G.L. van Meer, and P.S. Gartside. 1977. Sewage worker's syndrome. Lancet 1: 1009.

Clark, C.S., H.S. Bjornson, J.W. Holland, V.J. Elia, V.A. Majeti, C.R. Meyer, W.F. Balistreri, G.L. van Meer, P.S. Gartside, B.L. Specker, C.C. Linnemann, Jr., R. Jaffa, P.V. Scarpino, K. Brenner, W.J. Davis-Hoover, G.W. Barrett, T.S. Anderson, and D.L. Alexander. 1981. Evaluation of the Health Risks Associated with the Treatment and Disposal of Municipal Wastewater and Sludge. EPA-600/S1-81-030, U.S. Environmental Protection Agency, Cincinnati, OH.

Clark, C.S., H.S. Bjornson, J. Schwartz-Fulton, J.W. Holland, and P.S. Gartside. 1984. Biological health risks associated with the composting of wastewater treatment plant sludge. J. Water Pollut. Control Fed. 56: 1269–1276.

Clark, D.L., B.B. Milner, M.H. Stewart, R.L. Wolfe, and B.H. Olson. 1991. Comparative study of commercial 4-methylumbelliferyl-β-D-glucuronide preparations with the Standard Methods membrane filtration fecal coliform test for the detection of *Escherichia coli* in water samples. Appl. Environ. Microbiol. 57: 1528–1534.

Clark, J.A., C.A. Burger, and L.E. Sabatinos. 1982. Characterization of indicator bacteria in municipal raw water, drinking water, and new main water. Can. J. Microbiol. 28: 1002–1013.

Clark, R.M., E.J. Read, and J.C. Hoff. 1989. Analysis of inactivation of *Giardia lamblia* by chlorine. J. Environ. Eng. Div. 115: 80–90.

Clarke, N.A., and S.L. Chang. 1975. Removal of enteroviruses from sewage by bench-scale rotary-tube trickling filters. Appl. Microbiol. 30: 223–228.

Cleasby, J.L., D.J. Hilmoe, and C. J. Dimitracopoulos. 1984. Slow sand and direct in-line filtration of a surface water. J. Amer. Water Works Assoc. 76: 44–55.

Clements, K.D., and S. Bullivant. 1991. An unusual symbiont from the gut of surgeonfish may be the largest known prokaryote. J. Bacteriol. 173: 5359–5362.

Cleuziat, P., and J. Robert-Baudouy. 1990. Specific detection of *Escherichia coli* and *Shigella* species using fragments of genes coding for β-glucuronidase. FEMS Microbiol. Lett. 72: 315–322.

Cliver, D.O. 1984. Significance of water and environment in the transmission of virus disease. Monog. Virol. 15: 30–42.

Cliver, D.O. 1985. Vehicular transmission of hepatitis A. Public Health. Rev. 13: 235–292.

Cloete, T.E., and V.S. Brözel. 2002. Biofouling: Chemical control of biofouling in water systems, pp. 601–609, In: *Encyclopedia of Environmental Microbiology*, G. Bitton, editor-in-chief, Wiley-Interscience, N.Y.

Cloete, T.E., and D.J. Oosthuizen. 2001. The role of extracellular exopolymers in the removal of phosphorus from activated sludge. Water Res. 35: 3595–3598.

Cloete, T.E., and P.L. Steyn. 1988. The role of *Acinetobacter* as a phosphorus removing agent in activated sludge. Water Res. 22: 971–976.

Close, M., L.R. Hodgson, and G. Tod. 1989. Field evaluation of fluorescent whitening agents and sodium tripolyphosphate as indicators of septic tank contamination in domestic wells. N. Z. J. Mar. Freshwater Res. 23: 563–568.

Cobb, G.D., and E.J. Bouwer. 1991. Effect of electron acceptors on halogenated organic compounds biotransformations in a biofilm column. Environ. Sci. Technol. 25: 1068–1074.

Codony, F., P. Domenico, and J. Mas. 2003. Assessment of bismuth thiols and conventional disinfectants on drinking water biofilms. J. Al. Microbiol. 95: 288–293.

De Coen, W.M., K. Van Campenhout, G. Vandenbergh, B. Denayer, C.R. Janssen, and J.P. Giesy. 1999. The occurrence of estrogenic effects in Flanders, Belgium: A screening of resident fish populations. Abstract, *9th. Intern. Symp. Toxicity Testing*, Pretoria, South Africa (Sept. 26–Oct. 1, 1999)

Coetzee, J.N. 1987. Bacteriophage taxonomy, pp. 45–86, In: *Phage Ecology*, S.M. Goyal, C.P. Gerba, and G. Bitton, Eds., John Wiley & Sons, New York.

Cohen. A., A.M. Breure, J.G. van Andel, and A. van Deursen. 1980. Influence of phase separation on the anaerobic digestion of glucose. I. Maximum COD-turnover rate during continuous operation. Water Res. 14: 1439–1448.

Cohen, M.L. 1992. Epidemiology of drug resistance: Implications for a post-antimicrobial era. Science 257: 1050–1055.

Colbourne, J.S., P.J. Dennis, R.M. Trew, C. Berry, and G. Vesey. 1988. *Legionella* and public water supplies. *Proceedinds of the International Conference on Water and Wastewater Microbiology*, Vol. 1, Newport Beach, CA.

Cole, C.A., J.B. Stamberg, and D.F. Bishop. 1973. Hydrogen peroxide cures filamentous growth in activated sludge. J. Water Pollut. Control Fed. 45: 829–836.

Colignon, A., M.N. Fortin, and G. Martin. 1986. Action de l'ozone sur le foisonnement filamenteux. Sci. de l'Eau 5: 137–142.

Collins, M.R., T.T. Eightmy, J. Fenstermacher, and S.K. Spanos. 1992. Removing natural organic matter by conventional slow sand filtration. J. Amer. Water Works. Assoc. 84: 80–90.

Colleran, E., S. Finnegan, and P. Lens. 1995. Anaerobic treatment of sulphate-containing waste streams. Antonie van Leeuwenhoek 67: 29–46.

Colwell, R.R.,and D.J. Grimes. 1986. Evidence of genetic modification of microorganisms occurring in natural aquatic environments, pp. 222–230, In: *Aquatic Toxicology and Environmental Fate*, Vol. 9, T.M. Poston, and R. Purdy, Eds., ASTM Publication No. 921, ASTM, Philadelphia.

Colwell, R.R., and G.S. Sayler. 1978. Microbial degradation of industrial chemicals, pp. 111–134, In: *Water Pollution Microbiology*, Vol. 2, R. Mitchell, Ed., John Wiley & Sons, New York.

Comeau, Y., K.J. Hall, R.E.W. Hancock, and W.K. Oldham. 1986. Biochemical model for enhanced biological phosphorus removal. Water Res. 20: 1511–1521.

Comeau, Y., B. Rabinowitz, K.J. Hall, and W.K. Oldham. 1987. Phosphate release and uptake in enhanced biological phosphorus removal from wastewater. J. Water Pollut. Control Fed. 59: 707–715.

Compeau, G., and R. Bartha. 1985. Sulfate reducing bacteria: Principal methylators of mercury in anoxic estuarine sediments. Appl. Environ. Microbiol. 50: 498–502.

Compeau, G., and R. Bartha. 1987. Effect of salinity on mercury-methylating activity of sulfate-reducing bacteria in estuarine sediments. Appl. Environ. Microbiol. 53: 261–265.

Compeau, G.C., W.D. Mahaffey, and L. Patras. 1991. Full-scale bioremediation of contaminated soil and water, pp. 91–109, In: *Environmental Biotechnology for Waste Treatment*, G.S. Sayler, R. Fox, and J.W. Blackburn. Eds., Plenum, New York.

Condie, L.W. 1986. Toxicological problems associated with chlorine dioxide. J. Amer. Water Works Assoc. 78: 73–78.

Connery, N., A.S. Thompson, S. Patrick, and M.J. Larkin. 2002. Studies of *Microthrix parvicella* in situ and in laboratory culture: production and use of specific antibodies. Water Sci. & Technol. 46 (1–2): 115–118.

Cook, D., G. Newcombe, and P. Sztajnbok. 2001. The application of powdered activated carbon for mib and geosmin removal: predicting pac doses in four raw waters. Water Res. 35: 1325–1333.

Cooper, P.F., and D.H.V. Wheeldon. 1981. Fluidized-and expanded-bed reactors for wastewater treatment. Water Pollut. Control 79: 286–306.

Cooper, R.C. 1991. Public health concerns in wastewater reuse. Water Sci. Technol. 24: 55–65.

Cooper R.C., A.W. Olivieri, R.E. Danielson, P.G. Badger, R.C. Spear, and S. Selvin. 1986. Evaluation of military field-water Quality. Vol 5, Infectious organisms of military concern associated with consumption: Assessment of health risks, and recommendations for establishing related standards. APO 82PP2817, U.S. Army Medical Research and Development Command Ft. Detrick, MD.

Cornax, R., M.A. Morinigo, I.G. Paez, M.A. Munoz, and J.J. Borrego. 1990. Application of direct plaque assay for detection and enumeration of bacteriophages of *Bacteroides fragilis* from contaminated water samples. Appl. Environ. Microbiol. 56: 3170–3173.

Correa, I.E., N. Harb, and M. Molina. 1989. Incidence and prevalence of *Giardia* spp. in Puerto Rican waters: Removal of cysts by conventional sewage treatment plants, In: 89th Annual Meeting, American Society Microbiology, New Orleans, LA, 14–18, 1989.

Corso, P.S., M.H. Kramer, K.A. Blair, D.G. Addiss, J.P. Davis, and A.C. Haddix. 2003. Cost of illness in the 1993 waterborne *Cryptosporidium* outbreak, Milwaukee, Wisconsin. Emerg. Infect. Dis. 9: 426–431.

Coskuner, G., and T.P. Curtis. 2002. *In situ* characterization of nitrifiers in an activated sludge plant: detection of *Nitrobacter Spp.* J. Appl. Microbiol. 93: 431–437.

Costan, G., N. Bermingham, C. Blaise, and J.F. Ferard. 1993. Potential ecotoxic effects probe (PEEP): A novel index to assess and compare the toxic potential of industrial effluents. Environ. Toxicol. Water Qual. 8: 115–140.

Costerton, J.W. 1980. Some techniques involved in study of adsorption of microorganisms to surfaces, pp. 403–423, In: *Adsorption of Microorganisms to Surfaces*, G. Bitton, and K.C. Marshall, Eds., John Wiley & Sons, New York.

Costerton, J.W., and G.C. Geesey. 1979. Microbial contamination of surfaces, pp. 211–221, In: *Surface Contamination*, K.L. Mittal, Ed., Plenum, New York.

Costerton, J.W., J. Boivin, E.J. Laishley, and R.D. Bryant. 1989. A new test for microbial corrosion, pp. 20–25, In: *Sixth Asian-Pacific Corrosion Control Conference*, Corrosion Association of Singapore Asian-Pacific Materials and Corrosion Association, Singapore.

Costerton, J.W., Z. Lewandowski, D.E. Caldwell, D.R. Korber, and H.M. Lappin-Scott. 1995. Microbial biofilms. Annu. Rev. Microbiol. 49: 711–745.

Cotte, L., M. Rabodonirina, F. Chapuis, F. Bailly, F. Bissuel, C.l Raynal, P. Gelas, F. Persat, M.-A. Piens, and C. Trepo. 1999. Waterborne outbreak of intestinal microsporidiosis in persons with and without human immunodeficiency virus infection. J. Infect. Dis. 180: 2003–2008.

Council on Dental Materials, Instruments and Equipment, Council on Dental Practice, and Council on Dental Therapeutics. 1988. Infection control recommendations for the dental office and dental laboratory. J. Am. Dent. Assoc. 116: 241–248.

Covert, T.C., L.C. Shadix, E.W. Rice, J.R. Haines, and R.W. Freyberg. 1989. Evaluation of the Autoanalysis Colilert test for detection and enumeration of total coliforms. Appl. Environ. Microbiol. 55: 2443–2447.

Covert, T.C., E.W. Rice, S.A. Johnson, D. Berman, C.H. Johnson, and P.J. Mason. 1992. Comparing defined-substrate tests for the detection of *Escherichia coli* in wastewater. J. Am. Water Works Assoc. 84: 98–105.

Covert, T.C., M.R. Rodgers, A.L. Reyes, and G.N. Stelma, Jr. 1999. Occurrence of nontuberculous mycobacteria in environmental samples. Appl. Environ. Microbiol. 65: 2492–2496.

Cowan, D.A., and S.G. Burton. 2002. Archaea in Biotechnology, pp. 259–276, In: *Encyclopedia of Environmental Microbiology*, G. Bitton, editor-in-chief, Wiley-Interscience, N.Y.

Cox, C.S. 1987. *The Aerobiological Pathway of Microorganisms*. John Wiley & Sons, Chichester, U.K.

Cox, C.S., and C.M. Mathes, Eds. 1995. *Bioaerosols Handbook*. Lewis, Boca Raton, FL.

Cox, D.P. 1978. The biodegradation of polyethylene glycols. Adv. Appl. Microbiol. 23: 173–194.

Craig, D.L., H.J. Fallowfield, and N.J. Cromar. 2003. Effectiveness of guideline faecal indicator organism values in estimation of exposure risk at recreational coastal sites. Water Sci. Technol. 47 (3): 191–198.

Crance, J.M., C. Gantzer, L. Schwartzbrod, and R. Deloince. 1998. Effect of temperature on the survival of hepatitis A virus and its capsidal antigen in synthetic seawater. Environ. Toxicol. Water Qual. 13: 89–92.

Craun, G.F. 1979. Waterborne giardiasis in the United States. Am. J. Public Health 69: 817–820.

Craun, G.F. 1984a. Health aspects of groundwater pollution, pp. 135–179, In: *Groundwater Pollution Microbiology*, G. Bitton, and C.P. Gerba, Eds., John Wiley & Sons, New York.

Craun, G.F. 1984b. Waterborne outbreaks of giardiasis: Current status, pp. 243–261, In: *Giadia and Giardiasis*, S.L. Erlandsen and E.A. Meyers, Eds., Plenum, New York.

Craun, G.F. 1986a. Statistics of waterborne outbreaks in the U.S. (1920–1980), pp. 73–159, In: *Water Diseases in the United States*, G.F. Craun, Ed., CRC Press, Boca Raton, FL.

Craun, G.F., Ed. 1986b. *Waterborne Diseases in the United States*. CRC Press, Boca Raton, FL. pp. 295.

Craun, G.F. 1988. Surface water supplies and health. J. Am. Water Works Assoc. 80: 40–52.

Craun, G.F. 2001. Waterborne disease outbreaks caused by distribution system deficiencies. J. AWWA 93 (9): 64–75.

Craun, G.F., and R. Calderon. 1996. Microbial risks in groundwater systems: Epidemilogy of water-borne outbreaks, In: *Under the Microscope. Examining Microbes in Groundwater.* Amer. Water Wks, Assoc. Res. Foundation, Dever, CO.

Crawford, R.L., and K.T. O'Reilly. 1989. Bacterial decontamination of agricultural wastewaters, pp. 73–89, In: *Biotreatment of Agricultural Wastewater*, M.E. Huntley, Ed., CRC Press, Boca Raton, FL.

Crawford, R.L., K.T. O'Reilly, and H.-L. Tao. 1989. Microorganism stabilization for *in situ* degradation of toxic chemicals, pp. 203–211, In: *Biotechnology and Biodegradation*, D. Kamely, A. Chakrabarty, and G.S. Omenn, Eds., Gulf, Houston, TX.

Créach, V., A.-C. Baudoux, G. Bertru, and B. Le Rouzic. 2003. Direct estimate of active bacteria: CTC use and limitations, J. Microbiol. Meth. 52: 19–28.

Criddle, C.S., J.T. DeWitt, D. Grbic-Galic, and P.L. McCarty. 1990. Transformation of carbon tetrachloride by *Pseudomonas sp.*, strain KC under denitrification conditions. Appl. Environ. Microbiol. 56: 3240–3246.

Crites, R.W., and G. Tchobanoglous. 1998. *Small and Decentralized Wastewater Management Systems.* McGraw-Hill Book Co, New York.

Croci, L., D. De Medici, C. Scalfaro, A. Fiore, M. Divizia, D. Donia, A.M. Cosentino, P. Moretti, and G. Costantini. 2000. Determination of enteroviruses, hepatitis A virus, bacteriophages and *Escherichia coli* in Adriatic Sea mussels. J. Appl. Microbiol. 88: 293–298.

Crook, J. 1985. Water reuse in California. J. Am. Water Works Assoc. 77: 60–71.

Crook, J. 1998. Findings of NRC report on the viability of augmenting drinking water supplies with reclaimed water, pp. 291–305, In: *Water Reuse*, Conference Proceedings Feb. 1–4, 1998, Lake Buena Vista, FL, AWWA-WEF.

Crowley, D.E., and R.S. Dungan. 2002. Metals: Microbial processes affecting metals, pp. 1878–1893, In: *Encyclopedia of Environmental Microbiology*, G. Bitton, editor-in-chief, Wiley-Interscience, N.Y.

Cruz, J.R., P. Caceres, F. Cano, J. Flores, A. Bartlett, and B. Torun. 1990. Adenovirus type 40 and Ead 41 and rotavirus associated with diarrhea in Children from Guatemala. J. Clin. Microbiol. 28: 1780–1784.

Cullen, T.R., and R.D. Letterman. 1985. The effect of slow sand filter maintainance on water quality. J. Am. Water Works Assoc. 77: 48–55.

Cunningham, J.A., H. Rahme, G.D. Hopkins, C. Lebron, and M. Reinhard. 2001. Enhanced *in situ* bioremediation of BTEX-contaminated groundwater by combined injection of nitrate and sulfate. Envron. Sci. Technol. 35: 1663–1670.

Curds, C.R. 1975. Protozoa, pp. 203–268, In: *Ecological Aspects of Used-Water Treatment.* Vol. 1. C.R. Curds and H.A. Hawkes, Eds., Academic Press, London.

Curds, C.R. 1982. The ecology and role of protozoa in aerobic sewage treatment processes. Annu. Rev. Microb. 36: 27–46.

Curds, C.R., and H.A. Hawkes, Eds. 1975. *Ecological Aspects of Used-Water Treatment.* Vol. 1, Academic Press, London.

Curds, C.R., and H.A. Hawkes, Eds. 1983. *Ecological Aspects of Used-Water Treatment*, Vol. 2, Academic Press, London.

Current, W.L. 1987. *Cryptosporidium*: Its biology and potential for environmental transmission. Crit. Rev. Env. Control 17: 21.

Current, W.L. 1988. The biology of *Cryptosporidium.* ASM News 54: 605–611.

Current, W.L., and L.S. Garcia. 1991. Cryptosporidiosis. Clin. Microbiol. Rev. 4: 325–358.

Curtis C., A. Lima, S.J. Lorano, and G.D. Veith. 1982. Evaluation of a bacterial bioluminescence bioassay as a method for predicting acute toxicity of organic chemicals to fish, pp. 170–178, In: *Aquatic Toxicity and Hazard Assessment*, J.G. Pearson, R.B. Foster, and W.E. Bishop, Eds., STP No. 766. ASTM, Philadelphia, PA.

Curtis, T.P., D.D. Mara, and S.A. Silva. 1992a. The effect of sunlight on faecal coliforms: Implications for research and design. Water Sci. Technol. 26: 1729–1738.

Curtis, T.P., D.D. Mara, and S.A. Silva. 1992b. Influence of pH, oxygen, and humic substances on ability of sunlight to damage fecal coliforms in waste stabilization pond water. Appl. Environ. Microbiol. 58: 1335–1343.

Cypionka, H., F. Widdel, and N. Pfennig. 1985. Survival of sulfate reducing bacteria after oxygen stress, and growth in sulfate-free oxygen-sulfide gradients. FEMS Microbiol. Ecol. 31: 39–45.

Dabert, P., B. Sialve, J.-P. Delgenes, R. Moletta, and J.J. Godon. 2001. Characterization of the microbial 16S rDNA diversity of an aerobic phosphorus-removal ecosystem and monitoring of its transition to nitrate respiration. Appl. Microbiol. Bitechnol. 55: 500–509.

Dagues, R.E. 1981. Inhibition of nitrogenous BOD and treatment plant performance evaluation. J. Water Pollut. Control Fed. 53: 1738–1741.

Dahab, M.F., and R.Y. Surampalli. 2002. Effects of aerobic and anaerobic digestion systems on pathogen and pathogen indicator reduction in municipal sludge. Water Sci. Technol. 46 (10): 181–187.

Daigger, G.T., M.H. Robbins, Jr., and B.R. Marshall. 1985. The design of a selector to control low-F/M filamentous bulking. J. Water Pollut. Control Fed. 57: 220–226.

Daims, H., J.L. Nielsen, P.H. Nielsen, K. Schleifer, and M. Wagner. 2001. *In situ* characterization of *Nitrospira*-like nitrite-oxidizing bacteria active in wastewater treatment plants. Appl. Environ. Microbiol. 67: 5273–5284.

Damgaard, L.R., L. P. Nielsen, and N. P. Revsbech. 2001. Methane microprofiles in a sewage biofilm determined with a microscale biosensor. Water Res. 35: 1379–1386.

Damgaard, L.R., and N.P. Revsbech. 1997. A microscale biosensor for methane containing methanotrophic bacteria and an internal oxygen reservoir. Anal. Chem. 69: 2262–2267.

Damgaard-Larsen, S., K.O. Jensen, E. Lund, and B. Nisser. 1977. Survival and movement of enterovirus in connection with land disposal of sludges. Water Res. 11: 503–508.

Darnall, D.W., B. Greene, M.T. Henzl, J.M. Hosea, R.A. McPherson, J. Sneddon, and M.D. Alexander. 1986. Selective recovery of gold and other metal ions from an algal biomass. Env. Sci. Technol. 20: 206–210.

Daniell, T.J., M.L. Davy, and R.J. Smith, R.J. 2000. Development of a genetically modified bacteriophage for use in tracing sources of pollution. J. Appl. Microbiol. 88: 860–869.

Dart, R.K., and R.J. Stretton. 1980. *Microbiological Aspects of Pollution Control*. Elsevier Amsterdam.

D'Ascenzo, G., A. Di Corcia, A. Gentili, R. Mancini, R. Mastropasqua, M. Nazzari, and R. Samperi. 2003. Fate of natural estrogen conjugates in municipal sewage transport and treatment facilities. Sci. Total Environ. 302: 199–209.

Davenport, R.J., T.P. Curtis, M. Goodfellow, F.M. Stainsby, and M. Bingley. 2000. Quantitative use of fluorescent *In situ* hybridization to examine relationships between mycolic acid-containing actinomycetes and foaming in activated sludge plants. Appl. Environ. Microbiol. 66: 1158–1166.

Davenport, R.J., and T.P. Curtis. 2002. Are filamentous mycolata important in foaming? Water Sci. & Technol. 46 (1–2): 529–533.

Davies-Colley, R.J., R.G. Bell, and A.M. Donnison. 1994. Sunlight inactivation of enterococci and fecal coliform in sewage effluent diluted in seawater. Appl. Environ. Microbiol. 60: 2049–2058.

Davis, J.W., and C.L. Carpenter. 1990. Aerobic biodegradation of vinyl chloride in groundwater sample. Appl. Environ. Microbiol. 56: 3878–3880.

Davis, M.L., and D.A. Cornwell. 1985. *Introduction to Environmental Engineering*. PWS Engineering, Boston, MA.

Day, H.R., and G.T. Felbeck, Jr. 1974. Production and analysis of a humic-like exudate from the aquatic fungus *Aureobasidium pullulans*. J. Am. Water Works Assoc. 66: 484–489.

Deakyne, C.W., M.A. Patel, and D.J. Krichten. 1984. Pilot plant demonstration of biological phosphorus removal. J. Water Poll. Contr. Fed. 56: 867–873.

Dean, R.B., and E. Lund. 1981. *Water Reuse: Problems and Solutions*. Academic Press, London, U.K.

Debartolomeis, J., and V.J. Cabelli. 1991. Evaluation of an *Escherichia coli* host strain for enumeration of F male-specific bacteriophages. Appl. Environ. Microbiol. 57: 1301–1305.

DeBoer, J., and K.D. Linstedt. 1985. Advances in water reuse applications. Water Res. 19: 1455–1461.

Decamp, O., A. Warren, and R. Sanchez. 1999. The role of ciliated protozoa in subsurface flow wetlands and their potential as bioindicators. Water Sci. Technol. 40 (3): 91–97.

DeFlaun, M. 2002. Bioaugmentation, pp. 434–442, In: *Encyclopedia of Environmental Microbiology*, G. Bitton, editor-in-chief, Wiley-Interscience, N.Y.

De Laat, J., F. Bouanga, M. Dore, and J. Mallevialle. 1985. Influence du developpement bacterien au sein des filtres de charbon actif en grains sur l'elimination de composes organiques biodegradables et non biodegradables. Water Res. 19: 1565–1578.

Delahaye, E., B. Welté, Y. Levi, G. Leblon, and A. Montiel. 2003. An ATP-based method for monitoring the microbiological drinking water quality in a distribution network. Water Res. 37: 3689–3696.

De Leon, R., S.M. Matsui, R.S. Baric, J.E. Herrmann, N.R. Blacklow, H.B. Greenberg, and M.D. Sobsey. 1993. Detection of Norwalk virus in stool specimens by reverse transcriptase-polymerase chain reaction and nonradioactive oligoprobes. J. Clin. Microbiol. 30: 3151–3157.

DeLeon, R., M.D. Sobsey, R.M. Matsui, and R.S. Baric. 1992. Detection of Norwalk virus by reverse transcriptase-polymerase chain reaction and non-reactive oligoprobes (RT-PCR-OP). IWPRC Int. Symp., Washington, D.C., May 26–29, 1992.

DeLong, E.F. 1993. Single cell identification using fluorescently labeled, ribosomal RNA-specific probes, pp. 285–294, In: *Handbook of Methods of Aquatic Microbial Ecology*, P.F. Kemp, B.F. Sherr, E.B. Sherr, and J.J. Cole, Eds., Lewis, Boca Raton, FL.

Delwiche, C.C. 1970. The nitrogen cycle. Sci. Amer. 223: 137–146.

Demain, A.L. 1984. Capabilities of microorganisms (and microbiologists), pp. 277–299, In: *Genetic Control of Environmental Pollutants*, G.S. Omenn, and A. Hollaender, Eds., Plenum New York.

De Mik, G., and I. De Groot. 1977. Mechanisms of inactivation of bacteriophage fX174 and its DNA in aerosols by ozone and ozonized cyclohexene. J. Hyg. 78: 199–211.

Deng, M.Q., D.O. Cliver, and T.W. Mariam. 1997. Immunomagnetic capture PCR to detect viable *Cryptosporidium parvum* oocysts from environmental samples. Appl. Environ. Microbiol. 63: 3134–3138.

Denileon, G.P. 2001. The who, what, why and how of counter terrorism issues. J. Amer. Water Works Assoc. 93 (5): 78–85.

Denis, F.A., E. Blanchovin, A. DeLigneres, and P. Flamen. 1974. Coxsackie A16 infection from lake water. J. Amer. Med. Assoc. 228: 1370.

Dennis, P.J., D. Green, and B.P. Jones. 1984. A note on the temperature tolerance of *Legionella*. J. Appl. Bacteriol. 56: 349–350.

Dennis, W.H., V.P. Olivieri, and C.W. Kruse. 1979. Mechanism of disinfection: Incorporation of Cl-36 into f2 virus. Water Res. 13: 363–369.

Dermer, O.C, V.S. Curtis, and F.R Leach. 1980. *Biochemical Indicators of Subsurface Pollution*. Ann Arbor Scientific Publications, Ann Arbor, MI.

Desbrow, C., E.J. Routledge, G.C. Brighty, J.P. Sumpter, and M. Waldock. 1998. Identification of estrogenic chemicals in STW effluent. 1. Chemical fractionation and *in vitro* biological screening. Environ. Sci. Technol. 32: 1549–1558.

De Serres, G., T.L. Cromeans, B. Levesque, N. Brassard, C. Barthe, M. Dionne, H. Prud'homme, D. Paradis, C.N. Shapiro, O.V. Naiman, and H.S. Margolis. 1999. Molecular confirmation of hepatis A virus from well water: Epidemiology and public health implications. J. Infect. Dis. 179: 37–43.

Deshusses, M.A., and H.H.J. Cox. 2002. Biotrickling filters for air pollution control, pp. 782–795, In: *Encyclopedia of Environmental Microbiology*, G. Bitton, editor-in-chief, Wiley-Interscience, Hoboken, N.J.

Devinny, J.S., and D.E. Chitwood. 2002. Biofiltration and bioodors, pp. 593–601, In: *Encyclopedia of Environmental Microbiology*, G. Bitton, editor-in-chief, Wiley-Interscience, Hoboken, N.J.

DeWaters, J.E., and F.A. DiGiano. 1990. The influence of ozonated natural organic matter on the biodegradation of a micropollutant in a GAC bed. J. Am. Water Works Assoc. 82: 69–75.

Dice, J.C. 1985. Denver's seven decades of experience with chloramination. J. Am. Water Works Assoc. 77: 34–37.

Diehl, J.D., Jr. 1991. Improved method for coliform verification. Appl. Environ. Microbiol. 57: 604–605.

Diekert, G., U. Konheiser, K. Piechulla, and R.K. Thauer. 1981. Nickel requirement and factor F_{430} content of methanogenic bacteria. J. Bacteriol. 148: 459–465.

DiGiano, F.A., C. Clarkin, M.J. Charles, M.J. Maerker, D.E. Francisco, and C. LaRocca. 1992. Testing of the EPA toxicity identification evaluation protocol in the textile dye manufacturing industry. Water Sci. Technol. 25: 55–63.

DiGiano, F.A., G. Andreottola, S. Adham, C. Buckley, P. Cornel, G.T. Daigger, A.G. (Tony) Fane, N. Galil, J.G. Jacangelo, A. Pollice, B.E. Rittmann, A. Rozzi, T. Stephenson, and Z. Ujang. 2004. Safe water for everyone. Water Environ. Technol. 16: 31–35.

Dionisi, H.M., A.C. Layton, G. Harms, I.R. Gregory, K.G. Robinson, and G. S. Sayler. 2002. Quantification of *Nitrosomonas oligotropha*-like ammonia-oxidizing bacteria and *Nitrospira* spp. from full-scale wastewater treatment plants by competitive PCR. Appl. Environ. Microbiol. 68: 245–253.

DiStefano, T.D., J.M. Gossett, and S.H. Zinder. 1991. Reductive dechlorination of high concentrations of pentachloroethene to ethene by an anaerobic enrichment culture in the absence of methanogenesis. Appl. Environ. Microbiol. 57: 2287–2292.

Divizia, M., C. Gnesivo, R.A. Bonapasta, G. Morace, G. Pisani, and A. Pana. 1993. Virus isolation and identification by PCR in an outbreak of hepatitis A: Epidemiological investigation. Water Sci. Technol. 27: 199–205.

Dixon, C., M. Meselson, T. Guillemin, and P. Hanna. 1999. Anthrax. New Engl. J. Med. 341: 815–826.

Dizer, H., A. Nasser, and J.M. Lopez. 1984. Penetration of different human pathogenic viruses into sand columns percolated with distilled water, groundwater, or wastewater. Appl. Environ. Microbiol. 47: 409–415.

Dizer, H., W. Bartocha, H. Bartel, K. Seidel, J.M. Lopez-Pila, and A. Grohmann. 1993. Use of ultraviolet radiation for inactivation of bacteria and coliphages in pretreated wastewater. Water Res. 27: 397–403.

Dmochewitz, S., and K. Ballschmiter. 1988. Microbial transformation of technical mixtures of polychlorinated biphenyls (PCB) by the fungus *Aspergillus niger*. Chemosphere 17: 111–121.

Doggett, M.S. 2000. Characterization of fungal biofilms within a municipal water distribution system. Appl. Environ. Microbiol. 66: 1249–1251.

Domek, M.J., M.W. LeChevallier, S.C. Cameron, and G.A. McFeters. 1984. Evidence for the role of copper in the injury process of coliform bacteria in drinking water. Appl. Environ. Microbiol. 48: 289–293.

Dodgson, K.S., G.F. White, J.A. Massey, J. Shapleigh, and W.J. Payne. 1984. Utilization of sodium dodecyl sulphate by denitrifying bacteria under anaerobic conditions. FEMS Microbiol. Lett. 24: 53–56.

Dolfing, J., and J.W. Mulder. 1985. Comparison of methane production rate and coenzyme F_{420} content of methanogenic consortia in anaerobic granular sludge. Appl. Environ. Microbiol. 49: 1142–1145.

Donlan, R.M. 2002. Biofilms: Microbial life on surfaces. Emerging Infect. Dis. 8: 880–890.

Donlon, B.A., E. Razo-Flores, J.A. Field, and G. Lettinga. 1995. Toxicity of N-substituted aromatics to acetoclastic methanogenic bacteria in granular sludge. Appl. Environ. Microbiol. 61: 3889–3893.

Donovan, J.F., J.E. Bates, and C.H. Rowell. 1980. *Guidelines for Water Reuse*. Camp, Dresser and McKee, Boston, MA, 106 pp.

Doohan, M. 1975. Rotifera, pp. 289–304, In: *Ecological Aspects of Used-Water Treatment*, Vol. 1: *The Organisms and Their Ecology*, C.R. Curds, and H.A. Hawkes, Eds., Academic Press, London.

Doré, W.J., and D.N. Lees. 1995. Behavior of *Escherichia coli* and male-specific bacteriophage in environmentally contaminated bivalve mollusks before and after depuration. Appl. Environ. Microbiol. 61: 2830–2834.

Dott, W., and P. Kampfer. 1988. Biochemical methods for automated bacterial identification and testing metabolic activities in water and wastewater. In: *International Conference on Water and Wastewater Microbiology*. Vol. 1. Newport Beach, CA., Feb. 8–11, 1988.

Dott, W., D. Feidierker, P. Kampfer, H. Schleibinger, and S. Strechel. 1989. Comparison of autochthonous bacteria and commercially available cultures with respect to their effectiveness in fuel oil degradation. J. Ind. Microbiol. 4: 365–374.

Dowd, S.E. 2002. Wastewater and biosolids as sources of airborne microorganisms, pp. 3320–3330, In: *Encyclopedia of Environmental Microbiology*, G. Bitton, editor-in-chief, Wiley-Interscience, N.Y.

Dowd, S.E., and S.D. Pillai. 1999. Identifying the sources of biosolids derived pathogens indicator organisms in aerosols by ribosomal DNA fingerprinting. J. Env. Sci. Hlth. A34: 1061–1074.

Doyle, M.P., and J.L. Schoeni. 1987. Isolation of *Escherichia coli* 0157:H7 from retail fresh meats and poultry. Appl. Environ. Microbiol. 53: 2394–2396.

Drakides, C. 1980. La microfaune des boues activees. Etude d'une methode d'observation et application au suivi d'un pilote en phase de demarrage. Water Res. 14: 1199–1207.

Driedger, A.M., J.L. Rennecker, and B.J. Mariñas. 2000. Sequential inactivation of *Cryptosporidium parvum* oocysts with ozone and free chlorine. Water Res. 34: 3591–3597.

Driedger, A., E. Staub, U. Pinkernell, B. Mariñas, W. Köster, and U. von Gunten 2001. Inactivation of *Bacillus subtilis* spores and formation of bromate during ozonation. Water Res. 35: 2950–2960.

Drikas, M., C.W.K. Chow, J. House, and M.D. Burch. 2001. Using coagulation, flocculation and settling to remove toxic cyanobacteria. J. AWWA 93 (2): 100–111.

Dubey, J.P. 2002. *Toxoplasma gondii*, pp. 3176–3183, In: *Encyclopedia of Environmental Microbiology*, G. Bitton, editor-in-chief, Wiley, N.Y.

Dubois, E., F. Le Guyader, L. Haugarreau, H. Kopecka, M. Cormier, and M. Pommepuy. 1997. Molecular epidemiological survey of rotaviruses in sewage by reverse transcriptase seminested PCR and restriction fragment length polymorphism assay. Appl. Environ. Microbiol. 63: 1794–1800.

Duboise, S.M., B.E. Moore, and B.P. Sagik. 1976. Poliovirus survival and movement in a sandy forest soil. Appl. Environ. Microbiol. 31: 536–543.

Dufour, A.P. 1986. Diseases caused by water contact, pp. 23–41, In: *Waterborne Diseases in the United States*, G.F. Craun, Ed., CRC Press, Boca Raton, FL.

Dugan, P.R. 1987a. Prevention of formation of acid drainage from high-sulfur coal refuse by inhibition of iron- and sulfur-oxidizing microorganisms. I. Preliminary experiments in controlled shaken flasks. Biotech. Bioeng. 29: 41–48.

Dugan, P.R. 1987b. Prevention of formation of acid drainage from high-sulfur coal refuse by inhibition of iron- and sulfur- oxidizing microorganisms. II. Inhibition in "run of mine refuse under simulated field conditions". Biotech. Bioeng. 29: 49–54.

Dumètre, A., and M.-L. Dardé. 2003. How to detect *Toxoplasma gondii* oocysts in environmental samples? FEMS Microbiol. Rev. 27: 651–661.

duMoulin, G.C., I. Sherman, and K.D. Stottmeier. 1981. *Mycobacterium intracellulare*: An emerging pathogen. Ann. Meeting Amer. Soc. Microbiol., Washington, D.C.

duMoulin, G.C., K.D. Stottmeier, P.A. Pelletier, A.Y. Tsang, and J. Hedley-Whyte. 1988. Concentration of *Mycobacterium avium* by hospital hot water systems. JAMA 260: 1599–1601.

DuPont, H.L., C.L. Chappell, C.R. Sterling, P.C. Okhuysen, J.B. Rose, and W. Jakubowski. 1995. The infectivity of *Crypyosporidium parvum* in healthy volunteers. N. Engl. J. Med. 332: 855–859.

Duncan, A. 1988. The ecology of slow sand filters, pp. 163–180, In: *Slow Sand Filtration: Recent Development in Water Treatment Technology*, N.J.D. Graham, Ed., Ellis Horwood, Chichester, U.K.

Duran, A.E., M. Muniesa, X. Méndez, F. Valero, F. Lucena, and J. Jofre. 2002. Removal and inactivation of indicator bacteriophages in fresh waters. J. Appl. Microbiol. 92: 338–347.

Durán, A.E., M. Muniesa, L. Mocé-Llivina, C. Campos, J. Jofre, and F. Lucena. 2003. Usefulness of different groups of bacteriophages as model micro-organisms for evaluating chlorination. J. Appl. Microbiol. 95: 29–37.

Dutka B.J. and G. Bitton, Eds. 1986. *Toxicity Testing using Microorganisms.* Vol 2. CRC Press, Boca Raton, FL.

Dutka, B.J., A. El Shaarawi, M.T. Martins, and P.S. Sanchez. 1987. North and south American studies on the potential of coliphage as a water quality indicator. Water Res. 21: 1127–1134.

Dutton, R.J., G. Bitton, and B. Koopman 1988. Enzyme biosynthesis versus enzyme activity as a basis for microbial toxicity testing. Toxicity Assess. 3: 245–253.

Dutton, R.J, G. Bitton, B. Koopman, and O. Agami. 1990. Inhibition of β-galactosidase biosynthesis in *Escherichia coli*: Effect of alterations of the outer membrane permeability to environmental toxicants. Toxicity Assess. 5: 253–264.

Duvoort-van Engers, L.E., and S. Coppola. 1986. State of the Art on sludge composting, pp. 59–75, In: *Processing and Use of Organic Sludge and Liquid Agricultural Wastes*, P. l'Hermite Ed., D. Reidel, Dordrecht, the Netherlands.

Dvorak, D.H., R.S. Hedin, H.M. Edenborn, and P.E. McIntire. 1992. Treatment of metal-contaminated water using bacterial sulfate reduction: Results from pilot-scale reactors. Biotechnol. Bioeng. 40: 609–616.

Dwyer, D.F., and J.M. Tiedje. 1983. Degradation of ethylene glycol and polyethylene glycols by methanogenic consortia. Appl. Environ. Microbiol. 46: 185–190.

Dwyer, D.F., M.L. Krumme, S.A. Boyd, and J.M. Tiedje. 1986. Kinetics of phenol biodegradation by an immobilized methanogenic consortium. Appl. Environ. Microbiol. 52: 345–351.

Dwyer, D.F., F. Rojo, and K.N. Timmis. 1988. Fate and behaviour in an activated sludge of a genetically-engineered microorganism designed to degrade substituted aromatic compounds, pp. 77–88, In: *The Release of Genetically-Engineered Microorganisms*, M. Sussman, C.H. Collins, F.A. Skinner, and D.E. Stewart-Tull, Eds., Academic Press, London.

Van Dyk, T.K., W.R. Majarian, K.B. Konstantinov, R.M. Young, P.S. Dhurjati, and R.A. LaRossa. 1994. Rapid and sensitive pollutant detection by induction of heat shock gene-bioluminescence gene fusions. Appl. Environ. Microbiol. 60: 1414–1420.

Eaton, J.W., C.F. Kolpin, and H.S. Swofford. 1973. Chlorinated urban water: A cause of dyalysis-induced hemolytic anemia. Science 181: 463–464.

Eberl, L. R. Schulze, A. Ammendola, O. Geisenberger, R. Erhart, C. Sternberg, S. Molin, and R. Amann. 1997. Use of green fluorescent protein as a marker for ecological studies of activated sludge communities. FEMS Microbiol. 149: 77–83.

EC 80/179. 1980. Directive of July 1980 relating to the quality of water intended for human consumption. Off. J. European Community 23, 11–29.

Eccles, H., and S. Hunt, Eds. 1986. *Immobilization of Ions by Bio-Sorption*, Ellis Horwood, Chichester, U.K.

Edberg, S.C., M.J. Allen, D.B. Smith, and the National Collaborative Study. 1988. National field evaluation of a defined substrate method for the simultaneous enumeration of total coliforms and *Escherichia coli* from drinking water: Comparison with the standard multiple-tube fermentation method. Appl. Environ. Microbiol. 54: 1595–1601.

Edberg, S.C., M.J. Allen, and D.B. Smith. 1989. Rapid, specific autoanalytical method for the simultaneous detection of total coliforms and *E. coli* from drinking water. Water Sci. Technol. 21: 173–177.

Edberg, S.C., M.J. Allen, D.B. Smith, and N.J. Kriz. 1990. Enumeration of total coliforms and *Escherichia coli* from source water by the defined substrate technology. Appl. Environ. Microbiol. 56: 366–369.

Edeline, F. 1988. *L'epuration biologique des eaux residuaires: Theorie et technologie*. Editions CEDEBOC, Liege, Belgium.

Edgehill, R.U., and R.K. Finn. 1983a. Activated sludge treatment of synthetic wastewater containing pentachlorophenol. Biotechnol. Bioeng. 25: 2165–2176.

Edgehill, R.R.U., and R.K. Finn. 1983b. Microbial treatment of soil to remove pentachlorophenol. Appl. Environ. Microbiol. 45: 1122–1125.

Edwards T., and B.C. McBride. 1975. New method for the isolation and identification of methanogenic bacteria. Appl. Microbiol. 29: 540–545.

Egli, C., T. Tschan, R. Schlotz, A.M. Cook, and T. Leisinger. 1988. Transformation of tetrachloromethane to dichloromethane and carbon dioxide by *Acetobacterium woodii*. Appl. Environ. Microbiol. 54: 2819–2823.

Eguchi, M., M. Kitagawa, Y. Suzuki, M. Nakamuara, T. Kawai, K. Okamura, S. Sasaki, and Y. Miyake. 2001. A field evaluation of *in situ* biodegradation of trichloroethylene through methane injection. Water Res. 35: 2145–2152.

Ehrlich, H.L. l98l. *Geomicrobiology*. Marcel Dekker, New York.

Eighmy, T.T., D. Maratea, and P.L. Bishop. 1983. Electron microscopic examination of wastewater biofilm formation and structural components. Appl. Environ. Microbiol. 45: 1921–1931.

Eighmy, T.T., M.R. Collins, S.K. Spanos, and J. Fenstermacher. 1992. Microbial populations, activities and carbon metabolism in slow sand filters. Water Res. 26: 1319–1328.

Eikelboom, D.H. 1975. Filamentous organisms observed in activated sludge. Water Res. 9: 365–388.

Eikelboom, D.H., and H.J.J. van Buijsen. 1981. *Microscopic Sludge Investigation Manual*. Report No. A94a, TNO Research Institute, Delft, The Netherlands.

Eikelboom, D.H., and B. Geurkink. 2002. Filamentous micro-organisms observed in industrial activated sludge plants. Water Sci. & Technol. 46 (1–2): 535–542.

Eisenberg, T.N., E.J Middlebrooks, and V.D. Adams. 1987. Sensitizer photooxidation for wastewater disinfection and detoxification. Water Sci. Technol. 19: 1225–1258.

Eisenhardt, A., E. Lund, and B. Nissen. 1977. The effect of sludge digestion on virus infectivity. Water Res. 11: 579–581.

Eitzen, E., J. Pavlin, T. Cieslak, G. Christopher, and R. Culpepper, Eds. 1998. *Medical Management of Biological Casualties Handbook*. Fort Dietrick, MD, USAMRIID.

Elhmmali, M.M., D.J. Roberts, and R.P. Evershed. 1997. Bile acids as a new class of sewage pollution. Environ. Sci. Technol. 31: 3663–3666.

Elhmmali, M.M., D.J. Roberts, and R.P. Evershed. 2000. Combined analysis of bile acids and serols/ stanols from riverine particulates to assess sewage discharges and other fecal sources. Environ. Sci. Technol. 34: 39–46.

Ellis, J., and W. Korth. 1993. Removal of geosmin and methylisoborneol from drinking water by adsorption on ultrastable zeolite-Y. Water Res. 27: 535–539.

Van Elsas, J.D., G.F. Duarte, A.S. Rosado, and K. Smalla. 1998. Microbiological and molecular biological methods for monitoring microbial inoculants and their effects in the soil environment. J. Microbiol. Meth. 32: 133–154.

El-Sharkawi, F., L. El-Attar, A. Abdel Gawad, and S. Molazem. 1989. Some environmental factors affecting survival of fecal pathogens and indicator organisms in seawater. Water Sci. Technol. 21: 115–120.

Emelko, M.B., and P.M. Huck. 2004. Microspheres as surrogates for *Cryptosporidium* filtration. J. Amer. Water Works Assoc. 96 (3): 94–105.

Engelbrecht, R.S. 1983. Source, testing and distribution, In: *Assessment of Microbiology and Turbidity Standards for Drinking Water*, P.S. Berger, and Y. Argaman, Eds., EPA Report No. EPA570-9-83001. U.S. Environmental protection Agency, Washington, D.C.

Engelbrecht, R.S., and E.O. Greening. 1978. Chlorine-resistant indicators, pp. 243–265, In: *Indicators of Viruses in Water and Food*. G. Berg, Ed., Ann Arbor Scientific Publications, Ann Arbor, MI., 424 pp.

Engelbrecht, R.S., D.H. Foster, E.O. Greening, and S.H. Lee. 1974. New microbial indicators of wastewater chlorination efficiency. Report No. EPA-670/2-73/082, Cincinnati, OH.

Engstrand, L. 2001. *Helicobacter* in water and waterborne routes of transmission. J. Appl. Microbiol. 90: 80S–84S.

Enriquez, C.E., C.J. Hurst, and C.P. Gerba. 1995. Survival of the enteric adenoviruses 40 and 41 in tap, sea, and wastewater. Water Res. 29: 2548–2553.

Enriquez, C.A. Alum, E.M. Suarez-Rey, C.Y. Choi, G. Oron, and C.P. Gerba. 2003. Bacteriophages MS2 and PRD1 in Turfgrass by Subsurface Drip Irrigation. J. Environ. Eng. 129: 852–857.

Enzinger, R.M., and R.C. Cooper. 1976. Role of bacteria and protozoa in the removal of *E. coli* from estuarine waters. Appl. Environ. Microbiol. 31: 758–763.

Epstein, E. 1979. In: *Workshop on the Health and Legal Implications of Sewage Sludge Composting*. Vol. 2, Energy Resources Co., Cambridge, MA.

Ercolani, G.L. 1976. Bacteriological Quality assessment of fresh marketed lettuce and fennel. Appl. Environ. Microbiol. 31: 847–852.

Erdal, U.G., Z.K. Erdal, and C.W. Randall. 2003. The competition between PAOs (phosphorus accumulating organisms) and GAOs (glycogen accumulating organisms) in EBPR (enhanced biological phosphorus removal) systems at different temperatures and the effects on system performance. Water Sci. Technol. 47 (11): 1–8.

Ericksen, T.H., and A.P. Dufour. 1986. Methods to identify water pathogens and indicator organisms, pp. 195–214, In: *Waterborne Diseases in the United States*, G.F. Craun, Ed., CRC Press, Boca Raton, FL.

Eschenhagen, M., M. Schupplerb, and I. Röskea. 2003. Molecular characterization of the microbial community structure in two activated sludge systems for the advanced treatment of domestic effluents. Water Res. 37: 3224–3232.

Escobar, I.C., and A.A. Randall. 2001. Case study: Ozonation and distribution system biostability. J. AWWA 93 (10): 77–89.

Escobar, I.C., A.A. Randall, and J.S. Taylor. 2001. Bacterial growth in distribution systems: Effect of assimilable organic carbon and biodegradable dissolved organic carbon. Environ. Sci. Technol. 35: 3442–3447.

Estes, M.K., and M.E. Hardy. 1995. Norwalk virus and other enteric caliciviruses, p. 1009–1034, In: *Infections of the Gastrointestinal Tract*, M.J. Blaser, P.D. Smith, J.I. Ravdin, H.B. Greenberg, and R.L. Guerrant, Eds., Raven Press, New York.

van Etten, J.L., L.C. Lane, and R.H. Meints. 1991. Viruses and viruslike particles of eucaryotic algae. Microbiol. Rev. 55: 586–620.

European Union (EU). 1998. Council Directive 98.83/EC on the quality of water intended for human consumption. Official J. of the European Communitities. 5.12.98, L330/32-L330/53.

Evans, P.J., D.T. Mang, K.S. Kim, and L.Y. Young. 1991. Anaerobic degradation of toluene by a denitrifying bacterium. Appl. Environ. Microbiol. 57: 1139–1145.

Evans, T.M., J.E. Schillinger, and D.G. Stuart. 1978. Rapid determination of bacteriological water quality by using *Limulus* lysate. Appl. Environ. Microbiol. 35: 376–382.

Fair, G.M., and W.F. Wells. 1934. Measurement of atmospheric pollution and contamination from wastewater by sewage treatment works. Proc. 19th annual meeting New Jersey Sew. Works Assoc., Trenton, N.J.

Falconer, I.R. 1989. Effect on human health of some toxic cyanobacteria (blue-green algae) in reservoirs, lakes and rivers. Toxicity Assess. 4: 175–184.

Falconer, I.R., and T.H. Buckley. 1989. Tumour promotion by *Microcystis* sp., a blue-green alga occurring in water supplies. Med. J. Aust. 150: 351.

Falih, A.M.K., and M. Wainwright. 1995. Nitrification *in vitro* by a range of filamentous fungi and yeasts. Lett. Appl. Microbiol. 21: 18–19.

Falkinham III, J.O. 2002. *Mycobacterium avium complex*, pp. 2112–2120, In: *Encyclopedia of Environmental Microbiology*, G. Bitton, editor-in-chief, Wiley-Interscience, N.Y.

Falkinham III, J.O., C.D. Norton, and M.W. LeChevallier. 2001. Factors influencing numbers of *Mycobacterium avium, Mycobacterium intracellulare*, and other mycobacteria in drinking water distribution systems. Appl. Environ. Microbiol. 67: 1225–1231.

Fane, S,A., N. Ashbolt, and S.B. White. 2002. Decentralized urban water reuse: The implications of system scale for cost and pathogen risk. Water Sci. Technol. 46 (6–7): 281–288.

Fang, H.H.P., I.W.C. Lau, and D.W.C. Chung. 1997. Inhibition of methanogenic activity of starch-degrading granules by aromatic compounds. Water Sci. Technol. 35: 247–253.

Fannin, K.F., S.C. Vana, and W. Jakubowski. 1985. Effect of an activated sludge wastewater treatment plant on airborne air densities of aerosols containing bacteria and viruses. Appl. Environ. Microbiol. 49: 1191–1196.

Farahbakhsh, K., and D.W. Smith. 2004. Removal of coliphages in secondary effluent by microfiltration – mechanisms of removal and impact of operating parameters. Water Res. 38: 585–592.

Farooq, S., and S. Akhlaque. 1983. Comparative response of mixed cultures of bacteria and virus to ozonation. Water Res. 17: 809–812.

Farooq, S., C.S. Kurucz, T.D. Waite, W.J. Cooper, S.R. Mane, and J.H. Greenfield. 1992. Treatment of wastewater with high enegy electron beam irradiation. Water Sci. Technol. 26: 1265–1274.

Farquhar, G.J., and W.C. Boyle. 1972. Control of *Thiothrix* in activated sludge. J. Water Pollut. Control Fed. 44: 14–19.

Farrah, S.R., and G. Bitton. 1982. Methods (other than microporous filters) for concentration of viruses from water, pp. 117–149, In: *Methods in Environmental Virology*, C.P. Gerba, and S.M. Goyal, Eds., Marcel Dekker, New York.

Farrah, S.R., and G. Bitton. 1984. Enteric bacteria in aerobically digested sludge. Appl. Environ. Microbiol. 47: 831–834.

Farrah, S.R., and G. Bitton. 1990. Viruses in the soil environment, pp. 529–556, In: *Soil Biochemistry*, Vol. 6, J.-M. Bollag, and G. Stotzky, Eds., Marcel Dekker, New York.

Farrah, S.R., G. Bitton, E.M. Hoffmann, O. Lanni, O.C. Pancorbo, M.C. Lutrick, and J.E. Bertrand. 1981. Survival of enteroviruses and coliform bacteria in a sludge lagoon. Appl. Environ. Microbiol. 41: 459–465.

Farrah, S.R., D.R. Preston, G.A. Toranzos, M. Girard, G.A. Erdos, and V. Vasuhdivan. 1991. Use of modified diatomaceous earth for removal and recovery of viruses in water. Appl. Environ. Microbiol. 57: 2502–2506.

Farrel, J.B., J. Smith, S. Hathaway, and R. Dean. 1974. Lime stabilization of primary sludges. J. Water Pollut. Control Fed. 46: 113–122.

Farrel, J.B., A.E. Erlap, J. Rickabaugh, D. Freedman, and S. Hayes. 1988. Influence of feeding procedure on microbial reductions and performance on anaerobic digestion. J. Water Pollut. Control Fed. 60: 635–644.

Faruque, S.M., R. Khan, M. Kamruzzaman, S. Yamasaki, Q.S. Ahmad, T. Azim, G.B. Nair, Y. Takeda, and D.A. Sack. 2002. Isolation of *Shigella dysenteriae* Type 1 and *S. flexneri* strains from surface waters in Bangladesh: Comparative molecular analysis of environmental *Shigella* isolates versus clinical strains. Appl. Environ. Microbiol. 68: 3908–3913.

Fathepure, B.Z., J.P. Nengu, and S.A. Boyd. 1987. Anaerobic bacteria that dechlorinate perchloroethene. Appl. Environ. Microbiol. 53: 2671–2674.

Fathepure, B.Z., J.M. Tiedje, and S.A. Boyd. 1988. Reductive dechlorination of hexachlorobenzene to tri- and dichlorobenzenes in anaerobic sewage sludge. Appl. Environ. Microbiol. 54: 327–330.

Fattal, B., M. Margalith, H.I. Shuval, Y. Wax, and A. Morag. 1987. Viral antibodies in agricultural populations exposed to aerosols from wastewater irrigation during a viral disease outbreak. Amer. J. Epidemiol. 125: 899–906.

Fattal, B., R.J. Vasl, E. Katzenelson, and H.I. Shuval. 1983. Survival of bacterial indicator organisms and enteric viruses in the mediterranean coastal waters off Tel-Aviv. Water Res. 17: 397–402.

Fattal, B., Y. Wax, M. Davies, and H.I. Shuval. 1986. Health risks associated with wastewater irrigation: An epidemiological study. Am. J. Public Health 76: 977–979.

Fattal, B., M. Margalith, H.I. Shuval, and A. Morag. 1984. Community exposure to wastewater and antibody prevalence to several enteroviruses, In: *Proceedings Water Reuse Symposium III*, American Water Works Association, Denver, CO.

Fayer, R., and B.L.P. Ungar. 1986. *Cryptosporidium spp.* and cryptosporidiosis. Microbiol. Rev. 50: 458–483.

Fayer, R., C.A. Farley, E.J. Lewis, J.M. Trout, and T.K. Graczyk. 1997. Potential role of the Eastern oyster, *Crassostrea virginica*, in the epidemiology of *Cryptosporidium parvum*. Appl. Environ. Microbiol. 63: 2086–2088.

Feachem, R.G., D.J. Bradley, H. Garelick, and D.D. Mara. 1983. *Sanitation and Disease: Health Aspect of Excreta and Wastewater Management*. John Wiley & Sons, Chichester.

Federal Register. 1987. National Primary Drinking Water Regulations: Filtration, disinfection, turbidity, *Giardia lamblia*, viruses, *Legionella* and heterotrophic bacteria. Proposed rule, 40 CFR parts 141 and 142. Fed. Register 52: 212: 42718 (Nov. 3, 1987).

Federal Register. 1993. Standards for the use and disposal of sewage sludge. Federal Register, Subpart D of the Part 503 Regulation. Vol. 58, No 32, February 19, 1993.

Fedorak, P.M., and S.E. Hrudey. 1984. The effects of phenols and some alkyl phenolics on batch anaerobic methanogenesis. Water Res. 18: 361–367.

Fedorak, P.M., and S.E. Hrudey. 1988. Anaerobic degradation of phenolic compounds with applications to treatment of industrial wastewaters, pp. 169–225, In: *Biotreatment Systems*, Vol. 1, D.L. Wise, Ed., CRC Press, Boca Raton, FL.

Fedorak, P.M., D.J. Roberts, and S.E. Hrudey. 1986. The effects of cyanide on the methanogenic degradation of phenolic compounds. Water Res. 20: 1315–1320.

Fedorak, P.M., and D.W.S. Westlake. 1980. Airborne bacterial densities at an activated sludge treatment plant. J. Water Pollut. Control Fed. 52: 2185–2192.

Feiler, H. 1980. *Fate of Priority Pollutants in Publicly Owned Treatment Works*, EPA-440/1-80-301, U.S. Environmental Protection Agency, Washington, D.C.

Fenchel, T.M., and B.B. Jorgensen. 1977. Detritus food chain in aquatic ecosystems. Adv. Microb. Ecol. 1: 1–58.

Feng, P.C.S., and P.A. Hartman. 1982. Fluorogenic assay for immediate confirmation of *Escherichia coli*. Appl. Environ. Microbiol. 43: 1320–1329.

Ferguson, D.W., M.J. McGuire, B. Koch, R.L. Wolfe, and E.M. Aieta. 1990. Comparing PEROXONE and ozone for controlling taste and odor compounds, disinfection by-products and microorganisms. J. Am. Water Works Assoc. 82: 181–191.

Fernandez, A., C. Tejedor, and A. Chordi. 1992. Effect of different factors on the die-off of fecal bacteria in a stabilization pond purification plant. Water Res. 26: 1093–1098.

Field, J.A., and G. Lettinga. 1987. The methanogenic toxicity and anaerobic degradability of a hydrolyzable tannin. Water Res. 21: 367–374.

Field, J.A., E. de Jong, G.F. Costa, and J.A.M. de Bont. 1992. Biodegradation of polycyclic aromatic hydrocarbons by new isolates of white rot fungi. Appl. Environ. Microbiol. 58: 2219–2226.

Fields, B.S., E.B. Shotts, Jr., J.C. Feeley, G.W. Gorman, and W.T. Martin. 1984. Proliferation of *Legionella pneumophila* as an intracellular parasite of the ciliated protozoan *Tetrahymena pyriformis*. Appl. Environ. Microbiol. 47: 467–471.

Fiksdal, L., J.S. Maki, S.J. LaCroix, and J.T. Staley. 1985. Survival and detection of *Bacteroides spp.*, prospective indicator bacteria. Appl. Environ. Microbiol. 49: 148–150.

Fiksdal, L., M. Pommepuy, M.-P. Caprais, and I. Midttun. 1994. Monitoring of fecal pollution in coastal waters by use of rapid enzymatic techniques. Appl. Environ. Microbiol. 60: 1581–1584.

Filipe, C.D.M., G.T. Daigger, and C.P.L. Grady. 2001a. Effects of pH on the rates of aerobic metabolism of phosphate-accumulating and glycogen-accumulationg organisms. Water Environ. Res. 73: 213–222.

Filipe, C.D.M., G.T. Daigger, and C.P.L. Grady. 2001b. pH as a key factor in the competition between glycogen-accumulationg organisms and phosphate-accumulating organisms. Water Environ. Res. 73: 223–232.

Finance, C.F. Lucena, M. Briguaud, M. Aymard, R. Pares, and L. Schwartzbrod. 1982. Etude quantitative et qualitative de la pollution virale de l'eau de mer a Barcelone. Rev. Fr. Sci. Eau 1: 139–149.

Finch, G.R., and N. Fairbairn. 1991. Comparative inactivation of poliovirus type 3 and MS2 coliphage in demand-free phosphate buffer by using ozone. Appl. Environ. Microbiol. 57: 3121–3126.

Finch, G.R., N. Neumann, L.L. Gyürék, L.L. Bradbury, and M. Belosevic M. 1998. Sequential chemical disinfection for the control of *Giardia* and *Cryptosporidium* in drinking water. *Proceedings of the Water Quality Technology Conference*, AWWA, Denver, CO.

Finlay, B.B., and S. Falkow. 1989. Common themes in microbial pathogenicity. Microbiol. Rev. 53: 210–230.

Finlay, B.J., and T. Fenchel. 1991. An anaerobic protozoan, with symbiotic methanogens, living in municipal landfill material. FEMS Microb. Ecol. 85: 169–180.

Finstein, M.S., J. Cirello, D.J. Suler, M.L. Morris, and P.F. Strom. 1980. Microbial ecosystems responsible for anaerobic digestion and composting. J. Water Pollut. Control Fed. 52: 2675–2685.

Finstein, M.S. 1992. Composting in the context of municipal solid waste management, pp. 355–374, In: *Environmental Microbiology*, R. Mitchell, Ed., Wiley-Liss, New York.

Fitch, M.W., G.E. Speitel, Jr., and G. Georgiou. 1996. Degradation of trichloroethylene by methanol-grown cultures of *Methylosinus trichosporium* OB3b PP358. Appl. Environ. Microbiol. 62: 1124–1128.

Fitzgerald, P.R. 1982. *Proceedings on Microbial Health Considerations of Soil Disposal of Wastewaters*. University of Oklahoma, Norman, OK. pp. 101–120, EPA-600/9-83-017.

Fitzmaurice, G.D., and N.F. Gray. 1989. Evaluation of manufactured inocula for use in the BOD test. Water Res. 23: 655–657.

Flemming, H.-C., and J. Wingender. 2002. Extracellular polymeric substances (EPS): Structural, ecological and technical aspects, pp. 1223–1231, In: *Encyclopedia of Environmental Microbiology*, G. Bitton, editor-in-chief, Wiley-Interscience, N.Y.

Flewett, T.H. 1982. Clinical features of rotavirus infections, pp. 125–137, In: *Virus Infections of the Gastrointestinal Tract*, D.A. Tyrell, and A.Z. Kapikian, Eds., Marcel Dekker, New York.

Fliermans, C.B., and R.S. Harvey. 1984. Effectiveness of 1-bromo-3-chloro-5,5-dimethylhydantoin against *Legionella pneumophila* in a cooling tower. Appl. Environ. Microbiol. 47: 1307–1310.

Fliermans, C.B., W.B. Cherry, L.H. Orrison, S.J. Smith, and L. Thacker. 1979. Isolation of *Legionella pneumophila* from nonepidemic aquatic habitats. Appl. Environ. Microbiol. 37: 1239–1242.

Fliermans, C.B., W.B. Cherry, L.H. Orrison, S.J. Smith, and D.H. Pope. 1981. Ecological distribution of *Legionella pneumophila*. Appl. Environ. Microbiol. 41: 9–16.

Florentz, M., and P. Granger. 1983. Phosphorus-31 nuclear magnetic resonance of activated sludge: Use for the study of the biological removal of phosphates from wastewater. Environ. Technol. Lett. 4: 9–12.

Focht, D.D., and W. Verstraete. 1977. Biochemical ecology of nitrification and denitrification. Adv. Microb. Ecol. 1: 135–214.

Fogarty, A.M., and O.H. Tuovinen. 1991. Microbiological degradation of pesticides in yard waste composting. Microbiol. Rev. 55: 225–233.

Fogel, M.M., A.R. Taddeo, and S. Fogel. 1986. Biodegradation of chlorinated ethenes by a methane-utilizing mixed culture. Appl. Environ. Microbiol. 50: 720–724.

Folmar, L.C., N.D. Denslow, V. Rao, M. Chow, D.A. Crain, J. Enblom, J. Marcino, and L.J. Guillette, Jr. 1996. Vitellogenin induction and reduced serum testosterone concentrations in feral male carp (*Cyprinus carpio*) captured near a major metropolitan sewage treatment plant. Environ. Health Persp. 104: 1096–1101.

Foran, J.A. 2000. Early warning systems for hazardous biological agents in potable water. Environ. Health Perspec. 108: 993–996.

Ford, T., and R. Mitchell. 1990. The ecology of microbial corrosion. Adv. Microb. Ecol. 11: 231–262.

Forster, C.F., and J. Dallas-Newton. 1980. Activated sludge settlement- some suppositions and suggestions. Water Pollut. Control 79: 338–351.

Forster, C.F., and D.W.M. Johnston. 1987. Aerobic processes, pp. 15–56, In: *Environmental Biotechnology*, C.F. Forster, and D.A.J. Wase, Eds., Ellis Horwood, Chichester, U.K.

Forster, C.F., and E. Senior. 1987. Solid waste, pp. 176–233, In: *Environmental Biotechnology*, C.F. Forster, and D.A.J. Wase, Eds., Ellis Horwood, Chichester, U.K.

Forster, C.F., and D.A.J. Wase, Eds. 1987. *Environmental Biotechnology*. Ellis Horwood, Chichester, U.K.

Forsyth, J.V., Y.M. Tsao, and R.O. Bleam. 1995. Bioremediation: when is bioaugmentation needed?, pp. 1–15, In: *Bioaugmentation for Site Remediation*, R.E. Hinchee, J. Frederickson, and B.C. Allerman, Eds., Batelle Press, Columbus, OH.

Foster, D.H., and R.S. Engelbrecht. 1973. Microbial hazards in disposing of wastewater on soil, pp. 247–259, In: *Recycling Treated Municipal Wastewater and Sludge Through Forest and Cropland*, W.E. Sopper, and L.T. Kardos, Eds., Pennsylvania State University Press, University Park, PA.

Foster, D.M., D.S. Walsh, and O.J. Sproul. 1980. Ozone inactivation of cell-and fecal-associated viruses and bacteria. J. Water Pollut. Control Fed. 52: 2174–2184.

Foster, S.O., E.L. Palmer, G.W. Gary, M.L. Martin, K.L. Herrmann, P. Beasley, and J. Sampson. 1980. Gastroenteritis due to rotavirus in an isolated Pacific Island group: An epidemic of 3,439 cases. J. Infect. Dis. 141: 32–35.

Fox, A., G.C. Stewart, L.N. Waller, K.F. Fox, W.M. Harley, and R. L. Price. 2003. Carbohydrates and glycoproteins of *Bacillus anthracis* and related bacilli: targets for biodetection. J. Microbiol. Meth. 54: 143–152.

Fox, A.B., and G.W. Hambricks. 1984. Recreationally associated *Pseudomonas aeruginosa* folliculitis: Report of an epidemic. Arch. Dermatol. 120: 1304–1307.

Fox, J.L. 1989. Contemplating suicide genes in the environment. ASM News 55: 259–261.

Fox, K.R., R.J. Miltner, G.S. Logsdon, D.L. Dicks, and L.F. Drolet. 1984. Pilot-plant studies of slow-rate filtration. J. Am. Water Works Assoc. 76: 62–68.

Fox, J. 1997. Antibiotic resistance on rise globally. Am. Soc. Microbiol. News 63: 655.

Francis, C.A., A.C. Lockley, D.P. Sartory, and J. Watkins. 2001. A simple modified membrane filtration medium for the enumeration of aerobic spore-bearing bacilli in Water. Water Res. 35: 3758–3761.

Franco, E., L. Toti, R. Gabrieli, L. Croci, D. De Medici, and A. Panà. 1990. Depuration of *Mytilus galloprovincialis* experimentally contaminated with hepatitis A virus. Int. J. Food Microbiol. 11: 321–328.

Franco, R. M. B., and R. Cantusio Neto. 2002. Occurrence of cryptosporidial oocysts and *Giardia* cysts in bottled mineral water commercialized in the City of Campinas, State of São Paulo, Brazil. Mem. Inst. Oswaldo Cruz 97: 205–207.

Francy, D.S., and R.A. Darner. 2000. Comparison of methods for determining *Escherichia coli* concentrations in recreational waters. Water Res. 34: 2770–2778.

Franz, D.R., P.B. Jahrling, A.M. Friedlander, D.J. McClain, D.L. Hoover, W.R. Bryne, J.A. Pavlin, G.W. Christopher, and E.M. Eitzen. 1997. Clinical recognition and management of patients exposed to biological warfare agents. J. Amer. Med. Assoc. 278: 399–411.

Fraser, J., and N. Pan. 1998. Algae laden pond effluents-Tough duty for reclamation filters, pp. 593–607, In: *Water Reuse Conference Proceedings*, Feb. 1–4, 1998, Lake Buena Vista, FL., American Water Works Association & Water Environment Federation.

Frederickson, J.K., F.J. Brockman, D.J. Workman, S.W. Li, and T.O. Stevens. 1991. Isolation and characterization of a subsurface bacterium capable of growth on toluene, naphtalene, and other aromatic compounds. Appl. Environ. Microbiol. 55: 796–803.

Frias, J., F. Ribas, and F. Lucena. 1992. A method for the measurement of biodegradable organic carbon in waters. Water Res. 26: 255–258.

Friedman, B.A., P.R. Dugan, R.A. Pfister, and C.C. Remsen. 1969. Structure of exocellular polymers and their relationship to bacterial flocculation. J. Bacteriol. 98: 1328–1334.

Friello, D.A., J.R. Mylroie, and A.M. Chakrabarty. 1976. Use of genetically engineered microorganisms for rapid degradation of fuel hydrocarbons, pp. 205–213, In: *Proceedings International Biodegradation Symposium*, J.M. Sharpley, and A.M. Kaplan, Eds., Applied Science Publications, London.

Friis, L., C. Edling, and L. Hagmar. 1993. Mortality and incidence of cancer among sewage workers: a retrospective cohort study. Br. J. Ind. Med. 50: 563–567.

Frostell, B. 1981. Anaerobic treatment in a sludge bed system compared with a filter system. J. Water Pollut. Control Fed. 53: 216–222.

Fuhs, G.W., and M. Chen. 1975. Microbiological basis of phosphate removal in the activated sludge process for the treatment of wastewater. Microb. Ecol. 2: 119–138.

Fujioka, R.S. 1997. Indicators of marine recreational water quality, pp. 176–183, In: *Manual of Environmental Microbiology*, C.J. Hurst, G.R. Knudsen, M.J. McInerney, L.D. Stet zenbach, and M.V. Walter, Eds., ASM Press, Washington, D.C.

Fujioka, R.S., H.H. Hashimoto, E.B. Siwak, and R.H.F. Young. 1981. Effect of sunlight on survival of indicator bacteria in seawater. Appl. Environ. Microbiol. 41: 690–696.

Fujioka, R.S., and B.S. Yoneyama. 2002. Sunlight inactivation of human enteric viruses and fecal bacteria. Water Sci. & Technol. 46 (11–12): 291–295.

Funderburg, S.W., and C.A. Sorber. 1985. Coliphages as indicators of enteric viruses in activated sludge. Water Res. 19: 547–555.

Funderburg, S.W., B.E. Moore, C.A. Sorber, and B.P. Sagik. 1978. Survival of poliovirus in model wastewater holding pond. Prog. Water Technol. 10: 619–629.

Furness, B. W., M.J. Beach, and J.M. Roberts. 2000. Giardiasis surveillance – United States, 1992–1997. MMWR. CDC Surveillance Summaries: Morbidity and Mortality Weekly Report. CDC Surveillance Summaries/Centers for Disease Control 49 (7): 1–13.

Gainesville Regional Utilities. 1985. Onsite Systems for Wastewater Treatment in the Gainesville Urban Area. Report Prepared for the Water Manag. Adv. Comm. to the Gainesville City Commission.

Gajardo, R., N. Bouchriti, R.M. Pinto, and A. Bosch. 1995. Genotyping of rotaviruses isolated from sewage. Appl. Environ. Microbiol. 61: 3460–3462.

Galun, M., P. Keller, D. Malki, H. Feldstein, E. Galun, S.M. Siegel, and B.Z. Siegel. 1982. Removal of uranium (VI) from solution by fungal biomass and fungal wall-related biopolymers. Science 219: 285–286.

Gameson, A.L.H., and D.J. Gould. 1975. Effects of solar radiation on the mortality of some terrestrial bacteria in seawater, pp. 209–219, In: *Discharge of Sewage from Sea Outfalls*, A.L.H. Gameson, Ed., Pergamon Press, London.

Gantzer, C., P. Gaspard, L. Galvez, A. Huyard, N. Dumouthier, and J. Schwartzbrod. 2001. Monitoring of bacterial and parasitological contamination during various treatment of sludge. Water Res. 35: 3763–3770.

Garcia, J.L., S. Roussos, and M. Bensoussan. 1981. Etudes taxonomiques des bacteries denitrifiantes isolees sur benzoate dans les sols de rizieres du Senegal. Cahiers ORSTOM Ser. Biol. 43: 13–25.

Garcia-Fulgueiras, A., C. Navarro, D. fenoll, J. Garcia, P. Gonzalez-Diego, T. Jimenez-Bunuales, M. Rodriguez, R. Lopez, F. Pacheco, J. Ruiz, M. Segovia, B. Baladron, and C. Pelaz. 2003. Legionnaires' disease outbreak in Murcia, Spain. Emerg. Infect. Dis. 9: 915–921.

Garrigues, R.M., D.K. Ammerman, L.P. Wiseman, and W.E. Olson. 1998. City of Palm Beach reuse feasibility study, pp. 165–176, In: *Water Reuse Conference Proceedings*, Feb. 1–4, 1998, Lake Buena Vista, FL., American Water Works Association & Water Environment Federation.

Gaspard, P.G., J. Wiart, and J. Schwartzbrod. 1997. Sludge hygienization: helminth eggs (ascaris ova) destruction by lime treatment. Recent Res. Dev. Microbiol. 1: 77–83.

Gaudet, I.D., L.Z. Florence, and R.N. Coleman. 1996. Evaluation of test media for routine monitoring of *Escherichia coli* in nonpotable waters. Appl. Environ. Microbiol. 62: 4032–4035.

Gaudy, A.F., Jr. 1972. Biochemical oxygen demand, pp. 305–332, In: *Water Pollution Microbiology*. R. Mitchell, Ed., John Wiley & Sons, New York.

Gaudy, A.F., Jr., and E.T. Gaudy. 1988. *Elements of Bioenvironmental Engineering*. Engineering Press, San Jose, CA.

Gauthier, M.J., and V.A. Breittmayer. 1990. Regulation of gene expression in *E. coli* cells starved in seawater: Influence on their survival in marine environments. Int. Symp. Health-Related Water Microbiol., Tubingen, W. Germany, April 1–6, 1990.

Gauthier, M.J., G.N. Flatau, D. Le Rudulier, R.L. Clement, and M.-P. Combarro. 1991. Intracellular accumulation of potassium and glutamate specifically enhances survival of *Escherichia coli* in seawater. Appl. Environ. Microbiol. 57: 272–276.

Gauthier, M.J., G.N. Flatau, R.L. Clement, and P.M. Munro. 1992. Sensitivity of *Escherichia coli* cells to seawater closely depends on their growth stage. J. Appl. Bacteriol. 73: 257–262.

Gavaghan, P.D., J.L. Sykora, W. Jakubowski, C.A. Sorber, L.W. Casson, A.M. Sninsky, M.D. Lichte, and G. Keleti. 1992. Viability and infectivity of *Giardia* and *Cryptosporidium* in digested sludge. IWPRC Int. Symp., Washington, D.C., May 26–29, 1992.

Gealt, M.A., M.D. Chai, K.B. Alpert, and J.C. Boyer. 1985. Transfer of plasmid pBR322 and pBR325 in wastewater from laboratory strains of *Escherichia coli* to bacteria indigenous to the waste disposal system. Appl. Environ. Microbiol. 49: 836–841.

Geenens, D., C. Jonkers, and C. Thoeye. 2002. Go or no go for gel entrapped nitrifiers? A Belgian case study. Water Sci. & Technol. 46 (1–2): 465–471.

Geldenhuys, J.C., and P.D. Pretorius. 1989. The occurrence of enteric viruses in polluted water, correlation to indicators organisms and factors influencing their numbers. Water Sci. Technol. 21: 105–109.

Geldreich, E.E. 1980. Microbiological processes in water supply distribution (seminar), Annual Meeting of the American Society of Microbiology, Miami Beach, FL, May 14, 1980.

Geldreich, E.E. 1990. Microbiological quality of source waters for water supply, pp. 3–31, In: *Drinking Water Microbiology*, G.A. McFeters, Ed., Springer-Verlag, New York.

Geldreich, E.E., M.J. Allen, and R.H. Taylor. 1978. Interferences to coliform detection in potable water supplies, pp. 13–20, In: *Evaluation of the Microbiology Standards for Drinking Water*, C.W. Hendricks, Ed., U.S. Environmental Protection Agency, Washington, D.C.

Geldreich, E.E., K.R. Fox, J.A. Goodrich, E.W. Rice, R.M. Clark, and D.L. Swerdlow. 1992. Searching for a water supply connection in the Cabool, Missouri disease outbreak of *Escherichia coli* 0157:H7. Water Res. 26: 1127–1137.

Geldreich E.E., and B.A. Kenner. 1969. Concepts of fecal streptococci in stream pollution. J. Water Pollut. Control Fed. 41: R336–R341.

Geldreich, E.E., and D.J. Reasoner. 1990. Home treatment devices and water quality, pp. 147–167, In: *Drinking Water Microbiology*, G.A. McFeters, Ed., Springer Verlag, New York.

Geldreich, E.E., and E.W. Rice. 1987. Occurrence, significance, and detection of *Klebsiella* in water systems. J. Am. Water Works Assoc. 79: 74–80.

Geldreich, E.E., R.H. Taylor, J.C. Blannon, and D.J. Reasoner. 1985. Bacterial colonization of point-of-use water treatment devices. J. Am. Water Works Assoc. 77: 72–80.

Gennaccaro, A.L., M.R. McLaughlin, W. Quintero-Betancourt, D.E. Huffman, and J.B. Rose. 2003. Infectious *Cryptosporidium parvum* oocysts in final reclaimed effluent. Appl. Environ. Microbiol. 69: 4983–4984.

Gerba, C.P. 1984. Applied and theoretical aspects of virus adsorption to surfaces. Adv. Appl. Microbiol. 30: 133–168.

Gerba, C.P. 1987a. Phage as indicators of fecal pollution, pp. 197–209, In: *Phage Ecology*, S.M. Goyal, C.P. Gerba, and G. Bitton, Eds., Wiley Interscience, New York.

Gerba, C.P. 1987b. Recovering viruses from sewage, effluents, and water, pp. 1–23, In: *Methods for Recovering Viruses from the Environment*, G. Berg, Ed., CRC Press, Boca Raton, FL.

Gerba, C.P. 2000a. Assessment of enteric pathogen shedding by bathers during recreational activity and its impact on water quality. Quant. Microbiol. 2: 55–68.

Gerba, C.P. 2000b. Indicator organisms, pp. 491–503, In: *Environmental Microbiology*, R.M. Maier, I.L. Pepper, and C.P. Gerba, Eds., Academic Press, San Diego, CA, 585 pp.

Gerba, C.P., and G. Bitton. 1984. Microbial pollutants: Their survival and transport pattern to groundwater, In: *Groundwater Pollution Microbiology*, G. Bitton, and C.P. Gerba, Eds., John Wiley & Sons, New York.

Gerba, C.P., S.M. Goyal, C.J. Hurst, and R.L. LaBelle. 1980. Type and strain dependence of enterovirus adsorption to activated sludge, soils and estuarine sediments. Water Res. 14: 1197–1198.

Gerba, C.P., D. Gramos, and N. Nwachuku. 2002. Comparative inactivation of enteroviruses and adenovirus 2 by UV light. Appl. Environ. Microbiol. 68: 5167–5169.

Gerba, C.P., and C.N. Haas. 1986. Risks associated with enteric viruses in drinking water, pp. 460–468, In: *Progress in Chemical Disinfection*, G.E. Janauer, Ed., State University of New York, Binghamton, N.Y.

Gerba, C.P., M.S. Huber, J. Naranjo, J.B. Rose, and S. Bradford. 1995. Occurrence of enteric pathogens in composted domestic solid waste containing disposable diapers. Waste Manag. Res. 13: 315–324.

Gerba, C.P., and J.S. McLeod. 1976. Effect of sediments on the survival of *Escherichia coli* in marine waters. Appl. Environ. Microbiol. 32: 114–120.

Gerba, C.P., I.L. Pepper, and L.F. Whitehead III. 2002. A risk assessment of emerging pathogens of concern in the land application of biosolids. Water Sci. Technol. 46 (10): 225–230.

Gerba, C.P., and G. Schaiberger. 1975. Effect of particulates on virus survival in seawater. J. Water Pollut. Control Fed. 47: 93–103.

Gerba, C.P., S.N. Singh, and J.B. Rose. 1985. Waterborne gastroenteritis and viral hepatitis. CRC Crit. Rev. Environ. Cont. 15: 213–236.

Gerba, C.P., Stagg, C.H., and M.G. Abadie. 1978. Characterization of sewage solids-associated viruses and behavior in natural waters. Water Res. 12: 805–812.

Gerba, C.P., and R.B. Thurman. 1986. Evaluation of the efficacy of Microdyn against waterborne bacteria and viruses. Monograph, University of Arizona, Tucson, AZ.

Gerba, C.P., J.A. Thurston, J.A. Falabi, P.M. Watt, and M.M. Karpiscak. 1999. Optimization of artificial wetland design for removal of indicator microorganisms and pathogenic protozoa. Water Sci. Technol. 40 (4–5): 363–368.

Gerba, C.P. C. Wallis, and J.L. Melnick. 1975a. Viruses in water: the problem, some solutions. Environ. Sci. Technol. 9: 1122–1126.

Gerba, C.P., C. Wallis, and J.L. Melnick. 1975b. Fate of wastewater bacteria and viruses in soils. J. Irrig. Drainage Div., ASCE 3: 157–168.

Gerba, C.P., C. Wallis, and J.L. Melnick. 1977. Disinfection of wastewater by photodynamic action. J. Water Pollut. Control Fed. 49: 578–583.

Gerba, C.P., M.V. Yates, and S.R. Yates. 1991. Quantitation of factors controlling viral and bacterial transport in the subsurface, pp. 77–88, In: *Modeling the Environmental Fate of Microorganisms*, C.J. Hurst, Ed., American Society Microbiology, Washington, D.C.

Gerber, N.N. 1979. Volatile substances from actinomycetes: their role in the odor pollution of water. Crit. Rev. Microbiol. 7: 191–214.

Ghosh, S., and D.L. Klass. 1978. Two-phase anaerobic digestion. Proc. Biochem. 13: 15–24.

Ghosh, S., J.P. Ombregt, and P. Pipyn. 1985. Methane production from industrial wastes by two-phase anaerobic digestion. Water Res. 19: 1083–1088.

Gibson, D.T., and V. Subramanian. 1988. Microbial degradation of aromatic hydrocarbons, pp. 181–193, In: *Microbial Degradation of Organic Compounds*, D.T. Gibson, Ed., Marcel Dekker, New York.

Gieseke, A., U. Purkhold, M. Wagner, R. Amann, and A. Schramm. 2001. Community structure and activity dynamics of nitrifying bacteria in a phosphate-removing biofilm. Appl. Environ. Microbiol. 67: 1351–1362.

Giesy, J.P. R.L. Craney, J.L. Newsted, C.J. Rosiu, A. Benda, R.G. Kreis, and F.J. Horvath. 1988. Comparison of three sediments bioassay methods using Detroit River sediments. Environ. Toxicol. Chem. 7: 483–498.

Giger, W., and P.V. Roberts. 1978. Characterization of persistent organic carbon, pp. 135–175, In: *Water Pollution Microbiology*. Vol. 2, R. Mitchell, Ed., John Wiley & Sons, New York.

Giger, W., M. Ahel, M. Koch, H.U. Laubscher, C. Schaffner, and J. Schneider. 1987. Behaviour of alkylphenols polyethoxylate surfactants and of nitriloacetate in sewage treatment. Water Sci. Technol. 19: 449–460.

Gilpin, R.W. 1984. Laboratory and field applications of U.V light disinfection on six species of *Legionella* and other bacteria in water. Proceedings of the Second International Symposium of the American Society for Microbiology, Washington, D.C.

van Ginkel, C.G., J. Tramler, K.Ch.A.M. Luyben, and A. Klapwijk. 1983. Characterization of *Nitrosomonas europaea* immobilized in calcium alginate. Enzyme Microb. Technol. 5: 297–303.

Girones, R., A. Allard, G. Wadell, and J. Jofre. 1993. Application of PCR for the detection of adenovirus in polluted waters. Water Sci. Technol. 27: 235–241.

Glass, J.S., and R.T. O'Brien. 1980. Enterovirus and coliphage inactivation during activated sludge treatment. Water Res. 14: 877–882.

Glatzer, M.B. 1998. Shellfish-borne disease outbreaks in the U.S., 1992–1998. Internal technical report. U.S. Food and Drug Administration, Southeast Regional Office, Atlanta, GA.

Gloyna, E.F. 1971. *Waste Stabilization Ponds*. WHO Monograph Series No. 60. World Health Organization, Geneva, Switzerland.

Glynn, W., C. Baker, A. LoRe, and A. Quaglieri. 1987. *Mobile Waste Processing Systems and Treatment Technologies*. Noyes Data Corp., Park Ridge, N.J., 136 pp.

Goddard, A.J., and C.F. Forster. 1987a. Stable foams in activated sludge plants. Enz. Microbial Technol. 9: 164–168.

Goddard, A.J., and C.F. Forster. 1987b. A further examination into the problem of stable foams in activated sludge plants. Microbios 50: 29–42.

Goddard, M.R., J. Bates, and M. Butler. 1981. Recovery of indigenous enteroviruses from raw and digested sludges. Appl. Environ. Microbiol. 42: 1023–1028.

Godfrey, A.J., and L.E. Bryan. 1984. Intrinsic resistance and whole cell factors contributing to antibiotic resistance, pp. 113–145, In: *Antibiotic Drug Resistance*, L.E. Bryan, Ed., Academic Press, Orlando, FL.

Goldstein, N. 1988. Steady growth for sludge composting. Biocycle 29 (10): 29–43.

Gomez-Bautista, M., L.M. Ortega-Mora, E. Tabares, V. Lopez-Rodas, and E. Costas. 2000. Detection of infectious *Cryptosporidium parvum* oocysts in mussels (*Mytilus galloprovincialis*) and cockles (*Cerastoderma edule*). Appl. Environ. Microbiol. 66: 1866–1870.

Gomez-Suarez, C., H.C. van der Mei, and H.J. Busscher. 2002. Adhesion, immobilization and retention of microorganisms on solid substrata, pp. 100–113, In: *Encyclopedia of Environmental Microbiology*, G. Bitton, editor-in-chief, Wiley-Interscience, N.Y.

Gostelow, P., S.A. Parsons, and R.M. Stuetz. 2001. Odour measurements for sewage treatment works. Water Res. 35: 579–597.

Gostelow, P., and H.B. Gotaas. 1956. *Composting: Sanitary Disposal and Reclamation of Organic Wastes*. WHO Monograph No. 31, World Health Organization, Geneva.

Goyal, S.M. 1987. Methods in phage ecology, pp. 267–287, In: *Phage Ecology*, S.M. Goyal, C.P. Gerba, and G. Bitton, Eds., Wiley-Interscience, New York.

Goyal, S.M., and C.P. Gerba. 1982a. Concentration of viruses from water by membrane filters, pp. 59–116, In: *Methods in Environmental Virology*, C.P. Gerba, and S.M. Goyal, Eds., Marcel Dekker, New York.

Goyal, S.M., and C.P. Gerba. 1982b. Occurrence of endotoxins in groundwater during land application of wastewater. J. Environ. Sci. Health. A17: 187–196.

Goyal, S.M., and C.P. Gerba. 1983. Viradel method for detection of rotavirus from seawater. J. Virol. Methods 7: 279–285.

Goyal, S.M., C.P. Gerba, and J.L. Melnick. 1977. Occurrence and distribution of bacterial indicators and pathogens in canal communities along the Texas coast. Appl. Environ. Microbiol. 34: 139–149.

Goyal, S.M., C.P. Gerba, and J.L. Melnick. 1979. Human enteroviruses in oysters and their overlying water. Appl. Environ. Microbiol. 37: 572–581.

Goyal, S.M., K.S. Zerda, and C.P. Gerba. 1980. Concentration of coliphage from large volumes of water and wastewater. Appl. Environ. Microbiol. 39: 85–91.

Goyal, S.M., W.N. Adams, M.L. O'Malley, and D.W. Lear. 1984. Human pathogenic viruses at sewage disposal sites in the middle Atlantic region. Appl. Environ. Microbiol. 48: 758–763.

Goyal, S.M., C.P. Gerba, and G. Bitton, Eds. 1987. *Phage Ecology*. Wiley-Interscience, New York, 321 pp.

Grabow, W.O.K. 1986. Indicator systems for assessment of the virological safety of treated drinking water. Water Sci. Technol. 18: 159–165.

Grabow, W.O.K. 1990. Microbiology of drinking water treatment: Reclaimed wastewater, pp. 185–203, In: *Drinking Water Microbiology*, G.A. McFeters, Ed., Springer-Verlag, New York.

Grabow, W.O.K., J.S. Burger, and E.M. Nupen. 1980. Evaluation of acid-fast bacteria, *Candida albicans*, enteric viruses and conventional indicators for monitoring wastewater reclamation systems. Prog. Water Technol. 12: 803–817.

Grabow, W.O.K., and P. Coubrough. 1986. Practical direct plaque assay for coliphages in 100 ml samples of drinking water. Appl. Environ. Microbiol. 52: 430–433.

Grabow, N.A., and R. Kfir. 1990. Growth of Legionella bacteria in activated carbon filters. Presented at the International Symposium on Health-Related Microbiology, Tubingen, W. Germany, April 1–6, 1990.

Grabow, W.O.K., and O.W. Prozesky. 1973. Drug resistance of coliform bacteria in hospital and city sewage. Antimicrob. Agents Chemother. 3: 175–180.

Grace, R.D., N.E. Dewar, W.G. Barnes, and G.R. Hodges. 1981. Susceptibility of *Legionella pneumophila* to three cooling towers microbicides. Appl. Environ. Microbiol. 41: 233–236.

Graczyk, T.K., R. Fayer, M.R. Cranfield, and D.B. Conn. 1997a. In vitro interactions of Asian freshwater clam (*Corbicula fluminea*) hemocytes and *Cryptosporidium parvum* oocysts. Appl. Environ. Microbiol. 63: 2910–2912.

Graczyk, T.K., R. Fayer, E.J. Lewis, C.A. Farley, and J.M. Trout. 1997b. In vitro interactions between hemocytes of the Eastern oyster, *Crassostrea virginica*, Gmelin 1791, and *Cryptosporidium parvum* oocysts. J. Parasitol. 83: 949–952.

Graczyk, T.K., R. Fayer, M.R. Cranfield, and R. Owens. 1997c. *Cryptosporidium parvum* oocysts recovered from water by the membrane filter dissolution method retain their infectivity. J. Parasitol. 83: 111–114.

Grady, C.P.L. 1986. Biodegradation of hazardous wastes by conventional biological treatments. Haz. Wastes Haz. Mater. 3: 333–365.

Grady, C.P.L., Jr., and H.C. Lim. 1980. *Biological Waste Treatment: Theory and Applications*. Marcel Dekker, New York, 963 pp.

Grant, S.B., C.P. Pendroy, C.L. Mayer, J.K. Bellin, and C.J. Palmer. 1996. Prevalence of enterohemorrhagic *Escherichia coli* in raw and treated municipal sewage. Appl. Environ. Microbiol. 62: 3466–3469.

Grant, W.D., and P.E. Long. 1981. *Environmental Microbiology*. Halsted Press, New York.

Gratacap-Cavallier, B., O. Genoulaz, K. Brengel-Pesce, H. Soule, P. Innocenti-Francillard, M. Bost, L. Gofti, D. Zmirou, and J.M. Seigneurin. 2000. Detection of human and animal rotavirus sequences in drinking water. Appl. Environ. Microbiol. 66: 2690–2692.

Gray, N.F. 1994. *Drinking Water Quality: Problems and Solutions.* John Wiley & Sons, Chichester, 315 pp.

Grbic-Galic, D., and T.M. Vogel. 1987. Transformation of toluene and benzene by mixed methanogenic cultures. Appl. Environ. Microbiol. 53: 254–260.

Gregersen P., K. Grunnet, S.A. Uldum, B.H. Andersen, and H. Madsen. 1999. Pontiac fever at a sewage treatment plant in the food industry. Scan. J. Work Environ. Hlth. 25 (3): 291–305.

Grenier, D. 1995. Quantitative analysis of bacterial aerosols in two different dental clinic environments. Appl. Environ. Microbiol. 61: 3165–3168.

Griffin, P.M. 1995. *Escherichia coli* O157:H7 and other enterohemorrhagic *Escherichia coli*, pp. 739–761, In: *Infections of the Gastrointestinal Tract*, M.J. Blaser, P.D. Smith, J.I. Ravdin, H.B. Greenberg, and R.L. Guerrant, Eds., Raven Press, New York.

Grimason, A.M., S. Wiandt, B. Baleux, W.N. Thitai, J. Bontoux, and H.V. Smith. 1996. Occurrence and removal of Giardia sp. cysts by Kenyan and French waste stabilization pond systems. Water Sci. Technol. 33 (7): 83–89.

Grimes, D.J., F.L. Singleton, J. Stemmler, L.M. Palmer, P. Brayton, and R.R. Colwell. 1984. Microbiological effects of wastewater effluent discharge into coastal waters of Puerto Rico. Water Res. 18: 613–619.

van Groenestijn, J.W., G.J.F.M. Vlekke, D.M.E. Anink, M.H. Deinema, and A.J.B. Zehnder. 1988a. Role of cations in accumulation and release of phosphate by *Acinetobacter* strain 210A. Appl. Environ. Microbiol. 54: 2894–2901.

van Groenestijn, J.W., M.M.A. Bentvelsen, M.H. Deinema, and A.J.B. Zehnder. 1988b. Polyphosphate-degrading enzymes in *Acinetobacter spp.* and activated sludge. Appl. Env. Microbiol. 55: 219–223.

Groethe, D.R., and J.G. Eaton. 1975. Chlorine-induced mortality in fish. Trans. Am. Fish Soc. 104: 800–805.

Grotenhuis, J.T.C., M. Smit, C.M. Plugge, X. Yuansheng, A.A.M. van Lammeren, A.J.M. Stams, and A.J.B. Zehnder. 1991. Bacteriological composition and structure of granular sludge adapted to different substrates. Appl. Environ. Microbiol. 57: 1942–1949.

Grubbs, R.B. 1984. Panel discussion: Emerging industrial applications, pp. 331–349, In: *Genetic Control of Environmental Pollutants*, G.S. Omenn, and A. Hollaender, Eds., Plenum, New York.

Grubbs, T.R., and F.W. Pontius. 1992. USEPA releases draft ground water disinfection rule. J. Am. Water Works Assoc. 84: 25–31.

Guerrant, R.L., and N.M. Thielman. 1995. Types of *Escherichia coli* enteropathogens, pp. 687–690, In: *Infections of the Gastrointestinal Tract*, M.J. Blaser, P.D. Smith, J.I. Ravdin, H.B. Greenberg, and R.L. Guerrant, Eds., Raven Press, New York.

Guest, H. 1987. *The World of Microbes*, Science Tech. Publishers, Madison, WI.

Guest, R.K., and D.W. Smith. 2003. A potential new role for fungi in a wastewater MBR biological nitrogen reduction system. J. Environ. Eng. Sci. 1: 433–437.

Gui, G.P.H., P.R.S. Thomas, M.L.V. Tizard, J. Lake, J.D. Sanderson, and J. Hermon-Taylor. 1997. Two-year outcomes analysis of Crohn's disease treated with rifabutin and macrolide antibiotics. J. Antimicrobial Chemotherapy 39: 393–400.

Gujer, W., and A.J.B. Zehnder. 1983. Conversion processes in anaerobic digestion. Water Sci. Technol. 15: 127–134.

Gullick, R.W., W.M. Grayman, R.A. Deininger, and R.M. Males. 2003. Design of early warning systems for source waters. J. Amer. Water Works Assoc. 95 (11): 58–72.

Gunnerson, C.G., H.I. Shuval, and S. Arlosoroff. 1984. Health effects of wastewater irrigation and their control in developing countries. In: *Proc. Water Reuse Symposium III*, American Water Works Association, Denver, CO.

von Gunten, U. 2003. Ozonation of drinking water: Part II. Disinfection and by-product formation in presence of bromide, iodide or chlorine. Water Res. 37: 1469–1487.

Guthrie, M.A., E.J. Kirsch, R.F. Wukasch, and C.P.L. Grady, Jr. 1984. Pentachlorophenol biodegradation. II. Anaerobic. Water Res. 18: 451–461.

Haas, C.N., and R.S. Engelbrecht. 1980. Physiological alterations of vegetative microorganisms resulting from chlorination. J. Water Pollut. Control Fed. 52: 1976–1989.

Haas, C.N., M.G. Kerallus, D.M. Brncich, and M.A. Zapkin. 1986. Alteration of chemical and disinfectant properties of hypochlorite by sodium, potassium and lithium. Environ. Sci. Technol. 20: 822–826.

Haas, C.N., M.A. Meyer, M.S. Paller, and M.A. Zapkin. 1983. The utility of endotoxins as surrogate indicator in potable water microbiology. Water Res. 17: 803–807.

Haas, C.N., J.B. Rose, and C.P. Gerba. 1999. *Quantitative Microbial Risk Assessment.* Wiley, New York.

Haas, C.N., B.F. Severin, D. Roy, R.S. Engelbrecht, A. Lalchandani, and S. Farooq. 1985. Field observations on the occurrence of new indicators of disinfection efficiency. Water Res. 19: 323–329.

de Haas, D.W. 1989. Fractionation of bioaccumulated phosphorus compounds in activated sludge. Water Sci. Technol. 21: 1721–1725.

Habbari, K., A. Tifnouti, G. Bitton, and A. Mandil. 2000. Geohelminthic infections associated with raw wastewater reuse for agricultural purposes in Beni-Mellal, Morocco. Parasitol. Intern. 48: 249–254.

Haber, C.L., L.N. Allen, S. Zhao, and R.S. Hanson. 1983. Methylotrophic bacteria: Biochemical diversity and genetics. Science 221: 1147–1153.

Hackel, U., J. Klein, R. Megnet, and F. Wagner. 1975. Immobilization of cells in polymeric matrices. Eur. J. Appl. Microbiol. 1: 291–294.

Häfliger, D., P. Hübner, and J. Lüthy. 2000. Outbreak of viral gastroenteritis due to sewage-contaminated drinking water. Int. J. Food Microbiol. 54: 123–126.

Hagedorn, C. 1984. Microbiological aspects of groundwater pollution due to septic tanks, pp. 181–195, In: *Groundwater Pollution Microbiology*, G. Bitton, and C.P. Gerba, Eds., John Wiley & Sons, New York.

Hagedorn, C., S.L. Robinson, J.R. Filtz, S.M. Grubbs, T.A. Angier, and R.B. Reneau, Jr. 1999. Determining sources of fecal pollution in a rural Virginia watershed with antibiotic resistance patterns in fecal streptococci. Appl. Environ. Microbiol. 65: 5522–5531.

Hagedorn, C., J.B. Crozier 1, K.A. Mentz, A.M. Booth, A.K. Graves, N.J. Nelson, and R.B. Reneau Jr. 2003. Carbon source utilization profiles as a method to identify sources of faecal pollution in water. J. Appl. Microbiol. 94: 792–799.

Häggblom, M.M., and L.Y. Young. 1990. Chlorophenol degradation coupled to sulfate reduction. Appl. Environ. Microbiol. 56: 3255–3260.

Häggblom, M.M., and R.J. Valo. 1995. Bioremediation of chlorophenol wastes, pp. 389–434, In: *Microbial Transformation and Degradation of Toxic Organic Chemicals*, L.Y. Young, and C.E. Cerniglia, Eds., Wiley-Liss, New York.

Hain, K.E., and R.T. O'Brien. 1979. The survival of enteric viruses in septic tanks and septic tank drain fields. Water Resources Res. Inst. Report No. 108, New Mexico Water Resources Res. Inst., New Mexico State University., Las Cruces, N.M.

Haines, J.R., T.C. Covert, and C.C. Rankin. 1993. Evaluation of indoxyl-beta-D-glucuronide as a chromogen in media specific for Escherichia coli. Appl. Environ. Microbiol. 59: 2758–2759.

Hajji, K.T., F.F. Lepine, J.G. Bisaillon, R. Beaudet, J. Hawari, and S.R. Guiot. 2000. Effects of bioaugmentation strategies in UASB reactors with a methanogenic consortium for removal of phenolic compounds. Bioeng. Biotechnol. 67: 417–423.

Hall, J.C., and R.J. Foxen. 1983. Nitrification in BOD_5 test increases POTW noncompliance. J. Water Pollut. Control Fed. 55: 1461–1469.

Hall, R.M., and M.D. Sobsey. 1993. Inactivation of hepatitis A virus (HAV) and MS-2 by ozone and ozone-hydrogen peroxide in buffered water. Water Sci. Technol. 27: 371–378.

Hallam, N.B., J.R. West, C.F. Forster, and J. Simms. 2001. The potential for biofilm growth in water distribution systems. Water Res. 35: 4063–4071.

Hallam, N.B., J.R. West, C.F. Forster, J.C. Powell, and I. Spencer. 2002. The decay of chlorine associated with the pipe wall in water distribution systems. Water Res. 36: 3479–3488.

Hallas, L.E., and M.A. Heitkamp. 1995. Microbiological treatment of chemical process wastewater. pp. 349–387. In: *Microbial Transformation and Degradation of Toxic Organic Chemicals*, L.Y. Young, and C.E. Cerniglia, Eds., Wiley-Liss, New York.

Hallier-Soulier, S., and E. Guillot. 2000. Detection of cryptosporidia and *Cryptosporidium parvum* oocysts in environmental water samples by immunomagnetic separation-polymerase chain reaction. J. Appl. Microbiol. 89: 5–10.

Hallier-Soulier, S., and E. Guillot. 2003. An immunomagnetic separation-reverse transcription polymerase chain reaction (IMS-RT-PCR) test for sensitive and rapid detection of viable waterborne *Cryptosporidium parvum*. Environ. Microbiol. 5: 592–598.

Halling-Sorensen, B., S. Nors Nielsen, P.F. Lanzky, F. Ingerslev, H.C.H. Lützheft, and S.E. Jorgensen. 1998. Occurrence, fate and effects of pharmaceutical substances in the environment – A review. Chemosphere 36: 357–393.

Halvorson, H.O., D. Pramer, and M. Rogul, Eds. 1985. *Engineered Organisms in the Environment: Scientific Issues*. American Society Microbiology, Washington, D.C.

Hamelin, C., F. Sarhan, and Y.S. Chung. 1978. Induction of deoxyribonucleic acid degradation in *Escherichia coli* by ozone. Experientia 34: 1578–1579.

Hamilton, W.A. 1985. Sulphate anaerobic bacteria and anaerobic corrosion. Annu. Rev. Microbiol. 39: 195–217.

Hamilton, W.A. 1987. Biofilms: Microbial interactions and metabolic activities, pp. 361–385, In: *Ecology of Microbial Communities*, M. Fletcher, T.R.G. Gray, and J.G. Jones. Eds., Cambridge University Press, Cambridge, U.K.

Hammer, M.J. 1986. *Water and Wastewater Technology*. John Wiley & Sons, New York, 536 pp.

Hanaki, K., and C. Polprasert. 1989. Contribution of methanogenesis to denitrification in an upflow filter. J. Water Pollut. Control Fed. 61: 1604–1611.

Hanaki, K., Z. Hong, and T. Matsuo. 1992. Production of nitrous oxide gas during denitrification of wastewater. Water Sci. Technol. 26: 1027–1036.

Hancock, R.E.W. 1984. Alterations in outer membrane permeability. Annu. Rev. Microbiol. 38: 237–264.

Handwerker, J., J.G. Fox, and D.B. Schauer. 1995. Detection of *Helicobacter pylori* in drinking water using polymerase chain reaction amplification, Abstr. O-203, p. 435, In: Abstracts of the 95th General Meeting of the American Society of Microbiology, Washington, D.C.

Handzel, T.R., R.M. Green, C. Sanchez, H. Chung, and M.D. Sobsey. 1993. Improved specifity in detecting F-specific coliphages in environmental samples by suppression of somatic phages. Water Sci. Technol. 27: 123–131.

Hanel, L. 1988. *Biological Treatment of Sewage by the Activated Sludge Process*. Ellis Horwood, Chichester, U.K.

Hannah, S.A., B.M. Austern, A.E. Eralp, and R.H Wise. 1986. Comparative removal of toxic pollutants by six wastewater treatment processes. J. Water Pollut. Control Fed. 58: 27–34.

Hannah, S.A., B.M. Austern, A.E. Eralp, and R.A. Dobbs. 1988. Removal of organic toxic pollutants by trickling filter and activated sludge. J. Water Pollut. Control Fed. 60: 1281–1283.

Hänninen, M.-L., H. Haajanen, T. Pummi, K. Wermundsen, M.-L. Katila, H. Sarkkinen, I. Miettinen, and H. Rautelin. 2003. Detection and typing of *Campylobacter jejuni* and *Campylobacter coli* and analysis of indicator organisms in three waterborne outbreaks in Finland. Appl. Environ. Microbiol. 69: 1391–1396.

Hanstveit, A.O., W.J.Th. van Gemert, D.B. Janssen, W.H. Rulkens, and H.J. van Veen. 1988. Literature study on the feasibility of microbiological decontamination of polluted soils, pp. 63–128, In: *Biotreatment Systems*, Vol. 1, D.L. Wise, Ed., CRC Press, Boca Raton, FL.

Harakeh, M. 1985. Factors influencing chlorine disinfection of wastewater effluent contaminated by rotaviruses, enteroviruses, and bacteriophages, pp. 681–690, In: *Water Chlorination: Chemistry, Environmental Impact and Health Effects*, Vol. 5, R.L. Joley, Ed., Lewis, Chelsea, MI.

Harb, O.S., and Y. Abu Kwaik. 2002. Legionella in the environment: Persistence, evolution and pathogenicity, pp. 1796–1806, In: *Encyclopedia of Environmental Microbiology*, G. Bitton, editor-in-chief, Wiley-Interscience, N.Y.

Harb, O.S., L.Y. Gao, and Y. Abu Kwaik. 2000. From protozoa to mammalian cells: a new paradigm in the life cycle of intracellular bacterial pathogens. A minreview. Environ. Microbiol. 2: 251–265.

Hardman, D.J. 1987. Microbial control of environmental pollution: The use of genetic techniques to engineer organisms with novel catabolic capabilities, pp. 295–317, In: *Environmental Biotechnology*, C.F. Forster, and D.A.J. Wase, Eds., Ellis Horwood, Chichester, U.K.

Hardoyo, K. Yamada, H. Shinjo, J. Kato, and H. Ohtake. 1994. Production and release of polyphosphate by a genetically engineered strain of *Escherichia coli*. Appl. Environ. Microbiol. 60: 3485–3490.

Harp, J.A., R. Fayer, B.A. Pesch, and G.J. Jackson. 1996. Effect of pasteurization on infectivity of *Cryptosporidium parvum* oocysts in water and milk. Appl. Environ. Microbiol. 62: 2866–2868.

Harmsen, H.J.M., H.M.P. Kengen, A.D.L. Akkermans, A.J.M. Stams, and W.M. De Vos. 1996. Detection and localization of syntrophic propionate-oxidizing bacteria in granular sludge by *in situ* hybridization using 16S rRNA-based oligonucleotide probes. Appl. Environ. Microbiol. 62: 1656–1663.

Harremoes, P. 1978. Biofilm kinetics, pp. 71–109, In: *Water Pollution Microbiology*, Vol. 2, R. Mitchell, Ed., John Wiley & Sons, New York.

Harremoes, P., and M.H. Christensen. 1971. Denitrifikation med methan. Vand 1: 7–11.

Harries, J., D.A. Sheahan, S. Jobling, P. Matthiesen, P. Neall, E.J. Routledge, R. Rycroft, J.P. Sumpter, and T. Taylor. 1996. A survey of estrogenic activity in United Kingdom inland waters. Environ. Toxicol. Chem. 15: 1993–2002.

Harries, J., D.A. Sheahan, S. Jobling, P. Matthiesen, P. Neall, R. Rycroft, J.P. Sumpter, and T. Taylor, and N. Zaman. 1997. Estrogenic activity in five United Kingdom rivers detected by measurement of vitellogenesis in caged male trout. Environ. Toxicol. Chem. 16: 534–542.

Harrington, G.W., D.R. Noguera, A.I. Kandou, and D.J. Vanhoven. 2002. Pilot-scale evaluation of nitrification control strategies. J. Amer. Water Works Assoc. 94 (11): 78–89.

Harris, G.D., V.D. Adams, D.L. Sorensen, and R.R. Dupont. 1987a. The influence of photoreactivation and water quality on ultraviolet disinfection of secondary municipal wastewater. J. Water Pollut. Control Fed. 59: 781–787.

Harris, G.D., V.D. Adams, D.L. Sorenson, and M.S. Curtis. 1987b. Ultraviolet inactivation of selected bacteria and viruses with photoreactivation of the bacteria. Water Res. 21: 687–692.

Harris, R.H., and R. Mitchell. 1973. The role of polymers in microbial aggregation. Annu. Rev. Microbiol. 27: 27–50.

Harrison, J.R., and G.T. Daigger. 1987. A comparison of trickling filter media. J. Water Pollut. Control Fed. 59: 679–685.

Harshman, V. 2003. Follow your nose. Water Environ. Technol. 15: 3135.

Hartmann, L., and G. Laubenberger. 1968. Toxicity measurements in activated sludge. J. San. Eng. Div. 94: 247–252.

Hartmans, S., J.A.M. de Bont, J. Tramper, and K.C.A.M. Luyben. 1985. Bacterial degradation of vinyl chloride. Biotechnol. Lett. 7: 383–388.

Harvey, R.W. 1991. Parameters involved in modeling movement of bacteria in groundwater, pp. 89–114, In: *Modeling the Environmental Fate of Microorganisms*, C.J. Hurst, Ed., American Society of Microbiology, Washington, D.C.

Harvey, R.W. 1997a. *In situ* and laboratory methods to study subsurface microbial transport, pp. 586–599, In: *Manual of Environmental Microbiology*, C.J. Hurst, G.R. Knudsen, M.J. McInermey, L.J. Stetzenback, and M.V. Walters, Eds., ASM Press, Washington, DC.

Harvey, R.W. 1997b. Microorganisms as tracers in groundwater injection and recovery experiments: a review. FEMS Microbiology 20: 461–472.

Harvey, R.W., and S.P. Garabedian. 1991. Use of colloid filtration theory in modeling movement of bacteria though a contaminated sandy aquifer. Environ. Sci. Technol. 25: 178–185.

Harvey, R.W., and H. Harms. 2002. Tracers in groundwater: Use of microorganisms and microspheres, pp. 3194–3202, In: *Encyclopedia of Environmental Microbiology*, G. Bitton, editor-in-chief, Wiley-Interscience, N.Y.

Harvey, R.W., D.W. Metge, N. Kinner, and N. Mayberry. 1997. Physiological considerations in applying laboratory-determined buoyant densities to prediction of bacterial and protozoan transport in groundwater: Results of *in-situ* and laboratory tests. Environ. Sci. Technol. 31: 289–295.

Harvey, R.W., and J.N. Ryan. 2004. Use of PRD1 bacteriophage in groundwater viral transport, inactivation and attachment studies. FEMS Ecol. 49: 3–16.

Harwood, V.J., M. Brownell, W. Perusek, and J.E. Whitlock. 2001. Vancomycin-Resistant *Enterococcus* spp. isolated from wastewater and chicken feces in the United States. Appl. Environ. Microbiol. 67: 4930–4933.

Hashimoto, A., S. Kunikane, and T. Hirata. 2002. Prevalence of *Cryptosporidium* oocysts and *Giardia* cysts in the drinking water supply in Japan. Water Res. 36: 519–526.

Haufele, A., and H.V. Sprockhoff. 1973. Ozone for disinfection of water contaminated with vegetative and spore forms of bacteria, fungi and viruses. Zentralbl. Bakteriol. Hyg. Abt. 1, Orig. Reihe B 175: 53–70.

Havelaar, A.H., M. Bosman, and J. Borst. 1983. Otitis externa by *Pseudomonas aeruginosa* associated with whirlpools. J. Hyg. 90: 489–498.

Havelaar, A.H., and W.M. Hogeboom 1983. Factors affecting the enumeration of coliphage in sewage and sewage-polluted waters. Antonie van Leeuwenhoek J. Microbiol. 49: 387–397.

Havelaar, A.H., W.M. Hogeboom, K. Furuse, R. Pot, and M.P. Hormann. 1990a. F-specific RNA bacteriophages and sensitive host strains in faeces and wastewater of human and animal origin. J. Appl. Bacteriol. 69: 30–37.

Havelaar, A.H., C.C.E. Meulemans, W.M. Pot-Hogeboom, and J. Koster. 1990b. Inactivation of bacteriophage MS2 in wastewater effluent with monochromatic and polychromatic ultraviolet light. Water Res. 24: 1387–1391.

Havelaar, A.H., M. van Olphen, and J.F. Schijven. 1995. Removal and inactivation of viruses by drinking water treatment processes under full-scale conditions. Water Sci. Technol. 31 (5–6): 55–68.

Hawkes, H.A. 1983a. Activated sludge, pp. 77–162, In: *Ecological Aspects of Used Water Treatment*, Vol.2, C.R. Curds, and H.A. Hawkes, Eds., Academic Press, London.

Hawkes, H.A. 1983b. Stabilization ponds, pp. 163–217, In: *Ecological Aspects of Used-Water Treatment*, Vol.2, C.R. Curds and H.A. Hawkes, Eds., Academic Press, London.

Hawley, H.B., D.P. Morin, M.E. Geraghty, J. Tomkow, and C.A. Phillips. 1973. Coxsackievirus B epidemic at a boys summer camp: Isolation of virus from swimming water. J. Amer. Med. Assoc. 226: 33–36.

Hawley, R.L., and E.M. Eitzen Jr. 2001. Biological weapons-A primer for microbiologists. Ann. Rev. Microbiol. 55: 235–253.

Hayes, E.B., T.D. Matte, T.R. O'Brien, T.W. McKinley, G.S. Logsdon, J.B. Rose, B.L.P. Ungar, D.M. Word, P.F. Pinsky, M.L. Cummings, M.A. Wilson, E.G. Long, E.S. Hurwittz, and D.D. Juranek. 1989. Large community outbreak of cryptosporidiosis due to contamination of a filtered public water supply. N. England J. Med. 320: 1372–1376.

Hayes, K.P., and M.D. Burch. 1989. Odorous compounds associated with algal blooms in south australian waters. Water Res. 23: 115–121.

Hazen, T.C. 1988. Fecal coliforms as indicators in tropical waters. Tox. Assess. 3: 461–477.

Head, I,M., W.D. Hiorns, T.M. Embley, A.J. McCarthy, and J.R. Saunders. 1993. The phylogeny of autotrophic ammonia-oxidizing bacteria as determined by analysis of 16S ribosomal RNA gene sequences. J. Gen. Microbiol. 139: 1147–1153.

Head, I.M., J.R. Saunders, and R.W. Pickup. 1998. Microbial evolution, diversity, and ecology: A decade of ribosomal RNA analysis of uncultivated microorganisms. Microbial Ecol. 35: 1–21.

Hegarty, J.P., M.T. Dowd, and K.H. Baker. 1999. Occurrence of *Helicobacter pylori* in surface water in the United States. J. Appl. Microbiol. 87: 697–701.

Heijnen, J.J., and J.A. Roels. 1981. A macroscopic model describing yield and maintainance relationships in aerobic fermentation. Biotech. Bioeng. 23: 739–741.

Heitkamp, M.A., V. Camel, T.J. Reuter, and W.J. Adams. 1990. Biodegradation of *p*-nitrophenol in an aqueous waste stream by immobilized bacteria. Appl. Environ. Microbiol. 56: 2967–2973.

Heitkamp, M.A., and W.P. Stewart. 1996. A novel porous nylon biocarrier for immobilized bacteria. Appl. Environ. Microbiol. 62: 4659–4662.

Hejkal, T.W., F.M. Wellings, P.A. LaRock, and A.L. Lewis. 1979. Survival of poliovirus within organic solids during chlorination. Appl. Environ. Microbiol. 38: 114–118.

Hejzlar, J., and J. Chudoba. 1986. Microbial polymers in the aquatic environment: I. Production by activated sludge microorganisms under different conditions. Water Res. 20: 1209–1216.

Hemmes, J.H., K.C. Winkler, and S.M. Kool. 1960. Virus survival as a seasonal factor in influenza and poliomyelitis. Nature 188: 430–431.

Henderson, D.A., T.V. Inglesby, J.G. Bartlet, M.S. Ascher, and E. Eitzen. 1999. Smallpox as a biological weapon: medical and public health management. J. Amer. Med. Assoc. 281: 2127–2137.

Heng, B.H. 1994. Prevalence of hepatitis A virus among sewage workers in Singapore. Infect. Epidemiol. 113: 121–128.

Henney, R.C., M.C. Fralish, and W.V. Lacina. 1980. Shock load of chromium (VI). J. Water Pollut. Control Fed. 52: 2755–2760.

Henry, J.G., and R. Gehr. 1980. Odor control: An operator's guide. J. Water Poll. Contr. Fed. 52: 2523–2537.

Hensel, A., and K. Petzoldt. 1995. Biological and biochemical analysis of bacteria and viruses, pp. 335–360, In: *Bioaerosols Handbook*, C.S. Cox, and C.M. Mathes, Eds., Lewis, Boca Raton, FL.

Henson, J.M., P.H. Smith, and D.C. White. 1989. Examination of thermophilic methane-producing digesters by analysis of bacterial lipids. Appl. Environ. Microbiol. 50: 1428–1433.

Henze, M. 1997. Trends in advanced wastewater treatment. Water Sci. Technol. 35: 1–4.

Henze, M. and the Group on Mathematical Modelling for Design and Operation of Biological Wastewater Treatment. 2000. Activated sludge models ASM1, ASM2, ASM2d and ASM3. IWA Pub., London.

Herbst, E., I. Scheunert, W. Klein, and F. Korte. 1977. Fate of PCB-^{14}C in sewage treatment-laboratory experiments with activated sludge. Chemosphere 6: 725–730.

Hermon-Taylor, J., T.J. Bull, J.M. Sheridan, J. Cheng, M.L. Stellakis, and N. Sumar. 2000. Causation of Crohn's disease by *Mycobacterium avium subspecies paratuberculosis*. Can. J. Gastroenterol. 14: 521–539.

Hernandez, J.F., J.M. Guibert, J.M. Delattre, C. Oger, C. Charriere, B. Hugues, R. Serceau, and F. Sinegre. 1990. A miniaturized fluorogenic assay for enumeration of *E. coli* and enterococci in marine waters. Abstract, Int. Symp. on Health-Related Water Microbiology, Tubingen, West Germany, April 1–6, 1990.

Hernandez, J.F., J.M. Guibert, J.M. Delattre, C. Oger, C. Charriere, B. Hugues, R. Serceau, and F. Sinegre. 1991. Evaluation d'une methode miniaturisee de denombrement des *Escherichia coli* en eau de mer, fondee sur l'hydrolyse du 4-methylumbelliferyl β-D-glucuronide. Water Res. 25: 1073–1078.

Hernandez, J.F., A.M. Pourcher, J.M. Delattre, C. Oger, and J.L. Loeuillard. 1993. MPN miniaturized procedure for the enumeration of faecal streptococci in fresh and marine waters: The MUST procedure. Water Res. 27: 597–606.

Hernroth, B. E., A.-C. Conden-Hansson, A.-S. Rehnstam-Holm, R. Girones, and A.K. Allard. 2002. Environmental factors influencing human viral pathogens and their potential indicator organisms in the blue mussel, *Mytilus edulis*: the first Scandinavian report. Appl. Environ. Microbiol. 68: 4523–4533.

Herrmann, J.E., and N.R. Blacklow. 1995. Enteric adenoviruses, pp. 1047–1053, In: *Infections of the Gastrointestinal Tract*, M.J. Blaser, P.D. Smith, J.I. Ravdin, H.B. Greenberg, and R.L. Guerrant, Eds., Raven Press, New York.

Herndl, G.H., G. Müller-Niklas, and J. Frick. 1993. Major role of ultraviolet-B in controlling bacterioplankton growth in the surface layer of the ocean. Nature 362: 717–719.

Herson, D.S., B. McGonigle, M.A. Payer, and K.H. Baker. 1987. Attachment as a factor in the protection of *Enterobacter cloacae* from chlorination. Appl. Environ. Microbiol. 53: 1178–1180.

Hervio-Heath, D., R. Colwell, A. Derrien, A. Robert-Pillot, J. Fournier, and M. Pommepuy. 2002. Occurrence of pathogenic vibrios in coastal areas of France. J. Appl. Microbiol. 92: 1123–1135.

Herwaldt, B.L., G.F. Craun, S.L. Stokes, and D.D. Juranek. 1992. Outbreaks of waterborne disease in the United States: 1989–90. J. Am. Water Works Assoc. 84: 129–135.

Herwaldt, B.L., M.-L. Ackers, and the *Cyclospora* Working Group. 1997. An outbreak in 1996 of cyclosporiasis associated with imported raspberries. N. Engl. J. Med. 336: 1548–1556.

Hespanhol, I., and A.M.E. Prost. 1994. WHO guidelines and national standards for reuse and water quality. Water Res. 28: 119–124.

Hesselmann, R.P.X., C. Werlen, D. Hahn, J.R. van der Meer, and A.J.B. Zehnder. 1999. Enrichment, phylogenetic analysis and detection of a bacterium that performs enhanced biological phosphate removal in activated sludge. Syst. Appl. Microbiol. 22: 454–465.

Hetrick, F.M. 1978. Survival of human pathogenic viruses in estuarine and marine waters. ASM News 44: 300–303.

Hibler, C.P., C.M. Hancock, M. Perger, G. Wegrzyn, and K.D. Swabby. 1987. Inactivation of *Giardia* cysts with chlorine at 0.5°C to 5°C. Report to the American Waterworks Association Research Foundation. Denver, CO.

Hibler, C.P., and C.M. Hancock. 1990. Waterborne giardiasis, pp. 271–293, In: *Drinking Water Microbiology*, G.A. McFeters, Ed., Springer Verlag, New York.

Hickey, C.W., C. Blaise, and G. Costan. 1991. Microtesting appraisal of ATP and cell recovery toxicity end points after acute exposure of *Selenastrum capricornutum* to selected chemicals. Environ. Toxicol. Water Qual. 6: 383–403.

Hickey, R.F., J. Vanderwielen, and M.S. Switzenbaum. 1987. The effect of organic toxicants on methane production and hydrogen gas levels during the anaerobic digestion of waste activated sludge. Water Res. 21: 1417–1427.

Hijnen, W.A.M., and D. van der Kooij. 1992. The effect of low concentrations of assimilable organic carbon (AOC) in water on biological clogging of sand beds. Water Res. 26: 963–972.

Hilborn, E.D., M.O. Royster, and D.J. Drabkowski. 2002. Survey of US public helath laboratories: Microbial pathogens on the CCL. J. A.W.W.A. 94: 88–96.

Hill, R.T., I.T. Knight, M.S. Anikis, and R.R. Colwell. 1993. Benthic distribution of sewage sludge indicated by *Clostridium perfringens* at a deep-ocean dump site. Appl. Environ. Microbiol. 59: 47–51.

Hill, R.T., W.L. Staube, A.C. Palmisano, S.L. Gibson, and R.L. Colwell. 1996. Distribution of sewage indicated by *Clostridium perfringens* at a deep-water disposal site after cessation of sewage disposal. Appl. Environ. Microbiol. 62: 1741–1746.

Himberg, K., A.M. Keijola, L. Hiisvirta, H. Pyysalo, and K. Sivonen. 1989. The effect of water treatment processes on the removal of hepatoxins from *Microcystis* and *Oscillatoria* cyanobacteria: A laboratory study. Water Res. 23: 979–984.

Hinzelin, F., and J.C. Block. 1985. Yeast and filamentous fungi in drinking water. Environ. Lett. 6: 101–106.

Hiraishi. A., K. Masamune, and H. Kitamura. 1989. Characterization of the bacterial population structure in an anaerobic-aerobic activated sludge system on the basis of respiratory quinone profiles. Appl. Environ. Microbiol. 55: 897–901.

Hiraishi. A., Y. Ueda, and J. Ishihara. 1998. Quinone profiling of bacterial communities in natural and synthetic sewage activated sludge for enhanced phosphate removal. Appl. Environ. Microbiol. 64: 992–998.

Hitdlebaugh, J.A., and R.D. Miller. 1981. Operational problems with rotating biological contactors. J. Water Pollut. Control Fed. 53: 1283–1293.

Ho, C.-F., and D. Jenkins. 1991. The effect of surfactants on *Nocardia* foaming in activated sludge. Water Sci. Technol. 23: 879–887.

Hoehn, R.C. 1965. Biological methods for the control of tastes and odors. Southwest Water Works J. 47: 26.

Hoehn, R.C., D.B. Barnes, B.C. Thompson, C.W. Randall, T.J. Grizzard, and P.T.B. Shaffer. 1980. Algae as sources of trihalomethane precursors. J. Am. Water Works Assoc. 72: 344–350.

Hoff, J.C. 1978. The relationship of turbidity to disinfection of potable water, p. 103–117, In: *Evaluation of the Microbiology Standards for Drinking Water*, C.W. Hendricks, Ed., EPA-570/9-78/00C, U.S. Environmental Protection Agency, Washington, DC.

Hoff, J.C. 1986. Inactivation of microbial agents by chemical disinfectants. EPA Report No. 600/2-86/067, U.S Environmental Protection Agency, Water Engineering Research Laboratory, Cincinnati, OH.

Hoff, J.C., and E.W. Akin. 1986. Microbial resistance to disinfectants: Mechanisms and significance. Environ. Health Perspect. 69: 7–13.

Hoffman, R.L., and R.M. Atlas. 1977. Measurement of the effect of cadmium stress on protozoan grazing of bacteria (bacterivory) in activated sludge by fluorescence microscopy. Appl. Environ. Microbiol. 53: 2440–2444.

Hofle, M.G. 1990. RNA chemotaxonomy of bacterial isolates and natural microbial communities, pp. 129–159, In: *Aquatic Microbial Ecology: Biochemical and Molecular Approaches*, Springer-Verlag, New York.

Holdeman, L.V., I.J. Good, and W.E.C. Moore. 1976. Human fecal flora: variation in human fecal composition within individuals and a possible effect of emotional stress. Appl. Environ. Microbiol. 31: 359–375.

Holm, H.W., and J.W. Vennes. 1970. Occurrence of sulfur purple bacteria in a sewage treatment lagoon. Appl. Microbiol. 19: 988–996.

Holm, N.C., C.G. Gliesche, and P. Hirsch. 1996. Diversity and structure of *Hyphomicrobium* populations in a sewage treatment plant and its adjacent receiving lake. Appl. Environ. Microbiol. 62: 522–528.

Hong, S.M., J.K. Park, and Y.O. Lee. 2004. Mechanisms of microwave irradiation involved in the destruction of fecal coliforms from biosolids. Water Res. 38: 1615–1625.

van Hoof, F., J.G. Janssens, and H. van Dyck. 1985. Formation of mutagenic activity during surface water preozonation and their behaviour in water treatment. Chemosphere 14: 501–510.

Hooper, S.W. 1987. Characterization of pSS50, the 4-chlorobiphenyl mineralization plasmid, Doctoral dissertation, University of Tennessee, Knoxville, TN.

Hopkins, G.D., L. Semprini, and P.L. McCarty. 1993. Microcosm and *in situ* field studies of enhanced biotransformation of trichloroethylene by phenol-utilizing microorganisms. Appl. Environ. Microbiol. 59: 2277–2285.

Hopkins, R.J., P.A. Vial, C. Ferreccio, J. Ovalle, P. Prado, V. Sotomayor, R.G. Russel, S.S. Wasserman, and J.G. Morris. 1993. Seroprevalence of Helicobacter pylori in Chile: Vegetables may serve as one route of transmission. J. Infect. Dis. 168: 222–226.

Horan, N.J., and C.R. Eccles. 1986. Purification and characterization of extracellular polysaccharide from activated sludges. Water Res. 20: 1427–1432.

Horan, N.J., A.M. Bu'Ali, and C.R. Eccles. 1988. Isolation, identification and characterisation of filamentous and floc-forming bacteria from activated sludge flocs. Environ. Technol. Lett. 9: 449–457.

Horn, J.B., D.W. Hendricks, J.M. Scanlan, L.T. Rozelle, and W.C. Trnka. 1988. Removing *Giardia* cysts and other particles from low-turbidity waters using dual-stage filtration. J. Am. Water Works Assoc. 80: 68–77.

Horrocks, H.W. 1907. Experiments made to determine the conditions under which specific bacteria derived from sewage may be present in the air of ventilating pipes, drains, inspection chambers and sewers. Proc. Royal Soc. London, Ser. B7: 531.

Horvath, R.S. 1972. Microbial co-metabolism and the degradation of organic compounds in nature. Bacteriol. Rev. 36: 146–155.

Hosetti, B.B., and S. Frost. 1994. Catalase activity in wastewater. Water Res. 28: 497–500.

Hoshino, T., N. Noda, S. Tsuneda, A. Hirata, and Y. Inamori. 2001. Direct detection by *in situ* PCR of the *amoA* gene in biofilm resulting from a nitrogen removal process. Appl. Environ. Microbiol. 67: 5261–5266.

Hotez, P.J., F. Zheng, L.G. Xu, M.-G. Chen, S.-H. Xiao, S.-X. Liu, D. Blair, D.P. McManus, and G.M. Davis. 1997. Emerging and reemerging helminthiases and the public health of China. Emerg. Infect. Dis. 3: 303–317.

Houghton, S.R., and D.D. Mara. 1992. The effect of sulphide generation in waste stabilization ponds on photosynthetic populations and effluent quality. Water Sci. Technol. 26: 1759–1768.

Houndt, T., and H. Ochman. 2000. Long-term shifts in patterns of antibiotic resistance in enteric bacteria. Appl. Environ. Microbiol. 66: 5406–5409.

Houston, J., B.N. Dancer, and M.A. Learner. 1989a. Control of sewage filter flies using *Bacillus thuringiensis* var *israelensis*. I. Acute toxicity tests and pilot scale trial. Water Res. 23: 369–378.

Houston, J., B.N. Dancer, and M.A. Learner. 1989b. Control of sewage filter flies using *Bacillus thuringiensis* var *israelensis*. II. Full scale trials. Water Res. 23: 379–385.

Howgrave-Graham, A.R., and P.L. Steyn. 1988. Application of the fluorescent-antibody technique for the detection of *Sphaerotilus natans* in activated sludge. Appl. Environ. Microbiol. 54: 799–802.

Hoyana, Y., V. Bacon, R.E. Summons, W.E. Pereira, B. Helpern, and A.M. Duffield. 1973. Chlorination studies: IV. Reactions of aqueous hypochlorous acid with pyrimidine and purine bases. Biochem. Biophys. Res. Commun. 53: 1195–2001.

Hoyer, O. 2000. The status of UV technology in Europe. IUVA News 2: 22–27.

Hozalski, R.M., S. Goel, and E.J. Bouwer. 1992. Use of biofiltration for removal of natural organic matter to achieve biologically stable drinking water. Water Sci. Technol. 26: 2011–2014.

Hozalski, R.M., and E.J. Bouwer. 1998. Deposition and retention of bacteria in backwashed filters. J. Am. Water Works Assoc. 90: 71–85.

Hsu, S.C., R. Martin, and B.B. Wentworth. 1984. Isolation of *Legionella* species from drinking water. Appl. Environ. Microbiol. 48: 830–832.

Hu, M., L. Kang, and G. Yao. 1989. An outbreak of viral hepatitis A in Shangai, pp. 361–372, In: *Infectious Diseases of the Liver*, Proceedings of the 54th Falk Symposium, Oct. 12–14, 1989, Basel, Swizerland.

Hu, P., and P.F. Strom. 1991. Effect of pH on fungal growth and bulking in laboratory activated sludges. Res. J. Water Pollut. Control Fed. 63: 276–277.

Huang, C.-H., and D. Sedlak. 2001. Analysis of estrogenic hormones in municipal wastewater effluent and surface water using enzyme-linked immunosorbent assay and gas chromatography/mass spectroscopy. Environ. Tox. Chem. 20: 133–139.

Huang, F., G. Bitton, and I.-C. Kong.1999. Determination of the heavy metal binding capacity of aquatic samples using MetPLATE: a preliminary study. Sci. Total Environ. 234: 139–145.

Huang, J.C., and V.T. Bates. 1980. Comparative performance of rotating biological contactors using air and pure oxygen. J. Water Pollut. Control Fed. 52: 2686–2703.

Huang, J.Y.C., G.E. Wilson, and T.W. Schroepfer. 1979. Evaluation of activated carbon adsorption for sewer odor control. J. Water Pollut. Control Fed. 51: 1054–1062.

Huang, L.N., H. Zhou, Y.Q. Chen, S. Luo, C.Y. Lan, and L.H. Qu. 2002. Diversity and structure of the archaeal community in the leachate of a full-scale recirculating landfill as examined by direct 16S rRNA gene sequence retrieval. FEMS Microbiol. Lett. 214: 235–240.

Huang, P., J.T. Weber, D.M. Sosin, P.M. Griffin, E.G. Long, J.J. Murphy, F. Kocka, C. Peters, and C. Kallick. 1996. The first reported outbreak of diarreal illness associated with *Cyclospora* in the United States. Ann. Intern. Med. 123: 409–414.

Huck, P.M. 1990. Measurement of biodegradable organic matter and bacterial growth potential in drinking water. J. Am. Water Works Assoc. 82: 78–86.

Huertas, A., B. Barbeau, C. Desjardins, A. Galarza, M.A. Figueroa, and G.A. Toranzos. 2003. Evaluation of *Bacillus subtilis* and coliphage MS2 as indicators of advanced water treatment efficiency. Water Sci. Technol. 37 (3): 255–259.

Huffman, D.E., T.R. Slifko, K. Salisbury, and J.B. Rose. 2000. Inactivation of bacteria, virus and *Cryptosporidium* by a point-of-use device using pulsed broad spectrum white light. Water Res. 34: 2491–2498.

Huffman, D.E., A. Gennaccaroa, J.B. Rose, and B.W. Dussert. 2002. Low- and medium-pressure UV inactivation of microsporidia *Encephalitozoon intestinalis*. Water Res. 36: 3161–3164.

Hugues, B., A. Cini, M. Plissier, and J.R. Lefebvre. 1980. Recherche des virus dans le milieu marin a partir d'echantillon de volumes differents. Eau du Quebec 13: 199–203.

Hugues, B., J.R. Lefebvre, M. Plissier, and A. Cini. 1981. Distribution of viral and bacterial densities in seawater near a coastal discharge of treated domestic sewage. Zbl. Bakt. Hyg. I. Abt. Orig. B 173: 509–516.

Huisman, L., and W.E. Wood. 1974. Slow Sand Filtration. World Health Organization, Geneva.

Huixian, Z., Y. Zirui, L. Junhe, X. Xu, and Z. Jinqi. 2002. A possible new disinfection by-product – 2-chloro-5-oxo-3-hexene diacyl chloride (COHC) – in formation of MX by chlorinating model compounds. Water Res. 36: 4535–4542.

Huk, A., R.R. Colwell, R. Rahman, A. Ali, M.A.R. Chowdhury, S. Parveen, D.A. Sack, and E. Russek-Cohen. 1990. Detection of *Vibrio cholerae* 01 in the aquatic environment by fluorescent-monoclonal antibody and culture methods. Appl. Environ. Microbiol. 56: 2370–2373.

Huk, A., E. Lipp, and R. Colwell. 2002. Cholera, pp. 853–861, In: *Encyclopedia of Environmental Microbiology*, G. Bitton, editor-in-chief, Wiley-Interscience, N.Y.

Hulshoff Pol, L.W., W.J. de Zeeuw, C.T.M. Velzeboer, and G. Lettinga. 1982. Granulation in UASB reactors. Water Sci. Technol. 15: 291–305.

Hulshoff Pol, L.W., J. Dolfing, W.J. de Zeeuw, and G. Lettinga. 1983. Cultivation of well adapted pelletized methanogenic sludge. Biotechnol. Lett. 5: 329–332.

Hunen, W.A.M, and D. van der Kooij. 1992. The effect of low concentrations of assimilable organic carbon (AOC) in water on biological clogging of sand beds. Water Res. 26: 963–972.

Hunter, G.V., G.R. Bell, and C.N. Henderson. 1966. Coliform organism removal by diatomite filtration. J. Am. Works Assoc. 58: 1160–1169.

Hurst, C.J., and D.A. Brashear. 1987. Use of a vacuum filtration technique to study leaching of indigenous viruses from raw wastewater sludge. Water Res. 21: 809–812.

Hurst, C.J., C.P. Gerba, and I. Cech. 1980. Effects of environmental variables and soil characteristics on virus survival in soil. Appl. Environ. Microbiol. 40: 1067–1079.

Hurst, C.J., M.C. Roman, J.L. Garland, D.C. Obenhuber, and A.M. Brittain. 1997. Microbiological aspects of space exploration. Am. Soc. Microbiol. News 63: 611–614.

Huser, B.A., K. Wuhrmann, and A.J.B. Zehnder. 1982. *Methanothrix soehngenii* gen. nov. sp. nov., a new acetotrophic non-hydrogen-oxidizing bacterium. Arch. Microbiol. 132: 1–9.

Hussong, D., W.G. Burge, and N.K. Enkiri. 1985. Occurrence, growth and suppression of salmonellae in composted sewage sludge. Appl. Environ. Microbiol. 50: 887–893.

Hutchins, S.R. 1989. Biorestoration of fuel-contaminated aquifer using nitrate: Laboratory studies. Paper presented at the tenth. Annual Meeting, Society Environmental Toxicology and Chemistry, Toronto, Canada, Oct. 28–Nov. 2, 1989.

Hutchins, S.R. 1991. Biodegradation of monoaromatic hydrocarbons by aquifer microoorganisms using oxygen, nitrate, or nitrous oxide as the terminal electron acceptor. Appl. Environ. Microbiol. 57: 2403–2407.

Hutchins, S.R., G.W. Sewell, D.A. Kovacs, and G.A. Smith. 1991. Biodegradation of aromatic hydrocarbons by aquifer microorganisms under denitrifying conditions. Environ. Sci. Technol. 25: 68–76.

Hwang, Y., T. Matsuo, K. Hanaki, and N. Suzuki. 1994. Removal of odorous compounds in wastewater by using activated carbon, ozonation and aerated biofilter. Water Res. 28: 2309–2319.

Hyun, C.-K., E. Tamiya, T. Takeuchi, and I. Karube. 1993. A novel BOD sensor based on bacterial luminescence. Biotechnol. Bioeng. 41: 1107–1111.

Ichimura, K., and K. Watanabe. 1982. Preparation and characteristics of photo-crosslinkable polyvinyl alcohol. J. Polym. Sci. Polym. Chem. Ed. 20: 1419–1432.

Ida, S., and M. Alexander. 1965. Permeability of *Nitrobacter agilis* to organic compounds. J. Bacteriol. 90: 151–155.

Ijaz, M.K., and S.A Sattar. 1985. Comparison of airborne survival of calf rotavirus and poliovirus type 1 (Sabin) aerosolized as a mixture. Appl. Environ. Microbiol. 49: 289–295.

Ijzerman, M.M., and C. Hagedorn. 1992. Improved method for coliphage detection based on beta-galactosidase induction. J. Virol. Methods 40: 31–36.

Ijzerman, M.M., J.O. III Falkinham, R.B. Jr. Reneau, and C. Hagedorn. 1994. Field evaluation of two colorimetric coliphage detection methods. Appl. Environ. Microbiol. 60: 826–830.

Inamori, Y., K. Matsusige, R. Sudo, K. Chiba, H. Kikuchi, and T. Ebisuno. 1989. Advanced wastewater treatment using an immobilized microorganism/biofilm two-step process. Water Sci. Technol. 21: 1755–1758.

Inamori, Y., Y. Kuniyasu, R. Sudo, and M. Koga. 1991. Control of the growth of filamentous microorganisms using predacious ciliated protozoa. Water Sci. Technol. 23: 963–971.

International Reference Centre for Waste Disposal (IRCWD). 1985. Health Aspects of Wastewater and Excreta Use in Agriculture and Aquaculture: The Engelberg Report. IRCWD News No. 23: 11–18, Dubendorf, Switzerland.

Isaac-Renton, J.L., C.P.J. Fung, and A. Lochan. 1986. Evaluation of tangential-flow multiple-filter technique for detection of *Giardia lamblia* cysts in water. Appl. Environ. Microbiol. 52: 400–402.

Isaac-Renton, J., W.R. Bowie, A. King, G.S. Irwin, C.S. Ong, C.P. Fung, M.O. Shokeir, and J.P. Dubey. 1998. *Toxoplasma gondii* oocysts in drinking water. Appl. Environ. Microbiol. 63: 2278–2280.

Isaacson, M., A.R. Sayed, and W.H.J. Hattingh. 1986. *Studies on Health Aspects of Water Reclamation during 1974 to 1983 in Windhoek, South West Africa/Namibia.* WRC Report No. 38/1/86, Water Research Commission, Pretoria, 82 pp.

Ishii, K., M. Fukui, and S. Takii. 2000. Microbial succession during a composting process as evaluated by denaturing gradient gel electrophoresis analysis. J. Appl. Microbiol. 89: 768–777.

Ishizaki, K., N. Shinriki, and T. Ueda. 1984. Degradation of nucleic acids with ozone. V. Mechanism of action of ozone on deoxyribonucleoside 5′-monophosphates. Chem. Pharm. Bull. 32: 3601–3606.

Ishizaki, K., K. Sawadaishi, K. Miura, and N. Shinriki. 1987. Effect of ozone on plasmid DNA of *Escherichia coli in situ.* Water Res. 21: 823–827.

Isobe, K.O., M. Tarao, N.H. Chiem, L.Y. Minh, and H. Takada. 2004. Effect of environmental factors on the relationship between concentrations of coprostanol and fecal indicator bacteria in tropical (Mekong Delta) and temperate (Tokyo) freshwaters. Appl. Environ. Microbiol. 70: 814–821.

Iversen, Aina, Inger Kühn, Anders Franklin, and Roland Möllby. 2002. High Prevalence of Vancomycin-Resistant Enterococci in Swedish Sewage. Appl. Environ. Microbiol. 68: 2838–2842.

Iverson, W.P., and F.E. Brinckman. 1978. Microbial metabolism of heavy metals, pp. 201–232, In: *Water Pollution Microbiology*, Vol. 2, R. Mitchell, Ed., John Wiley & Sons, New York.

Izaguirre, G., C.J. Hwang, S.W. Krasner, and M.J. McGuire. 1982. Geosmin and 2-methylisoborneol from cyanobacteria in three water supply systems. Appl. Environ. Microbiol. 43: 708–714.

Jacobsen, B.N., N. Nyholm, B.M. Pedersen, O. Poulsen, and P. Ostfeldt. 1991. Microbial degradation of pentachlorophenol and lindane in laboratory-scale activated sludge reactors. Water Sci. Technol. 23: 349–356.

Jacobson, S.N., N.L. O'Mara, and M. Alexander. 1980. Evidence of cometabolism in sewage. Appl. Environ. Microbiol. 40: 917–921.

Jacongelo, J.G., S.A. Adham, and J.M. Laine. 1995. Mechanism of Cryptosporidium, Giadria and MS2 virus removal by MF and UF. J. AWWA 87 (9): 107–114.

Jagger, J. 1958. Photoreactivation. Bacteriol. Rev. 22: 99–114.

Jain, R.K., R.S. Burlage, and G.S. Sayler. 1988. Methods for detecting recombinant DNA in the environment. Crit. Rev. Biotechnol. 8: 33–47.

Jakubowski, W. 1986. USEPA-sponsored epidemiological studies of health risks associated with the treatment and disposal of wastewater and sewage sludge, pp. 140–153, In: *Epidemiological Studies of Risks Associated With Agricultural Use of Sewage Sludge: Knowledge and Needs*, J.C. Block, A.H. Havelaar, and P. l'Hermite, Eds., Elsevier Applied Science Pub., London, 168 pp.

Jakubowski, W., and T.H. Ericksen. 1979. Methods of detection of *Giardia* cysts in water supplies, In: *Waterborne Transmission of Giardiasis*, Report No. EPA-600/9-79-001, U.S. Environmental Protection Agency, Cincinnati, OH.

Jakubowski, W., and J.C. Hoff, Eds., 1979. Waterborne transmission of giardiasis. Proceedings of a symposium, EPA Report No. EPA-600/9-79-001, U.S. Environmental Protection Agency, Cincinnati, OH.

Jakubowski, W., J.L. Sykora, C.A. Sorber, L.W. Casson, and P.D. Cavaghan. 1990. Determining giardiasis prevalence by examination of sewage. International Symposium Health-Related Water Microbiology, Tubingen, W. Germany, April 1–6, 1990.

James, A.L., J.D. Perry, M. Ford, L. Armstrong, and F.K. Gould. 1996. Evaluation of cyclohexeno-esculetinβ-D-galactoside and 8-hydroxyquinolineβ-D-galactoside as substrates for the detection of β-galactosidase. Appl. Environ. Microbiol. 62: 3868–3870.

Jansons, J., L.W. Edmonds, B. Speight, and M.R. Bucens. 1989a. Survival of viruses in groundwater. Water Res. 23: 301–306.

Jansons, J., L.W. Edmonds, B. Speight, and M.R. Bucens. 1989b. Movement of viruses after artificial recharge. Water Res. 23: 293–299.

Janssen, C. 1997. Alternative assays for routine toxicity assessments: A review, pp. 813–839, In: *Ecotoxicology: Ecological Fundamentals, Chemical Exposure, and Biological Effects.* G. Schüürmann, and B. Markert, Eds., John Wiley & Sons, New York; Spektrum, Heidelberg, Germany.

Janssens, I., T. Tanghe, and W. Verstraete. 1997. Micropollutants: A bottleneck in sustainable waste-water treatment. Water Res. 35: 13–26.

Jarroll, E.F., J.C. Hoff, and E.A. Meyer. 1984. Resistance of cysts to disinfection agents, pp. 311–328, In: *Giardia and Giardiasis*, S.L. Erlandsen, and E.A. Meyer, Eds., Plenum, New York.

Jarvis, A.S., M.E. Honeycutt, V.A. McFarland, A.A. Bulich, and H.C. Bonds. 1996. A comparison of the Ames assay and Mutatox in assessing the mutagenic potential of contaminated dredged sediments. Ecotoxicol. Environ. Safety 33: 193–200.

Jawetz, E., J.L. Melnick, and E.A. Adelberg. 1984. *Review of Medical Microbiology*, 16th Ed., Lange, Los Altos, CA.

Jefferson, B., A.L. Laine, S.J. Judd, and T. Stephenson. 2000. Membrane bioreactors and their role in wastewater reuse. Water Sci. Technol. 41(1): 197–204.

Jeffrey, H.C., and R.M. Leach. 1972. *Atlas of Medical Helminthology and Protozoology*, Churchill Livingstone, Edinburgh.

Jehl-Pietri, C. 1992. Detection des virus enteriques dans le milieu hydrique: Cas du virus de l'hepatite dans l'environnement marin et les coquillages. Doctoral dissertation, Universite de Nancy 1, Nancy, France.

Jenkins, D. 1992. Towards a comprehensive model of activated sludge bulking and foaming. Water Sci. Technol. 25: 215–230.

Jenkins, D., M.G. Richard, and G.T. Daigger. 1984. *Manual on the Causes and Control of Activated Sludge Bulking and Foaming*, Water Research Commission, Pretoria, South Africa.

Jenkins, D., and M.G. Richard. 1985. The causes and control of activated-sludge bulking. Tappi 68: 73–76.

Jenkins, M.B., D.D. Bowman, and W.C. Ghiorse. 1998. Inactivation of *Cryptosporidium parvum* oocysts by ammonia. Appl. Environ. Microbiol. 64: 784–788.

Jenkins, R.L., J.P. Gute, S.W. Krasner, and R.B. Baird. 1980. The analysis and fate of odorous sulfur compounds in wastewaters. Water Res. 14: 441–448.

Jenkins, S.H., D.G. Keight, and A. Ewins. 1964. The solubility of heavy metal hydroxides in water, sewage and sewage sludge. II. The precipitation of metals by sewage. Int. J. Air Water Pollut. 8: 679–693.

Jensen, S.E., C.L. Anders, L.J. Goatcher, T. Perley, S. Kenefick, and S.E. Hrudey. 1994. Actinomycetes as a factor in odour problems affecting drinking water from the North Saskatchewan River. Water Res. 28: 1393–1401.

Jeon, C.O., D.S. Lee, and J.M. Park. 2003. Microbial communities in activated sludge performing enhanced biological phosphorus removal in a sequencing batch reactor. Water Res. 37: 2195–2205.

Jetten, M.S.M., S.J. Horn, and M.C.M. van Loosdrecht. 1997. Towards a more sustainable municipal wastewater treatment plant system. Water Sci. Technol. 35: 171–180.

Jewell, W.J. 1987. Anaerobic sewage treatment. Environ. Sci. Technol. 21: 14–20.

Jewell, W.J., and R.M. Kabrick. 1980. Autoheated aerobic thermophilic digestion with aeration. J. Water Pollut. Control Fed. 52: 512–523.

Jewell, W.J., Y.M. Nelson, and M.S. Wilson. 1992. Methanotrophic bacteria for nutrient removal from wastewater: Attached film system. Water Environ. Res. 64: 756–765.

Jiang, X., J. Wang, D.Y. Graham, and M.K. Estes. 1992. Detection of Norwalk virus in stool by polymerase chain reaction. J. Clin. Microbiol. 30: 2529–2534.

Jimenez, B., and A. Chavez, 2000. Chlorine disinfection of advanced primary effluent for reuse in irrigation in Mexico. AWWA Water Reuse Conf. Proc., Jan. 30–Feb. 2, 2000, San Antonio, TX.

Jofre, J., M. Blasi, A. Bosch, and F. Lucena. 1989. Occurrence of bacteriophages infecting *Bacteroides fragilis* and other viruses in polluted marine sediments. Water Sci. Technol. 21: 15–19.

Jofre, J., E. Olle, F. Ribas, A. Vidal, and F. Lucena. 1995. Potential usefulness of bacteriophages that infect *Bacteroides fragilis* as model organisms for monitoring virus removal in drinking water treatment plants. Appl. Environ. Microbiol. 61: 3227–3231.

Johnson, A.C., A. Belfroid, and A. Di Corcia. 2000. Estimating steroid estrogens inputs into activated sludge treatment works and observation on their removal from the effluent. Sci. Total Environ. 256: 163–173.

Johnson, A.C., and J.P. Sumpter. 2001. Removal of endocrine-disrupting chemicals in activated sludge treatment works. Environ. Sci. Technol. 35: 4697–4703.

Johnson, B.T. 1998. Microtox toxicity test system: New developments and applications, pp. 201–218, In: *Microscale Testing in Aquatic Toxicology: Advances, Techniques, and Practice*, P.G., Wells, K. Lee, and C. Blaise, Eds., CRC Press, Boca Raton, FL.

Johnson, C.H., E.W. Rice, and D.J. Reasoner. 1997. Inactivation of *Helicobacter pylori* by chlorination. Appl. Environ. Microbiol. 63: 4949–4971.

Johnson, D.C., C.E. Enriquez, I.J. Pepper, T.L. Davis, C.P. Gerba, and J.B. Rose. 1997. *Giardia* cysts and Cryptosporidium oocysts at sewage treatment works in Scotland. Water Sci. Technol. 35 (11–12): 261–268.

Johnson, D.E., Camann, J.W. Register, R.E. Thomas, C.A. Sorber, M.N. Guentzel, J.M. Taylor, and H.J. Harding. 1980. The evaluation of microbiological aerosols associated with the application of wastewater to land: Pleasanton, CA. Report No. EPA-600/1-80-015. U.S. Environmental Protection Agency, Cincinnati, OH.

Johnson, D.W., N.J. Pienazek, and J.B. Rose. 1993. DNA probe hybridization and PCR detection of *Cryptosporidium* compared to immunofluorescence assay. Water Sci. Technol. 27: 77–84.

Johnson, D.W., N.J. Pienazek, D.W. Griffin, L. Misener, and J.B. Rose. 1995a. Development of a PCR protocol for sensitive detection of *Cryptosporidium* ocysts in water samples. Appl. Environ. Microbiol. 61: 3849–3855.

Johnson, D.W., K.A. Reynolds, C.P. Gerba, I.L. Pepper, and J.B. Rose. 1995b. Detection of *Giardia* and *Cryptosporidium* in marine waters. Water Sci. Technol. 31: 439–442.

Johnson, L.M., C.S. McDowell, and M. Krupta. 1985. Microbiology in pollution control: From bugs to biotechnology. Dev. Ind. Microbiol. 26: 365–376.

Johnson, W.D. 1991. Dual distribution systems: The public utility perspective. Water Sci. Technol. 24: 343–352.

Johnston, J.B., and S.G. Robinson. 1982. Opportunities for development of new detoxification processes through genetic engineering, pp. 301–314, In: *Detoxification of Hazardous Wastes*, J.H. Exner, Ed., Ann Arbor Scientific Publications, Ann Arbor, MI.

Johnston, J.B., and S.G. Robinson. 1984. *Genetic Engineering and the Development of New Pollution Control Technologies*. Report No. EPA 600/2-84-037, U.S. Environmental Protection Agency, Cincinnati, OH.

Jolley, R.L., W.A. Brungs, and R.B. Cummings, Eds. 1985. *Water Chlorination: Chemistry, Environmental Impact, and Health Effects*. Lewis, Chelsea, MI.

Jolis, D., C. Lam and P. Pitt. 2001. Particle effects on ultraviolet disinfection of coliform bacteria in recycled water. Water Environ. Res. 73: 233–236.

Jonassen, T.O., E. Kjeldsberg, and B. Grinde. 1993. Detection of human astrovirus serotype 1 by polymerase chain reaction. J. Virol. Methods 44: 83–88.

Jones, B.R. 1956. Studies of pigmented non-sulfur purple bacteria in relation to cannery waste lagoon odors. Sewage Ind. Wastes 28: 883–893.

Jones, K. 2001. Campylobacters in water, sewage and the environment. J. Appl. Microbiol. 90: 68S–79S.

Jones, K., M. Betaieb, and D.R. Telford.199.) Seasonality of thermophilic campylobacters in surface waters: correlation with light. *1st UK Symposium on Health-Related Water Microbiology* (IAWPRC), Glasgow. pp. 124–130.

Jones, P.H., A.D. Tadwalkar, and C.L. Hsu. 1987. Enhanced uptake of phosphorus by activated sludge: Effect of substrate addition. Water Res. 21: 301–308.

Jones, W.L., and E.D. Schroeder. 1989. Use of cell-free extracts for the enhancement of biological wastewater treatment. J. Water Pollut. Control Fed. 61: 60–65.

Jordan, E.C. 1982. Fate of priority pollutants in publicly owned treatment works. 30-day study. Report No. EPA-440/1-82/302, U.S. Environmental Protection Agency, Washington, D.C.

Joret, J.-C., and Y. Levi. 1986. Methode rapide d'evaluation du carbone eliminable des eaux par voie biologique. Tribune CEBEDEAU 510: 3–9.

Joret, J.-C., A. Hassen, M.M. Bourbigot, F. Agbalika, P. Hartmann, and J.M Foliguet. 1986. Inactivation des virus dans l'eau sur une filiere de production a ozonation etagee. Water Res. 20: 871–876.

Joret, J.C., Y. Levi, T. Dupin, and M. Gibert. 1988. Rapid method for estimating bioeliminable organic carbon in water. Presented at the American Water Works Association Conference, Orlando, FL., June 19–23, 1988.

Joret, J.-C., P. Cervantes, Y. Levi, N. Dumoutier, L. Cognet, C. Hasley, M.O. Husson, and H. Leclerc. 1989. Rapid detection of *E. coli* in water using monoclonal antibodies. Water Sci. Technol. 21: 161–167.

Joret, J.C., Y. Levi, and C. Volk. 1990. Biodegradable dissolved organic carbon (BDOC) content of drinking water and potential regrowth of bacteria. Presented at the *International Symposium Health-Related Water Microbiology*, Tubingen, Germany, April 1–6, 1990.

Joret, J.-C., D. Perrine, and B. Langlais. 1992. Effect of temperature on the inactivation of *Cryptosporidium* oocysts by ozone. IWPRC International Symposium, Washington, D.C., May 26–29, 1992.

Jorgensen, J.H., J.C. Lee, G.A. Alexander, and H.W. Wolf. 1979. Comparison of *Limulus* assay, standard plate count and total coliform count for microbiological assessment of renovated wastewater. Appl. Environ. Microbiol. 37: 928–931.

Joyce, T.M., K.G. McGuigan, M. Elmore-Meegan, and R.M. Conroy. 1996. Inactivation of fecal bacteria in drinking water by solar heating. Appl. Environ. Microbiol. 62: 399–402.

Judd, S. 2004. A review of fouling of membrane bioreactors in sewage treatment. Water Sci. Technol. 49 (2): 229–235.

Jung, K., G. Bitton, and B. Koopman. 1996. Selective assay for heavy metal toxicity using a fluorogenic substrate. Environ. Toxicol. Chem. 15: 711–714.

Juretschko, S., G. Timmermann, M. Schmid, K.H. Schleifer, A. Pommerening-Röser, H.P. Koops, and M. Wagner. 1998. Combined molecular and conventional analyses of nitrifying bacterium diversity in activated sludge: *Nitrosococcus mobilis* and *Nitrospira*-like bacteria as dominant populations. Appl Environ Microbiol 64: 3042–3051.

Juttner, F. 1981. Detection of lipid degradation products in the water of a reservoir during a bloom of *Synura uvella*. Appl. Environ. Microbiol. 41: 100–106.

Kabrick, R.M., and W.J. Jewell. 1982. Fate of pathogens in thermophilic aerobic sludge digestion. Water Res. 16: 1051–1060.

Kabrick, R.M., W.J. Jewell, B.V. Salotto, and D. Berman. 1979. Inactivation of viruses, pathogenic bacteria and parasites in the autoheated thermophilic digestion of sewage sludges. Proceedings 34th. Industrial Waste Conference, Vol. 34: 771–789, Purdue University, Lafayette, IN.

Kadlec, R.H. 2002. Wastewater treatment – Wetlands and reedbeds, pp. 3389–3401, In: *Encyclopedia of Environmental Microbiology*, G. Bitton, editor-in-chief, Wiley-Interscience, N.Y.

Kahn, A.A., and G.E. Cerniglia. 1994. Detection of Pseudomonas aeruginosa from clinical and environmental samples by amplification of the exotoxin A gene using PCR. Appl. Environ. Microbiol. 60: 3739–3745.

Kakii, K., S. Kitamura, T. Shirakashi, and M. Kuriyama. 1985. Effect of calcium ion on sludge charateristics. J. Ferment. Technol. 63: 263–270.

Kalmbach, S., W. Manz, B. Bendinger, and U. Szewzyk. 2000. *In situ* probing reveals *Aquabacterium* commune as a widespread and highly abundant bacterial species in drinking water biofilms. Water Res. 34: 575–581.

Kämpfer, P., and M. Wagner. 2002. Filamentous bacteria in activated sludge: Current taxonomic status and ecology, pp. 1287–1306, In: *Encyclopedia of Environmental Microbiology*, G. Bitton, editor-in-chief, Wiley-Interscience, N.Y.

Kaminski, J.C. 1994. *Cryptosporidium* and the public water supply. N. Engl. J. Med. 331 1529.

Kanagawa, T., and D.P. Kelly. 1986. Breakdown of dimethylsulphide by mixed cultures and by *Thiobacillus thioparus*. FEMS Microbiol. Lett. 34: 13–19.

Kanagawa, T., and E. Mikami. 1989. Removal of methanethiol, dimethyl sulfide, dimethyl disulfide and hydrogen sulfide from contaminated air by *Thiobacillus thioparus* TK-m. Appl. Environ. Microbiol. 55: 555–558.

Kanagawa, T., Y. Kamagata, S. Aruga, T. Kohno, M. Horn, and M. Wagner. 2000. Phylogenetic analysis of and oligonucleotide probe development for Eikelboom type 021N filamentous bacteria isolated from bulking activated sludge. Appl. Environ. Microbiol. 66: 5043–5052.

Kaneko, M. 1989. Effect of suspended solids on inactivation of poliovirus and T2-phage by ozone. Water Sci. Technol. 21: 215–219.

Kaneko, M., K. Morimoto, and S. Nambu. 1976. The response of activated sludge to a polychlorinated biphenyl (KC-500). Water Res. 10: 157–163.

Kao, J.F., L.P. Hsieh, S.S. Cheng, and C.P. Huang. 1982. Effect of EDTA on cadmium in activated sludge systems. J. Water Pollut. Control Fed. 54: 1118–1126.

Kaplan, L.A., T.L. Bott, and D.J. Reasoner. 1993. Evaluation and simplification of the assimilable organic carbon nutrient bioassay for bacterial growth in drinking water. Appl. Environ. Microbiol. 59: 1532–1539.

Kaplan, L.A., D.J. Reasoner, E.W. Rice, and T.I. Bott. 1992. Rev. Sci. Eau 5: 69 (cited by van der Kooij, 1995).

Kapperud, G., T. Vardund, E. Skjerve, E. Hornes, and T.E. Michaelsen. 1993. Detection of pathogenic *Yersinia enterocolitica* in food and water by immunomagnetic separation, nested polymerase chain reactions, and colorimetric detection of amplified DNA. Appl. Environ. Microbiol. 59: 2938–2944.

Kapuschinski, R.B., and R. Mitchell. 1981. Solar radiation induces sublethal injury in *Escherichia coli* in seawater. Appl. Environ. Microbiol. 41: 670–674.

Kapuschinski, R.B., and R. Mitchell, 1982. Sunlight-induced mortality of viruses and *Escherichia coli* in coastal seawater. Environ. Sci. Technol. 17: 1–6.

Karhadkar, P.P., J.M. Audic, G.M. Faup, and P. Khanna. 1987. Sulfide and sulfate inhibition of methanogenesis. Water Res. 21: 1061–1066.

Karimi, A.A., and P.C. Singer. 1991. Trihalomethane formation in open reservoirs. J. Am. Water Works Assoc. 83: 84–88.

Karner, D.A., J.H. Standridge, G.W. Harrington, and R.P. Barnum. 2001. Microcystin algal toxins in source and finished drinking water. J. AWWA 93 (8): 72–81.

Karns, J.S., M.T. Muldoon, W.W. Mulbry, M.K. Derbyshire, and P.C. Kearney. 1987. Use of microorganisms and microbial systems in the degradation of pesticides. In: H.M. LeBaron, R.O. Mumma, R.C. Honeycutt, and J.H. Duesing, Eds., *Biotechnology in Agricultural Chemistry*, ACS Symp. Series 334, American Chemical Society, Washington, D.C.

Karim, M.R., F.D. Manshadi, M.M. Karpiscak, and C.P. Gerba. 2004. The persistence and removal of enteric pathogens in constructed wetlands. Water Res. 38: 1831–1837.

Karube. I. 1987. Microorganism based sensors, pp. 13–29, In: *Biosensors: Fundamentals and Applications*, A.P.F. Turner, I. Karube and G.S. Wilson, Eds., Oxford University Press, Oxford.

Karube, I., S. Kuriyama, T. Matsunaga, and S. Suzuki. 1980. Methane production from wastewaters by immobilized methanogenic bacteria. Biotech. Bioeng. 22: 847–857.

Karube, I., S. Mitsuda, T. Matsunaga, and S. Suzuki. 1977. A rapid method for estimation of BOD by using immobilized microbial cells. J. Ferment. Technol. 55: 243–246.

Karube, I., and E. Tamiya. 1987. Biosensors for environmental control. Pure Appl. Chem. 59: 545–554.

Kaspar, C.W., J.L. Burgess, I.T. Knight, and R.R. Colwell. 1990. Antibiotic resistance indexing of *Escherichia coli* to identify sources of fecal contamination in water. Can. J. Microbiol. 36: 891–894.

Kastner, J.R., J.S. Domingo, M. Denham, M. Molina, and R. Brigmon. 2000. Effect of chemical oxidation on subsurface microbiology and trichloroethene (TCE) biodegradation. Bioremediation J. 4: 219–236.

Katamay, M.M. 1990. Assessing defined-substrate technology for meeting monitoring requirements of the total coliform rule. J. Am. Water Works Assoc. 82: 83–87.

Kataoka, N., Y. Tokiwa, and K. Takeda. 1991. Improved technique for identification and enumeration of methanogenic bacterial colonies on roll tubes by epifluoresecnce microscopy. Appl. Environ. Microbiol. 57: 3671–3673.

Kato, J., K. Yamada, A. Muramatsu, Hardoyo, and H. Ohtake. 1993. Genetic improvement of *Escherichia coli* for enhanced biological removal of phosphate from wastewater. Appl. Environ. Microbiol. 59: 3744–3749.

Kato, K., and F. Kazama. 1991. Respiratory inhibition of *Sphaerotilus* by iron compounds and the distribution of the sorbed iron. Water Sci. Technol. 23. 947–954.

Kato, M., J.A. Field, and G. Lettinga. 1993. The high tolerance of methanogens in granular sludge to oxygen. Biotechnol. Bioeng. 42: 1360–1366.

Kato, M., J.A. Field, R. Kleerebezem, and G. Lettinga. 1994. Treatment of low strength wastewater in upflow anaerobic sludge blanket (UASB) reactors. J. Ferment. Bioeng. 77: 679–685.

Katz, A., and N. Narkis. 2001. Removal of chlorine dioxide disinfection by-products by ferrous salts. Water Res. 34: 101–108.

Katzenelson, E., I. Buium, and H.I. Shuval. 1976. Risk of communicable disease infection associated with wastewater irrigation in agricultural settlements. Science 194: 944–946.

Keene, W.E., J.M. McAnulty, F.C. Hoesly, L.P. Williams, K. Hedberg, G.L. Oxman, T.J. Barret, M.A. Pfaller, and D.W. Fleming. 1994. A swimming-associated outbreak of haemorrhagic colitis caused by *Escherichia coli* O157:H7 and *Shigella sonnei*. N. Eng. J. Med. 331: 579–584.

Keer, J.T., and L. Birch. 2003. Molecular methods for the assessment of bacterial viability. J. Microbiol. Meth. 53: 175–183.

Keevil, C.W. 2002. Pathogens in environmental biofilms, pp. 2339–2356, In: *Encyclopedia of Environmental Microbiology*, G. Bitton, editor-in-chief, Wiley-Interscience, N.Y.

Kelly, S.M., and W.W. Sanderson. 1958. The effect of chlorine in water on enteric viruses. Amer. J. Public Health 48: 1323–1327.

Kemmy, F.A., J.C. Fry, and R.A. Breach. 1989. Development and operational implementation of a modified and simplified method for determination of assimilable organic carbon (AOC) in drinking water. Water Sci. Technol. 21: 155–159.

Kemp, H.A., D.B. Archer, and M.R.A. Morgan. 1988. Enzyme-linked immunosorbent assays for the specific and sensitive quantification of *Methanosarcina mazei* and *Methanobacterium bryantii*. Appl. Environ. Micrbiol. 54: 1003–1008.

Keinänen, M.M., L.K. Korhonen, M.J. Lehtola, I.T. Miettinen, P.J. Martikainen, T. Vartiainen, and M.H. Suutari. 2002. The microbial community structure of drinking water biofilms can be affected by phosphorus availability. Appl. Environ. Microbiol. 68: 434–439.

Kennedy, J.E., Jr., G. Bitton, and J.L. Oblinger. 1985. Comparison of selective media for assay of coliphages in sewage effluent and lake water. Appl. Environ. Microbiol. 49: 33–36.

Kennedy, M.S., J. Grammas, and W.B. Arbuckle. 1990. Parachlorophenol degradation using bioaugmentation. J. Water Pollut. Control Fed. 62: 227–233.

Kepner, R.L., and J.R. Pratt. 1994. Use of fluorochromes for direct enumeration of total bacteria in environmental samples: Past and present. Microbiol. Rev. 58: 603–615.

Keswick, B.H., N.R. Blacklow, G.C. Cukor, H.L. DuPont, and J.L. Vollet. 1982. Norwalk virus and rotavirus in travellers' diarrhea in Mexico. Lancet i: 110–111.

Ketratanakul, A., and S. Ohgaki. 1989. Indigenous coliphages and RNA-F-specific coliphages associated with suspended solids in the activated sludge process. Water Sci. Technol. 21: 73–78.

Kfir, R., M. du Preez, and B. Genthe. 1993. The use of monoclonal antibodies for the detection of faecal bacteria in water. Water Sci. Technol. 27: 257–260.

Khan, A. L., D.L. Swerdlow, and D.D. Juranek. 2001. Precautions against biological and chemical terrorism directed at food and water supplies. Pub. Health Rep. 116: 3–14.

Khan, E., R.W. Babcock, S. Jongskul, F.A. Devadason, and S. Tuprakay. 2003. Determination of biodegradable dissolved organic carbon using entrapped mixed microbial cells. Water Res. 37: 4981–4991.

Khan, K.A., M.T. Suidan, and W.H. Cross. 1981. Anaerobic activated carbon filter for the treatment of phenol-bearing wastewater. J. Water Pollut. Control Fed. 53: 1519–1532.

Kiff, R.J., and D.R. Little. 1986. Biosorption of heavy metals by immobilized fungal biomass, pp. 71–80, In: *Immobilisation of Ions by Bio-Sorption*, H. Eccles, and S. Hunt, Eds., Ellis Horwood, Chichester, U.K.

Kilbanov, A.M., T.M. Tu, and K.P. Scott. 1983. Peroxidase catalyzed removal of phenols from coal-conversion wastewaters. Science 221: 259–261.

Kilvington, S., and J. Beeching. 1995a. Identification and epidemiological typing of *Naegleria foleri* with DNA probes. Appl. Environ. Microbiol. 61: 2071–2078.

Kilvington, S., and J. Beeching. 1995b. Development of a PCR for identification of *Naegleria fowleri* from the environment. Appl. Environ. Microbiol. 61: 3764–3767.

Kilvington, S., and D.G. White. 1985. Rapid identification of thermophilic *Naegleria* including *Naegleria fowleri* using API ZYM system. J. Clin. Pathol. 38: 1289–1292.

Kim, B.R., J.E. Anderson, S.A. Mueller, W.A. Gaines, and A.M. Kendall. 2002. Literature review – efficacy of various disinfectants against *Legionella* in water systems. Water Res. 36: 4433–4444.

Kim, S.B., M. Goodfellow, J. Kelly, G.S. Saddler, and A.C. Ward. 2002b. Application of oligonucleotide probes for the detection of *Thiothrix spp.* in activated sludge plants treating paper and board mill wastes. Water Sci. Technol. 46 (1–2): 559–564.

Kinzelman, J., C. Ng, E. Jackson, S. Gradus, and R. Bagley. 2003. Enterococci as indicators of Lake Michigan recreational water quality: Comparison of two methodologies and their impacts on public health regulatory events. Appl. Environ. Microbiol. 69: 92–96.

King, C.H., E.B. Shotts, Jr., R.E. Wooley, and K.G. Poter. 1988. Survival of coliform and bacterial pathogens within protozoa during chlorination. Appl. Environ. Microbiol. 54: 3023–3033.

King, E.F., and B.J. Dutka. 1986. Respirometric techniques, pp. 75–113, In: *Toxicity Testing Using Microorganisms*, Vol 1., G. Bitton, and B.J. Dutka, Eds., CRC Press, Boca Raton, FL.

King, G.M. 1988. Distribution and metabolism of quaternary amines in marine sediments, pp. 143–173, In: *Nitrogen Cycling in Costal Marine Environments*, T.H. Blackburn, and J. Sorensen, Eds., John Wiley & Sons, New York.

Kinner, N., D.L. Blackwill, and P.L. Bishop. 1983. Light and electron microscopic studies of microorganisms growing in rotating biological contactor biofilms. Appl. Environ. Microbiol. 45: 1659–1669.

Kinner, N.E., and C.R. Curds. 1989. Development of protozoan and metazoan communities in rotating biological contactor biofilms. Water Res. 23: 481–490.

Kirkland, K. B., R.A. Meriwether, J.K. Leiss, and W.R. MacKenzie. 1996. Steaming oysters does not prevent Norwalk-like gastroenteritis. Pub. Health Rep. 111: 527–530.

Kirchman, D.L., and H.W. Ducklow. 1993. Estimating conversion factors for the thymidine and leucine methods for measuring bacterial production, pp. 513–517, In: *Handbook of Methods of Aquatic Microbial Ecology*, P.F. Kemp et al. Eds., Lewis Pub., Boca Raton, FL.

Kirsop, B.H. 1984. Methanogenesis. Crit. Rev. Biotechnol. 1: 109–159.

Klamer, M., and E. Bååth. 1998. Microbial community dynamics during composting of straw material studied using phospholipid fatty acid analysis. FEMS Microbiol. Ecol. 27: 9–20.

Klatte, S., F.A. Rainey, and R.M. Kroppenstedt. 1994. Transfer of *Rhodococcus aichiensis* Tsukamura 1982 and *Nocardia amarae* and Lechevalier 1974 to the genus *Gordona* and *Gordona aichiensis* comb. Nov. and *Gordona amarae* comb. Nov. J. Syst. Bacteriol. 44: 769–773.

Klein, P.D., D.Y. Graham, A. Gailor, A.R. Opekun, and E. O'Brian Smith. 1991. Water source as risk factor for *Helicobacter pylori* infection in Peruvian children. Lancet 337: 1503–1506.

Knisely, R.F., 1966. Selective medium for *Bacillus anthracis*. J. Bacteriol. 92, pp. 784–786.

Knudsen, S., P. Saadbye, L.H. Hansen, A. Collier, B.L. Jacobsen, J. Schlundt, and O.H. Karlstrom. 1995. Development and testing of improved suicide functions for biological containment of bacteria Appl. Environ. Microbiol. 61: 985–991.

Knudson, G.B. 1985. Photorectivation of UV-irradiated *Legionella pneumphila* and other *Legionella* species. Appl. Environ. Microbiol. 49: 975–980.

Knudson, L.M., and P.A. Hartman. 1993. Antibiotic resistance among enterococcal isolates from environmental and clinical sources. J. Food Prot. 56: 489–492.

Kobayashi, H.A. 1984. Application of genetic engineering to industrial waste/wastewater treatment, pp. 195–214, In: *Genetic Control of Environmental Pollutants*, G.S. Omenn, and A. Hollaender, Eds., Plenum, New York.

Kobayashi, H.A., M. Stenstrom, and R.A. Mah. 1983. Use of photosymthetic bacteria for hydrogen sulfide removal from anaerobic waste treatment effluent. Water Res. 17: 579–587.

Koch, A.L. 2002. Viable but not culturable (VBNC) microorganisms, pp. 3246–3255, In: *Encyclopedia of Environmental Microbiology*, G. Bitton, editor-in-chief, Wiley-Interscience, N.Y.

Koch, G., M. Kühni, and H. Siegrist. 2001. Calibration and validation of an ASM3-based steady-state model for activated sludge systems. Part I. Prediction of nitrogen removal and sludge production. Water Res. 35: 2235–2245.

Kodikara, C.P., H.H. Crew, and, G.S.A.B. Stewart. 1991. Near on-line detection of enteric bacteria using *lux* recombinant bacteriophage. FEMS Microbiol. Lett. 83: 261–266.

Kofoed, A.D. 1984. Optimum use of sludge in agriculture, pp. 2–21, In: *Utilization of Sewage Sludge on Land: Rates of Application and Long-Term Effects of Metals*, S. Berglund, R.D. Davis, and P. L'Hermite, Eds., Reidel, Dordrecht.

Kogure, K., U. Simidu, and N. Taga. 1984. An improved direct viable count method for aquatic bacteria. Arch. Hydrobiol. 102: 117–122.

Koh, S.-C., J.P. Bowman, and G.S. Sayler. 1993. Soluble methane monooxygenase production and trichloroethylene degradation by a type I methanotroph, *Methylomonas methanica* 68-1. Appl. Environ. Microbiol. 59: 960–967.

Kohring, G.W., X. Zhang, and J. Wiegel. 1989. Anaerobic dechlorination of 2,4-dichlorophenol in freshwater sediments in the presence of sulfate. Appl. Environ. Microbiol. 55: 2735–2737.

Koide, M., A. Saito, N. Kusano, and F. Higa. 1993. Detection of *Legionella* spp. in cooling tower water by the polymerase chain reaction method. Appl. Environ. Microbiol. 59: 1943–1946.

Kong, I.-C., G. Bitton, B. Koopman, and K.-H. Jung. 1995. Heavy metal toxicity testing in environmental samples, Rev. Environ. Contam. Toxicol. 142, 119–147.

van der Kooij, D. 1983. Biological processes in carbon filters, pp. 119–152, In: *Activated Carbon in Drinking Water Technology*, Res. Report, AWWA Res. Foundation, Denver, CO.

van der Kooij, D. 1988. Properties of aeromonads and their occurrence and hygienic significance in drinking water. Zentralblatt fur Bakteriologie Mikrobiologie Hyg [B] 187: 1–17.

van der Kooij, D. 1990. Assimilable organic carbon (AOC) in drinking water, pp. 57–87, In: *Drinking Water Microbiology*, G.A. McFeters, Ed., Springer-Verlag, New York.

van der Kooij, D. 1992. Assimilable organic carbon as an indicator of bacterial regrowth. J. Am. Water Works Assoc. 84: 57–65.

van der Kooij, D. 1995. Significance and assessment of the biological stability of drinking water, pp. 89–102, In: *Water Pollution: Quality and Treatment of Drinking Water*, Springer-Verlag, New York.

van der Kooij, D. 2002. Assimilable organic carbon (AOC) in treated water: Determination and significance, pp. 312–327, In: *Encyclopedia of Environmental Microbiology*, G. Bitton, editor-in-chief, Wiley-Interscience, N.Y.

van der Kooij, D., and W.A.M. Hijnen. 1981. Utilization of low concentrations of starch by a *Flavobacterium* species isolated from tap water. Appl. Environ. Microbiol. 41: 216–221.

van der Kooij, D., and W.A.M. Hijnen. 1984. Substrate utilization by an oxalate-consuming *Spirillum* species in relation to its growth in ozonated water. Appl. Environ. Microbiol. 47: 551–559.

van der Kooij, D., and W.A.M. Hijnen. 1985a. Determination of the concentration of maltose and starch-like compounds in drinking water by growth measurements with a well defined strain of *Flavobacterium* species. Appl. Environ. Microbiol. 49: 765–771.

van der Kooij, D., and W.A.M. Hijnen. 1985b. Measuring the concentration of easily assimilable organic carbon in water treatment as a tool for limiting regrowth of bacteria in distribution systems.In: Proceedings of the American Water Works Association Technol. Conference.

van der Kooij, D., and W.A.M. Hijnen. 1988. Multiplication of a *Klebsiella pneumonae* strain in water at low concentration of substrate. *Proceedings International Conference on Water and Wastewater Microbiology*, Newport Beach, CA., Feb. 8–11, 1988.

van der Kooij, D., J.P. Oranje, and W.A.M. Hijnen. 1982a. Growth of *Pseudomonas aeruginosa* in tap water in relation to utilization of substrates at concentrations of a few microgram per liter. Appl. Environ. Microbiol. 44: 1086–1095.

van der Kooij, D., A. Visser, and W.A.M. Hijnen. 1982b. Determining the concentration of easily assimilable organic carbon in drinking water. J. Am. Water Works Assoc. 74: 540–545.

van der Kooij, D., and H.R. Veenendaal. 1994. Assessment of the biofilm formation potential of synthetic materials in contact with drinking water during distribution, pp. 1395–1407, In: Proc. Amer. Water Works Assoc. Water Qual. Technol. Conf., Miami, FL.

Kool, J.L., J.C. Carpenter, and B.S. Fields. 2000. Monochloramine and Legionnaires's disease. J. AWWA 92 (9): 88–96.

Koopman, B., and G. Bitton. 1986. Toxicant screening in wastewater systems, pp. 101–132, In: *Toxicity Testing Using Microorganisms*, Vol. 2, B.J. Dutka, and G. Bitton, Eds., CRC Press, Boca Raton, FL.

Koopman, B., G. Bitton, R.J. Dutton, and C.L. Logue. 1988. Toxicity testing in wastewater systems: Application of a short-term assay based on induction of the *lac* operon in *E. coli*. Wat Sci. Technol. 20: 137–143.

Koopman, B., G. Bitton, J.J. Delfino, C. Mazidji, G. Voiland, and D. Neita. 1989. Toxicity screening in wastewater systems. Final report (contract no. WM-222) to the Florida Department of Environmental Regulation, Tallahassee, FL.

Koopman, J.S., E.A. Eckert, H.B. Greenberg, B.C. Strohm, R.E. Isaacson, and A.S. Monto. 1982. Norwalk virus enteric illness aquired by swimming exposure. Am. J. Epidemiol. 115: 173.

Koramann, C., Bahnemann, D.W., and M.R. Hoffmann. 1991. Photolysis of chloroform and other molecules in aqueous TiO_2 suspensions. Environ. Sci. Technol. 25: 494–500.

Korich, D.G., J.R. Mead, M.S. Madore, and N.A. Sinclair. 1989. Effets of chlorine and ozone on *Cryptosporidium* oocyst viability (Abstr.), 89th Annual Meeting of the American Society of Microbiology, New Orleans, LA.

Korich, D.G., J.R. Mead, M.S. Madore, N.A. Sinclair, and C.R. Sterling. 1990. Effect of ozone, chlorine dioxide, chlorine, and monochloramine on *Cryptosporidium parvum* oocyst viability. Appl. Environ. Microbiol. 56: 1423–1428.

Kornberg, S.R. 1957. Adenosine triphosphate synthesis from polyphosphate by an enzyme from *E. coli*. Biochim. Biophys. Acta 26: 294–300.

Kornberg, A., N.N. Rao, and D. Ault-Riche. 1999. Inorganic polyphosphate: a molecule of many functions. Annu. Rev. Biochem. 68: 89–125.

Korsholm, E., and H. Søgaard. 1988. An evaluation of direct microscopical counts and endotoxin measurements as alternatives for total plate counts. Water Res 22: 783–788.

Koskinen, R., T. Ali-Vehmas, P. Kämpfer, M. Laurikkala, I. Tsitko, E. Kostyal, F. Atroshi, and M. Salkinoja-Salonen. 2000. Characterization of *Sphingomonas* isolates from Finnish and Swedish drinking water distribution systems. J. Appl. Microbiol. 89: 687–696.

Koster, I.W. 1988. Microbial, chemical and technological aspects of the anaerobic degradation of organic pollutants, pp. 285–316, In: *Biotreatment Systems*, Vol. 1, D.L. Wise, Ed., CRC Press, Boca Raton, FL.

Koster, I.W., and A. Cramer. 1987. Inhibition of methanogenesis from acetate in granular sludge by long-chain fatty acids. Appl. Environ. Microbiol. 53: 403–409.

Koster, L.W., A. Rinzema, A.L. Vegt, and G. Lettinga. 1986. Sulfide inhibition of the methanogenic activity of granular sludge at various pH-levels. Water Res. 20: 1561–1567.

Kott, Y. 1966. Estimation of low numbers of *Escherichia coli* bacteriophage by use of the most probable number method. Appl. Microbiol. 14: 141–144.

Kott, Y., and N. Betzer. 1972. The fate of *Vibrio cholerae* (El Tor) in oxidation pond effluents. Israel J. Med. Sci. 8: 1912.

Kott, Y., N. Roze, S. Sperber, and N. Betzer. 1974. Bacteriophages as viral pollution indicators. Water Res. 8: 165–171.

Kott, Y., L. Vinokur, and H. Ben-Ari.1980. Combined effects of disinfectants on bacteria and viruses, pp. 677–686, In: *Water Chlorination: Environmental Impact and Health Effects*. Vol. 3, Ann Arbor Science Publishers, Ann Arbor, MI.

Kowal, N.E. 1982. Health Effects of Land Treatment: Microbiological Report. No. EPA-600/1-82-007, U.S. Environmental Protection Agency, Cincinnati, OH.

Kramer, M.H., B.L. Herwaldt, G.F. Craun, R.L. Calderon, and D.D. Juranek. 1996. Waterborne disease: 1993 and 1994. J. Amer. Water Works Assoc. 88: 66–80.

Kreft, P., M. Umphres, J.-M. Hand, C. Tate, M.J. McGuire, and R.R. Trussel. 1985. Converting from chlorine to chloramines: A case study. J. Am. Water Works Assoc. 77: 38–45.

Krishnaswamy, R., and D.B. Wilson. 2000. Construction and characterization of an *Escherichia coli* strain genetically engineered for Ni(II) bioaccumulation. Appl. Environ. Microbiol. 66: 5383–5386.

Krumme, M.L., and S.A. Boyd. 1988. Reductive dechlorination of chlorinated phenols in anaerobic upflow bioreactor. Water Res. 22: 171–177.

Kuchenrither, R.D., and L.D. Benefield. 1983. Mortality patterns of indicator organisms during aerobic digestion. J. Water Pollut. Control Fed. 55: 76–80.

Kuchta, J.M., S.J. States, A.M. McNamara, R.M. Wadowsky, and R.B. Yee. 1983. Susceptibility of *Legionella pneumophila* to chlorine in.

Kuehn, W., and U. Mueller. 2000. Riverbank filtration: An overview. J. AWWA 92 (12): 60–69.

Kuhn, R.C., and K.H. Oshima. 2001. Evaluation and optimization of a reusable hollow fiber ultrafilter as a first step in concentrating *Cryptosporidium parvum* oocysts from water. Water Res. 35: 2779–2783.

Kuhn, S.P., and R.M. Pfister. 1990. Accumulation of cadmium by immobilized *Zooglea ramigera* 115. J. Ind. Microbiol. 6: 123–128.

Kühn, I., G. Allestam, G. Huys, P. Janssen, K. Kersters, K. Krovacek, and T.-A., Stenström. 1997. Diversity, persistence, and virulence of *Aeromonas* strains isolated from drinking water distribution systems in Sweden. Appl. Environ. Microbiol. 63: 2708–2715.

Kukkula, M., L. Maunula, E. Silvennoinen, and C.-H. von Bonsdorff. 1999. Outbreak of viral gastroenteritis due to drinking water contaminated by Norwalk-like viruses. J. Infect. Dis. 180: 1771–1776.

Kulikovsky, A., H.S. Pankratz, and S.L. Sadoff. 1975. Ultrastructural and chemical changes in spores of *Bacillus cereus* after action of disinfectants. J. Appl. Bacteriol. 38: 39–46.

Kumaran, P., and N. Shivaraman. 1988. Biological treatment of toxic industrial wastes, pp. 227–283, In: *Biotreament Systems*, Vol. 1, D.L. Wise Ed., CRC Press, Boca Raton, FL.

Kusters, J.G., and E.J. Kuipers. 2001. Antibiotic resistance of *Helicobacter pylori*. J. Appl. Microbiol. 90: 134S–144S.

Kwa, B.H., M. Moyad, M.A. Pentella, and J.B. Rose. 1993. A nude mouse model as an in vivo infectivity assay for cryptosporidiosis. Water Sci. Technol. 27: 65–68.

Kwan, K.K., and B.J. Dutka. 1990. Simple two-step sediment extraction procedure for use in genotoxicity and toxicity bioassays. Toxicity Assess. 5: 395–404.

Labatiuk, C.W., F.W. Schaefer III, G.R. Finch, and M. Belosevic. 1991. Comparison of animal infectivity, excystation, and fluorogenic dye as measures of *Giardia muris* cyst inactivation by ozone. Appl. Environ. Microbiol. 57: 3187–3192.

Lafleur, J., and J.E. Vena. 1991. Retrospective cohort mortality study of cancer among sewage plant workers. Amer. J. Ind. Med. 19: 75–86.

Laitinen, S., A. Nevalainen, M. Kotimaa, J. Liesivuori, and P.J. Marikainen. 1992. Relationship between bacterial counts and endotoxin concentration in the air of wastewater treatment plants. Appl. Environ. Microbiol. 58: 3474–3476.

Lajoie, C.A., G.S. Sayler, and C.J. Kelly. 2002. The Activated Sludge Biomolecular Database. Water Environ. Res. 74 (5): 1.

Lal, R., and D.M. Saxena. 1982. Accumulation, metabolism, and effects of organochlorine insecticides on microorganisms. Microbiol. Rev. 46: 95–127.

Lalezary, S., M. Pirbazari, and M.J. McGuire. 1986. Oxidation of five earthy-musty taste and odor compounds. J. Am. Water Works Assoc. 78: 62–69.

Lalezary-Craig, S., M. Pirbazari, M.S. Dale, T.S. Tanaka, and M.J. McGuire. 1988. Optimizing the removal of geosmin and 2-methylisoborneol by powdered activated carbon. J. Am. Water Works Assoc. 80: 73–80.

Lamar, R.T., and D.M. Dietrich. 1990. *In situ* depletion of pentachlorophenol from contaminated soil by *Phanerochaete* spp. Appl. Environ. Microbiol. 56: 3093–3100.

Lamar, R.T., M.J. Larsen, and T.K. Kirk. 1990. Sensitivity to and degradation of pentachlorophenol by *Phanerochaete* spp. Appl. Environ. Microbiol. 56: 3519–3526.

Lamb, A., and E.L. Tollefson. 1973. Toxic effects of cupric, chromate and chromic ions on biological oxidation. Water Res. 7: 599–613.

Lamka, K.G., M.W. Lechevallier, and R.J. Seidler. 1980. Bacterial contamination of drinking water supplies in a modern rural neighborhood. Appl. Environ. Microbiol. 39: 734–738.

La Motta, E.J. 1976. Internal diffusion and reaction in biological films. Environ. Sci. Technol. 19: 765–769.

Lampel, K.A., J.A. Jagow, M. Trucksess, and W.E. Hill. 1990. Polymerase chain reaction for detection of invasive *Shigella flexneri* in food. Appl. Environ. Microbiol. 56: 1536–1540.

Lance, J.C. 1972. Nitrogen removal by soil mechanisms. Water Pollut. Control Fed. 44: 1352–1361.

Lance, J.C., and C.P. Gerba. 1980. Poliovirus movement during high rate land filtration of sewage water. J. Environ. Qual. 9: 31–34.

Landeen, L.K., M.T. Yahya, and C.P. Gerba. 1989. Efficacy of copper and silver ions and reduced levels of free chlorine in inactivation of *Legionella pneumophila*. Appl. Environ. Microbiol. 55: 3045–3050.

Lane, S., and D. Lloyd. 2002. Current trends in research into the waterborne parasite *Giardia*. Crit. Rev. Microbiol. 28: 123–147.

Lange, J.L., P.S. Thorne, and N. Lynch. 1997. Application of flow cytometry and fluorescent *in situ* hybridization for assessment of exposures to airborne bacteria. Appl. Environ. Microbiol. 63: 1557–1563.

Lange, K.P., W.D. Bellamy, D.W. Hendricks, and G.S. Logsdon. 1986. Diatomaceous Earth Filtration of Giardia Cysts and Other Substances. J. Amer. Water Works Assoc. 78 (12): 76–84.

Langeland, G. 1982. *Salmonella* spp. in the working environment of sewage treatment plants in Oslo, Norway. Appl. Environ. Microbiol. 43: 1111–1115.

Langendijk, P.S., F. Schut, G.J. Jansen, G.C. Raangs, G.R. Kamphuis, M.H.F. Wilkinson, and G.W. Welling. 1995. Quantitative fluorescence in situ hybridization of *Bifidobacterium* spp. with genus-specific 16S rRNA-targeted probes and its application in fecal samples. Appl. Environ. Microbiol. 61: 3069–3075.

van Langenhove, H., K. Roelstraete, N. Schamp, and J. Houtmeyers. 1985. GC-MS identification of odorous volatiles in wastewater. Water Res. 19: 597–603.

Lapinski, J. and A. Tunnacliffe. 2003. Reduction of suspended biomass in municipal wastewater using bdelloid rotifers. Water Res. 37: 2027–2034.

Laquidara, M.J., F.C. Blanc, and J.C. O'Shaughnessy. 1986. Development of biofilm, operating characteristics and operational control in the anaerobic rotating biological contactor process. J. Water Pollut. Control Fed. 58: 107–114.

LaRiviere, J.W.M. 1977. Microbial ecology of liquid waste treatment. Adv. Microb. Ecol. 1: 215–259.

Larkin, J.M. 1980. Isolation of *Thiothrix* in pure culture and observation of a filamentous epiphyte on *Thiothrix*. Curr. Microbiol. 41: 155–158.

Larsen, L.H., N.P. Revsbech, and S.J. Binnerup. 1996. A microsensor for nitrate based on immobilized denitrifying bacteria. Appl. Environ. Microbiol. 62: 1248–1251.

Larsen, P.I., L.K. Sydnes, B. Landfald, and A.R. Strom. 1987. Osmoregulation in *Escherichia coli* by accumulation of organic osmolytes: betaines, glutamic acid and trehalose. Arch. Microbiol. 147: 1–7.

Larson, R.A., Ed. 1989. *Biohazards of Drinking Water Treatment*. Lewis, Chelsea, MI, 293 pp.

Larson, R.J., and S.L. Schaeffer. 1982. A rapid method for determining the toxicity of chemicals to activated sludge. Water Res. 16: 675–680.

Lau, A.O., P.F. Strom, and D. Jenkins. 1984a. Growth kinetics of *Sphaerotilus natans* and a floc former in pure and dual continuous culture. J. Water Pollut. Control Fed. 56: 41–51.

Lau, A.O., P.F. Strom, and D. Jenkins. 1984b. The competitive growth of floc-forming and filamentous bacteria: A model for activated sludge bulking. J. Water Pollut. Control Fed. 56: 52–61.

Lauer, W.C. 1991. Water quality for ptable reuse. Water Sci. Technol. 23: 2171–2180.

Lauer, W.C., S.E. Rogers, and J.M. Ray. 1985. The current status of Denver's potable water reuse project. J. Am. Water Works Assoc. 77: 52–59.

Lawrence, A.W., and P.L. McCarty. 1969. Kinetics of methane fermentation in anaerobic treatment. J. Water Pollut. Control Fed. 41: R1–R17.

Lawrence, A.W., P.L. McCarty, and F.J.A. Guerin. 1966. The effects of sulfides on anaerobic treatment. Air Water Int. J. 110: 2207–2210.

Layton, A.C., P.N. Karanth, C.A. Lajoie, A.J. Meyers, I.R. Gregory, R.D. Stapleton, D.E. Taylor, and G.S. Sayler. 2000. Quantification of *Hyphomicrobium* populations in activated sludge from an industrial wastewater treatment system as determined by 16S rRNA analysis. Appl. Environ. Microbiol. 66: 1167–1174.

Lazarova, V., and J. Manem. 1995. Biofilm characterization and activity analysis in water and wastewater treatment. Water. Res. 29: 2227–2245.

Lebaron, P., P. Catala, Celine Fajon, F. Joux, J. Baudart, and L. Bernard. 1997. A new sensitive whole-cell hybridization technique for the detection of bacteria involving a biotinylated oligonucleotide probe targeting rRNA and tyramide signal amplification. Appl. Environ. Microbiol. 63: 3274–3278.

LeChevalier, H.A., and D. Pramer. 1971. *The Microbes*, J.B. Lippincott, Philadelphia.

LeChevalier, H.A., and M.P. Lechevalier. 1975. *Actinomycetes of Sewage Treatment Plants*. EPA Report No. 600/2-75-031, U.S. Environmental Protection Agency, Cincinnati, OH.

LeChevalier, H.A., M.P. Lechevalier, and P.E. Wyszkowski. 1977. *Actinomycetes of Sewage Treatment Plants*. EPA Report No. 600/2-77-145, U.S. Environmental Protection Agency, Cincinnati, OH.

LeChevalier, M.P., and H.A. Lechevalier. 1974. *Nocardia amarae* sp. nov., an actinomycete common in foaming activated sludge. Int. J. Syst. Bacteriol. 24: 278–288.

LeChevallier, M.W. 2002. Microbial removal by pretreatment, coagulation and ion exchange, pp. 2012–2019, In: *Encyclopedia of Environmental Microbiology*, G. Bitton, editor-in-chief, Wiley-Interscience, N.Y.

LeChevallier, M.W., W.C. Becker, P. Schorr, and R.G. Lee. 1992. Evaluating the performance of biologically active rapid filters. J. Am. Water Works Assoc. 84: 136–146.

LeChevallier, M.W., S.C. Cameron, and G.A. McFeters. 1983. New medium for improved recovery of coliform bacteria from drinking water. Appl. Environ. Microbiol. 45: 484–492.

LeChevallier, M.W., C.D. Cawthon, and R.G. Lee. 1988. Inactivation of biofilm bacteria. Appl. Environ. Microbiol. 54: 2492–2499.

LeChevallier, M.W., G.D. Di Giovanni, J.L. Clancy, Z. Bukhari, S. Bukhari, J.S. Rosen, J. Sobrinho, and M.M. Frey. 2003. Comparison of Method 1623 and cell culture-PCR for detection of *Cryptosporidium* spp. in source waters. Appl. Environ. Microbiol. 69: 971–979.

LeChevallier, M.W., T.M. Evans, and R.J. Seidler. 1981. Effect of turbidity on chlorination efficiency and bacterial persistence in drinking water. Appl. Environ. Microbiol. 42: 159–167.

LeChevallier, M.W., T.S. Hassenauer, A.K. Camper, and G.A. McFeters. 1984a. Disinfection of bacteria attached to granular activated carbon. Appl. Environ. Microbiol. 48: 918–923.

LeChevallier, M.W., P.E. Jakanoski, A.K. Camper, and G.A. McFeters. 1984b. Evaluation of m-T7 as a fecal coliform medium. Appl. Environ. Microbiol. 48: 371–375.

LeChevallier, M.W., C.H. Lowry, and R.G. Lee. 1990. Disinfecting biofilm in a model distribution system. J. Am. Water Works Assoc. 82: 85–99.

LeChevallier, M.W., and G.A. McFeters. 1985a. Interactions between heterotrophic plate count bacteria and coliform organisms. Appl. Environ. Microbiol. 49: 1138–1141.

LeChevallier, M.W., and G.A. McFeters. 1985b. Enumerating injured coliforms in drinking water. J. Am. Water Works Assoc. 77: 81–87.

LeChevallier, M.W., and G.A. McFeters. 1990. Microbiology of activated carbon, pp. 104–119, In: *Drinking Water Microbiology*, G.A. McFeters, Ed., Springer Verlag, New York.

LeChevallier, M.W., W.D. Norton, and R.G. Lee. 1991a. *Giardia* and *Cryptosporidium* spp. in filtered drinking water supplies. Appl. Environ. Microbiol. 57: 2617–2621.

LeChevallier, M.W., W.D. Norton, and R.G. Lee. 1991b. Occurrence of *Giardia* and *Cryotosporidium* spp. in surface water supplies. Appl. Environ. Microbiol. 57: 2610–2616.

LeChevallier, M.W., W. Schulz, and R.G. Lee. 1991c. Bacterial nutrients in drinking water. Appl. Environ. Microbiol. 57: 857–862.

LeChevallier, M.W., R.J. Seidler, and T.M. Evans. 1980. Enumeration and characterization of standard plate count bacteria in chlorinated and raw water supplies. Appl. Environ. Microbiol. 40: 922–930.

LeChevallier, M.W., N.E. Shaw, L.A. Kaplan, and T.L. Bott. 1993. Development of a rapid assimilable organic carbon method in water. Appl. Environ. Microbiol. 59: 1526–1531.

LeChevallier, M.W., A. Singh, D.A. Schiemann, and G.A. McFeters. 1985. Changes in virulence of water enteropathogens with chlorine injury. Appl. Environ. Microbiol. 50: 412–419.

Leclerc, H., S. Edberg, V. Pierzo, and J.M. Delattre. 2000. Bacteriophages as indicators of enteric viruses and public health risk in groundwaters. J. Appl. Microbiol. 88: 5–21.

Leclerc, H., and A. Moreau. 2002. Microbiological safety of natural mineral water. FEMS Mirobiol. Rev. 26: 207–222.

Leclerc, H., L. Schwartzbrod, and E. Dei-Cas. 2002. Microbial agents associated with waterborne disease. Crit. Rev. Microbiol. 28: 371–409.

Le Dantec, C., J.-P. Duguet, A. Montiel, N. Dumoutier, S. Dubrou, and V. Vincent. 2002. Occurrence of mycobacteria in water treatment lines and in water distribution systems. Appl. Environ. Microbiol. 2002. 68: 5318–5325.

Lee, H.-B., and T.E. Paert. 1995. Determination of 4-nonylphenol in effluent and sludge from sewage treatment plants. Anal. Chem. 67: 1976–1980.

Lee, H.-B., T.E. Paert, D.T. Bennie, and R.J. Maguire. 1997. Determination of nonylphenol polyethoxylates and their carboxylic acid metabolites in sewage treatment plant sludge by supercritical carbon dioxide extraction. J. Chromatogr. A 785: 385–394.

Lee, K.M., C.A. Brunner, J.B. Farrel, and A.E. Eralp. 1989. Destruction of enteric bacteria and viruses during two-phase digestion. J. Water Pollut. Control Fed. 61: 1421–1429.

Lee, L.A., E.K. Maloney, N.H. Bean, and R.V. Tauxe. 1994. Increase in antimicrobial-resistant *Salmonella* infections in the United States. 1989–1990. J. Infect. Dis. 170: 18–34.

Lee, S. 2000. Landfill Gas Composition At Florida Construction And Demolition debris Facilities. Master of Engineering Thesis, University of Florida, Gainesville, Florida.

Lee, S.-H., and S.-J. Kim. 2002. Detection of infectious enteroviruses and adenoviruses in tap water in urban areas in Korea. Water Res. 36: 248–256.

Le Guyader, F., E. Dubois, D. Menard, and M. Pommepuy. 1994. Detection of hepatitis A virus, rotavirus, and enterovirus in naturally contaminated shellfish and sediment by reverse transcription-seminested PCR. Appl. Environ. Microbiol. 60: 3665–3671.

Le Guyader, F., F.H. Neill, M.K. Estes, S.S. Monroe, T. Ando, and R.L. Atmar. 1996. Detection and analysis of a small round-structured virus strain in oysters implicated in an outbreak of acute gastroenteritis. Appl. Environ. Microniol. 62: 4268–4272.

Le Guyader, F., L. Haugarreau, L. Miossec, E. Dubois, and M. Pommepuy. 2000. Three-year study to assess human enteric viruses in shellfish. Appl. Environ. Microbiol. 66: 3241–3248.

Lehtola, M.J., I.T. Miettinen, T. Vartiainen, and P.J. Martikainen. 2002. Changes in content of microbially available phosphorus, assimilable organic carbon and microbial growth potential during drinking water treatment processes. Water Res. 36: 3681–3690.

Leis, A., and H.-C. Flemming. 2002. Activity and carbon transformations in biofilms, pp. 81–92, In: *Encyclopedia of Environmental Microbiology*, G. Bitton, editor-in-chief, Wiley-Interscience, N.Y.

Leng, X., D.A. Mosier, and R.A. Oberst. 1996. Simplified method for recovery and PCR detection of *Cryptosporidium* DNA from bovine feces. Appl. Environ. Microbiol. 62: 643–647.

Leisinger, T. 1983. Microorganisms and xenobiotic compounds. Experientia 39: 1183–1191.

Leisinger, T., R. Hutter, A.M. Cook, and J. Nuesch, Eds. 1981. *Microbial Degradation of Xenobiotics and Recalcitrant Compounds.* Academic Press, Orlando, FL.

Lembke, L.L., R.N. Kniseley, R.C. van Nostrand, and M.D. Hale. 1981. Precision of the all-glass impinger and the Andersen microbial impactor for air sampling in solid waste handling facilities. Appl. Environ. Microbiol. 42: 222–225.

Lemmer, H. 1986. The ecology of scum causing actinomycetes in sewage treatment plants. Water Res. 20: 531–535.

Lemmer, H., and M. Baumann. 1988a. Scum actinomycetes in sewage treatment plants. Part 2: The effect of hydrophobic substrate. Water Res. 22: 761–763.

Lemmer, H., and M. Baumann. 1988b. Scum actinomycetes in sewage treatment plants. Part 3: Synergism with other sludge bacteria. Water Res. 22: 765–767.

Lemmer, H., and R.M. Kroppenstedt. 1984. Chemotaxonomy and physiology of some actinomycetes isolated from scumming activated sludge system. Appl. Microbiol. 5: 124–135.

Lens, P.N., M.-P. De Poorter, C.C. Cronenberg, and W.H. Verstraete. 1995. Sulfate reducing and methane producing bacteria in aerobic wastewater treatment systems. Water Res. 29: 871–880.

Leoni, E., P.P. Legnani, M.A. Bucci Sabattini, and F. Righi. 2001. Prevalence of *legionella* spp. in swimming pool environment. Water Res. 35: 3749–3753.

Lessard, E.J., and J.McN. Sieburth. 1983. Survival of natural sewage populations of enteric bacteria in diffusion and batch chambers in the marine environment. Appl. Environ. Microbiol. 45: 950–959.

Lestan, D., and R.T. Lamar. 1996. Development of fungal inocula for bioaugmentation of contaminated soils. Appl. Environ. Microbiol. 62: 2045–2052.

Letterman, R.D., A. Amirtharajah, and C.R. O'Melia. 1999. Coagulation and flocculation, pp. 6.1–6.66, In: *Water Quality and Treatment*, R.D. Letterman, Ed., McGraw-Hill, New York.

Lettinga, G. 1995. Anaerobic digestion and wastewater treatment systems. Antonie van Leeuwenhoek 67: 3–28.

Lettinga, G., J. Field, J. van Lier, G. Zeeman, and L.W. Hulshoff Pol. 1997. Advanced anaerobic wastewater treatment in the near future. Water Sci. Technol. 35 (10): 5–12.

Leung, K.T., R. Mackereth, Y.-C. Tien, and E. Topp. 2004. Comparison of AFLP and ERIC-PCR analyses for discriminating *Escherichia coli* from cattle, pig and human sources. FEMS Microbiol. Ecol. 47: 111–119.

Levi, Y., C. Henriet, J.P. Coutant, M. Lucas, and G. Leger. 1989. Monitoring acute toxicity in rivers with the help of the Microtox test. Water Supply 7: 25–31.

Levin, T.R., J.A. Schmittdiel, J.M. Henning, K. Kunz, C.J. Henke, C.J. Colby, and J.V. Selby. 1998. A cost analysis of a *Helicobacter pylori* eradication strategy in a large health maintenance organization. Am. J. Gastroenterol. 93: 743–747.

Levine, A.D., and T. Asano. 2004. Recovering sustainable water from wastewater. Environ. Sci. Technol. 38: 201A–208A.

Levine, M.M. 1987. *Escherichia coli* that cause diarrhea: Enterotoxigenic, enteropathogenic, enteroinvasive,enterohemorrhagic, and enteroadherent. J. Infect. Dis. 155: 377–389.

Levy, R.V. 1990. Invertebrates and associated bacteria in drinking water distribution lines, pp. 224–248, In: *Drinking Water Microbiology*, G.A. McFeters, Ed., Springer Verlag, New York.

Levy, R.V., R.D. Cheetham, J. Davis, G. Winer, and F.L. Hart. 1984. Novel method for studying the public health significance of macroinvertebrates in potable water. Appl. Environ. Microbiol. 47: 889–894.

Levy, R.V., F.L. Hart, and R.D. Cheetham. 1986. Occurrence and public health significance of invertebrates in drinking water systems. J. Am. Water Works Assoc. 78: 105–110.

Lewis, D.L., and D.K. Gattie. 2002. Pathogen risks from applying sewage sludge to land. Environ. Sci. Technol. 36: 287A–293A.

Lewis, C.M., and J.L. Mak. 1989. Comparison of membrane filtration and Autoanalysis Colilert presence-absence techniques for analysis of total coliforms and *Escherichia coli* in drinking water samples. Appl. Environ. Microbiol. 55: 3091–3094.

Lewis, G.D., F.J. Austin, M.W Loutit, and K. Sharples. 1986. Enterovirus removal from sewage: The effectiveness of four different treatment plants. Water Res. 20: 1291–1297.

Leyer, G.J., and E.A. Johnson. 1997. Acid adaptation sensitizes *Salmonella typhimurium* to hypochlorous acid. Appl. Environ. Microbiol. 63: 461–467.

Leyval, C., C. Arz, J.C. Block, and M. Rizet. 1984. *Escherichia coli* resistance to chlorine after successive chlorinations. Environ. Technol. Lett. 5: 359–364.

Li, B., and P.L. Bishop. 2004. Micro-profiles of activated sludge floc determined using microelectrodes. Water Res. 38: 1248–1258.

Li, G., W. Huang, D. N. Lerner, and X. Zhang. 2000. Enrichment of degrading microbes and bioremediation of petrochemical contaminants in polluted soil. Water Res. 34: 3845–3853.

Li, H., G.R. Finch, D.W. Smith, and M. Belosevic. 2001. Sequential inactivation of *cryptosporidium parvum using ozone* and chlorine. Water Res. 35: 4339–4348.

Liang, S., J.H. Min, M.K. Davis, J.F. Green, and D.S. Remer. 2003. Use of pulsed-UV processes to destroy NDMA. J. Amer. Water Works Assoc. 95 (9): 121–131.

Liao, B.Q., D.G. Allen, I.G. Droppo, G.G. Leppard and S.N. Liss. 2001. Surface properties of sludge and their role in bioflocculation and settleability. Water Res. 34: 339–350.

van Lieverloo, J.H.M., D. van der Kooij, and W. Hoogenboezem. 2002. Invertebrates and protozoa (free living) in drinking water distribution systems, pp. 1718–1733, In: *Encyclopedia of Environmental Microbiology*, G. Bitton, editor-in-chief, Wiley-Interscience, N.Y.

van Lieverloo, J.H.M., D.W. Bosboom, G. L. Bakker, A. J. Brouwer, R. Voogt, and J. E. M. De Roos. 2004. Sampling and quantifying invertebrates from drinking water distribution mains. Water Res. 38: 1101–1112.

Lighthart, B., and A.S. Frisch. 1976. Estimation of viable airborne microbes downwind from a point source. Appl. Environ. Microbiol. 31: 700–704.

Lighthart, B., and A.J. Mohr. 1987. Estimating downwind concentrations of viable airborne micro-organisms in dynamic atmospheric conditions. Appl. Environ. Microbiol. 53: 1580–1583.

Lim, B.R., K.-H. Ahn, P. Songprasert, J.W. Cho, and S.H. Lee. 2004. Microbial community structure of membrane fouling film in an intermittently and continuously aerated submerged membrane bioreactor treating domestic wastewater. Water Sci. Technol. 4 (2): 255–261.

Limsawat, S., and S. Ohgaki. 1997. Fate of liberated viral RNA in wastewater determined by PCR. Appl. Environ. Microbiol. 63: 2932–2933.

Lin, C.-Y. 1992. Effect of heavy metals on volatile fatty acid degradation in anaerobic digestion. Water Res. 26: 177–183.

Lin, S.D. 1985. *Giardia lamblia* and water supply. J. Am. Water Works Assoc. 77: 40–47.

Lin, Y.-F., and K.-C. Chen. 1993. Denitrification by immobilized sludge with polyvinyl alcohol gels. Water Sci. Technol. 28 (7): 159–164.

Linkfield, T.G., J.M. Suflita, and J.M. Tiedje. 1989. Characterization of the acclimation period prior to the anaerobic biodegradation of haloaromatic compounds. Appl. Environ. Microbiol. 55: 2773–2778.

Lippy, E.C. 1986. Chlorination to prevent and control waterborne diseases. J. Am. Water Works Assoc. 78: 49–52.

Lippy, E.C., and S.C. Waltrip. 1984. Waterborne disease outbreaks—1946–1980: a thirty-five-year perspective. J. Am. Water Works Assoc. 76: 60–67.

Lipson, S.M., and G. Stotzky. 1987. Interactions between viruses and clay minerals, pp. 197–230, In: *Human Viruses in Sediments, Sludges, and Soils*, V.C. Rao, and J.L. Melnick, Eds., CRC Press, Boca Raton, FL.

Liss, S.N., I.G. Droppo, D.T. Flannigan, and G.G. Leppard. 1996. Floc architecture in wastewater and natural riverine systems. Environ. Sci. Technol. 30: 680–686.

Little, C.D., A.V. Palumbo, S.E. Herbes, M.E. Lidstrom, R.L. Thyndall, and P.J. Gilmer. 1988. Trichloroethylene biodegradation by a methane-oxidizing bacterium. Appl. Environ. Microbiol. 54: 951–956.

Little, M.D. 1986. The detection and significance of pathogens in sludge: Parasites. In: *Control of Sludge Pathogens*, C.A. Sorber, Ed., Water Pollution Control Federation, Washington, D.C.

Liu, D. 1982. Effect of sewage sludge land disposal on the microbiological quality of groundwater. Water Res. 16: 957–961.

Liu D, and B.J. Dutka, Eds. 1984. *Toxicity Screening Procedures using Bacterial Systems.* Marcel Dekker, New York.

Liu, J., and B. Mattiasson. 2002. Microbial BOD sensors for wastewater analysis. Water Res. 36: 3786–3802.

Liu, W.T., O.-C. Chan, and H.H.P. Fang. 2002a. Characterization of microbial community in granular sludge treating brewery wastewater. Water Res. 36: 1767–1775.

Liu, W., H. Wu, Z. Wang, S.L. Ong, J.Y. Hu, and W.J. Ng. 2002b. Investigation of assimilable organic carbon (AOC) and bacterial regrowth in drinking water distribution system. Water Res. 36: 891–898.

Liu, X., and R.M. Slawson. 2001. Factors affecting drinking water biofiltration. J. AWWA 93 (12): 90–101.

Liu, Y. 2000. Effect of chemical uncoupler on the observed growth yield in batch culture of activated sludge. Water Res. 34: 2025–2030.

Liu, Y., and H.P. Fang. 2003. Influences of extracellular polymeric substances (EPS) on flocculation, settling, and dewatering of activated sludge. Crit. Rev. Environ. Sci. & Technol. 33: 237–264.

Livernoche, D., L. Jurasek, M. Desrochers, J. Dorica, and I.A. Veliky. 1983. Removal of colour from kraft mill waste waters with cultures of white-rot fungi and with immobilized mycelia of *Coriolus versicolor.* Biotech. Bioeng. 25: 2055–2065.

Loaec, M., R. Olier, and J. Guezennec. 1997. Uptake of lead, cadmium and zinc by a novel bacterial exopolysaccharide. Water Res. 31: 1171–1179.

Logan, K.B., G.E. Rees, N.D. Seeley, and S.B. Primrose. 1980. Rapid concentration of bacteriophages from large volumes of freshwater: Evaluation of positively-charged microporous filters. J. Virol. Methods 1: 87–97.

Logsdon, G.S., and J.C. Hoff. 1986. Barriers to the transmission of waterborne disease, pp. 255–274, In: *Waterborne Diseases in the United States*, G.F. Craun, Ed., CRC Press, Boca Raton, FL.

Logsdon, G.S., and E.C. Lippy. 1982. The role of filtration in preventing waterborne disease. J. Amer. Water Works Assoc. 74: 649–655.

Logsdon, G.S., J.M. Symons, R.L. Hoye, Jr., and M.M. Arozarena. 1981. Alternative filtration methods for removal of *Giardia* cysts and cysts models. J. Amer. Water Works Assoc. 73: 111–118.

Logue, C., B. Koopman, and G. Bitton. 1983. INT-reduction assays and control of sludge bulking. J. Environ. Eng. Sci., 109: 915–923.

Logue, C.L., B. Koopman, G.K. Brown, and G. Bitton. 1989. Toxicity screening in a large, municipal wastewater system. J. Water Pollut. Control Fed. 61: 632–640.

Longley, K.E., B.E. Moore, and C.A. Sorber. 1980. Comparison of chlorine and chlorine dioxide as disinfectants. J. Water Pollut. Control Fed. 52: 2098–2105.

Lopez, A.S., D.R. Dodson, M.J. Arrowood, P.A. Orlandi, Jr., A.J. da Silva, J.W. Bier, S.D. Hanauer, R.L. Kuster, S. Oltman, M.S. Baldwin, K.Y. Won, E.M. Nace, M.L. Eberhard, and B.L. Herwaldt. 2001. Outbreak of cyclosporiasis associated with basil in Missouri in 1999. Clin. Infet. Dis. 32: 1010–1017.

Lopman, B.A., D.W. Brown, and M. Koopmans. 2003. Caliciviruses: Occurrence in Europe, In: *Encyclopedia of Environmental Microbiology*, G. Bitton, editor-in-chief, Internet Edition, Wiley-Interscience, N.Y.

Lovley, D.R., J.D. Coates, J.C. Woodward, and E.J.F. Phillips. 1995. Benzene oxidation coupled to sulfate reduction. Appl. Environ. Microbiol. 61: 953–958.

Low, E.W., H.A. Chase, M.G. Milner, and T.P. Curtis. 2000. Uncoupling of metabolism to reduce biomass production in the activated sludge process. Water Res. 34: 3204–3212.

Lu, F., J. Lukasik, and S.R. Farrah. 2001. Immunological methods for the study of zoogloea strains in natural environments. Water Res. 35: 4011–4018.

Lu, Y., T.E. Redlinger, R. Avitia, A. Galindo, and K. Goodman. 2002. Isolation and genotyping of *Helicobacter pylori* from untreated municipal wastewater. Appl. Environ. Microbiol. 68: 1436–1439.

Lucena, F., C. Finance, J. Jofre, J. Sancho, and L. Schwartzbrod. 1982. Viral pollution determination of superficial waters (river water and sea water) from the urban area of Barcellona (Spain). Water Res. 16: 173–177.

Lucena, F., J. Lasobras, D. McIntosh, M. Forcadell, and P. Jofre. 1994. Effect of distance from the polluting focus on relative concentrations of Bacteroides fragilis phages and coliphages in mussels. Appl. Environ. Microbiol. 60: 2272–2277.

Luckiesh, M., and L.L Holladay. 1944. Disinfecting water by means of germicidal lamps. Gen. Electric Rev. 47: 45–54.

Luef, E., T. Prey, and C.P. Kubicek. 1994. Biosorption of zinc by fungal mycelial wastes. Appl. Microbiol. Biotechnol. 34: 688–692.

Lui J., L. Bjornsson, and B. Mattiasson (2000). Immobilised activated sludge based biosensor for biochemical oxygen demand measurement. Biosensors Bioelectron. 14: 883–893.

Lundgren, D.G., and W. Dean. 1979. Biogeochemistry of iron, pp. 211–223, In: *Biogeochemical Cycling of Mineral-Forming Elements*, P.A. Trudinger, and D.J. Swaine, Eds., Elsevier, Amsterdam.

Lydholm, B., and A.L. Nielsen. 1981. The use of soluble polyelectrolytes for the isolation of virus from sludge, pp. 85–90, In: *Viruses and Wastewater Treatment*, M. Goddard, and M. Butler, Eds., Pergamon Press, Oxford.

Lynch, P.A., B.J. Gilpin, L.W. Sinton, and M.G. Savill. 2002. The detection of *Bifidobacterium adolescentis* by colony hybridization as an indicator of human faecal pollution. J. Appl. Microbiol. 92: 526–533.

Ma, J.-F., T.M. Straub, I.L. Pepper, and C.P. Gerba. 1994. Cell culture and PCR determination of poliovirus inactivation by disinfectants. Appl. Environ. Microbiol. 60: 4203–4206.

Macario, A.J.L., and E.C. de Macario. 1988. Quantitative immunologic analysis of the methanogenic flora of digestors reveals a considerable diversity. Appl. Environ. Microbiol. 54: 79–86.

MacGregor, B.J. 2002. Phylogenetically based methods in microbial ecology, pp. 2489–2501, In: *Encyclopedia of Environmental Microbiology*, G. Bitton, editor-in-chief, Wiley-Interscience, N.Y.

Mach, P.A., and D.J. Grimes. 1982. R-plasmid transfer in a wastewater treatment plant. Appl. Environ. Microbiol. 44: 1395–1403.

Mackay, D.M., and J. A. Cherry. 1989. Groundwater contamination: Pump-and-treat remediation. Environ. Sci. Technol. 23: 630–636.

MacKenzie, W.R., N.J. Hoxie, M.E. Proctor, M.S. Gradus, K.A. Blair, D.E. Peterson, J.J. Kazmierezak, D.G. Addiss, K.R. Fox, J.B. Rose, and J.P. Davis. 1994. A massive outbreak in Milwaukee of *Cryptosporidium* infection transmitted through the public water supply. N. Engl. J. Med. 331: 161–167.

Mackey, E.D., T.M. Hargy, H.B. Wright, J.P. Malley Jr., and R.S. Cushing. 2002. Comparing Cryptosporidium and MS2 bioassays: Implications for UV reactor validation. J. AWWA 94 (2): 62–69.

Mackie, R.I., and M.P. Bryant. 1981. Metabolic activity of fatty acid-oxidizing bacteria and the contribution of acetate, propionate, butyrate and CO_2 to methanogenesis in cattle waste at 40 and 60°C. Appl. Environ. Microbiol. 41: 1363–1373.

MacLeod, F.A., S.R. Guiot, and J.W. Costerton. 1990. Layered structure of bacterial aggregates in an upflow anaerobic sludge bed and filter reactor. Appl. Environ. Microbiol. 56: 1598–1607.

MacRae, J.D., and J. Smit. 1991. Characterization of Caulobacters isolated from wastewater treatment systems. Appl. Environ. Microbiol. 55: 751–758.

Madge, B.A., and J.N. Jensen. 2002. Disinfection of wastewater using a 20-kHz ultrasound unit. Water Environ. Res. 74: 159–169.

Madigan, M.T. 1988. Microbiology, physiology, and ecology of phototrophic bacteria, pp. 39–111, In: *Biology of Anaerobic Microorganisms* A.J.B. Zehnder, Ed., Wiley-Interscience, New York.

Madoni, P., D. Davoli, and E. Chierici. 1993. Comparative analysis of the activated sludge microfauna in several sewage treatment plants. Water Res. 27: 1485–1491.

Madoni, P., D. Davoli, and G. Gibin. 2000. Survey of filamentous microorganisms from bulking and foaming activated-sludge plants in Italy. Water Res. 34: 1767–1772.

Madore, M.S., J.B. Rose, C.P. Gerba, M.J. Arrowood, and C.R. Sterling. 1987. Occurrence of *Cryptosporidium* oocysts in sewage effluents and select surface waters. J. Parasitol. 73: 702–705.

Magnusson, S., K. Gundersen, A. Brandberg, and E. Lycke. 1967. Marine bacteria and their possible relation to the virus inactivation capacity of seawater. Acta Pathol. Microbiol. Scand. 71: 274–280.

Mahbubani, M.H., A.K. Bej, M. Perlin, F.W. Schaefer III, W. Jakubowski, and R.M. Atlas. 1991. Detection of *Giardia* cysts by using the polymerase chain reaction and distinguishing live from dead cysts. Appl. Environ. Microbiol. 57: 3456–3461.

Maier, R.M. 2000a. Biogeochemical cycling, pp. 319–346, In: *Environmental Microbiology*, R.M. Maier, I.L. Pepper, and C.P. Gerba, Eds., Academic Press, San Diego, CA, 585 pp.

Maier, R.M. 2000b. Microorganisms and organic pollutants, pp. 363–402, In: *Environmental Microbiology*, R.M. Maier, I.L. Pepper, and C.P. Gerba, Eds., Academic Press, San Diego, 585 pp.

Maier, R.M., I.L. Pepper, and C.P. Gerba. 2000. *Environmental Microbiology*, Acad. Press, San Diego, CA. 585 pp.

Majeti, V.A., and C.S. Clark. 1981. Health risk of organics in land application. J. Environ. Eng. Div. ASCE 107: 339–357.

Makino, S.-I., and H.-I. Cheun. 2003. Application of the real-time PCR for the detection of airborne microbial pathogens in reference to the anthrax spores. J. Microbiol. Meth. 53: 141–147.

Makintubee, S., J. Mallonee, and G.R. Istre. 1987. Shigellosis outbreak associated with swimming. Am. J. Publ. Health 77: 166–171.

Mahoney, F.J., T.A. Farley, K.Y. Kelso, S.A. Wilson, J.M. Horan, and L.M. McFarland. 1992. An outbreak of hepatitis A associated with swimming in a public pool. J. Infect. Dis. 165: 613–618.

Maloney, S.W., J. Maneim, J. Mallevialle, and F. Fiesssinger. 1986. Transformation of trace organic compounds in drinking water by enzymatic oxidative coupling. Environ. Sci. Technol. 20: 249–253.

Manafi, M., and R. Sommer. 1993. Rapid identification of enterococci with a new fluorogenic-chromogenic assay. Water Sci. Technol. 27: 271–274.

Mandi, L., J. Darley, J. Barbe, and B. Baleux. 1992. Essais d'epuration des eaux usees de Marrakech par la jacinthe d'eau (charge organique, bacterienne et parasitologique). Rev. Sci. Eau 5: 313–333.

Mandi, L., N. Ouazzani, K. Bouhoum, and A. Boussaid. 1994. Wastewater treatment by stabilization ponds with and without macrophytes under arid climate. Water Sci. Technol. 28 (10): 177–181.

Manem, J.A., and B.E. Rittmann. 1992. Removing trace-level organic pollutants in a biological filter. J. Amer. Water Works Assoc. 84: 152–157.

Manero, A., X. Vilanova, M. Cerdà-Cuéllar, and A.R. Blanch. 2002. Characterization of sewage waters by biochemical fingerprinting of Enterococci. Water Res. 36: 2831–2835.

Mangione, E.J., G. Huitt, D. Lenaway, J. Beebe, A. Bailey, M. Figoski, M.P. Rau, K.D. Albrecht, and M.A. Yakrus. 2001. Nontuberculous mycobacterial disease following hot tub exposure. Emerg. Infect. Dis. 7: 1039–1042.

Manka, J., M. Rebhun, A. Mandelbaum, and A. Bortinger. 1974. Characterization of organics in secondary effluents. Environ. Sci. Technol. 8: 1017–1020.

Manning, J.F., and R.L. Irvine. 1985. The biological removal of phosphorus in a sequencing batch reactor. J. Water Pollut. Control Fed. 57: 87–94.

Mansfield, L.A., P.B. Melnyk, and G.C. Richardson. 1992. Selection and full-scale use of a chelated iron adsorbent for odor control. Water Environ. Res. 64: 120–127.

Manz, W., M. Wagner, R. Amann, and K.-H. Schleifer. 1994. In situ characterization of the microbial consortia active in two wastewater treatment plants. Water Res. 28: 1715–1723.

Manz, W., M. Eisenbrecher, T.R. Neu, and U. Szewzyk. 1998. Abundance and spatial organization of Gram-negative sulfate-reducing bacteria in activated sludge investigated by in situ probing with specific 16S rRNA targeted oligonucleotides. FEMS Microbiol. Ecol. 25: 43–61.

Mara, D.D., and S. Cairncross. 1989. Guidelines for the Safe Use of Wastewater and Excreta in Agriculture and Aquaculture. World Health Organization, Geneva, Switzerland.

Mara, D.D. 2002. Waste stabilization ponds, pp. 3330–3337, In: Encyclopedia of Environmental Microbiology, G. Bitton, editor-in-chief, Wiley-Interscience, N.Y.

Marais, G.V.R. 1974. Fecal bacterial kinetics in stabilization ponds. J. San. Eng. Div. ASCE 100: 119–139.

March, D., L. Benefield, E. Bennett, D. Linstedt, and R. Hartman. 1981. Coupled trickling filter-rotating biological contactor nitrification process. J. Water Pollut. Control Fed. 53: 1469–1480.

Marciano-Cabral, F. 1988. Biology of Naegleria spp. Microbiol. Rev. 52: 114–133.

Marison, L.W. 1988a. Growth kinetics, pp. 184–217, In: Biotechnology for Engineers: Biological Systems in Technological Processes, A. Scragg, Ed., Ellis Horwood, Chichester, U.K.

Marison, L.W. 1988b. Enzyme kinetics, pp. 96–119, In: Biotechnology for Engineers: Biological Systems in Technological Processes, A. Scragg, Ed., Ellis Horwood, Chichester, U.K.

Markovic, M.J., and A.L. Kroeger. 1989. Saprophitic actinomycetes as a causative agent of musty odor in drinking water (Abst). In: 89th Ann. Meeting, American Society of Microbiology, New Orleans, LA.

Marques, E., and M.T. Martins. 1983. Enterovirus isolation from seawater from beaches of Baixada Santista. In: 9th Latin Amer. Congress for Microbiol. Sau Paulo, Brazil (In Portuguese).

Marshall, K.C. 1976. Interfaces in Microbial Ecology, Harvard University Press, Cambridge, MA.

Marshall, T.H. 1998. Emerging electrotechnology for disinfection: Benchtest results and evaluation of the new Phoenix water system technology, pp. 41–53, In: Water Reuse Conference Proceedings, Feb. 1–4, 1998, Lake Buena Vista, FL., American Water Works. Assoc. & Water Environment Federation.

Marthi, B., V.P. Fieland, M. Walter, and R.J. Seidler. 1990. Survival of bacteria during aerosolization. Appl. Environ. Microbiol. 56: 3463–3467.

Marthi, B., and B. Lighthart. 1990. Effect of betaine on enumeration of airborne bacteria. Appl. Environ. Microbiol. 56: 1286–1289.

Marthi, B., B.T. Shaffer, B. Lighthart, and L. Ganio. 1991. Resuscitaion effects of catalase on airborne bacteria. Appl. Environ. Microbiol. 57: 2775–2776.

Martin, J.H., Jr., H.E. Bostian, and G. Stern. 1990. Reductions of enteric microorganisms during aerobic sludge digestion. Water Res. 24: 1377–1385.

Martins, A.M.P., K. Pagilla, J.J. Heijnena, and M.C.M. van Loosdrecht. 2004. Filamentous bulking sludge – a critical review. Water Res. 38: 793–817.

Martins, M.T., I.G. Rivera, D.L. Clark, and B.H. Olson. 1992. Detection of virulence factors in culturable *Escherichia coli* isolates from water samples by DNA probes and recovery of toxin-bearing strains in minimal *o*-nitrophenol-b-D-galactopyranoside-4-methylumbelliferyl-β-D-glucuronide media.Appl. Environ. Microbiol. 58: 3095–3100.

Martins, M.T., I.G. Rivera, D.L. Clark, M.H. Stewart, R.L. Wolfe, and B.H. Olson. 1993. Distribution of uidA gene sequences in Escherichia coli isolates in water sources and comparison with the expression of beta-glucuronidase activity in 4-methylumbelliferyl-beta-D-glucuronide media. Appl. Environ. Microbiol. 59: 2271–2276.

Martz, R.F., D.I. Sebacher, and D.C. White. 1983. Biomass measurement of methane forming bacteria in environmental samples. J. Microbiol. Methods 1: 53–61.

Mascher, F., S. Deller, F.P. Pichler-Semmelrock, S. Roehm, and E. Marth. 2003. The significance of sunlight for the elimination of indicator bacteria in small-scale bathing ponds in central Europe. Water Sci. Technol. 4& (3): 211–213.

Maszenan, A.-M., R.J. Seviour, B.K.C. Patel, and J. Wanner. 2000. A fluorescently-labelled r-RNA targeted oligonucleotide probe for the *in situ* detection of G-bacteria of the genus *Amaricoccus* in activated sludge. J. Appl. Microbiol. 88: 826–835.

Mateju, V., S. Cizinska, J. Krejei, and T. Janoch. 1992. Biological water denitrification: a review. Enzyme Microbiol Technol. 14: 170–183.

Matin, A., and S. Harakeh. 1990. Effect of starvation on bacterial resistance to disinfectants, pp. 88–103, In: *Drinking Water Microbiology*, G.A. McFeters, Ed., Springer Verlag, New York.

Matin, A., E. Auger, P. Blum, and J. Schultz. 1989. Genetic basis of starvation survival in non-differentiating bacteria. Annu. Rev. Microbiol. 43: 293–316.

Matsuda, H., H. Yamamori, T. Sato, Y. Ose, H. Nagase, H. Kito, and K. Sumida. 1992. Mutagenicity of ozonation products from humic substances and their components. Water Sci. Technol. 25: 363–370.

Matsui, S., H. Takigami, T. Matsuda, N. Taniguchi, J. Adachi, H. Kawami, and Y. Shimizu. 2000. Estrogens and estrogen mimics contamination in water and the role of sewage treatment. Water Sci. Technol. 42 (12): 173–179.

Matsui, S.M. 1995. Astroviruses, pp. 1035–1045, In: *Infections of the Gastrointestinal Tract*, M.J. Blaser, P.D. Smith, J.I. Ravdin, H.B. Greenberg, and R.L. Guerrant, Eds., Raven Press, New York.

Matsui, T., S. Kyosai, and M. Takahashi. 1991. Application of biotechnology to municipal waste-water treatment. Water Sci. Technol. 23: 1723–1732.

Matsuki, T., K. Watanabe, J. Fujimoto, Y. Miyamoto, T. Takada, K. Matsumoto, H. Oyaizu, and R. Tanaka. 2002. Development of 16S rRNA-gene-targeted group-specific primers for the detection and identification of predominant bacteria in human feces. Appl. Environ. Microbiol. 68: 5445–5451.

Matsunaga, T., I. Karube, and S. Suzuki. 1980. A specific microbial sensor for formic acid. European J. Appl. Microbiol. Biotechnol. 10: 235–243.

Matsunaga, T., S. Nakasono, Y. Kitajima, and K. Horiguchi. 1994. Electrochemical disinfection of bacteria in drinking water using activated carbon fibers. Biotechnol. Bioeng. 43: 429–433.

May, K.R. 1966. Multistage liquid impinger. Bacteriol. Rev. 30: 559–570.

Mayer, C.L., and C.J. Palmer. 1996. Evaluation of PCR, nested PCR, and fluorescent antibodies for detection of *Giardia* and *Cryptosporidium* species in wastewater. Appl. Environ. Microbiol. 62: 2081–2085.

Mazidji, C.N., B. Koopman, G. Bitton, and G. Voiland. 1990. Use of Microtox and *Ceriodaphnia* bioassays in wastewater fractionation. Toxicity Assess 5: 265–277.

Mazidji, C.N., B. Koopman, and G. Bitton. 1992. Chelating resin versus ion-exchange resin for heavy metal removal in toxicity fractionation. Water Sci. Technol. 26: 189–196.

Mbeunkui, F., C. Richaud, A.-L. Etienne, R.D. Schmid, and T.T. Bachmann. 2002. Bioavailable nitrate detection in water by an immobilized luminescent cyanobacterial reporter strain. Appl. Microbiol. Biotechnol. 60: 306–312.

McCambridge, J., and T.A. McMeekin. 1980. Relative effects of bacterial and protozoan predators on survival of *Escherichia coli* in estuarine water samples. Appl. Environ. Microbiol. 40: 907–911.

McCambridge, J., and T.A. McMeekin. 1981. Effect of solar radiation and predacious microorganisms on survival of fecal and other bacteria. Appl. Environ. Microbiol. 41: 1083–1087.

McCarty, P.L., B.E. Rittmann, and E.J. Bouwer. 1984. Microbiological processes affecting chemical transformations in groundwater, pp. 89–115, In: *Groundwater Pollution Microbiology*, G. Bitton, and C.P. Gerba, Eds., John Wiley & Sons, New York.

McCarty, S.C., J.H. Standridge, and M.C. Stasiak. 1992. Evaluating a commercially available defined-substrate test for recovery of *E. coli*. J. Am. Water Works Assoc. 84: 91–97.

McClure, N.C., J.C. Fry, and A.J. Weightman. 1990. Gene transfer in activated sludge, pp. 111–129, In: *Bacterial Genetics in Natural Environments*, J.C. Fry, and M.J. Day, Eds., Chapman & Hall, London.

McCuin, R.M., and J.L. Clancy. 2003. Modifications to United States Environmental Protection Agency Methods 1622 and 1623 for Detection of *Cryptosporidium* oocysts and *Giardia* Cysts in Water. Appl. Environ. Microbiol. 69: 267–274.

McDonald, L.C., C.R. Hackney, and B. Ray. 1983. Enhanced recovery of injured *Escherichia coli* by compounds that degrade hydrogen peroxide or block its formation. Appl. Environ. Microbiol. 45: 360–365.

McDonnell, S., K.B. Kirkland, W.G. Hlady, C. Aristeguieta, R.S. Hopkins, S.S. Monroe, and R.I. Glass. 1997. Failure of cooking to prevent shellfish-associated viral gastroenteritis. Arch. Int. Med. 157: 111–116.

McFarland, M.J., and W.J. Jewell. 1990. The effect of sulfate reduction on the thermophilic (55°C) methane fermentation process. J. Ind. Microbiol. 5: 247–258.

McFeters, G.A., Ed. 1990. *Drinking Water Microbiology*. Springer Verlag, New York, 502 pp.

McFeters, G.A., and A.K. Camper. 1988. Microbiological analysis and testing, pp. 73–95, In: *Bacterial Regrowth in Distribution Systems*, Characklis, W.G., Ed., Research Report, AWWA Research Foundation, Denver, CO.

McFeters, G.A., S.C. Cameron, and M.W. LeChevallier. 1982. Influence of diluents, media, and membrane filters on detection of injured waterborne coliform bacteria. Appl. Environ. Microbiol. 43: 97–103.

McGowan, K.L., E. Wickersham, and N.A. Strockbine. 1989. *Escherichia coli* O157:H7 from water (letter). Lancet i: 967–968.

McGrath, J.W., S. Cleary, A. Mullan, and J.P. Quinn. 2001. Acid-stimulated phosphate uptake by activated sludge microorganisms under aerobic laboratory conditions. Water Res. 35: 4317–4322.

McInernay, M.J., M.P. Bryant, R.B. Hespell, and J.W. Costerton. 1981. *Syntrophomonas wolfei*, gen. nov. sp. nov., an anaerobic syntrophic, fatty acid-oxidizing bacterium. Appl. Environ. Microbiol. 41: 1029–1039.

McKay, G. 1996. Design of adsorption contacting systems, pp. 99–132, In: *Use of Adsorbents for the Removal of Pollutants from Wastewater*, G. McKay, Ed., CRC Press, Boca Raton, FL.

McKay, L.D., J.A. Cherry, R.C. Bales, M.T. Yahya, and C.P. Gerba. 1993. A field example of bacteriophage as tracers of fracture flow. Env. Sci. Technol. 27: 1075–1079.

McKeegan, K.S., M.I. Borges-Walmsley, and A.R. Walmsley. 2002. Microbial and viral drug resistance mechanisms. Trends Microbiol. 10: s8–s14.

McKinley, V.L., and J.R. Vestal. 1985. Physical and chemical correlates of microbial activity and biomass in composting municipal sewage sludge. Appl. Environ. Microbiol. 50: 1395–1403.

McKinley, V.L., J.R. Vestal, and A.E. Eralp. 1985. Microbial activity in composting. Biocycle 26: 47–50.

McLean, R.J.C., D. Beauchemin, L. Clapham, and T.J. Beveridge. 1990. Metal-binding characteristics of the gamma-glutamyl capsular polymer of *Bacillus licheniformis* ATTC 9945. Appl. Environ. Microbiol. 56: 3671–3677.

McLean R.J., M. Whiteley, D.J. Stickler, and W.C. Fuqua. 1997. Evidence of autoinducer activity in naturally occurring biofilms. FEMS Microbiol. Lett. 154: 259–63.

McLoughlin, A.J. 1994. Controlled release of immobilized cells as a strategy to regulate ecological competence of inocula. Adv. Biochem. Eng. Biotech. 51: 2–45.

McMahon, K.D., M.A. Dojka, N.R. Pace, D. Jenkins, and J.D. Keasling. 2002. Polyphosphate kinase from activated sludge performing enhanced biological phosphorus removal. Appl. Environ. Microbiol. 68: 4971–4978.

McPherson, P., and M.A. Gealt. 1986. Isolation of indigenous wastewater bacterial strains capable of mobilizing plasmid pBR325. Appl. Environ. Microbiol. 51: 904–909.

Meadows, C.A., and B.H. Snudden. 1982. Prevalence of *Yersinia enterocolitica* in waters at the lower Chippewa River basin, Wisconsin. Appl. Environ. Microbiol. 43: 953–954.

Means, E.G., and B.H. Olson. 1981. Coliform inhibition by bacteriocin-like substances in drinking water distribution systems. Appl. Environ. Microbiol. 42: 506–512.

Mechsner, K., T. Fleischmann, C.A. Mason, and G. Hamer. 1990. UV disinfection: Short-term inactivation and revival. (Abstr.), In: *International Symposium Health-Related Microbiology*, Tubingen, Germany, April 1–6, 1990.

Mechsner, K., and G. Hamer. 1985. Denitrification by methanotrophic/methylotrophic bacterial associations in aquatic environments, pp. 257–271, In: *Denitrification in the Nitrogen Cycle*. Plenum, New York.

Medsker, L.L., D. Jenkins, and J.F. Thomas. 1968. Odorous compounds in natural waters: An earthy-smelling compound associated with blue-green algae and actinomycetes. Environ. Sci. Technol. 2: 461–464.

Meganck, M.T.J., and G.M. Faup. 1988. Enhanced biological phosphorus removal from waste waters, pp. 111–203, In: *Biotreatment Systems*, Vol. 3, D.L. Wise, Ed., CRC Press, Boca Raton, FL.

Meganck, M.T.J., D. Malnou, P. Le Flohic, G.M. Faup, and J.M. Rovel. 1984. The importance of the acidogenic microflora in biological phosphorus removal, pp. 254–263, In: Proceedings IAWPRC Conference *Enhanced Biological Phosphorus Removal from Wastewater*, Vol. 1, Paris, Sept. 1984.

Meier, J.R., R.B. Knohl, W.E. Coleman, H.R. Ringhand, J.W. Munch, W.H. Kaylor, R.P. Streicher, and F.C. Kopfler. 1987. Studies on the potent bacterial mutagen: aqueous stability, XAD-recovery and analytical determination in drinking water and in chlorinated humic acid solution. Mutation Res. 189: 363–370.

Melnick, J.L. 1976. Taxonomy of viruses. Prog. Med. Virol. 22: 211–221.

Mena,, K.D., C.P. Gerba, C.N. Haas, and J.B. Rose. 2003. Risk assessment of waterborne coxsackievirus. J. Amer. Water. Wks. Assoc. 95 (7): 122–129.

Menard, A.B., and D. Jenkins. 1970. Fate of phosphorus in wastewater treatment processes: Enhanced removal of phosphate by activated sludge. Environ. Sci. Technol. 4: 1115–1119.

Meng, Q.S., and C.P. Gerba. 1996. Comparative inactivation of enteric adenoviruses, polioviruses and coliphages by ultraviolet irradiation. Water Res. 30: 2665–2668.

Menon, P., G. Billen, and P. Servais. 2003. Mortality rates of autochthonous and fecal bacteria in natural aquatic ecosystems. Water Res. 37: 4151–4158.

Mentzing, L.O. 1981. Waterborne outbreaks of *Campylobacter enteritis* in central Sweden. Lancet ii: 352–354.

Meschke, J.S., and M.D. Sobsey. 2002. Norwalk-like viruses: Detection methodologies and environmental fate, pp. 2221–2235, In: *Encyclopedia of Environmental Microbiology*, G. Bitton, editor-in-chief, Wiley-Interscience, N.Y.

Meschke, J.S., and M.D. Sobsey. 2003. Comparative reduction of Norwalk virus, poliovirus type 1, F^+ RNA coliphage MS2 and *Escherichia coli* in miniature soil columns. Water Sci. Technol. 47 (3): 85–90.

Messer, J.W., and A.P. Dufour. 1998. A rapid, specific membrane filtration procedure for enumeration of enterococci in recreational water. Appl. Environ. Microbiol. 64: 678–680.

Messing, R. 1988. Immobilized cells in anaerobic waste treatment, pp. 311–316, In: *Bioreactor Immobilized Enzymes and Cells: Fundamentals and Applications*, M. Moo-Young, Ed., Elsevier Appl. Sci., New York.

Metcalf and Eddy, Inc. 1991. *Wastewater Engineering: Treatment, Disposal and Reuse.* 3rd Ed., McGraw-Hill, New York.

Metcalf, T.G., J.L. Melnick, and M.K. Estes. 1995. Environmental virology: from detection of virus in sewage and water by isolation to identification by molecular biology. A trip of over 50 years. Annu. Rev. Microbiol. 49: 461–487.

Miana, P., L. Grando, G. Caravello, and M. Fabris. 2002. *Microthrix parvicella* foaming at the Fusina WWTP. Water Sci. & Technol. 46 (1–2): 499–502.

Michal, G. 1978. Determination of Michaelis constant and inhibitor constants, pp. 29–42, In: *Principles of Enzymatic Analysis*, H.U. Bergmeyer, Ed., Verlag Chemie, Wienheim.

Michel, O., H. Ginanni, J. Duchateau, F. Vertongen, B. Le Bon, and R. Sergysels. 1991. Domestic endotoxin exposure and clinical severity of asthma. Clin. Exp. Allergy 21: 441–448.

Miettinen, I.T., T. Vartiainen, and P.J. Martikainen. 1997. Phophorus and bacterial growth in drinking water. Appl. Environ. Microbiol. 63: 3242–3245.

Miettinen, I.T., O. Zacheus, C.-H. von Bonsdorff, and T. Vartiainen. 2001. Waterborne epidemics in Finland 1998–1999. Water Sci. Technol. 43: 67–71.

Mignotte-Cadiergues, B., C. Gantzer, and L. Schwartzbrod. 2002. Evaluation of bacteriophages during the treatment of sludge. Water Sci. & Technol. 46 (10): 189–194.

Mikesell, M.D., and S.A. Boyd. 1985. Reductive dechlorination of the herbicides 2,4-D, 2,4,5-T and pentachlorophenol in anaerobic sewage sludges. J. Environ. Qual. 14: 337–340.

Mikesell, M.D., and S.A. Boyd. 1986. Complete reductive dechlorination and mineralization of pentachlorophenol by anaerobic microorganisms. Appl. Environ. Microbiol. 52: 861–865.

Mills, S.W., G.P. Alabaster, D.D. Mara, H.W. Pearson, and W.N. Thitai. 1992. Efficiency of faecal bacterial removal in waste stabilization ponds in Kenya. Water Sci. Technol. 26: 1739–1748.

Miller, M.B., and B.L. Bassler. 2001. Quorum sensing in bacteria. Annu. Rev. Microbiol. 55: 165–199.

Miller, R.A., M.A. Bronsdon, and W.R. Morton. 1986. Determination of the infectious dose of *Cryptosporidium* and the influence of inocculum size on disease severity in a primate model (Abst.), In: *Annual Meeting American Society Microbiology*, Washington, D.C.

Mills, A.L., J.S. Herman, G.M. Hornberger, and T.H. DeJesus. 1994. Effect of solution ionic strength and iron coatings on mineral grains on the sorption of bacterial cells to quartz sand. Appl. Environ. Microbiol. 60: 3300–3306.

Milstein, C., B. Nicklas, and A. Huettermann. 1989. Oxidation of aromatic compounds in organic solvents with laccase from *Trametes versicolor*. Appl. Microbiol. Technol. 31: 70–74.

Mino, T. 2002. Activated sludge models: Microbiological basis, pp. 14–26, In: *Encyclopedia of Environmental Microbiology*, G. Bitton, editor-in-chief, Wiley-Interscience, N.Y.

Mirenda, R.J., and W.S. Hall. 1992. The application of effluent characterization procedures in toxicity identification evaluations. Water Sci. Technol. 25: 39–44.

Mitchell, R. 1971. Destruction of bacteria and viruses in seawater. J. San. Eng. Div. ASCE 97: 425–432.

Mitchell, R., and Z. Nevo. 1964. Effect of bacterial polysaccharide accumulation on infiltration of water through sand. Appl. Microbiol. 12: 219–223.

Mobarry, B.K., M. Wagner, V. Urbain, B.E. Rittmann, and D.A. Stahl. 1996. Phylogenetic probes for analyzing abundance and spatial organization of nitrifying bacteria. Appl. Environ. Microbiol. 62: 2156–2162.

Mocé-Llivina, L., M. Muniesa, H. Pimenta-Vale, F. Lucena, and J. Jofre. 2003. Survival of bacterial indicator species and bacteriophages after thermal treatment of sludge and sewage. Appl. Environ. Microbiol. 69: 1452–1456.

Moe, C.L. 1997. Waterborne transmission of infectious agents, pp. 136–152, In: *Manual of Environmental Microbiology*, C.J. Hurst, G.R. Knudsen, M.J. McInerney, L.D. Stetzenbach, and M.V. Walter, Eds., ASM Press, Washington, D.C.

Moeller, J.R., and J. Calkins. 1980. Bactericidal agents in wastewater lagoons and lagoon design. J. Water Pollut. Control Fed. 52: 2442–2450.

Mofidi, A.A., H. Baribeau, P.A. Rochelle, R. De Leon, B.M. Coffey, and J.F. Green. 2001. Disinfection of *Cryptosporidium parvum* with polychromatic UV light. J. AWWA 93 (6): 95–109.

Mofidi, A.A., E.A. Meyer, P.M. Wallis, C.I. Chou, B.P. Meyer, S. Ramalingam, and B.M. Coffey. 2002. The effect of UV light on the inactivation of *Giardia lamblia* and *Giardia muris* cysts as determined by animal infectivity assay (P-2951-01). Water Res. 36: 2098–2108.

Mohr, A.J. 1991. Development of models to explain the survival of viruses and bacteria in aerosols, pp. 160–190, In: *Modeling the Environmental Fate of Microorganisms*, C.J. Hurst, Ed., American Society of Microbiology, Washington, D.C.

Mohr, A.J. 1997. Fate and transport of microorganisms in air, pp. 640–650, In: *Manual of Environmental Microbiology*, C.J. Hurst, G.R. Knudsen, M.J. McInerney, L.D. Stetzenbach, and M.V. Walter, Eds., ASM Press, Washington, D.C.

Monarca, S., D. Feretti, C. Collivignarelli, L. Guzzella, I. Zerbini, G. Bertanza, and R. Pedrazzani. 2000. The influence of different disinfectants on mutagenicity and toxicity of urban wastewater. Water Res. 34: 4261–4269.

Monsen, R.M., and E.M. Davis. 1984. Microbial responses to selected organic chemicals in industrial waste treatment units, pp. 233–249, In: *Toxicity Screening Procedures Using Bacterial Systems*, D. Liu, and B.J. Dutka, Eds., Marcel Dekker, New York.

Santegoeds, C.M., L.R. Damgaard, G. Hesselink, J. Zopfi, P. Lens, G. Muyzer, and D. de Beer. 1999. Distribution of sulfate-reducing and methanogenic bacteria in anaerobic aggregates determined by microsensor and molecular analysis. Appl. Environ. Microbiol. 65: 4618–4629.

Moore, A.C., B.L. Herwaldt, G.F. Craun, R.L. Calderon, A.K. Highsmith, and D.D. Juranek. 1994. Waterborne diseases in the United States, 1991 and 1992. J. Am. Water Works Assoc. 86: 87–99.

Moore, A.T., A. Vira, and S. Fogel. 1989. Biodegradation of *trans*-1,2-dichloroethylene by methane-utilizing bacteria in an aquifer simulator. Environ. Sci. Technol. 23: 403–406.

Moore B.E., B.P. Sagik, and C.A. Sorber. 1976. An assessment of potention health risks associated with land disposal of residual sludges, pp. 108–112, Presented at the *3rd Nat. Conf. on Sludge Management, Disposal and Reuse*, Miami, FL.

Moore, B.E., B.P. Sagik, and C.A. Sorber. 1978. Land application of sludge: Minimizing the impact of viruses on water resources, pp. 154–167, In: *Risk Assessment and Health Effects of Land Application of Municipal Wastewater and Sludges*, B.P. Sagik, and C.A. Sorber, Eds., University of Texas, San Antonio, TX.

Moore, B.E., B.P. Sagik, and C.A. Sorber. 1979. Procedure for recovery of airborne human enteric viruses during spray irrigation of treated wastewater. Appl. Environ. Microbiol. 38: 688–693.

Moore, B.E., B.P. Sagik, and C.A. Sorber. 1981. Viral transport to groundwater at a wastewater land application site. J. Water Pollut. Control Fed. 53: 1492–1502.

Moore, B.E., D.E. Camman, C.A. Turk, and C.A. Sorber. 1988. Microbial characterization of municipal wastewater at a spray irrigation site: The Lubbock infection surveillance study. J. Water Pollut. Control Fed. 60: 1222–1230.

Moore, N.J., and A.B. Margolin. 1994. Efficacy of nucleic acid probes for detection of poliovirus in water disinfected by chlorine, chlorine dioxide, ozone, and UV radiation. Appl. Environ. Microbiol. 60: 4189–4191.

Moos, L.P., E.J. Kirsch, R.F. Wukasch, and C.P.L Grady, Jr. 1983. Pentachlorophenol biodegradation. I. Aerobic. Water Res. 17: 1575–1584.

Morace, G., F.A. Aulicino, C. Angelozzi, L. Costanzo, F. Donadio, and M. Rapicetta. 2002. Microbial quality of wastewater: detection of hepatitis A virus by reverse transcriptase-polymerase chain reaction. J. Appl. Microbiol. 92: 828–836.

Moran, M.A., V.L. Torsvik, T. Torsvik, and R.E. Hodson. 1993. Direst extraction and purification of rRNA for ecological studies. Appl. Environ. Microbiol. 59: 915–918.

Moreno, Y., S. Botella, J.L. Alonso, M.A. Ferrús, M. Hernández, and J. Hernández. 2003. Specific detection of Arcobacter and Campylobacter strains in water and sewage by PCR and fluorescent In situ hybridization. Appl. Environ. Microbiol. 69: 1181–1186.

Moreno, Y., M.A. Ferrús, J.L. Alonso, A. Jiménez, and J. Hernández. 2003. Use of fluorescent *in situ* hybridization to evidence the presence of *Helicobacter pylori* in water. Water Res. 37: 2251–2256.

Mori, T., K. Itokazu, Y. Ishikura, F. Mishina, Y. Sakai, and M. Koga. 1992. Evaluation of control strategies for actinomycete scum in full-sale treatment plants. Water Sci. Technol 25: 231–237.

Morin, P., A. Camper, W. Jones, D. Gatel, and J.C. Goldman. 1996. Colonization and disinfection of biofilms hosting coliform-colonized carbon fines. Appl. Environ. Microbiol. 62: 4428–4432.

Morinigo, M.A., M.A. Munoz, R. Cornax, E. Martinez-Manzanares, and J.J. Borrego. 1992a. Presence of indicators and *Salmonella* in natural waters affected by outfall wastewater discharges. Water Sci. Technol. 25: 1–8.

Morinigo, M.A., D. Wheeler, C. Berry, C. Jones, M.A. Munoz, R. Cornax, and J.J. Borrego. 1992b. Evaluation of different bacteriophage groups as faecal indicators in contaminated natural waters in southern England. Water Res. 26: 267–271.

Morisada, S., N. Miyata, and K. Iwahori. 2002. Immunomagnetic separation of scum-forming bacteria using polyclonal antibody that recognizes mycolic acids. J. Microbiol. Meth. 51: 141–148.

Morris, J.C. 1975. Aspects of the quantitative assessment of germicidal efficiency. In: *Disinfection of Water and Wastewater.* J.D. Johnson, Ed., Ann Arbor Scientific Publications, Ann Arbor, MI.

Morita, S., A. Namikoshi, T. Hirata, K. Oguma, H. Katayama, S. Ohgaki, N. Motoyama, and M. Fujiwara. 2002. Efficacy of UV irradiation in inactivating *Cryptosporidium parvum* Oocysts. Appl. Environ. Microbiol. 68: 5387–5393.

Moslemy, P., R.J. Neufeld, and S.R. Guiot. 2002. Biodegradation of gasoline by gellan gum-encapsulated bacterial cells. Biotech. Bioeng. 80: 175–184.

Mota, P.Q.F., and S.C. Edberg. 2002. Nosocomial infections, pp. 2235–2250, In: *Encyclopedia of Environmental Microbiology*, G. Bitton, editor-in-chief, Wiley-Interscience, N.Y.

da Motta, M., M.N. Pons, and N. Roche. 2002. Study of filamentous bacteria by image analysis and relation with settleability. Water Science & Technology 46 (1–2): 363–369.

Mueller, R.F., and A. Steiner. 1992. Inhibition of anaerobic digestion by heavy metals. Water Sci. Technol. 26: 835–846.

Mulbry, W.W., and J.S. Karns. 1989. Purification and characterization of three parathion hydrolases from gram-negative bacterial strains. Appl. Environ. Microbiol. 55: 289–293.

Mullen, M.D., D.C. Wolf, F.G. Ferris, T.J. Beveridge, C.A. Flemming, and G.W. Bayley. 1989. Bacterial sorption of heavy metals. Appl. Environ. Microbiol. 55: 3143–3149.

Mullis, K.B., and F.A. Fallona. 1987. Specific synthesis of DNA *in vitro* via a polymerase catalyzed chain reaction. Methods Enzymol. 155: 335–350.

Munkittrick, K.R, E.A. Power, and G.A. Sergy. 1991. The relative sensitivity of Microtox, daphnid, rainbow trout and fathead minnow acute lethality tests. Environ. Toxicol. Water Qual. 6: 35–62.

Munnecke, D.M. 1981. The use of microbial enzymes for pesticides detoxification. In: *Microbial Degradation of Xenobiotics and Recalcitrant Compounds*, T. Leisinger et al. Eds., Academic Press, San Diego, CA.

Munro, P.M., F. Laumond, and M.J. Gauthier. 1987. Previous growth of enteric bacteria on a salted medium increases their survival in seawater. Lett. Appl. Microbiol. 4: 121–124.

Munro, P.M., M.J. Gauthier, V.A. Breittmayer, and J. Bongiovanni. 1989. Influence of osmoregulation on starvation survival of *Escherichia coli* in seawater. Appl. Environ. Microbiol. 55: 2017–2024.

Muraca, P., J.E. Stout, and V.L. Yu. 1987. Comparative assessment of chlorine, heat, ozone, and U.V light for killing *Legionella pneumophila* within a model plumbing system. Appl. Environ. Microbiol. 53: 447–453.

Muraca, P., V.L. Yu, and J.E. Stout. 1988. Environmental aspects of Legionnaires' Disease. J. Am. Water Works Assoc. 80: 78–86.

Murray, G.E., R.S. Tobin, B. Junkins, and D.J. Kushner. 1984. Effect of chlorination on antibiotic resistance profiles of sewage-related bacteria. Appl. Environ. Microbiol. 48: 73–77.

Murray, W.D., and L. van den Berg. 1981. Effect of nickel, cobalt, and molybdenum on performance of methanogenic fixed-film reactors. Appl. Environ. Microbiol. 42: 502–505.

Murthy, S. and J.T. Novak. 2001. Influence of cations on activated sldge effluent quality. Water Environ. Res. 73: 30–36.

Musial, C.E., M.J. Arrowood, C.R. Sterling, and C.P. Gerba. 1987. Detection of *Cryptosporidium* in water by using propylene cartridge filters. Appl. Environ. Microbiol. 53: 687–692.

Myint, K.S.A., J.R. Campbell, and A.L. Corwin. 2002. Hepatitis viruses (HAV-HEV), pp. 1530–1540, In: *Encyclopedia of Environmental Microbiology*, G. Bitton, editor-in-chief, Wiley-Interscience, N.Y.

Myler, C.A., and W. Sisk. 1991. Bioremediation of explosive contaminated soils, pp. 137–146, In: *Environmental Biotechnology for Waste Treatment*, G.S. Sayler, R. Fox, and J.W. Blackburn. Eds., Plenum, New York.

Myoga, H., H. Asano, Y. Nomura, and H. Yoshida. 1991. Effect of immobilization on the nitrification treatability of entrapped cell reactors using the PVA freezing method. Water Sci. Technol. 23: 1117–1124.

Nadan, S., J.E. Walter, W.O. K.Grabow, D.K. Mitchell, and M.B. Taylor. 2003. Molecular characterization of astroviruses by reverse transcriptase PCR and sequence analysis: Comparison of clinical and environmental isolates from South Africa. Appl. Environ. Microbiol. 69: 747–753.

Nagy L.A., and B.H. Olson. 1982. The occurrence of filamentous fungi in drinking water distribution systems. Can. J. Microbiol. 28: 667–671.

Najm, I., and R.R. Trussel. 2001. NDMA formation in water and wastewater. J. AWWA 93 (2): 92–99.

Najm, I.M., V.M. Snoeyink, B.W. Lykins, Jr., and J.Q. Adams. 1991. Using powdered activated carbon: A critical review. J. Am. Water Works. Assoc. 83: 65–76.

Nakae, T. 1986. Outer membrane permeability of bacteria. Crit. Rev. Microbiol. 13: 1–62.

Nakamoto, S., and N. Machida. 1992. Phenol removal from aqueous solutions by peroxidase-catalyzed reaction using additives. Water Res. 26: 49–54.

Nakamura, K., M. Shibata, and Y. Miyaji. 1989. Substrate affinity of oligotrophic bacteria in biofilm reactors. Water Sci. Technol. 21: 779–790.

Nakasaki, K., M. Shoda, and H. Kubota. 1985a. Effect of temperature on composting of sewage sludge. Appl. Environ. Microbiol. 50: 1526–1530.

Nakhforoosh, N., and J.B. Rose. 1989. Detection of *Giardia* with a gene probe (Abstr.), In: 89th Annual Meeting American Society Microbiology, New Orleans, LA, 14–18, 1989.

Namkung, E., R.G. Stratton, and B.E. Rittmann. 1983. Predicting removal of trace organic compounds by biofilms. J. Water Pollut. Control Fed. 55: 1366–1372.

Namkung, E., and B.E. Rittmann. 1987. Removal of taste- and odor-causing compounds by biofims grown on humic substances. J. Am. Water Works Assoc. 79: 107–112.

Narkis, N., and Y. Kott. 1992. Comparison between chlorine dioxide and chlorine for use as a disinfectant of wastewater effluents. Water Sci. Technol. 26: 1483–1492.

Narkis, N., A. Katz, F. Orshansky, Y. Kott, and Y. Friedland. 1995. Disinfection of effluents by combinations of chlorine dioxide and chlorine. Water Sci. Technol. 31 (5–6): 105–114.

Nasser, A.M., Y. Tchorch, and B. Fattal. 1993. Comparative survival of *E. coli*, F^+ bacteriophages, HAV and poliovirus 1 in wastewaters and groundwaters. Water Sci. Technol. 27: 401–407.

Nasser, A.M., N. Zaruk, L. Tenenbaum, and Y. Netzan. 2003. Comparative survival of *Cryptosporidium*, coxsackievirus A9 and *Escherichia coli* in stream, brackish and sea waters. Water Sci. Technol. 47 (3): 91–96.

Nataro, J.P. and J.B. Kaper. 1998. Diarrheagenic *Escherichia coli*. Clin. Microbiol. Rev. 11: 142–201.

Nathanson, J.A. 1986. *Basic Environmental Technology: Water Supply, Waste Disposal and Pollution Control*, John Wiley & Sons, New York.

National Research Council. 1979. *Hydrogen Sulfide*. Report by Committee on Medical and Biologic Effects of Environmental Pollutants, Division of Medical Sciences., National Research Council, Washington, D.C.

National Research Council. 1998. *Issues in Potable Reuse: The viability of Augmenting Drinking Water Supplies with Reclaimed Water*. National Academy Press, Washington, D.C.

National Water Research Institute/American Water Works Association Research Foundation. 2000. Ultraviolet disinfection guidelines for drinking water and water reuse. National Water Research Institute, Fountain Valley, CA.

Nazaly, N., and C.J. Knowles. 1981. Cyanide degradation by immobilized fungi. Biotech. Lett. 3: 363–368.

Neef, A., A. Zaglauer, H. Meier, R. Amann, H. Lemmer, and K.-H. Schleifer. 1996. Population analysis in a denitrifying sand filter: Conventional and *in situ* identification of *Paracoccus* spp. in methanol-fed biofilms. Appl. Environ. Microbiol. 62: 4329–4339.

Negulescu, M. 1985. *Municipal Wastewater Treatment*. Elsevier, Amsterdam.

Neidhardt, F.R. Van Bogelen, and V. Vaughn. 1984. The genetics and regulation of heat-shock proteins. Annu. Rev. Genet. 18: 295–329.

Neilson, A.H., A.S. Allard, and M. Remberger. 1985. Biodegradation and transformation of recalcitrant compounds, In: *Handbook of Environmental Chemistry*, O. Hutzinger, Ed., Springer, New York.

Nellor, M.H., R.B. Baird, and J.R. Smyth. 1985. Health effects of indirect potable water reuse. J. Am. Water Works Assoc. 77: 88–96.

Nelson, K.L., B.J. Cisneros, G. Tchobanoglous, and J.L. Darby. 2004. Sludge accumulation, characteristics, and pathogen inactivation in four primary waste stabilization ponds in central Mexico. Water Res. 38: 111–127.

Nelson, P.O., and A.W. Lawrence. 1980. Microbial viability measurements and activated sludge kinetics. Water Res. 14: 217–225.

Nelson, P.O., A.K. Chung, and M.C. Hudson. 1981. Factors affecting the fate of heavy metals in the activated sludge process. J. Water Pollut. Control Fed. 53: 1323–1333.

Nelson, S.M., and R.A. Roline. 1998. Evaluation of the sensitivity of rapid toxicity tests relative to daphnid acute lethality tests. Bull. Environ. Contam. Toxicol. 60: 292–299.

Nelson, T.C., J.Y.C. Huang, and D. Ramaswami. 1988. Decomposition of exopolysaccharide slime by a bacteriophage enzyme. Water Res. 22: 1185–1188.

Nerenberg, R., B.E. Rittmann, and W.J. Soucie. 2000. Ozone/biofiltration for removing MIB and Geosmin. J. AWWA 92 (12): 85–95.

Nercessian, D.M. Upton, D. Loyd, and C. Edwards. 1999. Phylogenetic analysis of peat bog methanogen populations. FEMS Microbiol. Lett. 173: 425–429.

Neu, H.C. 1992. The crisis of antibiotic resistance. Science 257: 1064–1072.

Neu, T.R., and J.R. Lawrence. 2002. Laser scanning microscopy in combination with fluorescence techniques for biofilm study, pp. 1772–1788, In: *Encyclopedia of Environmental Microbiology*, G. Bitton, editor-in-chief, Wiley-Interscience, N.Y.

Neufield, R.D. 1976. Heavy metal induced deflocculation of activated sludge. J. Water Pollut. Control Fed. 48: 1940–1947.

Neufield, R.D., and E.R. Hermann. 1975. Heavy metal removal by activated sludge. J. Water Pollut. Control Fed. 47: 310–329.

Neumeister, B., S. Schoniger, M. Faigle, M. Eichner, and K. Dietz. 1997. Multiplication of different *Legionella* species in mono Mac 6 cells and in Acanthamoeba castellani. Appl. Environ. Microbiol. 63: 1219–1224.

Nevo, Z., and R. Mitchell. 1967. Factors affecting biological clogging of sand associated with ground water recharge. Water Res. 1: 231–236.

Newman, D.K., T.J. Beveridge, and F.M.M. Morel. 1997. Precipitation of arsenic trisulfide by *Desulfotomaculum auripigmentum*. Appl. Environ. Microbiol. 63: 2022–2028.

Newsome, A.L., R.L. Baker, R.D. Miller, and R.R. Arnold. 1985. Interactions between *Naegleria fowleri* and *Legionella pneumophila*. Infect. Immun. 50: 449–452.

Nicell, J.A., J.K. Bewtra, K.E. Taylor, N. Biswas, and C. St. Pierre. 1992. Enzyme catalyzed polymerization and precipitation of aromatic compounds from wastewater. Water Sci. Technol. 25: 157–164.

Nichols, A.B. 1988. Water reuse closes water-wastewater loop. J. Water Pollut. Control Fed. 60: 1931–1937.

Nichols, P.D., and C.A. Mancuso Nichols. 2002. Archaea: Detection methods, pp. 246–259, In: *Encyclopedia of Environmental Microbiology*, G. Bitton, editor-in-chief, Wiley-Interscience, Hoboken, N.J.

Nicholson, D.K., S.L. Woods, J.D. Istok, and D.C. Peek. 1992. Reductive dechlorination of chlorophenols by a pentachlorophenol-acclimated methanogenic consortium. Appl. Environ. Microbiol. 58: 2280–2286.

Nicholson, W.L., and B. Galeano. 2003. UV resistance of *Bacillus anthracis* spores revisited: Validation of *Bacillus subtilis* spores as UV surrogates for spores of *B. anthracis* sterne. Appl. Environ. Microbiol. 69: 1327–1330.

Nicks, O.W. 1986. Conceptual design for a food production, water, and waste processing, and gas regeneration module. Progress Report, Contract No. NAG-9-161. Regenerative Concept Group, NASA, Johnson Space Center, Houston.

Niederwohrmeier, B., R. Bohm, and D. Strauch. 1985. Microwave treatment as an alternative pasteurization process for the disinfection of sewage sludge: Experiments with the treatment of liquid manure, pp. 135–147, In: *Inactivation of Microorganisms in Sewage Sludge by Stabilization Processes*, Strauch, D., A.H. Havelaar, and P. L'Hermite, Eds., Elsevier, London.

van Niekerk, A.M., D. Jenkins, and M.G. Richard. 1987. The competitive growth of *Zooglea ramigera* and Type 021N in activated sludge and pure culture. A model for low F/M bulking. J. Water Pollut. Control Fed. 59: 262–273.

Nielsen, P.H. 2002. Activated sludge – The floc, In: *Encyclopedia of Environmental Microbiology*, G. Bitton, editor-in-chief, Wiley-Interscience, Hoboken, N.J.

Nielsen, K.M., M.D. van Weerelt, T.N. Berg, A.M. Bones, A.N. Hagler, and J.D. van Elsas. 1997. Natural transformation and natural availability of transforming DNA to *Acinetobacter calcoaceticus* in soil microcosms. Appl. Environ. Microbiol. 63: 1945–1952.

Niemi, R.M., S. Knuth, and K. Lundstrom. 1982. Actinomycetes and fungi in surface waters and in potable water. Appl. Environ. Microbiol. 43: 378–388.

Nieminski, E.C. 2002. Aerobic endospores, pp. 140–148, In: *Encyclopedia of Environmental Microbiology*, G. Bitton, editor-in-chief, Wiley-Interscience, N.Y.

Nieminski, E.C., W.D. Bellamy, and L.R. Moss. 2000. Using surrogates to improve plant performance. J. Am. Water Works Assoc. 92 (3): 67–78.

Nilsson, I., and S. Ohlson. 1982. Columnar denitrification of water by immobilized *Pseudomonas denitrificans* cells. Eur. J. Appl. Microbiol. Biotechnol. 14: 86–90.

Niquette, P., P. Servais, and R. Savoir. 2000. Impacts of pipe materials on densities of fixed bacterial biomass in a drinking water distribution system. Water Res. 34: 1952–1956.

Nitisoravut, S., and P.Y. Yang. 1992. Denitrification of nitrate-rich water using entrapped-mixed-microbial cells immobilization technique. Water Sci. Technol. 26: 923–931.

Noble, P.A., H.F. Ridgway, and B.H. Olson. 1994. Incorporation of the luciferase genes into *Pseudomonas fluorescens* strain P17: Development of a bioluminescent sensor for assimilable organic carbon (Abstr. Q-351), In: 94th. Meeting American Society Microbiology, Las Vegas, Nevada, May 23–27, 1994.

Noble, R.T., and J.A. Fuhrman. 1997. Virus decay and its causes in coastal waters. Appl. Environ. Microbiol. 63: 77–83.

Noble, R.T., I.M. Lee, and K.C. Schiff. 2004. Inactivation of indicator microorganisms from various sources of faecal contamination in seawater and freshwater. J. Appl. Microbiol. 96: 464–472.

Nogueira, R., L.F. Melo, U. Purkhold, S. Wuertz, and M. Wagner. 2002. Nitrifying and heterotrophic population dynamics in biofilm reactors: effects of hydraulic retention time and the presence of organic carbon. Water Res. 36: 469–481.

Norberg, A.B., and S.-O. Enfors. 1982. Production of extracellular polysaccharide by *Zooglea ramigera*. Appl. Environ. Microbiol. 44: 1231–1237.

Norberg, A.B., and H. Persson. 1984. Accumulation of heavy metals ions by *Zooglea ramigera*. Biotech. Bioeng. 26: 239–246.

Norberg, A.B., and S. Rydin. 1984. Development of a continuous process for metal accumulation by *Zooglea ramigera*. Biotech. Bioeng. 26: 265–268.

Norton, C.D., and M.W. LeChevallier. 1997. Chloramination: its effect on distribution system water quality. J. Am. Water Works Assoc. 89: 66–77.

Norton, C.F. 1986. *Microbiology*. 2nd Ed., Addison-Wesley, Reading, MA, 860 pp.

Noss, C.I., and V.P. Olivieri. 1985. Disinfecting capabilities of oxychlorine compounds. Appl. Environ. Microbiol. 50: 1162–1164.

Noss, C.I., F.S. Hauchman, and V.P. Olivieri. 1986. Chlorine dioxide reactivity with proteins. Water Res. 20: 351–356.

Novak, J.T. 2001. The effect of ammonium ion on activated sludge settling properties. Water Environ. Res. 73: 409–414.

Nowak, G., and G.D. Brown. 1990. Characteristics of *Nostocoida lumicola* and its activity in activated sludge suspension. J. Water Pollut. Control Fed. 62: 137–142.

Nowak, G., G. Brown, and A. Yee. 1986. Effect of feed pattern and dissolved oxygen on growth of filamentous bacteria. J. Water Pollut. Control Fed. 58: 978–984.

Nuanualsuwan, S., and D.O. Cliver. 2003. Infectivity of RNA from inactivated Poliovirus. Appl. Environ. Microbiol. 69: 1629–1632.

Nuhoglu, A., T. Pekdemir, E. Yildiz, B. Keskinler, and G. Akay. 2002. Drinking water denitrification by a membrane bio-reactor. Water Res. 36: 1155–1166.

Oates, P.M., P. Shanahan, and M.F. Polz. 2003. Solar disinfection (SODIS): simulation of solar radiation for global assessment and application for point-of-use water treatment in Haiti. Water Res. 37: 47–54.

Obiri-Danso, K., and K. Jones. 2000. Intertidal sediments as reservoirs for hippurate negative campylobacters, salmonellae and faecal indicators in three EU recognised bathing waters in North West England. Water Res. 34: 519–527.

O'Brien, R.T., and J. Newman. 1979. Structural and compositional changes associated with chlorine inactivation of polioviruses. Appl. Environ. Microbiol. 38: 1034–1039.

O'Brien, S.J., R.T. Mitchell, I.A. Gillespie, and G.K. Adak. 2000. The microbiological status of ready to eat fruit and vegetables. Discussion paper: Advisory Committee on the Microbiological Safety of Food. ACM/476, pp. 1–34.

Obst, U., A. Holzapfel-Pschorn, and M. Wiegand-Rosinus. 1988. Application of enzyme assays for toxicological water testing. Toxicity Assess. 3: 81–91.

O'Connor, J.T., L. Hash, and A.B. Edwards. 1975. Deterioration of water quality in distribution systems. J. Am. Water Works Assoc. 67: 113–116.

Oda, M., M. Morita, H. Unno, and Y. Tanji. 2004. Rapid Detection of *Escherichia coli* O157:H7 by Using Green Fluorescent Protein-Labeled PP01 Bacteriophage. Appl. Environ. Microbiol. 70: 527–534.

Odeymi, O. 1990. Use of solar radiation for drinking water disinfection in West Africa (Abstr.), In: International Symposium Health-Related Microbiology, Tubingen, Germany, April 1–6, 1990.

Odom, J.M. 1990. Industrial and environmental concerns with sulfate reducing bacteria. ASM News 56: 473–476.

Office of Technology Assessment. 1984. Protecting the nation's groundwater from contamination. OTA-0-233. U.S. Congress, Washington, D.C.

O'Grady, D.P., P.H. Howard, and A.F. Werner. 1985. Activated sludge biodegradation of 12 commercial phtalate esters. Appl. Environ. Microbiol. 49: 443–445.

Ohtake, H., and Hardoyo. 1992. New biological method for detoxification and removal of hexavalent chromium. Water Sci. Technol. 25: 395–402.

Ohtake, H., K. Takahashi, Y. Tsuzuki, and K. Toda. 1985. Uptake and release of phosphate by a pure culture of *Acinetobacter calcoaceticus.* Water Res. 19: 1587–1594.

Okabe, S., H. Naitoh, H. Satoh, and Y. Watanabe. 2002. Structure and function of nitrifying biofilms as determined by molecular techniques and the use of microelectrodes. Water Sci. & Technol. 46 (1–2): 233–241.

Okamoto, K., Y. Yamamoto, H. Tanaka, M. Tanaka, and A. Itaya. 1985. Heterogenous photocatalytic decomposition of phenol over TiO_2 powder. Bull. Chem. Soc. Jpn. 58: 2015–2022.

O'Keefe, B., and J. Green. 1989. Coliphages as indicators of fecal pollution at three recreational beaches on the firth and forth. Water Res. 23: 1027–1030.

Okun, D.A. 1997. Distributing reclaimed water through dual systems. J. Am. Water Works Assoc. 89: 52–64.

Oldenhuis, R., J.Y. Oedzes, J.J. van der Waarde, and D.B. Janssen. 1991. Kinetics of chlorinated hydrocarbon degradation by *Methylosinus trichosporium* OB3b and toxicity of trichloroethylene. Appl. Environ. Microbiol. 57: 7–14.

Olenchock, S.A. 1997. Airborne endotoxin, pp. 661–665, In: *Manual of Environmental Microbiology*, C.J. Hurst, G.R. Knudsen, M.J. McInerney, L.D. Stetzenbach, and M.V. Walter, Eds., ASM Press, Washington, D.C.

Olive, D.M. 1989. Detection of enterotoxigenic *Escherichia coli* after polymerase chain reaction amplification with a thermostable DNA polymerase. J. Clin. Microbiol. 27: 261–265.

Oliver, B.G., and J.H. Carey. 1976. Ultraviolet disinfection, an alternative to chlorination. J. Water Pollut. Control Fed. 48: 2619–2627.

Oliver, B.G., and E.G. Cosgrove. 1977. The disinfection of sewage treatment plant effluents using ultraviolet light. Can J. Chem. Eng. 53: 170–174.

Oliver, B.G., and D.B. Shindler. 1980. Trihalomethanes from the chlorination of aquatic algae. Environ. Sci. Technol. 14: 1502–1505.

Olivieri, V.P. 1983. Measurement of microbial quality, In: *Assessment of Microbiology and Turbidity Standards For Drinking Water*, P.S. Berger, and Y. Argaman, Eds., EPA Report No. EPA 570-9-83-001, Office of Drinking Water, Washington, D.C.

Olivieri, V.P., W.H. Dennis, M.C. Snead, D.R. Richfield, and C.W. Kruse. 1980. Reaction of chlorine and chloramines with nucleic acids under disinfection conditions, In: *Water Chlorination: Environmental Impacts and Health Effects*, Vol. 3, R.L. Jolley, W.A. Brungs, and R.B. Cumming, Eds., Ann Arbor Science Publishers, Ann Arbor, MI.

Ollis, D.F. 1985. Contaminant degradation in water. Environ. Sci. Technol. 19: 480–484.

Ollos, P.J., P.M. Huck, and R.M. Slawson. 2003. Factors affecting biofilm accumulation in model distribution systems. J. Amer. Water Works Assoc. 95: 87–97.

O'Malley, M.L., D.W. Lear, W.N. Adams, J. Gaines, T.K. Sawyer, and E.J. Lewis. 1982. Microbial contamination of continental shelf sediments by wastewater. J. Water Pollut. Control Fed. 54: 1311–1317.

Olson, B.H. 1991. Tracking and using genes in the environment. Environ. Sci. Technol. 25: 604–611.

Olson, B.H., and R.A. Goldstein. 1988. Applying genetic ecology to environmental management. Environ. Sci. Technol. 22: 370–372.

Olson, B.H., R. McCleary, and J. Meeker. 1991. Background and models for bacterial biofilm formation and function in water distribution systems, pp. 255–285, In: *Modeling the Environmental Fate of Microorganisms*, C.J. Hurst, Ed., American Society of Microbiology, Washington, D.C.

Olson, B.H., and L.A. Nagy. 1984. Microbiology of potable water. Adv. Appl. Microbiol. 30: 73–132.

Olofsson, A.-C., A. Zita, and M. Hermansson. 1998. Floc stability and adhesion of green-fluorescent-protein-marked bacteria to flocs in activated sludge. Microbiology 144: 519–528.

O'Malley, M.L., D.W. Lear, W.N. Adams, J. Gaines, T.K. Sawyer, and E.J. Lewis. 1982. Microbial contamination of continental shelf sediments by wastewater sludge. J. Water Poll. Control Fed. 54: 1311–1317.

Omura, T., M. Onuma, J. Aizawa, T. Umita, and T. Yagi. 1989. Removal efficiencies of indicator microorganisms in sewage treatment plants. Water Sci. Technol. 21: 119–124.

Ongerth, J.E. and H.H. Stibbs. 1987. Identification of *Cryptosporidium* oocygsts in river water. Appl. Environ. Microbiol. 53: 672–676.

Ongerth, H.J., and W.F. Jopling. 1977. Water reuse in California, pp. 219–256, In: *Water Renovation and Reuse*, H.I. Shuval, Ed., Academic Press, New York.

Ongerth, J.E., 1990. Evaluation of treatment for removing *Giardia* cysts. J. Am. Water Works Assoc. 82: 85–96.

Ongerth, J.E., and P.E. Hutton. 2001. Testing diatomaceous earth for removal of *Cryptosporidium* oocycts. J. AWWA 93 (12): 54–63.

Ongerth, J.E., and S. Khan. 2004. Drug residuals: How xenobiotics can affect water supply sources. J. Amer. Water. Works Assoc. 96 (5): 94–101.

Ongerth, J.E., and H.H. Stibb. 1987. Identification of Cryptosporidium oocysts in river water. Appl. Environ. Microbiol. 53: 672–676.

Ono, K., H. Tsuji, S.K. Rai, A. Yamamoto, K. Masuda, T. Endo, H. Hotta, T. Kawamura, and S.I Uga. 2001. Contamination of river water by *Cryptosporidium parvum* oocysts in western Japan. Appl. Environ. Microbiol. 67: 3832–3836.

Oppenheimer, J.A., J.G. Jacangelo, J.-M. Laine, and J.E. Hoagland. 1997. Testing the equivalency of ultraviolet light and chlorine for disinfection of wastegater to reclamation standards. Water Environ. Res. 69: 14–24.

O'Reilly, K.T., R. Kadakia, R.A. Korus, and R.L Crawford. 1988. Utilization of immobilized bacteria to degrade aromatic compounds common to wood-treatment wastewaters, In: *Internatinal Conference Water and Wastewater Microbiology*, Newport Beach, CA, Feb. 8–11, 1988.

Oremland, R.S. 1988. Biogeochemistry of methanogenic bacteria, pp. 641–705, In: *Biology of Anaerobic Microorganisms*, A.J.B. Zehnder, Ed., John Wiley & Sons, New York.

Oremland, R.S., and S. Polcin. 1982. Methanogenesis and sulfate reduction: competitive and noncompetitive substrates in estuarine sediments. Appl. Environ. Microbiol. 44: 1270–1276.

Oron, G., Y. DeMalach, Z. Hoffman, and R. Cibotaru. 1991a. Subsurface microirrigation with effluent. J. Irrig. Drain. Eng. ASCE 117: 25–36.

Oron, G., Y. DeMalach, Z. Hoffman, Y. Keren, H. Hartmann, and N. Plazner. 1991b. Wastewater disposal by subsurface trickle irrigation. Water Sci. Technol. 23: 2149–2158.

Oron, G., M. Goemans, Y. Manor, and J. Feyen. 1995. Poliovirus distribution in the soil-plant system under reuse of secondary wastewater. Water Res. 29: 1069–1078.

Orsini, M., P. Laurenti, F. Boninti, D. Arzani, A. Lanni, and V. Romano-Spica. 2002. A molecular typing approach for evaluating bioaerosol exposure in wastewater treatment plant workers. Water Res. 36: 1375–1378.

Ortega, Y.R., V.A. Cama, and A.B. Prisma. 2002. Cyclospora: Basic biology, occurrence, fate and methodologies, pp. 995–1001, In: *Encyclopedia of Environmental Microbiology*, G. Bitton, editor-in-chief, Wiley-Interscience, Hoboken, N.J.

Ortiz-Roque, C.M., and T.C. Hazen. 1987. Abundance and distribution of legionellaceae in puerto rican water. Appl. Environ. Microbiol. 53: 2231–2236.

Oskam, G. 1995. Main principles of water quality improvement in reservoirs. Aqua 44: 23–29.

Oste, C. 1988. Polymerase chain reaction. Biotechniques 6: 162–167.

Ottolenghi, A.C., and V.V. Hamparian. 1987. Multiyear study of sludge application to farmland: Prevalence of bacterial enteric pathogens and antibody status of farm families. Appl. Environ. Microbiol. 53: 1118–1124.

Ou, C.Y., S. Kwok, S.W. Mitchell, D.H. Mack, J.J. Sninsky, J.W. Krebs, P. Feorino, D. Warfield, and G. Schochetman. 1988. DNA amplification for direct detection of HIV-1 in DNA of peripheral blood mononuclear cells. Science 239: 295–297.

Ouellette, R.P. 1991. A perspective on water pollution. Nat. Environ. J. 1: 20–24.

Oufdou, K. 1994. Etude de la dynamique et de la survie de *V. cholerae*, *E. coli* et de *P. aeruginosa* au cours d'un traitement des eaux usees par lagunage naturel sous climat aride a Marrakech. DES Thesis, Universite Cadi Ayyad, Marrakech, Morocco.

van Overbeek. 1998. *Responses of Bacterial Inoculants to Soil Conditions*. Ph.D. Thesis, Leiden, The Netherlands.

van Overbeek, L.S., and D. van Elsas. 2002. Genetically modified microorganisms (GMM) in soil environments, pp. 1429–1440, In: *Encyclopedia of Environmental Microbiology*, G. Bitton, editor-in-chief, Wiley-Interscience, N.Y.

Owen, W.F., D.C. Stuckey, J.B. Healy, Jr., L.Y. Young, and P.L. McCarty. 1979. Bioassay for monitoring biochemical potential and anaerobic toxicity. Water Res. 13: 485–492.

Ozaki, H., Z. Liu, and Y. Terashima. 1991. Utilization of microorganisms immobilized with magnetic particles for sewage and wastewater treatment. Water Sci. Technol. 23: 1125–1136.

Pagilla, K.R., A. Sood, and H. Kim. 2002. *Gordonia* (*Nocardia*) *amarae* foaming due to biosurfactant production. Water Sci. & Technol. 46 (1–2): 519–524.

Painter, H.A. 1970. A review of literature on inorganic nitrogen metabolism in microorganisms. Water Res. 4: 393–450.

Painter, H.A., and J.E. Loveless. 1983. Effect of temperature and pH value on the growth-rate constants of nitrifying bacteria in the activated sludge process. Water Res. 17: 237–248.

Painter H.A., and M. Viney. 1959. Composition of domestic sewage. J. Biochem. Microbiol. Technol. 1: 143–162.

Palchak, R.B., R. Cohen, M. Ainslie, and C. Lax Hoerner. 1988. Airborne endotoxin associated with industrial-scale production of protein products in gram-negative bacteria. Amer. Ind. Hyg. Assoc. 49: 420–421.

Palm, J.C., D. Jenkins, and D.S. Parker. 1980. Relationship between organic loading, dissolved oxygen concentration and sludge settleability in the completely-mixed activated sludge process. J. Water Pollut. Control Fed. 52: 2484–2506.

Palmer, C.J., Y.-L. Tsai, A.L. Lang, and L.R. Sangermano. 1993. Evaluation of Colilert-Marine Water for detection of total coliforms and *Escherichia coli* in the marine environment. Appl. Environ. Microbiol. 59: 786–790.

Palmer, S.R., P.R. Gully, J.M. White, A.D. Pearson, W.G. Suckling, D.M. Jones, J.C.L. Rawes, and J.L. Penner. 1983. Waterbborne outbreak of *Campylobacter* gastroenteritis. Lancet i: 287–290.

Palmgren, U., G. Strom, G. Blomquist, and P. Malmberg. 1986. The Nucleopore filter method: A technique for enumeration of viable and nonviable airborne microorganisms. Am. J. Ind. Med. 10: 325–327.

Pancorbo, O.C., G. Bitton, S.R. Farrah, G.E. Gifford, and A.R. Overman. 1988. Poliovirus retention in soil columns after application of chemical- and polyelectrolyte-conditioned dewatered sludges. Appl. Environ. Microbiol. 54: 118–123.

Panicker, P.V., and K.P. Krishnamoorthi. 1978. Elimination of enteric parasites during sewage treatment. Indian Assoc. Water Pollut. Control Tech. Annu. 5: 130–138 (cited by Feachem et al., 1983).

Panicker, P.V., and K.P. Krishnamoorthi. 1981. Parasite egg and cyst reduction in oxidation ditches and aerated lagoons. J. Water Pollut. Control Fed. 53: 1413–1419.

Parhad, N.M., and N.U. Rao. 1974. Effect of pH on survival of *Escherichia coli*. J. Water Pollut. Control Fed. 46: 980–986.

Park., N., T.N. Blandford, M.Y. Corapcioglu, and P.S. Huyakorn. 1990. VIRALT: A modular semi-analytical and numerical model for simulating viral transport in ground water. Office of Drinking Water, U.S. Environmental Protection Agency, Washington, D.C.

Park, S. 2002. Campylobacter jejuni and other enteric campylobacters, pp. 803–810, In: *Encyclopedia of Environmental Microbiology*, G. Bitton, editor-in-chief, Wiley-Interscience, N.Y.

Park, S.R., W.G. Mackay, and D.C. Reid. 2001. *Helicobacter* sp. recovered from drinking water biofilm sampled from a water distribution system. Water Res. 35: 1624–1626.

Parker, D.S., D. Jenkins, and W.J. Kaufman. 1971. Physical conditioning of the activated sludge floc. J. Water Pollut. Control Fed. 43: 1897.

Parker, D.S., and T. Richards. 1986. Nitrification in trickling filters. J. Water Pollut. Control Fed. 58: 896–902.

Parker, W.J., D.J. Thompson, J.P. Bell, and H. Melcer. 1993. Fate of volatile organic compounds in municipal activated sludge plants. Water Environ. Res. 65: 58–65.

Parkin, G.F., and R.E. Speece. 1982. Modeling toxicity in methane fermentation systems. J. Environ. Eng. Div. ASCE 108: 515–531.

Parkin, G.F., R.E. Speece, C.H.J. Yang, and W.M. Kocher. 1983. Response of methane fermentation systems to industrial toxicants. J. Water Pollut. Control Fed. 55: 44–53.

Parkinson, A., M.J. Barry, A., F.A. Roddick, and M.D. Hobday. 2001. Preliminary toxicity assessment of water after treatment with UV-irradiation and UVC/H_2O_2. Water Res. 35: 3656–3664.

Parrotta, M.J., and F. Bekdash. 1998. UV disinfection of small groundwater supplies. J. Am. Water Works Assoc. 90: 71–81.

Parshionikar, P.U., S. Willian-True, G.S. Fout, D.E. Robbins, S.A. Seys, J.D. Cassady, and R. Harris. 2003. Waterborne Outbreak of Gastroenteritis Associated with a Norovirus Appl. Environ. Microbiol. 69: 5263–5268.

Parveen, S., R.L. Murphree, L. Edmiston, C.W. Kaspar, K.M. Portier, and M.L. Tamplin. 1997. Association of multiple-antibiotic-resistance profiles with point and nonpoint sources of *Escherichia coli* in Apalachicola Bay. Appl. Environ. Microbiol. 63: 2607–2612.

Pasquill, F. 1961. The estimation of the dispersion of windborne material. Meteor. Mag. 90: 33–49.

Patel, G.B., B.J. Agnew, and C.J. Dicaire. 1991. Inhibition of pure culture of methanogens by benzene ring compounds. Appl. Environ. Microbiol. 57: 2969–2974.

Patrick, M.E., P.M. Adcock, T.M. Gomez, S.F. Altekruse, B.H. Holland, R.V. Tauxe, and D.L. Swerdlow. 2004. *Salmonella enteriditis* infections, United Sates, 1985–1999. Emerg. Infect. Dis. 10: 1–7.

Patterson, J.W. 1984. Perspectives on opportunities for genetic engineering applications in industrial pollution control, pp. 187–193, In: *Genetic Control of Environmental Pollutants*, G.S. Ommen, and A. Hollaender, Eds., Plenum, New York.

Patti, A.M., A.L. Santi, R. Gabrielli, S. Fiamma, M. Cauletti, and A. Pana. 1987. Hepatitis A virus and poliovirus 1 inactivation in estuarine water. Water Res. 21: 1335–1338.

Patureau, D., J. Davison, N. Bernet, and R. Moletta. 1994. Denitrification under various aeration conditions in *Comamonas* sp., strain SGLY2. FEMS Microbiol. Ecol. 14: 71–78.

Paul, E.A., and F.E. Clark. 1989. *Soil Microbiology and Biochemistry.* Academic, San Diego, CA.

Paul, J.H. 1993. The advances and limitations of methodology, pp. 15–46, In: *Aquatic Microbiology: An Ecological Approach*, T.E. Ford, Ed., Blackwell Sci. Pub., Oxford.

Paul, J.H., J.B. Rose, J. Brown, E.A. Shinn, S. Miller, and S.R. Farrah. 1995. Viral tracer studies indicate contamination of marine waters by sewage disposal practices in Key Largo, FL. Appl. Environ. Microbiol. 61: 2230–2234.

Pavlostathis, S.G., and S.K. Maeng. 2000. Fate and effect of silver on the anaerobic digestion process. Water Res. 34: 3957–3966.

Pavoni, J.L., M.W. Tenney, and W.F. Echelberger. 1972. Bacterial exocellular polymers and biological flocculation. J. Water Pollut. Control Fed. 44: 414–431.

Payment, P. 1989a. Elimination of viruses and bacteria during drinking water treatment: Review of 10 years of data from the Montreal metropolitan area, pp. 59–65, In: *Biohazards of Drinking Water Treatment*, R.A. Larson, Ed., Lewis, Chelsea, MI.

Payment, P. 1989b. Bacterial colonization of domestic reverse-osmosis filtration units. Can. J. Microbiol. 35: 1065–1067.

Payment, P. 1991. Fate of human enteric viruses, coliphages, and *Clostridium perfringens* during drinking-water treatment. Can. J. Microbiol. 37: 154–157.

Payment, P., and E. Franco. 1993. Clostridium perfringens and somatic coliphages as indicators of the efficiency of drinking water treatment for viruses and protozoan cysts. Appl. Environ. Microbiol. 59: 2418–2424.

Payment, P., E. Franco, L. Richardson, and J. Siemiatycki. 1991. Gastrointestinal health effects associated with the consumption of drinking water produced by point-of-use domestic reverse-osmosis filtration units. Appl. Environ. Microbiol. 57: 945–948.

Payment, P., E. Franco, and J. Siemiatycki. 1993. Absence of relationship between health effects due to tapwater consumption and drinking water quality parameters. Water Sci. Technol. 27: 137–143.

Payment, P., F. Gamache, and G. Paquette. 1989. Comparison of microbiological data from two water filtration plants and their distribution system. Water Sci. Technol. 21: 287–289.

Payment, P., A. Godfree, and D. Sartory. 2002. Clostridium, pp. 861–871, In: *Encyclopedia of Environmental Microbiology*, G. Bitton, editor-in-chief, Wiley-Interscience, N.Y.

Payment, P., J. Siemiatycki, L. Richardson, G. Renaud, E. Franco, and M. Prevost. 1997. A prospective epidemiological study of gastrointestinal health effects due to the consumption of drinking water. Int. J. Environ. Health Res. 7: 5–31.

Paxeus, N., P. Robinson, and P. Balmer. 1992. Study of organic pollutants in municipal wastewater in Goteborg, Sweden. Water Sci. Technol. 25: 249–256.

Pearson, H.W., D.D. Mara, S.W. Mills, and D.J. Sallman. 1987. Physicochemical parameters influencing faecal bacterial survival in waste stabilization ponds. Water Sci. Technol. 19: 145–152.

Peck, M.W. 1989. Changes in concentration of coenzyme F_{420} analogs during batch growth of *Metanosarcina barkeri* and *Methanosarcina mazei*. Appl. Environ. Microbiol. 55: 940–945.

Pedersen, D.C. 1981. *Density levels of pathogenic organisms in municipal wastewater sludge: A literature review.* Report No. EPA 600/2- 81-170, U.S Environmental Protection Agency, Cincinnati, OH.

Pedersen, D.C. 1983. Effectiveness of sludge treatment processes in reducing levels of bacteria, viruses, and parasites, pp. 9–31, In: *Biological Health Risks of Sludge Disposal to Land in Cold Climates*, P.M. Wallis, and D.L. Lehmann, Eds., University of Calgary Press, Calgary, Canada.

Pedersen, K. 1990. Biofilm development on stainless steel and PVC surfaces in drinking water. Water Res. 24: 239–243.

Peeters, J.E., E.A. Mazas, W.J. Masschelein, I.V. Martinez de Maturana, and E. Debacker. 1989. Effect of disinfection of drinking water with ozone or chlorine dioxide on survival of *Cryptosporidium parvum* oocysts. Appl. Environ. Microbiol. 55: 1519–1522.

Pelletier, P.A., and G.C. DuMoulin. 1988. Comparative resistance of mycobacteria to chloramine (Abst.), In: Annual Meeting American Society of Microbiology, Washington, D.C.

Peltier, W.H., and C.I. Weber. 1985. *Methods for Measuring the Acute Toxicity of Effluents to Freshwater and Marine Organisms* (3rd Ed.). Report No. EPA-600/4-85/013, U.S. Environmental Protection Agency, Cincinnati, OH.

Peng, M.M., L. Xiao, A.R. Freeman, M.J. Arrowood, A.A. Escalante, A.C. Weltman, C.S.L. Ong, W.R. MacKenzie, A.A. Lal, and C.B. Beard. 1997. Genetic polymorphism amomg *Cryptosporidium parvum* isolates: Evidence of two distinct human transmission cycles. Emerg. Infect. Dis. 3: 567–573.

Peng, Y., C. Gao, S. Wang, M. Ozaki, and A. Takigawa. 2003. Non-filamentous sludge bulking caused by a deficiency of nitrogen in industrial wastewater treatment. Water Sci. Technol. 47 (11): 289–295.

Perkins, J., and C. Hunter. 2000. Removal of enteric bacteria in a surface flow constructed wetland in Yorkshire, England. Water Res. 34: 1941–1947.

Persson, F., T.W., F. Sörensson, and M. Hermansson. 2002. Distribution and activity of ammonia oxidizing bacteria in a large full-scale trickling filter. Water Res. 36: 1439–1448.

Persson, P.E. 1979. Notes on muddy odour. III. Variability of sensory response to 2-methylisoborneol. Aqua Fenn. 9: 48–52.

Pescod, M.B., and J.V. Nair. 1972. Biological disk filtration for tropical waste treatment: Experimental studies. Water Res. 6: 1509–1523.

Petrilli, F.L., G.P. DeRenzi, P. Orlando, and S. DeFlora. 1980. Microbiological evaluation of coastal water in the Tyrrhenian Sea. Prog. Water Technol. 12: 129–136.

Petterson, N.S.R., N.J. Ashbolt, and A. Sharma. 2001. Microbial risks from wastewater irrigation of salad crops: A screening-level risk assessment. Water Environ. Res. 73: 667–672.

Pfenning, N. 1978. General physiology and ecology of photosynthetic bacteria, pp. 3–18, In: *The Photosynthetic Bacteria*, R.K. Clayton, and W.R. Sistrom, Eds., Plenum, New York.

Pfiffner, S.M., A.V. Palumbo, T.J. Phelps, and T.C. Hazen. 1997. Effects of nutrient dosing on subsurface methanotrophic populations and trichloroethylene degradation. J. Ind. Microbiol. & Biotechnol. 18: 204–212.

Pfuderer, G. 1985. Influence of lime treatment of raw sludge on the survival of pathogens, on the digestability of the sludge and on the production of methane: Hygienic investigations, pp. 85–97, In: *Inactivation of Microorganisms in Sewage Sludge by Stabilization Processes.* Strauch, D., A.H. Havelaar, and P. L'Hermite, Eds., Elsevier, London.

Phae, C.-G., and M. Shoda. 1991. A new fungus which degrades hydrogen sulfide, methanethiol, dimethyl sulfide and dimethyl disulfide. Biotechnol. Let. 11: 375–380.

Phelps, P.A., S.K. Agarwal, G.E. Speitel, Jr., and G. Giorgiou. 1992. *Methylosinus trichosporium* OB3b mutants having constitutive expression of soluble methane monooxygenase in the presence of high levels of copper. Appl. Environ. Microbiol. 58: 3701–3708.

Phelps, T.J., J.J. Niedzielski, R.M. Schram, S.E. Herbes, and D.C. White. 1990. Biodegradation of trichloroethylene in continuous-recycle expanded-bed biorectors. Appl. Environ. Microbiol. 56: 1701–1709.

Phelps, T.J., J.J. Niedzielski, K. Malachowski, R.M. Schram, S.E. Herbes, and D.C. White. 1991. Biodegradation of mixed-organic wastes by microbial consortia in continuous-recycle expanded-bed bioreactors. Environ. Sci. Technol. 25: 1461–1465.

Pietronave, S., L. Fracchia, M. Rinaldi, and M.G. Martinotti. 2004. Influence of biotic and abiotic factors on human pathogens in a finished compost. Water Res. 38: 1963–1970.

Phillips, S.J., D.S. Dalgarn, and S.K. Young. 1989. Recombinant DNA in wastewater: pBR322 degradation kinetics. J. Water Pollut. Control Fed. 61: 1588–1595.

Phillipsa, C.J., E.A. Paul, and J.I. Prossera. 2000. Quantitative analysis of ammonia oxidizing bacteria using competitive PCR. FEMS Microbiol. Ecol. 32: 167–175.

Pietri, Ch., and J.-Ph. Breittmayer. 1976. Etude de la survie d'un enterovirus en eau de mer. Rev. Int. Oceanog. Med. 42: 77–86.

Pickup, R.W. 1991. Development of molecular methods for the detection of specific bacteria in the environment. J. Gen. Microbiol. 137: 1009–1019.

Pike, E.B., E.G. Carrington, and S.A. Harman. 1988. Destruction of Salmonellae, enteroviruses and ova of parasites by pasteurization and anaerobic digestion. In: *International Conference on Water and Wastewater Microbiology*, Newport Beach, CA, Feb. 8–11, 1988.

Pilly, E. 1990. *Maladies infectieuses* (11th Ed). Editions C&R, La Madeleine, France, 645 pp.

Pintar, K.D.M., and Robin M. Slawson. 2003. Effect of temperature and disinfection strategies on ammonia-oxidizing bacteria in a bench-scale drinking water distribution system. Water Res. 37: 1805–1817.

Pinto, R.M., F.X. Abad, R. Gajardo, and A. Bosch. 1996. Detection of infectious astroviruses in water. Appl. Environ. Microbiol. 62: 1811–1813.

Piotrowski, M.R. 1989. Bioremediation: Testing the waters. Civil Eng. 59 (8): 51–53.

Pipes, W.O. 1974. Control bulking with chemicals. Water and Waste Engineering (Nov. 1974).

Pipes, W.O. 1978. Actinomycetes scum formation in activated sludge processes. J. Water Pollut. Control Fed. 5: 628–634.

Pipes, W.O., and W.B. Cooke. 1969. Purdue University #132 Proc. 23rd Ind. Conf. 53: 170–182.

Pipes, W.O., and J.T. Zmuda. 1997. Assessing the efficiency of wastewater treatment, pp. 231–242, In: *Manual of Environmental Microbiology*, C.J. Hurst, G.R. Knudsen, M.J. McInerney, L.D. Stetzenbach, and M.V. Walter, Eds., ASM Press, Washington, D.C.

Pipyn, P., W. Verstraete, and J.P. Ombregt. 1979. A pilot scale anaerobic upflow reactor treating distillery wastewaters. Biotech. Lett. 1: 495–500.

Pisarczyk, K.S., and L.A. Rossi. 1982. Sludge odor control and improved dewatering with potassium permanganate, In: 55th Annual Conference of the Water Pollution Control Federation, St. Louis, MO (Oct. 5, 1982).

Pitt, P. and D. Jenkins. 1990. Causes and control of *Nocardia* in activated sludge. J. Water Pollut. Control Fed. 62: 143–150.

Pizzi, N.G. 2002. *Water Treatment Operator Handbook*. Amer. Water Works Assoc., Denver, CO, 241 pp.

Placencia, A.M., J.T. Peeler, G.S. Oxborrow, and J.W. Danielson. 1982. Comparison of bacterial recovery by Reuter centrifugal air sampler and Slit-to-Agar sampler. Appl. Environ. Microbiol. 44: 512–513.

Plachy, P., I. Placha, and M. Vargova. 1995. Effect of physico-chemical parameters of sludge aerobic exothermic stabilization on the viability of *Ascaris suum* eggs. Helminthologia 32: 233–237.

Plissier, M., and P. Therre. 1961. Recherches sur l'inactivation *in vitro* du poliovirus dans l'eau de mer. Ann. Inst. Pasteur 101: 840–844.

Plovins, A., A.M. Alvarez, M. Ibanez, M. Molina, and C. Nombeda. 1994. Use of fluorescein-di-beta-D-galactopyranoside (FDG) and C sub(12)-FDG as substrates for beta-galactosidase detection by flow cytometry in animal, bacterial, and yeast cells. Appl. Environ. Microbiol. 60: 4638–4641.

Plummer, J.D., and J.K. Edzwald. 2001. Effect of ozone on algae as precursors for trihalomethane and haloacetic acid production. Environ. Sci. Technol. 35: 3661–3668.

Poggi, R. 1990. Impacts sanitaires des contaminations microbiologiques, In: *La mer et les rejets urbains*, IFREMER Proc. 11: 115–132 (Bendor, 13–15 Juin, 1990).

Pollard, P.C., and P.F. Greenfield. 1997. Measuring in situ bacterial specific growth rates and population dynamics in wastewater. Water Res. 31: 1074–1082.

Polprasert, C. 1989. *Organic Wastes Recycling*. John Wiley & Sons, Chichester, U.K. 357 pp.

Polprasert, C., M.G. Dissanayake, and N.C. Thanh. 1983. Bacterial die-off kinetics in waste stabilization ponds. J. Water Pollut. Control Fed. 55: 285–296.

Pomeroy, R.D. 1982. Biological treatment of odorous air. J. Water Pollut. Control Fed. 54: 1541–1545.

Pommepuy, M., M. Butin, A. Derrien, M. Gourmelon, R.R. Colwell, and M. Cormier. 1996. Retention of enteropathogenicity by viable but nonculturable *Escherichia coli* exposed to seawater and sunlight. Appl. Environ. Microbiol. 62: 4621–4626.

Pontius, F.W. 1992. A current look at the federal drinking water regulations. J. Am. Water Works Assoc. 84: 36–50.

Pontius, F.W. 2002. Regulatory compliance planning to ensure water supply safety. J. Water Works. Assoc. 94: 52–64.

Pontius, F.W. 2003. Update on USEPA's drinking water regulations. J. Water Works. Assoc. 95: 57–68.

Pope, R.J., and J.M. Lauria. 1989. Odors: The other effluent. Civil Eng. 59 (8): 42–44.

Pope, D.H., R.J. Soracco, H.K. Gill, and C.B. Fliermans. 1982. Growth of *Legionella pneumophila* in two-membered cultures with green algae and cyanobacteria. Curr. Microbiol. 7: 319–322.

Portier, R.J. 1986. Chitin immobilization systems for hazardous waste detoxification and biodegradation, pp. 229–244, In: *Immobilisation of Ions by Bio-Sorption*, H. Eccles, and S. Hunt, Eds., Ellis Horwood, Chichester, U.K.

Posch, T., J. Pernthaler, A. Alfreider, and R. Psenner. 1997. Cell-specific respiratory activity of aquatic bacteria studied with the tetrazolium reduction method, cyto-clear slides, and image analysis. Appl. Environ. Microbiol. 63: 867–873.

van Poucke, S.O., and H.J. Nelis. 1995. Development of a sensitive chemiluminometric assay for the detection of β-galactosidase in permeabilized coliform bacteria and comparison with fluorometry and colorimetry. Appl. Environ. Microbiol. 61: 4505–4509.

van Poucke, S.O., and H.J. Nelis. 1997. Limitations of highly sensitive presence-absence tests for detection of waterborne coliforms and *Escherichia coli*. Appl. Environ. Microbiol. 63: 771–774.

van Poucke, S.O., and H.J. Nelis. 2000. A 210-min solid phase cytometry test for the enumeration of *Escherichia coli* in drinking water. J. Appl. Microbiol. 89: 390–396.

Pougnard, C., Ph. Catala, J.-L. Drocourt, S. Legastelois, P. Pernin, E. Pringuez, and Ph. Lebaron. 2002. Rapid detection and enumeration of *Naegleria fowleri* in surface waters by solid-phase cytometry. Appl. Environ. Microbiol. 68: 3102–3107.

Pourcher, A.-M., L.A. Devriese, J.F. Hernandez, and J.M. Delattre. 1991. Enumeration by a miniaturized method of *Escherichia coli*, *Streptococcus bovis* and enterococci as indicators of the origin of faecal pollution of waters. J. Appl. Bacteriol. 70: 525–530.

Poynter, S.F.B., and J.S. Slade. 1977. The removal of viruses by slow sand filtration. Prog. Water Technol. 9: 75–78.

van Praagh, A.D., P.D. Gavaghan, and J.L. Sykora. 1993. *Giardia muris* cyst inactivation in anaerobic digester sludge. Water Sci. Technol. 27: 105–109.

Pretorius, W.A. 1971. Some operational characteristic of a bacterial disk unit. Water Res. 5: 1141–1146.

Prevot, J., S. Dubrou, and J. Marechal. 1993. Detection of human hepatitis A virus in environmental waters by an antigen-capture polymerase chain reaction method. Water Sci. Technol. 27: 227–233.

Prévost, M., A. Rompré, H. Baribeau, J. Coallier, and P. Lafrance. 1997. Service lines: their effect on microbiological quality. J. Am. Water Works. Assoc. 89: 78–91.

Price, G.J. 1982. Use of an anoxic zone to improve activated sludge settleability, In: *Bulking of Activated Sludge: Preventive and Remedial Methods*, B. Chambers, and E.J. Tomlinson, Eds., Ellis Horwood, Chichester, U.K.

ProMed-Mail. 2000. www.promedmail.org. *E. coli*, EHEC-Canada (ONT). 16 August, 2000.

Proulx, D., and J. de la Noue. 1988. Removal of macronutrients from wastewater by immobilized microalgae, pp. 301–310, In: *Bioreactor Immobilized Enzymes and Cells: Fundamentals and Applications*, M. Moo-Young, Ed., Elsevier, New York.

Pude, R.A., G.J. Jackson, J.W. Bier, T.K. Sawyer, and N.G. Risty. 1984. Survey of fresh vegetables for nematodes, amoebae and *Salmonella*. J. Assoc. Off. Anal. Chem. 67: 613–617.

van Puffelen, J. 1983. The importance of activated carbon, pp. 1–8, In: *Activated Carbon in Drinking Water Technology*, Res. Report, AWWA Res. Foundation, Denver, CO.

Pujol, R., Ph. Duchene, S. Schetrite, and J.P. Canler. 1991. Biological foams in activated sludge plants: Characterization and situation. Water Res. 25: 1399–1404.

Purdom, C.E., P.A. Hardiman, V.J. Bye, N.C. Eno, C.R. Tyler, and J.P. Sumper. 1994. Estrogenic effects of effluents from sewage treatment works. Chem. Ecol. 8: 275–285.

Pyle, B.H., S.C. Broadaway, and G.A. McFeters. 1995a. Factors affecting the determination of respiratory activity on the basis of cyanoditolyl tetrazolium chloride reduction with membrane filtration. Appl. Environ. Microbiol. 61: 4304–4309.

Pyle, B.H., S.C. Broadaway, and G.A. McFeters. 1995b. A rapid, direct method for enumerating respiring enterohemorrhagic *Escherichia coli* O157:H7 in water. Appl. Environ. Microbiol. 61: 2614–2619.

Qian, X., P.Y. Yang, and T. Maekawa. 2001. Evaluation of direct removal of nitrate with entrapped mixed microbial cell technology using ethanol as the carbon source. Water Environ. Res. 73: 584–589.

Qin, D., P.J. Bliss, D. Barnes, and P.A. Fitzgerald. 1991. Bacterial (total coliform) die off in maturation ponds. Water Sci. Technol. 23: 1525–1534.

Qualls, R.G., M.P. Flynn, and J.D. Johnson. 1983. The role of suspended particles in ultraviolet irradiation. J. Water Pollut. Control Fed. 55: 1280–1285.

Qualls, R.G., and J.D. Johnson. 1983. Bioassay and dose measurement in U.V. disinfection. Appl. Environ. Microbiol. 45: 872–877.

Qualls, R.G., S.F. Ossoff, J.C.H. Chang, M.H. Dorfman, C.M. Dumais, D.C. Lobe, and J.D. Johnson. 1985. Factors controlling sensitivity in ultraviolet disinfection of secondary effluents. J. Water Pollut. Control Fed. 57: 1006–1011.

Qualls, R.G., M.H. Dorfman, and J.D. Johnson. 1989. Evaluation of the efficiency of ultraviolet disinfection systems. Water Res. 23: 317–325.

Quan, X., H. Shi, H.J. Wang, and Y. Qian. 2003. Biodegradation of 2,4-dichlorophenol in sequencing batch reactors augmented with immobilized mixed culture. Chemosphere 50: 1069–1074.

Quan, X., H. Shi, H. Liu, P. Lv, and Y. Qian. 2004. Enhancement of 2,4-dichlorophenol degradation in conventional activated sludge systems bioaugmented with mixed special culture. Water Res. 38: 245–253.

Queiroz, A.P.S., F.M. Santos, A. Sassaroli, C.M. Hársi, T.A. Monezi, and D.U. Mehnert. 2001. Electropositive filter membrane as an alternative for the elimination of PCR inhibitors from sewage and water samples. Appl. Environ. Microbiol. 67: 4614–4618.

Quigley, C.J., and R.L. Corsi. 1995. Emissions of VOCs from a municipal sewer J. Air & Waste Management Assoc. 45: 395–403.

Quignon, F., M. Sardin, L. Kiene, and L. Schwartzbrod. 1997. Poliovirus 1 inactivation and interaction with biofilm: A pilot-scale study. Appl. Environ. Microbiol. 63: 978–982.

Qureshi, A.A., A.A. Bulich, and D.L. Isenberg. 1998. Microtox toxicity test systems: where they stand today, pp. 185–199, In: *Microscale Testing in Aquatic Toxicology: Advances, Techniques, and Practice*, P.G. Wells, K. Lee, and C. Blaise, Eds., CRC Press, Boca Raton, FL.

Rabold, J.G., C.W. Hoge, D.R. Shlim, C. Kefford, R. Rajah, and P. Echeverria. 1994. *Cyclospora* outbreak associated with chlorinated drinking water (letter). Lancet 344: 1360–1361.

Radehaus, P.M., and S.K. Schmidt. 1992. Characterization of a novel *Pseudomonas* sp. that mineralizes high concentrations of pentachlorophenol. Appl. Environ. Microbiol. 58: 2879–2885.

Radziminski, C., L. Ballantyne, J. Hodson, R. Creason, R.C. Andrews, and C. Chauret. 2002. Disinfection of *Bacillus subtilis* spores with chlorine dioxide: a bench-scale and pilot-scale study. Water Res. 36: 1629–1639.

Ramanathan, S., M. Ensor, and S. Daunert. 1997. Bacterial biosensors for monitoring toxic metals. Trends Biotechnol. 15: 500–506.

Ramsing, N.B., M. Kuhl, and B.B. Jorgensen. 1993. Distribution of sulfate-reducing bacteria, O_2 and H_2S in photosynthetic biofilms determined by oligonucleotide probes and microelectrodes. Appl. Environ. Microbiol. 59: 3840–3847.

Ramirez, G.W., J.L. Alonso, A. Villanueva, R. Guardino, J.A. Basiero, I. Bernecer, and J.J. Morenilla. 2000. A rapid, direct method for assessing chlorine effect on filamentous bacteria in activated sludge. Water Res. 34: 3894–3898.

Randall, A.A., L.D. Benefield, and W.E. Hill. 1997a. Enhanced biological phosphorus removal: The variations in location and form of intracellular phosphate induced by different substrates and observed with ^{31}P-NMR. Adv. Environ. Res. 1: 58–73.

Randall, A.A., L.D. Benefield, W.E. Hill, J.-P. Nicol, G.K. Boman, and S.-R. Jing. 1997b. The effect of volatile fatty acids on enhanced biological phosphorus removal and population structure in anaerobic/aerobic sequencing batch reactors. Water Sci. Technol 35: 153–160.

Randall, A.A., L.D. Benefield, and W.E. Hill. 1997c. Induction of phosphorus removal in an enhanced biological phosphorus removal bacterial population. Water Res. 31: 2869–2877.

Randall, A.A., and Y.-H. Liu. 2002. Polyhydroxyalkanoates form potentially a key aspect of aerobic phosphorus uptake in enhanced biological phosphorus removal. Water Res. 36: 3473–3478.

Rands, M.B., D.E. Cooper, C.P. Woo, G.C. Fletcher, and K.A. Rolfe. 1981. Compost filters for H_2S removal from anaerobic digestion and rendering exhausts. J. Water Pollut. Control Fed. 53: 185–189.

Rangsayatorn, N., P. Pokethitiyook, E.S. Upatham, and G.R. Lanza. 2004. Cadmium biosorption by cells of *Spirulina platensis* TISTR 8217 immobilized in alginate and silica gel. Environ. Intern. 30: 57–63.

Rao, V.C., S.B. Lakhe, S.V. Waghmare, and P. Dube. 1977. Virus removal in activated sludge sewage treatment. Prog. Water Technol 9: 113–127.

Rao, V.C., T.G. Metcalf, and J.L. Melnick. 1986. Removal of pathogens during wastewater treatment, pp. 531–554, In: *Biotechnology*, Vol. 8, H.J. Rehm, and G. Reed, Eds., VCH, Germany.

Rao, V.C., K.M. Seidel, S.M. Goyal, T.G. Metcalf, and J.L. Melnick. 1984. Isolation of enteroviruses from water, suspended solids and sediments from Galveston Bay: Survival of poliovirus and rotavirus adsorbed to sediments. Appl. Environ. Microbiol. 48: 404–409.

Rao, V.C., J.M. Symons, A. Ling, P. Wang, T.G. Metcalf, J.C. Hoff, and J.L. Melnick. 1988. Removal of hepatitis A virus and rotavirus by drinking water treatment. J. Am. Water Works Assoc. 80: 59–67.

Rapala, J., K. Lahtia, L.A. Räsänen, A.-L. Esala, S.I. Niemelä, and K. Sivonen. 2002. Endotoxins associated with cyanobacteria and their removal during drinking water treatment. Water Res. 36: 2625–2637.

Raskin, L., L.K. Poulsen, D.R. Noguera, B.E. Rittmann, and D.A. Stahl. 1994. Quantification of methanogenic groups in anaerobic biological reactors by oligonucleotide probe hybridization. Appl. Environ. Microbiol. 60: 1241–1248.

Ratto, A., B.J. Dutka, C. Vega, C. Lopez, and A. El-Shaarawi. 1989. Potable water safety assessed by coliphage and bacterial tests. Water Res. 23: 253–255.

Raunkjer, K., T. Hvitved-Jacobsen, and P.H. Nielsen. 1994. Measurement of pools of protein, carbohydrate and lipid in domestic wastewater. Water Res. 28: 251–262.

Rawat, K.P., A. Sharma, and S.M. Rao. 1998. Microbiological and physicochemical analysis of radiation disinfected municipal sewage. Water Res. 32: 737–740.

Rawlings, D.E. 2002. Heavy metal mining using microbes. Annual Rev. Microbiol. 56: 65–91.

Ray, C., T. Grischek, J. Schubert, J.Z. Wang, and T.F. Speth. 2002. A perspective of riverbank filtration. J. AWWA 94 (4): 49–159.

Ray, R., R. Aggarwal, P.N. Salunke, N.N. Mehrotra, G.P. Talwar, and S.R. Naik. 1991. Hepatitis E virus genome in stools of hepatitis patients during large epidemic in north India. Lancet 2: 438–442.

Reading, N.S., M.D. Cameron, and S.D. Aust. 2002. Fungi for biotechnology, pp. 1383–1394, In: *Encyclopedia of Environmental Microbiology*, G. Bitton, editor-in-chief, Wiley-Interscience, N.Y.

Reasoner, D.J. 1990. Monitoring heterotrophic bacteria in potable water, pp. 452–477, In: *Drinking Water Microbiology*. G.A. McFeters, Ed., Springer-Verlag, New York.

Reasoner, D.J. 2002. Home treatment devices- Microbiology of point of use and point of entries devices, pp. 1563–1575, In: *Encyclopedia of Environmental Microbiology*, G. Bitton, editor-in-chief, Wiley-Interscience, N.Y.

Reasoner, D.J., and E.E. Geldreich. 1985. A new medium for the enumeration and subculture of bacteria from potable water. Appl. Environ. Microbiol. 49: 1–7.

Reasoner, D.J., J.C. Blannon, and E.E. Geldreich. 1979. Rapid seven-hour fecal coliform test. Appl. Environ. Microbiol. 38: 229–236.

Reasoner, D.J., J.C. Blannon, and E.E. Geldreich. 1987. Microbiological characteristics of third-faucet point-of-use devices. J. Am. Water Works Assoc. 79: 60–66.

Rebac, S., J. Ruskova, S. Gerbens, J. van Lier, A.J.M. Stams, and G. Lettinga. 1995. High rate anaerobic treatment of wastewater under psychrophilic conditions. J. Ferment. Bioeng. 80: 499–506.

Rebuhn, M., and G. Engel. 1988. Reuse of wastewater for industrial cooling systems. J. Water Pollut. Control Fed. 60: 237–241.

Reddy, R.K. 1984. Use of aquatic macrophyte filters for water purification, pp. 660–678, In: *Proceedings of the Third AWWA Water Reuse Symposium III*, San Diego, CA, Aug. 26–31, 1984, Denver, CO.

Reed, S.C., E.J. Middlebrooks, and R.W. Crites. 1988. *Natural Systems for Wastewater Management and Treatment*. McGraw-Hill, New York.

Regan, J.M., G.W. Harrington, and D.R. Noguera. 2002. Ammonia- and nitrite-oxidizing bacterial communities in a pilot-scale chloraminated drinking water distribution system. Appl. Environ. Microbiol. 68: 73–81.

Reif, J.S., M.C. Hatch, M. Bracken, L.B. Holmes, B.A. Schwetz, and P.C. Singer. 1996. Reproductive and developmental effects of disinfection by-products in drinking water. Environ. Health Perspect. 104: 1056–1061.

Reimers, R.S., D.B. McDonell, M.D. Little, T.G. Ackers, and W.D. Henriques. 1986. Chemical inactivation of pathogens in municipal sludges, In: *Control of Sludge Pathogens*, C.A. Sorber, Ed., Water Pollution Control Federation, Washington, D.C.

Reines, H.D., and F.V. Cook. 1981. Pneumonia and bacteremia due to Aeromonas *hydrophila*. Chest 80: 264–268.

Reinhartz, A., I. Lampert, M. Herzberg, and F. Fish. 1987. A new short-term, sensitive bacterial assay kit for the detection of toxicants. Toxicity Assess. 2: 193–206.

Reinthaler, F.F., J. Posch, G. Feierl, G. Wüst, D. Haas, G. Ruckenbauer, F. Mascher, and E. Marth. 2003. Antibiotic resistance of *E. coli* in sewage and sludge. Water Res. 37: 1685–1690.

Ren, S., and P.D. Frymier. 2003. The use of a genetically engineered *Pseudomonas* species (Shk1) as a bioluminescent reporter for heavy metal toxicity screening in wastewater treatment plant influent. Water Environ. Res. 75: 21–29.

Rendtorff, R.C 1979. The experimental transmission of Giardia lamblia among volunteer subjects, pp. 64–81, In: *Waterborne Transmission of Giardiasis*, W.W. Jakubowski, and J.C. Hoff, Eds., Environmnetal Protection Agency, Office of R&D. Env. Res. Center, Cincinnati, Ohio, EPA-600/9-79-001.

Rennecker, J.L., A.M. Driedger, S.A. Rubin, and B.J. Mariñas. 2000. Synergy in sequential inactivation of *Cryptosporidium parvum* with ozone/free chlorine and ozone/monochloramine. Water Res. 34: 4121–4130.

Reuther, C.G. 1996. Brighter light better water. Environ. Health Perspect. 104: 1046–1048.

de los Reyes, F.L., W. Ritter, and L. Raskin. 1997. Group-specific small-subunits rRNA hybridization probes to characterize filamentous foaming in activated sludge systems. Appl. Environ. Microbiol. 63: 1107–1117.

de los Reyes, F.L., and L. Raskin. 2002. Role of filamentous microorganisms in activated sludge foaming: relationship of mycolata levels to foaming initiation and stability. Water Res. 36: 445–459.

Ribas, F., J. Frias, and F. Lucena. 1991. A new dynamic method for the rapid determination of the biodegradable dissolved organic carbon in drinking water. J. Appl. Bacteriol. 71: 371–378.

Ribas, F., J. Perramon, A. Terradillos, J. Frias, and F. Lucena. 2000. The *Pseudomonas* group as an indicator of potential regrowth in water distribution systems. J. Appl. Microbiol. 88: 704–710.

Rice, E.W., M.J. Allen, D.J. Brenner, and S.C. Edberg. 1991. Assay for β-glucuronidase in species of the genus *Escherichia* and its application for drinking-water analysis. Appl. Environ. Microbiol. 57: 592–593.

Rice, E.W., and J.C. Hoff. 1981. Inactivation of *Giardia lamblia* cysts by ultraviolet irradiation. Appl. Environ. Microbiol. 42: 546–547.

Rice, E.W., D.J. Reasoner, and P.V. Scarpino. 1988. Determining biodegradable organic matter in drinking water: A progress report. Water Quality Technology Conference of the American Water Works Association, St. Louis, MO, Nov. 13–17.

Rice, E.W., M.J. Allen, T.C. Covert, J. Langewis, and J. Standridge. 1993. Identifying *Escherichia* species with biochemical test kits and standard bacteriological tests. J. Amer. Water Works Assoc. 85: 74–76.

Rice, E.W., M.J. Allen, and S.C. Edberg. 1990. Efficacy of β-glucuronidase assay for identification of *Escherichia coli* by the defined-substrate technology. Appl. Environ. Microbiol. 56: 1203–1205.

Rice, E.W., K.R. Fox, R.J. Miltner, D.A. Lytle, and C.H. Johnson. 1996. Evaluating plant performance with endospores. J. Am. Water Works Assoc. 88: 122–136.

Rice, R.G. 1989. Ozone oxidation products – Implications for drinking water treatment, pp. 153–170, In: *Biohazards of Drinking Water Treatment*, R.A. Larson, Ed., Lewis, Chelsea, MI.

Richard, M.G., D. Jenkins, O. Hao, and G. Shimizu. 1982. *The Isolation and Characterization of Filamentous Microorganisms from Activated Sludge Bulking*. Report No. 81-2, Sanitary Engineering and Environmental Health Research Laboratory, University of California, Berkeley, CA.

Richard, M.G., O. Hao, and D. Jenkins. 1985a. Growth kinetics of Sphaerotilus species and their significance in activated sludge bulking. J. Water Pollut. Control Fed. 57: 68–81.

Richard, M.G., G.P. Shimizu, and D. Jenkins. 1985b. The growth physiology of the filamentous organism type 021N and its significance to activated sludge bulking. J. Water Pollut. Control Fed. 57: 1152–1162.

Rickert, D.A., and J.V. Hunter. 1971. General nature of soluble and particulate organics in sewage and secondary effluent. Water Res. 5: 421–436.

Ridgley, S.M., and D.V. Calvin. 1982. Household hazardous waste disposal project. Metro toxicant Program No. 1. Toxicant Control Planning Section, Seattle, WA.

Ridgway, H.F., and B.H. Olson. 1981. Scanning electron microscope evidence for bacterial colonization of a drinking-water distribution system. Appl. Environ. Microbiol. 41: 274–287.

Ridgway, H.F., and B.H. Olson. 1982. Chlorine resistance patterns of bacteria from two drinking water distribution systems. Appl. Environ. Microbiol. 44: 972–987.

Ridgway, H.F., E.G. Means, and B.H. Olson. 1981. Iron bacteria in drinking-water distribution systems: Elemental analysis of *Gallionella* stalks, using x-ray energy-dispersive microanalysis. Appl. Environ. Microbiol. 41: 288–297.

Riedel, K., K-P. Lange, H.J. Stein, M. Khun, P. Ott, and F. Scheller. 1990. A microbial sensor for BOD. Water Res. 24: 883–887.

Riehl, M.L., H.H. Wieser, and B.T. Rheins. 1952. Effect of lime-treated water upon survival of bacteria. J. Am. Water Works Assoc. 44: 466–470.

Riesser, V.W., J.R. Perrich, B.B. Silver, and J.R. McCammon. 1977. Possible mechanisms of poliovirus inactivation by ozone, pp. 186–192, In: *Forum on Ozone Disinfection*, E.G. Fochtman, R.G. Rice, and M.E. Browning, Eds., International Ozone Institute, New York.

Rinzema, A., and G. Lettinga. 1988. Anaerobic treatment of sulfate-containing waste water, pp. 65–109, In: *Biotreatment Systems*, Vol. 3, D.L. Wise, Ed., CRC Press, Boca Raton, FL.

Ripp, S., and G.S. Sayler. 2002. Field release of genetically engineered microorganisms, pp. 1278–1287, In: *Encyclopedia of Environmental Microbiology*, G. Bitton, editor-in-chief, Wiley-Interscience, N.Y.

Rippey, S.R., and W.D. Watkins. 1992. Comparative rates of disinfection of microbial indicator organisms in chlorinated sewage effluents. Water Sci. Technol. 26: 2185–2189.

Rippon, J.W. 1974. *Medical Mycology: The pathogenic fungi and pathogenic actinomycetes*. W.B. Saunders, Philadelphia, PA.

Rising, M.L., and A.-L. Reysenbach. 2002. Thermophiles, diversity of, pp. 3140–3147, In: *Encyclopedia of Environmental Microbiology*, G. Bitton, editor-in-chief, Wiley-Interscience, N.Y.

Ritchie, D.A., C. Edwards, I.R. McDonald, and J.C. Murrell. 1997. Detection of methanogens in natural environments. Global Change Biol. 3: 339–350.

Rittmann, B.E. 1984. Needs and strategies for genetic control: Municipal wastes, pp. 215–228, In: *Genetic Control of Environmental Pollutants*, G.S. Omenn, and A. Hollaender, Eds., Plenum, New York.

Rittmann, B.E. 1987. Aerobic biological treatment. Environ. Sci. Technol. 21: 128–136.

Rittmann, B.E. 1989. Biodegradation processes to make drinking water biologically stable, pp. 257–263, In: *Biohazards of Drinking Water Treatment*, R.A. Larson, Ed., Lewis, Chelsea, MI.

Rittmann, B.E. 1995a. Transformations of organic micropollutants by biological processes, pp. 31–42, In: *Water Pollution: Quality and Treatment of Drinking Water*, Springer, New York.

Rittmann, B.E. 1995b. Fundamentals and application of biofilm processes in drinking-water treatment, pp. 61–87, In: *Water Pollution: Quality and Treatment of Drinking Water*, Springer, New York.

Rittmann, B.E. 1996. How input active biomass affects sludge age and process stability. J. Environ. Eng. ASCE 122: 4–8.

Rittmann, B.E. 2002. The role of molecular methods in evaluating biological treatment processes. Water Environ. Res. 74: 421–427.

Rittmann, B.E., and C.W. Brunner. 1984. The nonsteady-state biofilm process for advanced organic removal. J. Water Pollut. Control Fed. 56: 874–880.

Rittmann, B.E., L.A. Crawford, C.K. Tuck, and E. Namkung. 1986. *In situ* determination of kinetic parameters for biofilms: Isolation and characterization of oligotrophic biofilms. Biotech. Bioeng. 28: 1753–1760.

Rittmann, B.E., D.E. Jackson, and S.L Storck. 1988. Potential for treatment of hazardous organic chemicals with biological processes, pp. 15–64, In: *Biotreatment Systems*, Vol. 3, D.L. Wise, Ed., CRC Press, Boca Raton, FL.

Rittmann, B.E., and C.S. Laspidou. 2002. Biofilm detachment, pp. 544–550, In: *Encyclopedia of Environmental Microbiology*, G. Bitton, editor-in-chief, Wiley-Interscience, N.Y.

Rittmann, B.E., C.S. Laspidou, J. Flax, D.A. Stahl, V. Urbain, H. Harduin, J.J. van der Waarde, B. Geurkink, M.J.C. Henssen, H. Brouwer, A. Klapwijk, and M. Wetterauw. 1999. Molecular and modeling analyses of the structure and function of nitrifying activated sludge. Water Sci. Technol. 39 (1): 51–59.

Rittmann, B.E., and P.L. McCarty. 1980a. Model of steady-state biofilm kinetics. Biotech. Bioeng. 22: 2243–2357.

Rittmann, B.E., and P.L. McCarty. 1980b. Evaluation of steady-state biofilm kinetics. Biotech. Bioeng. 22: 2359–2373.

Rittmann, B.E., and P.L. McCarty. 1981. Substrate flux into biofilms of any thickness. J. Environ. Eng. Div. ASCE 107: 831–849.

Rittmann, B., and P.L. McCarty. 2001. *Environmental Biotechnology: Principles and applications*. McGraw-Hill, New York.

Rittmann, B.E., B.F. Smets, and D.A. Stahl. 1990. The role of genes in biological treatment processes. Environ. Sci. Technol. 24: 23–29.

Rittmann, B.E., and V.L. Snoeyink. 1984. Achieving biologically stable drinking water. J. Am. Water Works Assoc. 76: 106–114.

Ro, K.S., R.W. Babcock, and M.K. Stenstrom. 1997. Demonstration of bioaugmentation in a fluidized-bed process treating 1-naphthylamine. Water Res. 31: 1687–1693.

Roane, T.M., and I.L. Pepper. 2000. Microscopic Techniques, pp. 195–211, In: *Environmental Microbiology*, R.M. Maier, I.L. Pepper, and C.P. Gerba, Eds., Academic Press, San Diego, CA, 585 pp.

Robeck, G.G., N.E. Clark, and K.A. Dostal. 1962. Effectiveness of water treatment processes in virus removal. J. Amer. Water Works Assoc. 54 (10): 1275–1292.

Roberton, A.M., and R.S. Wolfe. 1970. ATP pools in *Methanobacterium*. J. Bacteriol. 102: 43–51.

Roberts, P.V., L. Semprini, G.D. Hopkins, D. Grbic-Galic, P.L. McCarty, and M. Reinhard. 1989. *In situ* aquifer restoration of chlorinated aliphatics by methanotrophic bacteria. Report No. EPA/600/S2-89/033, U.S. Environmental Protection Agency, Ada, OK.

Robertson, J.B., and S.C. Edberg. 1997. Natural protection of spring and well drinking water against surface microbial contamination. I. Hydrogeological parameters. Crit. Rev. Microbiol. 23: 143–178.

Robertson, L.J., A.T. Campbell, and H.V. Smith. 1992. Survival of *Cryptosporidium parvum* oocysts under various environmental pressures. Appl. Environ. Microbiol. 58: 3494–3500.

Robertson, L.J., C.A. Paton, A.T. Campbell, P.G. Smith, M.H. Jackson, R.A. Gilmour, S.E. Black, D.A. Stevenson, and H.V. Smith. 2000. *Giardia* cysts and *Cryptosporidium* oocysts at sewage treatment works in Scotland, UK. Water Res. 34: 2310–2322.

Robison, B.J. 1984. Evaluation of a fluorogenic assay for detection of *E. coli* in foods. Appl. Environ. Microbiol. 48: 285–288.

Rochelle, P.A. 2002. *Giardia*: Detection and occurrence in the environment, pp. 1477–1489, In: *Encyclopedia of Environmental Microbiology*, G. Bitton, editor-in-chief, Wiley-Interscience, N.Y.

Rochelle, P.A., D.M. Fergusson, T.J. Handojo, R. De Leon, M.H. Stewart, and R.L. Wolfe. 1997a. An assay combining cell culture with reverse transcriptase PCR to detect and determine the infectivity of waterborne *Cryptosporidium parvum*. Appl. Environ. Microbiol. 63: 2029–2037.

Rochelle, P.A., R. De Leon, M.H. Stewart, and R.L. Wolfe. 1997b. Comparison of primers and optimization of PCR conditions for detection of *Cryptosporidium parvum* and *Giardia lamblia* in water. Appl. Environ. Microbiol. 63: 106–114.

Rochelle, P.A., M.M. Marshall, J.R. Mead, A.M. Johnson, D.G. Korich, J.S. Rosen, and R. De Leon. 2002. Comparison of *in vitro* cell culture and a mouse assay for measuring infectivity of *Cryptosporidium parvum*. Appl. Environ. Microbiol. 68: 3809–3817.

Rochkind-Dubinsky, M.L., G.S. Sayler, and J.W. Blackburn. 1987. *Microbiological Decomposition of Chlorinated Aromatic Compounds*. Marcel Dekker, New York.

Rodgers, F.G., P. Hufton, E. Kurzawska, C. Molloy, and S. Morgan. 1985. Morphological response of human rotavirus to ultraviolet radiation, heat and disinfectants. J. Med. Microbiol. 20: 123–130.

Rodgers, J., and C.W. Keevil. 1995. Survival of *Cryptosporidium parvum* oocysts in biofilms and planktonic samples in a model system, pp. 209–213, In: *Protozoan Parasites and Water*, W.B. Betts, D. Casemore, C. Fricker, H. Smith, and J. Watkins, Eds., The Royal Society of Chemistry, Cambridge, U.K.

Rodgers, M.R., C.M. Bernardino, and W. Jakubowski. 1992. A comparison of methods for extracting amplifiable *Giardia* DNA from various environmental samples. Presented at the International Water Pollution Research Conference, Washington, D.C., May 26–29, 1992.

Rodgers-Gray, T.P., S. Jobling, S. Morris, C. Kelly, S. Kirby, A. Janbakhsh, J.E. Harries, M.J. Waldock, J.P. Sumpter, and C.R. Tyler. 2000. Long-term temporal changes in the estrogenic composition of treated sewage effluent and its biological effects on fish. Environ. Sci. Technol. 34: 1524–1528.

Rodriguez, R.L., and R.C. Tait. 1983. *Recombinant DNA Techniques: An Introduction*. Addison-Wesley, Reading, MA.

Rogers, J., A.B. Dowsett, P.J. Dennis, J.V. Lee, and C.W. Keevil. 1994. Influence of plumbing materials on biofilm formation and growth on *Legionella pneumophila* in potable water systems. Appl. Environ. Microbiol. 60: 1842–1851.

Rogers, S.E., and W.C. Lauer. 1986. Disinfection for potable reuse. J. Water Pollut. Control Fed. 58: 193–198.

Rogers, S.E., and W.C. Lauer. 1992. Denver's demonstration of potable water reuse: Water quality and health effects testing. Water Sci. Technol. 26: 1555–1564.

Rollinger, Y., and W. Dott. 1987. Survival of selected bacterial species in sterilized activated carbon filters and biological activated carbon filters. Appl. Environ. Microbiol. 53: 77–781.

Rook, J.J. 1974. Formation of haloforms during chlorination of natural waters. Water Treat. Exam. 23: 234–243.

Rosas, I., A. Baez, and M. Coutino. 1984. Bacteriological quality of crops irrigated with wastewater in the Xochimilco plots, Mexico City, Mexico. Appl. Environ. Microbiol. 47: 1074–1079.

Rose, J.B. 1986. Microbial aspects of wastewater reuse for irrigation. CRC Crit. Rev. Environ. Control. 16: 231–256.

Rose, J.B. 1988. Occurrence and significance of *Cryptosporidium* in water. J. Am. Water Works Assoc. 80: 53–58.

Rose, J.B. 1990. Occurrence and Control of *Cryptosporidium* in drinking water, pp. 294–321, In: *Drinking Water Microbiology*, G.A. McFeters, Ed., Springer Verlag, New York.

Rose, J.B., A. Cifrino, M.S. Madore, C.P. Gerba, C.R. Sterling, and M.J. Arrowood. 1986. Detection of *Cryptosporidium* from wastewater and freshwater environments. Water Sci. Technol. 18: 233–237.

Rose, J.B., R. De Leon, and C.P. Gerba. 1989a. *Giardia* and virus monitoring of sewage effluents in the state of Arizona. Water Sci. Technol. 21: 43–47.

Rose, J.B., L.K. Landeen, K.R. Riley, and C.P. Gerba. 1989b. Evaluation of immunofluoresecence techniques for detection of *Cryptosporidium* oocysts and *Giardia* cysts from environmental samples. Appl. Environ. Microbiol. 55: 3189–3196.

Rose, J.B., L.J. Dickson, S.R. Farrah, and R.P. Carnahan. 1996. Removal of pathogenic and indicator microorganisms by a full-scale water reclamation facility. Water Res. 30: 2785–2797.

Rose, J.B., S.R. Farrah, D. Friedman, K. Riley, C.L. Hamann, and M. Robbins. 1999. Public health evaluation of advanced reclaimed water for potable applications. Water Sci. Technol. 40 (4–5): 247–252.

Rose, J.B., and C.P. Gerba. 1991. Assessing potential health risks from viruses and parasites in reclaimed water in Arizona and Florida. Water Sci. Technol. 23: 2091–2098.

Rose, J.B., C.P. Gerba, and W. Jakubowski. 1991. Survey of potable water supplies for *Cryptosporidium* and *Giardia*. Environ. Sci. Technol. 25: 1393–1399.

Rose, J.B., C.E. Musial, M.J. Arrowood, C.R. Sterling, and C.P. Gerba. 1985. Development of a method for the detection of *Cryptosporidium* in drinking water. Presented at the Water Technology Conference, In: Proceedings of the Water Technology Conference, AWWA, Houston, TX, Dec. 8–11, 1985.

Rose, J.B., X. Zhou, D.W. Griffin, and J.H. Paul. 1997. Comparing PCR and plaque assay for detection and enumeration of coliphage in polluted marine waters. Appl. Environ. Microbiol. 63: 4564–4566.

Rosello-Mora, R.A., M. Wagner, R. Amann, and K.-H. Schleifer. 1995. The abundance of *Zooglea ramigera* in sewage treatment plants. Appl. Environ. Microbiol. 61: 702–707.

Rosenberg, F.A. 2002. Bottled water, microbiology of, pp. 795–802, In: *Encyclopedia of Environmental Microbiology*, G. Bitton, editor-in-chief, Wiley-Interscience, N.Y.

Rosenberg, M.L., K.K. Hazlet, J. Schaefer, J.G. Wells, and R.C. Pruneda. 1976. Shigellosis from swimming. J. Am. Med. Assoc. 236: 1849–1852.

Rosenszweig, W.D., H.A. Minnigh, and W.O. Pipes. 1983. Chlorine demand and inactivation of fungal propagules. Appl. Environ. Microbiol. 45: 182–186.

Rosenszweig, W.D., H.A. Minnigh, and W.O. Pipes. 1986. Fungi in potable distribution systems. J. Am. Water Works Assoc. 78: 53–55.

Rosenszweig, W.D., and W.O. Pipes. 1989. Presence of fungi in drinking water, pp. 85–93, In: *Biohazards of Drinking Water Treatment*, R.A. Larson, Ed., Lewis, Chelsea, MI.

Rosenszweig, W.D., G. Ramirez-Toro, H. Minnigh, and W.O. Pipes. 1989. Fungi in potable water (Abst.), In: 89th Ann. Meeting, American Society of Microbiology, New Orleans, LA, May 14–18, 1989.

Rosenthal, S.L., J.H. Zuger, and E. Apollo. 1975. Respiratory colonization with *Pseudomonas putrefaciens* after near-drowning in salt water. J. Clin. Path. 64: 382.

Ross, I.S., and C.C. Townsley. 1986. The uptake of heavy metals by filamentous fungi, pp. 49–58, In: *Immobilisation of Ions by Bio-Sorption*, H. Eccles, and S. Hunt, Eds., Ellis Horwood, Chichester, U.K.

Rossetti, S., M.C. Tomei, C. Levantesi, R. Ramador, and V. Tandoi. 2002. *Microthrix parvicella*: A new approach for kinetic and physiological characterization. Water Sci. & Technol. 46 (1–2): 65–72.

Roszak, D.B., and R.R. Colwell. 1987. Survival strategies of bacteria in the natural environment. Microbiol. Rev. 51: 365–379.

Rotz, L.D., A.S. Khan, S.R. Lillibridge, S.M. Ostroff, and J.M. Hughes. 2002. Public health assessment of potential biological terrorism agents. Emerg. Infect. Dis. 8 (2): 225–229.

Rouf, M.A., and J.L. Stokes. 1962. Isolation and identification of the sudanophilic granules of *Sphaerotilus natans*. J. Bacteriol. 83: 343–347.

Roy, D., P.K.Y. Wong, R.S. Engelbrecht, and E.S.K. Chian. 1981. Mechanism of enteroviral inactivation by ozone. Appl. Environ. Microbiol. 41: 718–723.

Roy-Arcand, J., and J.S. Archibald. 1991. Direct dechlorination of chloro-phenolic compounds by laccases from *Trametes versicolor*. Enzyme Microb. Technol. 13: 194–203.

Rozen, Y., and S. Belkin. 2001. Survival of enteric bacteria in seawater. FEMS Microbiol. Rev. 25: 513–529.

Rozich, A.F., R.M. Sykes, G.B. Walkenshaw, and D.R. Rodgers. 1982. Control of algal filamentous bulking at the Southerly wastewater treatment plant. J. Water Pollut. Control Fed. 54: 231–237.

Rubin, A.J. 1988. Factors affecting the inactivation of *Giardia* cysts by monochloramine and comparison with other disinfectants, pp. 224–229, In: *Proceedings: Conference on Current Research in Drinking Water Treatment*. Report No. EPA-600/9-88/004, Environmental Protection Agency, Cincinnati, OH.

Rubin, A.J., J.P. Engel, and O.J. Sproul. 1983. Disinfection of amoebic cysts in water with free chlorine. J. Water Pollut. Control Fed. 55: 1174–1182.

Rudd, T., R.M. Sterritt, and J.N. Lester. 1984. Complexation of heavy metals by extracellular polymers in the activated sludge process. J. Water Pollut. Control Fed. 56: 1260–1268.

Rush, B.A., R.A. Chapman, and R.W. Ineson. 1990. A probable waterborne outbreak of cryptosporidiosis in the Sheffield area. J. Med. Microbiol. 32: 239–242.

Rusin, P., C.E. Enriquez, D. Johnson, and C.P. Gerba. 2000. Environmentally transmitted pathogens, pp. 447–489, In: *Environmental Microbiology*, R.M. Maier, I.L. Pepper, and C.P. Gerba, Eds., Academic Press, San Diego, CA, 585 pp.

Russ, C.F., and W.A. Yanko. 1981. Factors affecting Salmonellae repopulation in composted sludge. Appl. Environ. Microbiol. 41: 597–602.

Russel, A.D. 2002. Antibiotic and biocide resistance in bacteria: Introduction. J. Appl. Microbiol. 92: 1S–3S.

Russel, A.D., J.R. Furr, and J.-Y. Maillard. 1997. Microbial susceptibility and resistance to biocides. Am. Soc. Microbiol. News 63: 481–487.

Rylander, R., K. Andersson, L. Belin, G. Berglund, R. Bergstrom, L.-A. Hanson, M. Lundholm, and I. Mattsby. 1976. Sewage worker's syndrome. Lancet 2: 478–479.

Sabatini, D.A., J.W. Smith, and L.W. Moore. 1990. Treatment of chlordane-contaminated water by the activated rotating biological contactor. J. Environ. Qual. 19: 334–338.

Saber, D.L., and R.L. Crawford. 1985. Isolation and characterization of *Flavobacterium* strains that degrade pentachlorophenol. Appl. Environ. Microbiol. 50: 1512–1518.

Saber, D.L., and R.L. Crawford. 1989. Isolation and characterization of *Flavobacterium* strains that degrade pentachlorophenol. Appl. Environ. Microbiol. 55: 1512–1518.

Sabra, W., A.-P. Zeng, H. Lünsdorf, and W.-D. Deckwer. 2000. Effect of oxygen on formation and structure of *Azotobacter vinelandii* alginate and Its role in protecting nitrogenase. Appl. Environ. Microbiol. 66: 4037–4044.

Sachon, G., J. Wiart, and J.L. Martel. 1997. Le plan d'épandage agricole des boues d'épuration: une spécificité française. Tech. Sci. Méth. 2: 43–51.

Saeger, V.W., and E.S. Tucker. 1976. Biodegradation of phtalic acid esters in river water and activated sludge. Appl. Environ. Microbiol. 31: 29–34.

De Saeger, S., and C. van Peteghem. 1996. Dipstick enzyme immunoassay to detect *Fusarium* T-2 toxin in wheat. Appl. Environ. Microbiol. 62: 1880–1884.

Safferman, R.S., and M.E. Morris. 1976. Assessment of virus removal by a multi-stage activated sludge process. Water Res. 10: 413–420.

Safferman, R.S., A.A. Rosen, C.I. Mashni, and M.E. Morris. 1967. Earthy-smelly substance from a blue-green alga. Environ. Sci. Technol. 1: 429–430.

Sagy, M., and Y. Kott. 1990. Efficiency of rotating biological contactors in removing pathogenic bacteria from domestic sewage. Water Res. 24: 1125–1128.

Sahasrabudhe, A., A. Pande, and V. Modi. 1991. Dehalogenation of a mixture of chloroaromatics by immobilized *Pseudomonas* sp. US1 excells. Appl. Microbiol. Biotech. 35: 830–832.

Sahm, H. 1984. Anaerobic wastewater treatment. Adv. Biochem. Eng. Biotech. 29: 84–115.

Saier, M., K. Koch, and J. Wekerle. 1985. Influence of thermophilic anaerobic digestion (55°C) and subsequent mesophilic digestion of sludge on the survival of viruses with and without pasteurization of the digested sludge, pp. 28–37, In: *Inactivation of Microorganisms in Sewage Sludge by Stabilization Processes*, Strauch, D., A.H. Havelaar, and P. L'Hermite, Eds., Elsevier, London.

Sails, A.D., F.J. Bolton, A.J. Fox, D.R.A. Wareing, and D.L.A. Greenway. 2002. Detection of *Campylobacter jejuni* and *Campylobacter coli* in environmental waters by PCR enzyme-linked immunosorbent assay. Appl. Environ. Microbiol. 68: 1319–1324.

Salanitro, J.P., P.C. Johnson, G.E. Spinnler, P.M. Maner, H.L. Wisniewski, and C. Bruce. 2000. Field-scale demonstration of enhanced MTBE bioremediation through aquifer bioaugmentation and oxygenation. Environ. Sci. Technol. 34: 4152–4162.

Salem, S., D.H.J.G. Berends, J.J. Heijnen, and M.C.M. Van Loosdrecht. 2003. Bio-augmentation by nitrification with return sludge. Water Res. 37: 1794–1804.

Salkinoja-Salonen, M., P. Middeldorp, M. Briglia, R. Valo, M. Haggblom, and A. McBain. 1989. Cleanup of old industrial sites, pp. 203–211, In: *Biotechnology and Biodegradation*, D. Kamely, A. Chakrabarty, and G.S. Omenn, Eds., Gulf, Houston, TX.

Salmen, P., D.M. Dwyes, H. Vorse, and W. Kruse. 1983. Whirlpool-associated *Pseudomonas aeruginosa* urinary tract infections. J. Am. Med. Assoc. 250: 2025–2026.

Sanders, W.T.M., G. Zeeman, and A.J.M. Stams. 2002. Biosolids: Anaerobic digestion of, pp. 738–750, In: *Encyclopedia of Environmental Microbiology*, G. Bitton, Ed., Wiley-Interscience, N.Y.

Sandt, C., and D. Herson. 1989. Plasmid mobilization from a genetically engineered microorganism to an environmentally isolated strain of *Enterobacter cloacae* in sterile drinking water, In: 89th Annual Meeting of the American Society Microbiology, New Orleans, LA, May 14–18, 1989.

Santegoeds, C.M., L.R. Damgaard, G. Hesselink, J. Zopfi, P. Lens, G. Muyzer, and D. De Beer. 1999. Distribution of sulfate-reducing and methanogenic bacteria in anaerobic aggregates determined by microsensor and molecular analyses. Appl. Environ. Microbiol. 65: 4618–4629.

Saqqar, M.M., and M.B. Pescod. 1992. Modelling coliform reduction in wastewater stabilization ponds. Water Sci. Technol. 26: 1667–1677.

Saqqar, M.M., and M.B. Wood. 1991. Microbiological performance of multistage stabilization ponds for effluent use in agriculture. Water Sci. Technol. 23: 1517–1524.

Sarhan, H.R., and H.A. Foster. 1991. A rapid fluorogenic method for the detection of *Escherichia coli* by the production of b-glucuronidase. J. Appl. Bacteriol. 70: 394–400.

Sarikaya, H.Z., and A.M. Saarci. 1987. Bacterial die-off in waste stabilization ponds. J. Environ. Eng. Div. ASCE 113: 366–382.

Sarikaya, H.Z., A.M. Saarci, and A.F. Abdulfattah. 1987. Effect of pond depth on bacterial die-off. J. Environ. Eng. Div. 113: 1350–1362.

Sarner, E. 1986. Removal of particulate and dissolved organics in aerobic fixed-film biological processes. J. Water Pollut. Control Fed. 58: 165–172.

Sasikala, Ch., and Ch.V. Ramana. 1995. Biotechnological potential of anoxygenic phototrophic bacteria. I. Production of single-cell protein, vitamins, ubiquinone, hormones, and enzymes, and use in waste treatment. Adv. Appl. Microbiol. 41: 173–226.

Sathasivan, A., and S. Ohgaki. 1999. Application of new bacterial regrowth potential method for water distribution systems: a clear evidence of phosphorus limitation. Water Res. 33: 137–144.

Sato, C., S.W. Leung, and J.L. Schnoor. 1988. Toxic response of *Nitrosomonas europea* to copper in inorganic medium and wastewater. Water Res. 22: 1117–1127.

Satoh, H., S. Okabe, Y. Yamaguchi, and Y. Watanabe. 2003. Evaluation of the impact of bioaugmentation and biostimulation by *in situ* hybridization and microelectrode. Water Res. 37: 2206–2216.

Sattar, S.A., S. Ramia, and J.C.N. Westwood. 1976. Calcium hydroxide (lime) and the elimination of human pathogenic viruses from sewage: Studies with experimentally contaminated (poliovirus type 1, Sabin) and pilot plant samples. Can. J. Public Health. 67: 221–225.

Sattar, S.A., M.K. Ijaz, C.M. Johnson-Lussenburg, and V.S. Springthorpe. 1984. Effect of relative humidity on the airborne survival of rotavirus SA-11. Appl. Environ. Microbiol. 47: 879–881.

Sauch, J.F. 1989. Use of immunofluorescence and phase-contrast microscopy for detection and identification of *Giardia* cysts in water samples. Appl. Environ. Microbiol. 50: 1434–1438.

Sauch, J.F., and D. Berman. 1991. Immunofluorescence and morphology of *Giardia lamblia* cysts exposed to chlorine. Appl. Environ. Microbiol. 57: 1573–1575.

Sauch, J.F., D. Flanigan, M.L. Galvin, D. Berman, and W. Jakubowski. 1991. Propidium iodide as an indicator of *Giardia* cyst viability. Appl. Environ. Microbiol. 57: 3243–3247.

Saunders, A.M., A. Oehmen, L.L. Blackall, Z. Yuan, and J. Keller. 2003. The effect of GAOs (glycogen accumulating organisms) on anaerobic carbon requirements in full-scale Australian EBPR (enhanced biological phosphorus removal) plants. Water Sci. Technol. 47 (11): 37–43.

Saunders, V.A., and J.R. Saunders. 1987. *Microbial Genetics Applied to Biotechnology: Principles and Techniques of Gene Transfer and Manipulation.* MacMillan, New York.

Savenhed, R., H. Boren, A. Grimwall, B.V. Lundgren. P. Balmer, and T. Hedberg. 1987. Removal of individual off-flavour compounds in water during artificial groundwater recharge and during treatment by alum coagulation/sand filtration. Water Res. 21: 277–283.

Sawyer, C.N., and P.L. McCarty, 1967. *Chemistry for Sanitary Engineers.* McGraw-Hill, New York.

Sawyer, T.K., E.J. Lewis, M. Galassa, D.W. Lear, M.L. O'Malley, W.N. Adams, and J. Gaines. 1982. Pathogenic amoebas in ocean sediments near wastewater sludge disposal sites. J. Water Pollut. Control Fed. 54: 1318–1323.

Sayama, N., and Y. Itokawa. 1980. Treatment of cresol, phenol and formalin using fixed-film reactors. J. Appl. Bacteriol. 9: 395–403.

Sayler, G.S., and J.W. Blackburn. 1989. Modern biological methods: The role of biolotechnology, pp. 53–71, In: *Biotreatment of Agricultural Wastewater*, M.E. Huntley, Ed., CRC Press, Boca Raton, FL.

Sayler, G.S., and A.C. Layton. 1990. Environmental applications of nucleic acid hybridization. Ann. Rev. Microbiol. 44: 625–648.

Sayler, G.S., A. Breen, J.W. Blackburn, and O. Yagi. 1984. Predictive assessment of priority pollutant bio-oxidation kinetics in activated sludge. Environ. Prog. 3: 153–163.

Sayler, G.S., R. Fox, and J.W. Blackburn. Eds. 1991. *Environmental Biotechnology for Waste Treatment.* Plenum, New York.

Scalf, M.R., W.J. Dunlap, and J.F. Kreissel. 1977. *Environmental Effects of Septic Tank Systems.* Report No. EPA-600/3-77-096.

Schachter, B. 2003. Slimy business: the biotechnology of biofilms. Nature Biotechnol. 21: 361–365.

Schaefer, F.W. 1997. Detection of protozoan parasites in source and finished drinking waters, pp. 153–175, In: *Manual of Environmental Microbiology*, C.J. Hurst, G.R. Knudsen, M.J. McInerney, L.D. Stetzenbach, and M.V. Walter, Eds., ASM Press, Washington, D.C.

Schaefer, F.W. III, C.H. Johnson, C.H. Hsu, and E.W. Rice. 1991. Determination of *Giardia lamblia* infective dose for the mongolian gerbil (*Meriones unguiculatus*). Appl. Environ. Microbiol. 57: 2408–2409.

Schaiberger, G.E., T.D. Edmond, and C.P. Gerba. 1982. Distribution of enteroviruses in sediments contiguous with a deep marine sewage outfall. Water Res. 16: 1425–1428.

Schaper, M., J. Jofre, M. Uys, and W.O.K. Grabow. 2002. Distribution of genotypes of F-specific RNA bacteriophages in human and non-human sources of faecal pollution in South Africa and Spain. J. Appl. Microbiol. 92: 657–667.

Schaub, S.A., and C.A. Sorber. 1977. Virus and bacteria removal from wastewater by rapid infiltration through soil. Appl. Environ. Microbiol. 33: 609–619.

Scheuerman, P.R. 1984. Fate of Viruses During Aerobic Digestion of Wastewater Sludges. Doctoral Dissertation, University of Florida, Gainesville, FL.

Scheuerman, P.R., G. Bitton, A.R. Overman, and G.E. Gifford. 1979. Transport of viruses through organic soils and sediments. J. Environ. Eng. Div. ASCE 105: 629–640.

Scheuerman, P.R., S.R. Farrah, and G. Bitton. 1991. Laboratory studies of virus survival during aerobic and anaerobic digestion of sewage sludge. Water Res. 25: 241–245.

Schiemann, D.A. 1990. *Yersinia enterocolitica* in drinking water, pp. 322–339, In: *Drinking Water Microbiology*, G.A. McFeters, Ed., Springer Verlag, New York.

Schiewer, S., and B. Volesky. 2000. Biosorption processes for heavy metal removal. pp. 329–362, In: *Environmental Microbe-Metal Interactions*, D.R. Lovley, Ed., ASM Press, Washington.

Schiff, G.M., G.M. Stefanovic, E.C. Young, G.S. Sander, J.K. Pennekamp, and R.L. Ward. 1984a. Studies of echovirus 12 in volunteers: Determination of minimal infectious dose and the effect of previous infection on infectious dose. J. Infect. Dis. 150: 858–866.

Schiff, G.M., G.M. Stefanovic, E.C. Young, and J.K. Pennekamp. 1984b. Minimum human infective dose of enteric virus (echovirus 12) in drinking water. Monog. Virol. 15: 222–228.

Schimel, J. 2002. Traces gases, soil, pp. 3183–3194, In: *Encyclopedia of Environmental Microbiology*, G. Bitton, editor-in-chief, Wiley-Interscience, N.Y.

Schink, B. 1988. Principles and limits of anaerobic degradation: Environmental and technological aspects, pp. 771–846, In: *Biology of Anaerobic Microorganisms*, A.J.B. Zehnder, Ed., John Wiley & Sons, New York.

von Schirnding, Y.E.R., N. Strauss, P. Robertson, R. Kfir, B. Fattal, A. Mathee, M. Franck, and V.J. Cabelli. 1993. Bather's morbidity from recreational exposure to sea water. Water Sci. Technol. 27: 183–186.

Schmidt, M., U. Panne, J. Adams, and R. Niessner. 2004. Investigation of biocide efficacy by photo-acoustic biofilm monitoring. Water Res. 38: 1189–1196.

Schmidt, I.O., Sliekers, M. Schmid, E. Bock, J. Fuerst, J.G. Kuenen, M.S.M. Jetten, and M. Strous. 2003. New concepts of microbial treatment processes for the nitrogen removal in wastewater. FEMS Microbiol. Rev. 27: 481–492.

Schmidt, S.K., and M. Alexander. 1985. Effect of dissolved organic carbon and second substrates on the biodegradation of organic compounds at low concentrations. Appl. Environ. Microbiol. 49: 822–827.

Schneider, R.P., and A. Leis. 2002. Conditioning films in aquatic environments, pp. 928–941, In: *Encyclopedia of Environmental Microbiology*, G. Bitton, Ed., Wiley-Interscience, N.Y.

Schönhuber, W., B. Fuchs, S. Juretschko, and R. Amann. 1997. Improved sensitivity of whole cell hybridization by the combination of horseradish peroxidase-labeled oligonucleotides and thyramine signal amplification. Appl. Environ. Microbiol. 63: 3268–3273.

Schramm, A., L.H. Larsen, N.P. Revsbech, N.B. Ramsing, R. Amann, and K.-H. Schleifer. 1996. Structure and function of a nitrifying biofilm determined by *in situ* hybridization and use of microelectrodes. Appl. Environ. Microbiol. 62: 4641–4647.

Schuler, P.F., and M.M. Ghosh. 1990. Diatomaceous earth filtration of cysts and other particulates using chemical additives. J. Am. Water Works Assoc. 82: 67–75.

Schulze-Robbecke, R., B. Janning, and R. Fischeder. 1992. Occurrence of mycobacteria in biofilm samples. Tuber. Lung Dis. 73: 141–144.

Schumacher, G., and I. Sekoulov. 2002. Polishing of secondary effluent by an algal biofilm process. Water Sci. Technol. 46 (8): 83–90.

Schupp, D.G., and S.L. Erlandsen. 1987. A new method to determine *Giardia* cyst viability: correlation of fluorescein diacetate and propidium iodide staining with animal infectivity. Appl. Environ. Microbiol. 53: 704–707.

Schuppler, M., M. Wagner, G. Schön, and U.B. Göbel. 1998. *In situ* identification of nocardioform actinomycetes in activated sludge using fluorescent rRNA-targeted oligonucleotide probes. Microbiology 144: 249–259.

Schwab, K.J., F.H. Neill, F. Le Guyader, M.K. Estes, and R.L. Atmar1. 2001. Development of a reverse transcription-PCR-DNA enzyme immunoassay for detection of "Norwalk-Like" viruses and hepatitis A virus in stool and shellfish. Appl. Environ. Microbiol. 67: 742–749.

Schwartzbrod, J., J.L. Stien, K. Bouhoum, and B. Baleux. 1989. Impact of wastewater treatment on helminth eggs. Water Sci. Technol. 21: 295–297.

Schwartzbrod, J., M. Maux, and T. Chesnot. 2002. Parasitic protozoa: Fate in wastewater treatment plants, pp. 2327–2337 In: *Encyclopedia of Environmental Microbiology*, G. Bitton, editor-in-chief, Wiley-Interscience, N.Y.

Schwartzbrod, L. Ed. 1991. *Virologie des milieux hydriques*. TEC & DOC Lavoisier, Paris, France, 304 pp.

Schwartzbrod, L., C. Jehl-Pietri, S. Boher, B. Hugues, M. Albert, and C. Beril. 1990. Les contaminations par les virus, In: La mer et les rejets urbains, IFREMER Proceedings 11: 101–114.

Schwartzbrod, L., and C. Mattieu. 1986. Virus recovery from wastewater treatment plant sludges. Water Res. 20: 1011–1013.

Schwartzbrod, L., P. Vilagines, J. Schwartzbrod, B. Sarrette, R. Vilagines, and J. Collomb. 1985. Evaluation of the viral population in two wastewater treatment plants: Study of different sampling techniques. Water Res. 19: 1353–1356.

Schwebach, G.H., D. Cafaro, J. Egan, M. Grimes, and G. Michael. 1988. Overhauling health effects perspectives. J. Water Pollut Control Fed. 60: 473–479.

Scragg, A., Ed. 1988. *Biotechnology for Engineers: Biological Systems in Technological Processes*. Ellis Horwood, Chichester, U.K., 390 pp.

Seetharam, G.B., and B.A. Saville. 2003. Degradation of phenol using tyrosinase immobilized on siliceous supports. Water Res. 37: 436–440.

Segall, B.A., and C.R. Ott. 1980. Septage treatment at a municipal plant. J. Water Pollut. Control Fed. 52: 2145–2157.

Segel, I.H. 1975. *Enzyme Kinetics*, John Wiley & Sons, New York, 955 pp.

Seidler, R. J., and T.M. Evans. 1983. Analytical methods for microbial water quality, In: *Assessment of Microbiology and Turbidity Standards For Drinking Water*, P.S. Berger, and Y. Argaman, Eds., Report No. EPA-570-9-83-001. Environmental Protection Agency, Wahington, D.C.

Seidler, R.J., J.E. Morrow, and S.T. Bagley. 1977. Klebsielleae in drinking water emanating from redwood tanks. Appl. Environ. Microbiol. 33: 893–900.

Séka, A.M., S. Carbooter, and W. Verstraete. 2001a. A test for predicting propensity of activated sludge to acute filamentous bulking. Water Environ. Res. 73: 237–242.

Séka, A.M., Y. Kalogo, F. Hammes, J. Kielemoes, and W. Verstraete. 2001a. Chlorine-susceptible and chlorine-resistant Type 021N bacteria occurring in bulking activated sludges. Appl. Environ. Microbiol. 67: 5303–5307.

Seka, A.M., T. Van de wiele, and W. Verstraete. 2001. Feasibility of a multi-component additive for efficient control of activated sludge filamentous bulking. Water Res. 35: 2995–3003.

Sekiguchi, Y., Y. Kamagata, K. Nakamura, A. Ohashi, and H. Harada. 1999. Fluorescence *in situ* hybridization using 16S rRNA-targeted oligonucleotides reveals localization of methanogens and selected uncultured bacteria in mesophilic and thermophilic sludge granules. Appl. Environ. Microbiol. 65: 1280–1288.

Sekla, L., D. Gemmill, J. Manfreda, M. Lysyk, W. Stackiw, C. Kay, C. Hooper, L. van Buckenhout, and G. Eibisch. 1980. Sewage treatment plant workers and their environment: A health study, pp. 281–293, In: *Proceedings of the Symposium on Wastewater Aerosols and Disease*, H. Pahren, and W. Jakubowski, Eds., Report No. EPA-600/9-80-028, U.S. Environmental Protection Agency, Cincinnati, OH.

Sekla, L.H., W. Stackiw, A.G. Buchanan, and S.E. Parker. 1982. *Legionella pneumophila* pneumonia. Can. Med. Assoc. J. 126: 116–119.

Selvaratnam, C., B.A. Schoedel, B.L. McFarland, and C.F. Kulp. 1997. Application of the polymerase chain reaction (PCR) and the reverse transcriptase PCR for determine the fate of phenol-degrading *Pseudomonas putida* ATCC 11172 in a bioaugmented sequencing batch reactor. Appl. Microbiol. Biotechnol. 47: 236–240.

Sembries, S., and R.L. Crawford. 1997. Production of *Clostridium bifermentans* spores as inoculum-for bioremediation of nitroaromatic contaminants. Appl. Environ. Microbiol. 63: 2100–2104.

Semprini, L., P.V. Roberts, G.D. Hopkins, and P.L. McCarty. 1990. A field evaluation of in-situ biodegradation of chlorinated ethenes: Part 2, results of biostimulation and biotransformation experiments. Ground Water 28: 715–727.

Serafim, L.S., P.C. Lemos, C. Levantesi, V. Tandoi, H. Santos, and M.A.M. Reis. 2002. Methods for detection and visualization of intracellular polymers stored by polyphosphate-accumulating microorganisms. J. Microbiol. Meth. 51: 1–18.

De Serres, G., T.L. Cromeans, B. Levesque, N. Brassard, C. Barthe, M. Dionne, H. Prud'homme, D. Paradis, C.N. Shapiro, O.V. Nainan, and H.S. Margolis. 1999. Molecular confirmation of hepatitis A virus from well water: epidemiology and public health implications. J. Infect. Dis. 179: 37–43.

Servais, P. 1996. Rôle du chlore et de la matière organique biodégradable dans le contrôle de la croissance bactérienne en réseaux de distribution de l'eau potable. Tribune Eau 96: 3–12.

Servais, P., G. Billen, and M.-C. Hascoet. 1987. Determination of the biodegradable fraction of dissolved organic matter in waters. Water Res. 21: 445–452.

Servais, P., A. Anzil, and C. Ventresque. 1989. Simple method for determination of biodegradable dissolved organic carbon in water. Appl. Environ. Microbiol. 55: 2732–2734.

Servais, P., G. Billen, C. Ventresque, and G.P. Bablon. 1991. Microbial activity in GAC filters at the Choisy-le-Roi treatment plant. J. Am. Water Works. Assoc. 83: 62–68.

Severin, B.F. 1980. Disinfection of municipal wastewater effluents with ultraviolet light. J. Water Pollut. Control Fed. 52: 2007–2018.

Seviour, E.M., C.J. Williams, R.J. Seviour, J.A. Soddell, and K.C. Lindrea. 1990. A survey of filamentous bacterial populations from foaming activated sludge plants in eastern states of Australia. Water Res. 24: 493–498.

Seviour, E.M., C. Williams, B. DeGrey, J.A. Soddell, R.J. Seviour, and K.C. Lindrea. 1994. Studies on filamentous bacteria from australian activated sludge plants. Water Res. 28: 2335–2342.

Seviour, R.J. 2002. Activated sludge-The "G-bacteria", pp. 61–68, In: *Encyclopedia of Environmental Microbiology*, G. Bitton, editor-in-chief, Wiley-Interscience, N.Y.

Seviour, R.J., T. Mino, and M. Onuki. 2003. The microbiology of biological phosphorus removal in activated sludge systems. FEMS Microbiol. Rev. 27: 99–127.

Seymour, I.J., and H. Appleton. 2001. Foodborne viruses and fresh produce. J. Appli. Microbiol. 91: 759–773.

Sezgin, M. 1982. Variation of sludge volume index with activated sludge characteristics. Water Res. 16: 83–88.

Sezgin, M., D. Jenkins, and D.S. Parker. 1978. A unified theory of activated sludge bulking. J. Water Pollut. Control Fed. 50: 362–381.

Sezgin, M., D. Jenkins, and J.C. Palm. 1980. Floc size, filament length and settling properties of prototype activated sludge plants. Prog. Water Technol. 12: 171–182.

Sezgin, M., and P.R. Karr. 1986. Control of actinomycete scum on aeration basins and clarifiers. J. Water Pollut. Control Fed. 58: 972–977.

Sezgin, M., M.P. Lechevalier, and P.R. Karr. 1988. Isolation and identification of actinomycetes present in activated sludge scum. In: *Proceedings of the International Conference on Water and Wastewater Microbiology*, Newport Beach, CA, Feb. 8–11, 1988.

Shamat, N.A., and W.J. Maier. 1980. Kinetics of biodegradation of chlorinated organics. J. Water Pollut. Control Fed. 52: 2158–2166.

Shands, K.N., J.L. Ho, R.D. Meyer, G.W. Gorman, P.H. Edelstein, G.F. Mallison, S.M. Finegold, and D.W. Fraser. 1985. Potable water as a source of Legionnaire's disease. J. Am. Med. Assoc. 253: 1412–1416.

Shao, Y.J., M. Starr, K. Kaporis, H.S. Kim, and D. Jenkins. 1997. Polymer addition as a solution to *Nocardia* foaming problems. Water. Environ. Res. 69: 25–27.

Shapiro, J. 1967. Induced rapid release and uptake of phosphate by microorganisms. Science 155: 1269–1271.

Shapiro, J., G.V. Levin, and H.G. Zea. 1967. Anoxically induced release of phosphate in wastewater treatment. J. Water Pollut. Control Fed. 39: 1810–1818.

Sharma, P.K., D.L. Balkwill, A. Frenkel, and M.A. Vairavamurthy. 2000. A New Klebsiella planticola strain (Cd-1) grows anaerobically at high cadmium concentrations and precipitates cadmium sulfide. Appl. Environ. Microbiol. 66: 3083–3087.

Sharp, D.G., R. Floyd, and J.D. Johnson. 1976. Initial fast reaction of bromine on reovirus in turbulent flowing water. Appl. Environ. Microbiol. 31: 171–181.

Sharp, D.G., D.C. Young, R. Floyd, and J.D. Johnson. 1980. Effect of ionic environment on the inactivation of poliovirus in water by chlorine. Appl. Environ. Microbiol. 39: 530–534.

Shaw, K., S. Walker, and B. Koopman. 2000. Improving filtration of Cryptosporium. J. AWWA 92 (11): 103–111.

Sheehan, T. 2002. Bioterrorism, pp. 771–782, In: *Encyclopedia of Environmental Microbiology*, G. Bitton, editor-in-chief, Wiley-Interscience, N.Y.

Shelton, D.R., and J.M. Tiedje. 1984. General method for determining anaerobic biodegradation potential. Appl. Environ. Microbiol. 47: 850–857.

Sherr, B.F., E.B. Sherr, and R.D. Fallon. 1987. Use of monodispersed, fluorescently labeled bacteria to estimate *in situ* protozoan bacterivory. Appl. Environ. Microbiol. 53: 958–965.

Shiaris, M.P., A.C. Rex, G.W. Pettibone, K. Keay, P. McNamus, M.A. Rex, J. Ebersole, and E. Gallagher. 1987. Distribution of indicator bacteria and *Vibrio parahaemolyticus* in sewage-polluted intertidal sediments. Appl. Environ. Microbiol. 53: 1756–1761.

Shields, J.M., and B.H. Olson. 2003. PCR-restriction fragment length polymorphism method for detection of *Cyclospora cayetanensis* in environmental waters without microscopic confirmation. Appl. Environ. Microbiol. 69: 4662–4669.

Shields, P.A., and S.R. Farrah. 2002. Characterization of virus adsorption by using DEAE-sepharose and octyl-sepharose. Appl. Environ. Microbiol. 6: 3965–3968.

Shieh, Y.-C., R.S. Baric, J.W. Woods, and K.R. Calci. 2003. Molecular surveillance of enterovirus and Norwalk-like virus in oysters relocated to a municipal-sewage-impacted Gulf estuary. Appl. Environ. Microbiol. 69: 7130–7136.

Shieh, Y.-S.C., S.S. Monroe, R.L. Fankhauser, G.W. Langlois, W. Burkhardt III, and R.S. Baric. Detection of Norwalk like virus in shellfish implicated in illness. J. Infect. Dis. (In Press).

Shih, J.L., and J. Lederberg. 1974. Effect of chloramine on *Bacillus subtilis* deoxyribonucleic acid. J. Bacteriol. 125: 934–945.

Shilo, M. 1979. *Strategies of Microbial Life in Extreme Environments.* Verlag Chemie, New York.

Shim, S.S., and K. Kawamoto. 2002. Enzyme production activity of Phanerochaete chrysosporium and degradation of pentachlorophenol in a bioreactor. Water Res. 36: 4445–4454.

Shimshony, A. 1997. Epidemiology of emerging zoonoses in Israel. Emerg. Infect. Dis. 3: 229–238.

Shin, G.-A., K.G. Linden, M.J. Arrowood, and M.D. Sobsey. 2001. Low-pressure UV inactivation and DNA repair potential of *Cryptosporidium parvum* oocysts. Appl. Environ. Microbiol. 67: 3029–3032.

Shonheit, P., J. Moll, and R.K. Thauer. 1979. Nickel, cobalt, and molybdenum requirement for growth of *Methanobacterium thermoautotrophicum.* Arch. Microbiol. 123: 105–107.

Shonheit, P., J.K. Kristjansson, and R.K. Thauer. 1982. Kinetic mechanism for the ability of sulfate reducers to outcompete methanogens for acetate. Arch. Microbiol. 132: 285–288.

Shoop, D.S., L.L. Myers, and J.B. LeFever. 1990. Enumeration of enterotoxigenic *Bacteroides fragilis* in municipal sewage. Appl. Environ. Microbiol. 56: 2243–2244.

Shugatt, R.H., D.P. O'Grady, S. Banerjee, P.H. Howard, and W.E. Gledhill. 1984. Shake flask biodegradation of 14 commercial phthalate esters. Appl. Environ. Microbiol. 47: 601–606.

Shuval, H.I. Ed. 1977. *Water Renovation and Reuse.* Academic Press, New York.

Shuval, H. 1991. Health Guidelines and standards for wastewater reuse in agriculture: Historical perpectives. 1991. Water Sci. Technol. 23: 2073–2080.

Shuval, H.I. 1992. Investigation of cholera and typhoid fever transmission by raw wastewater irrigated in Santiago, Chile. Presented at the International Water Pollution Research International Symposium, Washington, D.C., May 26–29, 1992.

Shuval, H.I., A. Thompson, B. Fattal, S. Cymbalista, and Y. Wiener. 1971. Natural virus inactivation processes in seawater. J. San. Eng. Div. ASCE 97: 587–600.

Shuval, H.I., Yekutiel, P., and B. Fattal. 1986. An epidemiological model of the potential health risk associated with various pathogens in wastewater irrigation. Water Sci. Technol. 18: 191–198.

Shuval, H.I., N. Guttman-Bass, J. Applebaum, and B. Fattal. 1989. Aerosolized enteric bacteria and viruses generated by spray irrigation of wastewater. Water Sci. Technol. 21: 131–135.

Sias, S.R., A.S. Stouthamer, and J.L. Ingraham. 1980. The assimilatory and dissimilatory nitrate reductases of *Pseudomonas aeruginosa* are encoded by different genes. J. Gen. Microbiol. 118: 229–234.

Sibille, I., T. Sime-Ngando, L. Mathieu, and J.C. Block. 1998. Protozoan bacterivory and *Escherichia coli* survival in drinking water distribution systems. Appl. Environ. Microbiol. 64: 197–202.

Sidhu, J., R.A. Gibbs, G.E. Ho, and I. Unkovich. 2001. The role of indigenous microorganisms in suppression of salmonella regrowth in composted biosolids. Water Res. 35: 913–920.

Siefert, E., R.L Irgens, and N. Pfenning. 1978. Phototrophic purple and green bacteria in a sewage treatment plant. Appl. Environ. Microbiol. 35: 38–44.

Silver, S., and T.K. Misra. 1988. Plasmid-mediated heavy metal resistance. Annu. Rev. Microbiology 42: 717–743.

Simon, H.M., and R.M. Goodman. 2002. Archaea in soil habitats, pp. 293–305, In: *Encyclopedia of Environmental Microbiology*, G. Bitton, editor-in-chief, Wiley-Interscience, N.Y.

Simmons III, O.D., M.D. Sobsey, C.D. Heaney, F.W. Schaefer II, and D.S. Francy. 2001. Concentration and detection of *Cryptosporidium* oocysts in surface water samples by method 1622 using ultrafiltration and capsule filtration. Appl. Environ. Microbiol. 67: 1123–1127.

Sims, J.L., J.M. Suflita, and H.H. Russell. 1991. *Reductive Dehalogenation of Organic Contaminants in Soils and Ground Water.* Report No. EPA-540/4-90/054, U.S. Environmental Protection Agency, R.S. Kerr Environmental Research Laboratory, Ada, OK.

Singh, A., and G.A. McFeters. 1986. Recovery, growth, and production of heat-stable enterotoxin by *Escherichia coli* after copper-induced injury. Appl. Environ. Microbiol. 51: 738–742.

Singh, A., and G.A. McFeters. 1987. Survival and virulence of copper- and chlorine-stressed *Yersinia enterocolitica* in experimentally infected mice. Appl. Environ. Microbiol. 53: 1768–1774.

Singh, A., and G.A. McFeters. 1990. Injury of enteropathogenic bacteria in drinking water, pp. 368–379, In: *Drinking Water Microbiology*, G.A. McFeters, Ed., Springer Verlag, New York.

Singh, A., M.W. LeChevallier, and G.A. McFeters. 1985. Reduced virulence of *Yersinia enterocolitica* by copper-induced injury. Appl. Environ. Microbiol. 50: 406–411.

Singh, A., R. Yeager, and G.A. McFeters. 1986a. Assessment of *in vivo* revival, growth and pathogenicity of *E. coli* strains after copper- and chlorine-induced injury. Appl. Environ. Microbiol. 52: 832–837.

Singh, S.N., M. Bassous, C.P. Gerba, and L.M. Kelley. 1986b. Use of dyes and proteins as indicators of virus adsorption to soils. Water Res. 20: 267–272.

Sinton, L.W. 1986. Microbial contamination of alluvial gravel aquifers by septic tank effluents. Water Air Soil Pollut. 28: 407–425.

Sivela, S., and V. Sundman. 1975. Demonstration of *Thiobacillus* type bacteria which utilize methyl sulfides. Arch. Microbiol. 103: 303–304.

Sivonen, K. 1999. Toxins produced by cyanobacteria. Vesitalous 5: 11–18.

Sivonen, K., M. Namikoshi, W.R. Evans, W.W. Carmichael, F. Sun, L. Rouhiainen, R. Luukkainen, and K.L. Rinehart. 1992. Isolation and characterization of a variety of microcystins from seven strains of the cyanobacterial genus *Anabaena.* Appl. Environ. Microbiol. 58: 2495–2500.

Skaliy P., T.A. Thompson, G.W. Gorman, G.K. Morris, H.V. McEachern, and D.C. Mackel. 1980. Laboratory studies of disinfectants against *Legionella pneumophila.* Appl. Environ. Microbiol. 40: 697–700.

Skinhoj, P., F.B. Hollinger, K. Hovind-Hougen, and P. Lous. 1981. Infectious liver diseases in three groups of Copenhagen workers: Correlations of hepatitis A infections to sewage exposure. Arch. Environ. Health 36: 139–143.

Skraber, S., C. Gantzer, A. Maul, and L. Schwartzbrod. 2002. Fates of bacteriophages and bacterial indicators in the Moselle river (France). Water Res. 36: 3629–3637.

Slezak, L.A., and R.C. Sims. 1984. The application and effectiveness of slow sand filtration in the United States. J. Am. Water Works Assoc. 76: 38–43.

Slijkhuis, H., and M.H. Deinema. 1988. Effect of environmental conditions on the occurrence of *Microthrix parvicella* in activated sludge. Water Res. 22: 825–828.

Slomczynski, D., J.P. Nakas, and S.W. Tanenbaum. 1995. Production and characterization of laccase from *Botrytis cinerea* 61-34. Appl. Environ. Microbiol. 61: 907–912.

Small, I.C., and G.F. Greaves. 1968. A survey of animals in distribution systems. Water Treat. Exam. 19: 150–183.

Smith, A.J., and D.S. Hoare. 1968. Acetate assimilation by *Nitrobacter agilis* in relation to it's "obligate autotrophy". J. Bacteriol. 95: 844–855.

Smith, C.A., C.B. Phiefer, S.J. Macnaughton, A. Peacock, R.S. Burkhalter, R. Kirkegaard, and D.C. White. 2000. Quantitative lipid biomarker detection of unculturable microbes and chlorine exposure in water distribution system biofilms. Water Res. 34: 2683–2688.

Smith, D.B. 2002. Coliform bacteria: Control in drinking water distribution systems, pp. 905–914, In: *Encyclopedia of Environmental Microbiology*, G. Bitton, editor-in-chief, Wiley-Interscience, N.Y.

Smith, H.V. 1996. Detection of *Giardia* and *Cryptosporidium* in water: Current status and future prospects, pp. 195–225, In: *Molecular Approaches to Environmental Microbiology*, R.W. Pickup, and J.R. Saunders, Eds., Ellis Horwood, Chichester, U.K.

Smith, M.R., and R.A. Mah. 1978. Growth and methanogenesis by *Methanosarcina* strain 227 on acetate and methanol. Appl. Environ. Microbiol. 36: 870–879.

Smith, M.S., M.K. Firestone, and J.M. Tiedje. 1978. Acetylene inhibition method for short-term measurement of soil denitrification and its evaluation using nitrogen-13. Soil Sci Soc. Am. J. 42: 611–615.

Smith-Somerville, H.E., V.B. Huryn, C. Walker, and A.L. Winters. 1991. Survival of *Legionella pneumophila* in the cold-water ciliate *Tetrahymena vorax*. Appl. Environ. Microbiol. 57: 2742–2749.

Snaidr, J., R. Amann, I. Huber, W. Ludwig, and K.-H. Schleifer. 1997. Phylogenetic analysis and *in situ* identification of bacteria in activated sludge. Appl. Environ. Microbiol. 63: 2884–2896.

Snoeyink, V.L., and D. Jenkins. 1980. *Water Chemistry*. John Wiley & Sons, New York.

Snyder, J.W., Jr., C.N. Mains, R.E. Anderson, and G.K. Bissonnette. 1995. Effect of point-of-use, activated carbon filters on the bacteriological quality of rural groundwater supplies. Appl. Environ. Microbiol. 61: 4291–4295.

Snyder, S.A., D.L. Villeneuve, E.M. Snyder, and J.P. Giesy. 2001. Identification and quantification of estrogen receptors agonists in wastewater effluents. Environ. Sci. Technol. 35: 3620–3625.

Soave, R. 1996. *Cyclospora*: An overview. Clin. Infect. Dis. 23: 429–437.

Sobsey, M.D. 1989. Inactivation of health-related microorganisms in water by disinfection processes. Water Sci. Technol. 21: 179–195.

Sobsey, M.D., C.W. Dean, M.E. Knuckles, and R.A. Wagner. 1980. Interactions and survival of enteric viruses in soil materials. Appl. Environ. Microbiol. 40: 92–101.

Sobsey, M.D., and B. Olson. 1983. Microbial agents of waterborne disease, In: *Assessment of Microbiology and Turbidity Standards For Drinking Water*, P.S. Berger, and Y. Argaman, Eds., Report No. EPA-570-9-83-001, U.S. Environmental Protection Agency, Office of Drinking Water, Washington, D.C.

Sobsey, M.D., P.A. Shields, F.H. Hauchman, R.L. Hazard, and L.W. Canton III. 1986. Survival and transport of hepatitis A virus in soils, groundwater and wastewater. Water Sci. Technol. 18: 97–106.

Sobsey, M.D., T. Fuji, and P.A. Shields. 1988. Inactivation of hepatitis A virus and model viruses in water by free chlorine and monochloramine, In: *Proceedings of the International Conference on Water and Wastewater Microbiology*, IAWPRC, Pergamon Press, New York.

Soddell, J.A., and R.J. Seviour. 1990. Microbiology of foaming in activated sludge foams. J. Appl. Bacteriol. 69: 145–176.

Soddell, J.A., and R.J. Seviour. 1994. Incidence and morphological variability of *Nocardia pinensis* in Australian activated sludge plants. Water Res. 28: 2343–2351.

Sofer, S.S., G.A. Lewandowski, M.P. Lodaya, F.S. Lakhwala, K.C. Yang, and M. Singh. 1990. Biodegradation of 2-chlorophenol using immobilized activated sludge. J. Water Pollut. Control Fed. 62: 73–80.

Sole, M., M.J. Lopez de Alda, M. Castillo, C. Porte, K. Ladegaard-Pedersen, and D. Barcelo. 2000. Estrogenicity determination in sewage treatment plants and surface waters from Catalonian area (NE Spain). Environ. Sci. Technol. 34: 5076–5083.

Sommer, R., G. Weber, A. Cabaj, J. Wekerle, G. Keck, and G. Schauberger. 1993. UV inactivation of microorganisms in water. Zbl. Hyg. 189:. 214–224.

Sommer, R., G. Weber, A. Cabaj, J. Wekerle, and G. Schauberger. 1993. UV-inactivation of microorganisms in water. Zentralbl. Hyg. Umweltmed 189: 214–224.

Sommer, R., W. Pribil, S. Appelt, P. Gehringer, H. Eschweiler, H. Leth, A. Cabaj, and T. Haider. 2001. Inactivation of bacteriophages in water by means of non-ionizing (UV-253.7 nm) and ionizing (gamma) radiation: a comparative approach. Water Res. 35: 3109–3116.

Sonnenberg, A., and J.E. Everhart. 1997. Health impact of peptic ulcer in the United States. Am. J. Gastroenterol. 92: 614–620.

Soracco, R.J., H.K. Gill, C.B. Fliermans, and D.H. Pope. 1983. Susceptibility of algae and *Legionella pneumophila* to cooling tower biocides. Appl. Environ. Microbiol. 45: 1254–1260.

Sorber, C.A., B.E. Moore, D.E. Johnson, H.J. Harding, and R.E. Thomas. 1984. Microbiological aerosols from the application of liquid sludge to land. J. Water Pollut. Control Fed. 56: 830–836.

Sorensen, J., D. Christensen, and B.B. Jorgensen. 1981. Volatile fatty acids and hydrogen as substrates for sulfate-reducing bacteria in anaerobic marine sediments. Appl. Environ. Microbiol. 42: 5–11.

Sorvillo, F.J., S.H. Waterman, J.K. Vogt, and B. England. 1988. Shigellosis associated with recreational water contact in Los Angeles County. Am. J. Trop. Med. Hyg. 38: 613–617.

Sorvillo, F.J., K. Fujioka, B. Nahlen, M.P. Tormey, R. Kebabjian, and L. Mascola. 1992. Swimming-associated crytosporidiosis. Am. J. Public Health 82: 742–744.

Speece, R.E. 1983. Anaerobic biotechnology for industrial wastewater treatment. Environ. Sci. Technol. 17: 416A–427A.

Speece, R.E., G.F. Parkin, and D. Gallagher. 1983. Nickel stimulation of anaerobic digestion. Water Res. 17: 677–683.

Spendlove, J.C., and K.F. Fannin. 1983. Source, significance and control of indoor microbial aerosols: Human health effects. Public Health. Rep. 98: 229–244.

Spotts, W.E.A., M.E. Beatty, T.H. Taylor Jr., R. Weyant, J. Sobel, M.J. Arduino, and D.A. Ashford. 2003. Inactivation of *Bacillus anthracis* spores. Emerg. Infect. Dis. 9: 623–627.

Sproul, O.J., R.M. Pfister, and C.K. Kim. 1982. The mechanism of ozone inactivation of water borne viruses. Water Sci. Technol. 14: 303–314.

Staley, J.T., J. Crosa, F. DeWalle, and D. Carlson. 1988. Effect of wastewater disinfectants on survival of R-factor coliform bacteria. Report No. EPA-600/S2-87/092, U.S. Environmental protection Agency, Cincinnati, OH.

Stampi, S., G. De Luca, and F. Zanetti. 2001. Evaluation of the efficiency of peracetic acid in the disinfection of sewage effluents. J. Appl. Microbiol. 91: 833–838.

Stanfield, G., and P.H. Jago. 1987. The development and use of a method for measuring the concentration of assimilable organic carbon in water. Report PRU 1628-M, Water Res. Centre, Medmenham, UK.

Stanier, R.Y., E.A. Adelberg, J.L. Ingraham, and M.L. Wheelis. 1979. *Introduction to the Microbial World*. Prentice-Hall, Englewood Cliffs, N.J., 468 pp.

Stanley, S., S.D. Prior, D.J. Leak, and H. Dalton. 1983. Copper stress underlies the fundamental change in intracellular location of methane monooxygenase in methane-oxidizing organisms: Studies in batch and continuous cultures. Biotechnol. Lett. 5: 487–492.

States, S., J. Newberry, J. Wichterman, J. Kuchta, M. Scheuring, and L. Casson. 2004. Rapid analytical techniques for drinking water security investigations. J. Amer. Water Wks. Assoc. 96 (1): 52–64.

States, S., M. Scheuring, R. Evans, E. Buzza, B. Movahed, T. Gigliotti, and L. Casson. 2000. Membrane filtration as posttreatment. J. AWWA 92 (8): 59–68.

States, S.J., L.F. Conley, M. Ceraso, T.E. Stephenson, R.S. Wolford, R.M. Wadosky, A.M. McNamara, and R.B. Yee. 1985. Effect of metals on Legionella pneumophila growth in drinking water plumbing systems. Appl. Environ. Microbiol. 50: 1149–1154.

States, S.J., L.F. Conley, S.G. Towner, R.S. Wolford, T.E. Stephenson, A.M. McNamara, R.M. Wadowsky, and R.B. Yee. 1987. An alkaline approach to treating cooling waters for control of *legionella pneumophila*. Appl. Environ. Microbiol. 53: 1775–1779.

States, S.J., J.M. Kuchta, L.F. Conley, R.S. Wolford, R.M. Wadowsky, and R.B. Yee. 1989. Factors affecting the occurrence of the legionnaires' disease bacterium in public water supplies, pp. 67–83, In: *Biohazards of Drinking Water Treatment*, R.A. Larsen, Ed., Lewis, Chelsea, MI.

States, S.J., R.M. Wadowsky, J.M. Kuchta, R.S. Wolford, L.F. Conley, and R.B. Yee. 1990. *Legionella* in drinking water, pp. 340–367, In: *Drinking Water Microbiology*, G.A. McFeters, Ed., Springer Verlag, New York.

Stathopoulos, G.A., and T. Vayonas-Arvanitidou. 1990. Detection of *Campylobacter* and *Yersinia* species in waters and their relatioship to indicator microorganisms, In: *International Symposium on Health-Related Water Microbiology*, Tubingen, Germany, April 1–6, 1990.

Steele, M., S. Unger, and J. Odumeru. 2003. Sensitivity of PCR detection of Cyclospora cayetanensis in raspberries, basil, and mesclun lettuce. J. Microbiol. Meth. 54: 277–280.

Steinmann, C.R., S. Weinhart, and A. Melzer. 2003. A combined system of lagoon and constructed wetland for an effective wastewater treatment. Water Res. 37: 2035–2042.

Steger, K., Å. Jarvis, S. Smårs, and I. Sundh. 2003. Comparison of signature lipid methods to determine microbial community structure in compost. J. Microbiol. Meth. 55: 371–382.

Stelzer, W. 1990. Detection and spread of *Campylobacter* in water (Abst.), In: *International Symposium on Health-Related Water Microbiology*, Tubingen, Germany, April 1–6, 1990.

Stender, H., A.J. Broomer, K. Oliveira, H. Perry-O'Keefe, J.J. Hyldig-Nielsen, A. Sage, and J. Coull. 2001. Rapid detection, identification, and enumeration of *Escherichia coli* cells in municipal water by chemiluminescent *in situ* hybridization. Appl. Environ. Microbiol. 67: 142–147.

Stenquist, R.J., D.S. Parker, and T.J. Dosh. 1974. Carbon oxidation-nitrification in synthetic media trickling filters. J. Water Pollut. Control Fed. 46: 2327–2339.

Stenstrom, M.K., and S.S. Song. 1991. Effect of oxygen transport limitation on nitrification in the activated sludge process. Res. J. Water Pollut. Control Fed. 63: 208–219.

Sterritt, R.M., and J.N. Lester. 1986. Heavy metals immobilisation by bacterial extracellular polymers, pp. 121–134, In: *Immobilisation of Ions by Bio-Sorption*, H. Eccles, and S. Hunt, Eds., Ellis Horwood, Chichester, U.K.

Sterrit, R.M., and J.N. Lester. 1988. *Microbiology for Environmental and Public Health Engineers*. E. & F.N. Spon, London.

Stetzenbach, L.D. 1997. Introduction to aerobiology, pp. 619–628, In: *Manual of Environmental Microbiology*, C.J. Hurst, G.R. Knudsen, M.J. McInerney, L.D. Stetzenbach, and M.V. Walter, Eds., ASM Press, Washington, D.C.

Stewart, M.H. 2002. Spa and hot tub microbiology, pp. 2978–2984, In: *Encyclopedia of Environmental Microbiology*, G. Bitton, editor-in-chief, Wiley-Interscience, N.Y.

Stewart, M.H., and B.H. Olson. 1992a. Impact of growth conditions on resistance of *Klebsiella pneumoniae* to chloramines. Appl. Environ. Microbiol. 58: 2649–2653.

Stewart, M.H., and B.H. Olson. 1992b. Physiological studies of chloramine resistance developed by *Klebsiella pneumonae* under low-nutrient growth conditions. Appl. Environ. Microbiol. 58: 2918–2927.

Stewart, M.H., R.L. Wolfe, and E.G. Means. 1990. Assessment of the bacteriological activity associated with granular activated carbon treatment of drinking water. Appl. Environ. Microbiol. 56: 3822–3829.

Stewart, M.H., M.V. Yates, M.A. Anderson, C.P. Gerba, J.B. Rose, R. De Leon, and R.L. Wolfe. 2002. Predicted public helath consequences of body-contact recreation on a potable water reservoir. J. AWWA 94 (5): 84–97.

St. Laurent, D., C. Blaise, P. Macquarrie, R. Scroggins, and R. Trottier. 1992. Comparative assessment of herbicide phytotoxicity to *Selenastrum capricornutum* using microplate and flask bioassay procedures. Environ. Toxicol. Water Qual. 7: 35–48.

Stites, F.L. 2002. Bioaerosols: Transport and Fate, pp. 425–434, In: *Encyclopedia of Environmental Microbiology*, G. Bitton, editor-in-chief, Wiley-Interscience, N.Y.

Stoodley, P., K. Sauer, D.G. Davies, and J.W. Costerton. 2002. Biofilms as complex differentiated communities. Annu. Rev. Microbiol. 56: 187–209.

Stormo, K.E., and R.L. Crawford. 1992. Preparation of encapsulated microbial cells for environmental applications. Appl. Environ. Microbiol. 58: 727–730.

Stott, R., E. May, and D.D. Mara. 2003a. Parasite removal by natural wastewater treatment systems: performance of waste stabilisation ponds and constructed wetlands. Water Sci. Technol. 48: 97–104.

Stott, R., E. May, E. Ramirez, and A. Warren. 2003b. Predation of *Cryptosporidium* oocysts by protozoa and rotifers: implications for water quality and public health. Water Sci. & Technol. 47 (3): 77–83.

Stott, R., May, E., Matsushita, E., and Warren, A. 2001. Protozoan predation as a mechanism for the removal of *Cryptosporidium* oocysts from wastewaters in constructed wetlands. Wat. Sci. Tech. 44 (11–12): 194–198.

Stout, J.E., M.G. Best, and V.L. Yu. 1986. Susceptibility of members of the family *Legionellaceae* to thermal stress: Implications for heat eradication methods in water distribution systems. Appl. Environ. Microbiol. 52: 396–399.

Stout, J.E., V.L. Yu, and M.G. Best. 1985. Ecology of *Legionella pneumophila* within water distribution systems. Appl. Environ. Microbiol. 49: 221–228.

Stramer, S.L., and D.O. Cliver. 1981. Poliovirus removal from septic tank systems under conditions of saturated flow, (Abst.), In: American Society of Microbiology Annual Meeting, Dallas, TX.

Stramer, S.L., and D.O. Cliver. 1984. Septage treatments to reduce the numbers of bacteria and polioviruses. Appl. Environ. Microbiol. 48: 566–572.

Strand, S.E., J.V. Wodrich, and H.D. Stensel. 1991. Biodegradation of chlorinated solvents in a sparged, methanotrophic biofilm reactor. Res. J. Water Pollut. Control Fed. 63: 859–867.

Stratton, R.G., E. Namkung, and B.E. Rittmann. 1983. Secondary utilization of trace organics by biofilms on porous media. J. Am. Water Works Assoc. 75: 463–469.

Straub, T.M., I.L. Pepper, and C.P. Gerba. 1992. Persistence of viruses in desert soils amended with anaerobically digested sewage sludge. Appl. Environ. Microbiol. 58: 636–641.

Straub, T.M., I.L. Pepper, and C.P. Gerba. 1995. Removal of PCR inhibiting substances in sewage sludge amemded soil. Water Sci. Technol. 31: 311–315.

Strauch, D., A.H. Havelaar, and P. L'Hermite, Eds. 1985. *Inactivation of Microorganisms in Sewage Sludge by Stabilization Processes.* Elsevier, London.

Stringer, R., and C.W. Kruse. 1970. Amoebic cystidal properties of halogens in water. In: Proceedings of the National Specialty Conference on Disinfection, American Society of Civil Engineers, New York.

Strom, A.R., P. Falkenberg, and B. Landfald. 1986. Genetics of osmoregulation in *Escherichia coli*: uptake and biosynthesis of organic solutes. FEMS Microbiol. Rev. 39: 79–86.

Strom, P.F. 1985a. Effect of temperature on bacterial species diversity in thermophilic solid-waste composting. Appl. Environ. Microbiol. 50: 899–905.

Strom, P.F. 1985b. Identification of thermophilic bacteria in solid-waste composting. Appl. Environ. Microbiol. 50: 906–913.

Strom, P.F., and D. Jenkins. 1984. Identification and significance of filamentous microorganisms in activated sludge. J. Water Pollut. Control Fed. 49: 584–589.

Strous, M., J.A. Fuerst, E.H.M. Kramer, S. Logemann, G. Muyzer, K.T. van de Pas-Schoonen, R. Webb, J.G. Kuenen, and M.S.M. Jetten. 1999. Missing lithotroph identified as new planctomycete. Nature 400: 446–449.

Stumm, W., and E. Stumm-Zollinger. 1972. The role of phosphorus in eutrophication, pp. 11–42, In: *Water Pollution Microbiology*, R. Mitchell, Ed., Wiley-Interscience, New York, 416 pp.

Sturbaum, G.D., P.T. Klonicki, M.M. Marshall, B.H. Jost, B.L. Clay, and C.R. Sterling. 2002. Immunomagnetic separation (IMS)-fluorescent antibody detection and IMS-PCR detection of seeded *Cryptosporidium parvum* oocysts in natural waters and their limitations. Appl. Environ. Microbiol. 68: 2991–2996.

Sudo, R., and S. Aiba. 1984. Role and function of protozoa in the biological treatment of polluted waters. Adv. Biochem. Eng. Biotechnol. 29: 117–141.

Suffet, I.H., A. Corado, D. Chou, M.J. McGuire, and S. Butterworth. 1996. Taste and odor survey. J. Amer. Water Works Assoc. 88 (4): 168–180.

Suflita, J.M., and G.W. Sewell. 1991. Anaerobic biotransformation of contaminants in the subsurface. Report No. EPA-600-M-90/024, U.S. Environmental Protection Agency, R.S. Kerr Environmental Research Laboratory, Ada, O.K.

Suidan, M.T., W.H. Gross, M. Fong, and J.W. Calvert. 1981. Anaerobic carbon filter for degradation of phenols. J. Environ. Eng. Div. ASCE 107: 563–579.

Sujarittanonta, S., and J.H. Sherrard. 1981. Activated sludge nickel toxicity studies. J. Water Pollut. Control Fed. 53: 1314–1322.

Sullivan, J.P. Jr. 2004. The WateISAC – One year later. J. Amer. Water Works Assoc. 96 (1): 29–32.

Sun, D.D., J.H. Tay, and K.M. Tan. 2003. Photocatalytic degradation of *E. coliform* in water. Water Res. 37: 3452–3462.

Sunada, K., Y. Kikuchi, K. Hashimoto, and A. Fujishima. 1998. Bactericidal and detoxification effects of TiO_2 thin film photocatalysis. Environ. Sci. Technol. 32: 726–728.

Suriyawattanakul, L., W. Surareungchai, P. Sritongkam, M. Tanticharoen, and K. Kirtikara. 2002. The use of co-immobilization of *Trichosporon cutaneum* and Bacillus licheniformis for a BOD sensor. Appl. Microbiol. Biotechnol. 59: 40–44.

Sutherland, I.W. 1977. Enzyme action on bacterial surface carbohydrates, pp. 209–245, In: *Surface Carbohydrates of the Prokaryotic Cell*, I.W. Sutherland, Ed., Academic Press, New York.

Suttle, C.A., and F. Chen. 1992. Mechanisms and rates of decay of marine viruses in seawater. Appl. Environ. Microbiol. 58: 3721–3729.

Sutton, L.D., W.W. Wilke, N.A. Lynch, and R.N. Jones. 1995. *Helicobacter pylori*-containing sewage detected by an automated polymerase chain reaction amplication procedure (Abst. No. C-395), In: 95th General Meeting of the American Society of Microbiology, Washington, D.C.

Sutton, R. 1992. Use of biosurfactants produced by *Nocardia amarae* for removal and recovery of non-ionic organics from aqueous solutions. Water Sci. Technol. 26: 2393–2396.

Sutton, S.D. 2002. Quatification of microbial biomass, pp. 2652–2660, In: *Encyclopedia of Environmental Microbiology*, G. Bitton, Ed., Wiley-Interscience, N.Y.

Suylen, G.M.H., and J.G. Kuenen. 1986. Chemostat enrichment and isolation of *Hyphomicrobium* HG, a dimethyl sulfide oxidizing methylotroph and reevaluation of *Thiobacillus* MS1. Antonie van Leeuwenhoek J. Microbiol. Serol. 52: 281–293.

Swartz, M.N. 1997. Use of antimicrobial agents and drug resistance. N. Engl. J. Med. 337: 491–492.

Swerdlow, D.L., B.A. Woodruff, R.C. Brady, P.M. Griffin, S. Tippen, H.D. Donnell, Jr., E.E. Geldreich, B.J. Payne, A. Meyer, Jr., J.S. Wells, K.D. Greene, M. Bright, N.H. Bean, and P.A. Blake. 1992. A waterborne outbreak in Missouri of *Escherichia coli* 0157:H7 associated with bloody diarrhea and death. Ann. Intern. Med. 117: 812–819.

Switzenbaum, M.S. 1983. Anaerobic treatment of wastewater: Recent developments. ASM News 49: 532–536.

Sykes, G., and F.A Skinner, Eds. 1973. *Actinomycetales: Characteristics and practical importance*. Academic Press, London.

Sykes, J.C. 1989. The use of biological selector technology to minimize sludge bulking, pp. C1–C25, In: *Biological Nitrogen and Phosphorus Removal: The Florida Experience*, TREEO Center, Univ. of Florida, Gainesville, FL., March 22–23, 1989.

Sykora, J.L., G. Keleti, R. Roche, D.R. Volk, G.P. Kay, R.A. Burgess, M.A. Shapiro, and E.C. Lippy. 1980. Endotoxins, algae and *Limulus* amoebocyte lysate test in drinking water. Water Res. 14: 829–839.

Sykora, J.L., C.A. Sorber, W. Jakubowski, L.W. Casson, P.D. Cavaghan, and M.A. Shapiro. 1990. Distribution of *Giardia* cysts in wastewater, In: *International Symposium on Health-Related Water Microbiology*, Tubingen, Germany, April 1–6, 1990.

Szymona, M., and W. Ostrowski. 1964. Inorganic polyphosphate glucokinase of *Mycobacterium phlei*. Biochim. Biophys. Acta 85: 283–295.

Tago, Y., and K. Aida. 1977. Exocellular micropolysaccharide closely related to bacterial floc formation. Appl. Environ. Microbiol. 34: 308–314.

Takada, S., M. Nakamura, R. Kondo, and K. Sakai. 1996. Degradation of polychlorinated dibenzo-*p*-dioxins and polychlorinated dibenzofurans by the white rot fungus *Phanerochaete sordida* YK-624. Appl. Environ. Microbiol. 62: 4323–4328.

Takase, L., T. Omori, and Y. Minoda. 1986. Microbial degradation products from biphenyl-related compounds. Agric. Biol. Chem. 50: 681–686.

Talbot, H.W., J.E. Morrow, and R.J. Seidler. 1979. Control of coliform bacteria in finished drinking water stored in redwood tanks. J. Am. Water Works. Assoc. 71: 349–353.

Tanaka H., M. Minamiyama, T. Toyoda, K. Komori, Y. Tanaka, and K. Kaguchi. 1998. Development of atoxicity monitor using bacteria. *Proceedings of WEFTEC '98 Water Environment Federation*, 71st Vol. 4, 165–180.

Tanaka, J.-I., K. Syutsubo, K. Watanabe, H. Izumida, and S. Harayama. 2003. Activity and population structure of nitrifying bacteria in an activated-sludge reactor containing polymer beads. Environ. Microbiol. 5: 278–286.

Tanaka, K., M. Tada, T. Kimata, S. Harada, Y. Fujii, T. Mizugushi, N. Mori, and H. Emori. 1991. Development of new nitrogen system using nitrifying bacteria in synthetic resin pellets. Water Sci. Technol. 23: 681–690.

Tanaka, K., T. Sumino, H. Nakamura, T. Ogasawara, and H. Emori. 1996. Application of nitrification by cells immobilised in polyethylene glycol. Prog. Biotechnol. 11: 622–632.

Tartera, C., and J. Jofre. 1987. Bacteriophage active against *Bacteroides fragilis* in sewage-polluted waters. Appl. Environ. Microbiol. 53: 1632–1637.

Tartera, C., F. Lucena, and J. Jofre. 1989. Human origin of *Bacteroides fragilis* bacteriophages present in the environment. Appl. Environ. Microbiol. 55: 2696–2701.

Tawfik, A., B. Klapwijk, F. El-Gohary, and G. Lettinga. 2002. Treatment of anaerobically treated domestic wastewater using rotating biological contactor. Water Sci. Technol. 45 (10): 371–376.

Tawfik, A., B. Klapwijk, J. Van Buuren, F. El-Gohary, and G. Lettinga. 2004. Physico-chemical factors affecting the *E. coli* removal in a rotating biological contactor (RBC) treating UASB effluent. Water Res. 38: 1081–1088.

Taylor, D.N., K.T. McDermott, J.R. Little, J.G. Wells, and M.J. Blaser. 1983. *Campylobacter* enteritis from untreated water in the Rocky Mountains. Ann. Intern. Med. 99: 38–40.

Taylor, G.R., and M. Butler. 1982. A comparison of the virucidal properties of chlorine, chlorine dioxide, bromine chloride and iodine. J. Hyg. 89: 321–328.

Taylor, R.H., M.J. Allen, and E.E. Geldreich. 1979. Testing of home carbon filters. J. Am. Water Works Assoc. 71: 577–581.

Taylor, R.H., J.O. Falkinham III, C.D. Norton, and M.W. LeChevallier. 2000. Chlorine, chloramine, chlorine dioxide, and ozone susceptibility of *Mycobacterium avium*. Appl. Environ. Microbiol. 66: 1702–1705.

Teeuw, K.B., C.M.J.E. Vandenbroucke-Grauls, and J. Verhoef. 1994. Airborne gram-negative bacteria and endotoxin in sick building syndrome. A study in Dutch governmental office buildings. Arch. Internal Med. 154: 2339–2349.

Teitzel, G.M., and M.R. Parsek. 2003. Heavy metal resistance of biofilm and planktonic *Pseudomonas aeruginosa*. Appl. Environ. Microbiol. 69: 2313–2320.

Teltsch, B., and E. Katzenelson. 1978. Airborne enteric bacteria and viruses from spray irrigation with wastewater. Appl. Environ. Microbiol. 35: 290–296.

Teltsch, B., S. Kedmi, Y. Bonnet, Borenzstajn-Rotem, and E. Katzenelson. 1980. Isolation and identification of pathogenic microorganisms at wastewater irrigation fields: Ratios in air and wasterwater. Appl. Environ. Microbiol. 39: 1183–1190.

Tenney, M.W., and W. Stumm. 1965. Chemical flocculation of microorganisms in biological waste treatment. J. Water Pollut. Control Fed. 37: 1370–1388.

Ternes, T.A., M. Stumpf, J. Mueller, K. Haberer, R.D. Wilken, and M. Servos. 1999. Behaviour and occurrence of estrogens in municipal sewage treatment plants. I. Inverstigations in Germany, Canada and Brazil. Sci. Total Environ. 225: 81–90.

Tetreault, M.J., A.H. Benedict, C. Kaempfer, and E.F. Barth. 1986. Biological phosphorus removal: A technology evaluation. J. Water Pollut. Control Fed. 58: 823–837.

Theron, J., and T.E. Cloete. 2002. Emerging waterborne infections: Contributing factors, agents, and detection tools. Crit. Rev. Microbiol. 28: 1–26.

Thomas, J.A., J.A. Soddell, and D.I. Kurtböke. 2002. Fighting foam with phages? Water Sci. & Technol. 46 (1–2): 511–518.

Thomas, J.M., and C.H. Ward. 1989. *In situ* biorestoration of organic contaminants in the subsurface. Environ. Sci. Technol. 23: 760–766.

Thompson, S.S., J.L. Jackson, M. Suva-Castillo, W.A. Yanko, Z. El Jack, J. Kuo, C.-L. Chen, F.P. Williams, and D.P. Schnurr. 2003. Detection of infectious adenoviruses in tertiary-treated and untraviolet-disinfected wastewater. Water Environ. Res. 75: 163–170.

Thorne, P.S., M.S. Kiekhaefer, P. Whitten, and K.J. Donham. 1992. Comparison of bioaerosol sampling methods in barns housing swine. Appl. Environ. Microbiol. 58: 2543–2551.

Thyndall, R.L., and E.L. Domingue, 1982. Cocultivation of *Legionella pneumophila* and free-living amoebae. Appl. Environ. Microbiol. 44: 954–959.

Tiedje, J.M. 1988. Ecology of denitrification and dissimilatory nitrate reduction to ammonium, pp. 179–244, In: *Biology of Anaerobic Microorganisms*, A.J.B. Zehnder, Ed., John Wiley & Sons, New York.

Tiedje, J.M., S.A. Boyd, and B.Z. Fathepure. 1987. Anaerobic biodegradation of aromatic hydrocarbons. Dev. Ind. Microbiol. 27: 117–127.

Tiedje, J.M., R. Colwell, Y.L. Grossman, R.E. Hodson, R.E. Lenski, R.N. Mack, and P.J. Regal. 1989. The planned introduction of genetically engineered organisms: Ecological considerations and recommendations. Soc. Ind. Microbiol. News 39: 149–165.

Tillet, D.M., and A.S. Myerson. 1987. The removal of pyritic sulfur from coal employing *Thiobacillus ferrooxidans* in a packed column reactor. Biotech. Bioeng. 29: 146–150.

Tim, U.S., and S. Mostaghimi. 1991. Model for predicting virus movement through soils. Ground Water 29: 251–259.

Timmis, K.N., R.J. Steffan, and R. Unterman. 1994. Designing microorganisms for the treatment of toxic wastes. Annu. Rev. Microbiol. 48: 525–557.

Tison, D.L., D.H. Pope, W.B. Cherry, and C.B. Fliermans. 1980. Growth of *Legionella pneumophila* in association with blue-green algae. Appl. Environ. Microbiol. 39: 456–459.

Tobien, T., W.J. Cooper, M.G. Nickelsen, E. Pernas, K.E. O'Shea, and K.-D. Asmus. 2000. Odor control in wastewater treatment: The removal of thioanisole from water-A model case study by pulse radiolysis and electron beam treatment. Environ. Sci. Technol. 34: 1286–1291.

Tobin, R.S., D.K. Smith, and J.A. Lindsay. 1981. Effects of activated carbon and bacteriostatic filters on microbiological quality of drinking water. Appl. Environ. Microbiol. 41: 646–651.

Tobin, R.S., P. Ewan, K. Walsh, and B. Dutka. 1986. A survey of *Legionella pneumophila* in water in 12 canadian cities. Water Res. 20: 495–501.

Toerien, D.F., A. Gerber, L.H. Lotter, and T.E. Cloete. 1990. Enhanced biological phosphorus removal in activated sludge systems. Adv. Microb. Ecol. 11: 173–230.

Tokuz, Y. 1989. Biodegradation and removal of phenols in rotating biological contactors. Water Sci. Technol. 21: 1751–1754.

Tomlinson, T.G., and I.L. Williams. 1975. Fungi, pp. 93–152, In: *Ecological Aspects of Used Water Treatment*, Vol. 1, C.R. Curds, and H.A. Hawkes, Eds., Academic Press, London.

Topping, B. 1987. The biodegradability of *para*-dichlorobenzene and its behaviour in model activated sludge plants. Water Res. 21: 295–300.

Toranzos, G.A., and G.A. McFeters. 1997. Detection of indicator microorganisms in environmental freshwaters and drinking waters, pp. 184–194, In: *Manual of Environmental Microbiology*, C.J. Hurst, G.R. Knudsen, M.J. McInerney, L.D. Stetzenbach, and M.V. Walter, Eds., ASM Press, Washington, D.C.

Torok, T.J., R.V. Tauxe, R.P. Wise, J.R. Livengood, R. Sokolow, and S. Mauvais. 1997. A large community outbreak of Salmonellosis caused by intentional contamination of restaurant salad bars. J. Amer. Med. Assoc. 278: 389–395.

Torpey, W.N., H. Heukelekian, A.J. Kaplovski, and R. Epstein. 1971. Rotating disks with biological growths prepare wastewater for disposal or reuse. J. Water Pollut. Control Fed. 43: 2181–2188.

Torres, A.G. 2002. Shigella, pp. 2865–2871, In: *Encyclopedia of Environmental Microbiology*, G. Bitton, editor-in-chief, Wiley-Interscience, N.Y.

Tortora, G.J., B.R. Funke, and C.L. Case. 1989. *Microbiology: An Introduction.* Benjamin/Cummings, Redwood City, CA, 810 pp.

Torvinen, E., S. Suomalainen, M.J. Lehtola, I.T. Miettinen, O. Zacheus, L. Paulin, M.-L. Katila, and P.J. Martikainen. 2004. Mycobacteria in water and loose deposits of drinking water distribution systems in Finland. Appl. Environ. Microbiol. 70: 1973–198.

Toze, S., L.I. Sly, I.C. MacRae, and J.A. Fuerst. 1990. Inhibition of growth of *Legionella* species by heterotrophic plate count bacteria isolated from chlorinated drinking water. Curr. Microbiol. 21: 139–143.

Travers, S.M., and D.A. Lovett. 1984. Activated sludge treatment of abattoir wastewater. II. Influence of dissolved oxygen concentration. Water Res. 18: 435–439.

Travis, R.J., and N.D. Rosenberg. 1997. Modeling in situ bioremediation of TCE at Savannah River: Effect of product toxicity and microbial interactions on TCE degradation. Environ. Sci. Technol. 31: 3093–3102.

Trepeta, R.W., and S.C. Edberg. 1984. Methylumbelliferyl-β-D-glucuronide-based medium for rapid isolation and identification of *E. coli*. J. Clin. Microbiol. 19: 172–174.

Trevors, J.T. 1986. Bacterial growth and activity as indicators of toxicity, pp. 9–25, In: *Toxicity Testing Using Microorganisms*, Vol. 1, G. Bitton, and B.J. Dutka, Eds., CRC Press, Boca Raton, FL.

Trevors, J.T. 1989. The role of microbial metal resistance and detoxification mechanisms in environmental bioassay research. Hydrobiology 188: 143 147.

Trevors, J.T., K.M. Oddie, and B.H. Beliveau. 1985. Metal resistance in bacteria. FEMS Microbiol. Rev. 32: 39–54.

Trulear, M.G., and W.G. Characklis. 1982. Dynamics of biofilm processes. J. Water Pollut. Control Fed. 54: 1288–1301.

Tsai, W.-L., C.E. Miller, and E.R. Richter. 2000. Determination of the sensitivity of a rapid *Escherichia coli* O157:H7 assay for testing 375-gram composite samples. Appl. Environ. Microbiol. 66: 4149–4151.

Tsai, Y.-L., and B.H. Olson. 1992. Rapid method for separation of bacterial DNA from humic substances in sediments for polymerase chain reaction. Appl. Environ. Microbiol. 58: 2292–2295.

Tsai, Y.-L., C.J. Palmer, and L.R. Sangermano. 1993. Detection of *Escherichia coli* in sewage and sludge by polymerase chain reaction. Appl. Environ. Microbiol. 59: 353–357.

Tsai, Y.-L., B. Tran, L.R. Sangermano, and C.J. Palmer. 1994. Detection of poliovirus, hepatitis A virus, and rotavirus from sewage and ocean water by triplex reverse transcriptase PCR. Appl. Environ. Microbiol. 60: 2400–2407.

Tsien, H-C, G.A. Brusseau, R.S. Hanson, and L.P. Wackett. 1989. Biodegradation of trichloroethylene by *Methylococcus trichosporium* OB3b. Appl. Environ. Microbiol. 55: 3155–3161.

Tsuneda, S., T. Nagano, T. Hoshino, Y. Ejiri, N. Noda, and A. Hirata. 2003. Characterization of nitrifying granules produced in an aerobic upflow fluidized bed reactor. Water Res. 37: 4965–4973.

Tucker, E.S., V.W. Saeger, and O. Hicks. 1975. Activated sludge primary biodegradation of polychlorinated biphenyls. Bull. Environ. Contam. Toxicol. 14: 705–713.

Tunlid, A. 2002. Lipid biomarkers in environmental microbiology, pp. 1817–1824, In: *Encyclopedia of Environmental Microbiology*, G. Bitton, editor-in-chief, Wiley-Interscience, N.Y.

Tuovinen, O.H., and J.C. Hsu. 1982. Aerobic and anaerobic microorganisms in tubercles of the Columbus, Ohio, water distribution system. Appl. Environ. Microbiol. 44: 761–764.

Turner, A.P.F., I. Karube, and G.S. Wilson, Eds. 1987. *Biosensors: Fundamentals and Applications.* Oxford University Press, Oxford.

Uberol, V., and S.K. Bhattacharya. 1997. Toxicity and degradability of nitrophenols in anaerobic systems. Water Environ. Res. 69: 146–161.

Uchiyama H., T. Nakajima, O. Yagi, and T. Nakahara. 1992. Role of heterotrophic bacteria in complete mineralization of trichloroethylene by *Methylocystis* sp. strain M. Appl. Environ. Microbiol. 58: 3067–3071.

Ueda, T., and N.J. Horan. 2000. Fate of indigenous bacteriophage in a membrane bioreactor. Water Res. 34: 2151–2159.

Uhlmann, D. 1979. *Hydrobiology: A Text for Engineers and Scientists.* John Wiley & Sons, New York.

Ulitzur, S. 1986. Bioluminescence test for genotoxic agents. Methods Enzymol. 133: 264–274.

Unz, R.F., and S.R. Farrah. 1976. Observations on the formation of wastewater zoogleae. Water Res. 10: 665–671.

Unz, R.F., and T.M. Williams. 1988. Effect of controlled pH on the development of rosette-forming bacteria in axenic culture and bulking activated sludge, In: *International Conference on Water and Wastewater Microbiology*, Newport Beach, CA, Feb. 8–11, 1988.

Urbain, V., J.C. Block, and J. Maneim. 1993. Bioflocculation in activated sludge: An analytic approach. Water Res. 27: 829–838.

U.S. Census Bureau. 1993. 1990 Census of population. Characteristics of the population, Vol. 1. U.S. Government Printing Office, Washington, D.C.

U.S. DHEW. 1967. Policy Statement on the Use of Ultraviolet Process for Disinfection of Water, U.S. Department Health, Education, and Walfare, Washington. D.C.

U.S. DHEW. 1969. Manual of Septic Tank Practices, Public Health Service Publication No. 528, Washington, D.C.

U.S. EPA. 1975. *Process Design Manual for Nitrogen Control*, Office of Technology Transfer, Washington, D.C.

U.S. EPA. 1977. *Wastewater Treatment Facilities for Sewered Small Communities*. Report No. EPA-625/1-77-009, U.S. Environmental Protection Agency, Cincinnati, OH.

U.S. EPA. 1980. *Onsite Wastewater Treatment and Disposal Systems*. Report No. EPA-625/1-80-012. U.S. Environmental Protection Agency, Cincinnati, OH.

U.S. EPA. 1981. *Process Design Manual for Land Treatment of Municipal Wastewater*. Report No. EPA-625/1-81-013. U.S. Environmental Protection Agency, Cincinnati, OH.

U.S. EPA. 1982. *Estimating Microorganisms Densities in Aerosols from Spray Irrigation of Wastewater*. Report No. EPA 600/9-82-003. U.S. Environmental Protection Agency, Center for Environmental Research Information, Cincinnati, OH.

U.S. EPA. 1983. *Process Design Manual for Land Application of Municipal Sludge*. Report No. EPA-625/1-83-016, U.S. Environmental Protection Agency, Cincinnati, OH.

U.S. EPA. 1984a. *Environmental Regulations and Technology: Use and Disposal of Municipal Wastewater Sludge*. Report No. EPA-625/10-84-003, U.S. Environmental Protection Agency, Cincinnati, OH.

U.S. EPA. 1984b. *Guidelines for deriving numerical aquatic site-specific water quality criteria by modifying national criteria*. EPA/600/384009, Environ. Res. Lab., Duluth, Minn.

U.S. EPA 1985. *Odor and Corrosion Control in Sanitary Systems and Treatment Plants*. Report No. EPA-625/1-85-018. U.S. Environmental Protection Agency, Cincinnati, OH.

U.S. EPA. 1986. *Septic Systems and Groundwater Protection: An Executive Guide*. Office of Groundwater Protection, Washington, D.C.

U.S. EPA. 1987a. *The Causes and Control of Activated Sludge Bulking and Foaming*. Report No. EPA/625/8-87/012, U.S. Environmental Protection Agency, Cincinnati, OH.

U.S EPA. 1987b. *Design Manual: Phosphorus Removal*. Report # EPA/625/1-87/001, U.S. Environmental Protection Agency, Cincinnati, OH.

U.S. EPA. 1988. *Methods for Aquatic Toxicity Identification Evaluation – Phase I Toxicity Characterization Procedures*. Report No. EPA 600/3-88/034, U.S. Environmental Protection Agency, Duluth, MN.

U.S. EPA. 1989a. *Methods for Aquatic Toxicity Identification Evaluation- Phase II Toxicity Identification Procedures*. Report No. EPA 600/3-88/035, U.S. Environmental Protection Agency, Duluth, MN.

U.S. EPA. 1989b. *Methods for Aquatic Toxicity Identification Evaluation- Phase III Toxicity Confirmation Procedures*. Report No. EPA 600/3-88/036, U.S. Environmental Protection Agency, Duluth, MN.

U.S. EPA. 1989c. Algal (*Selenastrum capricornutum*) growth test, pp. 147–174, In: *Short-Term Methods for Estimating the Chronic Toxicity of Effluents and Receiving Waters to Freshwater Organisms*. Report No. EPA-600/4-89/001, U.S. Environmental Protection Agency, Environmental Monitoring System Laboratory, Cincinnati, OH.

U.S EPA. 1989d. National primary drinking water regulations; filtration and disinfection; turbidity; *Giardia lamblia*; viruses, *Legionella*, and heterotrophic bacteria. Fed. Regist. 54: 27486–27541.

U.S. EPA. 1989e. Drinking Water Health Effects Task Force. *Health Effects of Drinking Water Treatment Technologies*. Lewis, Chelsea, MI.

U.S. EPA. 1989f. *Assessment of Needed Publicly Owned Wastewater Treatment Facilities in the United States*. 1988 Needs Survey Report to Congress, Report No. EPA-430/09-89-001. U.S. Environmental Protection Agency, Cincinnati, OH.

U.S. EPA. 1989g. National Primary Drinking Water Regulations: Total Coliforms (including fecal coliforms and *E. coli*); Final Rule; Fed Reg. 54: 124: 27544–27568 (June 29, 1989).

U.S. EPA. 1992a. *Control of Pathogens and Vector Attraction in Sewage Sludge (including Domestic Septage) Under 40 CFR Part 503*. Report No. EPA/625R-92/013. U.S. Environmental Protection Agency, Cincinnati, OH.

U.S. EPA. 1992b. *Guidelines for Water Reuse*. Report No. EPA/625/R-92/004. U.S. Environmental Protection Agency, Cincinnati, OH.

U.S. EPA. 1994. *Interim Guidance on determination and use of water-effect ratios for metals*. EPA-823-B-94-001, Washington, DC.

U.S. EPA. 1998. Announcement of the drinking water candidates contaminant list. Fed. Reg. 63 (40): 10273 (march 2, 1998).

U.S. EPA. 1999. *Alternative Disinfectants and Oxidants Guidance Manual.* Report # 815-R-99-014, U.S. EPA, Office of Water, Washington, DC.

U.S. EPA. 2000. *Bacterial water quality standards for recreational waters (fresh and marine).* Status report. Office of Science and Technology, Standards and Applied Science Division [Online.] http://www.epa.gov/OST/beaches/.

U.S. EPA. 2001a. Method 1623: *Cryptosporidium* and *Giardia* in water by filtration/IMS/FA. Publication EPA-821-R-99-006. United States Environmental Protection Agency, Washington, D.C.

U.S. EPA. 2001b. Method 1601: Male-specific (F+) and somatic coliphage in water by two-step enrichment procedure: Washington, D.C., U.S. Environmental Protection Agency, EPA-821-R-01-030.

U.S. EPA. 2001c. Method 1602: Male-specific (F+) and somatic coliphage in water by single agar layer (SAL) procedure: Washington, D.C., U.S. Environmental Protection Agency, EPA-821-R-01-029.

U.S. DHEW. 1992. Current estimates from the National Health Interview Survey, 1991. Vital and Health Statistics, Washington, D.C.

Valcke, D., and W. Verstraete. 1983. A practical method to estimate the acetoclastic methanogenic biomass in anaerobic sludges. J. Water Pollut. Control Fed. 55: 1191–1195.

Vallom, J.K., and A.J. McLoughlin. 1984. Lysis as a factor in sludge flocculation. Water Res. 18: 1523–1528.

Valls, M., S. Atrian, V. de Lorenzo, and L.A. Fernández. 2000. Engineering a mouse metallothionein on the cell surface of *Ralstonia eutropha* CH34 for immobilization of heavy metals in soil. Nature Biotechnol. 18: 661–665.

Valo, R.J., M.M. Haggblom, and M.S. Salkinoja-Salonen. 1990. Bioremediation of chlorophenol containing simulated ground water by immobilized bacteria. Water Res. 24: 253–258.

Vance, D.B. 1991. Onsite bioremediation of oil and grease contaminated soil. Natl. Environ. J. 1: 26–30.

Vandenbergh, P.A., A.M. Wright, and A.K. Vidaver. 1985. Partial purification and characterization of a polysaccharide depolymerase associated with phage-infected *Erwinia amylovora.* Appl. Environ. Microbiol. 49: 994–996.

Vander, A.J., J.H. Sherman, and D.S. Luciano. 1985. *Human Physiology: The Mechanisms of Body Function.* McGraw-Hill, New York.

Vandevivere, P., and P. Baveye. 1992. Effect of bacterial extracellular polymers on the saturated hydraulic conductivity of sand columns. Appl. Environ. Microbiol. 58: 1690–1698.

Varma, M.M., W.A. Thomas, and C. Prasad. 1976. Resistance to inorganic salts and antibiotics among sewage-borne Enteriobacteriaceae and Achromobacteriaceae. J. Appl. Bacteriol. 41: 347–349.

Varma, M., J.D. Hestera, F.W. Schaefer, III, M.W. Warea, and H.D.A. Lindquist. 2003. Detection of *Cyclospora cayetanensis* using a quantitative real-time PCR assay. J. Microbiol. Methods 53: 27–36.

Vaughn, J.M., E.F. Landry, T.J. Vicale, and W.F. Penello. 1979. Survey of human enteroviruses occurring in fresh and marine surface waters on Long Island. Appl. Environ. Microbiol. 38: 290–296.

Vaughn, J.M., E.F. Landry, C.A. Beckwith, and M.Z. Thomas. 1981. Virus removal during ground-water recharge: Effects of infiltration rate on adsorption of poliovirus to soil. Appl. Environ. Microbiol. 41: 139–143.

Vaughn, J.M., E.F. Landry, and M.Z. Thomas. 1983. Entrainment of viruses from septic tank leach fields through a shallow, sandy soil aquifer. Appl. Environ. Microbiol. 45: 1474–1480.

Vaughn, J.M., and J.F. Novotny. 1991. Virus inactivation by disinfectants, pp. 217–241, In: Modeling the Environmental Fate of Microorganisms. C.J. Hurst, Ed., American Society of Microbiology, Washington, D.C.

Venczel, L.V., M. Arrowood, M. Hurd, and M.D. Sobsey. 1997. Inactivation of *Cryptosporidium parvum* oocysts by a mixed-oxidant disinfectant and by free chlorine. Appl. Environ. Microbiol. 63: 1598–1601.

Venkobachar, C., Z. Invegar, and A.V.S. Prabhakara Raj. 1977. Mechanism of disinfection: Effect of chlorine on cell membrane functions. Water. Res. 11: 727–729.

Venosa, A., J.R. Haines, W. Nisamannepong, R. Gouind, S. Pradhan, and B. Siddique. 1992. Efficacy of commercial products in enhancing oil biodegradation in closed laboratory reactors. J. Ind. Microbiol. 10: 13–23.

Venosa, A.D. 1986. Detection and significance of pathogens in sludge, pp. 1–14, In: *Control of Sludge Pathogens*, C.A. Sorber, Ed., Water Pollution Control Federation, Washington, D.C.

Venosa, A.D., A.C. Petrasek, D. Brown, H.L. Sparks, and D.M. Allen. 1984. Disinfection of secondary effluent with ozone/UV. J. Water Pollut. Control Fed. 56: 137–142.

Verstraete, W., and M. Alexander. 1972. Heterotrophic nitrification by *Arthrobacter* sp. J. Bacteriol. 110: 955–961.

Verstraete, W., and E. van Vaerenbergh. 1986. Aerobic activated sludge, In: *Biotechnology.* H.J. Rehm, and G. Reed, Eds., VCH, Weinheim, Germany.

Vesey, G., J.S. Slade, M. Byrne, K. Sheppard, P.J. Dennis, and C.R. Fricker. 1993. Routine monitoring of *Cryptosporidium* oocysts in water using flow cytometry. J. Appl. Bacteriol. 75: 87–90.

Vesper, S.J., W. Davis-Hoover, and L.C. Murdoch. 1992. Oxygen sources for *in situ* subsurface bioremediation (Abst. No. Q-185), In: 92nd meeting of American Society of Microbiology, May 26–30, New Orleans, LA.

Vidales, J.A., C.P. Gerba, and M.M. Karpiscak. 2003. Virus removal from wastewater in a multispecies subsurface-flow constructed wetland. Water Environ. Res. 75: 238–244.

Villacorta-Martinez de Maturana, I., M.E. Ares-Mazas, D. Duran-Oreiro, and M.J. Lorenzo-Lorenzo. 1992. Efficacy of activated sludge in removing *Cryptosporidium parvum* from sewage. Appl. Environ. Microbiol. 58: 3514–3516.

Vissier, F.A., J.B. van Lier, A.J.L. Macario, and E. Conway de Macario. 1991. Diversity and population dynamics of bacteria in a granular consortium. Appl. Environ. 57: 1728–1734.

Visvesvara, G.S., M.J. Peralta, F.H. Brandt, M. Wilson, C. Aloisio, and E. Franko. 1987. Production of monoclonal antibodies to *Naegleria fowleri*, agent of primary amoebic meningoencephalitis. J. Clin. Microbiol. 25: 1629–1634.

Vogel, T.M., and P.L. McCarty. 1985. Biotransformation of tetrachloroethylene to trichloroethylene, dichloroethylene, vinyl chloride and carbon dioxide under methanogenic conditions. Appl. Environ. Microbiol. 49: 1080–1083.

Vogels, G.D., J.T. Keltjens, and C. van der Drift. 1988. Biochemistry of methane production, pp. 707–770, In: *Biology of Anaerobic Microorganisms*, A.J.B. Zehnder, Ed., John Wiley & Sons, New York.

Vogelsang, C., A. Husby, and K. Østgaard. 1997. Functional stability of temperature-compensated nitrification in domestic wastewater treatment obtained with PVA-SbQ alginate gel entrapment. Water Res. 31: 1659–1664.

Volk, C.J., and M.W. LeChevallier. 2000. Assessing biodegradable organic matter. J. AWWA 92 (5): 64–76.

Volkmann, H., T. Schwartz, P. Bischoff, S. Kirchen, and U. Obst. 2004. Detection of clinically relevant antibiotic-resistance genes in municipal wastewater using real-time PCR (TaqMan). J. Microbiol. Meth. 56: 277–286.

Volskay, V.T., Jr., and C.P.L. Grady, Jr. 1988. Toxicity of selected RCRA compounds to activated sludge microorganisms. J. Water Pollut. Control Fed. 60: 1850–1856.

Volskay, V.T., Jr., and C.P.L Grady, Jr. 1990. Respiration inhibition kinetics analysis. Water Res. 24: 863–874.

Volskay, V.T., Jr., C.P.L Grady, Jr., and H.H. Tabak. 1990. Effect of selected RCRA compounds on activated sludge activity. J. Water Pollut. Control Fed. 62: 655–664.

Voordouw, G. 1992. Evolution of hydrogenase genes. Adv. Inorg. Chem. 38: 397–422.

Voordouw, G., V. Niviere, F.G. Ferris, P.M. Fedorak, and D.W.S. Westlake. 1990. Distribution of hydrogenase genes in *Desulfovibrio* spp. and their use in identification of species from the oil-field environment. Appl. Environ. Microbiol. 56: 3748–3754.

Van der Waarde, J., J. Krooneman, B. Geurkink, A. van der Werf, D. Eikelboom, C. Beimfohr, J. Snaidr, C. Levantesi, and V. Tandoi. 2002. Microorganisms in Activated Sludge and Biofilm Processes. Water Sci. Technol. 46 (1–2): 99–104.

Wachinski, A.M. 1988. Waste management in U.S space program. J. Water Pollut. Control Fed. 60: 1790–1797.

Wada, S., H. Ichikawa, and K. Tatsumi. 1992. Removal of phenols with tyrosinase immobilized on magnetite. Water Sci. Technol. 26: 2057–2059.

Wada, S., H. Ichikawa, and K. Tatsumi. 1993. Removal of phenols from wastewater by soluble and immobilized tyrosinase. Biotech. Bioeng. 42: 854–858.

Wadowsky, R.M., and R.B. Yee. 1985. Effect on non-legionellaceae bacteria on the multiplication of *Legionella pneumophila* in potable water. Appl. Environ. Microbiol. 49: 1206–1210.

Wagner, M., and R. Amann. 1996. In *Microbial Community Analysis*, T.E. Cloete, and N.Y.O. Muyima, Eds., IAWQ, London, pp. 61–72.

Wagner, M., R. Amann, H. Lemmer, and K.-H. Schleifer. 1993. Probing activated sludge with oligonucleotides specific for proteobacteria: inadequacy of culture-dependent methods for describing microbial community structure. Appl. Environ. Microbiol. 59: 1520–1525.

Wagner, M., R. Amann, P. Kampfer, B. Assmus, A. Hartmann, P. Hutzler, N. Springer, and K.-H. Schleifer. 1994. Identification and *in situ* detection of Gram-negative filamentous bacteria in activated sludge. System. Appl. Microbiol. 17: 405–417.

Wagner, M., G. Rath, R. Amann, H.-P. Koops, and K.-H. Schleifer. 1995. *In situ* identification of ammonia-oxidizing bacteria. Syst. Appl. Microbiol. 18: 251–264.

Wagner-Wiening, C., and P. Kimmig. 1995. Detection of viable *Cryptosporidium parvum* oocysts by PCR. Appl. Environ. Microbiol. 61: 4514–4516.

Wahlen, 1993. The global methane cycle. Ann. Rev. Earth & Planet. Sci. 21: 407–426.

Walker, D.C., S.-V. Len, and B. Sheehan. 2004. Development and evaluation of a reflective solar disinfection pouch for treatment of drinking water. Appl. Environ. Microbiol. 70: 2545–2550.

Walker, G.S., F.P. Lee, and E.M. Aieta. 1986. Chlorine dioxide for taste and odor control. J. Am. Water Works Assoc. 78: 84–93.

Walker, J.M., and G.B. Wilson. 1973. Composting sewage sludge: Why? Compost Sci. 14: 10–12.

Wallis, C., C.H. Stagg, and J.L. Melnick. 1974. The hazards of incorporating charcoal filters into domestic water systems. Water Res. 8: 111–113.

Wallis, P.M., and D.L. Lehmann, Eds. 1983. *Biological Health Risks of Sludge Disposal to Land in Cold Climates*, University of Calgary Press, Calgary, Canada.

Wallis, P.M., S.L. Erlandsen, J.L. Isaac-Renton, M.E. Olson, W.J. Robertson, and H. van Keulen. 1996. Prevalence of *Giardia* cysts and *Cryptosporidium* oocysts and characterization of *Giardia* spp. Isolated from drinking water in Canada. Appl. Environ. Microbiol. 62: 2789–2797.

Walsh, G.E., and R.L. Garnas. 1983. Determination of bioactivity of chemical fractions of liquid wastes using freshwater and saltwater algae and crustaceans. Environ. Sci. Technol. 17: 180–182.

Walter, M.V. 1997. Bioaugmentation, pp. 753–757, In: *Manual of Environmental Microbiology*, C.J. Hurst, G.R. Knudsen, M.J. McInerney, L.D. Stetzenbach, and M.V. Walter, Eds., ASM Press, Washington, D.C.

Walter, M.V., and J.W. Vennes. 1985. Occurrence of multiple-antibiotic-resistant enteric bacteria in domestic sewage and oxidation lagoons. Appl. Environ. Microbiol. 50: 930–933.

Wang, X., X. Yu, and R. Bartha. 1990. Effect of bioremediation on polycyclic aromatic hydrocarbon residues in soil. Environ. Sci. Technol. 24: 1086–1089.

Wang, Y.-T., H.D. Gabbard, and P.-C. Pai. 1991. Inhibition of acetate methanogenesis by phenols. J. Environ. Eng. 117: 487–496.

Wanner, J. 1994. Activated Sludge Bulking And Foaming Control. Technomic Publishing Co. Lancaster, PA, 327 pp.

Wanner, J. 2002. Filamentous bulking in activated sludge, control of, pp. 1306–1315, In: *Encyclopedia of Environmental Microbiology*, G. Bitton, editor-in-chief, Wiley-Interscience, N.Y.

Wanner, J., and P. Grau. 1989. Identification of filamentous microorganisms from activated sludge: A compromise between wishes and possibilities. Water Res. 23: 883–891.

Wanner, J., J. Chudoba, K. Kuckman, and L. Proske. 1987a. Control of activated sludge filamentous bulking. VII. Effect of anoxic conditions. Water Res. 21: 1447–1451.

Wanner, J., K. Kuckman, V. Ottova, and P. Grau. 1987b. Effect of anaerobic conditions on activated sludge filamentous bulking in laboratory systems. Water Res. 21: 1541–1546.

Wanner, O., and W. Gujer. 1984. Competition in biofilms. Presented at the 12th Conference of International Association on Water Pollution, Amsterdam.

Ward, B.B. 2002. Nitrification in aquatic systems, pp. 2144–2167, In: *Encyclopedia of Environmental Microbiology*, G. Bitton, editor-in-chief, Wiley-Interscience, N.Y.

Ward, M.L. 2003. *Toxicity and hormonal activity in municipal solid waste (MSW) leachates from Florida landfills.* Ph.D. Dissertation, University of Florida, Gainesville. 288 pp.

Ward, N.R., R.L. Wolfe, and B.H. Olson. 1984. Effect of pH, application technique and chlorine-to-nitrogen ratio on disinfectant activity of inorganic chloramines with pure culture bacteria. Appl. Environ. Microbiol. 48: 508–514.

Ward, R.L., and C.S. Ashley. 1977a. Identification of the viricidal agent in wastewater sludge. Appl. Environ. Microbiol. 33: 860–864.

Ward, R.L., and C.S. Ashley. 1977b. Discovery of an agent in wastewater sludge that reduces the heat required to inactivate reovirus. Appl. Environ. Microbiol. 34: 681–688.

Ward, R.L., C.S. Ashley, and R.H. Moseley. 1976. Heat inactivation of poliovirus in wastewater sludge. Appl. Environ. Microbiol. 32: 339–346.

Ward, R.L., J.G. Yeager, and C.S. Ashley. 1981. Response of bacteria in wastewater sludge to moisture loss by evaporation and effect of moisture content on bacterial inactivation by ionizing radiation. Appl. Environ. Microbiol. 41: 1123–1127.

Ward, R.L., D.R. Knowlton, J. Stober, W. Jakubowski, T. Mills, P. Graham, and D.E. Camann. 1989. Effect of wastewater spray irrigation on rotavirus infection rates in an exposed population. Water Res. 23: 1505–1509.

Ware, M.W., L. Wymer, H.D.A. Lindquist, and F.W. Schaefer III. 2003. Evaluation of an alternative IMS dissociation procedure for use with Method 1622: detection of *Cryptosporidium* in water. J. Microbiol. Meth. 55: 575–583.

Washington, B., C. Lue-Hing, D.R. Zenz, K.C. Rao, and A.W. Kobayashi. 1983. Exertion of 5-day nitrogenous oxygen demand in nitrifying wastewater. J. Water Pollut. Control Fed. 55: 1196–1200.

Watanabe, I. 1973. Isolation of pentachlorophenol decomposing bacteria from soil. Soil Sci. Plant Nutr. 19: 109–116.

Water Pollution Control Federation. 1979. *Odor Control for Wastewater Facilities.* Manual of Practice No 22, WPCF, Washington, D.C.

Wattie, E., and C.T. Butterfield. 1944. Relative resistance of *Escherichia coli* and *Escherichia typhosa* to chlorine and chloramines. Public Health Rep. 59: 1661–1665.

Wattie, E., and C.W. Chambers. 1943. Relative resistance of coliforms organisms and certain enteric pathogens to excess-lime treatment. J. Am. Water Works Assoc. 35: 709–720.

Watkins, W.D., S.R. Rippey, C.R. Clavet, D.J. Kelley-Reitz, and W. Burkhardt III. 1988. Novel conpound for identifying *E. coli.* Appl. Environ. Microbiol. 54: 1874–1875.

Watts, R.J., S. King, M.P. Orr, G.C. Miller, and B.E. Henry. 1995. Photocatalytic inactivation of coliform bacteria and viruses in secondary wastewater effluent. Water Res. 29: 95–100.

Way, J.S., K.L. Josephson, S.D. Pillai, M. Abbaszadegan, C.P. Gerba, and I.L. Pepper. 1993. Specific detection of *Salmonella* spp. by multiplex polymerase chain reaction. Appl. Environ. Microbiol. 59: 1473–1479.

Webb, C. 1987. Cell immobilization, pp. 347–376, In: *Environmental Biotechnology*, C.F. Forster, and D.A.J. Wase, Eds., Ellis Horwood, Chichester, U.K.

Webb, J.S., S. McGinness, and H.M. Lappin-Scott. 1998. Metal removal by sulfate reducing bacteria from natural and constructed wetlands. J. Appl. Microbiol. 84: 240–248.

Weber, W.J., M. Pirbazari, and G.L. Melson. 1978. Biological growth on activated carbon: An investigation by scanning electron microscopy. Environ. Sci. Technol. 12: 817–819.

Webster, K.A., J.D.E. Pow, M. Giles, J. Catchpole, and M.J. Woodward. 1993. Detection of *Cryptosporidium parvum* using a specific polymerase chain reaction. Vet. Parasitol. 50: 35–44.

Weddle, C.L., and D. Jenkins, 1971. The viability and activity of activated sludge. Water Res. 5: 621–640.

von Wedel, R.J., J.F. Mosquera, C.D. Goldsmith, G.R. Hater, A. Wong, T.A. Fox, W.T. Hunt, M.S. Paules, J.M. Quiros, and J.W. Wiegand. 1988. Bacterial biodegradation of petroleum hydrocarbons in groundwater: *In situ* augmented bioreclamation with enrichment isolates in California, In: *International Conference on Water and Wastewater Microbiology*, Newport Beach, CA, Feb. 8–11, 1988.

Wegelin, M., S. Canonica, K. Mechsner, F. Pesaro, and A. Metzler. 1994. Solar water disinfection: scope of the process and analysis of radiation experiments. J Water SRT-Aqua 43: 154–169.

Wegelin, M., and B. Sommer. 1998. Solar water disinfection (SODIS) – destined for worldwide use? Waterlines 16: 30–32.

Weinbauer, M.G., S.W. Wilhelm, C.A. Suttle, and D.R. Garza. 1997. Photoreactivation compensates for U.V. damage and restores infectivity to natural marine virus communities. Appl. Environ. Microbiol. 63: 2200–2205.

Weiner, R.M. 1997. Biopolymers from marine prokaryotes. Trends Biotechnol. 15: 390–394.

Wells, P.G., K. Lee, and C. Blaise, Eds. 1998. *Microscale Testing in Aquatic Toxicology: Advances, Techniques, and Practice.* CRC Press, Boca Raton, FL.

van der Waarde, J., J. Krooneman, B. Geurkink, A. van der Wer, D. Eikelboom, F. Beimfohr, J. Snaidr, C. Levantesi, and V. Tandoi. 2002. Molecular monitoring of bulking sludge in industrial wastewater treatment plants. Water Sci. Technol. 46 (1–2): 551–558.

van der Wende, E., and W.G. Characklis. 1990. Biofilms in potable water distribution systems, pp. 249–268, In: *Drinking Water Microbiology*, G.A. McFeters, Ed., Springer Verlag, New York.

Weng, C.N., and A.H. Molof. 1974. Nitrification in the biological fixed film RBC. J. Water Pollut. Control Fed. 46: 1674–1685.

Wentzel, M.C., L.H. Lotter, R.E. Loewenthal, and G.V.R. Marais. 1986. Metabolic behaviour of *Acinetobacter* spp. in enhanced biological phosphorus removal: A biochemical model. Water SA 12: 209–224.

Werner, M., and R. Kayser. 1991. Denitrification with biogas as external carbon source. Water Sci. Technol. 23: 701–708.

Werner, P. 1985. Eine methode zur bestimmung der veirkemungsneigung von trinkwasser. Von Wasser 65: 257–262.

Westphal, P.A., and G.L. Christensen. 1983. Lime stabilization: Effectiveness of two process modifications. J. Water Pollut. Control Fed. 55: 1381–1386.

Wetzler, T.F., J.R. Rea, G.J. Ma, and M. Glass. 1979. Non-association of *Yersinia* with traditional coliform indicators, In: Proceedings Annual Meeting American Water Works Association, Denver, CO.

Wheeler, D., J. Bartram, and B.J. Lloyd. 1988. The removal of viruses by filtration through sand, pp. 207–229, In: *Slow Sand Filtration: Recent Development in Water Treatment Technology*, N.J.D. Graham, Ed., Ellis Horwood Pub., Chichester, U.K.

Whitby, G.E., G. Palmateer, W.G. Cook, J. Maarschalkerweerd, D. Huber, and K. Flood. 1984. Ultraviolet disinfection of secondary effluent. J. Water Pollut. Control Fed. 56: 844–850.

White, C., J.A. Sayer, and G.M. Gadd. 1997. Microbial solubilization and immobilization of toxic metals: Key biogeochemical processes for treatment of contamination. FEMS Microbiol. Rev. 20: 503–516.

White, D.C., R.J. Bobbie, J.D. King, J. Nickels, and P. Amoe. 1979. Lipid analysis of sediments for microbial biomass and community structure, pp. 87–103, In: *Methodology for Biomass Determinations and Microbial Activities in Sediments*, C.D. Litchfield, and P.L Seyfried, Eds., ASTM STP 673, American Society for Testing and Materials, Philadelphia.

White, J.M., D.P. Labeda, M.P. LeChevallier, J.R. Owens, D.D. Jones, and J.L. Gauthier. 1986. Novel actinomycete isolated from bulking industrial sludge. Appl. Environ. Microbiol. 52: 1324–1330.

Whitman, W.B., and R.S. Wolfe. 1980. Presence of nickel in factor F_{430} from *Methanobacterium bryantii.* Biochem. Biophys. Res. Commun. 92: 1196–1201.

Whitmore, T.N., and S. Denny. 1992. The effect of disinfectants on a geosmin-producing strain of *Streptomyces griseus.* J. Appl. Bacteriol. 72: 160–165.

WHO/UNICEF. 2000. Global water supply and sanitation assessment 2000. www.who.int/water_sanitation_health.

WHO. 2003. *The Right to Water*. WHO Pub. Geneva, Switzerland.

Wickramanayake, G.B., A.J. Rubin, and O.J. Sproul. 1985. Effect of ozone and storage temperature on *Giardia* cysts. J. Am. Water Works Assoc. 77: 74–77.

Widada, F., H. Nojiri, and T. Omori. 2002. Recent developments in molecular techniques for identification and monitoring of xenobiotic-degrading bacteria and their catabolic genes in bio-remediation. Appl. Microbiol. Biotechnol. 60: 45–59.

Widdel, F. 1988. Microbiology and ecology of sulfate- and sulfur-reducing bacteria, pp. 469–585, In: *Biology of Anaerobic Microorganisms*, A.J.B. Zehnder, Ed., John Wiley & Sons, New York.

Wiedenmann, A., B. Fischer, U. Sraub, C.-H. Wang, B. Flehmig, and D. Schoenen. 1993. Disinfection of hepatitis A virus and MS-2 coliphage in water by ultraviolet irradiation: Comparison of UV-susceptibility. Water Sci. Technol. 27: 335–338.

Wiedmann-Al-Ahmad, M., H.V. Tichy, and G. Schön. 1994. Characterization of *Acinetobacter* type strains and isolates obtained from wastewater treatment plants by PCR fingerprinting. Appl. Environ. Microbiol. 60: 4066–4071.

Wiggins, B.A. 1996. Discriminant analysis of antibiotic resistance patterns in fecal streptococci, a method to differentiate human and animal sources of fecal pollution in natural waters. Appl. Environ. Microbiol. 62: 3997–4002.

Wilcox, D.P., E. Chang, K.L. Dickson, and K.R. Johansson. 1983. Microbial growth associated with granular activated carbon in a pilot water treatment facility. Appl. Environ. Microbiol. 45: 406–416.

Wilczak, A., J.G. Jacangelo, J.P. Marcinko, L.H. Odell, G.J. Kirmeyer, and R.L. Wolfe. 1996. Occurrence of nitrification in chloraminated distribution systems. J. Am. Water Works Assoc. 88: 74–85.

Willcocks, M.M., J.B. Kurtz, T.W. Lee, and M.J. Carter. 1995. Prevalence of human astrovirus serotype 4: Capsid protein sequence and comparison with other strains. Epidemiology and Infection 114: 385–391.

Willcomb, G.E. 1923. Twenty years of filtration practice at Albany. J. Am. Water Works Assoc. 10: 97–103.

Wilderer, P.A., H.-J. Bungartz, H. Lemmer, M. Wagner, J. Keller, and S. Wuertz. 2002. Modern scientific methods and their potential in wastewater science and technology. Water Res. 36: 370–393.

Wilén, B.-M., B. Jina, and P. Lant. 2003. The influence of key chemical constituents in activated sludge on surface and flocculating properties. Water Res. 37: 2127–2139.

Wiles, S., A.S. Whiteley, J.C. Philp, and M.J. Bailey. 2003. Development of bespoke bioluminescent reporters with the potential for in situ deployment within a phenolic-remediating wastewater treatment system. J. Microbiol. Meth. 55: 667–677.

Williams, C.M., J.C.H. Shih, and J.W. Spears. 1986. Effect of nickel on biological methane generation from a laboratory poultry waste digester. Biotech. Bioeng. 28: 1608–1610.

Williams, F.P., and E.W. Akin. 1986. Waterborne viral gastroenteritis. J. Am. Water Works Assoc. 78: 34–39.

Williams, R.B., and G.L. Culp. 1986. *Handbook of Public Water Systems*. Van Nostrand Reinhold, New York.

Williams, R.T., P.S. Ziegenfuss, G.B. Mohrman, and W.E. Sisk. 1989. Composting of explosives and propellant contaminated sediments, pp. 269–281, In: *Hazardous Waste Treatment: Biosystems for Pollution Control*, Air and Waste Management Association, Pittsburgh, PA.

Williams, T.M., and R.F. Unz. 1983. Environmental distribution of *Zoogloea* strains. Water Res. 17: 779–787.

Williams, T.M., and R.F. Unz. 1985a. Filamentous sulfur bacteria of activated sludge characterization of *Thiothrix, Beggiatoa*, and Eikelboom "type 021N" strains. Appl. Environ. Microbiol. 49: 887–898.

Williams, T.M., and R.F. Unz. 1985b. Isolation and characterization of filamentous bacteria present in bulking activated sludge. Appl Microbiol. Biotechnol. 22: 273–282.

Williams, T.M., and R.F. Unz. 1989. The nutrition of *Thiothrix*, Type 021N, *Beggiatoa* and *Leucothrix* strains. Water Res. 23: 15–22.

Williams, T.M., R.F. Unz, and J.T. Doman. 1987. Ultrastructure of *Thiothrix* spp. and "type 021N" bacteria. Appl. Environ. Microbiol. 53: 1560–1570.

Williamson, K.J., and D.G. Johnson. 1981. A bacterial bioassay for assessment of wastewater toxicity. Water Res. 15: 383–390.

Williamson, K., and P.L. McCarty. 1976. A model for substrate utilization by bacterial films. J. Water Pollut. Control Fed. 48: 9–24.

Wilson, B., P. Roessler, M. Abbaszadegan, C.P. Gerba, and E. van Dellen. 1992. UV dose bioassay using coliphage MS-2 (Abst.), In: Annual meeting of American Society of Microbiology. May 26–30, New Orleans, LA.

Wilson, J.T., et al. 1986. *In situ* biorestoration as a groundwater remediation technique. Ground Water Monit. Rev. 6: 56–64.

Wilson, J.T., D.H. Kampbell, S.R. Hutchins, D.A. Kovacs, W. Korreck, R.H. Douglass, and D.J. Hendrix. 1989. Performance of two demonstration projects for the bioremediation of an aquifer contaminated by fuel spills from underground storage tanks. Presented at the Tenth Annual Meeting of the Society of Environmental Toxicology and Chemistry, Toronto, Canada, Oct. 28–Nov. 2, 1989.

Wilson, K.H., J.S. Ikeda, and R.B. Blitchington. 1997. Phylogenic placement of community members of human colonic biota. Clin. Infect. Dis. 25 (Suppl. 2): S114–S116.

Winzer, K., K.R. Hardie, and P. Williams. 2002. Bacterial cell-to-cell communication: sorry, can't talk now – gone to lunch! Current Opinion Microbiol. 5 (2): 216–222.

Witherell, L.E., R.W. Duncan, K.M. Stone, L.J. Stratton, L. Orciani, S. Kappel, and D.A. Jillson. 1988. Investigation of *Legionella pneumophila* in drinking water. J. Am. Water Works Assoc. 80: 87–93.

Withers, H., S. Swift, and P. Williams. 2001. Quorum sensing as an integral component of gene regulatory networks in Gram-negative bacteria. Curr. Opin. Microbiol. 4: 186–193.

Woese, C.R. 1987. Bacterial evolution. Microbiol. Rev. 51: 221–271.

Wolfe, R.L. 1990. Ultraviolet disinfection of potable water, Env. Sci. Technol. 24: 768–773.

Wolfe, R.L., and N.I. Lieu. 2002. Nitrifying bacteria in drinking water. pp. 2167–2176, In: *Encyclopedia of Environmental Microbiology*, G. Bitton, editor-in-chief, Wiley-Interscience, N.Y.

Wolfe, R.L., E.G. Means III, M.K. Davis, and S.E. Barrett. 1988. Biological nitrification in covered reservoirs containing chloraminated water. J. Amer. Water Works Assoc. 80 (9): 109–114.

Wolfe, R.L., and B.H. Olson. 1985. Inability of laboratory models to accurately predict field performance of disinfectants, pp. 555–573, In: *Water Chlorination: environmental impact and health effects*, Vol. 5, R.L. Jolley, R.J. Bull, W.P. Davis, S. Katz, M.H. Roberts, and V.A. Jacobs, Eds., Ann Arbor Science Publishers, Ann Arbor, MI.

Wolfe, R.L., N.R. Ward, and B.H. Olson. 1984. Inorganic chloramines as drinking water disinfectants: A review. J. Am. Water Works Assoc. 76: 74–88.

Wolinsky, E. 1979. Nontuberculous mycobacteria and associated diseases. Amr. Rev. Resp. Dis. 119: 107–159.

von Wolzogen-Kuhr, C.A.H., and A.S. van der Vlugt. 1934. The graphitization of cast iron as an electrobiochemichal process in anaerobic soils. Water 18: 147–165.

Won, W.D., and H. Ross. 1973. Persistence of virus and bacteria in seawater. J. Environ. Eng. Div. ASCE 99: 205–211.

Wood, D.J., K. Bijlsma, J.C. de Jong, and C. Tonkin. 1989a. Evaluation of a commercial monoclonal antibody-based immunoassay for detection of adenovirus types 40 and 41 in stool specimens. J. Clin. Microbiol. 27: 1155–1158.

Woods, S.L., J.F. Ferguson, and M.M. Benjamin. 1989b. Characterization of chlorophenol and chloromethoxybenzene biodegradation during anaerobic treatment. Environ. Sci. Technol. 23: 62–68.

World Health Organization. 1979. *WHO International Reference Center for Community Water Supply Annual Report.* Rijswijk, The Netherlands.

World Health Organization (WHO). 1989. *Guidelines for Use of Wastewater in Agriculture and Aquaculture.* WHO Technical Report Series 778, Geneva, Switzerland.

World Health Organization (WHO). 1996a. The World Health Report 1996: *Fighting disease, Fostering Development.* World Health Organization, Geneva, Switzerland.

World Health Organization (WHO). 1996b. *Guidelines for drinking water quality*, Vol. 2: Recommendations, World Health Organization, Geneva, Switzerland.

World Health Organization (WHO). 1996c. *Water and sanitation*: WHO fact sheet no. 112, World Health Organization, Geneva.

Wu, L., D.K. Thompson, G. Li, R.A. Hurt, J.M. Tiedje, and J. Zhou. 2001. Development and evaluation of functional gene arrays for detection of selected genes in the environment. Appl. Environ. Microbiol. 67: 5780–5790.

Wu, W., J. Hu, X. Gu, Y. Zhao, and H. Zhang. 1987. Cultivation of anaerobic granular sludge with aerobic activated sludge as seed. Water Res. 21: 789–799.

Wu, W.W., M.M. Benjamin, and G.V. Korshin. 2001. Effects of thermal treatment on halogenated disinfection by-products in drinking water. Water Res. 35: 3545–3550.

Wuertz, S. 2002. Gene exchange in biofilms, pp. 1408–1420, In: *Encyclopedia of Environmental Microbiology*, G. Bitton, editor-in-chief, Wiley-Interscience, N.Y.

Wyn-Jones, A.P., and J. Sellwood. 2001. Enteric viruses in the aquatic environment. J. Appl. Microbiol. 91: 945–962.

Xagoraraki. I., G.W. Harrington, P. Assavasilavasukul, and J.H. Standridge. 2004. Removal of emerging waterborne pathogens and pathogen indicators. J. Amer. Water Works Assoc. 96 (5): 102–113.

Xu, H., and B.J. Dutka. 1987. ATP-TOX system: A new rapid sensitive bacterial toxicity screening system based on the determination of ATP. Toxicity Assess. 2: 149–166.

Xu, L.-Q., S.-H. Yu, Z.-X Jiang, J.-L. Yang, C.-Q. Lai, and X.-J. Zhang. 1995. Soil-transmitted helminthiases: Nationwide survey in China. Bull. WHO 73: 507–513.

Yadav, J.S., J.F. Quensen III, J.M. Tiedje, and C.A. Reddy. 1995. Degradation of polychlorinated biphenyl mixtures (Aroclors 1242, 1254, and 1260) by the white rot fungus *Phanerochaete chrysosporium* as evidenced by congener-specific analysis. Appl. Environ. Microbiol. 61: 2560–2565.

Yaguchi, J., K. Chigusa, and Y. Ohkubo. 1991. Isolation of microorganisms capable of lysing the filamentous bacterium, type 021N (studies on lytic enzyme against the filamentous bacterium, type 021N, screening studies). Water Sci. Technol. 23: 955–962.

Yahya, M.T., and C.P. Gerba. 1990. Inactivation of bacteriophage MS-2 and poliovirus by copper, silver and low levels of free chlorine (Abst.), In: International Symposium on Health-Related Microbiology, Tubingen, Germany, April 1–6, 1990.

Yahya, M.T., L.K. Landeen, S.M. Kutz, and C.P. Gerba. 1989. Swimming pool disinfection: an evaluation of the efficacy of copper:silver ions. J. Environ. Health 51: 282–287.

Yahya, M.T., and W.A Yanko. 1992. Comparison of a long-term enteric virus monitoring data base with bacteriophage reduction in full scale reclamation plants. Presented at the International Water Pollution Research Symposium., Washington, D.C., May 26–29, 1992.

Yakub, G.P., D.A. Castric, K.L. Stadterman-Knauer, M.J. Tobin, M. Blazina, T.N. Heineman, G.Y. Yee, and L. Frazier. 2002. Evaluation of Colilert and Enterolert defined substrate methodology for wastewater applications. Water Environ. Res. 74: 131–135.

Yamaguchi, T., S. Yamazaki, S. Uemura, I-C. Tseng, A. Ohashi, and H. Harada. 2001. Microbial-ecological significance of sulfide precipitation within anaerobic granular sludge revealed by micro-electrodes study. Water Res. 35: 3411–3417.

Yang, J., and R.E. Speece. 1985. Effects of engineering controls on methane fermentation toxicity response. J. Water Pollut. Control Fed. 57: 1134–1141.

Yang, J., and R.E. Speece. 1986. The effects of chloroform toxicity on methane fermentation. Water Res. 20: 1273–1279.

Yang, P.Y., K. Cao, and S.J. Kim. 2002. Entrapped mixed microbial cell process for combined secondary and tertiary wastewater treatment. Water Environ. Res. 74: 226–234.

Yates, M.V. 1985. Septic tank density and ground water contamination. Ground Water 23: 586–591.

Yates, M.V., and C.P. Gerba. 1984. Factors controlling the survival of virus in groundwater. Water Sci. Technol. 17: 681–687.

Yates, M.V., C.P. Gerba, and L.M. Kelley. 1985. Virus persistence in groundwater. Appl. Environ. Microbiol. 49: 778–781.

Yates, M.V., and Y. Ouyang. 1992. VIRTUS, a model of virus transport in unsaturated soils. Appl. Environ. Microbiol. 58: 1609–1616.

Yates, M.V., and S.R. Yates. 1987. A comparison of geostatistical methods for estimating virus inactivation rates in ground water. Water Res. 21: 1119–1125.

Yates, M.V., and S.R. Yates. 1989. Septic tank setback distances: A way to minimize virus contamination of drinking water. Ground Water 27: 202–208.

Yates, M.V., and S.R. Yates. 1991. Modeling microbial transport in the subsurface: A mathematical discussion, pp. 48–76, In: *Modeling the Environmental Fate of Microorganisms*, C.J. Hurst, Ed., Amer. Soc. Microbiol., Washington, D.C.

Yates, M.V., S.R. Yates, A.W. Warrick, and C.P. Gerba. 1986. Predicting virus fate to determine septic tank setback distances using geostatistics. Appl. Environ. Microbiol. 49: 479–483.

Yaziz, M.I., and B.J. Lloyd. 1979. The removal of Salmonellas in conventional sewage treatment plants. J. Appl. Bacteriol. 46: 131–142.

Yeager, J.G., and R.T. O'Brien. 1983. Irradiation as a means to minimize public health risks from sludge-borne pathogens. J. Water Pollut. Control Fed. 55: 977–983.

Yee, R.B., and R.M. Wadowsky. 1982. Multiplication of *Legionella pneumophila* in unsterilized tap water. Appl. Environ. Microbiol. 43: 1130–1134.

Yerushalmi, L., S. Rocheleau, R. Cimpoia, M. Sarrazin, G. Sunahara, A. Peisajovich, G. Leclair, and S.R. Guiot. 2003. Enhanced biodegradation of petroleum hydrocarbons in contaminated soil. Biodegrad. J. 7: 37–51.

Yoda, M., M. Kitagawa, and Y. Miyaji. 1987. Long term competition between sulfate-reducing and methane-producing bacteria for acetate in anaerobic biofilm. Water Res. 21: 1547–1556.

Yoda, M., M. Kitagawa, and Y. Miyaji. 1989. Granular sludge formation in the anaerobic expanded micro-carrier bed process. Water Sci. Technol. 21: 109–120.

York, D.W., and N.R. Burg. 1998. Protozoan pathogens: A comparison of reclaimed water and other irrigation waters, pp. 81–87, In: *Water Reuse Conference Proceedings*, Feb. 1–4, 1998, Lake Buena Vista, FL., American Water Works Association & Water Environment Federation.

York, D.W., and L. Wadsworth. 1998. Reuse in Florida: Moving towards the 21st century, pp. 1–13, In: *Water Reuse Conference Proceedings*, Feb. 1–4, 1998, Lake Buena Vista, FL., American Water Works Association & Water Environment Federation.

Yoshpe-Purer, Y., and S. Golderman. 1987. Occurrence of *Staphylococcus aureus* and *Pseudomonas aeruginosa* in Israeli coastal water. Appl. Environ. Microbiol. 53: 1138–1141.

You, I.S., and R. Bartha. 1982. Stimulation of 3,4-dichloroaniline mineralization by aniline. Appl. Environ. Microbiol. 44: 678–681.

Young, J.C. 1983. Comparison of three forms of 2-chloro-6-(trichloromethyl) pyridine as a nitrification inhibitor in BOD tests. J. Water Pollut. Control Fed. 55: 415–416.

Young, J.C., and P.L. McCarty. 1969. The anaerobic filter for waste treatment. J. Water Pollut. Control Fed. 41: R160–173.

Young, L.Y., and M.M. Haggblom. 1989. The anaerobic microbiology and biodegradation of aromatic compounds, pp. 3–17, In: *Biotechnology and Biodegradation*, D. Kamely, A. Chakrabarty, and G.S. Ommen, Eds., Gulf, Houston, TX.

Young, L.Y. 1997. Anaerobic biodegradability assay, pp. 802–805, In: *Manual of Environmental Microbiology*, C.J. Hurst, G.R. Knudsen, M.J. McInerney, L.D. Stetzenbach, and M.V. Walter, Eds., ASM Press, Washington, D.C.

Young, L.Y., and C.E. Cerniglia, Eds. 1995. *Microbial Transformation and Degradation of Toxic Organic Chemicals*, Wiley-Liss, New York.

Young, S.B., and P. Setlow. 2004. Mechanisms of *Bacillus subtilis* spore resistance to and killing by aqueous ozone. J. Appl. Microbiol. 96: 1133–1142.

Yu, T., and P.L. Bishop. 2001. Stratification and oxido-reduction potential change in an aerobic and sulfate-reducing biofilm studied using microelectrodes. Water Environ. Res. 73: 368–373.

Yu, T., C. de la Rosa, and R. Lu. 2004. Microsensor measurement of oxygen concentration in biofilms: from one dimension to three dimensions. Water Sci. Technol. 49 (11–12): 353–358.

Zajik, J.E. 1971. *Water Pollution: Disposal and Reuse*, Marcel Dekker, New York.

Zaoutis, T., and J.D. Klein. 1998. Enterovirus infections. Pediatrics Rev. 19: 183–191.

Zaske, S.K., W.S. Dockins, and G.A. McFeters. 1980. Cell envelope damage in *E. coli* caused by short-term stress in water. Appl. Environ. Microbiol. 41: 386–390.

Zehnder, A.J.B., Ed. 1988. *Biology of Anaerobic Microorganisms.* John Wiley & Sons, New York.

Zeikus, J.G. 1980. Chemical and fuel production by anaerobic bacteria. Annu. Rev. Microbiol. 34: 423–464.

Zeyer, J., P. Eicher, J. Dolfing, and R.P. Schwarzenbach. 1989. Anaerobic degradation of aromatic hydrocarbons. In: *Biotechnology and Biodegradation*, D. Kamely, A. Chakrabarty, and G.S. Omenn, Eds., Gulf, Houston, TX.

Zhang, L., M. Hirai, and M. Shoda. 1991. Removal characteristics of dimethyl sulfide, methanethiol, and hydrogen sulfide by *Hyphomicrobium* sp. I55 isolated from peat biofilter. J. Ferment. Bioeng. 72: 392–396.

Zhou, H., and Y.F. Xie. 2002. Using BAC for HAA removal. Part 1: Batch study. J. AWWA 94 (4): 194–200.

Zhou, J., and D. Thompson. 2002. Microarrays: Applications in environmental microbiology, pp. 1968–1979, In: *Encyclopedia of Environmental Microbiology*, G. Bitton, editor-in-chief, Wiley-Interscience, N.Y.

Ziebell, W.A., D.H. Nero, J.F. Deininger, and E. McCoy. 1975. Use of bacteria in assessing waste treatment and soil disposal systems. In: *Home Sewage Disposal.* Proceeding of the National Home Sewage Disposal Symposium, Dec. 9–10, 1974. American Society of Agricultural Engineering Publication No. 175, St. Joseph, MI.

Ziegler, M., M. Lange, and W. Dott. 1990. Isolation and morphological and cytological characterization of filamentous bacteria from bulking sludge. Water Res. 24: 1437–1451.

Zierler, S., R.A. Danley, and L. Feingold. 1987. Type of disinfectant in drinking water and patterns of mortality in Massachussetts. Environ. Health Perspect. 68: 275–287.

Ziglio, G., G. Andreottola, S. Barbesti, G. Boschetti, L. Bruni, P. Foladori, and R. Villa. 2002. Assessment of activated sludge viability with flow cytometry. Water Res. 36: 460–468.

Zilinskas, R.A. 1997. Iraq's biological weapons. The past as future? J. Amer. Med. Assoc. 278: 418–424.

Zilles, J.L., J. Peccia, M.-W. Kim, C.-H. Hung, and D.R. Noguera. 2002. Involvement of *Rhodocyclus*-related organisms in phosphorus removal in full-scale wastewater treatment plants. Appl. Environ. Microbiol. 68: 2763–2769.

Zimmer, J.L., and R.M. Slawson. 2002. Potential repair of *Escherichia coli* DNA following exposure to UV radiation from both medium- and low-pressure UV sources used in drinking water treatment. Appl. Environ. Microbiol. 68: 3293–3299.

Zimmer, J.L., R.M. Slawson, and P.M. Huck. 2003. Inactivation and potential repair of *Cryptosporidium parvum* following low- and medium-pressure ultraviolet Irradiation. Water Res. 37: 3517–3523.

Zimmermann, K., and J.W. Mannhalter. 1996. Technical aspects of quantitative competitive PCR. Bio/Techniques 21: 268–279.

Zimmerman, N.J., P.C. Reist, and A.G. Turner. 1987. Comparison of two biological aerosol sampling methods. Appl. Environ. Microbiol. 53: 99–104.

Zinder, S.H., S.C. Cardwell, T. Anguish, M. Yee, and M. Koch. 1984. Methanogenesis in a thermophilic (58°C) anaerobic digestor: *Methanothrix* sp. as an important aceticlastic methanogen. Appl. Environ. Microbiol. 47: 796–807.

Ziogou, K., P.W.W. Kirk, and J.N. Lester. 1989. Behaviour of phthalic acid esters during batch anaerobic digestion of sludge. Water Res. 23: 743–748.

Zita, A., and M. Hermansson. 1994. Effects of ionic strength on bacterial adhesion and stability of flocs in a wastewater activated sludge. Appl. Environ. Microbiol. 60: 3041–3048.

Zita, A., and M. Hermansson. 1997a. Effects bacterial cell surface structures and hydrophobicity on attachment to activated sludge flocs. Appl. Environ. Microbiol. 63: 1168–1170.

Zita, A., and M. Hermansson. 1997b. Determination of bacterial cell surface hydrophobicity of single cells in cultures and in wastewater *in situ.* FEMS Microbiol. Lett. 152: 299–306.

Zoeteman, B.C.J., G.J. Piet, and L. Postma. 1980. Taste as an indicator for drinking water quality. J. Am. Water Works Assoc. 72: 537–540.

Zuckermann, H., M. Staal, L.J. Stal, J. Reuss, S.L. Kekkert, F. Harren, and D. Parker. 1997. On-line monitoring of nitrogenase activity in cyanobacteria by sensitive laser photoacoustic detection of ethylene. Appl. Environ. Microbiol. 63: 4243–4251.

Zyman, J., and C.A. Sorber. 1988. Influence of simulated rainfall on the transport and survival of selected indicator organisms in sludge-amended soils. J. Water Pollut. Control Fed. 60: 2105–2110.

INDEX